978-1-62265-930-2 (online)   978-1-62265-931-9 (paper)

Integral Geometry and Fields Theorematic Proofs and Numerical Models

Yangke

Abstract

This work is divided into Divergence Theorem at Manifold (Difference、corresponding to traditional Остроградский-Gauss Theorem)、Green Theorem at Manifold (Difference、corresponding to traditional Green Theorem) and Curl Theorem at Manifold (Difference、corresponding to traditional Stokes Theorem) three parts;

From different perspectives of three new typical integral theorems, this work demonstrates that base on individualized geometric object (Manifold)coordinates, through matrixing operations, realize new formular conjunctions between different typical integrals; and provides corresponding numerical models;

New integral formular demonstrations and numerical models indicate: Base on individualized geometric object (Manifold) coordinates, through unified standardized、concise、matrixing integral operations, science explorers can obtain analytic integral values or float integral values in discretional precision of discretional complicated geometric objects (Manifold; Especially point irregular、asymmetrical geometric shapes in 2-Dimensional and 3-Dimensional Euclidean space of real world);

New typical numerical modelings possess vast mathematical、physical、engineering applicational field, involve that in 2-Dimensional and 3-Dimensional Euclidean space of real world, exact integral calculation of vector field [Electric field、magnetic field、hydromechanical field、gravitational field etc.] and scalar field [Electric potential field、temperature field、density field etc.] about discretional complicated geometric objects and their boundary regions,and then realize direct triple connection between calculus、topology and physical engineering calculation.

Keywords:

Calculus; Topology; Theoretical Physics; Poincare Conjecture; Divergence Theorem at Manifold; Green Theorem at Manifold; Curl Theorem at Manifold; Proof; Numerical Models

Mathematics Subject Classification (2000): 58C35

PACS numbers: 41.20.-q, 47.85.Np & 95.30.Sf

978-1-62265-930-2 (online) 978-1-62265-931-9 (paper)   Yang Ke

Introduction

Divergence Theorem(i.e. Остроградский-Gauss Theorem)、Curl Theorem (i.e. Stokes Theorem) and Green Theorem are hard cores in modern mathematical and physical system [1][2][3][4][5][6][7][8][9][10][11][12][13][14][15][16][17]. Logic system of Divergence Theorem and Curl Theorem's traditional proof, in 3-Dimensional Cartesian coordinates, established formular association between surface integral (Based on projective method) and triple integrals (or formular association between surface integral and spacial closed curve integral), radicated that projective method was primary method of surface integral. Basal logic of projective method is that, convert surface integral (in 3-Dimensional Cartesian coordinates) to double integrals (at certain Cartesian coordinates plane), achieves object by indirect fashion.

Defects of projective method are obvious:

1. Projective zone of integral surface at certain Cartesian coordinates plane can't be overlapped, it means integral surface must be a function surface (Very simple and crude); as complicated parametrized surfaces that exist extensively in reality world and physical、engineering fields, projective method is disabled to calculate;

2. Projective method usually requests that integral surfaces are symmetric(such as point symmetry, axis symmetry, face symmetry etc.[2]). As asymmetrical、irregular surface that exist extensively in reality world and physical、engineering field, projective method is very fussy, and be disabled to calculate in most instances [1][2][3][4][5][6][7][8][9]; So as to solutions of many important questions in physics、engineering field (For examples, instantiation of Maxwell Equations in electromagnetism, Navier-Stokes Equations in hydrodynamics etc.) are built on solving partial differential equations (e.g. finite difference method、finite element method、boundary element method etc.[18]) in Cartesian coordinates or other coordinates (e.g. Body-Fitted coordinates). For more than a century, countless mathematical、physical and engineering practices have proved: Depend on projective method、partial differential equations in Cartesian coordinates or other coordinates, it is difficult or disable to obtain

analytical solution or numerical solution about complicated geometric objects(Manifold);

3.As geometrical differences of unlimited quantitative individual surfaces, projective orient on surface、frequency of projection are different and different (Sectional projection especially). There are 100 integral examples, maybe there are 100 projective plans. Steps of calculation can't be standardized and can't form a model, be difficulty to computer programming;

Algorithm design of solving partial differential equations is a huge project and tedious process; a specific geometric shape corresponding to a specific design plan, can not form an unified、standardized method template, in the face of Protean nonlinear geometric shapes, poor to cope with; solving result is approximate value (existing convergence and stability problems, etc.), even in today, 21st century, in the case of highly developed electronic computer numerical processing technology, solution accuracy of partial differential equations is still difficult to achieve realistic physics and engineering requirements;

4.No matter complicated degree of integral surface, actual calculating course of projective method is generally complicated and tedious;

5.Crucial points of physics and mathematical analysis, Острогадский-Gauss theorem and Stokes theorem (Green theorem with certain meanings), there aren't direct calculational examples by projective method almost. (Even if they exist, be individual special examples, unrepresentative, such as closed integral 'Surface' like hexahedron(cuboid) exterior [12][13][15][16][17], closed 'Curve' made up of several straight lines on hexahedron's different exteriors [12][13][15][16], closed 'Curve' structured by side lines of intersectant triangle -- plane 'x+y+z=1' and three Cartesian coordinates planes [12][13][14][15][16], etc). It is regretted and perplexed that, there aren't plentiful magnificent and individual direct numerical models about magic Острогадский-Gauss、Stokes and Green theorem.

In spherical coordinates, there are parametrized surface integral about vector field [13]、triple integrals about scalar field (3 variables function) in spacial bounded closed region (Variable transform by 3th order Jaccobian determinant); In polar coordinates, there is double integrals about scalar field (2 variables function) in plane bounded closed region (Variable transform by 2th order Jaccobian determinant) [12][13][14][16][17]. But exist following problems:

1. In spherical coordinates, 'Surface integral about vector field at closed surface' and 'Triple integrals about scalar field in space bounded closed region' are isolated one another, not to be associated by Divergence Theorem. Furthermore, calculated results both sides can't check each other;

2. 'Surface integral about vector field at closed surface' and 'Triple integrals about scalar field in space bounded closed region' are parochially confined with orthogonal curve coordinates (viz. spherical coordinates、cylindrical coordinates [12][13][14][16][17]) and generalized spherical coordinates [14]、generalized cylindrical coordinates, not to spread to unlimited quantitative free surface coordinates;

3. In polar coordinates, 'Closed curve integral about vector field' and 'Double integrals about scalar field in plane bounded closed region' are isolated one another, not to be associated by Green' theorem. Furthermore, calculated results both sides can't check each other too;

4. 'Double integrals about scalar field in plane bounded closed region' is parochially confined with orthogonal curve coordinates (viz. polar coordinates)[11], not to spread to unlimited quantitative discretional curve coordinates.

Traditional manifold calculus deducts out Green Theorem、Остроградский-Gauss Theorem and Stokes Theorem by exterior differential form, and even generalized Stokes Theorem about n-Dimensional space integral [20], viz. $\int_{\partial\Sigma} \omega = \int_{\Sigma} d\omega$

But these theorems (deducted by exterior differential form), scantly possess abstract academic meaning, and can't reveal

978-1-62265-930-2 (online)   978-1-62265-931-9 (paper)   Yang Ke

idiographic realization procedures of integrals, much less
idiographic numerical models and their physical、engineering
applications.

No end about exploration and cognition.

Contents

Part I   Divergence Theorem at Manifold

Part II   Green Theorem at Manifold

978-1-62265-930-2 (online)   978-1-62265-931-9 (paper)

Part III   Curl Theorem at Manifold

978-1-62265-930-2 (online)   978-1-62265-931-9 (paper)

Part I   Divergence Theorem at Manifold

Chapter 1 Precondition of Proofs
一 Constitute Abstract Simply Connected Orientable Closed Parametrized Surface Coordinates

1.1 Introspection between Divergence Theorem and Poincare Conjecture

Review object of Proof -- Divergence Theorem:

Divergence Theorem

Suppose spacial bounded closed region 'Ω' is surrounded by smooth or piecewise smooth closed surface 'S'. Functions 'P(x,y,z),Q(x,y,z), R(x,y,z)'[Structure vector field 'A'] and its partial derivatives are continuous at spacial bounded closed region 'Ω', then:

$$\iint\limits_{S} A \cdot n \, dS = \iiint\limits_{\Omega} divA \cdot d\omega$$

thereinto, surface 'S' is outer side of spacial bounded closed region 'Ω's entire boundary surface, 'n' is unit normal vector of surface 'S's outer side, and 'divA' is divergence of vector field 'A'.

In definiens of theorem, it emphasizes that boundary closed surface of spacial bounded closed region 'Ω' must be orientable surface (Observer can discriminate its outer side or inner side).

In proof of traditional Остроградский-Gauss' theorem in Cartesian coordinates, 'Abstract orientable closed surface Σ' is described as that:

'Abstract closed surface Σ' is piecewise surrounded by its three sub-surfaces(Viz. $\Sigma_1 : z = z_1(x,y), \Sigma_2 : z = z_2(x,y), \& \Sigma_3$);

thereinto, sub-surfaces $\Sigma_1 : z = z_1(x,y), \Sigma_2 : z = z_2(x,y), \& \Sigma_3$ are 2-variable function togetherly.

(See also <Calculus [6th Edition]> Tongji University, Department of Mathematics; Higher Education Press, Beijing; ed.1, 1978.10; ed.6, 2007.6; 9th Printing, 2009.8; pp.168-170)

In other words, objectively, Divergence Theorem requests that no matter in Cartesian coordinates or in other coordinates, correlative surface must possess two attributes: (1)Closed; (2)Orientable.

Depart from traditional Cartesian coordinates, how to depict abstract universal 'Orientable Closed Surface', and constitute 'Orientable Closed Surface Coordinates' ? Ready-made solution is absence.

Poincare Conjecture[19] concludes that 'Every closed n-Manifold which is homotopy equivalent to n-Sphere is homeomorphic to n-Sphere'. In 3-Dimensional Euclidean space that Divergence or Curl theorem refers, corresponding opinion is that: 'Every simply connected 、 orientable closed 2-Manifold is homeomorphic to 2-Sphere'.

In other words, according as Poincare Conjecture, in 3-Dimensional Euclidean space that Divergence or Curl theorem refers, every simply connected、orientable closed surface(Although it is 'simply connected' scantly), no matter its Protean geometrical shape, 'Be homeomorphic to 2-Sphere' is universal attribute.

Naturally, next question is 'In 3-Dimensional Euclidean space, base on Poincare Conjecture, it is possible to define an abstract universal expression of simply connected、orientable closed surface ?' -- It is kernel content of this chapter.

In spacial analytic geometry, parametrized expression of '2-Sphere' above-mentioned is '[sin(u)cos(v),sin(u)sin(v), cos(u)]', thereinto, u[0,Pi],v[0,2*Pi] (In proper meaning, this parametrized expression is transform expression of '2-Sphere' between '3-Dimensional Cartesian coordinates' and 'Spherical coordinates'; expression of '2-Sphere' in Spherical coordinates is constant 1);

In topology field, define 'Homeomorphism' as 'Two manifolds, if it is possible to change one to another by operations of bending、 extending 、 cutting etc., then recognise that two manifolds above-mentioned are homeomorphous'.

Reconsider Poincare Conjecture by visual angle of topology and analytic geometry, since parametrized expression of '2-Sphere' is '[sin(u)cos(v), sin(u)sin(v), cos(u)], u[0,Pi], v[0,2*Pi]', then its transfiguration '[a*sin(u)cos(v),b*sin(u)sin(v),c*cos(u)], u[0,Pi], v[0,2*Pi]' (Thereinto, 'a,b,c' are discretional nonzero constants) is parametrized expression of discretional ellipsoid. In 3-Dimensional Euclidean space, discretional ellipsoid is homeomorphic to 2-Sphere, this is general knowledge of topology, dispense with discussion;

If 'a,b,c' are discretional '1th order derivable continuous functions', what instance appears possibly ?

See also below figures:

Figure(I)1.1.1 Non Simply Connected Orientable Closed Surface, Deduced from parametrized expression of '[a*sin(u)cos(v), b*sin(u)sin(v), c*cos(u)], u[0,Pi], v[0, 2*Pi]'

Suppose discretional pending coefficients

a = sin(u)+cos(v), b = cos(u), c = cos(v/2)

(viz. 'a,b,c' are discretional '1th order derivable continuous functions'), then target parametrized surface [a*sin(u)*cos(v), b*sin(u)*sin(v), c*cos(u)], $u \in [0,\pi]$, $v \in [0,2\pi]$ equals

[(sin(u)+cos(v))*sin(u)*cos(v), cos(u)*sin(u)*sin(v),

cos(v/2)*cos(u)], $u \in [0,\pi]$, $v \in [0,2\pi]$.

```
// Corresponding drawing instructions of Waterloo Maplesoft 17:
> restart;
> with(plots):with(linalg):
> a:=sin(u)+cos(v);  # Input discretional '1th order derivable
continuous functions' by pending coefficients 'a,b,c'
```

$$a := \sin(u) + \cos(v)$$

```
> b:=cos(u);
```

$$b := \cos(u)$$

```
> c:=cos(v/2);
```

$$c := \cos\left(\frac{v}{2}\right)$$

```
> CS:=[a*sin(u)*cos(v),b*sin(u)*sin(v),c*cos(u)];
 # Target parametrized expression
```

$$CS := \left[(\sin(u) + \cos(v))\sin(u)\cos(v), \cos(u)\sin(u)\sin(v), \cos\left(\frac{v}{2}\right)\cos(u) \right]$$

```
> rgu:=[0,Pi]; # Set range of 'u,v'
```

$$rgu := [0, \pi]$$

```
> rgv:=[0,2*Pi];
```

$$rgv := [0, 2\pi]$$

```
>plot3d(CS,u=rgu[1]..rgu[2],v=rgv[1]..rgv[2],scaling=constrained,
 projection=0.9,numpoints=6000);
 # Drawing instruction of Waterloo Maplesoft 17
```

Actual shape of target parametrized surface:

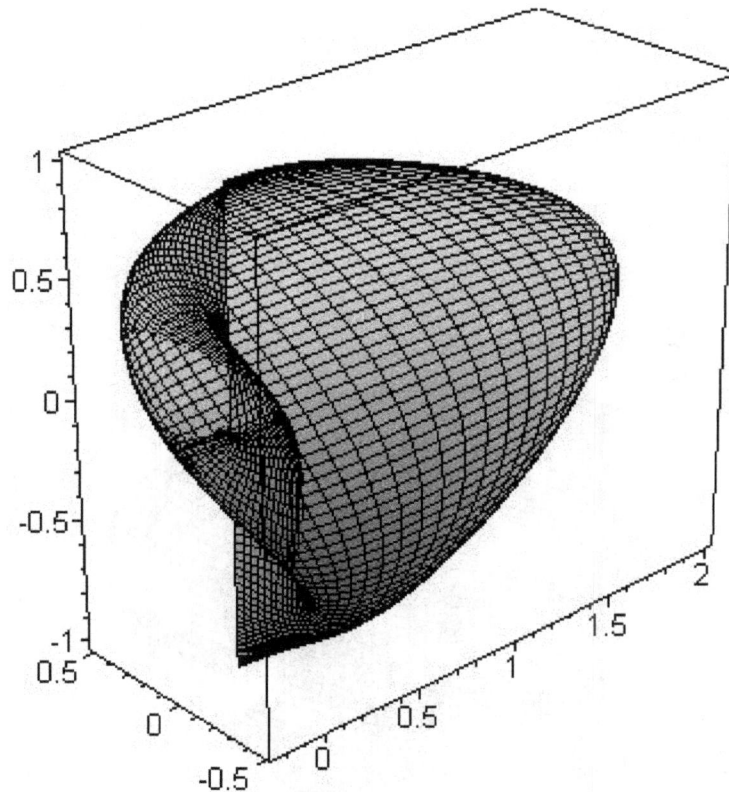

Figure(I)1.1.1 Non Simply Connected Orientable Closed Surface,
            Deduced from parametrized expression
`[a*sin(u)cos(v), b*sin(u)sin(v), c*cos(u)],u[0,Pi], v[0,2*Pi]`

Observe shape of Figure(I)1.1.1 by intuitionistic vision, parts
Of target parametrized surface above-mentioned show intersected、
overlapped、 fractured、 unclosed states; therefore target parametrized
surface belongs to non simply connected orientable closed surface,
be independent of 'Poincare Conjecture' and 'Divergence Theorem at
Manifold'

Figure(I)1.1.2 Non Simply Connected Orientable Closed Surface, Deduced from parametrized expression '[a*sin(u)cos(v),b*sin(u)sin(v), c*cos(u)],u[0,Pi],v[0,2*Pi]'

Suppose discretional pending coefficients

a = sin(u+v)+cos(v), b = cos(v), c = cos(v/2)

(viz. 'a,b,c' are discretional '1th order derivable continuous functions'), then target parametrized surface [a*sin(u)*cos(v), b*sin(u)*sin(v), c*cos(u)], u∈[0,π], v∈[0,2π]) equals

[(sin(u+v)+cos(v))*sin(u)*cos(v), cos(v)*sin(u)*sin(v),

cos(v/2)*cos(u)], u∈[0,π], v∈[0,2π].

Actual shape of target parametrized surface:

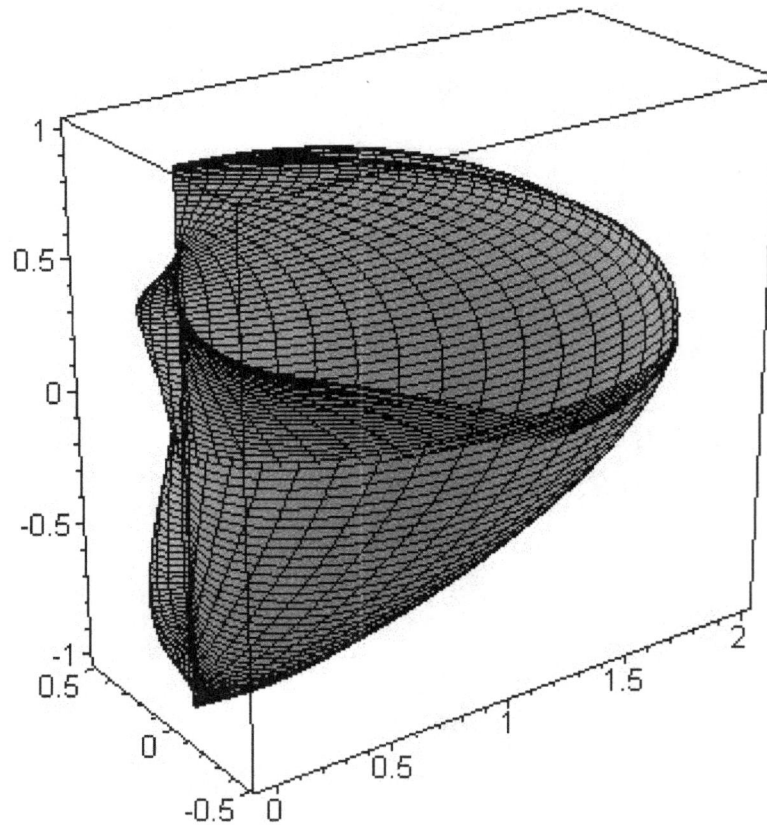

Figure(I)1.1.2 Non Simply Connected Orientable Closed Surface, Deduced from parametrized expression '[a*sin(u)cos(v), b*sin(u)sin(v),c*cos(u)], u[0,Pi], v[0,2*Pi]'

Observe shape of Figure(I)1.1.2 by intuitionistic vision, parts of target parametrized surface above-mentioned show intersected、 overlapped、fractured、unclosed states; therefore target parametrized surface belongs to non simply connected orientable closed surface, be independent of 'Poincare Conjecture' and 'Divergence Theorem at Manifold'.

Figure(I)1.1.3 Simply Connected Orientable Closed Surface, Deduced from parametrized expression '[a*sin(u)cos(v), b*sin(u)sin(v), c*cos(u)], u[0,Pi], v[0,2*Pi]'

Suppose discretional pending coefficients

a = sin(u), b = cos(u)+cos(u+3*v)/3, c = cos(u),

(viz. 'a,b,c' are discretional '1th order derivable continuous functions'), then target parametrized surface [a*sin(u)*cos(v), b*sin(u)*sin(v), c*cos(u)], u∈[0,π], v∈[0,2π] equals

[sin(u)*sin(u)*cos(v), (cos(u)+cos(u+3*v)/3)*sin(u)*sin(v), cos(u)*cos(u)], u∈[0,π], v∈[0,2π].

Actual shape of target parametrized surface:

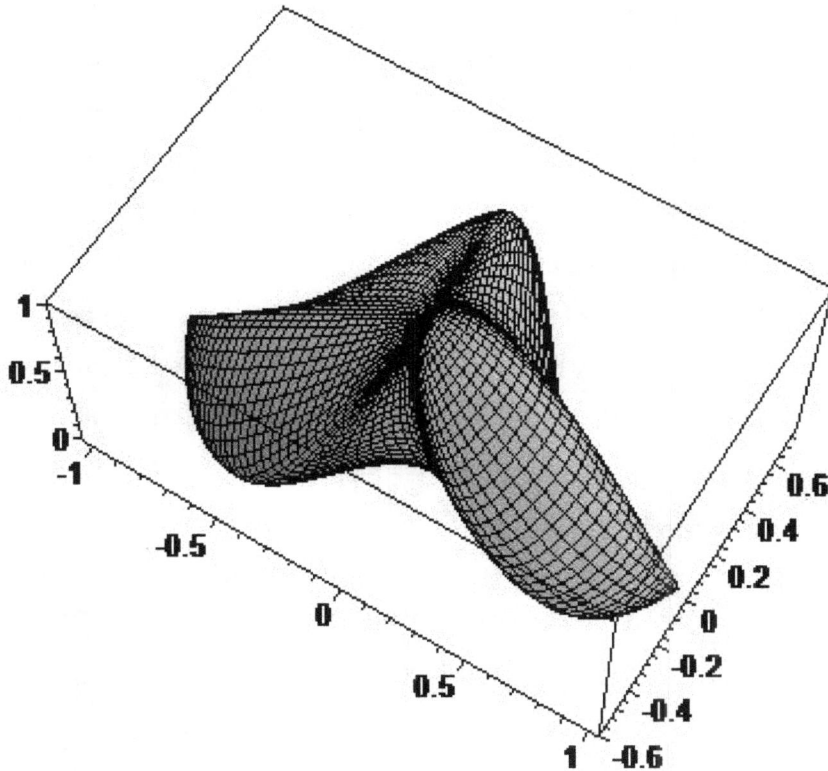

Figure(I)1.1.3 Simply Connected Orientable Closed Surface, Deduced from parametrized expression '[a*sin(u)cos(v),b*sin(u)sin(v), c*cos(u)],u[0,Pi],v[0,2*Pi]'

Observe shape of Figure(I)1.1.3 by intuitionistic vision, there aren't intersected、overlapped、fractured、unclosed states about target parametrized surface above-mentioned, and genus value of target parametrized surface is '0'; therefore target parametrized surface belongs to simply connected orientable closed surface, be relational to 'Poincare Conjecture' and 'Divergence Theorem at Manifold'.

In lay of actual operation, draw certain a parametrized surface by 'Plot3D'(Command of Waterloo Maplesoft), must judge that 'This parametrized surface is simply connected orientable closed surface or not' by intuitionistic vision at first, then decide that 'This parametrized surface is the same with numerical models of Divergence (or curl) theorem at manifold or not' at second. It is impossible to judge that 'Certain a parametrized surface is simply connected orientable closed surface or not' by parametrized expression itself.

Determinant diathesis of 'Parametrized surface is simply connected orientable closed surface or not' is in topology field and not in analytical geometry field; Only method of parametrized expression in analytical geometry is unable to conclude and deduce certain a surface as a simply connected orientable closed surface.

By original phenomenons, experimental data above-meitioned indicate that, similarly belong to parametrized surface '[a*sin(u)cos(v), b*sin(u)sin(v), c*cos(u)], u[0,Pi], v[0,2*Pi]', because of different values of pending coefficients 'a,b,c', a part of surfaces belong to simply connected orientable closed surface, and another part of surfaces are exceptional.

In other words, there are two instances about parametrized surface '[a*sin(u)cos(v),b*sin(u)sin(v),c*cos(u)],u[0,Pi],v[0,2*Pi]':

(1)In the case of pending coefficients 'a,b,c' are discretional nonzero constants, parametrized surface is ellipsoid (Be homeomorphic to 2-Sphere naturally);

(2)In the case of pending coefficients 'a,b,c' are discretional 1th order derivable continuous functions, parametrized surface is simply connected orientable closed surface possibly (Be homeomorphic to

2-Sphere),and parametrized surface isn't simply connected orientable closed surface possibly (Not to be homeomorphic to 2-Sphere).

Naturally, next question is "In the mode of parametrized surface [a*sin(u)cos(v), b*sin(u)sin(v), c*cos(u)], u[0,Pi], v[0,2*Pi], it is possible to exclude the instance of 'Parametrized surface is not simply connected orientable closed surface' by certain logical defining ?"

1.2 Base on Poincare Conjecture,
Define Abstract Simply Connected Orientable Closed Surface

Suppose 'Discretional Surfaces' as a set, then 'Discretional Simply Connected Orientable Closed Surfaces' is subset of former, Poincare Conjecture is attribute of this subset, current thesis 'Proof of Divergence or Curl Theorem at Manifold' discuss that Divergence or Curl Theorem is the same with this subset or not.

Poincare Conjecture supplies the realizing route of describing 'Discretional simply connected orientable closed surface's certain attribute (Viz. 'Be homeomorphic to 2-Sphere' this attribute) by parametrized expression.

Base on situation above-mentioned, abstract unlimited quantitative idiographic simply connected orientable closed surfaces as an uniform expression:
  [a*sin(u)cos(v),b*sin(u)sin(v),c*cos(u)],u[0,Pi],v[0,2*Pi]
(Thereinto, pending coefficients 'a,b,c' can't be designated discretionarily, 'a,b,c' must obey topology attribute of simply connected orientable closed surface)

In other words, if pending coefficients 'a,b,c' can be designated discretionarily, then target surface '[a*sin(u)cos(v),b*sin(u)sin(v), c*cos(u)],u[0,Pi],v[0,2*Pi]' may be simply connected orientable closed surface, or may be not;

If preestablish that 'Target surface [a*sin(u)cos(v), b*sin(u) sin(v), c*cos(u)], u[0,Pi], v[0,2*Pi] itself is simply connected orientable closed surface', then pending coefficients 'a,b,c' can't be designated discretionarily.

Explain phenomenons above-mentioned on geometrical meaning, in 3-Dimensional Cartesian coordinates, spherical surface (Viz. [sin(u)cos(v), sin(u)sin(v), cos(u)], u[0,Pi], v[0,2*Pi]) itself extends continuously discretionarily along three directions of axis 'x,y,z' (Viz. [a*sin(u)cos(v), b*sin(u)sin(v), c*cos(u)], u[0,Pi], v[0,2*Pi], thereinto, 'a,b,c' are discretional 1th order derivable continuous functions), not always comes into being simply connected orientable closed surface;

Contrarily, in 3-Dimensional Cartesian coordinates, discretional simply connected orientable closed surface, consequentially, based on spherical surface (Viz.[sin(u)cos(v),sin(u)sin(v),cos(u)],u[0,Pi], v[0,2*Pi]),extended continuously along three directions of axis 'x,y,z', came into being (Likewise consequentially, discretional simply connected orientable closed surface extends continuously along three directions of axis 'x,y,z', comes into being spherical surface) -- Base on Poincare Conjecture.

For example, surface of hexahedron can be regarded as simply connected orientable closed surface -- But surface of hexahedron is difficult or can't be described by parametrized expression -- But it is incontestable: Base on Poincare Conjecture, surface of hexahedron is homeomorphic to 2-Sphere consequentially, surface of hexahedron based on spherical surface(Viz.[sin(u)cos(v),sin(u)sin(v),cos(u)], u[0,Pi],v[0,2*Pi]),extended continuously along three directions of axis 'x,y,z', came into being (Likewise consequentially, surface of hexahedron extends continuously along three directions of axis 'x,y,z', comes into being spherical surface); Base on Poincare Conjecture, similarly, surface of hexahedron can be described by '[a*sin(u)cos(v), b*sin(u)sin(v),c*cos(u)],u[0,Pi],v[0,2*Pi]' Mode.

Describe abstract universal simply connected orientable closed surface by '[a*sin(u)cos(v), b*sin(u)sin(v), c*cos(u)], u[0,Pi], v[0,2*Pi]' Mode -- Actually, describe certain inner configuration and attribute (Viz. 'Be homeomorphic to 2-Sphere' this inner configuration and attribute) of simply connected orientable closed surface by Poincare Conjecture, define a felicitous precondition for proofs.

It is necessary to indicate expressly, in the case of 'Idiographic surfaces are multiply connected orientable closed surfaces' (For example, torus and its homeomorphous surfaces), abstraction is impossible yet, correlative theory is absence.

1.3 Constitute Abstract Simply Connected Orientable Closed Surface Coordinates

Base on Poincare Conjecture, '[a*sin(u)cos(v), b*sin(u)sin(v), c*cos(u)],u[0,Pi],v[0,2*Pi]' is just abstract universal expression of simply connected orientable closed surface, and it isn't coordinates; abstract simply connected orientable closed surface coordinates is '[r*a*sin(u)cos(v),r*b*sin(u)sin(v),r*c*cos(u)], r[0, ∞], u[0,Pi], v[0,2*Pi]', thereinto, 'r' is radius, 'a,b,c' is pending coefficients (Because of 'a,b,c' maybe nonzero constants, maybe 1th order derivable continuous functions), be uncertain.

Actually, expression of abstract simply connected orientable closed surface '[a*sin(u)cos(v),b*sin(u)sin(v),c*cos(u)],u[0,Pi], v[0,2*Pi]' is the same as expression of ellipsoid '[a*sin(u)cos(v), b*sin(u)sin(v), c*cos(u)], u[0,Pi], v[0,2*Pi]' formally, only explanations of pending coefficients 'a,b,c' are different: Former explains 'a,b,c' as 'Discretional nonzero constants or 1th order derivable continuous functions (Non Discretional, be restricted by attribute of simply connected orientable closed surface)'; and latter explains 'a,b,c' as 'Discretional nonzero constants' -- So that corresponding relation between abstract simply connected orientable closed surface coordinates and Cartesian coordinates is 'x=r*a*sin(u)cos(v),y=r*b*sin(u)sin(v),z=r*c cos(u)' (Be the same as transform expression of ellipsoid coordinates and Cartesian coordinates).

Chapter 2 Prove Divergence Theorem at Manifold

2.1 Prove Divergence Theorem at Manifold

Divergence Theorem

Suppose spacial bounded closed region 'Ω' is surrounded by smooth or piecewise smooth closed surface 'S'. Functions 'P(x,y,z),Q(x,y,z), R(x,y,z)' [Structure vector field 'A'] and its partial derivatives are continuous at spacial bounded closed region 'Ω', then:

$$\iint_S A \cdot n \, dS = \iiint_\Omega divA \cdot d\omega \qquad (1)$$

thereinto, surface 'S' is outer side of spacial bounded closed region 'Ω's entire boundary surface, 'n' is unit normal vector of surface 'S's outer side, and 'divA' is divergence of vector field 'A'.

Proof:

Define parametrized expression of abstract simply connected orientable closed surface 'S':

[a*sin(u)cos(v),b*sin(u)sin(v),c*cos(u)]   (2)

Thereinto, 'a,b,c' are nonzero constants or 1th order derivable continuous functions, simply connected orientable closed surface 'S' determines value of 'a,b,c'; set range of parameters 'u,v' in $[0,\pi],[0,2\pi]$, make surface 'S' closed. (See also Poincare Conjecture: 'Every closed n-Manifold which is homotopy equivalent to n-Sphere is homeomorphic to n-Sphere' [19], and Chapter 1 this Part 'Precondition of Proofs: Constitute Abstract Simply Connected Orientable Closed Parametrized Surface Coordinates')

According to surface parametrized expression (2), define and calculate first matrix of partial derivatives, obtain tangent plane normal vector of surface 'S':

$$\begin{bmatrix} i & j & k \\ \dfrac{\partial}{\partial u} a\sin(u)\cos(v) & \dfrac{\partial}{\partial u} b\sin(u)\sin(v) & \dfrac{\partial}{\partial u} c\cos(u) \\ \dfrac{\partial}{\partial v} a\sin(u)\cos(v) & \dfrac{\partial}{\partial v} b\sin(u)\sin(v) & \dfrac{\partial}{\partial v} c\cos(u) \end{bmatrix} =$$

$$i\,c\,\sin(u)^2\,b\,\cos(v) + a\,\cos(u)\,\cos(v)^2\,k\,b\,\sin(u) + a\,\sin(u)^2\,\sin(v)\,j\,c$$
$$+\,a\,\sin(u)\,\sin(v)^2\,k\,b\,\cos(u)$$

$$(3)$$

From expression(3), respectively pick up coefficients of item 'i,j,k', obtain tangent plane normal vector of surface 'S':

$$[\,c\sin(u)^2 b\cos(v)\,,\ \sin(u)^2 a\sin(v)c\,,\ \sin(u)ab\cos(u)\,]\qquad(4)$$

Surface Integral of Dot Product [Vetor field 'A' and Tangent Plane Normal Vector (4)] in range of 'u,v':

$$\int_0^{2\pi}\int_0^{\pi} P(x,y,z)\,c\,\sin(u)^2\,b\,\cos(v) + Q(x,y,z)\,\sin(u)^2\,a\,\sin(v)\,c$$
$$+\,R(x,y,z)\,\sin(u)\,a\,b\,\cos(u)\,du\,dv =$$

$$\int_0^{2\pi}\frac{1}{2}\,Q(x,y,z)\,a\,\sin(v)\,c\,\pi + \frac{1}{2}\,P(x,y,z)\,c\,b\,\cos(v)\,\pi\,dv = 0$$

$$(5)$$

// Relative to Protean given 3-Dimensional space vector fields (Composed of idiographic given 3-variable functions), abstract 3-Dimensional space vector field '[P(x,y,z),Q(x,y,z),R(x,y,z)]' is a sort of balanced、symmetrical data structure;

In Proof of Divergence Theorem at Manifold, objectively require a sort of balanced、symmetrical abstract orientable closed surface's expression, match with abstract 3-Dimensional space vector field '[P(x,y,z),Q(x,y,z),R(x,y,z)]';

In meaning of data structure, abstract simply connected orientable closed surface's expression '[a*sin(u)*cos(v), b*sin(u)*sin(v), c*cos(u)],u[0,Pi],v[0,2*Pi]' also possesses balanced、symmetrical attribute.

//Owing to universality and homogeneity of abstract vector field [P(x,y,z),Q(x,y,z),R(x,y,z)], Integral by Embedded Sub-Variable 'u,v' of Variable 'x,y,z', Result of Integral can be described as 'P(x,y,z)[or Q(x,y,z),or R(x,y,z)]'.

In other words, Integral by Embedded Sub-Variable 'u,v' of Variable 'x,y,z', Can't change structure of abstract function P(x,y,z) [or Q(x,y,z),or R(x,y,z)], so that abstract function structure P(x,y,z) [or Q(x,y,z),or R(x,y,z)] can preserve primary form after integral.

(See also Chapter 4 (This Part) Section 4.1 'Prove Divergence Theorem Finite Sums Limits at Manifold' and Section 4.3 'Discuss About Abstract Vector Field、Finite Sums Limit、Integral and Riemann Sums')

//'Integral value equals 0' possesses specific mathematical and physical meaning:

In mathematical meaning, 'Integral value equals 0' is inevitable result of logic illation, incarnates logic balanced state between various integral elements;

In physical meaning, 'Integral value equals 0' implies that flux of 'Abstract Vector Field at Abstract simply connected orientable closed surface' is constant for '0' (immobile and pending) identically; If integral value equals certain a positive/negative number or an expression, it implies that always there is effusive/influent flux or unknown flux about 'Abstract Vector Field at Abstract simply connected orientable closed surface', this instance will be unaccountable.

Entirely multiply radius vector 'r' (suppose r>0) by each item of surface 'S' parametrized expression(2), convert "Parametrized expression of surface 'S'" to "Parametrized expression of surface 'S' coordinates"

$$[r*a*sin(u)cos(v),r*b*sin(u)sin(v),r*c*cos(u)] \qquad (6)$$

According to parametrized expression of surface 'S' coordinates (6), define and calculate second matrix of partial derivatives, obtain general expression of surface 'S' coordinates volume element coefficient:

$$\begin{bmatrix} \dfrac{\partial}{\partial r} ra\sin(u)\cos(v) & \dfrac{\partial}{\partial u} ra\sin(u)\cos(v) & \dfrac{\partial}{\partial v} ra\sin(u)\cos(v) \\[2ex] \dfrac{\partial}{\partial r} rb\sin(u)\sin(v) & \dfrac{\partial}{\partial u} rb\sin(u)\sin(v) & \dfrac{\partial}{\partial v} rb\sin(u)\sin(v) \\[2ex] \dfrac{\partial}{\partial r} rc\cos(u) & \dfrac{\partial}{\partial u} rc\cos(u) & \dfrac{\partial}{\partial v} rc\cos(u) \end{bmatrix}$$

$$= abcr^2 \sin(u) \qquad\qquad\qquad (7)$$

//In proper meaning, surface 'S' coordinates volume element

coefficient is ratio of "Surface 'S' coordinates volume element" and 'Spacial Cartesian coordinates volume element'.

Calculate divergence of vector field 'A', and convert Divergence from Cartesian Coordinates expression(8) to surface 'S' coordinates expression(9):

$$\frac{\partial}{\partial x} P(x,y,z) + \frac{\partial}{\partial y} Q(x,y,z) + \frac{\partial}{\partial z} R(x,y,z) \tag{8}$$

$$\left(\frac{\partial}{\partial x} P(x,y,z) \right) \left(\frac{\partial}{\partial r} (r\,a\,\sin(u)\,\cos(v)) \right) \left(\frac{\partial}{\partial u} (r\,a\,\sin(u)\,\cos(v)) \right)$$

$$\left(\frac{\partial}{\partial v} (r\,a\,\sin(u)\,\cos(v)) \right) + \left(\frac{\partial}{\partial y} Q(x,y,z) \right) \left(\frac{\partial}{\partial r} (r\,b\,\sin(u)\,\sin(v)) \right)$$

$$\left(\frac{\partial}{\partial u} (r\,b\,\sin(u)\,\sin(v)) \right) \left(\frac{\partial}{\partial v} (r\,b\,\sin(u)\,\sin(v)) \right)$$

$$+ \left(\frac{\partial}{\partial z} R(x,y,z) \right) \left(\frac{\partial}{\partial r} (r\,c\,\cos(u)) \right) \left(\frac{\partial}{\partial u} (r\,c\,\cos(u)) \right) \left(\frac{\partial}{\partial v} (r\,c\,\cos(u)) \right) =$$

$$- \left(\frac{\partial}{\partial x} P(x,y,z) \right) a^3 \sin(u)^2 \cos(v)^2 r^2 \cos(u) \sin(v)$$

$$+ \left(\frac{\partial}{\partial y} Q(x,y,z) \right) b^3 \sin(u)^2 \sin(v)^2 r^2 \cos(u) \cos(v) \tag{9}$$

//In Spacial Cartesian Coordinates, Divergence of abstract vector field '[P(x,y,z),Q(x,y,z),R(x,y,z)]' is

$$\frac{\partial}{\partial x} P(x,y,z) + \frac{\partial}{\partial y} Q(x,y,z) + \frac{\partial}{\partial z} R(x,y,z)$$

In logic deduction of this Proof, it is necessary to import divergence of abstract vector field above-mentioned to abstract simply connected orientable closed surface coordinates.

Three elements of abstract vector field's divergence: $\frac{\partial}{\partial x} P(x,y,z)$, $\frac{\partial}{\partial y} Q(x,y,z)$, $\frac{\partial}{\partial z} R(x,y,z)$ are abstract differential function, and their variables 'x,y,z' contain sub-variables 'r,u,v'.

Transform Expression between 'Spacial Cartesian coordinates' and 'Abstract simply connected orientable closed surface coordinates' is:

x = r*a*sin(u)cos(v), y = r*b*sin(u)sin(v), z = r*c*cos(u).
--"Coordinates Transform Differential Functions" (Correspond to

"Three Differential Variables $\frac{\partial}{\partial x}$, $\frac{\partial}{\partial y}$, $\frac{\partial}{\partial z}$' of Differential Functions

$\frac{\partial}{\partial \mathrm{x}} \mathrm{P}(x,y,z)$, $\frac{\partial}{\partial y} Q(x,y,z)$, $\frac{\partial}{\partial z} R(x,y,z)$'") are:

$$\frac{\partial}{\partial r}(ra\sin(u)\cos()) \quad \frac{\partial}{\partial u} r\sin(\ )\cos(\ ))\ v\ (\ \frac{\partial}{\partial v}\sin(u)\cos(\ ))\quad v \quad ,$$

$$\frac{\partial}{\partial r}(rb\sin(u)\sin(v))\frac{\partial}{\partial u}(rb\sin(u)\sin(v))\frac{\partial}{\partial v}(rb\sin(u)\sin(v)) \quad \text{and}$$

$$\frac{\partial}{\partial r}(rc\cos(u))\frac{\partial}{\partial u}(rc\cos(u))\frac{\partial}{\partial v}(rc\cos(u))$$

Product of "Differential Functions $\frac{\partial}{\partial \mathrm{x}} \mathrm{P}(x,y,z)$, $\frac{\partial}{\partial y} Q(x,y,z)$,

$\frac{\partial}{\partial z} R(x,y,z)$'" and "Coordinates Transform Differential Functions" (Viz. Product of two different functions) constitute "Divergence in Abstract Simply Connected Orientable Closed Surface Coordinates"

'Chain differentiation' or 'Coordinates conversion' ?

If it is 'Chain differentiation', according to principle of 'Multiply in identical chain, Plus in subdivision chain', it ought to be:

$$\left(\frac{\partial}{\partial x} \mathrm{P}(x,y,z)\right)$$

$$\left(\left(\frac{\partial}{\partial r}(r\,a\,\sin(u)\,\cos(v))\right)+\left(\frac{\partial}{\partial u}(r\,a\,\sin(u)\,\cos(v))\right)+\left(\frac{\partial}{\partial v}(r\,a\,\sin(u)\,\cos(v))\right)\right)$$

$$+\left(\frac{\partial}{\partial y}\mathrm{Q}(x,y,z)\right)$$

$$\left(\left(\frac{\partial}{\partial r}(r\,b\,\sin(u)\,\sin(v))\right)+\left(\frac{\partial}{\partial u}(r\,b\,\sin(u)\,\sin(v))\right)+\left(\frac{\partial}{\partial v}(r\,b\,\sin(u)\,\sin(v))\right)\right)$$

$$+\left(\frac{\partial}{\partial z}\mathrm{R}(x,y,z)\right)\left(\left(\frac{\partial}{\partial r}(r\,c\,\cos(u))\right)+\left(\frac{\partial}{\partial u}(r\,c\,\cos(u))\right)+\left(\frac{\partial}{\partial v}(r\,c\,\cos(u))\right)\right)$$

No matter 'Chain differentiation' or 'Solve divergence/curl', question for discussion is 'How to solve differential coefficient' or 'method of differentiation' about abstract vector field 'P[x,y,z], Q[x,y,z],R[x,y,z]'; and this step, question for discussion is 'How to convert result of divergence from a coordinates to another'; character and hiberarchy of 'two questions' above-mentioned are disparate. It is 'multiply', not 'plus' - be determined by

3-Dimensional spacial attribute of coordinates.

Triple Integrals of Product [Divergence (9) and Surface 'S' Coordinates Volume Element] in range of 'r,u,v':   (10)

$$
\int_0^{2\pi}\int_0^{\pi}\int_0^1 \left(-\left(\frac{\partial}{\partial x}P(x,y,z)\right)a^3\sin(u)^2\cos(v)^2 r^2\cos(u)\sin(v)\right.
$$

$$
\left.+\left(\frac{\partial}{\partial y}Q(x,y,z)\right)b^3\sin(u)^2\sin(v)^2 r^2\cos(u)\cos(v)\right)a\sin(u)\,r^2\,b\,c\,dr\,du\,dv =
$$

$$
\int_0^{2\pi}\int_0^{\pi}\frac{1}{5}\left(-\left(\frac{\partial}{\partial x}P(x,y,z)\right)a^3\sin(u)^2\cos(v)^2\cos(u)\sin(v)\right.
$$

$$
\left.+\left(\frac{\partial}{\partial y}Q(x,y,z)\right)b^3\sin(u)^2\sin(v)^2\cos(u)\cos(v)\right)a\sin(u)\,b\,c\,du\,dv = 0
$$

//Owing to universality and homogeneity of abstract vector field's

divergence '$\frac{\partial}{\partial x}P(x,y,z)+\frac{\partial}{\partial y}Q(x,y,z)+\frac{\partial}{\partial z}R(x,y,z)$', (or their elements

'$\frac{\partial}{\partial x}P(x,y,z)$, $\frac{\partial}{\partial y}Q(x,y,z)$, $\frac{\partial}{\partial z}R(x,y,z)$'), Integral by Embedded

Sub-Variable 'r,u,v' of Variable 'x,y,z', character of integral can be regarded as "Integral about Product [Differential Function Elements

of Divergence (Viz. $\frac{\partial}{\partial x}P(x,y,z)$, $\frac{\partial}{\partial y}Q(x,y,z)$, $\frac{\partial}{\partial z}R(x,y,z)$) 、 Coordinates

Transform Differential Functions 、 Surface 'S' coordinates Volume Element]". Integral by Embedded Sub-Variable 'r,u,v' of Variable 'x,y,z', Can't change structure of abstract divergence (Viz.

'$\frac{\partial}{\partial x}P(x,y,z)+\frac{\partial}{\partial y}Q(x,y,z)+\frac{\partial}{\partial z}R(x,y,z)$') or their Differential Function

Elements (Viz. $\frac{\partial}{\partial x}P(x,y,z)$, or $\frac{\partial}{\partial y}Q(x,y,z)$, or $\frac{\partial}{\partial z}R(x,y,z)$), they can

preserve primary form after integral.

Three Coordinates Transform Differential Functions (Correspond to

Three Differential Variables '$\frac{\partial}{\partial x}$, $\frac{\partial}{\partial y}$, $\frac{\partial}{\partial z}$'), Viz:

$$\frac{\partial}{\partial r}(ra\sin(u)\cos(v))\quad \frac{\partial}{\partial u}(r\sin(v)\cos(v))\,v\;(\;\frac{\partial}{\partial v}\sin(u)\cos(v))\quad v\qquad,$$

$$\frac{\partial}{\partial r}(rb\sin(u)\sin(v))\frac{\partial}{\partial u}(rb\sin(u)\sin(v))\frac{\partial}{\partial v}(rb\sin(u)\sin(v))\quad \text{and}$$

$$\frac{\partial}{\partial r}(rc\cos(u))\frac{\partial}{\partial u}(rc\cos(u))\frac{\partial}{\partial v}(rc\cos(u))$$

will be changed after integral.

(See also Chapter 4 (This Part) Section 4.1 'Prove Divergence Theorem Finite Sums Limits at Manifold' and Section 4.3 'Discuss About Abstract Vector Field、Finite Sums Limit、Integral and Riemann Sums')

// Be different to ordinary triple integrals, in this position, first integral range is 'r∈[0,1]' always, and not 'r∈[0,n] or r∈[0,∞]'; Because of '1', it is possible to hold a correct ratio between 'a,b,c,r,u,v'

// About 'Integral value is unequal to 0', See also Chapter 6 (This Part) Section 6.1 'Divergence Theorem at Manifold and Klein Bottle'

Thereinto:

$$\int_0^{2\pi}\int_0^{\pi}\int_0^1 a\,b\,c\,r^2\sin(u)\,dr\,du\,dv \neq 0$$

Viz. Suppose triple integrals of surface 'S' coordinates volume element in range of 'r,u,v' can't be '0';

Explain visually ,it can be regarded as "Suppose spacial bounded closed region 'Ω' can't be '0 volume'"

Viz. expression(5) = expression(10):

$$\int_0^{2\pi}\int_0^{\pi} P(x,y,z)\,c\,\sin(u)^2\,b\,\cos(v) + Q(x,y,z)\,\sin(u)^2\,a\,\sin(v)\,c$$
$$+ R(x,y,z)\,\sin(u)\,a\,b\,\cos(u)\,du\,dv$$

$$=$$

$$\int_0^{2\pi}\int_0^{\pi}\int_0^1 \left(-\left(\frac{\partial}{\partial x}P(x,y,z)\right)a^3\sin(u)^2\cos(v)^2\,r^2\cos(u)\sin(v)\right.$$

$$\left.+\left(\frac{\partial}{\partial y}Q(x,y,z)\right)b^3\sin(u)^2\sin(v)^2\,r^2\cos(u)\cos(v)\right)a\sin(u)\,r^2\,b\,c\,dr\,du\,dv$$

Above-mentioned equation can be described as:

$$\iint\limits_S A \cdot n\, dS = \iiint\limits_\Omega \mathrm{div} A\, d\,\omega \qquad (1), \text{ Complete Proof.}$$

2.2 Prove Divergence Theorem at Manifold
   ### [Program Template of Waterloo Maplesoft, Optional]

Divergence Theorem

Suppose spacial bounded closed region 'Ω' is surrounded by smooth or piecewise smooth closed surface 'S'. Functions 'P(x,y,z),Q(x,y,z), R(x,y,z)' [Structure vector field 'A'] and its partial derivatives are continuous at spacial bounded closed region 'Ω', then:

$$\iint\limits_S A \cdot n\, dS = \iiint\limits_\Omega \mathrm{div} A \cdot d\omega \qquad (1)$$

thereinto, surface 'S' is outer side of special bounded closed region 'Ω's entire boundary surface, 'n' is unit normal vector of surface 'S's outer side, and 'divA' is divergence of vector field 'A'.

Symbol System:

Vector Field 'V',

Divergence 'diV1,diV2' of Vector Field 'V';

Abstract simply connected orientable closed parametrized surface 'CS' [Closed],

Tangent Plane Normal Vector '[A,B,C]' of Surface 'CS'

Spacial bounded closed region 'Ω' that Surface 'CS' surrounds,

General expression of surface 'CS' coordinates volume element coefficient 'J'

```
> restart; # Reset computer algebraic system
> with(linalg): # Load 'linear algebra' package
> CS:=[a*sin(u)*cos(v),b*sin(u)*sin(v),c*cos(u)];
```

Define parametrized expression of abstract simply connected orientable closed surface 'CS', 'a,b,c' are nonzero constants or 1th order derivable continuous functions, simply connected orientable closed surface 'CS' determines the value of 'a,b,c'. (See also Poincare Conjecture: 'Every closed n-Manifold which is homotopy equivalent to

n-Sphere is homeomorphic to n-Sphere'[19], and Chapter 1 this Part 'Constitute Abstract Simply Connected Orientable Closed Parametrized Surface Coordinates')

$$CS := [a\sin(u)\cos(v),\, b\sin(u)\sin(v),\, c\cos(u)]$$

> rgu:=[0,Pi];

$$rgu := [0, \pi]$$

> rgv:=[0,2*Pi];

Set range of parameters 'u,v', make surface 'CS' closed

$$rgv := [0, 2\pi]$$

> V:=[(P)(x,y,z),(Q)(x,y,z),(R)(x,y,z)];

 # Define abstract spacial vector field 'V' (Suppose Vector Field 'V' possesses 1th order continuous partial derivatives in spacial bounded closed region 'Ω' that Surface 'CS' surrounds)

$$V := [\mathrm{P}(x,y,z),\, \mathrm{Q}(x,y,z),\, \mathrm{R}(x,y,z)]$$

>Diff(V[1],x)+Diff(V[2],y)+Diff(V[3],z)=diff(V[1],x)+diff(V[2],y)
+diff(V[3],z);diV1:=rhs(%);

Calculate divergence 'diV1' of abstract vector field 'V'

$$\left(\frac{\partial}{\partial x}\mathrm{P}(x,y,z)\right)+\left(\frac{\partial}{\partial y}\mathrm{Q}(x,y,z)\right)+\left(\frac{\partial}{\partial z}\mathrm{R}(x,y,z)\right)=$$
$$\left(\frac{\partial}{\partial x}\mathrm{P}(x,y,z)\right)+\left(\frac{\partial}{\partial y}\mathrm{Q}(x,y,z)\right)+\left(\frac{\partial}{\partial z}\mathrm{R}(x,y,z)\right)$$
$$diV1 := \left(\frac{\partial}{\partial x}\mathrm{P}(x,y,z)\right)+\left(\frac{\partial}{\partial y}\mathrm{Q}(x,y,z)\right)+\left(\frac{\partial}{\partial z}\mathrm{R}(x,y,z)\right)$$

> x:=CS[1]:y:=CS[2]:z:=CS[3]:

>matrix(3,3,[i,j,k,Diff(CS[1],u),Diff(CS[2],u),Diff(CS[3],u),
Diff(CS[1],v),Diff(CS[2],v),Diff(CS[3],v)])=
matrix(3,3,[i,j,k,diff(CS[1],u),diff(CS[2],u),diff(CS[3],u),
diff(CS[1],v),diff(CS[2],v),diff(CS[3],v)]);m1:=rhs(%);

Define and calculate matrix of partial derivatives 'm1', obtain 'Tangent plane normal vector' of surface 'CS'

$$\begin{bmatrix} i & j & k \\[4pt] \frac{\partial}{\partial u}(a\sin(u)\cos(v)) & \frac{\partial}{\partial u}(b\sin(u)\sin(v)) & \frac{\partial}{\partial u}(c\cos(u)) \\[4pt] \frac{\partial}{\partial v}(a\sin(u)\cos(v)) & \frac{\partial}{\partial v}(b\sin(u)\sin(v)) & \frac{\partial}{\partial v}(c\cos(u)) \end{bmatrix} =$$

$$\begin{bmatrix} i & j & k \\ a\cos(u)\cos(v) & b\cos(u)\sin(v) & -c\sin(u) \\ -a\sin(u)\sin(v) & b\sin(u)\cos(v) & 0 \end{bmatrix}$$

$$m1 := \begin{bmatrix} i & j & k \\ a\cos(u)\cos(v) & b\cos(u)\sin(v) & -c\sin(u) \\ -a\sin(u)\sin(v) & b\sin(u)\cos(v) & 0 \end{bmatrix}$$

```
> det(m1);
```

$$i\,c\sin(u)^2\,b\cos(v) + a\cos(u)\cos(v)^2\,k\,b\sin(u) + a\sin(u)^2\sin(v)\,j\,c$$
$$+ a\sin(u)\sin(v)^2\,k\,b\cos(u)$$

```
> mn:=simplify(%);
```

$$mn := \sin(u)\,(i\,c\sin(u)\,b\cos(v) + a\sin(u)\sin(v)\,j\,c + a\,k\,b\cos(u))$$

```
> A:=coeff(mn,i);  # Obtain coefficient of 'i'
```

$$A := c\sin(u)^2\,b\cos(v)$$

```
> B:=coeff(mn,j);  # Obtain coefficient of 'j'
```

$$B := \sin(u)^2\,a\sin(v)\,c$$

```
> C:=coeff(mn,k);
# Obtain coefficient of 'k'
```

$$C := \sin(u)\,a\,b\cos(u)$$

```
> [A,B,C]; # [A,B,C] structure 'Tangent plane normal vector'
```

$$[c\sin(u)^2\,b\cos(v),\ \sin(u)^2\sin(v)\,a\,c,\ \sin(u)\,a\cos(u)\,b]$$

```
>Int(Int(V[1]*A+V[2]*B+V[3]*C,u=rgu[1]..rgu[2]),v=rgv[1]..rgv[2]);
# Integral of dot product (vector field'V' and tangent plane normal
vector [A,B,C] of surface 'CS') in range of parameters 'u,v'
```

$$\int_0^{2\pi}\int_0^{\pi} P(a\sin(u)\cos(v), b\sin(u)\sin(v), c\cos(u))\,c\sin(u)^2\,b\cos(v)$$
$$+ Q(a\sin(u)\cos(v), b\sin(u)\sin(v), c\cos(u))\sin(u)^2\,a\sin(v)\,c$$
$$+ R(a\sin(u)\cos(v), b\sin(u)\sin(v), c\cos(u))\sin(u)\,a\,b\cos(u)\,du\,dv$$

```
> value(%);
```

$$2\pi\int_0^{\pi} P(a\sin(\_X)\cos(\_X), b\sin(\_X)^2, c\cos(\_X))\,c\sin(\_X)^2\,b\cos(\_X)$$
$$+ \sin(\_X)\,Q(a\sin(\_X)\cos(\_X), b\sin(\_X)^2, c\cos(\_X))\,a\,c$$
$$- \sin(\_X)\,Q(a\sin(\_X)\cos(\_X), b\sin(\_X)^2, c\cos(\_X))\,a\,c\cos(\_X)^2$$

$$+ R(a \sin(\_X) \cos(\_X), b \sin(\_X)^2, c \cos(\_X)) \sin(\_X) \, a \, b \cos(\_X) \, d\_X$$

```
# Import 'x = a*sin(u)*cos(v),y = b*sin(u)*sin(v),z = c*cos(u)' to
abstract vector field '[P(x,y,z), Q(x,y,z), R(x,y,z)]', obtain
insoluble result, abnegate it


> x:='x':y:='y':z:='z':
> Int(Int(V[1]*A+V[2]*B+V[3]*C,u=rgu[1]..rgu[2]),v=rgv[1]..rgv[2])
=Int(int(V[1]*A+V[2]*B+V[3]*C,u=rgu[1]..rgu[2]),v=rgv[1]..rgv[2]);
# Integral of dot product (vector field'V' and tangent plane normal
vector [A,B,C] of  surface 'CS', reserve original form of abstract
vector field 'P(x,y,z),Q(x,y,z),R(x,y,z)') in range of 'u,v'
```

$$\int_0^{2\pi} \int_0^{\pi} P(x,y,z) \, c \sin(u)^2 \, b \cos(v) + Q(x,y,z) \sin(u)^2 \, a \sin(v) \, c$$

$$+ R(x,y,z) \sin(u) \, a \, b \cos(u) \, du \, dv =$$

$$\int_0^{2\pi} \frac{1}{2} Q(x,y,z) \, a \sin(v) \, c \, \pi + \frac{1}{2} P(x,y,z) \, c \, b \cos(v) \, \pi \, dv$$

```
> alpha:=rhs(value(%)); # Calculate value of surface integral
```
$$\alpha := 0$$ `# Obtain a constant '0'`
```
# Owing to universality and homogeneity of abstract vector field
[P(x,y,z),Q(x,y,z),R(x,y,z)],Integral by Embedded Sub-Variable 'u,v'
of Variable 'x,y,z', Result of Integral can be described as 'P(x,y,z)
[or Q(x,y,z),or R(x,y,z)]'.
```

In other words, Integral by Embedded Sub-Variable 'u,v' of Variable 'x,y,z', Can't change structure of abstract function P(x,y,z)[or Q(x,y,z),or R(x,y,z)], so that abstract function structure P(x,y,z) [or Q(x,y,z), or R(x,y,z)] can preserve primary form after integral

[See also Chapter 4 (This Part) Section 4.1 'Prove Divergence Theorem Finite Sums Limits at Manifold' and Section 4.3 'Discuss About Abstract Vector Field、Finite Sums Limit、Integral and Riemann Sums']

```
> x:='x':y:='y':z:='z':
> [r*CS[1],r*CS[2],r*CS[3]];
# Entirely multiply radius vector 'r' (suppose r>0) by each item of
```

surface 'CS' parametrized expression, Convert 'Parametrized expression of surface' to 'Parametrized expression of surface coordinates'

$$[\, r\, a\, \sin(u)\cos(v),\, r\, b\, \sin(u)\sin(v),\, r\, c\, \cos(u)\,]$$

```
> x:=r*CS[1]:y:=r*CS[2]:z:=r*CS[3]:
> matrix(3,3,[Diff(r*CS[1],r),Diff(r*CS[1],u),Diff(r*CS[1],v),
Diff(r*CS[2],r),Diff(r*CS[2],u),Diff(r*CS[2],v),
Diff(r*CS[3],r),Diff(r*CS[3],u),Diff(r*CS[3],v)])=
matrix(3,3,[diff(r*CS[1],r),diff(r*CS[1],u),diff(r*CS[1],v),
diff(r*CS[2],r),diff(r*CS[2],u),diff(r*CS[2],v),
diff(r*CS[3],r),diff(r*CS[3],u),diff(r*CS[3],v)]);m2:=rhs(%);
# Define and calculate matrix of partial derivatives 'm2', obtain
general expression of surface coordinates volume element coefficient
```

$$\begin{bmatrix} \dfrac{\partial}{\partial r}(r\,a\,\sin(u)\cos(v)) & \dfrac{\partial}{\partial u}(r\,a\,\sin(u)\cos(v)) & \dfrac{\partial}{\partial v}(r\,a\,\sin(u)\cos(v)) \\[2mm] \dfrac{\partial}{\partial r}(r\,b\,\sin(u)\sin(v)) & \dfrac{\partial}{\partial u}(r\,b\,\sin(u)\sin(v)) & \dfrac{\partial}{\partial v}(r\,b\,\sin(u)\sin(v)) \\[2mm] \dfrac{\partial}{\partial r}(r\,c\,\cos(u)) & \dfrac{\partial}{\partial u}(r\,c\,\cos(u)) & \dfrac{\partial}{\partial v}(r\,c\,\cos(u)) \end{bmatrix} =$$

$$\begin{bmatrix} a\sin(u)\cos(v) & r\,a\cos(u)\cos(v) & -r\,a\sin(u)\sin(v) \\ b\sin(u)\sin(v) & r\,b\cos(u)\sin(v) & r\,b\sin(u)\cos(v) \\ c\cos(u) & -r\,c\sin(u) & 0 \end{bmatrix}$$

$$m2 := \begin{bmatrix} a\sin(u)\cos(v) & r\,a\cos(u)\cos(v) & -r\,a\sin(u)\sin(v) \\ b\sin(u)\sin(v) & r\,b\cos(u)\sin(v) & r\,b\sin(u)\cos(v) \\ c\cos(u) & -r\,c\sin(u) & 0 \end{bmatrix}$$

```
> det(m2);
```

$$a\sin(u)^3\cos(v)^2 r^2 b\,c + b\sin(u)^3\sin(v)^2 r^2 a\,c + c\cos(u)^2 r^2 a\cos(v)^2 b\sin(u)$$
$$+ c\cos(u)^2 r^2 a\sin(u)\sin(v)^2 b$$

```
> J:=simplify(%);
# General expression of surface coordinates volume element
coefficient
```

$$J := a\sin(u)\,r^2\,b\,c$$

```
> x:='x':y:='y':z:='z':
> Diff(V[1],x)*Diff(r*CS[1],r)*Diff(r*CS[1],u)*Diff(r*CS[1],v)
+Diff(V[2],y)*Diff(r*CS[2],r)*Diff(r*CS[2],u)*Diff(r*CS[2],v)
```

```
+Diff(V[3],z)*Diff(r*CS[3],r)*Diff(r*CS[3],u)*Diff(r*CS[3],v);
diV2:=value(%);
# Convert 'diV1' from Cartesian Coordinates expression to Surface
Coordinates expression
# Reserve original form of abstract vector field '[P(x,y,z),Q(x,y,z),
R(x,y,z)]' and its partial derivatives (insoluble);Calculate partial
derivatives of 'r*a*sin(u)*cos(v), r*b*sin(u)*sin(v), r*c*cos(u)'
(Computable), obtain a new expression 'diV2' (Divergence)
```

$$\left(\frac{\partial}{\partial x}P(x,y,z)\right)\left(\frac{\partial}{\partial r}(r\,a\,\sin(u)\cos(v))\right)\left(\frac{\partial}{\partial u}(r\,a\,\sin(u)\cos(v))\right)$$

$$\left(\frac{\partial}{\partial v}(r\,a\,\sin(u)\cos(v))\right)+\left(\frac{\partial}{\partial y}Q(x,y,z)\right)\left(\frac{\partial}{\partial r}(r\,b\,\sin(u)\sin(v))\right)$$

$$\left(\frac{\partial}{\partial u}(r\,b\,\sin(u)\sin(v))\right)\left(\frac{\partial}{\partial v}(r\,b\,\sin(u)\sin(v))\right)$$

$$+\left(\frac{\partial}{\partial z}R(x,y,z)\right)\left(\frac{\partial}{\partial r}(r\,c\,\cos(u))\right)\left(\frac{\partial}{\partial u}(r\,c\,\cos(u))\right)\left(\frac{\partial}{\partial v}(r\,c\,\cos(u))\right)$$

$$diV2:=-\left(\frac{\partial}{\partial x}P(x,y,z)\right)a^3\sin(u)^2\cos(v)^2\,r^2\cos(u)\sin(v)$$

$$+\left(\frac{\partial}{\partial y}Q(x,y,z)\right)b^3\sin(u)^2\sin(v)^2\,r^2\cos(u)\cos(v)$$

```
> Int(Int(Int(diV2*J,r=0..1),u=rgu[1]..rgu[2]),v=rgv[1]..rgv[2]);
# Triple Integrals of Product (Divergence 'diV2' and Surface
Coordinates Volume Element) in range of 'r,u,v'
```

$$\int_0^{2\pi}\int_0^{\pi}\int_0^1\left(\left(-\left(\frac{\partial}{\partial x}P(x,y,z)\right)a^3\sin(u)^2\cos(v)^2\,r^2\cos(u)\sin(v)\right.\right.$$

$$\left.\left.+\left(\frac{\partial}{\partial y}Q(x,y,z)\right)b^3\sin(u)^2\sin(v)^2\,r^2\cos(u)\cos(v)\right)a\sin(u)\,r^2\,b\,c\,dr\,du\,dv\right.$$

```
> beta:=value(%); # Calculate value of volume element integral
```

$$\beta:=0 \quad \text{\# Obtain a constant '0'}$$

```
# Owing to universality and homogeneity of abstract vector field's
```

divergence ' $\dfrac{\partial}{\partial x}P(x,y,z)+\dfrac{\partial}{\partial y}Q(x,y,z)+\dfrac{\partial}{\partial z}R(x,y,z)$ ', (or their elements

' $\dfrac{\partial}{\partial x}P(x,y,z)$, $\dfrac{\partial}{\partial y}Q(x,y,z)$, $\dfrac{\partial}{\partial z}R(x,y,z)$ '), Integral by Embedded

Sub-Variable 'r,u,v' of Variable 'x,y,z', character of integral can be regarded as "Integral about Product [Differential Function Elements of Divergence (Viz. $\dfrac{\partial}{\partial x}P(x,y,z)$, $\dfrac{\partial}{\partial y}Q(x,y,z)$, $\dfrac{\partial}{\partial z}R(x,y,z)$) 、 Coordinates Transform Differential Functions 、 Surface 'S' coordinates Volume Element]". Integral by Embedded Sub-Variable 'r,u,v' of Variable 'x,y,z', Can't change structure of abstract divergence (Viz. '$\dfrac{\partial}{\partial x}P(x,y,z)+\dfrac{\partial}{\partial y}Q(x,y,z)+\dfrac{\partial}{\partial z}R(x,y,z)$') or their Differential Function Elements (Viz. $\dfrac{\partial}{\partial x}P(x,y,z)$, or $\dfrac{\partial}{\partial y}Q(x,y,z)$, or $\dfrac{\partial}{\partial z}R(x,y,z)$), they can preserve primary form after integral.

Three Coordinates Transform Differential Functions (Correspond to Three Differential Variables '$\dfrac{\partial}{\partial x}$, $\dfrac{\partial}{\partial y}$, $\dfrac{\partial}{\partial z}$'), Viz:

$$\frac{\partial}{\partial r}(ra\sin(u)\cos(\ )) \quad \frac{\partial}{\partial u}(r\sin(\ )\cos(\ ))\,v\;(\;\frac{\partial}{\partial v}\sin(\ )\cos(\ ))\quad v\quad,$$

$$\frac{\partial}{\partial r}(rb\sin(u)\sin(v))\frac{\partial}{\partial u}(rb\sin(u)\sin(v))\frac{\partial}{\partial v}(rb\sin(u)\sin(v)) \quad \text{and}$$

$$\frac{\partial}{\partial r}(rc\cos(u))\frac{\partial}{\partial u}(rc\cos(u))\frac{\partial}{\partial v}(rc\cos(u))$$

will be changed after integral.

[See also Chapter 4 (This Part) Section 4.1 'Prove Divergence Theorem Finite Sums Limits at Manifold' and Section 4.3 'Discuss About Abstract Vector Field、Finite Sums Limit、Integral and Riemann Sums']

Thereinto:

$$\int_0^{2\pi}\int_0^{\pi}\int_0^1 a\,b\,c\,r^2\sin(u)\,dr\,du\,dv \neq 0$$

Viz:

$$\int_0^{2\pi}\int_0^{\pi} P(x,y,z)\,c\,\sin(u)^2\,b\,\cos(v) + Q(x,y,z)\,\sin(u)^2\,a\,\sin(v)\,c$$

$$+ R(x,y,z)\,\sin(u)\,a\,b\,\cos(u)\,du\,dv$$

$$=$$

$$\int_0^{2\pi} \int_0^{\pi} \int_0^1 \left(\left(-\left(\frac{\partial}{\partial x}P(x,y,z)\right)a^3\sin(u)^2\cos(v)^2 r^2\cos(u)\sin(v)\right.\right.$$

$$\left.\left.+\left(\frac{\partial}{\partial y}Q(x,y,z)\right)b^3\sin(u)^2\sin(v)^2 r^2\cos(u)\cos(v)\right)a\sin(u)r^2 b\,c\,dr\,du\,dv\right.$$

Value of 'Closed Surface Integral' is equal to 'Volume Element Integral', Complete Proof.

2.3 Prove Divergence Theorem In Torus Coordinates
(Multiple Connected Orientable Closed Surface Coordinates)

Divergence Theorem

Suppose spacial bounded closed region 'Ω' is surrounded by smooth or piecewise smooth closed surface 'S'. Functions 'P(x,y,z),Q(x,y,z), R(x,y,z)'[Structure vector field 'A'] and its partial derivatives are continuous at spacial bounded closed region 'Ω', then:

$$\iint_S A\cdot n\,dS = \iiint_\Omega divA\cdot d\omega \qquad (1)$$

Thereinto, surface 'S' is outer side of spacial bounded closed region 'Ω''s entire boundary surface, 'n' is unit normal vector of surface 'S's outer side, and 'divA' is divergence of vector field 'A'.

Proof:

Define parametrized expression of torus surface 'S':

[(2+cos(u))cos(v),(2+cos(u))sin(v),sin(u)]   (2)

Set range of parameters 'u,v' in [0,2π],[0,2π], make torus surface 'S' closed.

According to torus surface parametrized expression (2),define and calculate first matrix of partial derivatives, obtain tangent plane normal vector of torus surface 'S':   (3)

$$\begin{bmatrix} i & j & k \\ \dfrac{\partial}{\partial u}(2+\cos(u))\cos(v) & \dfrac{\partial}{\partial u}(2+\cos(u))\sin(v) & \dfrac{\partial}{\partial u}\sin(u) \\ \dfrac{\partial}{\partial v}(2+\cos(u))\cos(v) & \dfrac{\partial}{\partial v}(2+\cos(u))\sin(v) & \dfrac{\partial}{\partial v}\sin(u) \end{bmatrix} =$$

$$-2\,i\cos(u)\cos(v) - i\cos(u)^2\cos(v) - 2\sin(u)\cos(v)^2\,k - \sin(u)\cos(v)^2\,k\cos(u)$$
$$-2\sin(v)\,j\cos(u) - 2\,k\sin(u)\sin(v)^2 - \sin(v)\,j\cos(u)^2 - \cos(u)\,k\sin(u)\sin(v)^2$$

From expression (3), respectively pick up coefficients of item 'i,j,k',obtain tangent plane normal vector of torus surface 'S':(4)

$$[-2\cos(u)\cos(v) - \cos(u)^2\cos(v),\, -2\sin(v)\cos(u) - \sin(v)\cos(u)^2,\, -2\sin(u) - \cos(u)\sin(u)]$$

Surface Integral of Dot Product [Vetor field 'A' and Tangent Plane Normal Vector (4)] in range of 'u,v':

$$\int_0^{2\pi}\int_0^{2\pi} P(x,y,z)\,(-2\cos(u)\cos(v) - \cos(u)^2\cos(v))$$
$$+ Q(x,y,z)\,(-2\sin(v)\cos(u) - \sin(v)\cos(u)^2)$$
$$+ R(x,y,z)\,(-2\sin(u) - \cos(u)\sin(u))\,du\,dv$$

$$=$$

$$\int_0^{2\pi} -P(x,y,z)\cos(v)\,\pi - Q(x,y,z)\sin(v)\,\pi\,dv = 0 \qquad (5)$$

Entirely multiply radius vector 'r' (suppose r>0) by each item of torus surface 'S' parametrized expression (2), convert "Parametrized expression of torus surface 'S'" to "Parametrized expression of torus surface 'S' coordinates":

[r*(2+cos(u))cos(v), r*(2+cos(u))sin(v), r*sin(u)] (6)

According to parametrized expression of torus surface 'S' coordinates (6), define and calculate second matrix of partial derivatives, obtain general expression of torus surface 'S' coordinates volume element coefficient:

$$
\begin{bmatrix}
\dfrac{\partial}{\partial r} r(2+\cos(u))\cos(v) & \dfrac{\partial}{\partial u} r(2+\cos(u))\cos(v) & \dfrac{\partial}{\partial v} r(2+\cos(u))\cos(v) \\[3mm]
\dfrac{\partial}{\partial r} r(2+\cos(u))\sin(v) & \dfrac{\partial}{\partial u} r(2+\cos(u))\sin(v) & \dfrac{\partial}{\partial v} r(2+\cos(u))\sin(v) \\[3mm]
\dfrac{\partial}{\partial r} r\sin(u) & \dfrac{\partial}{\partial u} r\sin(u) & \dfrac{\partial}{\partial v} r\sin(u)
\end{bmatrix}
$$

$$
= \; -r^2(5\cos(u)+2cos(u)^2+2) \tag{7}
$$

Calculate divergence of vector field 'A', and convert divergence from Cartesian Coordinates expression(8) to Torus Surface Coordinates expression(9):

$$
\frac{\partial}{\partial x}P(x,y,z)+\frac{\partial}{\partial y}Q(x,y,z)+\frac{\partial}{\partial z}R(x,y,z) \tag{8}
$$

$$
\left(\frac{\partial}{\partial x}P(x,y,z)\right)\left(\frac{\partial}{\partial r}(r(2+\cos(u))\cos(v))\right)\left(\frac{\partial}{\partial u}(r(2+\cos(u))\cos(v))\right)
$$

$$
\left(\frac{\partial}{\partial v}(r(2+\cos(u))\cos(v))\right)+\left(\frac{\partial}{\partial y}Q(x,y,z)\right)\left(\frac{\partial}{\partial r}(r(2+\cos(u))\sin(v))\right)
$$

$$
\left(\frac{\partial}{\partial u}(r(2+\cos(u))\sin(v))\right)\left(\frac{\partial}{\partial v}(r(2+\cos(u))\sin(v))\right)
$$

$$
+\left(\frac{\partial}{\partial z}R(x,y,z)\right)\left(\frac{\partial}{\partial r}(r\sin(u))\right)\left(\frac{\partial}{\partial u}(r\sin(u))\right)\left(\frac{\partial}{\partial v}(r\sin(u))\right)=
$$

$$
\left(\frac{\partial}{\partial x}P(x,y,z)\right)(2+\cos(u))^2\cos(v)^2 r^2\sin(u)\sin(v)
$$

$$
-\left(\frac{\partial}{\partial y}Q(x,y,z)\right)(2+\cos(u))^2\sin(v)^2 r^2\sin(u)\cos(v) \tag{9}
$$

Triple Integrals of Product [Divergence(9) and Torus Surface Coordinates Volume Element] in range of 'r,u,v':   (10)

$$
\int_0^{2\pi}\int_0^{2\pi}\int_0^1 -\left(\left(\frac{\partial}{\partial x}P(x,y,z)\right)(2+\cos(u))^2\cos(v)^2 r^2\sin(u)\sin(v)\right.
$$

$$
\left.-\left(\frac{\partial}{\partial y}Q(x,y,z)\right)(2+\cos(u))^2\sin(v)^2 r^2\sin(u)\cos(v)\right)r^2
$$

$$
(5\cos(u)+2\cos(u)^2+2)\;dr\,du\,dv
$$

$$
=
$$

$$
\int_0^{2\pi}\int_0^{2\pi} -\frac{1}{5}\left(\left(\frac{\partial}{\partial x}P(x,y,z)\right)(2+\cos(u))^2\cos(v)^2\sin(u)\sin(v)\right.
$$

$$-\left(\frac{\partial}{\partial y}Q(x,y,z)\right)(2+\cos(u))^2\sin(v)^2\sin(u)\cos(v)\Bigg)(5\cos(u)+2\cos(u)^2+2)$$
$$du\ dv=0$$

// Be different to ordinary triple integrals, in this position, first integral range is 'r∈[0,1]' always, and not 'r∈[0,n] or r∈[0,∞]'; Because of '1', it is possible to hold a correct ratio between 'r,u,v'.

// About 'Integral value is unequal to 0', See also Chapter 6 (This Part) Section 6.1'Divergence Theorem at Manifold and Klein Bottle'

Viz. expression(5) = expression(10):

$$\int_0^{2\pi}\int_0^{2\pi} P(x,y,z)\,(-2\cos(u)\cos(v)-\cos(u)^2\cos(v))$$
$$+\,Q(x,y,z)\,(-2\sin(v)\cos(u)-\sin(v)\cos(u)^2)$$
$$+\,R(x,y,z)\,(-2\sin(u)-\cos(u)\sin(u))\,du\ dv$$

$$=$$

$$\int_0^{2\pi}\int_0^{2\pi}\int_0^1 -\left(\left(\frac{\partial}{\partial x}P(x,y,z)\right)(2+\cos(u))^2\cos(v)^2\,r^2\sin(u)\sin(v)\right.$$

$$\left.-\left(\frac{\partial}{\partial y}Q(x,y,z)\right)(2+\cos(u))^2\sin(v)^2\,r^2\sin(u)\cos(v)\right)r^2$$
$$(5\cos(u)+2\cos(u)^2+2)\,dr\ du\ dv$$

Above-mentioned equation can be described as:

$$\iint_S A\cdot n\,dS=\iiint_\Omega divA\ d\ \omega\quad (1),\ \text{Complete Proof.}$$

2.4 Prove Divergence Theorem In Torus Coordinates
(Multiple Connected Orientable Closed Surface Coordinates)
[Program Template of Waterloo Maplesoft, Optional]

Divergence Theorem

Suppose spacial bounded closed region 'Ω' is surrounded by smooth or piecewise smooth closed surface 'S'. Functions 'P(x,y,z),Q(x,y,z), R(x,y,z)' [Structure vector field 'A'] and its partial derivatives are continuous at spacial bounded closed region 'Ω', then:

$$\iint\limits_{S} A \cdot n\, dS = \iiint\limits_{\Omega} divA \cdot d\omega \qquad (1)$$

Thereinto, surface 'S' is outer side of spacial bounded closed region 'Ω's entire boundary surface, 'n' is unit normal vector of surface 'S's outer side, and 'divA' is divergence of vector field 'A'.

Symbol System:

Vector Field 'V',

Divergence 'diV1,diV2' of Vector Field 'V';

Torus surface 'CS' [Closed],

Tangent Plane Normal Vector '[A,B,C]' of Torus Surface 'CS'

Spacial bounded closed region ' Ω ' that Torus Surface 'CS' surrounds,

General expression of Torus Surface 'CS' coordinates volume element coefficient 'J'

```
> restart;
> with(plots):with(linalg):
> CS:=[(2+cos(u))*cos(v),(2+cos(u))*sin(v),sin(u)];
# Define parametrized expression of torus surface 'CS'
```
$$CS := [(2 + \cos(u))\cos(v), (2 + \cos(u))\sin(v), \sin(u)]$$
```
> rgu:=[0,2*Pi];
```
$$rgu := [0, 2\pi]$$
```
> rgv:=[0,2*Pi];
# Set range of parameters 'u,v', make torus surface 'CS' closed
```
$$rgv := [0, 2\pi]$$
```
>plot3d(CS,u=rgu[1]..rgu[2],v=rgv[1]..rgv[2],scaling=constrained,
projection=0.9,numpoints=3000);
```

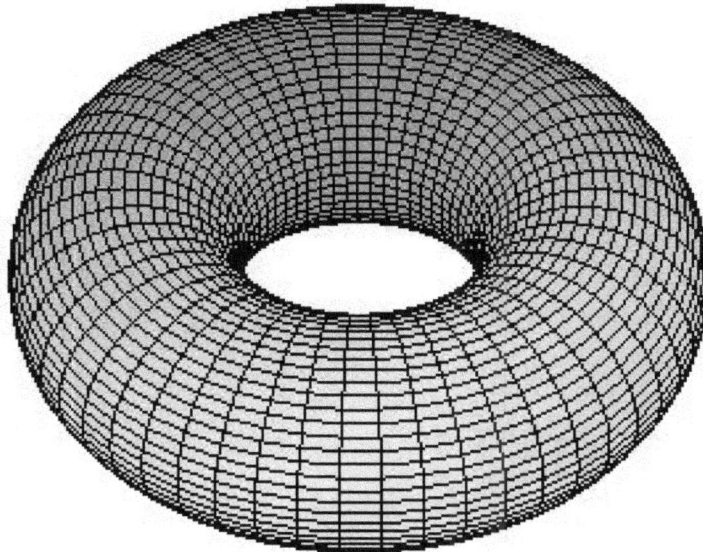

Figure(I)2.4 Torus Surface 'CS'

```
> V:=[(P)(x,y,z),(Q)(x,y,z),(R)(x,y,z)];
```
Define abstract spacial vector field 'V'(Suppose Vector Field 'V'
possesses 1th order continuous partial derivatives in spacial bounded
closed region 'Ω' that Torus Surface 'CS' surrounds)

$$V := [P(x, y, z), Q(x, y, z), R(x, y, z)]$$

```
> Diff(V[1],x)+Diff(V[2],y)+Diff(V[3],z)
=diff(V[1],x)+diff(V[2],y)+diff(V[3],z);diV1:=rhs(%);
```
Calculate divergence 'diV1 of abstract vector field 'V'

$$\left(\frac{\partial}{\partial x}P(x,y,z)\right)+\left(\frac{\partial}{\partial y}Q(x,y,z)\right)+\left(\frac{\partial}{\partial z}R(x,y,z)\right)=$$

$$\left(\frac{\partial}{\partial x}P(x,y,z)\right)+\left(\frac{\partial}{\partial y}Q(x,y,z)\right)+\left(\frac{\partial}{\partial z}R(x,y,z)\right)$$

$$diV1 := \left(\frac{\partial}{\partial x}P(x,y,z)\right)+\left(\frac{\partial}{\partial y}Q(x,y,z)\right)+\left(\frac{\partial}{\partial z}R(x,y,z)\right)$$

```
> x:=CS[1]:y:=CS[2]:z:=CS[3]:
> matrix(3,3,[i,j,k,Diff(CS[1],u),Diff(CS[2],u),Diff(CS[3],u),
Diff(CS[1],v),Diff(CS[2],v),Diff(CS[3],v)])=
```

```
matrix(3,3,[i,j,k,diff(CS[1],u),diff(CS[2],u),diff(CS[3],u),
diff(CS[1],v),diff(CS[2],v),diff(CS[3],v)]);m1:=rhs(%);
```

Define and calculate matrix of partial derivatives 'm1', obtain
'Tangent plane normal vector' of Torus surface 'CS'

$$
\begin{bmatrix}
i & j & k \\
\dfrac{\partial}{\partial u}((2+\cos(u))\cos(v)) & \dfrac{\partial}{\partial u}((2+\cos(u))\sin(v)) & \dfrac{d}{du}\sin(u) \\
\dfrac{\partial}{\partial v}((2+\cos(u))\cos(v)) & \dfrac{\partial}{\partial v}((2+\cos(u))\sin(v)) & \dfrac{\partial}{\partial v}\sin(u)
\end{bmatrix} =
$$

$$
\begin{bmatrix}
i & j & k \\
-\sin(u)\cos(v) & -\sin(u)\sin(v) & \cos(u) \\
-(2+\cos(u))\sin(v) & (2+\cos(u))\cos(v) & 0
\end{bmatrix}
$$

$$
m1 := \begin{bmatrix}
i & j & k \\
-\sin(u)\cos(v) & -\sin(u)\sin(v) & \cos(u) \\
-(2+\cos(u))\sin(v) & (2+\cos(u))\cos(v) & 0
\end{bmatrix}
$$

```
> det(m1);
```

$$
-2\,i\cos(u)\cos(v)-i\cos(u)^2\cos(v)-2\sin(u)\cos(v)^2 k - \sin(u)\cos(v)^2 k\cos(u)
$$
$$
-2\sin(v)j\cos(u)-2\,k\sin(u)\sin(v)^2 - \sin(v)j\cos(u)^2 - \cos(u)k\sin(u)\sin(v)^2
$$

```
> mn:=simplify(%);
```

$$
mn := -2\,i\cos(u)\cos(v)-i\cos(u)^2\cos(v)-2\sin(v)j\cos(u)-2\,k\sin(u)
$$
$$
-\sin(v)j\cos(u)^2 - \cos(u)k\sin(u)
$$

```
> A:=coeff(mn,i); # Obtain coefficient of 'i'
```

$$
A := -2\cos(u)\cos(v)-\cos(u)^2\cos(v)
$$

```
> B:=coeff(mn,j); # Obtain coefficient of 'j'
```

$$
B := -2\sin(v)\cos(u)-\sin(v)\cos(u)^2
$$

```
> C:=coeff(mn,k);
```

Obtain coefficient of 'k', [A,B,C] structure 'Tangent plane normal
vector'

$$
C := -2\sin(u)-\cos(u)\sin(u)
$$

```
> [A,B,C]; # [A,B,C] structure 'Tangent plane normal vector'
```

$$
[-\cos(u)^2\cos(v)-2\cos(u)\cos(v),\, -\cos(u)^2\sin(v)-2\cos(u)\sin(v),
$$
$$
-\sin(u)\cos(u)-2\sin(u)]
$$

```
> Int(Int(V[1]*A+V[2]*B+V[3]*C,u=rgu[1]..rgu[2]),v=rgv[1]..
rgv[2]);
```

Integral of dot product (vector field'V' and tangent plane normal

vector [A,B,C] of torus surface 'CS') in range of parameters 'u,v'

$$\int_0^{2\pi} \int_0^{2\pi} P((2+\cos(u))\cos(v),(2+\cos(u))\sin(v),\sin(u))$$

$$(-2\cos(u)\cos(v)-\cos(u)^2\cos(v))+$$
$$Q((2+\cos(u))\cos(v),(2+\cos(u))\sin(v),\sin(u))$$

$$(-2\sin(v)\cos(u)-\sin(v)\cos(u)^2)$$
$$+R((2+\cos(u))\cos(v),(2+\cos(u))\sin(v),\sin(u))(-2\sin(u)-\cos(u)\sin(u))$$
$$du\ dv$$

```
> value(%);
```

$$2\pi\int_0^{2\pi} -2\,P(2\cos(\_X)+\cos(\_X)^2,2\sin(\_X)+\cos(\_X)\sin(\_X),\sin(\_X))\cos(\_X)^2$$

$$-P(2\cos(\_X)+\cos(\_X)^2,2\sin(\_X)+\cos(\_X)\sin(\_X),\sin(\_X))\cos(\_X)^3-$$
$$2\,Q(2\cos(\_X)+\cos(\_X)^2,2\sin(\_X)+\cos(\_X)\sin(\_X),\sin(\_X))\cos(\_X)\sin(\_X)$$

$$-$$

$$Q(2\cos(\_X)+\cos(\_X)^2,2\sin(\_X)+\cos(\_X)\sin(\_X),\sin(\_X))\sin(\_X)\cos(\_X)^2$$
$$-2\,R(2\cos(\_X)+\cos(\_X)^2,2\sin(\_X)+\cos(\_X)\sin(\_X),\sin(\_X))\sin(\_X)$$
$$-R(2\cos(\_X)+\cos(\_X)^2,2\sin(\_X)+\cos(\_X)\sin(\_X),\sin(\_X))\cos(\_X)\sin(\_X)$$

$$d\_X$$

```
# Import 'x = (2+cos(u))*cos(v),y = (2+cos(u))*sin(v),z = sin(u)' to
abstract vector field '[P(x,y,z),  Q(x,y,z),  R(x,y,z)]', obtain
insoluble result, abnegate it
```

```
> x:='x':y:='y':z:='z':
> Int(Int(V[1]*A+V[2]*B+V[3]*C,u=rgu[1]..rgu[2]),v=rgv[1]..rgv[2])
=Int(int(V[1]*A+V[2]*B+V[3]*C,u=rgu[1]..rgu[2]),v=rgv[1]..rgv[2]);
# Integral of Dot Product (Vector field'V' and Tangent plane normal
vector [A,B,C] of torus surface 'CS', reserve original form of
abstract vector field 'P(x,y,z),Q(x,y,z),R(x,y,z)') in range of 'u,v'
```

$$\int_0^{2\pi} \int_0^{2\pi} P(x,y,z)(-2\cos(u)\cos(v)-\cos(u)^2\cos(v))$$

$$+Q(x,y,z)(-2\sin(v)\cos(u)-\sin(v)\cos(u)^2)$$
$$+R(x,y,z)(-2\sin(u)-\cos(u)\sin(u))\ du\ dv=$$
$$\int_0^{2\pi} -P(x,y,z)\cos(v)\pi-Q(x,y,z)\sin(v)\pi\ dv$$

```
> alpha:=rhs(value(%)); # Calculate value of surface integral
```

$$\alpha := 0$$

```
> x:='x':y:='y':z:='z':
```

\# Entirely multiply radius vector 'r' (suppose r>0) by each item of Torus surface 'CS' parametrized expression, convert 'Parametrized expression of Torus surface' to 'Parametrized expression of Torus surface coordinates'

$$[r(2+\cos(u))\cos(v), r(2+\cos(u))\sin(v), r\sin(u)]$$

```
> x:=r*CS[1]:y:=r*CS[2]:z:=r*CS[3]:
> matrix(3,3,[Diff(r*CS[1],r),Diff(r*CS[1],u),Diff(r*CS[1],v),
Diff(r*CS[2],r),Diff(r*CS[2],u),Diff(r*CS[2],v),Diff(r*CS[3],r),
Diff(r*CS[3],u),Diff(r*CS[3],v)])=
matrix(3,3,[diff(r*CS[1],r),diff(r*CS[1],u),diff(r*CS[1],v),
diff(r*CS[2],r),diff(r*CS[2],u),diff(r*CS[2],v),diff(r*CS[3],r),
diff(r*CS[3],u),diff(r*CS[3],v)]);m2:=rhs(%);
```

\# Define and calculate matrix of partial derivatives 'm2', obtain general expression of Torus surface coordinates volume element coefficient

$$\begin{bmatrix} \dfrac{\partial}{\partial r}(r(2+\cos(u))\cos(v)) & \dfrac{\partial}{\partial u}(r(2+\cos(u))\cos(v)) & \dfrac{\partial}{\partial v}(r(2+\cos(u))\cos(v)) \\ \dfrac{\partial}{\partial r}(r(2+\cos(u))\sin(v)) & \dfrac{\partial}{\partial u}(r(2+\cos(u))\sin(v)) & \dfrac{\partial}{\partial v}(r(2+\cos(u))\sin(v)) \\ \dfrac{\partial}{\partial r}(r\sin(u)) & \dfrac{\partial}{\partial u}(r\sin(u)) & \dfrac{\partial}{\partial v}(r\sin(u)) \end{bmatrix}$$

$$= \begin{bmatrix} (2+\cos(u))\cos(v) & -r\sin(u)\cos(v) & -r(2+\cos(u))\sin(v) \\ (2+\cos(u))\sin(v) & -r\sin(u)\sin(v) & r(2+\cos(u))\cos(v) \\ \sin(u) & r\cos(u) & 0 \end{bmatrix}$$

$$m2 := \begin{bmatrix} (2+\cos(u))\cos(v) & -r\sin(u)\cos(v) & -r(2+\cos(u))\sin(v) \\ (2+\cos(u))\sin(v) & -r\sin(u)\sin(v) & r(2+\cos(u))\cos(v) \\ \sin(u) & r\cos(u) & 0 \end{bmatrix}$$

```
> det(m2);
```

$$-4\cos(v)^2 r^2 \cos(u) - 4\cos(v)^2 r^2 \cos(u)^2 - \cos(v)^2 r^2 \cos(u)^3 - 4\sin(v)^2 r^2 \cos(u)$$
$$- 4\sin(v)^2 r^2 \cos(u)^2 - \sin(v)^2 r^2 \cos(u)^3 - 2 r^2 \sin(u)^2 \cos(v)^2$$
$$- r^2 \sin(u)^2 \cos(v)^2 \cos(u) - 2 r^2 \sin(v)^2 \sin(u)^2 - r^2 \sin(v)^2 \sin(u)^2 \cos(u)$$

```
> J:=simplify(%);
```

\# General expression of Torus surface coordinates volume element coefficient

$$J := -r^2 (5\cos(u) + 2\cos(u)^2 + 2)$$

```
> x:='x':y:='y':z:='z':
> Diff(V[1],x)*Diff(r*CS[1],r)*Diff(r*CS[1],u)*Diff(r*CS[1],v)
+Diff(V[2],y)*Diff(r*CS[2],r)*Diff(r*CS[2],u)*Diff(r*CS[2],v)
+Diff(V[3],z)*Diff(r*CS[3],r)*Diff(r*CS[3],u)*Diff(r*CS[3],v);
diV2:=value(%);
```

Convert 'diV1' from Cartesian Coordinates expression to Torus Surface Coordinates expression

Reserve original form of abstract vector field '[P(x,y,z),Q(x,y,z), R(x,y,z)]' and its partial derivatives (insoluble); Calculate partial derivatives of 'r*(2+cos(u))*cos(v),r*(2+cos(u))*sin(v),r*sin(u)' (Computable),obtain a new expression 'diV2' (Divergence)

$$\left(\frac{\partial}{\partial x}P(x,y,z)\right)\left(\frac{\partial}{\partial r}(r(2+\cos(u))\cos(v))\right)\left(\frac{\partial}{\partial u}(r(2+\cos(u))\cos(v))\right)$$

$$\left(\frac{\partial}{\partial v}(r(2+\cos(u))\cos(v))\right)+\left(\frac{\partial}{\partial y}Q(x,y,z)\right)\left(\frac{\partial}{\partial r}(r(2+\cos(u))\sin(v))\right)$$

$$\left(\frac{\partial}{\partial u}(r(2+\cos(u))\sin(v))\right)\left(\frac{\partial}{\partial v}(r(2+\cos(u))\sin(v))\right)$$

$$+\left(\frac{\partial}{\partial z}R(x,y,z)\right)\left(\frac{\partial}{\partial r}(r\sin(u))\right)\left(\frac{\partial}{\partial u}(r\sin(u))\right)\left(\frac{\partial}{\partial v}(r\sin(u))\right)$$

$$diV2:=\left(\frac{\partial}{\partial x}P(x,y,z)\right)(2+\cos(u))^2\cos(v)^2 r^2\sin(u)\sin(v)$$

$$-\left(\frac{\partial}{\partial y}Q(x,y,z)\right)(2+\cos(u))^2\sin(v)^2 r^2\sin(u)\cos(v)$$

```
> Int(Int(Int(diV2*J,r=0..1),u=rgu[1]..rgu[2]),v=rgv[1]..rgv[2]);
```

Triple Integrals of Product (Divergence 'diV2' and Torus Surface Coordinates Volume Element) in range of 'r,u,v'

$$\int_0^{2\pi}\int_0^{2\pi}\int_0^1 -\left(\left(\frac{\partial}{\partial x}P(x,y,z)\right)(2+\cos(u))^2\cos(v)^2 r^2\sin(u)\sin(v)\right.$$

$$\left.-\left(\frac{\partial}{\partial y}Q(x,y,z)\right)(2+\cos(u))^2\sin(v)^2 r^2\sin(u)\cos(v)\right)r^2$$

$$(5\cos(u)+2\cos(u)^2+2)\,dr\,du\,dv$$

```
> beta:=value(%); # Calculate value of volume element integral
```

$$\beta:=0$$

Viz.

$$\int_0^{2\pi} \int_0^{2\pi} P(x,y,z)\,(-2\cos(u)\cos(v) - \cos(u)^2 \cos(v))$$

$$+ Q(x,y,z)\,(-2\sin(v)\cos(u) - \sin(v)\cos(u)^2)$$

$$+ R(x,y,z)\,(-2\sin(u) - \cos(u)\sin(u))\,du\,dv$$

$$=$$

$$\int_0^{2\pi} \int_0^{2\pi} \int_0^{1} -\left(\left(\frac{\partial}{\partial x} P(x,y,z)\right)(2+\cos(u))^2 \cos(v)^2 r^2 \sin(u)\sin(v)\right.$$

$$\left. -\left(\frac{\partial}{\partial y} Q(x,y,z)\right)(2+\cos(u))^2 \sin(v)^2 r^2 \sin(u)\cos(v)\right) r^2$$

$$(5\cos(u) + 2\cos(u)^2 + 2)\,dr\,du\,dv$$

Above-mentioned equation can be described as:

$$\iint_S A\cdot n\,dS = \iiint_\Omega \operatorname{div}A\,d\,\omega \qquad (1),\ \text{Complete Proof.}$$

Chapter 3 Numerical Models of Divergence Theorem at Manifold

3.1 Numerical Model of Divergence Theorem at Manifold (I)

Known: Parametric expression of simply connected orientable closed surface (Irregular、Asymmetrical)

$$\left[9\cos(u) - 18\sin\left(\frac{u}{2}\right) - \frac{9}{2}\sin(2u) - 7u, \frac{3}{2}\sin(u)\cos(v) + \cos(6u),\right.$$

$$\left.\frac{3}{2}\sin(u)\sin(v) + \cos(5u)\right] \tag{1}$$

Thereinto, u∈ [0, π],v∈ [0,2π];

and Integral Vector Field

$$\left[\frac{\left(\frac{x}{3} + \frac{y}{6} - \frac{z}{5} - 3\right)^3}{2} - \frac{z^3}{7}, \frac{1}{6}xyz - \frac{1}{6}y^2, \frac{x^2}{5} + \frac{yz}{5}\right] \tag{2}$$

Calculate and Validate Divergence Theorem at Manifold

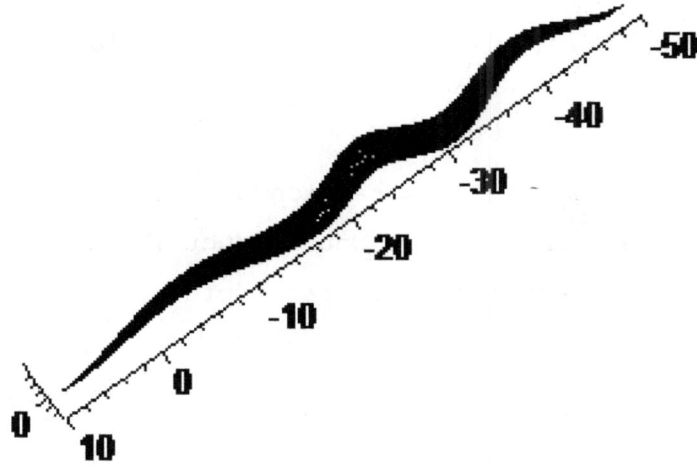

Figure(I)3.1 Simply connected orientable closed surface(1)

[Irregular、Asymmetrical]

Solution:

First, Free Surface Integral:

According to surface parametric expression (1), define and calculate First Matrix of Partial Derivatives, obtain tangent plane normal vector of surface:

$$
\begin{bmatrix}
i & j & k \\
\dfrac{\partial}{\partial u}(9\cos(u)-18\sin(\tfrac{u}{2})-\tfrac{9}{2}\sin(2u)-7u) & \dfrac{\partial}{\partial u}(\tfrac{3}{2}\sin(u)\cos(v)+\cos(6u)) & \dfrac{\partial}{\partial u}(\tfrac{3}{2}\sin(u)\sin(v)+\cos(5u)) \\
\dfrac{\partial}{\partial v}(9\cos(u)-18\sin(\tfrac{u}{2})-\tfrac{9}{2}\sin(2u)-7u) & \dfrac{\partial}{\partial v}(\tfrac{3}{2}\sin(u)\cos(v)+\cos(6u)) & \dfrac{\partial}{\partial v}(\tfrac{3}{2}\sin(u)\sin(v)+\cos(5u))
\end{bmatrix}
$$

$$
=
$$

$$
\frac{3}{4}\sin(u)\left(-12\,i\cos(v)\sin(6\,u)+3\,i\cos(u)-10\,i\sin(v)\sin(5\,u)+18\,j\sin(u)\cos(v)\right.
$$

$$
\left.+18\,k\sin(u)\sin(v)+18\cos\left(\frac{u}{2}\right)j\cos(v)+18\cos\left(\frac{u}{2}\right)k\sin(v)\right)
$$

$$+ 18 \cos(2\,u)\,j\,\cos(v) + 18 \cos(2\,u)\,k\,\sin(v) + 14\,j\,\cos(v) + 14\,k\,\sin(v) \bigg) \tag{3}$$

From expression(3), respectively pick up coefficients of item 'i,j,k', obtain tangent plane normal vector of surface (4):

$$\bigg[\frac{3}{4} \sin(u)\,(-12 \cos(v) \sin(6\,u) + 3 \cos(u) - 10 \sin(v) \sin(5\,u)),$$

$$\frac{3}{4} \sin(u) \bigg(18 \sin(u) \cos(v) + 18 \cos\!\left(\frac{u}{2}\right) \cos(v) + 18 \cos(2\,u) \cos(v) + 14 \cos(v) \bigg),$$

$$\frac{3}{4} \sin(u) \bigg(18 \sin(u) \sin(v) + 18 \cos\!\left(\frac{u}{2}\right) \sin(v) + 18 \cos(2\,u) \sin(v) + 14 \sin(v) \bigg) \bigg]$$

Input parametrized expression of surface (1) to Vector Field (2); and Calculate Integral of Dot Product [Vetor Field (2) and Tangent plane normal vector (4)] in range of 'u,v':

$$\int_0^{2\pi} \int_0^{\pi} \frac{3}{4} \bigg(\frac{1}{2} \bigg(3 \cos(u) - 6 \sin\!\left(\frac{u}{2}\right) - \frac{3}{2} \sin(2\,u) - \frac{7\,u}{3} + \frac{1}{4} \sin(u) \cos(v) + \frac{1}{6} \cos(6\,u)$$

$$- \frac{3}{10} \sin(u) \sin(v) - \frac{1}{5} \cos(5\,u) - 3 \bigg)^3 - \frac{1}{7} \bigg(\frac{3}{2} \sin(u) \sin(v) + \cos(5\,u) \bigg)^3 \bigg) \sin(u)$$

$$(-12 \cos(v) \sin(6\,u) + 3 \cos(u) - 10 \sin(v) \sin(5\,u)) + \frac{3}{4} \bigg(\frac{1}{6}$$

$$\bigg(9 \cos(u) - 18 \sin\!\left(\frac{u}{2}\right) - \frac{9}{2} \sin(2\,u) - 7\,u \bigg) \bigg(\frac{3}{2} \sin(u) \cos(v) + \cos(6\,u) \bigg)$$

$$\bigg(\frac{3}{2} \sin(u) \sin(v) + \cos(5\,u) \bigg) - \frac{1}{6} \bigg(\frac{3}{2} \sin(u) \cos(v) + \cos(6\,u) \bigg)^2 \bigg) \sin(u)$$

$$\bigg(18 \sin(u) \cos(v) + 18 \cos\!\left(\frac{u}{2}\right) \cos(v) + 18 \cos(2\,u) \cos(v) + 14 \cos(v) \bigg) + \frac{3}{4} \bigg($$

$$\frac{1}{5} \bigg(9 \cos(u) - 18 \sin\!\left(\frac{u}{2}\right) - \frac{9}{2} \sin(2\,u) - 7\,u \bigg)^2$$

$$+ \frac{1}{5} \bigg(\frac{3}{2} \sin(u) \cos(v) + \cos(6\,u) \bigg) \bigg(\frac{3}{2} \sin(u) \sin(v) + \cos(5\,u) \bigg) \bigg) \sin(u)$$

$$\bigg(18 \sin(u) \sin(v) + 18 \cos\!\left(\frac{u}{2}\right) \sin(v) + 18 \cos(2\,u) \sin(v) + 14 \sin(v) \bigg) du\,dv =$$

$$\frac{4554929006888647279}{13480992207540000} \pi + \frac{154491754161}{248934400} \pi^2 + \frac{16863}{160} \pi^3 + \frac{245}{96} \pi^4 \tag{5}$$

Surface Integral of 'Modulus of Tangent Plane Normal Vector [A,B,C]' in range of 'u,v', absolute value of result is area value of surface (6):

$$\int_0^{2\pi}\int_0^{\pi}\frac{3}{4}\Bigg(\sin(u)^2\left(-12\cos(v)\sin(6\,u)-10\sin(v)\sin(5\,u)+3\cos(u)\right)^2$$

$$+\sin(u)^2\left(18\sin(u)\cos(v)+18\cos\left(\frac{u}{2}\right)\cos(v)+18\cos(2\,u)\cos(v)+14\cos(v)\right)^2$$

$$+\sin(u)^2\left(18\sin(u)\sin(v)+18\cos\left(\frac{u}{2}\right)\sin(v)+18\cos(2\,u)\sin(v)+14\sin(v)\right)^2\Bigg)^{\wedge}$$

(1/2)

$$du\ dv\ =\ 330.6921620 \qquad (6)$$

Second, Free Volume Element Integral:

Entirely multiply radius vector 'r' (suppose r>0) by each item of surface parametrized expression (1), convert 'Parametrized expression of surface' to 'Parametrized expression of surface coordinates':

$$\left[r\left(9\cos(u)-18\sin\left(\frac{u}{2}\right)-\frac{9}{2}\sin(2\,u)-7\,u\right),r\left(\frac{3}{2}\sin(u)\cos(v)+\cos(6\,u)\right),\right.$$
$$\left.r\left(\frac{3}{2}\sin(u)\sin(v)+\cos(5\,u)\right)\right]$$

(7)

According to parametrized expression of surface coordinates (7), define and calculate Second Matrix of Partial Derivatives, obtain general expression of surface coordinates volume element coefficient(8):

$$\begin{bmatrix}\frac{\partial}{\partial r}(r(9\cos(u)-18\sin(\frac{u}{2})-\frac{9}{2}\sin(2u)-7u)) & \frac{\partial}{\partial u}(r(9\cos(u)-18\sin(\frac{u}{2})-\frac{9}{2}\sin(2u)-7u)) & \frac{\partial}{\partial v}(r(9\cos(u)-18\sin(\frac{u}{2})-\frac{9}{2}\sin(2u)-7u)) \\ \frac{\partial}{\partial r}(r(\frac{3}{2}\sin(u)\cos(v)+\cos(6u))) & \frac{\partial}{\partial u}(r(\frac{3}{2}\sin(u)\cos(v)+\cos(6u))) & \frac{\partial}{\partial v}(r(\frac{3}{2}\sin(u)\cos(v)+\cos(6u))) \\ \frac{\partial}{\partial r}(r(\frac{3}{2}\sin(u)\sin(v)+\cos(5u))) & \frac{\partial}{\partial u}(r(\frac{3}{2}\sin(u)\sin(v)+\cos(5u))) & \frac{\partial}{\partial v}(r(\frac{3}{2}\sin(u)\sin(v)+\cos(5u)))\end{bmatrix}$$

$$=$$

$$-\frac{3}{8}\sin(u)\left(-54-54\sin(u)\cos\left(\frac{u}{2}\right)-54\sin(u)\cos(2\,u)-432\sin\left(\frac{u}{2}\right)\cos(v)\sin(6\,u)\right.$$

$$+216\cos(u)\cos(v)\sin(6\,u)+180\cos(u)\sin(v)\sin(5\,u)$$

$$-360\sin\left(\frac{u}{2}\right)\sin(v)\sin(5\,u)-108\sin(2\,u)\cos(v)\sin(6\,u)$$

$$- 168\, u \cos(v) \sin(6\,u) - 90 \sin(2\,u) \sin(v) \sin(5\,u) - 36 \cos(v) \cos(6\,u) \cos(2\,u)$$

$$- 140\, u \sin(v) \sin(5\,u) - 36 \cos(v) \cos(6\,u) \cos\!\left(\frac{u}{2}\right) - 36 \sin(v) \cos(5\,u) \cos(2\,u)$$

$$- 36 \sin(v) \cos(5\,u) \cos\!\left(\frac{u}{2}\right) + 108 \sin\!\left(\frac{u}{2}\right) \cos(u) + 27 \sin(2\,u) \cos(u)$$

$$+ 42\, u \cos(u) - 36 \sin(u) \cos(v) \cos(6\,u) - 28 \cos(v) \cos(6\,u)$$

$$\left. - 36 \sin(u) \sin(v) \cos(5\,u) - 28 \sin(v) \cos(5\,u) - 42 \sin(u) \right) r^2 \tag{8}$$

Calculate Divergence of Vector Field (2):

$$\left(\frac{\partial}{\partial x}\left(\frac{\left(\frac{x}{3}+\frac{y}{6}-\frac{z}{5}-3 \right)^3}{2} - \frac{z^3}{7} \right) \right) + \left(\frac{\partial}{\partial y}\left(\frac{1}{6} x\,y\,z - \frac{1}{6} y^2 \right) \right) + \left(\frac{\partial}{\partial z}\left(\frac{x^2}{5} + \frac{y z}{5} \right) \right) =$$

$$\frac{\left(\frac{x}{3}+\frac{y}{6}-\frac{z}{5}-3 \right)^2}{2} + \frac{x z}{6} - \frac{2 y}{15} \tag{9}$$

Input parametrized expression of surface coordinates (7) to Divergence (9); and Calculate Triple Integrals of Product[Divergence (9) and Surface Coordinates Volume Element] in range of 'r,u,v' (10):

$$\int_0^{2\pi} \int_0^{\pi} \int_0^1 \; -\frac{3}{8}\left(\frac{1}{2}\left(\frac{1}{3} r \left(9 \cos(u) - 18 \sin\!\left(\frac{u}{2}\right) - \frac{9}{2} \sin(2\,u) - 7\,u \right) \right. \right.$$

$$+ \frac{1}{6} r \left(\frac{3}{2} \sin(u) \cos(v) + \cos(6\,u) \right) - \frac{1}{5} r \left(\frac{3}{2} \sin(u) \sin(v) + \cos(5\,u) \right) - 3 \Bigg)^2$$

$$+ \frac{1}{6} r^2 \left(9 \cos(u) - 18 \sin\!\left(\frac{u}{2}\right) - \frac{9}{2} \sin(2\,u) - 7\,u \right) \left(\frac{3}{2} \sin(u) \sin(v) + \cos(5\,u) \right)$$

$$\left. -\frac{2}{15} r \left(\frac{3}{2} \sin(u) \cos(v) + \cos(6\,u) \right) \right) \sin(u) \left(-54 - 36 \sin(u) \cos(v) \cos(6\,u) \right.$$

$$- 28 \cos(v) \cos(6\,u) - 36 \sin(u) \sin(v) \cos(5\,u) - 28 \sin(v) \cos(5\,u)$$

$$+ 27 \sin(2\,u) \cos(u) + 42\, u \cos(u) - 54 \sin(u) \cos(2\,u) - 54 \sin(u) \cos\!\left(\frac{u}{2}\right)$$

$$+ 108 \sin\!\left(\frac{u}{2}\right) \cos(u) - 42 \sin(u) + 216 \cos(u) \cos(v) \sin(6\,u)$$

$$+ 180 \cos(u) \sin(v) \sin(5\,u) - 432 \sin\!\left(\frac{u}{2}\right) \cos(v) \sin(6\,u)$$

$$- 360 \sin\left(\frac{u}{2}\right) \sin(v) \sin(5\,u) - 108 \sin(2\,u) \cos(v) \sin(6\,u)$$

$$- 168\,u \cos(v) \sin(6\,u) - 90 \sin(2\,u) \sin(v) \sin(5\,u) - 36 \sin(v) \cos(5\,u) \cos\left(\frac{u}{2}\right)$$

$$- 140\,u \sin(v) \sin(5\,u) - 36 \cos(v) \cos(6\,u) \cos(2\,u) - 36 \sin(v) \cos(5\,u) \cos(2\,u)$$

$$- 36 \cos(v) \cos(6\,u) \cos\left(\frac{u}{2}\right)\Bigg) r^2 \, dr \, du \, dv =$$

$$\frac{154491754161}{248934400} \pi^2 + \frac{4554929006888647279}{13480992207540000} \pi + \frac{16863}{160} \pi^3 + \frac{245}{96} \pi^4 \qquad (10)$$

Be different to ordinary triple integrals, in this position, first integral range is 'r∈[0,1]' always, and not 'r∈[0,n]' or 'r∈[0,∞]'; Because of '1', it is possible to hold a correct ratio between 'r,u,v'

//Be different to abstract divergence in 'Proof', and in 'Numerical Models', be able to input parametrized expression of idiographic surface coordinates (7) to idiographic divergence (9) directly, then calculate product of divergence (9) and surface coordinates volume element, complete triple integrals

Integral Precision Value (5) [Vector Field (2) on Target Surface (1)], is equal to Volume Element Integral Precision Value (10) [Divergence (9) in Spacial Bounded Closed Region 'Ω' that Target Surface (1) surrounds], Complete Calculation and Validation of Divergence Theorem at Manifold

Triple Integrals of 'Surface Coordinates Volume Element' in range of 'r,u,v', absolute value of result is volume value of space bounded closed region surrounded by closed surface (11):

$$\int_0^{2\pi} \int_0^{\pi} \int_0^1 -\frac{3}{8} \sin(u) \Bigg(216 \cos(u) \cos(v) \sin(6\,u) + 180 \cos(u) \sin(v) \sin(5\,u)$$

$$- 432 \sin\left(\frac{u}{2}\right) \cos(v) \sin(6\,u) - 108 \sin(2\,u) \cos(v) \sin(6\,u)$$

$$- 168\,u \cos(v) \sin(6\,u) - 36 \sin(u) \cos(v) \cos(6\,u) - 36 \sin(u) \sin(v) \cos(5\,u)$$

$$- 360 \sin\left(\frac{u}{2}\right) \sin(v) \sin(5\,u) - 90 \sin(2\,u) \sin(v) \sin(5\,u) - 140\,u \sin(v) \sin(5\,u)$$

$$- 36 \cos(v) \cos(6\,u) \cos\left(\frac{u}{2}\right) - 36 \cos(v) \cos(6\,u) \cos(2\,u)$$

$$- 36 \sin(v) \cos(5\,u) \cos\!\left(\frac{u}{2}\right) - 36 \sin(v) \cos(5\,u) \cos(2\,u) + 108 \sin\!\left(\frac{u}{2}\right) \cos(u)$$

$$+ 27 \sin(2\,u) \cos(u) + 42\,u \cos(u) - 54 \sin(u) \cos\!\left(\frac{u}{2}\right) - 54 \sin(u) \cos(2\,u)$$

$$\left. - 28 \cos(v) \cos(6\,u) - 28 \sin(v) \cos(5\,u) - 42 \sin(u) - 54 \right) r^2 \, dr \, du \, dv$$

$$= \ \frac{45}{16}\,\pi^2 + \frac{243}{5}\,\pi \tag{11}$$

3.2 Numerical Model of Divergence Theorem at Manifold (I)
[Program Template of Waterloo Maplesoft, Optional]

```
> restart; # Reset computer algebraic system
> with(plots):with(linalg): # Load 'plots' and 'linalg' package
> CS:=[9*(cos(u)-2*sin(u/2)),sin(u)*cos(v)+cos(6*u),sin(u)*sin(v)
+cos(5*u)];
# Define parametrized expression of discretional simply connected
orientable closed surface 'CS'
```

$$CS := \left[\, 9 \cos(u) - 18 \sin\!\left(\frac{u}{2}\right), \ \sin(u) \cos(v) + \cos(6\,u), \ \sin(u) \sin(v) + \cos(5\,u) \,\right]$$

```
> rgu:=[0,Pi];
```

$$rgu := [\,0, \pi\,]$$

```
> rgv:=[0,2*Pi];
# Define range of 'u,v', make surface 'CS' closed
```

$$rgv := [\,0, 2\,\pi\,]$$

```
>plot3d(CS,u=rgu[1]..rgu[2],v=rgv[1]..rgv[2],scaling=constrained,
projection=0.9,numpoints=1000);g1:=%:
# Draw discretional simply connected orientable closed surface 'CS'
```

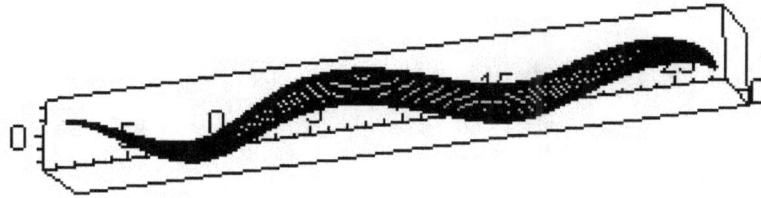

Figure(I)3.2.1 Asymmetrical Irregular Discretional
Simply Connected Orientable Closed Surface 'CS'

```
> V:=[(x/3+y/6-z/5)^3/12-y^2/5+z^2/7,x*y/7-z^2/9,x^2/6+y*z/5];
```
\# Define discretional spacial vector field 'V' (Suppose Vector Field
'V' possesses 1th order continuous partial derivatives in spacial
bounded closed region 'Ω' that Surface 'CS' surrounds)

$$V := \left[\frac{\left(\dfrac{x}{3}+\dfrac{y}{6}-\dfrac{z}{5}\right)^3}{12} - \frac{y^2}{5} + \frac{z^2}{7}, \frac{x\,y}{7} - \frac{z^2}{9}, \frac{x^2}{6} + \frac{y\,z}{5} \right]$$

```
>Diff(V[1],x)+Diff(V[2],y)+Diff(V[3],z)=diff(V[1],x)+diff(V[2],y)
  +diff(V[3],z);diV:=rhs(%);
```
\# Calculate Divergence 'diV' of vector field 'V'

$$\left(\frac{\partial}{\partial x}\left(\frac{\left(\dfrac{x}{3}+\dfrac{y}{6}-\dfrac{z}{5}\right)^3}{12} - \frac{y^2}{5} + \frac{z^2}{7} \right) \right) + \left(\frac{\partial}{\partial y}\left(\frac{x\,y}{7} - \frac{z^2}{9} \right) \right) + \left(\frac{\partial}{\partial z}\left(\frac{x^2}{6} + \frac{y\,z}{5} \right) \right) =$$

$$\frac{\left(\dfrac{x}{3}+\dfrac{y}{6}-\dfrac{z}{5}\right)^2}{12} + \frac{x}{7} + \frac{y}{5}$$

$$diV := \frac{\left(\dfrac{x}{3}+\dfrac{y}{6}-\dfrac{z}{5}\right)^2}{12} + \frac{x}{7} + \frac{y}{5}$$

```
> rgx:=[-30,10];
```

$$rgx := [-30, 10]$$

```
> rgy:=[-20,20];
```

$$rgy := [-20, 20]$$

```
> rgz:=[-20,20];
```

$$rgz := [-20, 20]$$

```
>fieldplot3d(V,x=rgx[1]..rgx[2],y=rgy[1]..rgy[2],z=rgz[1]..
rgz[2],arrows=SLIM,color=blue):g2:=%: # Draw vector field 'V'
>implicitplot3d(diV,x=rgx[1]..rgx[2],y=rgy[1]..rgy[2],z=rgz[1]..
rgz[2],style=wireframe,color=cyan,numpoints=3000):g3:=%:
# Draw level surface of divergence 'diV'
> display(g1,g2,g3); # Synthesize figures
```

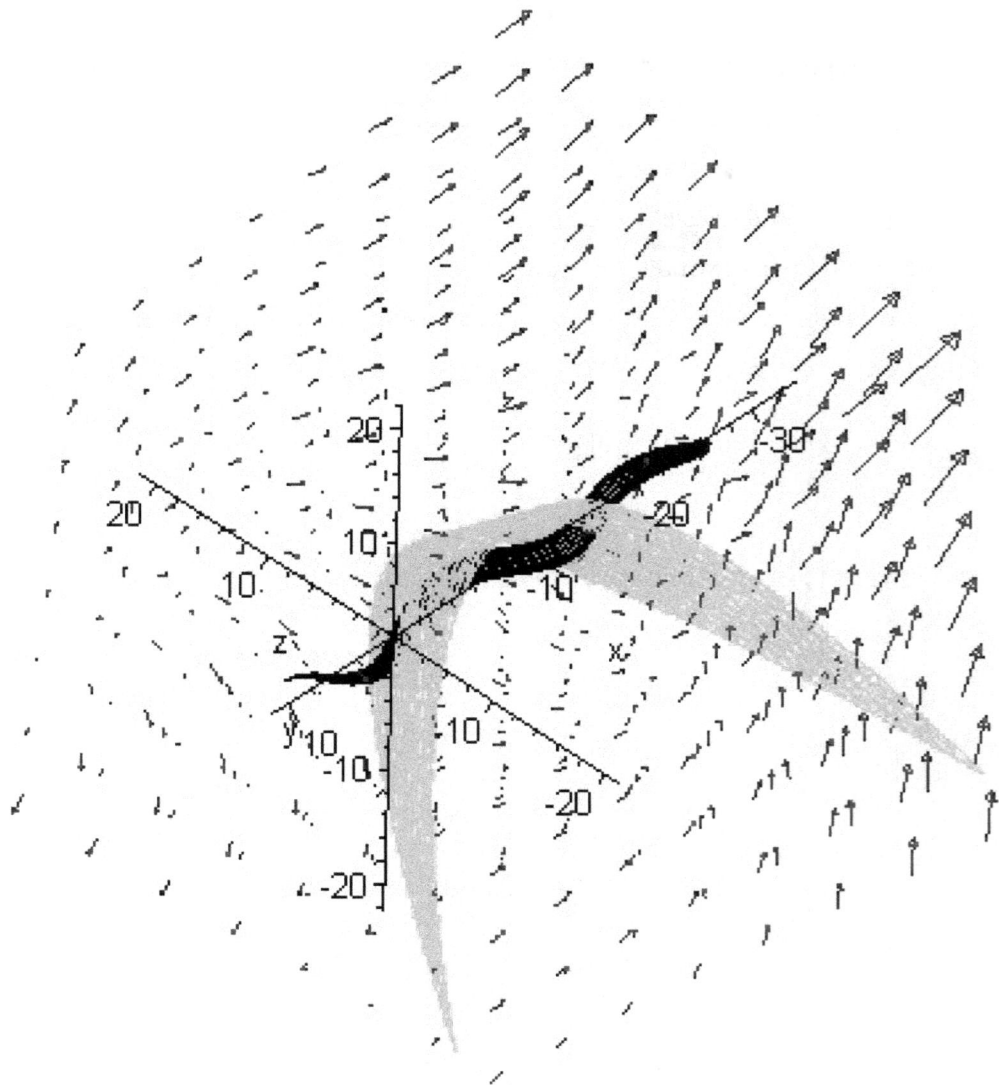

Figure(I)3.2.2 Surface 'CS'、Vector Field 'V' and
Level Surface of Divergence 'diV'

```
> x:=CS[1]:y:=CS[2]:z:=CS[3]:
> matrix(3,3,[i,j,k,Diff(x,u),Diff(y,u),Diff(z,u),Diff(x,v),
Diff(y,v),Diff(z,v)])=
matrix(3,3,[i,j,k,diff(x,u),diff(y,u),diff(z,u),diff(x,v),
diff(y,v),diff(z,v)]);m1:=rhs(%);
# Define and calculate matrix of partial derivatives 'm1'; obtain
'Tangent plane normal vector' of surface 'CS'
```

$[\,i\,,j\,,k\,]$

$$\left[\frac{d}{du}\left(9\cos(u)-18\sin\left(\frac{u}{2}\right)\right),\frac{\partial}{\partial u}\left(\sin(u)\cos(v)+\cos(6\,u)\right),\right.$$

$$\left.\frac{\partial}{\partial u}\left(\sin(u)\sin(v)+\cos(5\,u)\right)\right]$$

$$\left[\frac{\partial}{\partial v}\left(9\cos(u)-18\sin\left(\frac{u}{2}\right)\right),\frac{\partial}{\partial v}\left(\sin(u)\cos(v)+\cos(6\,u)\right),\right.$$

$$\left.\frac{\partial}{\partial v}\left(\sin(u)\sin(v)+\cos(5\,u)\right)\right]=$$

$$\begin{bmatrix} i & j & k \\ -9\sin(u)-9\cos\left(\frac{u}{2}\right) & \cos(u)\cos(v)-6\sin(6\,u) & \cos(u)\sin(v)-5\sin(5\,u) \\ 0 & -\sin(u)\sin(v) & \sin(u)\cos(v) \end{bmatrix}$$

$$m1 := \begin{bmatrix} i & j & k \\ -9\sin(u)-9\cos\left(\frac{u}{2}\right) & \cos(u)\cos(v)-6\sin(6\,u) & \cos(u)\sin(v)-5\sin(5\,u) \\ 0 & -\sin(u)\sin(v) & \sin(u)\cos(v) \end{bmatrix}$$

```
> det(m1):
> mn:=simplify(%);
```

$$mn := \sin(u)\left(-6\,i\cos(v)\sin(6\,u)+i\cos(u)-5\,i\sin(v)\sin(5\,u)+9\,j\sin(u)\cos(v)\right.$$

$$\left.+9\,k\sin(u)\sin(v)+9\cos\left(\frac{u}{2}\right)j\cos(v)+9\cos\left(\frac{u}{2}\right)k\sin(v)\right)$$

```
> A:=coeff(mn,i); # Obtain coefficient of 'i'
```

$$A := \sin(u)\left(-6\cos(v)\sin(6\,u)+\cos(u)-5\sin(v)\sin(5\,u)\right)$$

```
> B:=coeff(mn,j); # Obtain coefficient of 'j'
```

$$B := \sin(u)\left(9\sin(u)\cos(v)+9\cos\left(\frac{u}{2}\right)\cos(v)\right)$$

```
> C:=coeff(mn,k);
 # Obtain coefficient of 'k', [A,B,C] structure 'Tangent plane normal
vector'
```

$$C := \sin(u)\left(9\sin(u)\sin(v)+9\cos\left(\frac{u}{2}\right)\sin(v)\right)$$

```
> [A,B,C]: # [A,B,C] structure 'Tangent plane normal vector'
> Int(Int(V[1]*A+V[2]*B+V[3]*C,u=rgu[1]..rgu[2]),v=rgv[1]..rgv[2]);
# Integrals of Dot Product (Vector 'V' and Tangent plane normal vector
[A,B,C] of surface 'CS') in range of parameters 'u,v'
```

$$\int_0^{2\pi}\int_0^{\pi}\left(\frac{1}{12}\right.$$

$$\left(3\cos(u)-6\sin\left(\frac{u}{2}\right)+\frac{1}{6}\sin(u)\cos(v)+\frac{1}{6}\cos(6u)-\frac{1}{5}\sin(u)\sin(v)-\frac{1}{5}\cos(5u)\right)^{\wedge}$$

$$\left.^3-\frac{1}{5}(\sin(u)\cos(v)+\cos(6u))^2+\frac{1}{7}(\sin(u)\sin(v)+\cos(5u))^2\right)\sin(u)$$

$$(-6\cos(v)\sin(6u)+\cos(u)-5\sin(v)\sin(5u))+\left(\right.$$

$$\frac{1}{7}\left(9\cos(u)-18\sin\left(\frac{u}{2}\right)\right)(\sin(u)\cos(v)+\cos(6u))$$

$$\left.-\frac{1}{9}(\sin(u)\sin(v)+\cos(5u))^2\right)\sin(u)\left(9\sin(u)\cos(v)+9\cos\left(\frac{u}{2}\right)\cos(v)\right)+\left(\right.$$

$$\frac{1}{6}\left(9\cos(u)-18\sin\left(\frac{u}{2}\right)\right)^2$$

$$\left.+\frac{1}{5}(\sin(u)\cos(v)+\cos(6u))(\sin(u)\sin(v)+\cos(5u))\right)\sin(u)$$

$$\left(9\sin(u)\sin(v)+9\cos\left(\frac{u}{2}\right)\sin(v)\right)du\,dv$$

```
> alpha:=value(%);delta:=evalf(alpha);
# Calculate value of surface integral
```

$$\alpha:=\frac{269021701916899\pi}{152846502558750}$$

$$\delta:=5.529446786$$

```
> Int(Int(sqrt(A^2+B^2+C^2),u=rgu[1]..rgu[2]),v=rgv[1]..rgv[2]);
# Surface Integral of 'Modulus of Tangent Plane Normal Vector [A,B,C]'
in range of 'u,v', absolute value of result is area value of surface
```

$$\int_0^{2\pi}\int_0^{\pi}\left(\sin(u)^2(-6\cos(v)\sin(6u)-5\sin(v)\sin(5u)+\cos(u))^2\right.$$

$$+ \sin(u)^2 \left(9 \sin(u) \cos(v) + 9 \cos(v) \cos\left(\frac{u}{2}\right) \right)^2$$

$$+ \sin(u)^2 \left(9 \sin(u) \sin(v) + 9 \sin(v) \cos\left(\frac{u}{2}\right) \right)^2 \Bigg)^{(1/2)} \quad du \, dv$$

```
> evalf(%); # Area value of idiographic surface
```

$$172.1960019$$

```
> x:='x':y:='y':z:='z': # Reset variable 'x', 'y','z'
> [r*CS[1],r*CS[2],r*CS[3]];
```

Entirely multiply radius vector 'r' (suppose r>0) by each item of surface 'CS' parametrized expression, Convert 'Parametrized expression of surface' to 'Parametrized expression of surface coordinates'

$$\left[r\left(9\cos(u) - 18\sin\left(\frac{u}{2}\right)\right), r\left(\sin(u)\cos(v) + \cos(6\,u)\right), r\left(\sin(u)\sin(v) + \cos(5\,u)\right)\right]$$

```
> x:=r*CS[1]:y:=r*CS[2]:z:=r*CS[3]:
> matrix(3,3,[Diff(x,r),Diff(x,u),Diff(x,v),Diff(y,r),Diff(y,u),
Diff(y,v),Diff(z,r),Diff(z,u),Diff(z,v)])=
matrix(3,3,[diff(x,r),diff(x,u),diff(x,v),diff(y,r),diff(y,u),
diff(y,v),diff(z,r),diff(z,u),diff(z,v)]);m2:=rhs(%);
```

Define and calculate matrix of partial derivatives 'm2', obtain general expression of surface coordinates volume element coefficient

$$\left[\frac{\partial}{\partial r}\left(r\left(9\cos(u) - 18\sin\left(\frac{u}{2}\right)\right)\right), \frac{\partial}{\partial u}\left(r\left(9\cos(u) - 18\sin\left(\frac{u}{2}\right)\right)\right),\right.$$
$$\left.\frac{\partial}{\partial v}\left(r\left(9\cos(u) - 18\sin\left(\frac{u}{2}\right)\right)\right)\right]$$

$$\left[\frac{\partial}{\partial r}\left(r\left(\sin(u)\cos(v) + \cos(6\,u)\right)\right), \frac{\partial}{\partial u}\left(r\left(\sin(u)\cos(v) + \cos(6\,u)\right)\right),\right.$$
$$\left.\frac{\partial}{\partial v}\left(r\left(\sin(u)\cos(v) + \cos(6\,u)\right)\right)\right]$$

$$\left[\frac{\partial}{\partial r}\left(r\left(\sin(u)\sin(v) + \cos(5\,u)\right)\right), \frac{\partial}{\partial u}\left(r\left(\sin(u)\sin(v) + \cos(5\,u)\right)\right),\right.$$
$$\left.\frac{\partial}{\partial v}\left(r\left(\sin(u)\sin(v) + \cos(5\,u)\right)\right)\right] =$$

$$\begin{bmatrix} 9\cos(u) - 18\sin\left(\dfrac{u}{2}\right) & r\left(-9\sin(u) - 9\cos\left(\dfrac{u}{2}\right)\right) & 0 \\ \sin(u)\cos(v) + \cos(6\,u) & r\left(\cos(u)\cos(v) - 6\sin(6\,u)\right) & -r\sin(u)\sin(v) \\ \sin(u)\sin(v) + \cos(5\,u) & r\left(\cos(u)\sin(v) - 5\sin(5\,u)\right) & r\sin(u)\cos(v) \end{bmatrix}$$

$$m2 := \begin{bmatrix} 9\cos(u) - 18\sin\left(\dfrac{u}{2}\right) & r\left(-9\sin(u) - 9\cos\left(\dfrac{u}{2}\right)\right) & 0 \\[2mm] \sin(u)\cos(v) + \cos(6u) & r\,(\cos(u)\cos(v) - 6\sin(6u)) & -r\sin(u)\sin(v) \\[2mm] \sin(u)\sin(v) + \cos(5u) & r\,(\cos(u)\sin(v) - 5\sin(5u)) & r\sin(u)\cos(v) \end{bmatrix}$$

```
> det(m2):

> J:=simplify(%);
```

`# General expression of surface coordinates volume element coefficient`

$$J := 9\sin(u)\Big[-6\cos(u)\cos(v)\sin(6u) - 5\cos(u)\sin(v)\sin(5u)$$

$$+ 12\sin\left(\frac{u}{2}\right)\cos(v)\sin(6u) - 2\sin\left(\frac{u}{2}\right)\cos(u) + 10\sin\left(\frac{u}{2}\right)\sin(v)\sin(5u)$$

$$+ \sin(u)\cos(v)\cos(6u) + \cos(v)\cos(6u)\cos\left(\frac{u}{2}\right) + 1 + \sin(u)\cos\left(\frac{u}{2}\right)$$

$$+ \sin(u)\sin(v)\cos(5u) + \sin(v)\cos(5u)\cos\left(\frac{u}{2}\right)\Big] r^2$$

```
> Int(Int(Int(diV*J,r=0..1),u=rgu[1]..rgu[2]),v=rgv[1]..rgv[2]);
```

`# Triple Integrals of Product(Divergence 'diV' and Surface Coordinates`
`Volume Element) in range of 'r,u,v'`

`// Be different to abstract divergence 'diV1,diV2' in 'Proof',in`
`'Numerical Models', be able to input parametrized expression of`
`idiographic surface coordinates 'x=r*CS[1],y=r*CS[2],z=r*CS[3]' to`
`idiographic divergence 'diV' directly, then calculate product of`
`divergence and surface coordinates volume element, complete triple`
`integrals`

$$\int_0^{2\pi}\int_0^{\pi}\int_0^1 9\left(\frac{1}{12}\left(\frac{1}{3}r\left(9\cos(u) - 18\sin\left(\frac{u}{2}\right)\right)\right) + \frac{1}{6}r\,(\sin(u)\cos(v) + \cos(6u))\right.$$

$$\left. - \frac{1}{5}r\,(\sin(u)\sin(v) + \cos(5u))\right)^2 + \frac{1}{7}r\left(9\cos(u) - 18\sin\left(\frac{u}{2}\right)\right)$$

$$+ \frac{1}{5}r\,(\sin(u)\cos(v) + \cos(6u))\right)\sin(u)\Big(-6\cos(u)\cos(v)\sin(6u)$$

$$- 5\cos(u)\sin(v)\sin(5u) + 12\sin\left(\frac{u}{2}\right)\cos(v)\sin(6u) - 2\sin\left(\frac{u}{2}\right)\cos(u)$$

$$+ 10\sin\left(\frac{u}{2}\right)\sin(v)\sin(5u) + \sin(u)\cos(v)\cos(6u) + \cos(v)\cos(6u)\cos\left(\frac{u}{2}\right) + 1$$

$$+ \sin(u)\cos\left(\frac{u}{2}\right) + \sin(u)\sin(v)\cos(5u) + \sin(v)\cos(5u)\cos\left(\frac{u}{2}\right)\Big) r^2 \; dr \; du \; dv$$

```
> beta:=value(%);epsilon:=evalf(beta);
```

`# Calculate value of volume element integral`

$$\beta := \frac{269021701916899\pi}{152846502558750}$$

$$\varepsilon := 5.529446786$$

> alpha;beta; # Two analytic values are equal

$$\frac{269021701916899\pi}{152846502558750}$$

$$\frac{269021701916899\pi}{152846502558750}$$

> delta;epsilon; # Two float values are equal

$$5.529446786$$

$$5.529446786$$

> Int(Int(Int(J,r=0..1),u=rgu[1]..rgu[2]),v=rgv[1]..rgv[2]);
Triple Integrals of 'Surface Coordinates Volume Element' in range
of 'r,u,v', absolute value of result is volume value of space bounded
closed region surrounded by closed surface

$$\int_0^{2\pi}\int_0^{\pi}\int_0^1 9\sin(u)\left(12\cos(v)\sin\left(\frac{u}{2}\right)\sin(6\,u)+\cos(v)\cos(6\,u)\sin(u)\right.$$

$$+\cos(v)\cos(6\,u)\cos\left(\frac{u}{2}\right)-6\cos(v)\sin(6\,u)\cos(u)+10\sin(v)\sin\left(\frac{u}{2}\right)\sin(5\,u)$$

$$+\sin(v)\cos(5\,u)\sin(u)+\sin(v)\cos(5\,u)\cos\left(\frac{u}{2}\right)-5\sin(v)\sin(5\,u)\cos(u)$$

$$\left.-2\sin\left(\frac{u}{2}\right)\cos(u)+\sin(u)\cos\left(\frac{u}{2}\right)+1\right)r^2\,dr\,du\,dv$$

> value(%);evalf(%);
Volume value of space bounded closed region surrounded by closed
surface

$$\frac{108\,\pi}{5}$$

$$67.85840133$$

3.3 Numerical Model of Divergence Theorem at Manifold (II)

Known: Parametric parametric expression of Simply Connected
Orientable Closed Surface (Irregular、Asymmetrical)

$$\left[2\sin(u)\cos(v) + \frac{2}{7}\sin(u)\cos(v-2)\cos(7v) - \sin(u), \right.$$

$$2\sin(u)\sin(v) + \frac{1}{7}\sin(u)\sin(v-2)\cos(8v) - \sin(u),$$

$$\left. 3\cos(u) - \frac{1}{7}\cos\left(\frac{u}{2}\right)\cos(12u - v) \right] \qquad (1)$$

Thereinto, u∈[0,π],v∈[0,2π]; and Integral Vector Field

$$\left[\frac{\left(\dfrac{x}{3} + \dfrac{y}{6} - \dfrac{z}{5}\right)^3}{2} - \frac{y^2 z}{7}, \frac{1}{6}xyz - \frac{1}{6}y^2, \frac{1}{5}x^2 + \frac{1}{15}xyz \right] \qquad (2)$$

Calculate and Validate Divergence Theorem at Manifold

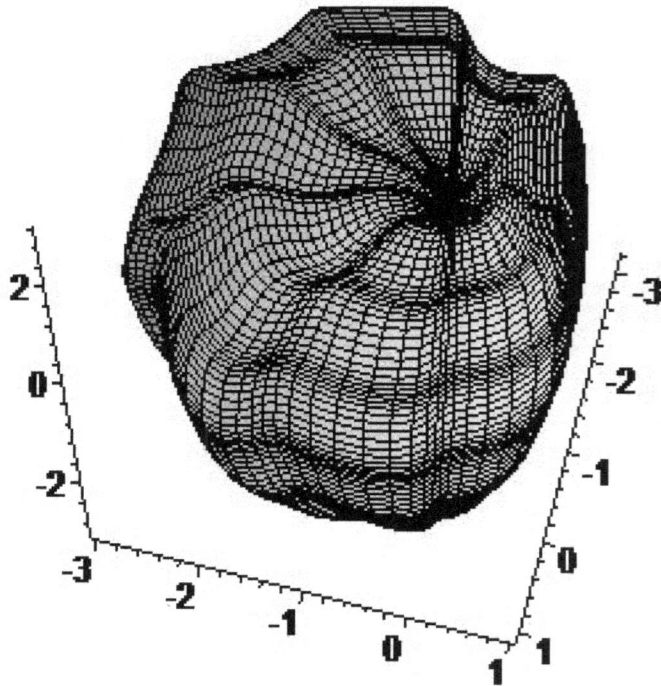

Figure(I)3.3 Simply Connected Orientable Closed Surface (1)

[Irregular、Asymmetrical]

Solution:

First, Free Surface Integral:

According to surface parametric expression (1), define and calculate First Matrix of Partial Derivatives, obtain Tangent Plane Normal Vector of surface:

$$
\begin{bmatrix}
i & j & k \\
\frac{\partial}{\partial u}(2\sin(u)cos(v)+\frac{2}{7}\sin(u)\cos(v\text{-}2)\cos(7v)\text{-}\sin(u)) & \frac{\partial}{\partial u}(2\sin(u)\sin(v)+\frac{1}{7}\sin(u)\sin(v\text{-}2)\cos(8v)\text{-}\sin(u)) & \frac{\partial}{\partial u}(3\cos(u)-\frac{1}{7}\cos(\frac{u}{2})\cos(12u-v)) \\
\frac{\partial}{\partial v}(2\sin(u)cos(v)+\frac{2}{7}\sin(u)\cos(v\text{-}2)\cos(7v)\text{-}\sin(u)) & \frac{\partial}{\partial v}(2\sin(u)\sin(v)+\frac{1}{7}\sin(u)\sin(v\text{-}2)\cos(8v)\text{-}\sin(u)) & \frac{\partial}{\partial v}(3\cos(u)-\frac{1}{7}\cos(\frac{u}{2})\cos(12u-v))
\end{bmatrix}
$$

$$= $$

$$
\frac{1}{7}\,i\,\cos\!\left(\frac{u}{2}\right)\sin(12\,u-v)\cos(u)+\frac{3}{7}\,i\,\sin(u)^2\cos(v-2)\cos(8\,v)
$$

$$
+\frac{2}{7}\cos(u)\cos(v)\,k\,\sin(u)\cos(v-2)\cos(8\,v)
$$

$$
-\frac{16}{7}\cos(u)\cos(v)\,k\,\sin(u)\sin(v-2)\sin(8\,v)
$$

$$
+\frac{2}{49}\cos(u)\cos(v-2)\cos(7\,v)\,j\,\cos\!\left(\frac{u}{2}\right)\sin(12\,u-v)
$$

$$
+\frac{4}{7}\cos(u)\cos(v-2)\cos(7\,v)\,k\,\sin(u)\cos(v)
$$

$$
+\frac{2}{49}\cos(u)\cos(v-2)^2\cos(7\,v)\,k\,\sin(u)\cos(8\,v)
$$

$$
+\frac{2}{7}\sin(u)\sin(v)\,k\,\cos(u)\sin(v-2)\cos(8\,v)
$$

$$
-\frac{1}{49}\sin(u)\sin(v-2)\cos(7\,v)\,j\,\sin\!\left(\frac{u}{2}\right)\cos(12\,u-v)
$$

$$
-\frac{24}{49}\sin(u)\sin(v-2)\cos(7\,v)\,j\,\cos\!\left(\frac{u}{2}\right)\sin(12\,u-v)
$$

$$
+\frac{4}{7}\sin(u)\sin(v-2)\cos(7\,v)\,k\,\cos(u)\sin(v)
$$

$$
+\frac{2}{49}\sin(u)\sin(v-2)^2\cos(7\,v)\,k\,\cos(u)\cos(8\,v)
$$

$$
-\frac{1}{7}\sin(u)\cos(v-2)\sin(7\,v)\,j\,\sin\!\left(\frac{u}{2}\right)\cos(12\,u-v)
$$

$$
-\frac{24}{7}\sin(u)\cos(v-2)\sin(7\,v)\,j\,\cos\!\left(\frac{u}{2}\right)\sin(12\,u-v)
$$

$$
+4\sin(u)\cos(v-2)\sin(7\,v)\,k\,\cos(u)\sin(v)
$$

$$
+\frac{96}{49}\,i\,\cos\!\left(\frac{u}{2}\right)\sin(12\,u-v)\sin(u)\sin(v-2)\sin(8\,v)
$$

$$-\frac{12}{49}\,i\,\cos\!\left(\frac{u}{2}\right)\sin(12\,u-v)\sin(u)\cos(v-2)\cos(8\,v)$$

$$+\frac{4}{49}\,i\,\sin\!\left(\frac{u}{2}\right)\cos(12\,u-v)\sin(u)\sin(v-2)\sin(8\,v)$$

$$-\frac{1}{49}\,i\,\cos\!\left(\frac{u}{2}\right)\sin(12\,u-v)\cos(u)\sin(v-2)\cos(8\,v)$$

$$-\frac{1}{98}\,i\,\sin\!\left(\frac{u}{2}\right)\cos(12\,u-v)\sin(u)\cos(v-2)\cos(8\,v)$$

$$-\frac{16}{49}\cos(u)\cos(v-2)\cos(7\,v)\,k\,\sin(u)\sin(v-2)\sin(8\,v)$$

$$+\frac{2}{7}\sin(u)\cos(v-2)\sin(7\,v)\,k\,\cos(u)\sin(v-2)\cos(8\,v)$$

$$-\frac{2}{7}\,i\,\cos\!\left(\frac{u}{2}\right)\sin(12\,u-v)\cos(u)\sin(v)-\frac{1}{7}\,i\,\sin\!\left(\frac{u}{2}\right)\cos(12\,u-v)\sin(u)\cos(v)$$

$$-\frac{24}{7}\,i\,\cos\!\left(\frac{u}{2}\right)\sin(12\,u-v)\sin(u)\cos(v)+\frac{2}{7}\cos(u)\cos(v)\,j\,\cos\!\left(\frac{u}{2}\right)\sin(12\,u-v)$$

$$-\frac{1}{7}\cos(u)\,k\,\sin(u)\cos(v-2)\cos(8\,v)+\frac{8}{7}\cos(u)\,k\,\sin(u)\sin(v-2)\sin(8\,v)$$

$$-\frac{1}{7}\sin(u)\sin(v)\,j\,\sin\!\left(\frac{u}{2}\right)\cos(12\,u-v)-\frac{24}{7}\sin(u)\sin(v)\,j\,\cos\!\left(\frac{u}{2}\right)\sin(12\,u-v)$$

$$-\frac{2}{7}\sin(u)\sin(v-2)\cos(7\,v)\,k\,\cos(u)-2\sin(u)\cos(v-2)\sin(7\,v)\,k\,\cos(u)$$

$$+6\,i\,\sin(u)^{2}\cos(v)+6\sin(u)^{2}\sin(v)\,j-\frac{24}{7}\,i\,\sin(u)^{2}\sin(v-2)\sin(8\,v)$$

$$-\frac{1}{7}\cos(u)\,j\,\cos\!\left(\frac{u}{2}\right)\sin(12\,u-v)-2\cos(u)\,k\,\sin(u)\cos(v)$$

$$+4\sin(u)\sin(v)^{2}\,k\,\cos(u)-2\sin(u)\sin(v)\,k\,\cos(u)$$

$$+\frac{6}{7}\sin(u)^{2}\sin(v-2)\cos(7\,v)\,j+6\sin(u)^{2}\cos(v-2)\sin(7\,v)\,j$$

$$+4\cos(u)\cos(v)^{2}\,k\,\sin(u) \qquad (3)$$

From expression(3), respectively pick up coefficients of item 'i,j,k', obtain tangent plane normal vector of surface(4):

$$\left[\frac{1}{7}\cos\!\left(\frac{u}{2}\right)\sin(12\,u-v)\cos(u)+\frac{96}{49}\cos\!\left(\frac{u}{2}\right)\sin(12\,u-v)\sin(u)\sin(v-2)\sin(8\,v)\right.$$

$$-\frac{12}{49}\cos\!\left(\frac{u}{2}\right)\sin(12\,u-v)\sin(u)\cos(v-2)\cos(8\,v)$$

$$+\frac{4}{49}\sin\!\left(\frac{u}{2}\right)\cos(12\,u-v)\sin(u)\sin(v-2)\sin(8\,v)$$

$$-\frac{1}{49}\cos\!\left(\frac{u}{2}\right)\sin(12\,u-v)\cos(u)\sin(v-2)\cos(8\,v)$$

$$-\frac{1}{98}\sin\left(\frac{u}{2}\right)\cos(12\,u-v)\sin(u)\cos(v-2)\cos(8\,v)$$

$$-\frac{2}{7}\cos\left(\frac{u}{2}\right)\sin(12\,u-v)\cos(u)\sin(v)-\frac{1}{7}\sin\left(\frac{u}{2}\right)\cos(12\,u-v)\sin(u)\cos(v)$$

$$-\frac{24}{7}\cos\left(\frac{u}{2}\right)\sin(12\,u-v)\sin(u)\cos(v)+\frac{3}{7}\cos(v-2)\cos(8\,v)$$

$$-\frac{24}{7}\sin(v-2)\sin(8\,v)+6\cos(v)-6\cos(v)\cos(u)^2$$

$$-\frac{3}{7}\cos(v-2)\cos(8\,v)\cos(u)^2+\frac{24}{7}\sin(v-2)\sin(8\,v)\cos(u)^2,$$

$$\frac{2}{49}\cos(u)\cos(v-2)\cos(7\,v)\cos\left(\frac{u}{2}\right)\sin(12\,u-v)$$

$$-\frac{1}{49}\sin(u)\sin(v-2)\cos(7\,v)\sin\left(\frac{u}{2}\right)\cos(12\,u-v)$$

$$-\frac{24}{49}\sin(u)\sin(v-2)\cos(7\,v)\cos\left(\frac{u}{2}\right)\sin(12\,u-v)$$

$$-\frac{1}{7}\sin(u)\cos(v-2)\sin(7\,v)\sin\left(\frac{u}{2}\right)\cos(12\,u-v)$$

$$-\frac{24}{7}\sin(u)\cos(v-2)\sin(7\,v)\cos\left(\frac{u}{2}\right)\sin(12\,u-v)$$

$$+\frac{2}{7}\cos(u)\cos(v)\cos\left(\frac{u}{2}\right)\sin(12\,u-v)-\frac{1}{7}\sin(u)\sin(v)\sin\left(\frac{u}{2}\right)\cos(12\,u-v)$$

$$-\frac{24}{7}\sin(u)\sin(v)\cos\left(\frac{u}{2}\right)\sin(12\,u-v)-\frac{1}{7}\cos\left(\frac{u}{2}\right)\sin(12\,u-v)\cos(u)$$

$$+\frac{6}{7}\sin(v-2)\cos(7\,v)+6\cos(v-2)\sin(7\,v)+6\sin(v)-6\sin(v)\cos(u)^2$$

$$-\frac{6}{7}\sin(v-2)\cos(7\,v)\cos(u)^2-6\cos(v-2)\sin(7\,v)\cos(u)^2,$$

$$\frac{2}{7}\cos(u)\cos(v)\sin(u)\cos(v-2)\cos(8\,v)$$

$$-\frac{16}{7}\cos(u)\cos(v)\sin(u)\sin(v-2)\sin(8\,v)$$

$$+\frac{4}{7}\cos(u)\cos(v-2)\cos(7\,v)\sin(u)\cos(v)$$

$$+\frac{2}{7}\sin(u)\sin(v)\cos(u)\sin(v-2)\cos(8\,v)$$

$$+\frac{4}{7}\sin(u)\sin(v-2)\cos(7\,v)\cos(u)\sin(v)$$

$$+4\sin(u)\cos(v-2)\sin(7\,v)\cos(u)\sin(v)$$

$$-\frac{16}{49}\cos(u)\cos(v-2)\cos(7\,v)\sin(u)\sin(v-2)\sin(8\,v)$$

$$+\frac{2}{7}\sin(u)\cos(v-2)\sin(7\,v)\cos(u)\sin(v-2)\cos(8\,v)$$

$$-\frac{1}{7}\cos(u)\sin(u)\cos(v-2)\cos(8\,v)+\frac{8}{7}\cos(u)\sin(u)\sin(v-2)\sin(8\,v)$$

$$-\frac{2}{7}\sin(u)\sin(v-2)\cos(7\,v)\cos(u)-2\sin(u)\cos(v-2)\sin(7\,v)\cos(u)$$

$$-2\cos(u)\sin(u)\cos(v)-2\sin(u)\sin(v)\cos(u)$$

$$\left.+\frac{2}{49}\cos(u)\cos(7\,v)\sin(u)\cos(8\,v)+4\cos(u)\sin(u)\right]$$

(4)

Input surface parametric expression (1) to Vector Field (2); and
Calculate Integral of Dot Product [Vetor Field (2) and Tangent Plane
Normal Vector (4)] in range of 'u,v':

$$\int_0^{2\,\pi}\int_0^{\pi}\left(\left[\frac{1}{2}\left(\frac{2}{3}\sin(u)\cos(v)+\frac{2}{21}\sin(u)\cos(v-2)\cos(7\,v)-\frac{1}{2}\sin(u)+\frac{1}{3}\sin(u)\sin(v)\right.\right.\right.$$

$$+\frac{1}{42}\sin(u)\sin(v-2)\cos(8\,v)-\frac{3}{5}\cos(u)+\frac{1}{35}\cos\left(\frac{u}{2}\right)\cos(12\,u-v)\right)^3-\frac{1}{7}$$

$$\left(2\sin(u)\sin(v)+\frac{1}{7}\sin(u)\sin(v-2)\cos(8\,v)-\sin(u)\right)^2$$

$$\left.\left(3\cos(u)-\frac{1}{7}\cos\left(\frac{u}{2}\right)\cos(12\,u-v)\right)\right]\left(\frac{1}{7}\cos\left(\frac{u}{2}\right)\sin(12\,u-v)\cos(u)\right.$$

$$+\frac{96}{49}\cos\left(\frac{u}{2}\right)\sin(12\,u-v)\sin(u)\sin(v-2)\sin(8\,v)$$

$$-\frac{12}{49}\cos\left(\frac{u}{2}\right)\sin(12\,u-v)\sin(u)\cos(v-2)\cos(8\,v)$$

$$+\frac{4}{49}\sin\left(\frac{u}{2}\right)\cos(12\,u-v)\sin(u)\sin(v-2)\sin(8\,v)$$

$$-\frac{1}{49}\cos\left(\frac{u}{2}\right)\sin(12\,u-v)\cos(u)\sin(v-2)\cos(8\,v)$$

$$-\frac{1}{98}\sin\left(\frac{u}{2}\right)\cos(12\,u-v)\sin(u)\cos(v-2)\cos(8\,v)$$

$$-\frac{2}{7}\cos\left(\frac{u}{2}\right)\sin(12\,u-v)\cos(u)\sin(v)-\frac{1}{7}\sin\left(\frac{u}{2}\right)\cos(12\,u-v)\sin(u)\cos(v)$$

$$-\frac{24}{7}\cos\left(\frac{u}{2}\right)\sin(12\,u-v)\sin(u)\cos(v)+\frac{3}{7}\cos(v-2)\cos(8\,v)$$

$$-\frac{24}{7}\sin(v-2)\sin(8\,v)+6\cos(v)-6\cos(v)\cos(u)^2$$

$$\left.-\frac{3}{7}\cos(v-2)\cos(8\,v)\cos(u)^2+\frac{24}{7}\sin(v-2)\sin(8\,v)\cos(u)^2\right)+\left(\frac{1}{6}\right.$$

$$\left(2\sin(u)\cos(v)+\frac{2}{7}\sin(u)\cos(v-2)\cos(7v)-\sin(u)\right)$$

$$\left(2\sin(u)\sin(v)+\frac{1}{7}\sin(u)\sin(v-2)\cos(8v)-\sin(u)\right)$$

$$\left(3\cos(u)-\frac{1}{7}\cos\left(\frac{u}{2}\right)\cos(12u-v)\right)$$

$$-\frac{1}{6}\left(2\sin(u)\sin(v)+\frac{1}{7}\sin(u)\sin(v-2)\cos(8v)-\sin(u)\right)^2\right)\Bigg($$

$$\frac{2}{49}\cos(u)\cos(v-2)\cos(7v)\cos\left(\frac{u}{2}\right)\sin(12u-v)$$

$$-\frac{1}{49}\sin(u)\sin(v-2)\cos(7v)\sin\left(\frac{u}{2}\right)\cos(12u-v)$$

$$-\frac{24}{49}\sin(u)\sin(v-2)\cos(7v)\cos\left(\frac{u}{2}\right)\sin(12u-v)$$

$$-\frac{1}{7}\sin(u)\cos(v-2)\sin(7v)\sin\left(\frac{u}{2}\right)\cos(12u-v)$$

$$-\frac{24}{7}\sin(u)\cos(v-2)\sin(7v)\cos\left(\frac{u}{2}\right)\sin(12u-v)$$

$$+\frac{2}{7}\cos(u)\cos(v)\cos\left(\frac{u}{2}\right)\sin(12u-v)-\frac{1}{7}\sin(u)\sin(v)\sin\left(\frac{u}{2}\right)\cos(12u-v)$$

$$-\frac{24}{7}\sin(u)\sin(v)\cos\left(\frac{u}{2}\right)\sin(12u-v)-\frac{1}{7}\cos\left(\frac{u}{2}\right)\sin(12u-v)\cos(u)$$

$$+\frac{6}{7}\sin(v-2)\cos(7v)+6\cos(v-2)\sin(7v)+6\sin(v)-6\sin(v)\cos(u)^2$$

$$-\frac{6}{7}\sin(v-2)\cos(7v)\cos(u)^2-6\cos(v-2)\sin(7v)\cos(u)^2\Bigg)+\Bigg($$

$$\frac{1}{5}\left(2\sin(u)\cos(v)+\frac{2}{7}\sin(u)\cos(v-2)\cos(7v)-\sin(u)\right)^2+\frac{1}{15}$$

$$\left(2\sin(u)\cos(v)+\frac{2}{7}\sin(u)\cos(v-2)\cos(7v)-\sin(u)\right)$$

$$\left(2\sin(u)\sin(v)+\frac{1}{7}\sin(u)\sin(v-2)\cos(8v)-\sin(u)\right)$$

$$\left(3\cos(u)-\frac{1}{7}\cos\left(\frac{u}{2}\right)\cos(12u-v)\right)\Bigg)\left(\frac{2}{7}\cos(u)\cos(v)\sin(u)\cos(v-2)\cos(8v)\right.$$

$$-\frac{16}{7}\cos(u)\cos(v)\sin(u)\sin(v-2)\sin(8v)$$

$$+\frac{4}{7}\cos(u)\cos(v-2)\cos(7v)\sin(u)\cos(v)$$

$$+\frac{2}{7}\sin(u)\sin(v)\cos(u)\sin(v-2)\cos(8v)$$

$$+ \frac{4}{7} \sin(u) \sin(v-2) \cos(7v) \cos(u) \sin(v)$$

$$+ 4 \sin(u) \cos(v-2) \sin(7v) \cos(u) \sin(v)$$

$$- \frac{16}{49} \cos(u) \cos(v-2) \cos(7v) \sin(u) \sin(v-2) \sin(8v)$$

$$+ \frac{2}{7} \sin(u) \cos(v-2) \sin(7v) \cos(u) \sin(v-2) \cos(8v)$$

$$- \frac{1}{7} \cos(u) \sin(u) \cos(v-2) \cos(8v) + \frac{8}{7} \cos(u) \sin(u) \sin(v-2) \sin(8v)$$

$$- \frac{2}{7} \sin(u) \sin(v-2) \cos(7v) \cos(u) - 2 \sin(u) \cos(v-2) \sin(7v) \cos(u)$$

$$- 2 \cos(u) \sin(u) \cos(v) - 2 \sin(u) \sin(v) \cos(u)$$

$$+ \frac{2}{49} \cos(u) \cos(7v) \sin(u) \cos(8v) + 4 \cos(u) \sin(u) \Bigg) du\, dv =$$

$$\frac{1905330274956 20399\, \pi}{48617598427138125} + \frac{2304571\, \pi^2}{1536640}$$

$$+ \frac{304753183422133444 0562983}{49935849797624790416016600} \cos(4)\, \pi$$

$$+ \frac{49743302900491179001 8901}{110152609847701743564 74250} \sin(4)\, \pi + \frac{11761}{384160} \sin(4)\, \pi^2$$

$$- \frac{3}{439040} \cos(4)\, \pi^2$$

(5)

Second, Free Volume Element Integral:

Entirely multiply radius vector 'r' (suppose r>0) by each item of surface parametrized expression (1), convert 'Parametrized expression of surface' to 'Parametrized expression of surface coordinates':

$$\Bigg[r\Bigg(2 \sin(u) \cos(v) + \frac{2}{7} \sin(u) \cos(v-2) \cos(7v) - \sin(u) \Bigg),$$

$$r\Bigg(2 \sin(u) \sin(v) + \frac{1}{7} \sin(u) \sin(v-2) \cos(8v) - \sin(u) \Bigg),$$

$$r\Bigg(3 \cos(u) - \frac{1}{7} \cos\Bigg(\frac{u}{2} \Bigg) \cos(12u-v) \Bigg) \Bigg]$$

(6)

According to parametrized expression of surface coordinates (6), define and calculate Second Matrix of Partial Derivatives, obtain general expression of surface coordinates volume element coefficient (7):

$$
\begin{bmatrix}
\frac{\partial}{\partial r}(r(2\sin(u)cos(v)+\frac{2}{7}\sin(u)\cos(v\text{-}2)\cos(7v)\text{-}\sin(u))) & \frac{\partial}{\partial u}(r(2\sin(u)cos(v)+\frac{2}{7}\sin(u)\cos(v\text{-}2)\cos(7v)\text{-}\sin(u))) & \frac{\partial}{\partial v}(r(2\sin(u)cos(v)+\frac{2}{7}\sin(u)\cos(v\text{-}2)\cos(7v)\text{-}\sin(u))) \\
\frac{\partial}{\partial r}(r(2\sin(u)\sin(v)+\frac{1}{7}\sin(u)\sin(v\text{-}2)\cos(8v)\text{-}\sin(u))) & \frac{\partial}{\partial u}(r(2\sin(u)\sin(v)+\frac{1}{7}\sin(u)\sin(v\text{-}2)\cos(8v)\text{-}\sin(u))) & \frac{\partial}{\partial v}(r(2\sin(u)\sin(v)+\frac{1}{7}\sin(u)\sin(v\text{-}2)\cos(8v)\text{-}\sin(u))) \\
\frac{\partial}{\partial r}(r(3\cos(u)-\frac{1}{7}\cos(\frac{u}{2})\cos(12u-v))) & \frac{\partial}{\partial u}(r(3\cos(u)-\frac{1}{7}\cos(\frac{u}{2})\cos(12u-v))) & \frac{\partial}{\partial v}(r(3\cos(u)-\frac{1}{7}\cos(\frac{u}{2})\cos(12u-v)))
\end{bmatrix}
$$

$=$

$$-\frac{2}{49}\sin(u)^2\sin(v)\,r^2\sin(v-2)\cos(7v)\sin\!\left(\frac{u}{2}\right)\cos(12u-v)$$

$$-\frac{1}{49}\sin(u)^2\cos(v)\,r^2\cos(v-2)\cos(8v)\sin\!\left(\frac{u}{2}\right)\cos(12u-v)$$

$$-\frac{24}{49}\sin(u)^2\cos(v)\,r^2\cos(v-2)\cos(8v)\cos\!\left(\frac{u}{2}\right)\sin(12u-v)$$

$$+\frac{8}{49}\sin(u)^2\cos(v)\,r^2\sin(v-2)\sin(8v)\sin\!\left(\frac{u}{2}\right)\cos(12u-v)$$

$$+\frac{192}{49}\sin(u)^2\cos(v)\,r^2\sin(v-2)\sin(8v)\cos\!\left(\frac{u}{2}\right)\sin(12u-v)$$

$$-\frac{48}{49}\sin(u)^2\sin(v)\,r^2\sin(v-2)\cos(7v)\cos\!\left(\frac{u}{2}\right)\sin(12u-v)$$

$$-\frac{2}{7}\sin(u)^2\sin(v)\,r^2\cos(v-2)\sin(7v)\sin\!\left(\frac{u}{2}\right)\cos(12u-v)$$

$$-\frac{2}{49}\sin(u)^2\cos(v-2)\cos(7v)\,r^2\cos(v)\sin\!\left(\frac{u}{2}\right)\cos(12u-v)$$

$$-\frac{48}{49}\sin(u)^2\cos(v-2)\cos(7v)\,r^2\cos(v)\cos\!\left(\frac{u}{2}\right)\sin(12u-v)$$

$$-\frac{1}{343}\sin(u)^2\cos(v-2)^2\cos(7v)\,r^2\cos(8v)\sin\!\left(\frac{u}{2}\right)\cos(12u-v)$$

$$-\frac{24}{343}\sin(u)^2\cos(v-2)^2\cos(7v)\,r^2\cos(8v)\cos\!\left(\frac{u}{2}\right)\sin(12u-v)$$

$$-\frac{24}{49}\sin(u)^2\sin(v-2)\cos(8v)\,r^2\sin(v)\cos\!\left(\frac{u}{2}\right)\sin(12u-v)$$

$$-\frac{1}{343}\sin(u)^2\sin(v-2)^2\cos(8v)\,r^2\cos(7v)\sin\!\left(\frac{u}{2}\right)\cos(12u-v)$$

$$-\frac{24}{343}\sin(u)^2\sin(v-2)^2\cos(8v)\,r^2\cos(7v)\cos\!\left(\frac{u}{2}\right)\sin(12u-v)$$

$$-\frac{48}{49}r^2\cos(u)^2\cos(v-2)\cos(7v)\sin(u)\sin(v-2)\sin(8v)$$

$$+\frac{6}{7}r^2\sin(u)\cos(v-2)\sin(7v)\cos(u)^2\sin(v-2)\cos(8v)$$

$$+\frac{1}{49}\cos\!\left(\frac{u}{2}\right)\cos(12u-v)\,r^2\cos(u)\sin(u)\cos(v-2)\cos(8v)$$

$$-\frac{8}{49}\cos\!\left(\frac{u}{2}\right)\cos(12u-v)\,r^2\cos(u)\sin(u)\sin(v-2)\sin(8v)$$

$$-\frac{48}{7}\sin(u)^2\sin(v)\,r^2\cos(v-2)\sin(7\,v)\cos\!\left(\frac{u}{2}\right)\sin(12\,u-v)$$

$$+\frac{2}{49}\cos\!\left(\frac{u}{2}\right)\cos(12\,u-v)\,r^2\sin(u)\sin(v-2)\cos(7\,v)\cos(u)$$

$$+\frac{2}{7}\cos\!\left(\frac{u}{2}\right)\cos(12\,u-v)\,r^2\sin(u)\cos(v-2)\sin(7\,v)\cos(u)$$

$$-\frac{1}{49}\sin(u)^2\sin(v-2)\cos(8\,v)\,r^2\sin(v)\sin\!\left(\frac{u}{2}\right)\cos(12\,u-v)$$

$$+\frac{12}{7}r^2\cos(u)^2\cos(v-2)\cos(7\,v)\sin(u)\cos(v)$$

$$-\frac{48}{7}r^2\cos(u)^2\cos(v)\sin(u)\sin(v-2)\sin(8\,v)$$

$$+\frac{6}{49}r^2\cos(u)^2\cos(v-2)^2\cos(7\,v)\sin(u)\cos(8\,v)$$

$$+\frac{6}{7}\sin(u)^3\sin(v-2)\cos(8\,v)\,r^2\sin(v)$$

$$-\frac{2}{343}\cos\!\left(\frac{u}{2}\right)\cos(12\,u-v)\,r^2\cos(u)\cos(v-2)^2\cos(7\,v)\sin(u)\cos(8\,v)$$

$$+\frac{2}{7}\cos\!\left(\frac{u}{2}\right)\cos(12\,u-v)\,r^2\cos(u)\sin(u)\cos(v)$$

$$+\frac{192}{343}\sin(u)^2\cos(v-2)\cos(7\,v)\,r^2\sin(v-2)\sin(8\,v)\cos\!\left(\frac{u}{2}\right)\sin(12\,u-v)$$

$$+\frac{6}{7}\sin(u)^3\sin(v-2)\cos(8\,v)\,r^2\cos(v-2)\sin(7\,v)$$

$$+\frac{8}{343}\sin(u)^2\cos(v-2)\cos(7\,v)\,r^2\sin(v-2)\sin(8\,v)\sin\!\left(\frac{u}{2}\right)\cos(12\,u-v)$$

$$+\frac{1}{98}r^2\sin(u)^2\cos(v-2)\cos(8\,v)\sin\!\left(\frac{u}{2}\right)\cos(12\,u-v)$$

$$+\frac{12}{7}\sin(u)^3\cos(v-2)\cos(7\,v)\,r^2\cos(v)$$

$$-\frac{48}{49}\sin(u)^3\cos(v-2)\cos(7\,v)\,r^2\sin(v-2)\sin(8\,v)$$

$$+\frac{24}{7}r^2\sin(u)^2\cos(v-2)\sin(7\,v)\cos\!\left(\frac{u}{2}\right)\sin(12\,u-v)$$

$$-\frac{4}{49}\cos\!\left(\frac{u}{2}\right)\cos(12\,u-v)\,r^2\sin(u)\sin(v-2)\cos(7\,v)\cos(u)\sin(v)$$

$$+\frac{24}{49}r^2\sin(u)^2\sin(v-2)\cos(7\,v)\cos\!\left(\frac{u}{2}\right)\sin(12\,u-v)$$

$$-\frac{1}{49}\sin(u)^2\sin(v-2)\cos(8\,v)\,r^2\cos(v-2)\sin(7\,v)\sin\!\left(\frac{u}{2}\right)\cos(12\,u-v)$$

$$+\frac{1}{49}r^2\sin(u)^2\sin(v-2)\cos(7\,v)\sin\!\left(\frac{u}{2}\right)\cos(12\,u-v)$$

$$-\frac{24}{49}\sin(u)^2\sin(v-2)\cos(8\,v)\,r^2\cos(v-2)\sin(7\,v)\cos\!\left(\frac{u}{2}\right)\sin(12\,u-v)$$

$$-\frac{96}{49}r^2\sin(u)^2\sin(v-2)\sin(8v)\cos\left(\frac{u}{2}\right)\sin(12u-v)$$

$$-\frac{2}{49}\cos\left(\frac{u}{2}\right)\cos(12u-v)r^2\cos(u)\cos(v)\sin(u)\cos(v-2)\cos(8v)$$

$$-\frac{4}{49}r^2\sin(u)^2\sin(v-2)\sin(8v)\sin\left(\frac{u}{2}\right)\cos(12u-v)$$

$$+\frac{16}{49}\cos\left(\frac{u}{2}\right)\cos(12u-v)r^2\cos(u)\cos(v)\sin(u)\sin(v-2)\sin(8v)$$

$$+\frac{6}{7}r^2\cos(u)^2\cos(v)\sin(u)\cos(v-2)\cos(8v)$$

$$-\frac{4}{49}\cos\left(\frac{u}{2}\right)\cos(12u-v)r^2\cos(u)\cos(v-2)\cos(7v)\sin(u)\cos(v)$$

$$-\frac{4}{7}\cos\left(\frac{u}{2}\right)\cos(12u-v)r^2\sin(u)\sin(v)^2\cos(u)$$

$$-\frac{4}{7}\cos\left(\frac{u}{2}\right)\cos(12u-v)r^2\sin(u)\cos(v-2)\sin(7v)\cos(u)\sin(v)$$

$$+\frac{1}{7}r^2\sin(u)^2\cos(v-2)\sin(7v)\sin\left(\frac{u}{2}\right)\cos(12u-v)$$

$$-\frac{2}{49}\cos\left(\frac{u}{2}\right)\cos(12u-v)r^2\sin(u)\cos(v-2)\sin(7v)\cos(u)\sin(v-2)\cos(8v)$$

$$+\frac{12}{7}\sin(u)^3\sin(v)r^2\sin(v-2)\cos(7v)$$

$$+\frac{16}{343}\cos\left(\frac{u}{2}\right)\cos(12u-v)r^2\cos(u)\cos(v-2)\cos(7v)\sin(u)\sin(v-2)\sin(8v)$$

$$-\frac{48}{7}\sin(u)^2\sin(v)^2r^2\cos\left(\frac{u}{2}\right)\sin(12u-v)$$

$$-\frac{2}{49}\cos\left(\frac{u}{2}\right)\cos(12u-v)r^2\sin(u)\sin(v)\cos(u)\sin(v-2)\cos(8v)$$

$$+\frac{2}{7}\cos\left(\frac{u}{2}\right)\cos(12u-v)r^2\sin(u)\sin(v)\cos(u)$$

$$-\frac{2}{343}\cos\left(\frac{u}{2}\right)\cos(12u-v)r^2\sin(u)\sin(v-2)^2\cos(7v)\cos(u)\cos(8v)$$

$$+\frac{12}{49}r^2\sin(u)^2\cos(v-2)\cos(8v)\cos\left(\frac{u}{2}\right)\sin(12u-v)+12\sin(u)^3\sin(v)^2r^2$$

$$-6r^2\sin(u)^3\cos(v)+12\sin(u)^3\cos(v)^2r^2-6r^2\sin(u)^3\sin(v)$$

$$+12r^2\sin(u)\cos(v-2)\sin(7v)\cos(u)^2\sin(v)$$

$$+\frac{6}{7}r^2\sin(u)\sin(v)\cos(u)^2\sin(v-2)\cos(8v)$$

$$+\frac{6}{49}r^2\sin(u)\sin(v-2)^2\cos(7v)\cos(u)^2\cos(8v)$$

$+ \dfrac{12}{7} r^2 \sin(u) \sin(v-2) \cos(7v) \cos(u)^2 \sin(v)$

$- \dfrac{4}{7} \cos\left(\dfrac{u}{2}\right) \cos(12u-v) r^2 \cos(u) \cos(v)^2 \sin(u) - 6 r^2 \sin(u)^3 \cos(v-2) \sin(7v)$

$- \dfrac{3}{7} r^2 \sin(u)^3 \cos(v-2) \cos(8v) + \dfrac{24}{7} r^2 \sin(u)^3 \sin(v-2) \sin(8v)$

$- \dfrac{6}{7} r^2 \sin(u)^3 \sin(v-2) \cos(7v) - 6 r^2 \cos(u)^2 \sin(u) \cos(v)$

$+ 12 r^2 \sin(u) \sin(v)^2 \cos(u)^2 - 6 r^2 \sin(u) \sin(v) \cos(u)^2$

$+ 12 r^2 \cos(u)^2 \cos(v)^2 \sin(u) + \dfrac{1}{7} r^2 \sin(u)^2 \cos(v) \sin\left(\dfrac{u}{2}\right) \cos(12u-v)$

$+ \dfrac{24}{7} r^2 \sin(u)^2 \cos(v) \cos\left(\dfrac{u}{2}\right) \sin(12u-v)$

$- \dfrac{2}{7} \sin(u)^2 \sin(v)^2 r^2 \sin\left(\dfrac{u}{2}\right) \cos(12u-v) - \dfrac{6}{7} r^2 \sin(u) \sin(v-2) \cos(7v) \cos(u)^2$

$- 6 r^2 \sin(u) \cos(v-2) \sin(7v) \cos(u)^2$

$- \dfrac{2}{7} \sin(u)^2 \cos(v)^2 r^2 \sin\left(\dfrac{u}{2}\right) \cos(12u-v)$

$- \dfrac{48}{7} \sin(u)^2 \cos(v)^2 r^2 \cos\left(\dfrac{u}{2}\right) \sin(12u-v)$

$+ \dfrac{6}{7} \sin(u)^3 \cos(v) r^2 \cos(v-2) \cos(8v) - \dfrac{48}{7} \sin(u)^3 \cos(v) r^2 \sin(v-2) \sin(8v)$

$+ \dfrac{6}{49} \sin(u)^3 \cos(v-2)^2 \cos(7v) r^2 \cos(8v)$

$+ 12 \sin(u)^3 \sin(v) r^2 \cos(v-2) \sin(7v)$

$+ \dfrac{6}{49} \sin(u)^3 \sin(v-2)^2 \cos(8v) r^2 \cos(7v)$

$+ \dfrac{1}{7} r^2 \sin(u)^2 \sin(v) \sin\left(\dfrac{u}{2}\right) \cos(12u-v)$

$+ \dfrac{24}{7} r^2 \sin(u)^2 \sin(v) \cos\left(\dfrac{u}{2}\right) \sin(12u-v)$

$- \dfrac{3}{7} r^2 \cos(u)^2 \sin(u) \cos(v-2) \cos(8v) + \dfrac{24}{7} r^2 \cos(u)^2 \sin(u) \sin(v-2) \sin(8v)$

Calculate Divergence of Vector Field (2):

$$\left(\dfrac{\partial}{\partial x}\left(\dfrac{\left(\dfrac{x}{3}+\dfrac{y}{6}-\dfrac{z}{5}\right)^3}{2}-\dfrac{y^2 z}{7}\right)\right) + \left(\dfrac{\partial}{\partial y}\left(\dfrac{1}{6}xyz-\dfrac{1}{6}y^2\right)\right) + \left(\dfrac{\partial}{\partial z}\left(\dfrac{1}{5}x^2+\dfrac{1}{15}xyz\right)\right) =$$

$$\frac{\left(\dfrac{x}{3}+\dfrac{y}{6}-\dfrac{z}{5}\right)^{2}}{2}+\frac{xz}{6}-\frac{y}{3}+\frac{xy}{15} \quad (8)$$

Input parametrized expression of surface coordinates (6) to Divergence (8); and Calculate Triple Integrals of Product [Divergence (8) and Surface Coordinates Volume Element] in range of 'r,u,v' (9):

$$\int_{0}^{2\pi}\int_{0}^{\pi}\int_{0}^{1} -\frac{1}{686}\left(\frac{1}{2}\left(\frac{1}{3}r\left(2\sin(u)\cos(v)+\frac{2}{7}\sin(u)\cos(v-2)\cos(7v)-\sin(u)\right)\right.\right.$$

$$+\frac{1}{6}r\left(2\sin(u)\sin(v)+\frac{1}{7}\sin(u)\sin(v-2)\cos(8v)-\sin(u)\right)$$

$$\left.-\frac{1}{5}r\left(3\cos(u)-\frac{1}{7}\cos\left(\frac{u}{2}\right)\cos(12u-v)\right)\right)^{2}+\frac{1}{6}r^{2}$$

$$\left(2\sin(u)\cos(v)+\frac{2}{7}\sin(u)\cos(v-2)\cos(7v)-\sin(u)\right)$$

$$\left(3\cos(u)-\frac{1}{7}\cos\left(\frac{u}{2}\right)\cos(12u-v)\right)$$

$$-\frac{1}{3}r\left(2\sin(u)\sin(v)+\frac{1}{7}\sin(u)\sin(v-2)\cos(8v)-\sin(u)\right)+\frac{1}{15}r^{2}$$

$$\left(2\sin(u)\cos(v)+\frac{2}{7}\sin(u)\cos(v-2)\cos(7v)-\sin(u)\right)$$

$$\left.\left(2\sin(u)\sin(v)+\frac{1}{7}\sin(u)\sin(v-2)\cos(8v)-\sin(u)\right)\right)\sin(u)\left(-8232\right.$$

$$-2352\cos\left(\frac{u}{2}\right)\sin(12u-v)\sin(u)\cos(v)+4116\cos(v)+4116\sin(v)$$

$$-7\sin\left(\frac{u}{2}\right)\cos(12u-v)\sin(u)\cos(v-2)\cos(8v)$$

$$+56\sin\left(\frac{u}{2}\right)\cos(12u-v)\sin(u)\sin(v-2)\sin(8v)$$

$$-168\cos\left(\frac{u}{2}\right)\sin(12u-v)\sin(u)\cos(v-2)\cos(8v)$$

$$+1344\cos\left(\frac{u}{2}\right)\sin(12u-v)\sin(u)\sin(v-2)\sin(8v)$$

$$-98\sin\left(\frac{u}{2}\right)\cos(12u-v)\sin(u)\cos(v)+294\cos(v-2)\cos(8v)$$

$$-14\sin(u)\sin(v-2)\cos(7v)\sin\left(\frac{u}{2}\right)\cos(12u-v)$$

$$- 336 \sin(u) \sin(v-2) \cos(7v) \cos\left(\frac{u}{2}\right) \sin(12u-v)$$

$$- 98 \sin(u) \cos(v-2) \sin(7v) \sin\left(\frac{u}{2}\right) \cos(12u-v)$$

$$- 2352 \sin(u) \cos(v-2) \sin(7v) \cos\left(\frac{u}{2}\right) \sin(12u-v)$$

$$- 384 \sin(u) \cos(v-2) \cos(7v) \sin(v-2) \sin(8v) \cos\left(\frac{u}{2}\right) \sin(12u-v)$$

$$+ 28 \cos\left(\frac{u}{2}\right) \cos(12u-v) \cos(v-2) \sin(7v) \cos(u) \sin(v-2) \cos(8v)$$

$$+ 336 \sin(u) \sin(v-2) \cos(8v) \cos(v-2) \sin(7v) \cos\left(\frac{u}{2}\right) \sin(12u-v)$$

$$+ 14 \sin(u) \sin(v-2) \cos(8v) \cos(v-2) \sin(7v) \sin\left(\frac{u}{2}\right) \cos(12u-v)$$

$$- 32 \cos\left(\frac{u}{2}\right) \cos(12u-v) \cos(u) \cos(v-2) \cos(7v) \sin(v-2) \sin(8v)$$

$$- 16 \sin(u) \cos(v-2) \cos(7v) \sin(v-2) \sin(8v) \sin\left(\frac{u}{2}\right) \cos(12u-v)$$

$$- 98 \sin(u) \sin(v) \sin\left(\frac{u}{2}\right) \cos(12u-v)$$

$$+ 48 \sin(u) \cos(7v) \cos(8v) \cos\left(\frac{u}{2}\right) \sin(12u-v) - 2352 \sin(v-2) \sin(8v)$$

$$+ 588 \sin(v-2) \cos(7v) + 4116 \cos(v-2) \sin(7v)$$

$$- 2352 \sin(u) \sin(v) \cos\left(\frac{u}{2}\right) \sin(12u-v) - 84 \cos(7v) \cos(8v)$$

$$- 2688 \sin(u) \cos(v) \sin(v-2) \sin(8v) \cos\left(\frac{u}{2}\right) \sin(12u-v)$$

$$+ 28 \sin(u) \sin(v) \sin(v-2) \cos(7v) \sin\left(\frac{u}{2}\right) \cos(12u-v)$$

$$+ 14 \sin(u) \cos(v) \cos(v-2) \cos(8v) \sin\left(\frac{u}{2}\right) \cos(12u-v)$$

$$+ 336 \sin(u) \cos(v) \cos(v-2) \cos(8v) \cos\left(\frac{u}{2}\right) \sin(12u-v)$$

$$- 112 \sin(u) \cos(v) \sin(v-2) \sin(8v) \sin\left(\frac{u}{2}\right) \cos(12u-v)$$

$$+ 196 \sin(u) \sin(v) \cos(v-2) \sin(7v) \sin\left(\frac{u}{2}\right) \cos(12u-v)$$

$$+ 28 \sin(u) \cos(v-2) \cos(7v) \cos(v) \sin\left(\frac{u}{2}\right) \cos(12u-v)$$

$$+ 672 \sin(u) \cos(v-2) \cos(7v) \cos(v) \cos\left(\frac{u}{2}\right) \sin(12u-v)$$

$$+ 4704 \sin(u) \sin(v) \cos(v-2) \sin(7v) \cos\left(\frac{u}{2}\right) \sin(12u-v)$$

$$+ 14 \sin(u) \sin(v-2) \cos(8v) \sin(v) \sin\left(\frac{u}{2}\right) \cos(12u-v)$$

$$+ 672 \sin(u) \sin(v) \sin(v-2) \cos(7v) \cos\left(\frac{u}{2}\right) \sin(12u-v)$$

$$+ 336 \sin(u) \sin(v-2) \cos(8v) \sin(v) \cos\left(\frac{u}{2}\right) \sin(12u-v)$$

$$+ 28 \cos\left(\frac{u}{2}\right) \cos(12u-v) \cos(u) \cos(v) \cos(v-2) \cos(8v)$$

$$- 196 \cos\left(\frac{u}{2}\right) \cos(12u-v) \cos(u) \cos(v)$$

$$- 224 \cos\left(\frac{u}{2}\right) \cos(12u-v) \cos(u) \cos(v) \sin(v-2) \sin(8v)$$

$$+ 28 \cos\left(\frac{u}{2}\right) \cos(12u-v) \sin(v) \cos(u) \sin(v-2) \cos(8v)$$

$$+ 56 \cos\left(\frac{u}{2}\right) \cos(12u-v) \cos(u) \cos(v-2) \cos(7v) \cos(v)$$

$$- 196 \cos\left(\frac{u}{2}\right) \cos(12u-v) \sin(v) \cos(u)$$

$$+ 56 \cos\left(\frac{u}{2}\right) \cos(12u-v) \sin(v-2) \cos(7v) \cos(u) \sin(v)$$

$$+ 392 \cos\left(\frac{u}{2}\right) \cos(12u-v) \cos(v-2) \sin(7v) \cos(u) \sin(v)$$

$$- 196 \cos\left(\frac{u}{2}\right) \cos(12u-v) \cos(v-2) \sin(7v) \cos(u)$$

$$- 28 \cos\left(\frac{u}{2}\right) \cos(12u-v) \sin(v-2) \cos(7v) \cos(u)$$

$$+ 112 \cos\left(\frac{u}{2}\right) \cos(12u-v) \cos(u) \sin(v-2) \sin(8v)$$

$$- 14 \cos\left(\frac{u}{2}\right) \cos(12u-v) \cos(u) \cos(v-2) \cos(8v)$$

$$+ 2 \sin(u) \cos(7v) \cos(8v) \sin\left(\frac{u}{2}\right) \cos(12u-v)$$

$$+ 4 \cos\left(\frac{u}{2}\right) \cos(12u-v) \cos(u) \cos(7v) \cos(8v)$$

$$+ 4704 \cos(v) \sin(v-2) \sin(8v) - 1176 \cos(v-2) \cos(7v) \cos(v)$$
$$- 588 \sin(v) \sin(v-2) \cos(8v) - 8232 \cos(v-2) \sin(7v) \sin(v)$$

$$- 1176 \sin(v-2) \cos(7v) \sin(v) + 392 \cos\left(\frac{u}{2}\right) \cos(12u-v) \cos(u)$$

$$+ 4704 \sin(u) \cos\left(\frac{u}{2}\right) \sin(12u-v) + 196 \sin(u) \sin\left(\frac{u}{2}\right) \cos(12u-v)$$

$$- 588 \cos(v) \cos(v - 2) \cos(8\,v) + 672 \cos(v - 2) \cos(7\,v) \sin(v - 2) \sin(8\,v)$$

$$- 588 \cos(v - 2) \sin(7\,v) \sin(v - 2) \cos(8\,v) \Bigg) r^2 \; dr \; du \; dv =$$

$$\frac{1905330274956203999\,\pi}{48617598427138125} + \frac{2304571\,\pi^2}{1536640}$$

$$+ \frac{30475318342213344440562983}{4993584979762479041601 6600} \cos(4)\,\pi$$

$$+ \frac{49743302900491179001 8901}{110152609847701743564742 50} \sin(4)\,\pi + \frac{11761}{384160} \sin(4)\,\pi^2$$

$$- \frac{3}{439040} \cos(4)\,\pi^2$$

$$(9)$$

Integral Precision Value (5) [Vector Field (2) on Target Surface (1)], is equal to Volume Element Integral Precision Value (9) [Divergence (8) in Spacial Bounded Closed Region 'Ω' that Target Surface (1) surrounds], Complete Calculation and Validation of Divergence Theorem at Manifold

3.4 Numerical Model of Divergence Theorem at Manifold (II) [Program Template of Waterloo Maplesoft, Optional]

```
> restart;
> with(plots):with(linalg):
>CS:=[2*sin(u)*cos(v)+sin(5*u)/5,2*sin(u)*sin(v),
cos(u)+sin(5*v)/5+sin(2*u)];
# Define parametrized expression of discretional simply connected
orientable closed surface 'CS'
```

$$CS := \left[2 \sin(u) \cos(v) + \frac{1}{5} \sin(5\,u),\; 2 \sin(u) \sin(v),\; \cos(u) + \frac{1}{5} \sin(5\,v) + \sin(2\,u) \right]$$

```
> rgu:=[0,Pi];
```

$$rgu := [0, \pi]$$

```
> rgv:=[0,2*Pi];
# Set range of parameters 'u,v', make surface 'CS' closed
```

$$rgv := [0, 2\,\pi]$$

```
> plot3d(CS,u=rgu[1]..rgu[2],v=rgv[1]..rgv[2],scaling=constrained,
projection=0.9,numpoints=6000);g1:=%:
```

\# Draw discretional simply connected orientable closed surface 'CS'

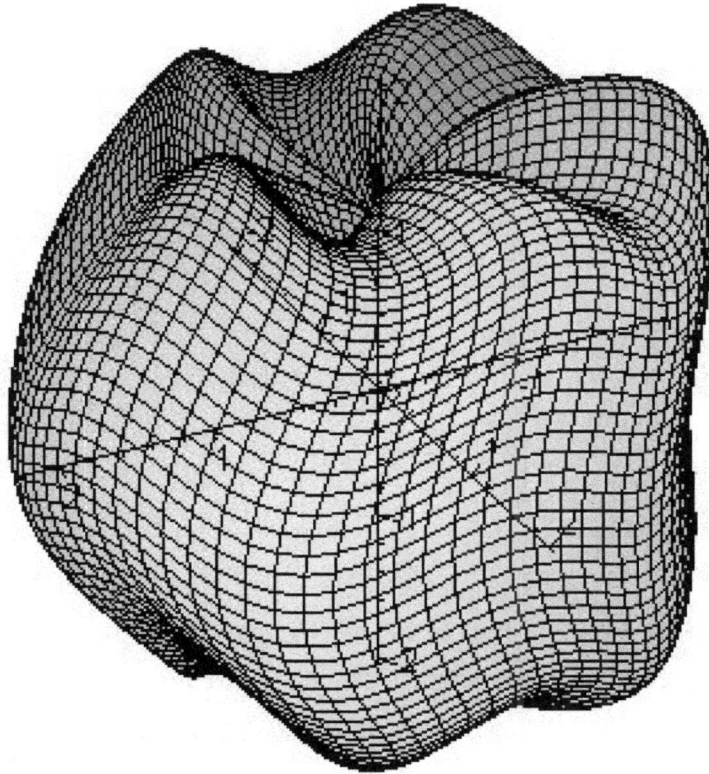

Figure(I)3.4.1 Asymmetrical、Irregular Discretional
Simply Connected、Orientable Closed Surface 'CS'

\> V:=[(x/3-y/6+z/5)^3/3-y^2/5-z^3/7,x*y*z/6-z^2/6,x^2*y/6-y*z/5];
\# Define discretional spacial vector field 'V' (Suppose Vector Field
'V' possesses 1th order continuous partial derivatives in spacial
bounded closed region 'Ω' that Surface 'CS' surrounds)

$$V:=\left[\frac{\left(\dfrac{x}{3}-\dfrac{y}{6}+\dfrac{z}{5}\right)^3}{3}-\frac{y^2}{5}-\frac{z^3}{7},\frac{1}{6}xyz-\frac{1}{6}z^2,\frac{1}{6}x^2y-\frac{1}{5}yz\right]$$

\> Diff(V[1],x)+Diff(V[2],y)+Diff(V[3],z)=diff(V[1],x)+diff(V[2],y)
+diff(V[3],z);diV:=rhs(%);
\# Calculate divergence 'diV' of vector field 'V'

$$\left(\frac{\partial}{\partial x}\left(\frac{\left(\frac{x}{3}-\frac{y}{6}+\frac{z}{5}\right)^3}{3}-\frac{y^2}{5}-\frac{z^3}{7}\right)\right)+\left(\frac{\partial}{\partial y}\left(\frac{1}{6}xyz-\frac{1}{6}z^2\right)\right)+\left(\frac{\partial}{\partial z}\left(\frac{1}{6}x^2y-\frac{1}{5}yz\right)\right)=$$

$$\frac{\left(\frac{x}{3}-\frac{y}{6}+\frac{z}{5}\right)^2}{3}+\frac{xz}{6}-\frac{y}{5}$$

$$diV:=\frac{\left(\frac{x}{3}-\frac{y}{6}+\frac{z}{5}\right)^2}{3}+\frac{xz}{6}-\frac{y}{5}$$

```
> rgx:=[-2,2];
```

$$rgx:=[-2,2]$$

```
> rgy:=[-2,2];
```

$$rgy:=[-2,2]$$

```
> rgz:=[-2,2];
```

$$rgz:=[-2,2]$$

```
> fieldplot3d(V,x=rgx[1]..rgx[2],y=rgy[1]..rgy[2],
z=rgz[1]..rgz[2],arrows=SLIM,color=blue):g2:=%:
> implicitplot3d(diV,x=rgx[1]..rgx[2],y=rgy[1]..rgy[2],
z=rgz[1]..rgz[2],style=wireframe,color=cyan,numpoints=3000):g3:=%
: > display(g1,g2,g3);
```

978-1-62265-930-2 (online)   978-1-62265-931-9 (paper)

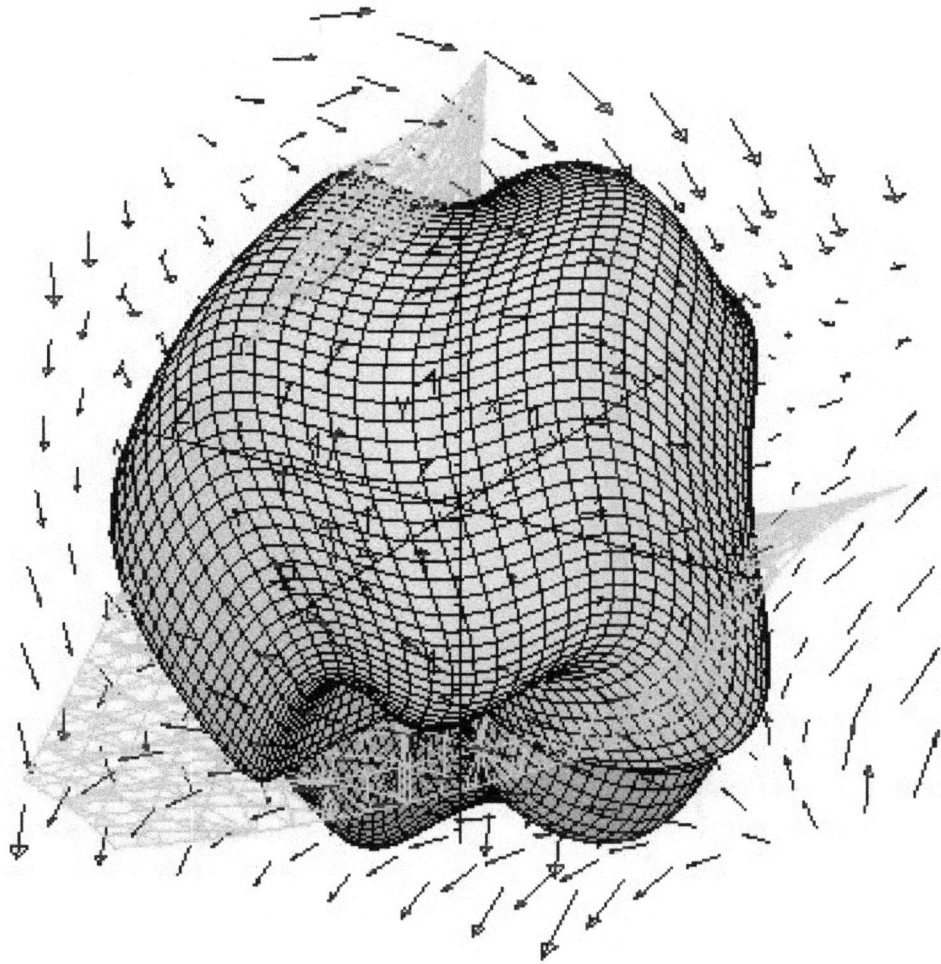

Figure(I)3.4.2 Surface[closed] 'CS'、Vector Field 'V' and
Level Surface of Divergence 'diV'

```
> x:=CS[1]:y:=CS[2]:z:=CS[3]:
> matrix(3,3,[i,j,k,Diff(x,u),Diff(y,u),Diff(z,u),Diff(x,v),
Diff(y,v),Diff(z,v)])=
matrix(3,3,[i,j,k,diff(x,u),diff(y,u),diff(z,u),diff(x,v),
diff(y,v),diff(z,v)]);m1:=rhs(%);
# Define and calculate matrix of partial derivatives 'm1', obtain
'Tangent plane normal vector' of surface 'CS'
```

$$[i,j,k]$$

$$\left[\frac{\partial}{\partial u}\left(2\sin(u)\cos(v)+\frac{1}{5}\sin(5\,u)\right),\frac{\partial}{\partial u}\left(2\sin(u)\sin(v)\right),\right.$$

$$\left.\frac{\partial}{\partial u}\left(\cos(u)+\frac{1}{5}\sin(5\,v)+\sin(2\,u)\right)\right]$$

$$\left[\frac{\partial}{\partial v}\left(2\sin(u)\cos(v)+\frac{1}{5}\sin(5\,u)\right),\frac{\partial}{\partial v}\left(2\sin(u)\sin(v)\right),\right.$$

$$\left.\frac{\partial}{\partial v}\left(\cos(u)+\frac{1}{5}\sin(5\,v)+\sin(2\,u)\right)\right]=$$

$$\begin{bmatrix} i & j & k \\ 2\cos(u)\cos(v)+\cos(5\,u) & 2\cos(u)\sin(v) & -\sin(u)+2\cos(2\,u) \\ -2\sin(u)\sin(v) & 2\sin(u)\cos(v) & \cos(5\,v) \end{bmatrix}$$

$$m1:=\begin{bmatrix} i & j & k \\ 2\cos(u)\cos(v)+\cos(5\,u) & 2\cos(u)\sin(v) & -\sin(u)+2\cos(2\,u) \\ -2\sin(u)\sin(v) & 2\sin(u)\cos(v) & \cos(5\,v) \end{bmatrix}$$

```
> det(m1):
> mn:=simplify(%);
```

$$mn:=2\,i\cos(u)\sin(v)\cos(5\,v)+2\,i\cos(v)-2\,i\cos(v)\cos(u)^2$$
$$-4\,i\sin(u)\cos(v)\cos(2\,u)-2\cos(u)\cos(v)\,j\cos(5\,v)-\cos(5\,u)\,j\cos(5\,v)$$
$$+2\cos(5\,u)\,k\sin(u)\cos(v)+2\sin(v)\,j-2\sin(v)\,j\cos(u)^2$$
$$-4\sin(u)\sin(v)\,j\cos(2\,u)+4\sin(u)\,k\cos(u)$$

```
> A:=coeff(mn,i); # Obtain coefficient of 'i'
```

$$A:=2\cos(u)\sin(v)\cos(5\,v)+2\cos(v)-2\cos(v)\cos(u)^2-4\sin(u)\cos(v)\cos(2\,u)$$

```
> B:=coeff(mn,j); # Obtain coefficient of 'j'
```

$$B:=-2\cos(u)\cos(v)\cos(5\,v)-\cos(5\,u)\cos(5\,v)+2\sin(v)-2\sin(v)\cos(u)^2$$
$$-4\sin(u)\sin(v)\cos(2\,u)$$

```
> C:=coeff(mn,k);
# Obtain coefficient of 'k'
```

$$C:=2\cos(5\,u)\sin(u)\cos(v)+4\sin(u)\cos(u)$$

```
> [A,B,C]: # [A,B,C] structure 'tangent plane normal vector'
> Int(Int(V[1]*A+V[2]*B+V[3]*C,u=rgu[1]..rgu[2]),v=rgv[1]..rgv[2]);
# Integrals of Dot Product (Vector field 'V' and Tangent plane normal
vector [A,B,C] of surface 'CS') in range of parameters 'u,v'
```

$$\int_0^{2\pi} \int_0^{\pi} \left(\frac{1}{3} \right($$

$$\frac{2}{3}\sin(u)\cos(v) + \frac{1}{15}\sin(5u) - \frac{1}{3}\sin(u)\sin(v) + \frac{1}{5}\cos(u) + \frac{1}{25}\sin(5v) + \frac{1}{5}\sin(2u)$$

$$\Big)^3 - \frac{4}{5}\sin(u)^2\sin(v)^2 - \frac{1}{7}\left(\cos(u) + \frac{1}{5}\sin(5v) + \sin(2u)\right)^3 \Big)($$

$$2\cos(u)\sin(v)\cos(5v) + 2\cos(v) - 2\cos(v)\cos(u)^2 - 4\sin(u)\cos(v)\cos(2u))$$

$$+ \left(\frac{1}{3}\left(2\sin(u)\cos(v) + \frac{1}{5}\sin(5u) \right)\sin(u)\sin(v)\left(\cos(u) + \frac{1}{5}\sin(5v) + \sin(2u) \right) \right.$$

$$\left. - \frac{1}{6}\left(\cos(u) + \frac{1}{5}\sin(5v) + \sin(2u) \right)^2 \right)(-2\cos(u)\cos(v)\cos(5v)$$

$$- \cos(5u)\cos(5v) + 2\sin(v) - 2\sin(v)\cos(u)^2 - 4\sin(u)\sin(v)\cos(2u)) + \Big($$

$$\frac{1}{3}\left(2\sin(u)\cos(v) + \frac{1}{5}\sin(5u) \right)^2 \sin(u)\sin(v)$$

$$- \frac{2}{5}\sin(u)\sin(v)\left(\cos(u) + \frac{1}{5}\sin(5v) + \sin(2u) \right) \Big)$$

$$(2\cos(5u)\sin(u)\cos(v) + 4\sin(u)\cos(u))\,du\,dv$$

```
> alpha:=value(%);delta:=evalf(alpha);
```

$$\alpha := \frac{13473832}{50675625}\pi + \frac{1934}{16875}\pi^2$$

$$\delta := 1.966428630$$

```
> x:='x':y:='y':z:='z':
> [r*CS[1],r*CS[2],r*CS[3]];
```

\# Entirely multiply radius vector 'r' (suppose r>0) by each item of surface 'CS' parametrized expression, Convert "Parametrized expression of surface 'CS'" to "Parametrized expression of surface 'CS' coordinates"

$$\left[r\left(2\sin(u)\cos(v) + \frac{1}{5}\sin(5u) \right), 2r\sin(u)\sin(v), r\left(\cos(u) + \frac{1}{5}\sin(5v) + \sin(2u) \right) \right]$$

```
> x:=r*CS[1]:y:=r*CS[2]:z:=r*CS[3]:
> matrix(3,3,[Diff(x,r),Diff(x,u),Diff(x,v),Diff(y,r),Diff(y,u),
Diff(y,v),Diff(z,r),Diff(z,u),Diff(z,v)])=
matrix(3,3,[diff(x,r),diff(x,u),diff(x,v),diff(y,r),diff(y,u),
diff(y,v),diff(z,r),diff(z,u),diff(z,v)]);m2:=rhs(%);
```

\# Define and calculate matrix of partial derivatives 'm2', obtain
general expression of surface coordinates volume element coefficient

$$\left[\frac{\partial}{\partial r}\left(r\left(2\sin(u)\cos(v) + \frac{1}{5}\sin(5\,u) \right) \right), \frac{\partial}{\partial u}\left(r\left(2\sin(u)\cos(v) + \frac{1}{5}\sin(5\,u) \right) \right), \right.$$
$$\left. \frac{\partial}{\partial v}\left(r\left(2\sin(u)\cos(v) + \frac{1}{5}\sin(5\,u) \right) \right) \right]$$
$$\left[\frac{\partial}{\partial r}\left(2\,r\sin(u)\sin(v) \right), \frac{\partial}{\partial u}\left(2\,r\sin(u)\sin(v) \right), \frac{\partial}{\partial v}\left(2\,r\sin(u)\sin(v) \right) \right]$$
$$\left[\frac{\partial}{\partial r}\left(r\left(\cos(u) + \frac{1}{5}\sin(5\,v) + \sin(2\,u) \right) \right), \frac{\partial}{\partial u}\left(r\left(\cos(u) + \frac{1}{5}\sin(5\,v) + \sin(2\,u) \right) \right), \right.$$
$$\left. \frac{\partial}{\partial v}\left(r\left(\cos(u) + \frac{1}{5}\sin(5\,v) + \sin(2\,u) \right) \right) \right]$$

=

$$\begin{bmatrix} 2\sin(u)\cos(v) + \frac{1}{5}\sin(5\,u) & r\left(2\cos(u)\cos(v) + \cos(5\,u) \right) & -2\,r\sin(u)\sin(v) \\ 2\sin(u)\sin(v) & 2\,r\cos(u)\sin(v) & 2\,r\sin(u)\cos(v) \\ \cos(u) + \frac{1}{5}\sin(5\,v) + \sin(2\,u) & r\left(-\sin(u) + 2\cos(2\,u) \right) & r\cos(5\,v) \end{bmatrix}$$

m2 :=

$$\begin{bmatrix} 2\sin(u)\cos(v) + \frac{1}{5}\sin(5\,u) & r\left(2\cos(u)\cos(v) + \cos(5\,u) \right) & -2\,r\sin(u)\sin(v) \\ 2\sin(u)\sin(v) & 2\,r\cos(u)\sin(v) & 2\,r\sin(u)\cos(v) \\ \cos(u) + \frac{1}{5}\sin(5\,v) + \sin(2\,u) & r\left(-\sin(u) + 2\cos(2\,u) \right) & r\cos(5\,v) \end{bmatrix}$$

> det(m2):

> J:=simplify(%);

\# General expression of surface coordinates volume element coefficient

$$J := \frac{2}{5}r^2\left(\sin(5\,u)\cos(u)\sin(v)\cos(5\,v) - 2\sin(5\,u)\sin(u)\cos(v)\cos(2\,u) \right.$$
$$- 5\sin(u)\sin(v)\cos(5\,v)\cos(5\,u) + 5\sin(2\,u)\sin(u)\cos(v)\cos(5\,u)$$
$$+ 5\cos(u)\sin(u)\cos(v)\cos(5\,u) + \sin(5\,v)\sin(u)\cos(v)\cos(5\,u)$$
$$+ 10\sin(2\,u)\sin(u)\cos(u) + 2\sin(5\,v)\sin(u)\cos(u) + 20\cos(2\,u)\cos(u)^2$$
$$\left. + \sin(5\,u)\cos(v) - \sin(5\,u)\cos(v)\cos(u)^2 + 10\sin(u) - 20\cos(2\,u) \right)$$

>Int(Int(Int(diV*J,r=0..1),u=rgu[1]..rgu[2]),v=rgv[1]..rgv[2]);
\# Triple Integrals of Product (Divergence 'diV' and Surface
Coordinates Volume Element) in range of 'r,u,v'

$$\int_0^{2\pi} \int_0^\pi \int_0^1 \frac{2}{5}\left(\frac{1}{3}\left(\frac{1}{3}r\left(2\sin(u)\cos(v) + \frac{1}{5}\sin(5\,u) \right) \right) - \frac{1}{3}r\sin(u)\sin(v) \right)$$

$$+ \frac{1}{5} r \left(\cos(u) + \frac{1}{5} \sin(5\,v) + \sin(2\,u) \right) \Big)^2$$

$$+ \frac{1}{6} r^2 \left(2 \sin(u) \cos(v) + \frac{1}{5} \sin(5\,u) \right) \left(\cos(u) + \frac{1}{5} \sin(5\,v) + \sin(2\,u) \right)$$

$$- \frac{2}{5} r \sin(u) \sin(v) \Big) r^2 \left(\sin(5\,u) \cos(u) \sin(v) \cos(5\,v) \right)$$

$$- 2 \sin(5\,u) \sin(u) \cos(v) \cos(2\,u) - 5 \sin(u) \sin(v) \cos(5\,v) \cos(5\,u)$$

$$+ 5 \sin(2\,u) \sin(u) \cos(v) \cos(5\,u) + 5 \cos(u) \sin(u) \cos(v) \cos(5\,u)$$

$$+ \sin(5\,v) \sin(u) \cos(v) \cos(5\,u) + 10 \sin(2\,u) \sin(u) \cos(u)$$

$$+ 2 \sin(5\,v) \sin(u) \cos(u) + 20 \cos(2\,u) \cos(u)^2 + \sin(5\,u) \cos(v)$$

$$- \sin(5\,u) \cos(v) \cos(u)^2 + 10 \sin(u) - 20 \cos(2\,u)) \, dr \, du \, dv$$

```
> beta:=value(%);epsilon:=evalf(beta);
```

$$\beta := \frac{13473832}{50675625} \pi + \frac{1934}{16875} \pi^2$$

$$\varepsilon := 1.966428630$$

```
> alpha;beta; # Two analytic values are equal
```

$$\frac{13473832}{50675625} \pi + \frac{1934}{16875} \pi^2$$

$$\frac{13473832}{50675625} \pi + \frac{1934}{16875} \pi^2$$

```
> delta,epsilon; # Two float values are equal
```

$$1.966428630, \ 1.966428630$$

3.5 Numerical Model of Divergence Theorem at Manifold (III)

```
Known:
A. Parametric expression of unclosed surface 1:
```

$$[v\,(5 + 2 \cos(u)) \cos(v),\, v\,(5 + 2 \cos(u)) \sin(v),\, 2\,v \sin(u)] \quad (1)$$

```
thereinto, u∈[0,2π],v∈[0,3π];
```

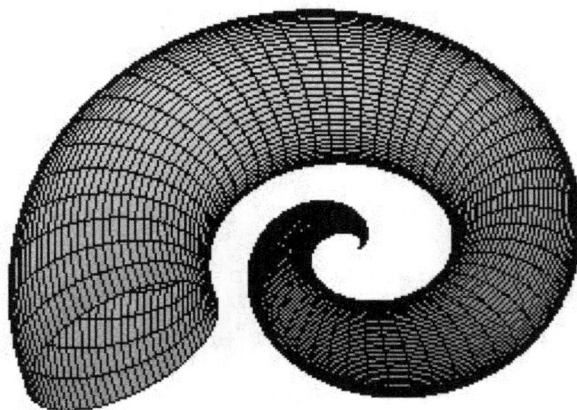

Figure(I)3.5.1 Unclosed Surface 1

B. Parametric expression of unclosed surface 2:

$$[-v(5+2\cos(u))\cos(v)-30\,\pi,\,-v(5+3\cos(u)+\sin(v))\sin(v),\,2\,v\sin(u)] \quad (2)$$

thereinto, $u \in [0, 2\pi]$, $v \in [0, 3\pi]$;

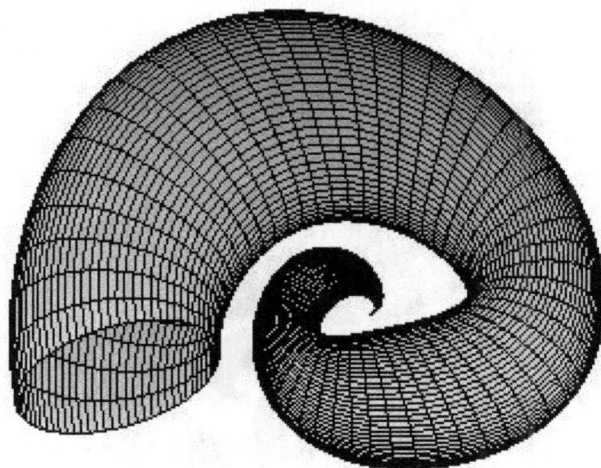

Figure(I)3.5.2 Unclosed Surface 2

C. Patchwork of unclosed surface 1 & 2, structures composite simply connected orientable closed surface 3:

Figure(I)3.5.3 Composite Simply Connected Orientable
Closed Surface 3; Patchwork of Unclosed Surface 1 & 2
[Irregular、Asymmetrical]

and Integral Vector Field

$$\left[\frac{\left(\frac{x}{3}+\frac{y}{6}-\frac{z}{5}-3\right)^3}{2}-\frac{z^3}{7},\frac{1}{6}xyz-\frac{1}{6}y^2,\frac{x^2}{5}+\frac{yz}{5}\right] \quad (3)$$

Calculate and Validate Divergence Theorem at Manifold.

Solution:

First, Free Surface Integral:

According to parametric expression (1) of unclosed surface 1,
define and calculate First Matrix of Partial Derivatives, obtain

tangent plane normal vector of unclosed surface 1:

$$
\begin{bmatrix}
i & j & k \\[4pt]
\dfrac{\partial}{\partial u}(v(5+2\cos(u))\cos(v)) & \dfrac{\partial}{\partial u}(v(5+2\cos(u))\sin(v)) & \dfrac{\partial}{\partial u}(2v\sin(u)) \\[8pt]
\dfrac{\partial}{\partial v}(v(5+2\cos(u))\cos(v)) & \dfrac{\partial}{\partial v}(v(5+2\cos(u))\sin(v)) & \dfrac{\partial}{\partial v}(2v\sin(u))
\end{bmatrix} =
$$

$$
\begin{aligned}
-2\,v\,(&2\sin(v)\cos(u)^2\,j\,v + 2\cos(u)^2\cos(v)\,i\,v + 2\sin(u)\cos(u)\,k\,v \\
&+ 5\sin(v)\cos(u)\,j\,v + 5\cos(u)\cos(v)\,i\,v + 5\sin(u)\,k\,v + 5\sin(v)\cos(u)\,i \\
&- 5\cos(u)\cos(v)\,j + 2\sin(v)\,i - 2\cos(v)\,j)
\end{aligned}
$$

$$(4)$$

From expression(4), respectively pick up coefficients of item
'i,j,k', obtain tangent plane normal vector of unclosed surface 1 (5):

$$
\begin{aligned}
[&-2\,v\,(2\cos(u)^2\cos(v)\,v + 5\cos(u)\cos(v)\,v + 5\sin(v)\cos(u) + 2\sin(v)), \\
&-2\,v\,(2\sin(v)\cos(u)^2\,v + 5\sin(v)\cos(u)\,v - 5\cos(u)\cos(v) - 2\cos(v)), \\
&-2\,v\,(2\sin(u)\cos(u)\,v + 5\,v\sin(u))]
\end{aligned}
$$

$$(5)$$

Input parametrized expression (1) of unclosed surface 1 to vector
field (3); and calculate Integral of Dot Product [Vetor Field (3) and
Tangent plane normal vector (5)] in range of 'u,v':   (6)

$$
\int_0^{3\pi}\int_0^{2\pi} -2\,v\,(2\cos(u)^2\cos(v)\,v + 5\cos(u)\cos(v)\,v + 5\sin(v)\cos(u) + 2\sin(v)) \Bigg(
$$

$$
\frac{1}{2}\left(\frac{1}{3}v(5+2\cos(u))\cos(v) + \frac{1}{6}v(5+2\cos(u))\sin(v) - \frac{2}{5}v\sin(u) - 3\right)^3
$$

$$
-\frac{8}{7}v^3\sin(u)^3 \Bigg) - 2\,v
$$

$$
(2\sin(v)\cos(u)^2\,v + 5\sin(v)\cos(u)\,v - 5\cos(u)\cos(v) - 2\cos(v))
$$

$$
\left(\frac{1}{3}v^3(5+2\cos(u))^2\cos(v)\sin(v)\sin(u) - \frac{1}{6}v^2(5+2\cos(u))^2\sin(v)^2\right) - 2\,v
$$

$$
(2\sin(u)\cos(u)\,v + 5\,v\sin(u))
$$

$$
\left(\frac{1}{5}v^2(5+2\cos(u))^2\cos(v)^2 + \frac{2}{5}v^2(5+2\cos(u))\sin(v)\sin(u)\right) du\,dv
$$

$$
=
$$

$$
\frac{15808}{5}\pi - \frac{24111}{10}\pi^7 + 1890\,\pi^6 + \frac{46629}{20}\pi^5 - 12282\,\pi^4 + 7663\,\pi^2 - \frac{134397}{20}\pi^3
$$

$$(6)$$

According to parametric expression(2) of unclosed surface 2, define
and calculate First Matrix of Partial Derivatives, obtain tangent plane
normal vector of unclosed surface 2:

$[i,j,k]$

$$\left[\frac{\partial}{\partial u}(-v(5+2\cos(u))\cos(v)-30\pi),\frac{\partial}{\partial u}(-v(5+3\cos(u)+\sin(v))\sin(v)),\right.$$

$$\left.\frac{\partial}{\partial u}(2v\sin(u))\right]$$

$$\left[\frac{\partial}{\partial v}(-v(5+2\cos(u))\cos(v)-30\pi),\frac{\partial}{\partial v}(-v(5+3\cos(u)+\sin(v))\sin(v)),\right.$$

$$\left.\frac{\partial}{\partial v}(2v\sin(u))\right]$$

$$=$$

$$-v(4\sin(u)\sin(v)\cos(v)^2\,k\,v-2\sin(u)\cos(v)^3\,k-5\sin(u)\cos(v)^2\,k\,v$$
$$-4\sin(v)\cos(u)^2\,j\,v-6\cos(u)^2\cos(v)\,i\,v-4\cos(u)\sin(v)\cos(v)\,i\,v$$
$$+6\sin(u)\cos(u)\,k\,v-5\sin(u)\sin(v)\cos(v)\,k-10\sin(v)\cos(u)\,j\,v$$
$$+2\cos(u)\cos(v)^2\,i-10\cos(u)\cos(v)\,i\,v+2\sin(u)\cos(v)\,k+15\sin(u)\,k\,v$$
$$-10\sin(v)\cos(u)\,i+10\cos(u)\cos(v)\,j-2\cos(u)\,i-6\sin(v)\,i+4\cos(v)\,j)\ \text{(7)}$$

From expression(7), respectively pick up coefficients of item 'i,j,k', obtain tangent plane normal vector of unclosed surface 2 (8):

$$[-v(-6\cos(u)^2\cos(v)\,v-4\cos(u)\sin(v)\cos(v)\,v+2\cos(u)\cos(v)^2$$
$$-10\cos(u)\cos(v)\,v-10\sin(v)\cos(u)-2\cos(u)-6\sin(v)),$$
$$-v(-4\sin(v)\cos(u)^2\,v-10\sin(v)\cos(u)\,v+10\cos(u)\cos(v)+4\cos(v)),-v($$
$$4\sin(u)\sin(v)\cos(v)^2\,v-2\sin(u)\cos(v)^3-5\sin(u)\cos(v)^2\,v+6\sin(u)\cos(u)\,v$$
$$-5\sin(u)\sin(v)\cos(v)+2\sin(u)\cos(v)+15\,v\sin(u))]\qquad\text{(8)}$$

Input parametrized expression (2) of unclosed surface 2 to vector field (3); and calculate Integral of Dot Product [Vetor Field (3) and Tangent plane normal vector (8)] in range of 'u,v':   (9)

$$\int_0^{3\pi}\int_0^{2\pi}-v(-6\cos(u)^2\cos(v)\,v-4\cos(u)\sin(v)\cos(v)\,v+2\cos(u)\cos(v)^2$$

$$-10\cos(u)\cos(v)\,v-10\sin(v)\cos(u)-2\cos(u)-6\sin(v))\left(\frac{1}{2}\right(\right.$$

$$-\frac{1}{3}v(5+2\cos(u))\cos(v)-10\pi-\frac{1}{6}v(5+3\cos(u)+\sin(v))\sin(v)-\frac{2}{5}v\sin(u)$$

$$\left.-3\right)^3-\frac{8}{7}v^3\sin(u)^3\right)-v$$

$$(-4\sin(v)\cos(u)^2\,v-10\sin(v)\cos(u)\,v+10\cos(u)\cos(v)+4\cos(v))\Big($$

$$-\frac{1}{3}(-v(5+2\cos(u))\cos(v)-30\pi)v^2(5+3\cos(u)+\sin(v))\sin(v)\sin(u)$$

$$-\frac{1}{6}v^2(5+3\cos(u)+\sin(v))^2\sin(v)^2\Big)-v(4\sin(u)\sin(v)\cos(v)^2v$$

$$-2\sin(u)\cos(v)^3-5\sin(u)\cos(v)^2v+6\sin(u)\cos(u)v-5\sin(u)\sin(v)\cos(v)$$

$$+2\sin(u)\cos(v)+15v\sin(u))$$

$$\Big(\frac{1}{5}(-v(5+2\cos(u))\cos(v)-30\pi)^2-\frac{2}{5}v^2(5+3\cos(u)+\sin(v))\sin(v)\sin(u)\Big)$$

$$du\,dv$$

$$=$$

$$-\frac{300623425081}{15360000}\pi^3-\frac{113939355534346549}{5227649280000}\pi^2-\frac{41256942056}{11390625}\pi-\frac{7629687}{256}\pi^7$$

$$-\frac{403919757}{11200}\pi^6+\frac{1723978081}{51200}\pi^5+\frac{839146995493}{19756800}\pi^4$$

(9)

Difference between (6) and (9) [viz.(6)-(9)] is surface integral
value of composite simply connected orientable closed surface. (10)

$$\frac{77269542056}{11390625}\pi+\frac{35062227}{1280}\pi^7+\frac{425087757}{11200}\pi^6-\frac{1604607841}{51200}\pi^5$$

$$-\frac{1081800013093}{19756800}\pi^4+\frac{153998831966986549}{5227649280000}\pi^2+\frac{197406529081}{15360000}\pi^3$$

(10)

Second, Free Volume Element Integral:

Entirely multiply radius vector 'r' (suppose r>0) by each item of
surface 1 parametrized expression (1), convert 'Parametrized
expression of surface 1' to 'Parametrized expression of surface 1
coordinates':

$$[rv(5+2\cos(u))\cos(v),rv(5+2\cos(u))\sin(v),2rv\sin(u)]$$ (11)

According to parametrized expression of surface 1 coordinates (11),
define and calculate Second Matrix of Partial Derivatives, obtain
general expression of surface 1 coordinates volume element coefficient

$$\Big[\frac{\partial}{\partial r}(rv(5+2\cos(u))\cos(v)),\frac{\partial}{\partial u}(rv(5+2\cos(u))\cos(v)),$$

$$\frac{\partial}{\partial v}(rv(5+2\cos(u))\cos(v))\Big]$$

$$\Big[\frac{\partial}{\partial r}(rv(5+2\cos(u))\sin(v)),\frac{\partial}{\partial u}(rv(5+2\cos(u))\sin(v)),$$

$$\frac{\partial}{\partial v}(rv(5+2\cos(u))\sin(v))\Big]$$

$$\left[\frac{\partial}{\partial r}(2\,r\,v\,\sin(u)),\frac{\partial}{\partial u}(2\,r\,v\,\sin(u)),\frac{\partial}{\partial v}(2\,r\,v\,\sin(u))\right]$$

$$=$$

$$-2\,r^2\,v^3\,(10\cos(u)^2+29\cos(u)+10) \qquad (12)$$

Calculate Divergence of Vector Field (3):

$$\left(\frac{\partial}{\partial x}\left(\frac{\left(\frac{x}{3}+\frac{y}{6}-\frac{z}{5}-3\right)^3}{2}-\frac{z^3}{7}\right)\right)+\left(\frac{\partial}{\partial y}\left(\frac{1}{6}x\,y\,z-\frac{1}{6}y^2\right)\right)+\left(\frac{\partial}{\partial z}\left(\frac{x^2}{5}+\frac{y\,z}{5}\right)\right)=$$

$$\frac{\left(\frac{x}{3}+\frac{y}{6}-\frac{z}{5}-3\right)^2}{2}+\frac{x\,z}{6}-\frac{2\,y}{15} \qquad (13)$$

Input parametrized expression of surface 1 coordinates (11) to Divergence (13); and Calculate Triple Integrals of Product [Divergence (13) and Surface 1 Coordinates Volume Element] in range of 'r,u,v' (14):

$$\int_0^{3\pi}\int_0^{2\pi}\int_0^1 -2\Bigg($$

$$\frac{1}{2}\left(\frac{1}{3}r\,v\,(5+2\cos(u))\cos(v)+\frac{1}{6}r\,v\,(5+2\cos(u))\sin(v)-\frac{2}{5}r\,v\,\sin(u)-3\right)^2$$

$$+\frac{1}{3}r^2\,v^2\,(5+2\cos(u))\cos(v)\sin(u)-\frac{2}{15}r\,v\,(5+2\cos(u))\sin(v)\Bigg)r^2\,v^3$$

$$(10\cos(u)^2+29\cos(u)+10)\,dr\,du\,dv$$

$$=$$

$$\frac{15808}{5}\pi+1890\,\pi^6+7663\,\pi^2+\frac{46629}{20}\pi^5-\frac{134397}{20}\pi^3-12282\,\pi^4-\frac{24111}{10}\pi^7 \qquad (14)$$

Entirely multiply radius vector 'r' (suppose r>0) by each item of surface 2 parametrized expression (2), convert 'Parametrized expression of surface 2' to 'Parametrized expression of surface 2 coordinates':

$$[r(-v(5+2\cos(u))\cos(v)-30\,\pi),-r\,v(5+3\cos(u)+\sin(v))\sin(v),2\,r\,v\,\sin(u)] \qquad (15)$$

According to parametrized expression of surface 2 coordinates (15), define and calculate Second Matrix of Partial Derivatives, obtain general expression of surface 2 coordinates volume element coefficient

$$\left[\frac{\partial}{\partial r}\left(r\left(-v\left(5+2\cos(u)\right)\cos(v)-30\,\pi\right)\right),\right.$$

$$\frac{\partial}{\partial u}\left(r\left(-v\left(5+2\cos(u)\right)\cos(v)-30\,\pi\right)\right),\frac{\partial}{\partial v}\left(r\left(-v\left(5+2\cos(u)\right)\cos(v)-30\,\pi\right)\right)$$

$$\left.\right]$$

$$\left[\frac{\partial}{\partial r}\left(-r\,v\left(5+3\cos(u)+\sin(v)\right)\sin(v)\right),\frac{\partial}{\partial u}\left(-r\,v\left(5+3\cos(u)+\sin(v)\right)\sin(v)\right),\right.$$

$$\left.\frac{\partial}{\partial v}\left(-r\,v\left(5+3\cos(u)+\sin(v)\right)\sin(v)\right)\right]$$

$$\left[\frac{\partial}{\partial r}\left(2\,r\,v\sin(u)\right),\frac{\partial}{\partial u}\left(2\,r\,v\sin(u)\right),\frac{\partial}{\partial v}\left(2\,r\,v\sin(u)\right)\right]$$

$$=$$

$$2\,r^2\,v\,(2\sin(v)\cos(u)^2\cos(v)^2\,v^2-5\sin(v)\cos(u)\cos(v)^2\,v^2-5\cos(u)^2\cos(v)^2\,v^2$$

$$-2\sin(v)\cos(u)^2\,v^2-60\sin(v)\cos(u)\cos(v)\,\pi\,v-4\,v^2\cos(v)^2\sin(v)$$

$$-90\cos(u)^2\cos(v)\,\pi\,v-5\cos(u)\,v^2\sin(v)-10\cos(u)^2\,v^2+30\cos(u)\cos(v)^2\,\pi$$

$$-150\cos(u)\cos(v)\,\pi\,v+5\,v^2\cos(v)^2-150\sin(v)\cos(u)\,\pi-31\cos(u)\,v^2$$

$$-90\sin(v)\,\pi-30\cos(u)\,\pi-15\,v^2)\qquad(16)$$

Input parametrized expression of surface 2 coordinates (15) to Divergence (13); and Calculate Triple Integrals of Product [Divergence (13) and Surface 2 Coordinates Volume Element] in range of 'r,u,v' (17):

$$\int_0^{3\pi}\int_0^{2\pi}\int_0^1 2\left(\frac{1}{2}\left(\frac{1}{3}r\left(-v\left(5+2\cos(u)\right)\cos(v)-30\,\pi\right)\right.\right.$$

$$\left.-\frac{1}{6}r\,v\left(5+3\cos(u)+\sin(v)\right)\sin(v)-\frac{2}{5}r\,v\sin(u)-3\right)^2$$

$$+\frac{1}{3}r^2\left(-v\left(5+2\cos(u)\right)\cos(v)-30\,\pi\right)v\sin(u)$$

$$\left.+\frac{2}{15}r\,v\left(5+3\cos(u)+\sin(v)\right)\sin(v)\right)r^2\,v\,(2\sin(v)\cos(u)^2\cos(v)^2\,v^2$$

$$-5\sin(v)\cos(u)\cos(v)^2\,v^2-5\cos(u)^2\cos(v)^2\,v^2-2\sin(v)\cos(u)^2\,v^2$$

$$-60\sin(v)\cos(u)\cos(v)\,\pi\,v-4\,v^2\cos(v)^2\sin(v)-90\cos(u)^2\cos(v)\,\pi\,v$$

$$-5\cos(u)\,v^2\sin(v)-10\cos(u)^2\,v^2+30\cos(u)\cos(v)^2\,\pi-150\cos(u)\cos(v)\,\pi\,v$$

$$+5\,v^2\cos(v)^2-150\sin(v)\cos(u)\,\pi-31\cos(u)\,v^2-90\sin(v)\,\pi-30\cos(u)\,\pi$$

$$-15\,v^2)\,dr\,du\,dv$$

$$=$$

$$-\frac{300623425081}{15360000}\pi^3 - \frac{113939355534346549}{5227649280000}\pi^2 - \frac{41256942056}{11390625}\pi - \frac{7629687}{256}\pi^7$$

$$-\frac{403919757}{11200}\pi^6 + \frac{1723978081}{51200}\pi^5 + \frac{839146995493}{19756800}\pi^4 \qquad (17)$$

Difference between (14) and (17) [viz.(14)-(17)] is volume element integral value of spacial bounded closed region that composite simply connected orientable closed surface 3 surrounds:

$$\frac{77269542056}{11390625}\pi + \frac{425087757}{11200}\pi^6 + \frac{153998831966986549}{5227649280000}\pi^2 - \frac{1604607841}{51200}\pi^5$$

$$+\frac{197406529081}{15360000}\pi^3 - \frac{1081800013093}{19756800}\pi^4 + \frac{35062227}{1280}\pi^7 \qquad (18)$$

Surface Integral Precision Value (10) [Vector Field (3) at composite Surface 3], is equal to Volume Element Integral Precision Value (18) [Divergence (13) in Spacial bounded Closed Region 'Ω' that composite Surface 3 surrounds], Complete Calculation and Validation of Divergence Theorem at Manifold.

3.6 Numerical Model of Divergence Theorem at Manifold (III)
[Program Template of Waterloo Maplesoft, Optional]

```
> restart;
> with(plots):with(linalg):
> CS1:=[v*(5+2*cos(u))*cos(v),v*(5+2*cos(u))*sin(v), 2*v*sin(u)];
```
$$CS1 := [v(5+2\cos(u))\cos(v), v(5+2\cos(u))\sin(v), 2v\sin(u)]$$
```
# Define discretional unclosed surface 'CS1'
> rgu1:=[0,2*Pi]; # Define range of 'u,v'
```
$$rgu1 := [0, 2\pi]$$
```
> rgv1:=[0,3*Pi];
```
$$rgv1 := [0, 3\pi]$$
```
> plot3d(CS1,u=rgu1[1]..rgu1[2],v=rgv1[1]..rgv1[2],
scaling=constrained,projection=0.9,numpoints=6000):g1:=%:
> CS2:=[-v*(5+2*cos(u))*cos(v)-30*Pi,
-v*(5+3*cos(u)+sin(v))*sin(v), 2*v*sin(u)];
# Define discretional unclosed surface'CS2'
```

$$CS2 := [-v(5+2\cos(u))\cos(v) - 30\,\pi, -v(5+3\cos(u)+\sin(v))\sin(v), 2\,v\sin(u)]$$

```
> rgu2:=[0,2*Pi];
```
$$rgu2 := [0, 2\,\pi]$$

```
> rgv2:=[0,3*Pi];
```
$$rgv2 := [0, 3\,\pi]$$

```
> plot3d(CS2,u=rgu2[1]..rgu2[2],v=rgv2[1]..rgv2[2],
scaling=constrained,projection=0.9,numpoints=6000):g2:=%:
> display(g1,g2):
```
Patchwork of Unclosed Surface 'CS1' & 'CS2', structures composite
simply connected orientable closed surface 'CS1/CS2':
```
> V:=[(x/3+y/6-z/5-3)^3/2-z^3/7,x*y*z/6-y^2/6,x^2/5+y*z/5];
```
Define discretional spacial vector field 'V' (Suppose Vector Field
'V' possesses 1th order continuous partial derivatives in spacial
bounded closed region 'Ω' that composite simply connected orientable
closed surface 'CS1/CS2' surrounds)

$$V := \left[\frac{\left(\frac{x}{3}+\frac{y}{6}-\frac{z}{5}-3\right)^3}{2} - \frac{z^3}{7}, \frac{1}{6}xyz - \frac{1}{6}y^2, \frac{x^2}{5} + \frac{yz}{5} \right]$$

```
> Diff(V[1],x)+Diff(V[2],y)+Diff(V[3],z)
= diff(V[1],x)+diff(V[2],y)+diff(V[3],z);diV:=rhs(%);
```
Calculate Divergence 'diV' of vector field 'V'

$$\left(\frac{\partial}{\partial x}\left(\frac{\left(\frac{x}{3}+\frac{y}{6}-\frac{z}{5}-3\right)^3}{2}-\frac{z^3}{7}\right)\right)+\left(\frac{\partial}{\partial y}\left(\frac{1}{6}xyz-\frac{1}{6}y^2\right)\right)+\left(\frac{\partial}{\partial z}\left(\frac{x^2}{5}+\frac{yz}{5}\right)\right)=$$

$$\frac{\left(\frac{x}{3}+\frac{y}{6}-\frac{z}{5}-3\right)^2}{2}+\frac{xz}{6}-\frac{2y}{15}$$

$$diV := \frac{\left(\frac{x}{3}+\frac{y}{6}-\frac{z}{5}-3\right)^2}{2}+\frac{xz}{6}-\frac{2y}{15}$$

```
> rgx:=[-140,40];
```
$$rgx := [-140, 40]$$

```
> rgy:=[-90,90];
```
$$rgy := [-90, 90]$$

```
> rgz:=[-90,90];
```

$$rgz := [-90, 90]$$

```
> fieldplot3d(V,x=rgx[1]..rgx[2],y=rgy[1]..rgy[2],
z=rgz[1]..rgz[2],arrows=SLIM,color=blue):g3:=%:
# Draw vector field 'V'
> implicitplot3d(diV,x=rgx[1]..rgx[2],y=rgy[1]..rgy[2],
z=rgz[1]..rgz[2],style=wireframe,color=cyan,numpoints=5000):
g4:=%: # Draw level surface of divergence 'diV'
> display(g1,g2,g3,g4); # Synthesize figures
```

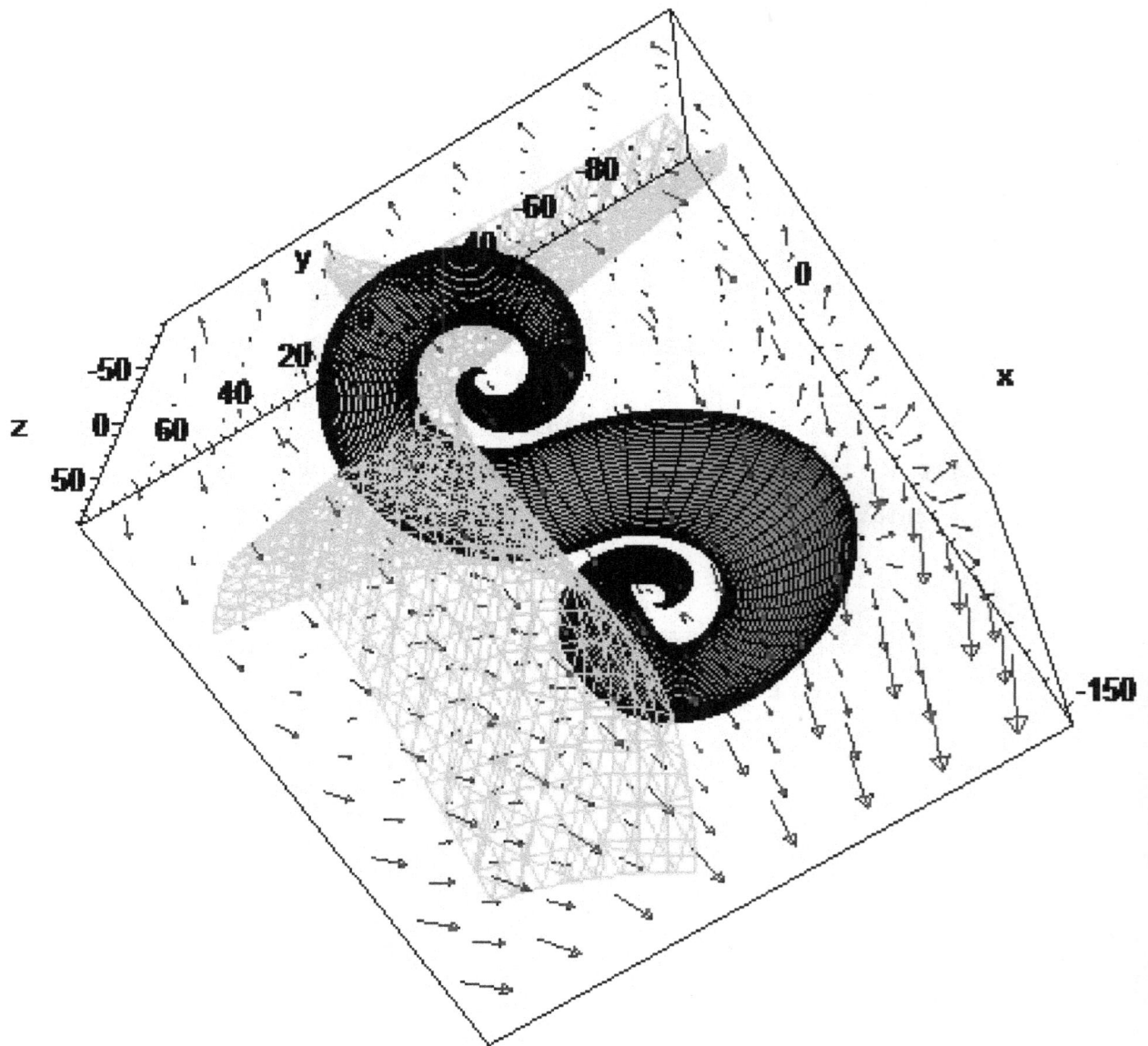

Figure(I)3.6 Composite Simply Connected Orientable
Closed Surface 'CS1/CS2'
Vector Field 'V' and Level Surface of Divergence 'diV'

```
> x:=CS1[1]:y:=CS1[2]:z:=CS1[3]:
> matrix(3,3,[i,j,k,Diff(x,u),Diff(y,u),Diff(z,u),Diff(x,v),
Diff(y,v),Diff(z,v)])=
matrix(3,3,[i,j,k,diff(x,u),diff(y,u),diff(z,u),diff(x,v),
diff(y,v),diff(z,v)]);m1:=rhs(%);
# Define and calculate matrix of partial derivatives 'm1'; obtain
'Tangent plane normal vector' of surface 'CS1'
```

$$
\begin{bmatrix}
i & j & k \\
\dfrac{\partial}{\partial u}(v(5+2\cos(u))\cos(v)) & \dfrac{\partial}{\partial u}(v(5+2\cos(u))\sin(v)) & \dfrac{\partial}{\partial u}(2\,v\sin(u)) \\
\dfrac{\partial}{\partial v}(v(5+2\cos(u))\cos(v)) & \dfrac{\partial}{\partial v}(v(5+2\cos(u))\sin(v)) & \dfrac{\partial}{\partial v}(2\,v\sin(u))
\end{bmatrix} =
$$

$[i,j,k]$
$[-2\,v\sin(u)\cos(v),-2\,v\sin(u)\sin(v),2\,v\cos(u)]$
$[(5+2\cos(u))\cos(v)-v(5+2\cos(u))\sin(v),$
$(5+2\cos(u))\sin(v)+v(5+2\cos(u))\cos(v),2\sin(u)]$

$m1 :=$
$[i,j,k]$
$[-2\,v\sin(u)\cos(v),-2\,v\sin(u)\sin(v),2\,v\cos(u)]$
$[(5+2\cos(u))\cos(v)-v(5+2\cos(u))\sin(v),$
$(5+2\cos(u))\sin(v)+v(5+2\cos(u))\cos(v),2\sin(u)]$

```
> det(m1):
> mn:=simplify(%);
```

$$
mn := -2\,v(2\sin(v)\cos(u)^2 j\,v+2\cos(u)^2\cos(v)\,i\,v+2\sin(u)\cos(u)\,k\,v
$$
$$
+5\sin(v)\cos(u)\,j\,v+5\cos(u)\cos(v)\,i\,v+5\sin(u)\,k\,v+5\sin(v)\cos(u)\,i
$$
$$
-5\cos(u)\cos(v)\,j+2\sin(v)\,i-2\cos(v)\,j
$$

```
> A:=coeff(mn,i); # Obtain coefficient of 'i'
```

$$
A := -2\,v(2\cos(u)^2\cos(v)\,v+5\cos(u)\cos(v)\,v+5\sin(v)\cos(u)+2\sin(v))
$$

```
> B:=coeff(mn,j); # Obtain coefficient of 'j'
```

$$
B := -2\,v(2\sin(v)\cos(u)^2 v+5\sin(v)\cos(u)\,v-5\cos(u)\cos(v)-2\cos(v))
$$

```
> C:=coeff(mn,k); # Obtain coefficient of 'k'
```

$$
C := -2\,v(2\sin(u)\cos(u)\,v+5\,v\sin(u))
$$

```
> [A,B,C]: # [A,B,C] structure 'Tangent plane normal vector'
> Int(Int(V[1]*A+V[2]*B+V[3]*C,u=rgu1[1]..rgu1[2]),
  v=rgv1[1]..rgv1[2]);
```

Integrals of Dot Product (Vector 'V' and Tangent plane normal vector
[A,B,C] of surface 'CS1') in range of parameters 'u,v'

$$\int_0^{3\pi}\int_0^{2\pi} -2\,v\,(2\cos(u)^2\cos(v)\,v + 5\cos(u)\cos(v)\,v + 5\sin(v)\cos(u) + 2\sin(v))\Bigg($$

$$\frac{1}{2}\left(\frac{1}{3}v(5+2\cos(u))\cos(v) + \frac{1}{6}v(5+2\cos(u))\sin(v) - \frac{2}{5}v\sin(u) - 3\right)^3$$

$$-\frac{8}{7}v^3\sin(u)^3\Bigg) - 2\,v$$

$$(2\sin(v)\cos(u)^2\,v + 5\sin(v)\cos(u)\,v - 5\cos(u)\cos(v) - 2\cos(v))$$

$$\left(\frac{1}{3}v^3(5+2\cos(u))^2\cos(v)\sin(v)\sin(u) - \frac{1}{6}v^2(5+2\cos(u))^2\sin(v)^2\right) - 2\,v$$

$$(2\sin(u)\cos(u)\,v + 5\,v\sin(u))$$

$$\left(\frac{1}{5}v^2(5+2\cos(u))^2\cos(v)^2 + \frac{2}{5}v^2(5+2\cos(u))\sin(v)\sin(u)\right)du\;dv$$

```
> v1:=value(%);
```

$$v1 := \frac{15808}{5}\pi - \frac{24111}{10}\pi^7 + 1890\,\pi^6 + \frac{46629}{20}\pi^5 - 12282\,\pi^4 + 7663\,\pi^2 - \frac{134397}{20}\pi^3$$

```
> x:=CS2[1]:y:=CS2[2]:z:=CS2[3]:
> matrix(3,3,[i,j,k,Diff(x,u),Diff(y,u),Diff(z,u),Diff(x,v),
Diff(y,v),Diff(z,v)])=
matrix(3,3,[i,j,k,diff(x,u),diff(y,u),diff(z,u),diff(x,v),
diff(y,v),diff(z,v)]);m1:=rhs(%);
```

Define and calculate matrix of partial derivatives 'm1'; obtain
'Tangent plane normal vector' of surface 'CS2'

$$[i,j,k]$$

$$\left[\frac{\partial}{\partial u}(-v(5+2\cos(u))\cos(v) - 30\,\pi),\frac{\partial}{\partial u}(-v(5+3\cos(u)+\sin(v))\sin(v)),\right.$$

$$\left.\frac{\partial}{\partial u}(2\,v\sin(u))\right]$$

$$\left[\frac{\partial}{\partial v}(-v(5+2\cos(u))\cos(v) - 30\,\pi),\frac{\partial}{\partial v}(-v(5+3\cos(u)+\sin(v))\sin(v)),\right.$$

$$\left.\frac{\partial}{\partial v}(2\,v\sin(u))\right] =$$

$$[i,j,k]$$

$[2\,v\sin(u)\cos(v)\,,3\,v\sin(u)\sin(v)\,,2\,v\cos(u)]$

$[-(5+2\cos(u))\cos(v)+v\,(5+2\cos(u))\sin(v)\,,$

$-(5+3\cos(u)+\sin(v))\sin(v)-v\cos(v)\sin(v)-v\,(5+3\cos(u)+\sin(v))\cos(v)\,,$

$2\sin(u)]$

$m1 :=$

   $[i\,,j\,,k]$

   $[2\,v\sin(u)\cos(v)\,,3\,v\sin(u)\sin(v)\,,2\,v\cos(u)]$

   $[-(5+2\cos(u))\cos(v)+v\,(5+2\cos(u))\sin(v)\,,$

   $-(5+3\cos(u)+\sin(v))\sin(v)-v\cos(v)\sin(v)-v\,(5+3\cos(u)+\sin(v))\cos(v)\,,$

   $2\sin(u)]$

```
> det(m1):
```

```
> mn:=simplify(%);
```

$mn := -v\,(4\sin(u)\sin(v)\cos(v)^2\,k\,v-2\sin(u)\cos(v)^3\,k-5\sin(u)\cos(v)^2\,k\,v$

$\qquad -4\sin(v)\cos(u)^2\,j\,v-4\sin(v)\cos(u)\cos(v)\,i\,v-6\cos(u)^2\cos(v)\,i\,v$

$\qquad -5\sin(u)\sin(v)\cos(v)\,k+6\sin(u)\cos(u)\,k\,v-10\sin(v)\cos(u)\,j\,v$

$\qquad +2\cos(u)\cos(v)^2\,i-10\cos(u)\cos(v)\,i\,v+2\sin(u)\cos(v)\,k+15\sin(u)\,k\,v$

$\qquad -10\sin(v)\cos(u)\,i+10\cos(u)\cos(v)\,j-6\sin(v)\,i-2\cos(u)\,i+4\cos(v)\,j)$

```
> A:=coeff(mn,i); # Obtain coefficient of 'i'
```

$A := -v\,(-4\sin(v)\cos(u)\cos(v)\,v-6\cos(u)^2\cos(v)\,v+2\cos(u)\cos(v)^2$

$\qquad -10\cos(u)\cos(v)\,v-10\sin(v)\cos(u)-6\sin(v)-2\cos(u))$

```
> B:=coeff(mn,j); # Obtain coefficient of 'j'
```

$B := -v\,(-4\sin(v)\cos(u)^2\,v-10\sin(v)\cos(u)\,v+10\cos(u)\cos(v)+4\cos(v))$

```
> C:=coeff(mn,k); # Obtain coefficient of 'k'
```

$C := -v\,(4\sin(u)\sin(v)\cos(v)^2\,v-2\sin(u)\cos(v)^3-5\sin(u)\cos(v)^2\,v$

$\qquad -5\sin(u)\sin(v)\cos(v)+6\sin(u)\cos(u)\,v+2\sin(u)\cos(v)+15\,v\sin(u))$

```
> [A,B,C]: # A,B,C] structure 'Tangent plane normal vector'
> Int(Int(V[1]*A+V[2]*B+V[3]*C,u=rgu2[1]..rgu2[2]),
v=rgv2[1]..rgv2[2]);
# Integrals of Dot Product (Vector 'V' and Tangent plane normal vector
[A,B,C] of surface 'CS2') in range of parameters 'u,v'
```

$$\int_0^{3\pi}\int_0^{2\pi} -v\,(-4\sin(v)\cos(u)\cos(v)\,v-6\cos(u)^2\cos(v)\,v+2\cos(u)\cos(v)^2$$

- 84 -

$$- 10 \cos(u) \cos(v) \, v - 10 \sin(v) \cos(u) - 6 \sin(v) - 2 \cos(u)) \Bigg(\frac{1}{2} \Bigg($$

$$- \frac{1}{3} v \, (5 + 2 \cos(u)) \cos(v) - 10 \, \pi - \frac{1}{6} v \, (5 + 3 \cos(u) + \sin(v)) \sin(v) - \frac{2}{5} v \sin(u)$$

$$- 3 \Bigg)^3 - \frac{8}{7} v^3 \sin(u)^3 \Bigg) - v$$

$$(-4 \sin(v) \cos(u)^2 \, v - 10 \sin(v) \cos(u) \, v + 10 \cos(u) \cos(v) + 4 \cos(v)) \Bigg($$

$$- \frac{1}{3} (-v \, (5 + 2 \cos(u)) \cos(v) - 30 \, \pi) \, v^2 \, (5 + 3 \cos(u) + \sin(v)) \sin(v) \sin(u)$$

$$- \frac{1}{6} v^2 \, (5 + 3 \cos(u) + \sin(v))^2 \sin(v)^2 \Bigg) - v \, (4 \sin(u) \sin(v) \cos(v)^2 \, v$$

$$- 2 \sin(u) \cos(v)^3 - 5 \sin(u) \cos(v)^2 \, v - 5 \sin(u) \sin(v) \cos(v) + 6 \sin(u) \cos(u) \, v$$

$$+ 2 \sin(u) \cos(v) + 15 \, v \sin(u))$$

$$\Bigg(\frac{1}{5} (-v \, (5 + 2 \cos(u)) \cos(v) - 30 \, \pi)^2 - \frac{2}{5} v^2 \, (5 + 3 \cos(u) + \sin(v)) \sin(v) \sin(u) \Bigg)$$

$$du \; dv$$

```
> v2:=value(%);
```

$$v2 := - \frac{300623425081}{15360000} \pi^3 - \frac{113939355534346549}{5227649280000} \pi^2 - \frac{41256942056}{11390625} \pi - \frac{7629687}{256} \pi^7$$

$$- \frac{403919757}{11200} \pi^6 + \frac{1723978081}{51200} \pi^5 + \frac{839146995493}{19756800} \pi^4$$

```
> alpha:=(v1-v2);delta:=evalf(alpha);
```

Surface integral value of composite simply connected orientable
closed surface 'CS1/CS2'

$$\alpha := \frac{77269542056}{11390625} \pi + \frac{35062227}{1280} \pi^7 + \frac{425087757}{11200} \pi^6 - \frac{1604607841}{51200} \pi^5$$

$$- \frac{1081800013093}{19756800} \pi^4 + \frac{153998831966986549}{5227649280000} \pi^2 + \frac{197406529081}{15360000} \pi^3$$

$$\delta := 0.1050079693 \; 10^9$$

```
> x:='x':y:='y':z:='z':
> [r*CS1[1],r*CS1[2],r*CS1[3]];
```

Entirely multiply radius vector 'r' (suppose r>0) by each item
of surface 'CS1' parametrized expression, convert "Parametrized
expression of surface 'CS1'" to "Parametrized expression of surface
'CS1' coordinates"

$$[r \, v \, (5 + 2 \cos(u)) \cos(v), \, r \, v \, (5 + 2 \cos(u)) \sin(v), \, 2 \, r \, v \sin(u)]$$

```
> x:=r*CS1[1]:y:=r*CS1[2]:z:=r*CS1[3]:
> matrix(3,3,[Diff(x,r),Diff(x,u),Diff(x,v),Diff(y,r),Diff(y,u),
Diff(y,v),Diff(z,r),Diff(z,u),Diff(z,v)])
= matrix(3,3,[diff(x,r),diff(x,u),diff(x,v),diff(y,r),diff(y,u),
diff(y,v),diff(z,r),diff(z,u),diff(z,v)]); m2:=rhs(%);
```

Define and calculate matrix of partial derivatives 'm2', obtain general expression of surface 'CS1' coordinates volume element coefficient

$$\left[\frac{\partial}{\partial r}(r\,v\,(5+2\cos(u))\cos(v)), \frac{\partial}{\partial u}(r\,v\,(5+2\cos(u))\cos(v)), \right.$$
$$\left. \frac{\partial}{\partial v}(r\,v\,(5+2\cos(u))\cos(v)) \right]$$
$$\left[\frac{\partial}{\partial r}(r\,v\,(5+2\cos(u))\sin(v)), \frac{\partial}{\partial u}(r\,v\,(5+2\cos(u))\sin(v)), \right.$$
$$\left. \frac{\partial}{\partial v}(r\,v\,(5+2\cos(u))\sin(v)) \right]$$
$$\left[\frac{\partial}{\partial r}(2\,r\,v\sin(u)), \frac{\partial}{\partial u}(2\,r\,v\sin(u)), \frac{\partial}{\partial v}(2\,r\,v\sin(u)) \right] =$$
$$[v\,(5+2\cos(u))\cos(v), -2\,r\,v\sin(u)\cos(v),$$
$$r\,(5+2\cos(u))\cos(v) - r\,v\,(5+2\cos(u))\sin(v)]$$
$$[v\,(5+2\cos(u))\sin(v), -2\,r\,v\sin(u)\sin(v),$$
$$r\,(5+2\cos(u))\sin(v) + r\,v\,(5+2\cos(u))\cos(v)]$$
$$[2\,v\sin(u), 2\,r\,v\cos(u), 2\,r\sin(u)]$$

$$m2 :=$$
$$[v\,(5+2\cos(u))\cos(v), -2\,r\,v\sin(u)\cos(v),$$
$$r\,(5+2\cos(u))\cos(v) - r\,v\,(5+2\cos(u))\sin(v)]$$
$$[v\,(5+2\cos(u))\sin(v), -2\,r\,v\sin(u)\sin(v),$$
$$r\,(5+2\cos(u))\sin(v) + r\,v\,(5+2\cos(u))\cos(v)]$$
$$[2\,v\sin(u), 2\,r\,v\cos(u), 2\,r\sin(u)]$$

```
> det(m2):
> J:=simplify(%);
```

General expression of surface 'CS1' coordinates volume element coefficient

$$J := -2\,r^2\,v^3\,(10\cos(u)^2 + 29\cos(u) + 10)$$

```
> Int(Int(Int(diV*J,r=0..1),u=rgu1[1]..rgu1[2]),
v=rgv1[1]..rgv1[2]);
```

Triple Integrals of Product (Divergence 'diV' and Surface 'CS1'

Coordinates Volume Element) in range of 'r,u,v'

$$\int_0^{3\pi} \int_0^{2\pi} \int_0^1 -2 \left(\frac{1}{2} \left(\frac{1}{3} r v (5 + 2 \cos(u)) \cos(v) + \frac{1}{6} r v (5 + 2 \cos(u)) \sin(v) - \frac{2}{5} r v \sin(u) - 3 \right)^2 \right.$$

$$\left. + \frac{1}{3} r^2 v^2 (5 + 2 \cos(u)) \cos(v) \sin(u) - \frac{2}{15} r v (5 + 2 \cos(u)) \sin(v) \right) r^2 v^3$$

$$(10 \cos(u)^2 + 29 \cos(u) + 10) \, dr \, du \, dv$$

> v3:=value(%);

$$v3 := \frac{15808}{5} \pi - \frac{24111}{10} \pi^7 + 1890 \pi^6 + \frac{46629}{20} \pi^5 - 12282 \pi^4 + 7663 \pi^2 - \frac{134397}{20} \pi^3$$

> x:='x':y:='y':z:='z':

> [r*CS2[1],r*CS2[2],r*CS2[3]];

$$[r(-v(5 + 2 \cos(u)) \cos(v) - 30\pi), -r v (5 + 3 \cos(u) + \sin(v)) \sin(v), 2 r v \sin(u)]$$

Entirely multiply radius vector 'r' (suppose r>0) by each item of surface 'CS2' parametrized expression, convert "Parametrized expression of surface 'CS2'" to "Parametrized expression of surface 'CS2' coordinates"

> x:=r*CS2[1]:y:=r*CS2[2]:z:=r*CS2[3]:

> matrix(3,3,[Diff(x,r),Diff(x,u),Diff(x,v),Diff(y,r),Diff(y,u), Diff(y,v),Diff(z,r),Diff(z,u),Diff(z,v)])

= matrix(3,3,[diff(x,r),diff(x,u),diff(x,v),diff(y,r),diff(y,u), diff(y,v),diff(z,r),diff(z,u),diff(z,v)]); m2:=rhs(%);

Define and calculate matrix of partial derivatives 'm2', obtain general expression of surface 'CS2' coordinates volume element coefficient

$$\left[\frac{\partial}{\partial r} (r(-v(5 + 2 \cos(u)) \cos(v) - 30\pi)), \right.$$

$$\frac{\partial}{\partial u} (r(-v(5 + 2 \cos(u)) \cos(v) - 30\pi)), \frac{\partial}{\partial v} (r(-v(5 + 2 \cos(u)) \cos(v) - 30\pi))$$

$$\left. \right]$$

$$\left[\frac{\partial}{\partial r} (-r v (5 + 3 \cos(u) + \sin(v)) \sin(v)), \frac{\partial}{\partial u} (-r v (5 + 3 \cos(u) + \sin(v)) \sin(v)), \right.$$

$$\frac{\partial}{\partial v}(-r\,v\,(5+3\cos(u)+\sin(v))\sin(v))\Bigg]$$

$$\left[\frac{\partial}{\partial r}(2\,r\,v\sin(u)),\frac{\partial}{\partial u}(2\,r\,v\sin(u)),\frac{\partial}{\partial v}(2\,r\,v\sin(u))\right]=$$

$$[-v\,(5+2\cos(u))\cos(v)-30\,\pi\,,2\,r\,v\sin(u)\cos(v),$$

$$r\,(-(5+2\cos(u))\cos(v)+v\,(5+2\cos(u))\sin(v))]$$

$$[-v\,(5+3\cos(u)+\sin(v))\sin(v),3\,r\,v\sin(u)\sin(v),$$

$$-r\,(5+3\cos(u)+\sin(v))\sin(v)-r\,v\cos(v)\sin(v)$$

$$-r\,v\,(5+3\cos(u)+\sin(v))\cos(v)]$$

$$[2\,v\sin(u),2\,r\,v\cos(u),2\,r\sin(u)]$$

$m2 :=$

$$[-v\,(5+2\cos(u))\cos(v)-30\,\pi\,,2\,r\,v\sin(u)\cos(v),$$

$$r\,(-(5+2\cos(u))\cos(v)+v\,(5+2\cos(u))\sin(v))]$$

$$[-v\,(5+3\cos(u)+\sin(v))\sin(v),3\,r\,v\sin(u)\sin(v),$$

$$-r\,(5+3\cos(u)+\sin(v))\sin(v)-r\,v\cos(v)\sin(v)$$

$$-r\,v\,(5+3\cos(u)+\sin(v))\cos(v)]$$

$$[2\,v\sin(u),2\,r\,v\cos(u),2\,r\sin(u)]$$

```
> det(m2):
> J:=simplify(%);
```

General expression of surface 'CS2' coordinates volume element
coefficient

$$J:=2\,r^2\,v\,(2\sin(v)\cos(u)^2\cos(v)^2\,v^2-5\sin(v)\cos(u)\cos(v)^2\,v^2$$

$$-5\cos(u)^2\cos(v)^2\,v^2-2\sin(v)\cos(u)^2\,v^2-60\sin(v)\cos(u)\cos(v)\,\pi\,v$$

$$-4\,v^2\cos(v)^2\sin(v)-90\cos(u)^2\cos(v)\,\pi\,v-5\cos(u)\,v^2\sin(v)-10\cos(u)^2\,v^2$$

$$+30\cos(u)\cos(v)^2\,\pi-150\cos(u)\cos(v)\,\pi\,v+5\,v^2\cos(v)^2-150\sin(v)\cos(u)\,\pi$$

$$-31\cos(u)\,v^2-90\sin(v)\,\pi-30\cos(u)\,\pi-15\,v^2)$$

```
> Int(Int(Int(diV*J,r=0..1),u=rgu2[1]..rgu2[2]),
v=rgv2[1]..rgv2[2]);
```

Triple Integrals of Product (Divergence 'diV' and Surface 'CS2'
Coordinates Volume Element) in range of 'r,u,v'

$$\int_0^{3\pi}\int_0^{2\pi}\int_0^1 2\left(\frac{1}{2}\left(\frac{1}{3}\,r\,(-v\,(5+2\cos(u))\cos(v)-30\,\pi)\right.\right.$$

$$\left.\left.-\frac{1}{6}\,r\,v\,(5+3\cos(u)+\sin(v))\sin(v)-\frac{2}{5}\,r\,v\sin(u)-3\right)^2\right.$$

$$\left.+\frac{1}{3}\,r^2\,(-v\,(5+2\cos(u))\cos(v)-30\,\pi)\,v\sin(u)\right.$$

$$+ \frac{2}{15} r\, v\, (5 + 3\cos(u) + \sin(v))\, \sin(v) \Big)\, r^2\, v\, (2\sin(v)\cos(u)^2\cos(v)^2\, v^2$$

$$- 5\sin(v)\cos(u)\cos(v)^2\, v^2 - 5\cos(u)^2\cos(v)^2\, v^2 - 2\sin(v)\cos(u)^2\, v^2$$

$$- 60\sin(v)\cos(u)\cos(v)\,\pi\, v - 4\, v^2\cos(v)^2\sin(v) - 90\cos(u)^2\cos(v)\,\pi\, v$$

$$- 5\cos(u)\, v^2\sin(v) - 10\cos(u)^2\, v^2 + 30\cos(u)\cos(v)^2\,\pi - 150\cos(u)\cos(v)\,\pi\, v$$

$$+ 5\, v^2\cos(v)^2 - 150\sin(v)\cos(u)\,\pi - 31\cos(u)\, v^2 - 90\sin(v)\,\pi - 30\cos(u)\,\pi$$

$$- 15\, v^2)\, dr\, du\, dv$$

```
> v4:=value(%);
```

$$v4 := -\frac{300623425081}{15360000}\,\pi^3 - \frac{113939355534346549}{5227649280000}\,\pi^2 - \frac{41256942056}{11390625}\,\pi - \frac{7629687}{256}\,\pi^7$$

$$- \frac{403919757}{11200}\,\pi^6 + \frac{1723978081}{51200}\,\pi^5 + \frac{839146995493}{19756800}\,\pi^4$$

```
> beta:=v3-v4;epsilon:=evalf(beta);
```

Volume element integral value of spacial bounded closed region that composite simply connected orientable closed surface 'CS1/CS2' surrounds

$$\beta := \frac{77269542056}{11390625}\,\pi + \frac{35062227}{1280}\,\pi^7 + \frac{425087757}{11200}\,\pi^6 - \frac{1604607841}{51200}\,\pi^5$$

$$- \frac{1081800013093}{19756800}\,\pi^4 + \frac{153998831966986549}{5227649280000}\,\pi^2 + \frac{197406529081}{15360000}\,\pi^3$$

$$\varepsilon := 0.1050079693\,10^9$$

```
> alpha;beta; # Two analytic values are equal
```

$$\frac{77269542056}{11390625}\,\pi + \frac{35062227}{1280}\,\pi^7 + \frac{425087757}{11200}\,\pi^6 - \frac{1604607841}{51200}\,\pi^5$$

$$- \frac{1081800013093}{19756800}\,\pi^4 + \frac{153998831966986549}{5227649280000}\,\pi^2 + \frac{197406529081}{15360000}\,\pi^3$$

$$\frac{77269542056}{11390625}\,\pi + \frac{35062227}{1280}\,\pi^7 + \frac{425087757}{11200}\,\pi^6 - \frac{1604607841}{51200}\,\pi^5$$

$$- \frac{1081800013093}{19756800}\,\pi^4 + \frac{153998831966986549}{5227649280000}\,\pi^2 + \frac{197406529081}{15360000}\,\pi^3$$

```
> delta;epsilon; # Two float values are equal
```

$$0.1050079693\,10^9$$

$$0.1050079693\,10^9$$

978-1-62265-930-2 (online)   978-1-62265-931-9 (paper)   Yang Ke

3.7 Numerical Model of Divergence Theorem in Torus Coordinates

Known: Parametric Expression of Torus
(Multiply connected Orientable Closed Surface)

$$[(2 + \cos(u)) \cos(v), (2 + \cos(u)) \sin(v), \sin(u)] \qquad (1)$$

Thereinto, u∈[0,2π],v∈[0,2π];

and Integral Vector Field

$$\left[\frac{\left(\frac{xy}{12} - \frac{z}{5} \right)^2}{6} - \frac{x^3}{12} - \frac{y^2}{3}, \frac{1}{3} xyz - \frac{1}{5} y^2, \frac{1}{5} x^3 - \frac{1}{5} xyz \right] \qquad (2)$$

Calculate and Validate Divergence Theorem at Torus

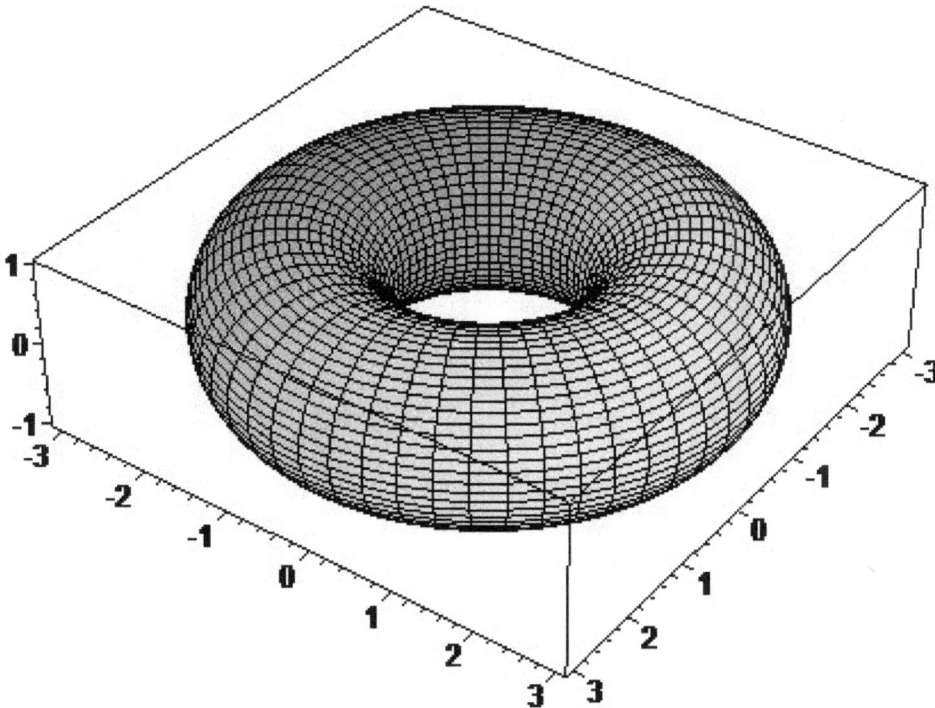

Figure(I)3.7 Torus (Multiply connected Orientable Closed Surface)

Solution:

First, Surface Integral about Torus:

According to Parametric Expression of Torus Surface(1), define and calculate First Matrix of Partial Derivatives, obtain tangent plane normal vector of Torus Surface:

$$
\begin{bmatrix}
i & j & k \\
\dfrac{\partial}{\partial u}\big((2+\cos(u))\cos(v)\big) & \dfrac{\partial}{\partial u}\big((2+\cos(u))\sin(v)\big) & \dfrac{d}{du}\sin(u) \\
\dfrac{\partial}{\partial v}\big((2+\cos(u))\cos(v)\big) & \dfrac{\partial}{\partial v}\big((2+\cos(u))\sin(v)\big) & \dfrac{\partial}{\partial v}\sin(u)
\end{bmatrix}
$$

$$
=
$$

$$
\begin{aligned}
& -i\cos(u)^2\cos(v) - \cos(u)^2\sin(v)\,j - 2\,i\cos(u)\cos(v) - \sin(u)\,k\cos(u) \\
& \quad - 2\cos(u)\sin(v)\,j - 2\sin(u)\,k
\end{aligned}
\tag{3}
$$

From expression (3), respectively pick up coefficients of item 'i,j,k', obtain tangent plane normal vector of Torus surface(4):

$$
\begin{aligned}
& [-\cos(u)^2\cos(v) - 2\cos(u)\cos(v),\ -\cos(u)^2\sin(v) - 2\cos(u)\sin(v), \\
& \quad -\sin(u)\cos(u) - 2\sin(u)]
\end{aligned}
\tag{4}
$$

Input expression (1) to Vector Field (2); and Calculate Integral of Dot Product [Vetor Field (2) and Tangent Plane Normal Vector (4)] in range of 'u,v' (5):

$$
\int_0^{2\pi}\int_0^{2\pi} \left(-\cos(u)^2\cos(v) - 2\cos(u)\cos(v)\right)\Bigg(
$$

$$
\frac{1}{6}\left(\frac{1}{12}(2+\cos(u))^2\cos(v)\sin(v) - \frac{1}{5}\sin(u)\right)^2 - \frac{1}{12}(2+\cos(u))^3\cos(v)^3
$$

$$
-\frac{1}{3}(2+\cos(u))^2\sin(v)^2\Bigg) + \left(-\cos(u)^2\sin(v) - 2\cos(u)\sin(v)\right)
$$

$$
\left(\frac{1}{3}(2+\cos(u))^2\cos(v)\sin(v)\sin(u) - \frac{1}{5}(2+\cos(u))^2\sin(v)^2\right) +
$$

$$
\left(-\sin(u)\cos(u) - 2\sin(u)\right)
$$

$$
\left(\frac{1}{5}(2+\cos(u))^3\cos(v)^3 - \frac{1}{5}(2+\cos(u))^2\cos(v)\sin(v)\sin(u)\right) du\,dv
$$

$$
= \frac{19\,\pi^2}{8}
\tag{5}
$$

Second, Volume Element Integral about Torus:

Entirely multiply radius vector 'r' (suppose r>0) by each item of Torus surface parametrized expression (1), convert 'Parametrized expression of Torus surface' to 'Parametrized expression of Torus surface coordinates':

$$
[r(2+\cos(u))\cos(v),\ r(2+\cos(u))\sin(v),\ r\sin(u)]
\tag{6}
$$

According to parametrized expression of Torus surface coordinates(6), define and calculate Second Matrix of Partial Derivatives, obtain general expression of Torus surface coordinates volume element coefficient:

$$\begin{bmatrix} \frac{\partial}{\partial r}(r(2+\cos(u))\cos(v)) & \frac{\partial}{\partial u}(r(2+\cos(u))\cos(v)) & \frac{\partial}{\partial v}(r(2+\cos(u))\cos(v)) \\ \frac{\partial}{\partial r}(r(2+\cos(u))\sin(v)) & \frac{\partial}{\partial u}(r(2+\cos(u))\sin(v)) & \frac{\partial}{\partial v}(r(2+\cos(u))\sin(v)) \\ \frac{\partial}{\partial r}(r\sin(u)) & \frac{\partial}{\partial u}(r\sin(u)) & \frac{\partial}{\partial v}(r\sin(u)) \end{bmatrix}$$

$$= -r^2(2\cos(u)^2+5\cos(u)+2) \tag{7}$$

Calculate Divergence of Vector Field (2):

$$\left(\frac{\partial}{\partial x}\left(\frac{\left(\frac{xy}{12}-\frac{z}{5}\right)^2}{6}-\frac{x^3}{12}-\frac{y^2}{3}\right)\right)+\left(\frac{\partial}{\partial y}\left(\frac{1}{3}xyz-\frac{1}{5}y^2\right)\right)+\left(\frac{\partial}{\partial z}\left(\frac{1}{5}x^3-\frac{1}{5}xyz\right)\right)=$$

$$\frac{\left(\frac{xy}{12}-\frac{z}{5}\right)y}{36}-\frac{x^2}{4}+\frac{xz}{3}-\frac{2y}{5}-\frac{xy}{5} \tag{8}$$

Input parametrized expression of Torus surface coordinates (6) to Divergence (8); and Calculate Triple Integrals of Product [Divergence (8) and Torus surface coordinates volume element] in range of 'r,u,v' (9):

$$\int_0^{2\pi}\int_0^{2\pi}\int_0^1 -\left(\frac{1}{36}\left(\frac{1}{12}(2+\cos(u))^2\cos(v)\sin(v)-\frac{1}{5}\sin(u)\right)(2+\cos(u))\sin(v)\right.$$

$$-\frac{1}{4}(2+\cos(u))^2\cos(v)^2+\frac{1}{3}(2+\cos(u))\cos(v)\sin(u)-\frac{2}{5}(2+\cos(u))\sin(v)$$

$$\left.-\frac{1}{5}(2+\cos(u))^2\cos(v)\sin(v)\right)r^2(2\cos(u)^2+5\cos(u)+2)\,dr\,du\,dv$$

$$=\frac{19\pi^2}{8} \tag{9}$$

Surface Integral Precision Value (5) [Vector Field (2) on Target Torus (1)], is equal to Volume Element Integral Precision Value (9) [Divergence (8) in Spacial Bounded Closed Region 'Ω' that Target Torus surface (1) surrounds],Complete Calculation and Validation of

Divergence Theorem at Torus Surface.

3.8 Numerical Model of Divergence Theorem In Torus Coordinates
   [Program Template of Waterloo Maplesoft, Optional]

```
> restart;
> with(plots):with(linalg):
> CS:=[(2+cos(u))*cos(v),(2+cos(u))*sin(v),sin(u)];
# Define parametrized expression of torus surface 'CS'
```
$$CS := [(2 + \cos(u)) \cos(v), (2 + \cos(u)) \sin(v), \sin(u)]$$
```
> rgu:=[0,2*Pi];
```
$$rgu := [0, 2\pi]$$
```
> rgv:=[0,2*Pi];
# Set range of parameters 'u,v', make torus surface 'CS' closed
```
$$rgv := [0, 2\pi]$$
```
>plot3d(CS,u=rgu[1]..rgu[2],v=rgv[1]..rgv[2],scaling=constrained,
projection=0.9,numpoints=3000):g1:=%:
> V:=[(x*y/12-z/5)^2/6-x^3/12-y^2/3,x*y*z/3-y^2/5,x^3/5-x*y*z/5];
# Define discretional spacial vector field 'V' (Suppose Vector Field
'V' possesses 1th order continuous partial derivatives in spacial
bounded closed region 'Ω' that Torus surface 'CS' surrounds)
```

$$V := \left[\frac{\left(\dfrac{xy}{12} - \dfrac{z}{5} \right)^2}{6} - \frac{x^3}{12} - \frac{y^2}{3}, \frac{1}{3}xyz - \frac{1}{5}y^2, \frac{1}{5}x^3 - \frac{1}{5}xyz \right]$$

```
> Diff(V[1],x)+Diff(V[2],y)+Diff(V[3],z)
  = diff(V[1],x)+diff(V[2],y)+diff(V[3],z);diV:=rhs(%);
# Calculate divergence 'diV' of vector field 'V'
```

$$\left(\frac{\partial}{\partial x} \left(\frac{\left(\dfrac{xy}{12} - \dfrac{z}{5} \right)^2}{6} - \frac{x^3}{12} - \frac{y^2}{3} \right) \right) + \left(\frac{\partial}{\partial y} \left(\frac{1}{3}xyz - \frac{1}{5}y^2 \right) \right) + \left(\frac{\partial}{\partial z} \left(\frac{1}{5}x^3 - \frac{1}{5}xyz \right) \right) =$$

$$\frac{\left(\dfrac{xy}{12} - \dfrac{z}{5} \right)y}{36} - \frac{x^2}{4} + \frac{xz}{3} - \frac{2y}{5} - \frac{xy}{5}$$

$$diV := \frac{\left(\dfrac{xy}{12} - \dfrac{z}{5} \right)y}{36} - \frac{x^2}{4} + \frac{xz}{3} - \frac{2y}{5} - \frac{xy}{5}$$

```
> rgx:=[-3,3];
```

$$rgx := [-3, 3]$$

```
> rgy:=[-3,3];
```

$$rgy := [-3, 3]$$

```
> rgz:=[-3,3];
```

$$rgz := [-3, 3]$$

```
> fieldplot3d(V,x=rgx[1]..rgx[2],y=rgy[1]..rgy[2],z=rgz[1]..
rgz[2],arrows=SLIM,color=blue):g2:=%:
> implicitplot3d(diV,x=rgx[1]..rgx[2],y=rgy[1]..rgy[2],z=rgz[1]..
rgz[2],style=wireframe,color=cyan,numpoints=3000):g3:=%:
> display(g1,g2,g3);
```

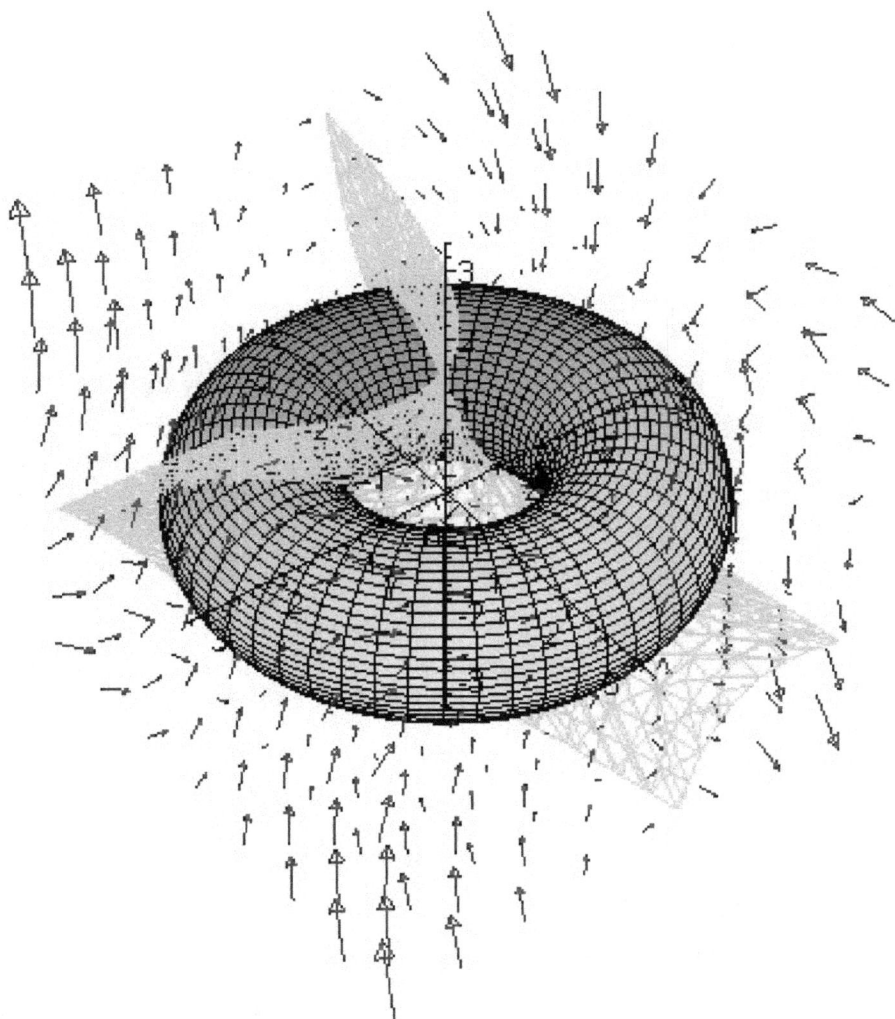

Figure(I)3.8 Torus Surface[closed] 'CS'、Vector Field 'V' and
Level Surface of Divergence 'diV'

```
> x:=CS[1]:y:=CS[2]:z:=CS[3]:
> matrix(3,3,[i,j,k,Diff(CS[1],u),Diff(CS[2],u),Diff(CS[3],u),
Diff(CS[1],v),Diff(CS[2],v),Diff(CS[3],v)])=
  matrix(3,3,[i,j,k,diff(CS[1],u),diff(CS[2],u),diff(CS[3],u),
diff(CS[1],v),diff(CS[2],v),diff(CS[3],v)]);m1:=rhs(%);
# Define and calculate matrix of partial derivatives 'm1'; obtain
'Tangent plane normal vector' of torus surface 'CS'
```

$$\begin{bmatrix} i & j & k \\ \dfrac{\partial}{\partial u}((2+\cos(u))\cos(v)) & \dfrac{\partial}{\partial u}((2+\cos(u))\sin(v)) & \dfrac{d}{du}\sin(u) \\ \dfrac{\partial}{\partial v}((2+\cos(u))\cos(v)) & \dfrac{\partial}{\partial v}((2+\cos(u))\sin(v)) & \dfrac{\partial}{\partial v}\sin(u) \end{bmatrix} =$$

$$\begin{bmatrix} i & j & k \\ -\sin(u)\cos(v) & -\sin(u)\sin(v) & \cos(u) \\ -(2+\cos(u))\sin(v) & (2+\cos(u))\cos(v) & 0 \end{bmatrix}$$

$$m1 := \begin{bmatrix} i & j & k \\ -\sin(u)\cos(v) & -\sin(u)\sin(v) & \cos(u) \\ -(2+\cos(u))\sin(v) & (2+\cos(u))\cos(v) & 0 \end{bmatrix}$$

```
> det(m1):
> mn:=simplify(%);
```

$$mn := -2\,i\cos(u)\cos(v) - i\cos(u)^2\cos(v) - 2\sin(v)\,j\cos(u) - 2\,k\sin(u)$$
$$- \sin(v)\,j\cos(u)^2 - \cos(u)\,k\sin(u)$$

```
> A:=coeff(mn,i); # Obtain coefficient of 'i';
```

$$A := -2\cos(u)\cos(v) - \cos(u)^2\cos(v)$$

```
> B:=coeff(mn,j); # Obtain coefficient of 'j';
```

$$B := -2\sin(v)\cos(u) - \sin(v)\cos(u)^2$$

```
> C:=coeff(mn,k); # Obtain coefficient of 'k'
```

$$C := -2\sin(u) - \cos(u)\sin(u)$$

```
> [A,B,C]: # [A,B,C] structure 'Tangent plane normal vector'
> Int(Int(V[1]*A+V[2]*B+V[3]*C,u=rgu[1]..rgu[2]),v=rgv[1]..rgv[2];
# Integrals of Dot Product (Spacial vector field 'V' and Tangent plane
normal vector [A,B,C] of torus surface 'CS') in range of parameters
'u,v'
```

$$\int_0^{2\pi}\int_0^{2\pi}\left(\frac{1}{6}\left(\frac{1}{12}(2+\cos(u))^2\cos(v)\sin(v)-\frac{1}{5}\sin(u)\right)^2-\frac{1}{12}(2+\cos(u))^3\cos(v)^3\right.$$

$$\left.-\frac{1}{3}(2+\cos(u))^2\sin(v)^2\right)(-2\cos(u)\cos(v)-\cos(u)^2\cos(v))+$$

$$\left(\frac{1}{3}(2+\cos(u))^2\cos(v)\sin(v)\sin(u)-\frac{1}{5}(2+\cos(u))^2\sin(v)^2\right)$$

$$(-2\sin(v)\cos(u)-\sin(v)\cos(u)^2)+$$

$$\left(\frac{1}{5}(2+\cos(u))^3\cos(v)^3-\frac{1}{5}(2+\cos(u))^2\cos(v)\sin(v)\sin(u)\right)$$

$$(-2\sin(u)-\cos(u)\sin(u))\,du\,dv$$

```
> alpha:=value(%);delta:=evalf(alpha);
```

$$\alpha:=\frac{19\pi^2}{8}$$

$$\delta:=23.44031046$$

```
> x:='x':y:='y':z:='z':
> [r*CS[1],r*CS[2],r*CS[3]];
```

\# Entirely multiply radius vector 'r' (suppose r>0) by each item of Torus surface 'CS' parametrized expression, Convert 'Parametrized expression of Torus surface' to 'Parametrized expression of Torus surface coordinates'

$$[r(2+\cos(u))\cos(v), r(2+\cos(u))\sin(v), r\sin(u)]$$

```
> x:=r*CS[1]:y:=r*CS[2]:z:=r*CS[3]:
> matrix(3,3,[Diff(r*CS[1],r),Diff(r*CS[1],u),Diff(r*CS[1],v),
Diff(r*CS[2],r),Diff(r*CS[2],u),Diff(r*CS[2],v),
Diff(r*CS[3],r),Diff(r*CS[3],u),Diff(r*CS[3],v)])=
matrix(3,3,[diff(r*CS[1],r),diff(r*CS[1],u),diff(r*CS[1],v),
diff(r*CS[2],r),diff(r*CS[2],u),diff(r*CS[2],v),
diff(r*CS[3],r),diff(r*CS[3],u),diff(r*CS[3],v)]);m2:=rhs(%);
```

\# Define and calculate matrix of partial derivatives 'm2', obtain general expression of torus surface coordinates volume element coefficient

$$\begin{bmatrix} \dfrac{\partial}{\partial r}\left(r\,(2+\cos(u))\cos(v)\right) & \dfrac{\partial}{\partial u}\left(r\,(2+\cos(u))\cos(v)\right) & \dfrac{\partial}{\partial v}\left(r\,(2+\cos(u))\cos(v)\right) \\[4pt] \dfrac{\partial}{\partial r}\left(r\,(2+\cos(u))\sin(v)\right) & \dfrac{\partial}{\partial u}\left(r\,(2+\cos(u))\sin(v)\right) & \dfrac{\partial}{\partial v}\left(r\,(2+\cos(u))\sin(v)\right) \\[4pt] \dfrac{\partial}{\partial r}\left(r\sin(u)\right) & \dfrac{\partial}{\partial u}\left(r\sin(u)\right) & \dfrac{\partial}{\partial v}\left(r\sin(u)\right) \end{bmatrix}$$

$$=\begin{bmatrix} (2+\cos(u))\cos(v) & -r\sin(u)\cos(v) & -r\,(2+\cos(u))\sin(v) \\ (2+\cos(u))\sin(v) & -r\sin(u)\sin(v) & r\,(2+\cos(u))\cos(v) \\ \sin(u) & r\cos(u) & 0 \end{bmatrix}$$

$$m2:=\begin{bmatrix} (2+\cos(u))\cos(v) & -r\sin(u)\cos(v) & -r\,(2+\cos(u))\sin(v) \\ (2+\cos(u))\sin(v) & -r\sin(u)\sin(v) & r\,(2+\cos(u))\cos(v) \\ \sin(u) & r\cos(u) & 0 \end{bmatrix}$$

```
> det(m2):
> J:=simplify(%);
```

General expression of torus surface coordinates volume element coefficient

$$J:=-r^2\left(5\cos(u)+2\cos(u)^2+2\right)$$

```
> Int(Int(Int(diV*J,r=0..1),u=rgu[1]..rgu[2]),v=rgv[1]..rgv[2]);
```

Triple Integrals of Product (Divergence 'diV' and Torus Surface
  Coordinates Volume Element) in range of 'r,u,v'

$$\int_0^{2\pi}\int_0^{2\pi}\int_0^1 -\left(\frac{1}{36}\left(\frac{1}{12}r^2(2+\cos(u))^2\cos(v)\sin(v)-\frac{1}{5}r\sin(u)\right)r\,(2+\cos(u))\sin(v)\right.$$

$$-\frac{1}{4}r^2(2+\cos(u))^2\cos(v)^2+\frac{1}{3}r^2(2+\cos(u))\cos(v)\sin(u)$$

$$\left.-\frac{2}{5}r\,(2+\cos(u))\sin(v)-\frac{1}{5}r^2(2+\cos(u))^2\cos(v)\sin(v)\right)r^2$$

$$(5\cos(u)+2\cos(u)^2+2)\,dr\,du\,dv$$

```
> beta:=value(%);epsilon:=evalf(beta);
```

$$\beta:=\frac{19\,\pi^2}{8}$$

$$\varepsilon:=23.44031046$$

```
> alpha,beta;  # Two analytic values are equal
```

$$\frac{19\,\pi^2}{8},\frac{19\,\pi^2}{8}$$

```
> delta,epsilon;  # Two float values are equal
```

978-1-62265-930-2 (online)   978-1-62265-931-9 (paper)

23.44031046, 23.44031046

Chapter 4 Prove Divergence Theorem Finite Sums Limits at Manifold

4.1 Prove Divergence Theorem Finite Sums Limits at Manifold

Divergence Theorem

Suppose spacial bounded closed region 'Ω' is surrounded by smooth or piecewise smooth closed surface 'S'. Functions 'P(x,y,z),Q(x,y,z), R(x,y,z)'[Structure vector field 'A'] and its partial derivatives are continuous at spacial bounded closed region 'Ω', then:

$$\iint\limits_{S} A \cdot n \, dS = \iiint\limits_{\Omega} divA \cdot d\omega \qquad (1)$$

Thereinto, surface 'S' is outer side of spacial bounded closed region'Ω's entire boundary surface, 'n' is unit normal vector of surface 'S's outer side, and 'divA' is divergence of vector field 'A'.

Proof (Finite Sums Limits):

Define parametrized expression of abstract simply connected orientable closed surface 'S':

[a*sin(u)cos(v),b*sin(u)sin(v),c*cos(u)]   (2)

Thereinto, 'a,b,c' are nonzero constants or 1th order derivable continuous functions, simply connected orientable closed surface 'S' determines value of 'a,b,c'; set range of parameters 'u,v' in [0,π],[0,2π], make surface 'S' closed. (See also Poincare Conjecture: 'Every closed n-Manifold which is homotopy equivalent to n-Sphere is homeomorphic to n-Sphere' [19], and Chapter 1 this Part 'Constitute Abstract Simply Connected Orientable Closed Parametrized Surface Coordinates')

According to surface parametrized expression (2), define and calculate first matrix of partial derivatives, obtain tangent plane normal vector of surface 'S':

$$\begin{bmatrix} i & j & k \\ \dfrac{\partial}{\partial u}a\sin(u)\cos(v) & \dfrac{\partial}{\partial u}b\sin(u)\sin(v) & \dfrac{\partial}{\partial u}c\cos(u) \\ \dfrac{\partial}{\partial v}a\sin(u)\cos(v) & \dfrac{\partial}{\partial v}b\sin(u)\sin(v) & \dfrac{\partial}{\partial v}c\cos(u) \end{bmatrix} =$$

$$i\,c\sin(u)^2\,b\cos(v)+a\cos(u)\cos(v)^2\,k\,b\sin(u)+a\sin(u)^2\sin(v)\,j\,c$$
$$+\,a\sin(u)\sin(v)^2\,k\,b\cos(u) \tag{3}$$

From expression(3), respectively pick up coefficients of item 'i,j,k', obtain tangent plane normal vector of surface 'S':

$$[\,c\sin(u)^2 b\cos(v)\,,\,\sin(u)^2 a\sin(v)c\,,\,\sin(u)ab\cos(u)\,] \tag{4}$$

Set Amount of surface 'S's segmental cells as 50
(This amount can be discretional natural number)   (5)

1.Microcosmic surface integral course at surface 'S's first segmental cell:

Segment range $[0,\pi]$ of 'u': du = $\dfrac{\pi}{50}$

Segment range $[0,2\pi]$ of 'v': dv = $\dfrac{2\pi}{50}=\dfrac{\pi}{25}$ (6)

Segment 'tangent plane normal vector'(7): [viz. Input (6) to (4)]

$$\left[c\sin\left(\frac{\pi}{50}\right)^2 b\cos\left(\frac{\pi}{25}\right),\,\sin\left(\frac{\pi}{50}\right)^2\sin\left(\frac{\pi}{25}\right)a\,c,\,\sin\left(\frac{\pi}{50}\right)a\cos\left(\frac{\pi}{50}\right)b\right] \tag{7}$$

Segment abstract vector field [P(x,y,z),Q(x,y,z),R(x,y,z)]:
       [P(x,y,z),Q(x,y,z),R(x,y,z)]   (8)

// Owing to universality and homogeneity of abstract vector field [P(x,y,z),Q(x,y,z),R(x,y,z)], its value at first segment cell is [P(x,y,z),Q(x,y,z),R(x,y,z)]; See also this Chapter Section 4.3 'Discuss About Abstract Vector Field、Finite Sums Limit、Integral and Riemann Sums'

Calculate microcosmic surface integral value at surface 'S's first segmental cell: (9)

Base on integral median theorem, dot product of abstract vector field (8) and tangent plane normal vector (7), times segment unit of 'u,v'(6), that is microcosmic surface integral value at first segmental cell(9)

$$\frac{1}{1250}\left(P(x,y,z)\, c\sin\left(\frac{\pi}{50}\right)^2 b\cos\left(\frac{\pi}{25}\right) + Q(x,y,z)\sin\left(\frac{\pi}{50}\right)^2 a\sin\left(\frac{\pi}{25}\right)c \right.$$
$$\left. + R(x,y,z)\sin\left(\frac{\pi}{50}\right) a\, b\cos\left(\frac{\pi}{50}\right) \right)\pi^2$$

(9)

2.Microcosmic surface integral course at surface 'S's all (viz.50) segmental cells:

Segment range $[0,\pi]$ of 'u': du $= \dfrac{s\pi}{50}$

Segment range $[0,2\pi]$ of 'v': dv $= \dfrac{2t\pi}{50} = \dfrac{t\pi}{25}$

(thereinto, 's' and 't' stand for natural number 1~50)   (10)

Segment 'tangent plane normal vector'(11): [viz. Input (10) to (4)]

$$\left[c\sin\left(\frac{\pi s}{50}\right)^2 b\cos\left(\frac{\pi t}{25}\right),\ \sin\left(\frac{\pi s}{50}\right)^2 \sin\left(\frac{\pi t}{25}\right) a\, c,\ \sin\left(\frac{\pi s}{50}\right) a\cos\left(\frac{\pi s}{50}\right) b \right]$$
(11)

// Expression (11) isn't that 'Single vector value', but that 'Set of finite quantitative (viz. 50) vector values'

Segment abstract vector field [P(x,y,z),Q(x,y,z),R(x,y,z)](12):
       [P(x,y,z),Q(x,y,z),R(x,y,z)]   (12)
// Owing to universality and homogeneity of abstract vector field [P(x,y,z),Q(x,y,z),R(x,y,z)],its values at all (viz.50) segment cells are [P(x,y,z),Q(x,y,z),R(x,y,z)]; See also this Chapter Section 4.3 'Discuss About Abstract Vector Field、Finite Sums Limit、Integral and Riemann Sums'

Calculate microcosmic surface integral value at surface'S's all segmental cells: (13)

Base on integral median theorem, dot product of abstract vector field (12) and tangent plane normal vector (11), times segment unit

of 'u,v'(10), these are microcosmic surface integral values at all
segmental cells(13)

$$\frac{1}{1250}\left(P(x,y,z)\, c \sin\left(\frac{s\,\pi}{50}\right)^2 b \cos\left(\frac{t\,\pi}{25}\right) + Q(x,y,z) \sin\left(\frac{s\,\pi}{50}\right)^2 a \sin\left(\frac{t\,\pi}{25}\right) c \right.$$
$$\left. + R(x,y,z) \sin\left(\frac{s\,\pi}{50}\right) a\, b \cos\left(\frac{s\,\pi}{50}\right) \right) \pi^2 \tag{13}$$

//Expression (13) isn't that 'Single value', but that 'Set of finite
quantitative (viz.50) values'

Structure sequence that is composed of finite quantitative (viz.
50) microcomosic surface integral values (14):
  [Lengthy expression of sequence is elided;
  In program template of Waterloo Maplesoft (See also Section 4.2):
  sqn:=seq(seq(((stdV[1]*stdA+stdV[2]*stdB+stdV[3]*stdC)*du*dv,
  s=1..dus),t=1..dus):
  change ':'(last) to ';', then obtain expression of sequence]

Accumulational sum of sequence (viz. Sum of integral values of dot
product [abstract vector field (12) and tangent plane normal vector
(11) at all segmental cells], obtain surface integral value at
manifold (15):
  [Lengthy expression of accumulation sum result is elided;
  In program template of Waterloo Maplesoft (See also Section 4.2):
  add(k,k=sqn):xi:=evalf(%);
  change ':'(middle) to ';', then obtain expression of result]
  Change expression of accumulational sum result to float value:

$$0.88\ 10^{-10}\, c\, b\, P(x,y,z) + 0.434\ 10^{-9}\, a\, c\, Q(x,y,z) + 0.630\ 10^{-9}\, a\, b\, R(x,y,z) \tag{15}$$

// Amount of surface 'S's segmental cells increases infinitely, sum
of integral values above-mentioned tends to '0' infinitely

  Set Amount of surface 'S's segmental cells as n
  (viz. uncertain natural number)   (16)

  3.Microcosmic surface integral course at surface 'S's first
segmental cell:

Segment range $[0,\pi]$ of 'u': du = $\dfrac{\pi}{n}$

Segment range $[0,2\pi]$ of 'v': dv = $\dfrac{2\pi}{n}$ (17)

Segment 'Tangent plane normal vector' (18): [viz. Input (17) to (4)]

$$\left[c\sin\left(\frac{\pi}{n}\right)^2 b\cos\left(\frac{2\pi}{n}\right),\ \sin\left(\frac{\pi}{n}\right)^2\sin\left(\frac{2\pi}{n}\right)a\,c,\ \sin\left(\frac{\pi}{n}\right)a\cos\left(\frac{\pi}{n}\right)b \right]$$ (18)

Segment abstract vector field [P(x,y,z),Q(x,y,z),R(x,y,z)](19):

$$[P(x,y,z),Q(x,y,z),R(x,y,z)]$$ (19)

// Owing to universality and homogeneity of abstract vector field [P(x,y,z),Q(x,y,z),R(x,y,z)],its value at first segment cell is [P(x,y,z),Q(x,y,z),R(x,y,z)]; See also this Chapter Section 4.3 'Discuss About Abstract Vector Field、Finite Sums Limit、Integral and Riemann Sums'

Calculate microcosmic surface integral value at surface 'S's first segmental cell: (20)

Base on integral median theorem, dot product of abstract vector field (19) and tangent plane normal vector (18), times segment unit of 'u,v'(17), that is microcosmic surface integral value at first segmental cell(20)

$$2\left(c\sin\left(\frac{\pi}{n}\right)^2 b\cos\left(\frac{2\pi}{n}\right)P(x,y,z) + \sin\left(\frac{\pi}{n}\right)^2\sin\left(\frac{2\pi}{n}\right)a\,c\,Q(x,y,z) \right.$$
$$\left. + \sin\left(\frac{\pi}{n}\right)a\cos\left(\frac{\pi}{n}\right)b\,R(x,y,z) \right)\pi^2 / n^2$$
                                                                (20)

4.Microcosmic surface integral course at surface 'S's all (viz. n cells) segmental cells:

Segment range $[0,\pi]$ of 'u': du = $\dfrac{s\pi}{n}$

Segment range $[0,2\pi]$ of 'v': dv = $\dfrac{2t\pi}{n}$

(thereinto, 's' and 't' stand for natural number 1~n)   (21)

Segment 'Tangent plane normal vector'(22): [viz. Input (21) to (4)]

$$\left[c \sin\left(\frac{\pi s}{n}\right)^2 b \cos\left(\frac{2\pi t}{n}\right), \sin\left(\frac{\pi s}{n}\right)^2 \sin\left(\frac{2\pi t}{n}\right) a\,c, \sin\left(\frac{\pi s}{n}\right) a \cos\left(\frac{\pi s}{n}\right) b \right] \quad (22)$$

// Expression (22) isn't that 'Single vector value', but that 'Set of finite quantitative (viz. n) vector values'

Segment abstract vector field [P(x,y,z),Q(x,y,z),R(x,y,z)](23):

$$[P(x,y,z),Q(x,y,z),R(x,y,z)] \qquad\qquad (23)$$

// Owing to universality and homogeneity of abstract vector field [P(x,y,z),Q(x,y,z),R(x,y,z)],its values at all (viz. n) segment cells are [P(x,y,z),Q(x,y,z),R(x,y,z)]; See also this Chapter Section 4.3 'Discuss About Abstract Vector Field、Finite Sums Limit、Integral and Riemann Sums'

Calculate microcosmic surface integral values at surface 'S's all segmental cells (24):

Base on integral median theorem, dot product of abstract vector field (23) and tangent plane normal vector (22), times segment unit of 'u,v' (21), these are microcosmic surface integral values at all segmental cells(24)

$$2\left(c \sin\left(\frac{\pi s}{n}\right)^2 b \cos\left(\frac{2\pi t}{n}\right) P(x,y,z) + \sin\left(\frac{\pi s}{n}\right)^2 \sin\left(\frac{2\pi t}{n}\right) a\,c\, Q(x,y,z) \right.$$
$$\left. + \sin\left(\frac{\pi s}{n}\right) a \cos\left(\frac{\pi s}{n}\right) b\, R(x,y,z) \right) \pi^2 / n^2$$
$$\qquad\qquad (24)$$

//Expression (24) isn't that 'Single value', but that 'Set of finite quantitative (viz.n) values'

Structure finite sums (25):

(In the case of "Amount 'n' of surface 'S's segmental cells is uncertain", sum of integral values [dot product about absract vector field (23) and tangent plane normal vector (22)] at all segmental cells):

$$\sum_{t=1}^{n}\left(\sum_{s=1}^{n}\left(2\left(c \sin\left(\frac{\pi s}{n}\right)^2 b \cos\left(\frac{2\pi t}{n}\right) P(x,y,z) + \sin\left(\frac{\pi s}{n}\right)^2 \sin\left(\frac{2\pi t}{n}\right) a\,c\, Q(x,y,z) \right.\right.\right.$$
$$\left.\left.\left. + \sin\left(\frac{\pi s}{n}\right) a \cos\left(\frac{\pi s}{n}\right) b\, R(x,y,z) \right) \pi^2 / n^2 \right)\right)$$

Infinitize finite sums, its limit operational value is 'Surface Integral Value at Manifold'(26):

(In the case of "Amount 'n' of surface 'S's segmental cells tends to infinity", limit value of integral values' sums [dot product about abstract vector field (23) and tangent plane normal vector (22)] at all segmental cells)

$$\lim_{n \to \infty} \sum_{t=1}^{n} \left(\sum_{s=1}^{n} \left[2\left(c \sin\left(\frac{\pi s}{n}\right)^2 b \cos\left(\frac{2\pi t}{n}\right) \right) P(x,y,z) \right. \right.$$

$$\left. \left. + \sin\left(\frac{\pi s}{n}\right)^2 \sin\left(\frac{2\pi t}{n}\right) a\, c\, Q(x,y,z) + \sin\left(\frac{\pi s}{n}\right) a \cos\left(\frac{\pi s}{n}\right) b\, R(x,y,z) \right] \pi^2 / n^2 \right) \right)$$

$$=$$

$$\lim_{n \to \infty} -\pi^2 c \left[P(x,y,z)\, b \cos\left(\frac{(n+1)\pi}{n}\right)^2 n \right.$$

$$+ Q(x,y,z)\, a \sin\left(\frac{(n+1)\pi}{n}\right) \cos\left(\frac{(n+1)\pi}{n}\right) \cos\left(\frac{\pi}{n}\right)^4 n$$

$$+ Q(x,y,z)\, a \sin\left(\frac{(n+1)\pi}{n}\right) \cos\left(\frac{(n+1)\pi}{n}\right) n$$

$$- P(x,y,z)\, b \cos\left(\frac{(n+1)\pi}{n}\right)^2 \cos\left(\frac{\pi}{n}\right)^2 n$$

$$+ \cos\left(\frac{(n+1)\pi}{n}\right)^2 \cos\left(\frac{\pi}{n}\right) \sin\left(\frac{\pi}{n}\right) Q(x,y,z)\, a\, n$$

$$- \sin\left(\frac{\pi}{n}\right)^2 \cos\left(\frac{\pi}{n}\right)^2 P(x,y,z)\, b \cos\left(\frac{(n+1)\pi}{n}\right)^2 n$$

$$- 2 Q(x,y,z)\, a \sin\left(\frac{(n+1)\pi}{n}\right) \cos\left(\frac{(n+1)\pi}{n}\right) \cos\left(\frac{\pi}{n}\right)^2 n$$

$$- \cos\left(\frac{(n+1)\pi}{n}\right)^2 \cos\left(\frac{\pi}{n}\right)^3 \sin\left(\frac{\pi}{n}\right) Q(x,y,z)\, a\, n$$

$$- \cos\left(\frac{(n+1)\pi}{n}\right) \sin\left(\frac{(n+1)\pi}{n}\right) \cos\left(\frac{\pi}{n}\right) \sin\left(\frac{\pi}{n}\right) P(x,y,z)\, b\, n$$

$$+ \cos\left(\frac{(n+1)\pi}{n}\right) \sin\left(\frac{(n+1)\pi}{n}\right) \cos\left(\frac{\pi}{n}\right)^3 \sin\left(\frac{\pi}{n}\right) P(x,y,z)\, b\, n$$

$$- \cos\left(\frac{\pi}{n}\right)^4 (n+1) P(x,y,z)\, b + 2 \cos\left(\frac{\pi}{n}\right)^6 (n+1) P(x,y,z)\, b$$

$$- \cos\left(\frac{\pi}{n}\right)^5 (n+1) \sin\left(\frac{\pi}{n}\right) Q(x,y,z)\, a + \cos\left(\frac{\pi}{n}\right)^7 (n+1) \sin\left(\frac{\pi}{n}\right) Q(x,y,z)\, a$$

$$- \cos\left(\frac{\pi}{n}\right)^8 (n+1)\, P(x,y,z)\, b + \cos\left(\frac{\pi}{n}\right)^5 (n+1) \sin\left(\frac{\pi}{n}\right)^3 Q(x,y,z)\, a$$

$$+ \cos\left(\frac{\pi}{n}\right)^4 (n+1) \sin\left(\frac{\pi}{n}\right)^2 b\, P(x,y,z) - \cos\left(\frac{\pi}{n}\right)^6 (n+1) \sin\left(\frac{\pi}{n}\right)^2 P(x,y,z)\, b$$

$$+ \sin\left(\frac{\pi}{n}\right)^2 \cos\left(\frac{\pi}{n}\right)^4 Q(x,y,z)\, a \sin\left(\frac{(n+1)\pi}{n}\right) \cos\left(\frac{(n+1)\pi}{n}\right)$$

$$- Q(x,y,z)\, a \sin\left(\frac{(n+1)\pi}{n}\right) \cos\left(\frac{(n+1)\pi}{n}\right) \cos\left(\frac{\pi}{n}\right)^4$$

$$- \sin\left(\frac{\pi}{n}\right)^2 \cos\left(\frac{\pi}{n}\right)^4 P(x,y,z)\, b \cos\left(\frac{(n+1)\pi}{n}\right)^2$$

$$+ P(x,y,z)\, b \cos\left(\frac{(n+1)\pi}{n}\right)^2 \cos\left(\frac{\pi}{n}\right)^4 - \cos\left(\frac{\pi}{n}\right)^6 P(x,y,z)\, b \cos\left(\frac{(n+1)\pi}{n}\right)^2$$

$$+ \cos\left(\frac{\pi}{n}\right)^6 Q(x,y,z)\, a \sin\left(\frac{(n+1)\pi}{n}\right) \cos\left(\frac{(n+1)\pi}{n}\right) \bigg) \Big/ \left(\left(-1 + \cos\left(\frac{\pi}{n}\right)^2\right)^2 \right.$$

$$n^2 \bigg) + \pi^2 c \left(P(x,y,z)\, b\, n \cos\left(\frac{\pi}{n}\right)^2 + \cos\left(\frac{\pi}{n}\right) \sin\left(\frac{\pi}{n}\right) Q(x,y,z)\, a\, n \right.$$

$$- P(x,y,z)\, b \cos\left(\frac{\pi}{n}\right)^4 n - \cos\left(\frac{\pi}{n}\right)^3 \sin\left(\frac{\pi}{n}\right) Q(x,y,z)\, a\, n$$

$$- \cos\left(\frac{\pi}{n}\right)^2 \sin\left(\frac{\pi}{n}\right)^2 P(x,y,z)\, b\, n - \cos\left(\frac{\pi}{n}\right)^4 P(x,y,z)\, b + 3 \cos\left(\frac{\pi}{n}\right)^6 P(x,y,z)\, b$$

$$- 2 \cos\left(\frac{\pi}{n}\right)^5 \sin\left(\frac{\pi}{n}\right) Q(x,y,z)\, a + 2 \cos\left(\frac{\pi}{n}\right)^7 \sin\left(\frac{\pi}{n}\right) Q(x,y,z)\, a$$

$$- 2 \cos\left(\frac{\pi}{n}\right)^8 P(x,y,z)\, b + 2 \cos\left(\frac{\pi}{n}\right)^5 \sin\left(\frac{\pi}{n}\right)^3 Q(x,y,z)\, a$$

$$+ \cos\left(\frac{\pi}{n}\right)^4 \sin\left(\frac{\pi}{n}\right)^2 b\, P(x,y,z) - 2 \cos\left(\frac{\pi}{n}\right)^6 \sin\left(\frac{\pi}{n}\right)^2 P(x,y,z)\, b \right) \Big/ \left(\vphantom{\frac{\pi}{n}} \right.$$

$$\left(-1 + \cos\left(\frac{\pi}{n}\right)^2 \right)^2 n^2 \right)$$

= 0   (26)

```
//'Integral value equals 0' possesses specific mathematical and
physical meaning:
    In mathematical meaning, 'Integral value equals 0' is inevitable
result of logic illation, incarnates logic balanced state between
various integral elements;
```

In physical meaning, 'Integral value equals 0' implies that flux of 'Abstract Vector Field at Abstract simply connected orientable closed surface' is constant for '0' (immobile and pending) identically; If integral value equals certain a positive/negative number or an expression, it implies that always there is effusive/influent flux or unknown flux about 'Abstract Vector Field at Abstract simply connected orientable closed surface', this instance will be unaccountable.

Entirely multiply radius vector 'r' (suppose r > 0) by each item of surface 'S' parametrized expression (2), convert "Parametrized expression of surface 'S'" to "Parametrized expression of surface 'S' coordinates":

[r*a*sin(u)cos(v),r*b*sin(u)sin(v),r*c*cos(u)]   (27)

Base on parametrized expression of surface 'S' coordinates (27), define and calculate second matrix of partial derivatives, obtain general expression of surface 'S' coordinates volume element coefficient (28):

$$\begin{bmatrix} \frac{\partial}{\partial r} ra\sin(u)\cos(v) & \frac{\partial}{\partial u} ra\sin(u)\cos(v) & \frac{\partial}{\partial v} ra\sin(u)\cos(v) \\ \frac{\partial}{\partial r} rb\sin(u)\sin(v) & \frac{\partial}{\partial u} rb\sin(u)\sin(v) & \frac{\partial}{\partial v} rb\sin(u)\sin(v) \\ \frac{\partial}{\partial r} rc\cos(u) & \frac{\partial}{\partial u} rc\cos(u) & \frac{\partial}{\partial v} rc\cos(u) \end{bmatrix}$$

$$= abcr^2 \sin(u) \qquad\qquad (28)$$

Calculate divergence of abstract vector field [P(x,y,z), Q(x,y,z),R(x,y,z)], and convert it from Cartesian Coordinates expression(29) to surface 'S' coordinates expression(30):

$$\frac{\partial}{\partial x} P(x,y,z) + \frac{\partial}{\partial y} Q(x,y,z) + \frac{\partial}{\partial z} R(x,y,z) \qquad (29)$$

$$\left(\frac{\partial}{\partial x} P(x,y,z) \right) \left(\frac{\partial}{\partial r} (r\,a\,\sin(u)\,\cos(v)) \right) \left(\frac{\partial}{\partial u} (r\,a\,\sin(u)\,\cos(v)) \right)$$

$$\left(\frac{\partial}{\partial v} (r\,a\,\sin(u)\,\cos(v)) \right) + \left(\frac{\partial}{\partial y} Q(x,y,z) \right) \left(\frac{\partial}{\partial r} (r\,b\,\sin(u)\,\sin(v)) \right)$$

$$\left(\frac{\partial}{\partial u}(r\,b\,\sin(u)\sin(v))\right)\left(\frac{\partial}{\partial v}(r\,b\,\sin(u)\sin(v))\right)$$

$$+\left(\frac{\partial}{\partial z}R(x,y,z)\right)\left(\frac{\partial}{\partial r}(r\,c\,\cos(u))\right)\left(\frac{\partial}{\partial u}(r\,c\,\cos(u))\right)\left(\frac{\partial}{\partial v}(r\,c\,\cos(u))\right)=$$

$$-\left(\frac{\partial}{\partial x}P(x,y,z)\right)a^3\sin(u)^2\cos(v)^2\,r^2\cos(u)\sin(v)$$

$$+\left(\frac{\partial}{\partial y}Q(x,y,z)\right)b^3\sin(u)^2\sin(v)^2\,r^2\cos(u)\cos(v)$$

(30)

//In Spacial Cartesian Coordinates, Divergence of abstract vector field [P(x,y,z),Q(x,y,z),R(x,y,z)] is

$$\frac{\partial}{\partial x}P(x,y,z)+\frac{\partial}{\partial y}Q(x,y,z)+\frac{\partial}{\partial z}R(x,y,z)$$

In logic deduction of this Proof, it is necessary to import divergence of abstract vector field above-mentioned to abstract simply connected orientable closed surface coordinates.

Three elements of abstract vector field's divergence: $\frac{\partial}{\partial x}P(x,y,z)$,

$\frac{\partial}{\partial y}Q(x,y,z)$, $\frac{\partial}{\partial z}R(x,y,z)$ are abstract differential function, and their

variables 'x,y,z' contain sub-variables 'r,u,v'.

Transform Expression between 'Spacial Cartesian coordinates' and 'Abstract simply connected orientable closed surface coordinates' is:

x = r*a*sin(u)cos(v), y = r*b*sin(u)sin(v), z = r*c*cos(u).

--"Coordinates Transform Differential Functions" (Correspond to

"Three Differential Variables '$\frac{\partial}{\partial x}$, $\frac{\partial}{\partial y}$, $\frac{\partial}{\partial z}$' of Differential Functions

'$\frac{\partial}{\partial x}P(x,y,z)$, $\frac{\partial}{\partial y}Q(x,y,z)$, $\frac{\partial}{\partial z}R(x,y,z)$'") are:

$$\frac{\partial}{\partial r}(ra\sin(u)\cos())\quad \frac{\partial}{\partial u}(r\sin(\;)\cos(\;))\,v\;(\;\frac{\partial}{\partial v}\sin(u)\cos(\;))\quad v$$

$$\frac{\partial}{\partial r}(rb\sin(u)\sin(v))\frac{\partial}{\partial u}(rb\sin(u)\sin(v))\frac{\partial}{\partial v}(rb\sin(u)\sin(v))\quad \text{and}$$

$$\frac{\partial}{\partial r}(rc\cos(u))\frac{\partial}{\partial u}(rc\cos(u))\frac{\partial}{\partial v}(rc\cos(u))$$

Product of "Differential Functions '$\frac{\partial}{\partial x}P(x,y,z)$, $\frac{\partial}{\partial y}Q(x,y,z)$,

$\frac{\partial}{\partial z} R(x,y,z)$ ' " and "Coordinates Transform Differential Functions" (Viz. Product of two different functions) constitute "Divergence in Abstract Simply Connected Orientable Closed Surface Coordinates"

'Chain differentiation' or 'Coordinates conversion' ?

If it is 'Chain differentiation', according to principle of 'Multiply in identical chain, Plus in subdivision chain', it ought to be:

$$\left(\frac{\partial}{\partial x} P(x,y,z) \right)$$

$$\left(\left(\frac{\partial}{\partial r} (r\,a\,\sin(u)\cos(v)) \right) + \left(\frac{\partial}{\partial u} (r\,a\,\sin(u)\cos(v)) \right) + \left(\frac{\partial}{\partial v} (r\,a\,\sin(u)\cos(v)) \right) \right)$$

$$+ \left(\frac{\partial}{\partial y} Q(x,y,z) \right)$$

$$\left(\left(\frac{\partial}{\partial r} (r\,b\,\sin(u)\sin(v)) \right) + \left(\frac{\partial}{\partial u} (r\,b\,\sin(u)\sin(v)) \right) + \left(\frac{\partial}{\partial v} (r\,b\,\sin(u)\sin(v)) \right) \right)$$

$$+ \left(\frac{\partial}{\partial z} R(x,y,z) \right) \left(\left(\frac{\partial}{\partial r} (r\,c\,\cos(u)) \right) + \left(\frac{\partial}{\partial u} (r\,c\,\cos(u)) \right) + \left(\frac{\partial}{\partial v} (r\,c\,\cos(u)) \right) \right)$$

No matter 'Chain differentiation' or 'Solve divergence/curl', question for discussion is 'How to solve differential coefficient' or 'Method of differentiation' about abstract vector field 'P[x,y,z], Q[x,y,z],R[x,y,z]'; and this step, question for discussion is 'How to convert result of solving divergence from a coordinates to another'; character and hiberarchy of 'Two questions' above-mentioned are disparate. It is 'Multiply',not 'Plus'– Be determined by 3-Dimensional spacial attribute of coordinates.

Set Amount of spacial bounded closed region 'Ω's (Surface 'S' surrounds) segmental cells as 20

(This amount can be discretional natural number) (31)

5.Microcosmic Triple Integrals Course in spacial bounded closed region 'Ω's first segmental cell:

Segment range [0,1] of 'r': dr = $\frac{1}{20}$

Segment range [0, π] of 'u': du = $\frac{\pi}{20}$

Segment range $[0, 2\pi]$ of 'v': $dv = \dfrac{2\pi}{20} = \dfrac{\pi}{10}$ (32)

Segment volume element coefficient 'J'(33)

[viz. Input (32) to (28)]

(Volume element coefficient (28), average value that corresponds to spacial bounded closed region 'Ω's first segmental cell)

$$\frac{1}{400} a \sin\left(\frac{\pi}{20}\right) b\, c$$

(33)

Segment abstract divergence (34): [viz. Input (32) to (30)]

(Abstract divergence (30), average value that corresponds to spacial bounded closed region 'Ω's first segmental cell)

$$-\frac{1}{400}\left(\frac{\partial}{\partial x}P(x,y,z)\right)a^3 \sin\left(\frac{\pi}{20}\right)^2 \cos\left(\frac{\pi}{10}\right)^2 \cos\left(\frac{\pi}{20}\right)\sin\left(\frac{\pi}{10}\right)$$

$$+\frac{1}{400}\left(\frac{\partial}{\partial y}Q(x,y,z)\right)b^3 \sin\left(\frac{\pi}{20}\right)^2 \sin\left(\frac{\pi}{10}\right)^2 \cos\left(\frac{\pi}{20}\right)\cos\left(\frac{\pi}{10}\right)$$

(34)

// Owing to universality and homogeneity of abstract scalar field

(Divergence) $\dfrac{\partial}{\partial x}P(x,y,z)+\dfrac{\partial}{\partial y}Q(x,y,z)+\dfrac{\partial}{\partial z}R(x,y,z)$, if this abstract

scalar field is continuous in spacial bounded closed region 'Ω', then this abstract scalar field still can be described as

$\dfrac{\partial}{\partial x}P(x,y,z)+\dfrac{\partial}{\partial y}Q(x,y,z)+\dfrac{\partial}{\partial z}R(x,y,z)$ in one or more segment cells of

spacial bounded closed region 'Ω'.

See also this Chapter Section 4.3 'Discuss About Abstract Vector Field、Finite Sums Limit、Integral and Riemann Sums'

Calculate microcosmic triple integrals value in spacial bounded closed region 'Ω's first segmental cell (35):

Base on integral median theorem, product of Abstract Divergence (34) and Volume Element Coefficient (33), times segment unit of 'r,u,v'(32), that is microcosmic triple integrals value in first segmental cell (35)

$$\frac{1}{1600000}\left(-\frac{1}{400}\left(\frac{\partial}{\partial x}P(x,y,z)\right)a^3\sin\left(\frac{\pi}{20}\right)^2\cos\left(\frac{\pi}{10}\right)^2\cos\left(\frac{\pi}{20}\right)\sin\left(\frac{\pi}{10}\right)\right.$$

$$\left.+\frac{1}{400}\left(\frac{\partial}{\partial y}Q(x,y,z)\right)b^3\sin\left(\frac{\pi}{20}\right)^2\sin\left(\frac{\pi}{10}\right)^2\cos\left(\frac{\pi}{20}\right)\cos\left(\frac{\pi}{10}\right)\right)a\sin\left(\frac{\pi}{20}\right)b\,c\,\pi^2$$

6.Microcosmic triple integrals course in spacial bounded closed region 'Ω's all (viz.20) segmental cells:

Segment range $[0,\pi]$ of 'u': du $=\dfrac{i\pi}{20}$

Segment range $[0,2\pi]$ of 'v': dv $=\dfrac{2j\pi}{20}=\dfrac{j\pi}{10}$

Segment range $[0,1]$ of 'r': dr $=\dfrac{k}{20}$

(thereinto,'i,j,k' stand for natural number 1~20)(36)

Segment volume element coefficient (37): [viz. Input (36) to (28)]

(Volume element coefficient (28),average values that correspond to spacial bounded closed region 'Ω's all segmental cells)

$$\frac{1}{400}a\sin\left(\frac{\pi\,i}{20}\right)k^2\,b\,c$$

(37)

// Expression (37) above mentioned isn't that 'Single value', but that 'Set of finite quantitative (viz. 20) values'

Segment abstract divergence (38): [viz. Input (36) to (30)]

(Abstract Divergence (30), average values that correspond to spacial bounded closed region 'Ω's all (viz.20) segmental cells)

$$-\frac{1}{400}\left(\frac{\partial}{\partial x}P(x,y,z)\right)a^3\sin\left(\frac{\pi\,i}{20}\right)^2\cos\left(\frac{\pi\,j}{10}\right)^2 k^2\cos\left(\frac{\pi\,i}{20}\right)\sin\left(\frac{\pi\,j}{10}\right)$$

$$+\frac{1}{400}\left(\frac{\partial}{\partial y}Q(x,y,z)\right)b^3\sin\left(\frac{\pi\,i}{20}\right)^2\sin\left(\frac{\pi\,j}{10}\right)^2 k^2\cos\left(\frac{\pi\,i}{20}\right)\cos\left(\frac{\pi\,j}{10}\right)$$

(38)

// Expression (38) isn't that 'Single value', but that 'Set of finite quantitative (viz.20) values'

// Owing to universality and homogeneity of abstract scalar field

(Divergence) $\dfrac{\partial}{\partial x}P(x,y,z)+\dfrac{\partial}{\partial y}Q(x,y,z)+\dfrac{\partial}{\partial z}R(x,y,z)$, if this abstract

scalar field is continuous in spacial bounded closed region 'Ω',then this abstract scalar field still can be described as

$$\frac{\partial}{\partial x}P(x,y,z)+\frac{\partial}{\partial y}Q(x,y,z)+\frac{\partial}{\partial z}R(x,y,z)$$ in one or more segment cells of

spacial bounded closed region 'Ω'.

   See also this Chapter Section 4.3 'Discuss About Abstract Vector Field、Finite Sums Limit、Integral and Riemann Sums'

   Calculate microcosmic triple integrals values in spacial bounded closed region 'Ω's all segmental cells (39):

   Base on integral median theorem, product of Abstract Divergence (38) and Volume Element Coefficient (37), times segment unit of 'r,u,v'(36), these are microcosmic triple integrals values in all (viz.20) segmental cells (39)

$$\frac{1}{1600000}\left(-\frac{1}{400}\left(\frac{\partial}{\partial x}\mathrm{P}(x,y,z)\right)a^3\sin\left(\frac{\pi\,i}{20}\right)^2\cos\left(\frac{\pi\,j}{10}\right)^2 k^2\cos\left(\frac{\pi\,i}{20}\right)\sin\left(\frac{\pi\,j}{10}\right)\right.$$

$$\left.+\frac{1}{400}\left(\frac{\partial}{\partial y}Q(x,y,z)\right)b^3\sin\left(\frac{\pi\,i}{20}\right)^2\sin\left(\frac{\pi\,j}{10}\right)^2 k^2\cos\left(\frac{\pi\,i}{20}\right)\cos\left(\frac{\pi\,j}{10}\right)\right)a\sin\left(\frac{\pi\,i}{20}\right)k^2\,b\,c$$

$$\pi^2 \tag{39}$$

//Expression (39) isn't that 'Single value', but that 'Set of finite quantitative (viz. 20) values'

   Structure sequence that is composed of finite quantitative (viz. 20) microcomosic triple integrals values (40):
   (Lengthy expression of sequence is elided;
   In program template of Waterloo Maplesoft (See also Section 4.2):
   sqn:=seq(seq(seq(ijkddiV2*ijkdJ*dr*du*dv,i=1..dus),j=1..dus),
   k=1..dus):
   change ':'(last) to ';', then obtain expression of sequence)

   Accumulational sum of sequence (viz. Sum of integral values of product [Abstract Divergence (38) and Volume Element at all (viz. 20) segmental cells], obtain triple integrals value at manifold (41):

   [Lengthy expression of accumulational sum result is elided;

In program template of Waterloo Maplesoft (See also Section 4.2):

add(k,k=sqn):omega:=evalf(%):

change ':'(middle and last) to ';', then obtain expression and float value of result]

// Amount of spacial bounded closed region 'Ω's (Surface 'S' surrounds) segmental cells increases infinitely, sum of integral values above-mentioned tends to '0' infinitely.

Set Amount of spacial bounded closed region 'Ω's (Surface 'S' surrounds) segmental cells as n (viz. uncertain natural number) (42)

7.Microcosmic Triple Integrals Course in spacial bounded closed region 'Ω's first segmental cell:

Segment range [0,1] of 'r': dr = $\dfrac{1}{n}$

Segment range [0,π] of 'u': du = $\dfrac{\pi}{n}$

Segment range [0,2π] of 'v': dv = $\dfrac{2\pi}{n}$ (43)

Segment volume element coefficient 'J' (44):

(viz. Input (43) to (28))

(Volume element coefficient (28),average value that corresponds to spacial bounded closed region 'Ω's first segmental cell)

$$\frac{a \sin\left(\dfrac{\pi}{n}\right) b\, c}{n^2}$$ (44)

Segment abstract divergence (45): [viz. Input (43) to (30)]

(Abstract divergence (30), average value that corresponds to spacial bounded closed region 'Ω's first segmental cell)

$$-\frac{\left(\dfrac{\partial}{\partial x}\mathrm{P}(x,y,z)\right) a^3 \sin\left(\dfrac{\pi}{n}\right)^2 \cos\left(\dfrac{2\pi}{n}\right)^2 \cos\left(\dfrac{\pi}{n}\right) \sin\left(\dfrac{2\pi}{n}\right)}{n^2}$$

$$+\frac{\left(\dfrac{\partial}{\partial y}\mathrm{Q}(x,y,z)\right) b^3 \sin\left(\dfrac{\pi}{n}\right)^2 \sin\left(\dfrac{2\pi}{n}\right)^2 \cos\left(\dfrac{\pi}{n}\right) \cos\left(\dfrac{2\pi}{n}\right)}{n^2}$$

      (45)

// Owing to universality and homogeneity of abstract scalar field

(Divergence) $\dfrac{\partial}{\partial x}P(x,y,z)+\dfrac{\partial}{\partial y}Q(x,y,z)+\dfrac{\partial}{\partial z}R(x,y,z)$, if this abstract

scalar field is continuous in spacial bounded closed region 'Ω',then

this abstract scalar field still can be described as

$\dfrac{\partial}{\partial x}P(x,y,z)+\dfrac{\partial}{\partial y}Q(x,y,z)+\dfrac{\partial}{\partial z}R(x,y,z)$ in one or more segment cells of

spacial bounded closed region 'Ω'.

See also this Chapter Section 4.3 'Discuss About Abstract Vector Field、Finite Sums Limit、Integral and Riemann Sums'

Calculate microcosmic triple integrals value in spacial bounded closed region 'Ω's first segmental cell (46):

Base on integral median theorem, product of Abstract Divergence (45) and Volume Element Coefficient (44), times segment unit of 'r,u,v'(43), that is microcosmic triple integrals value in first segmental cell (46)

$$2\left(-\dfrac{\left(\dfrac{\partial}{\partial x}\mathrm{P}(x,y,z)\right)a^3\sin\!\left(\dfrac{\pi}{n}\right)^2\cos\!\left(\dfrac{2\pi}{n}\right)^2\cos\!\left(\dfrac{\pi}{n}\right)\sin\!\left(\dfrac{2\pi}{n}\right)}{n^2}\right.$$

$$\left.+\dfrac{\left(\dfrac{\partial}{\partial y}\mathrm{Q}(x,y,z)\right)b^3\sin\!\left(\dfrac{\pi}{n}\right)^2\sin\!\left(\dfrac{2\pi}{n}\right)^2\cos\!\left(\dfrac{\pi}{n}\right)\cos\!\left(\dfrac{2\pi}{n}\right)}{n^2}\right)a\sin\!\left(\dfrac{\pi}{n}\right)b\,c\,\pi^2\,/\,n^5$$

8.Microcosmic triple integrals course in spacial bounded closed region 'Ω's all (viz.n) segmental cells:

Segment range [0, π] of 'u': du = $\dfrac{i\pi}{n}$

Segment range [0,2π] of 'v': dv = $\dfrac{2j\pi}{n}$

Segment range [0,1] of 'r': dr = $\dfrac{k}{n}$

(thereinto, 'i,j,k' stand for natural number 1~n)(47)

Segment volume element coefficient (48): [viz. Input (47) to (28)]

(Volume element coefficient (28),average values that correspond to spacial bounded closed region 'Ω's all (viz.n) segmental cells)

$$\frac{a \sin\left(\dfrac{\pi i}{n}\right) k^2 \, b \, c}{n^2}$$

(48)

//Expression (48) isn't that 'Single value', but that 'Set of finite quantitative (viz.n) values'

Segment abstract divergence (49): [viz. Input (47) to (30)]

(Abstract Divergence (30), average values that correspond to spacial bounded closed region 'Ω's all (viz.n) segmental cells)

$$-\frac{\left(\dfrac{\partial}{\partial x}P(x,y,z)\right)a^3 \sin\left(\dfrac{\pi i}{n}\right)^2 \cos\left(\dfrac{2\pi j}{n}\right)^2 k^2 \cos\left(\dfrac{\pi i}{n}\right)\sin\left(\dfrac{2\pi j}{n}\right)}{n^2}$$

$$+\frac{\left(\dfrac{\partial}{\partial y}Q(x,y,z)\right)b^3 \sin\left(\dfrac{\pi i}{n}\right)^2 \sin\left(\dfrac{2\pi j}{n}\right)^2 k^2 \cos\left(\dfrac{\pi i}{n}\right)\cos\left(\dfrac{2\pi j}{n}\right)}{n^2}$$

(49)

// Expression (49) isn't that 'Single value', but that 'Set of finite quantitative (viz. n) values'

// Owing to universality and homogeneity of abstract scalar field

(Divergence) $\dfrac{\partial}{\partial x}P(x,y,z)+\dfrac{\partial}{\partial y}Q(x,y,z)+\dfrac{\partial}{\partial z}R(x,y,z)$, if this abstract

scalar field is continuous in spacial bounded closed region 'Ω',then this abstract scalar field still can be described as

$\dfrac{\partial}{\partial x}P(x,y,z)+\dfrac{\partial}{\partial y}Q(x,y,z)+\dfrac{\partial}{\partial z}R(x,y,z)$ in one or more segment cells of

spacial bounded closed region 'Ω'.

See also this Chapter Section 4.3 'Discuss About Abstract Vector Field、Finite Sums Limit、Integral and Riemann Sums'

Calculate microcosmic triple integrals values in spacial bounded closed region 'Ω's all segmental cells (50):

Base on integral median theorem, product of Abstract Divergence (49) and Volume Element Coefficient (48), times segment unit of 'r,u,v' (47), that is microcosmic triple integrals values in all (viz.n) segmental cells (50)

$$2\left(-\frac{\left(\frac{\partial}{\partial x}P(x,y,z)\right)a^3\sin\left(\frac{\pi i}{n}\right)^2\cos\left(\frac{2\pi j}{n}\right)^2 k^2\cos\left(\frac{\pi i}{n}\right)\sin\left(\frac{2\pi j}{n}\right)}{n^2}\right.$$

$$\left.+\frac{\left(\frac{\partial}{\partial y}Q(x,y,z)\right)b^3\sin\left(\frac{\pi i}{n}\right)^2\sin\left(\frac{2\pi j}{n}\right)^2 k^2\cos\left(\frac{\pi i}{n}\right)\cos\left(\frac{2\pi j}{n}\right)}{n^2}\right)a\sin\left(\frac{\pi i}{n}\right)k^2 b\,c$$

$$\pi^2/n^5$$

//Expression (50) isn't that 'single value', but that 'Set of finite quantitative (viz.n) values'

Structure finite sums (51):

$$\sum_{k=1}^{n}\left(\sum_{j=1}^{n}\left(\sum_{i=1}^{n}\left(2\left(-\frac{\left(\frac{\partial}{\partial x}P(x,y,z)\right)a^3\sin\left(\frac{\pi i}{n}\right)^2\cos\left(\frac{2\pi j}{n}\right)^2 k^2\cos\left(\frac{\pi i}{n}\right)\sin\left(\frac{2\pi j}{n}\right)}{n^2}\right.\right.\right.\right.$$

$$\left.+\frac{\left(\frac{\partial}{\partial y}Q(x,y,z)\right)b^3\sin\left(\frac{\pi i}{n}\right)^2\sin\left(\frac{2\pi j}{n}\right)^2 k^2\cos\left(\frac{\pi i}{n}\right)\cos\left(\frac{2\pi j}{n}\right)}{n^2}\right)a\sin\left(\frac{\pi i}{n}\right)k^2 b\,c$$

$$\left.\left.\left.\pi^2/n^5\right)\right)\right)$$

(51)

Infinitize finite sums, its limit operational value is 'Triple Integrals Value at Manifold' (52):

(In the case of "Amount 'n' of spacial bounded closed region 'Ω's (Surface 'S' surrounds) segmental cells tends to infinity", limit value of integral values' sums [Product about Abstract Divergence(49) and Volume Element] at all segmental cells)

$$\lim_{n\to\infty}\sum_{k=1}^{n}\left(\sum_{j=1}^{n}\left(\sum_{i=1}^{n}\left(2\left(\vphantom{\frac{\pi}{n}}\right.\right.\right.\right.$$

$$-\frac{\left(\frac{\partial}{\partial x}P(x,y,z)\right)a^3\sin\left(\frac{\pi i}{n}\right)^2\cos\left(\frac{2\pi j}{n}\right)^2 k^2\cos\left(\frac{\pi i}{n}\right)\sin\left(\frac{2\pi j}{n}\right)}{n^2}$$

$$+ \frac{\left(\frac{\partial}{\partial y}Q(x,y,z)\right)b^3 \sin\left(\frac{\pi i}{n}\right)^2 \sin\left(\frac{2\pi j}{n}\right)^2 k^2 \cos\left(\frac{\pi i}{n}\right)\cos\left(\frac{2\pi j}{n}\right)}{n^2}\Bigg) a \sin\left(\frac{\pi i}{n}\right)k^2 b c$$

$$\pi^2/n^5 \Bigg)\Bigg)\Bigg)$$

$$= 0 \qquad\qquad\qquad\qquad\qquad\qquad\qquad\qquad (52)$$

// About 'Integral value is unequal to 0', See also Chapter 6 (This Part) Section 6.1 'Divergence Theorem at Manifold and Klein Bottle'

Thereinto, Suppose:

$$\lim_{n\to\infty}\sum_{k=1}^{n}\left(\sum_{j=1}^{n}\left(\sum_{i=1}^{n}\left(\frac{2\,a\sin\left(\frac{i\pi}{n}\right)k^2 b c \pi^2}{n^5}\right)\right)\right) \neq 0$$

Viz. Suppose sum of triple integrals values (Surface 'S' coordinates volume element itself) at all segmental cells can't be '0'; Explain visually, it can be regarded as "Suppose spacial bounded closed region 'Ω' can't be '0 volume'"

Viz. In the case of 'n → ∞', (25)=(51):

$$\sum_{t=1}^{n}\left(\sum_{s=1}^{n}\left(2\left(P(x,y,z)\,c\,\sin\left(\frac{s\pi}{n}\right)^2 b\cos\left(\frac{2t\pi}{n}\right)+Q(x,y,z)\sin\left(\frac{s\pi}{n}\right)^2 a\sin\left(\frac{2t\pi}{n}\right)c\right.\right.$$
$$\left.\left.+R(x,y,z)\sin\left(\frac{s\pi}{n}\right)a b\cos\left(\frac{s\pi}{n}\right)\right)\pi^2/n^2\right)\right)$$

$$=$$

$$\sum_{k=1}^{n}\left(\sum_{j=1}^{n}\left(\sum_{i=1}^{n}\left(2\left(-\frac{\left(\frac{\partial}{\partial x}P(x,y,z)\right)a^3\sin\left(\frac{i\pi}{n}\right)^2\cos\left(\frac{2j\pi}{n}\right)^2 k^2\cos\left(\frac{i\pi}{n}\right)\sin\left(\frac{2j\pi}{n}\right)}{n^2}\right.\right.\right.\right.$$

$$\left.+\frac{\left(\frac{\partial}{\partial y}Q(x,y,z)\right)b^3\sin\left(\frac{i\pi}{n}\right)^2\sin\left(\frac{2j\pi}{n}\right)^2 k^2\cos\left(\frac{i\pi}{n}\right)\cos\left(\frac{2j\pi}{n}\right)}{n^2}\right)a\sin\left(\frac{i\pi}{n}\right)k^2 b$$

$$c\,\pi^2/n^5 \Bigg)\Bigg)\Bigg)$$

Equation above-mentioned can be described as:

$$\iint\limits_{S} A \cdot n \, dS = \iiint\limits_{\Omega} divA \cdot d\omega$$ (1), Complete Proof.

4.2 Prove Divergence Theorem Finite Sums Limits at Manifold
[Program Template of Waterloo Maplesoft, Optional]

Divergence Theorem

Suppose spacial bounded closed region 'Ω' is surrounded by smooth or piecewise smooth closed surface 'S'. Functions 'P(x,y,z),Q(x,y,z), R(x,y,z)' [Structure vector field 'A'] and its partial derivatives are continuous at spacial bounded closed region 'Ω', then:

$$\iint\limits_{S} A \cdot n \, dS = \iiint\limits_{\Omega} divA \cdot d\omega$$ (1)

Thereinto, surface 'S' is outer side of spacial bounded closed region 'Ω's entire boundary surface, 'n' is unit normal vector of surface 'S's outer side, and 'divA' is divergence of vector field 'A'.

Symbol system:

Abstract spacial Vector Field 'V',

Divergence 'diV1,diV2' of Vector Field 'V';

Abstract simply connected orientable closed Parametrized Surface 'CS' [Closed],

Tangent Plane Normal Vector '[A,B,C]' of Surface 'CS',

Spacial bounded closed region 'Ω' that Surface 'CS' surrounds,

General expression of surface 'CS' coordinates volume element coefficient 'J';

Amount 'dus' of surface 'CS's segmental cells (It can be discretional natural number),

Segmental range 'du' of parameter 'u',

Segmental range 'dv' of parameter 'v';

'Tangent Plane Normal Vector [A,B,C]'s average value '[dA,dB,dC]' that corresponds to surface 'CS's first segmental cell,

Spacial vector field 'V's average value 'dV' that corresponds to surface 'CS's first segmental cell;

'Tangent Plane Normal Vector [A,B,C]'s average values '[stdA, stdB,stdC]' that correspond to surface 'CS's all segmental cells,

   (In actual expressions,'s,t' stand for natural number),

   Spacial vector field 'V's average values 'dV' that correspond to surface'CS's all segmental cells;

   Amount 'dus' of spacial bounded closed region 'Ω's segmental cells (It can be discretional natural number),

   Segmental range 'dr' of parameter 'r',

   Segmental range 'du' of parameter 'u',

   Segmental range 'dv' of parameter 'v';

   Volume element coefficient 'J's average value'dJ' that corresponds to spacial bounded closed region 'Ω's first segmental cell,

   Divergence 'diV2's average value 'ddiV2' that corresponds to spacial bounded closed region 'Ω's first segmental cell;

   Volume element coefficient 'J's average values 'ijkdJ' that correspond to special bounded closed region 'Ω's all segmental cells,

   Divergence 'diV2's average values 'ijkddiV2' that correspond to spacial bounded closed region 'Ω's all segmental cells

   (In actual expressions, 'i,j,k' stand for natural number)

> restart;

> with(linalg):

Define parametrized expression of abstract simply connected orientable closed surface 'CS':

> CS:=[a*sin(u)*cos(v),b*sin(u)*sin(v),c*cos(u)];   **(2)**

$$CS := [a \sin(u) \cos(v), b \sin(u) \sin(v), c \cos(u)]$$

Thereinto, 'a,b,c' are nonzero constants or 1th order derivable continuous functions, simply connected orientable closed surface 'CS' determines values of 'a,b,c'

> rgu:=[0,Pi];

$$rgu := [0, \pi]$$

> rgv:=[0,2*Pi];

Set range of parameters 'u,v' in $[0,\pi]$, $[0,2\pi]$, make surface 'CS' closed. (See also Poincare Conjecture: 'Every closed n-Manifold which

is homotopy equivalent to n-Sphere is homeomorphic to n-Sphere' and Chapter 1 this Part 'Constitute Abstract Simply Connected Orientable Closed Parametrized Surface Coordinates')

$$rgv := [\, 0, 2\,\pi\,]$$

```
> V:=[(P)(x,y,z),(Q)(x,y,z),(R)(x,y,z)];                    (3)
```

Define abstract spacial vector field 'V'

(Suppose Vector Field 'V' possesses 1th order continuous partial derivatives in spacial bounded closed region 'Ω' that Surface 'CS' surrounds)

$$V := [\,P(x, y, z), Q(x, y, z), R(x, y, z)\,]$$

```
> Diff(V[1],x)+Diff(V[2],y)+Diff(V[3],z)

= diff(V[1],x)+diff(V[2],y)+diff(V[3],z);diV1:=rhs(%);    (4)
```

Calculate divergence 'diV1'of spacial vector field 'V'

$$\left(\frac{\partial}{\partial x} P(x, y, z)\right) + \left(\frac{\partial}{\partial y} Q(x, y, z)\right) + \left(\frac{\partial}{\partial z} R(x, y, z)\right) =$$

$$\left(\frac{\partial}{\partial x} P(x, y, z)\right) + \left(\frac{\partial}{\partial y} Q(x, y, z)\right) + \left(\frac{\partial}{\partial z} R(x, y, z)\right)$$

$$diV1 := \left(\frac{\partial}{\partial x} P(x, y, z)\right) + \left(\frac{\partial}{\partial y} Q(x, y, z)\right) + \left(\frac{\partial}{\partial z} R(x, y, z)\right)$$

```
> x:=CS[1]:y:=CS[2]:z:=CS[3]:
> matrix(3,3,[i,j,k,Diff(CS[1],u),Diff(CS[2],u),Diff(CS[3],u),
Diff(CS[1],v),Diff(CS[2],v),Diff(CS[3],v)])=
matrix(3,3,[i,j,k,diff(CS[1],u),diff(CS[2],u),diff(CS[3],u),
diff(CS[1],v),diff(CS[2],v),diff(CS[3],v)]);m1:=rhs(%);
```

Define and calculate matrix of partial derivatives 'm1', obtain 'Tangent plane normal vector' of surface 'CS'

$$\begin{bmatrix} i & j & k \\ \dfrac{\partial}{\partial u}(a\sin(u)\cos(v)) & \dfrac{\partial}{\partial u}(b\sin(u)\sin(v)) & \dfrac{\partial}{\partial u}(c\cos(u)) \\ \dfrac{\partial}{\partial v}(a\sin(u)\cos(v)) & \dfrac{\partial}{\partial v}(b\sin(u)\sin(v)) & \dfrac{\partial}{\partial v}(c\cos(u)) \end{bmatrix} =$$

$$\begin{bmatrix} i & j & k \\ a\cos(u)\cos(v) & b\cos(u)\sin(v) & -c\sin(u) \\ -a\sin(u)\sin(v) & b\sin(u)\cos(v) & 0 \end{bmatrix}$$

$$m1 := \begin{bmatrix} i & j & k \\ a\cos(u)\cos(v) & b\cos(u)\sin(v) & -c\sin(u) \\ -a\sin(u)\sin(v) & b\sin(u)\cos(v) & 0 \end{bmatrix}$$

```
> det(m1);
```

$$i\, c \sin(u)^2\, b \cos(v) + a \cos(u) \cos(v)^2\, k\, b \sin(u) + a \sin(u)^2 \sin(v)\, j\, c$$
$$+ a \sin(u) \sin(v)^2\, k\, b \cos(u)$$

```
> mn:=simplify(%);
```

$$mn := \sin(u)\,(i\, c \sin(u)\, b \cos(v) + a \sin(u) \sin(v)\, j\, c + a\, k\, b \cos(u))$$

```
> A:=coeff(mn,i); # Obtain coefficient of 'i'
```

$$A := c \sin(u)^2\, b \cos(v)$$

```
> B:=coeff(mn,j); # Obtain coefficient of 'j'
```

$$B := \sin(u)^2\, a \sin(v)\, c$$

```
> C:=coeff(mn,k); # Obtain coefficient of 'k'
```

$$C := \sin(u)\, a\, b \cos(u)$$

```
> [A,B,C]; # [A,B,C] structure 'tangent plane normal vector'   (5)
```

$$[c \sin(u)^2\, b \cos(v),\ \sin(u)^2 \sin(v)\, a\, c,\ \sin(u)\, a \cos(u)\, b]$$

```
// Segment range '[0,Pi],[0,2*Pi]' of 'u,v':
> dus:=50;
# Set amount of surface 'CS's segmental cells, this amount can be
discretional natural number
```

$$dus := 50$$

```
> du:=(rgu[2]-rgu[1])/dus; # Segment range of 'u'
```

$$du := \frac{\pi}{50}$$

```
> dv:=(rgv[2]-rgv[1])/dus; # Segment range of 'v'
```

$$dv := \frac{\pi}{25}$$

 # Microcosmic surface integral course of surface 'CS's first segmental cell (50 segmental cells):

```
// Segment 'tangent plane normal vector'
> dA:=subs(v=rgv[1]+dv,subs(u=rgu[1]+du,A));
```

$$dA := c \sin\left(\frac{\pi}{50}\right)^2 b \cos\left(\frac{\pi}{25}\right)$$

```
> dB:=subs(v=rgv[1]+dv,subs(u=rgu[1]+du,B));
```

$$dB := \sin\left(\frac{\pi}{50}\right)^2 a \sin\left(\frac{\pi}{25}\right) c$$

```
> dC:=subs(v=rgv[1]+dv,subs(u=rgu[1]+du,C));
```

$$dC := \sin\left(\frac{\pi}{50}\right) a\, b\, \cos\left(\frac{\pi}{50}\right)$$

```
> [dA,dB,dC]; #`[dA,dB,dC]' are `tangent plane normal vector's average
```
value that corresponds to surface 'CS's first segmental cell

$$\left[c\, \sin\left(\frac{\pi}{50}\right)^2 b\, \cos\left(\frac{\pi}{25}\right),\ \sin\left(\frac{\pi}{50}\right)^2 \sin\left(\frac{\pi}{25}\right) a\, c,\ \sin\left(\frac{\pi}{50}\right) a\, \cos\left(\frac{\pi}{50}\right) b \right]$$

```
> x:='x':y:='y':z:='z':
```

// Transform variables, prevent that "Import 'x=a*sin(u)*cos(v),
y=b*sin(u)*sin(v),z=c*cos(u)' to abstract vector field '[P(x,y,z),
Q(x,y,z),R(x,y,z)]'"

// Segment spacial vector field 'V':
(This step possesses formal meaning in 'proof' only, and possesses
essential meaning in 'Numerical Model')

```
> dV:=subs(v=rgv[1]+dv,subs(u=rgu[1]+du,V));
```
Spacial vector field 'V's average value that corresponds to
surface 'CS's first segmental cell

$$dV := [\mathrm{P}(x,y,z),\ \mathrm{Q}(x,y,z),\ \mathrm{R}(x,y,z)]$$

// Calculate microcosmic surface integral value of surface 'CS's
first segmental cell:

```
> (dV[1]*dA+dV[2]*dB+dV[3]*dC)*du*dv;
```
Base on integral median theorem, integral expression of dot product
(Spacial vector field 'V' and '[dA,dB,dC]') at first segmental cell

$$\frac{1}{1250}\left(\mathrm{P}(x,y,z)\, c\, \sin\left(\frac{\pi}{50}\right)^2 b\, \cos\left(\frac{\pi}{25}\right) + \mathrm{Q}(x,y,z)\, \sin\left(\frac{\pi}{50}\right)^2 a\, \sin\left(\frac{\pi}{25}\right) c \right.$$
$$\left. + \mathrm{R}(x,y,z)\, \sin\left(\frac{\pi}{50}\right) a\, b\, \cos\left(\frac{\pi}{50}\right) \right) \pi^2$$

Microcosmic surface integral courses of surface 'CS's all segmental cells (50 segmental cells):
// Segment 'tangent plane normal vector'

```
> stdA:=subs(v=rgv[1]+t*dv,subs(u=rgu[1]+s*du,A));
```

$$stdA := c \sin\left(\frac{s\,\pi}{50}\right)^2 b \cos\left(\frac{t\,\pi}{25}\right)$$

```
> stdB:=subs(v=rgv[1]+t*dv,subs(u=rgu[1]+s*du,B));
```

$$stdB := \sin\left(\frac{s\,\pi}{50}\right)^2 a \sin\left(\frac{t\,\pi}{25}\right) c$$

```
> stdC:=subs(v=rgv[1]+t*dv,subs(u=rgu[1]+s*du,C));
```

$$stdC := \sin\left(\frac{s\,\pi}{50}\right) a\, b \cos\left(\frac{s\,\pi}{50}\right)$$

```
> [stdA,stdB,stdC]; #'[stdA,stdB,stdC]'
```
are 'tangent plane normal vector's average values that correspond to surface 'CS's all segmental cells

$$\left[c \sin\left(\frac{\pi\,s}{50}\right)^2 b \cos\left(\frac{\pi\,t}{25}\right), \sin\left(\frac{\pi\,s}{50}\right)^2 \sin\left(\frac{\pi\,t}{25}\right) a\,c, \sin\left(\frac{\pi\,s}{50}\right) a \cos\left(\frac{\pi\,s}{50}\right) b \right]$$

```
// In actual expressions, 's,t' stand for natural number 1~50
//'[stdA,stdB,stdC]' isn't that 'Single vector value', but that 'Set
of finite quantitative (Viz.50) vector values'


// Segment spacial vector field 'V'
```
(This step possesses formal meaning in 'Proof' only, and possesses essential meaning in 'Numerical Model')

```
> stdV:=subs(v=rgv[1]+t*dv,subs(u=rgu[1]+s*du,V));
```
\# Spacial vector field 'V's average values that correspond to surface'CS's all segmental cells

$$stdV := [\mathrm{P}(x,y,z), \mathrm{Q}(x,y,z), \mathrm{R}(x,y,z)]$$

```
// Calculate  microcosmosic surface integral value (All segmental
cells of surface 'CS')
> (stdV[1]*stdA+stdV[2]*stdB+stdV[3]*stdC)*du*dv;
```
\# Base on integral median theorem, integral expression of dot product (Spacial vector field 'V' and '[stdA,stdB,stdC]') at all segmental cells

$$\frac{1}{1250}\left(\mathrm{P}(x,y,z)\, c \sin\left(\frac{s\,\pi}{50}\right)^2 b \cos\left(\frac{t\,\pi}{25}\right) + \mathrm{Q}(x,y,z) \sin\left(\frac{s\,\pi}{50}\right)^2 a \sin\left(\frac{t\,\pi}{25}\right) c \right.$$

$$\left. + \mathrm{R}(x,y,z) \sin\left(\frac{s\,\pi}{50}\right) a\, b \cos\left(\frac{s\,\pi}{50}\right) \right) \pi^2$$

```
//Expression above-mentioned isn't that 'Single value', but that 'Set
of finite quantitative (viz.50) values'


// List of finite quantitative (viz.50) microcosmic surface integral
values:
> seq(seq([s*t,(stdV[1]*stdA+stdV[2]*stdB+stdV[3]*stdC)*du*dv,
evalf((stdV[1]*stdA+stdV[2]*stdB+stdV[3]*stdC)*du*dv)],s=1..dus),
t=1..dus):
# List of integral values of dot product(Spacial vector field 'V' and
'[stdA,stdB,stdC]') at all segmental cells, be elided
// Structure sequence that is composed of finite quantitative
microcomosic surface integral values:
> sqn:=seq(seq((stdV[1]*stdA+stdV[2]*stdB+stdV[3]*stdC)*du*dv,
s=1..dus),t=1..dus):
// Accumulational sum of sequence, obtain surface integral value  at
manifold:
> add(k,k=sqn):xi:=evalf(%);   (6)
# Sum of integral values of dot product(Spacial vector field 'V' and
'[stdA,stdB,stdC]')at all segmental cells;  Amount of surface 'CS's
segmental cells increases infinitely, Sum of integral values
above-mentioned tends to '0' infinitely
```

$$\xi := 0.5150 \, 10^{-9} \, Q(x,y,z) \, a \, c - 0.247 \, 10^{-9} \, R(x,y,z) \, a \, b - 0.263 \, 10^{-9} \, P(x,y,z) \, c \, b$$

```
// Renewedly segment range '[0,Pi],[0,2*Pi]' of 'u,v':
> dus:=n;
# Set amount of surface 'CS's segmental cells as natural number'n'
```
$$dus := n$$
```
> du:=(rgu[2]-rgu[1])/dus; # Segment range of 'u'
```
$$du := \frac{\pi}{n}$$
```
> dv:=(rgv[2]-rgv[1])/dus; # Segment range of 'v'
```
$$dv := \frac{2\,\pi}{n}$$

**# Microcosmic surface integral course of surface 'CS's first segmental
cell (n segmental cells):**
```
// Segment 'tangent plane normal vector'
```

```
> dA:=subs(v=rgv[1]+dv,subs(u=rgu[1]+du,A));
```

$$dA := c \sin\left(\frac{\pi}{n}\right)^2 b \cos\left(\frac{2\pi}{n}\right)$$

```
> dB:=subs(v=rgv[1]+dv,subs(u=rgu[1]+du,B));
```

$$dB := \sin\left(\frac{\pi}{n}\right)^2 a \sin\left(\frac{2\pi}{n}\right) c$$

```
> dC:=subs(v=rgv[1]+dv,subs(u=rgu[1]+du,C));
```

$$dC := \sin\left(\frac{\pi}{n}\right) a b \cos\left(\frac{\pi}{n}\right)$$

```
> [dA,dB,dC]; #'[dA,dB,dC]' is 'tangent plane normal vector's average
```
value that corresponds to surface 'CS's first segmental cell

$$\left[c \sin\left(\frac{\pi}{n}\right)^2 b \cos\left(\frac{2\pi}{n}\right), \sin\left(\frac{\pi}{n}\right)^2 \sin\left(\frac{2\pi}{n}\right) a c, \sin\left(\frac{\pi}{n}\right) a \cos\left(\frac{\pi}{n}\right) b \right]$$


```
// Segment spacial vector field 'V':
```
(This step possesses formal meaning in 'proof' only, and possesses
essential meaning in 'Numerical Model')
```
> dV:=subs(v=rgv[1]+dv,subs(u=rgu[1]+du,V));
```
Spacial vector field 'V's average value that corresponds to
surface 'CS's first segmental cell

$$dV := [\mathrm{P}(x,y,z), \mathrm{Q}(x,y,z), \mathrm{R}(x,y,z)]$$


```
// Calculate microcosmic surface integral value of surface 'CS's
first segmental cell:
> (dV[1]*dA+dV[2]*dB+dV[3]*dC)*du*dv;
```
Base on integral median theorem, integral value of dot product
(Spacial vector field 'V' and '[dA,dB,dC]') at first segmental cell

$$2\left(\mathrm{P}(x,y,z) c \sin\left(\frac{\pi}{n}\right)^2 b \cos\left(\frac{2\pi}{n}\right) + \mathrm{Q}(x,y,z) \sin\left(\frac{\pi}{n}\right)^2 a \sin\left(\frac{2\pi}{n}\right) c \right.$$
$$\left. + \mathrm{R}(x,y,z) \sin\left(\frac{\pi}{n}\right) a b \cos\left(\frac{\pi}{n}\right) \right) \pi^2 / n^2$$

Microcosmic surface integral course of surface 'CS's all segmental cells (n segmental cells):

```
// Segment 'tangent plane normal vector'
```

```
> stdA:=subs(v=rgv[1]+t*dv,subs(u=rgu[1]+s*du,A));
```

$$stdA := c\,\sin\!\left(\frac{s\,\pi}{n}\right)^{2} b\,\cos\!\left(\frac{2\,t\,\pi}{n}\right)$$

```
> stdB:=subs(v=rgv[1]+t*dv,subs(u=rgu[1]+s*du,B));
```

$$stdB := \sin\!\left(\frac{s\,\pi}{n}\right)^{2} a\,\sin\!\left(\frac{2\,t\,\pi}{n}\right) c$$

```
> stdC:=subs(v=rgv[1]+t*dv,subs(u=rgu[1]+s*du,C));
```

$$stdC := \sin\!\left(\frac{s\,\pi}{n}\right) a\,b\,\cos\!\left(\frac{s\,\pi}{n}\right)$$

```
> [stdA,stdB,stdC];
```

`#'[stdA,stdB,stdC]'` are `'tangent plane normal vector's average values`
that correspond to surface `'CS's` all segmental cells

$$\left[c\,\sin\!\left(\frac{\pi\,s}{n}\right)^{2} b\,\cos\!\left(\frac{2\,\pi\,t}{n}\right),\ \sin\!\left(\frac{\pi\,s}{n}\right)^{2}\sin\!\left(\frac{2\,\pi\,t}{n}\right) a\,c,\ \sin\!\left(\frac{\pi\,s}{n}\right) a\,\cos\!\left(\frac{\pi\,s}{n}\right) b\right]$$

```
// In actual expressions, 's,t' stand for natural number 1~n
//'[stdA,stdB,stdC]' isn't that 'single vector value', but that 'set
of finite quantitative(viz. n) vector values'
```

```
// Segment spacial vector field 'V'
```
(This step possesses formal meaning in `'Proof'` only, and possesses
essential meaning in `'Numerical Model'`)

```
> stdV:=subs(v=rgv[1]+t*dv,subs(u=rgu[1]+s*du,V));
```

`#` Spacial vector field `'V's` average values that correspond to
surface `'CS's` all segmental cells

$$stdV := [\,\mathrm{P}(x,y,z),\ \mathrm{Q}(x,y,z),\ \mathrm{R}(x,y,z)\,]$$

```
// Calculate microcosmosic surface integral values (All segmental
cells of surface 'CS'):
> (stdV[1]*stdA+stdV[2]*stdB+stdV[3]*stdC)*du*dv;
```

`#` Base on integral median theorem, integral expression of dot product
(Spacial vector field `'V'` and `'[stdA,stdB,stdC]'`) at all segmental
cells

$$2\left(P(x,y,z)\,c\,\sin\left(\frac{s\,\pi}{n}\right)^{2} b\,\cos\left(\frac{2\,t\,\pi}{n}\right) + Q(x,y,z)\,\sin\left(\frac{s\,\pi}{n}\right)^{2} a\,\sin\left(\frac{2\,t\,\pi}{n}\right) c\right.$$

$$\left. + R(x,y,z)\,\sin\left(\frac{s\,\pi}{n}\right) a\,b\,\cos\left(\frac{s\,\pi}{n}\right)\right)\pi^{2}\,/\,n^{2}$$

```
//Expression above-mentioned isn't that 'single value', but that 'set
of finite quantitative (viz. n) values'
```

```
// Structure finite sums:
> Sum(Sum((stdV[1]*stdA+stdV[2]*stdB+stdV[3]*stdC)*du*dv,
s=1..dus),t=1..dus);   (7)
```

In the case of "Amount 'n' of surface 'CS's segmental cells is pending", sum of integral values (dot product about Spacial vector field 'V' and '[stdA,stdB,stdC]') at all segmental cells

$$\sum_{t=1}^{n}\left(\sum_{s=1}^{n}\left(2\left(P(x,y,z)\,c\,\sin\left(\frac{s\,\pi}{n}\right)^{2} b\,\cos\left(\frac{2\,t\,\pi}{n}\right) + Q(x,y,z)\,\sin\left(\frac{s\,\pi}{n}\right)^{2} a\,\sin\left(\frac{2\,t\,\pi}{n}\right) c\right.\right.\right.$$

$$\left.\left.\left. + R(x,y,z)\,\sin\left(\frac{s\,\pi}{n}\right) a\,b\,\cos\left(\frac{s\,\pi}{n}\right)\right)\pi^{2}\,/\,n^{2}\right)\right)$$

```
> vs:=value(%):
> Limit(vs,n=infinity);
// Infinitize finite sums, its limit operational value is 'Surface
Integral Value at Manifold'
```

In the case of "Amount 'n' of surface 'CS's segmental cells tends to infinity", limit value of integral values' sums (dot product about Spacial vector field 'V' and '[stdA,stdB,stdC]') at all segmental cells

$$\lim_{n\to\infty} -\pi^{2}\,c\left(P(x,y,z)\,b\,\cos\left(\frac{(n+1)\,\pi}{n}\right)^{2} n\right.$$

$$+ Q(x,y,z)\,a\,\sin\left(\frac{(n+1)\,\pi}{n}\right)\cos\left(\frac{(n+1)\,\pi}{n}\right)\cos\left(\frac{\pi}{n}\right)^{4} n$$

$$+ Q(x,y,z)\,a\,\sin\left(\frac{(n+1)\,\pi}{n}\right)\cos\left(\frac{(n+1)\,\pi}{n}\right) n$$

$$- P(x,y,z)\,b\,\cos\left(\frac{(n+1)\,\pi}{n}\right)^{2}\cos\left(\frac{\pi}{n}\right)^{2} n$$

$$+ \cos\left(\frac{(n+1)\pi}{n}\right)^2 \cos\left(\frac{\pi}{n}\right) \sin\left(\frac{\pi}{n}\right) Q(x,y,z)\, a\, n$$

$$- \sin\left(\frac{\pi}{n}\right)^2 \cos\left(\frac{\pi}{n}\right)^2 P(x,y,z)\, b \cos\left(\frac{(n+1)\pi}{n}\right)^2 n$$

$$- 2\, Q(x,y,z)\, a \sin\left(\frac{(n+1)\pi}{n}\right) \cos\left(\frac{(n+1)\pi}{n}\right) \cos\left(\frac{\pi}{n}\right)^2 n$$

$$- \cos\left(\frac{(n+1)\pi}{n}\right)^2 \cos\left(\frac{\pi}{n}\right)^3 \sin\left(\frac{\pi}{n}\right) Q(x,y,z)\, a\, n$$

$$- \cos\left(\frac{(n+1)\pi}{n}\right) \sin\left(\frac{(n+1)\pi}{n}\right) \cos\left(\frac{\pi}{n}\right) \sin\left(\frac{\pi}{n}\right) P(x,y,z)\, b\, n$$

$$+ \cos\left(\frac{(n+1)\pi}{n}\right) \sin\left(\frac{(n+1)\pi}{n}\right) \cos\left(\frac{\pi}{n}\right)^3 \sin\left(\frac{\pi}{n}\right) P(x,y,z)\, b\, n$$

$$- \cos\left(\frac{\pi}{n}\right)^4 (n+1) P(x,y,z)\, b + 2\cos\left(\frac{\pi}{n}\right)^6 (n+1) P(x,y,z)\, b$$

$$- \cos\left(\frac{\pi}{n}\right)^5 (n+1) \sin\left(\frac{\pi}{n}\right) Q(x,y,z)\, a + \cos\left(\frac{\pi}{n}\right)^7 (n+1) \sin\left(\frac{\pi}{n}\right) Q(x,y,z)\, a$$

$$- \cos\left(\frac{\pi}{n}\right)^8 (n+1) P(x,y,z)\, b + \cos\left(\frac{\pi}{n}\right)^5 (n+1) \sin\left(\frac{\pi}{n}\right)^3 Q(x,y,z)\, a$$

$$+ \cos\left(\frac{\pi}{n}\right)^4 (n+1) \sin\left(\frac{\pi}{n}\right)^2 b\, P(x,y,z) - \cos\left(\frac{\pi}{n}\right)^6 (n+1) \sin\left(\frac{\pi}{n}\right)^2 P(x,y,z)\, b$$

$$+ \sin\left(\frac{\pi}{n}\right)^2 \cos\left(\frac{\pi}{n}\right)^4 Q(x,y,z)\, a \sin\left(\frac{(n+1)\pi}{n}\right) \cos\left(\frac{(n+1)\pi}{n}\right)$$

$$- Q(x,y,z)\, a \sin\left(\frac{(n+1)\pi}{n}\right) \cos\left(\frac{(n+1)\pi}{n}\right) \cos\left(\frac{\pi}{n}\right)^4$$

$$- \sin\left(\frac{\pi}{n}\right)^2 \cos\left(\frac{\pi}{n}\right)^4 P(x,y,z)\, b \cos\left(\frac{(n+1)\pi}{n}\right)^2$$

$$+ P(x,y,z)\, b \cos\left(\frac{(n+1)\pi}{n}\right)^2 \cos\left(\frac{\pi}{n}\right)^4 - \cos\left(\frac{\pi}{n}\right)^6 P(x,y,z)\, b \cos\left(\frac{(n+1)\pi}{n}\right)^2$$

$$+ \cos\left(\frac{\pi}{n}\right)^6 Q(x,y,z)\, a \sin\left(\frac{(n+1)\pi}{n}\right) \cos\left(\frac{(n+1)\pi}{n}\right) \Bigg) \Bigg/ \left(\left(-1 + \cos\left(\frac{\pi}{n}\right)^2\right)^2 \right.$$

$$n^2 \Bigg) + \pi^2 c \left(P(x,y,z)\, b\, n \cos\left(\frac{\pi}{n}\right)^2 + \cos\left(\frac{\pi}{n}\right) \sin\left(\frac{\pi}{n}\right) Q(x,y,z)\, a\, n \right.$$

$$- P(x,y,z)\, b \cos\left(\frac{\pi}{n}\right)^4 n - \cos\left(\frac{\pi}{n}\right)^3 \sin\left(\frac{\pi}{n}\right) Q(x,y,z)\, a\, n$$

978-1-62265-930-2 (online)   978-1-62265-931-9 (paper)   Yang Ke

$$- \cos\left(\frac{\pi}{n}\right)^2 \sin\left(\frac{\pi}{n}\right)^2 P(x,y,z)\,b\,n - \cos\left(\frac{\pi}{n}\right)^4 P(x,y,z)\,b + 3\cos\left(\frac{\pi}{n}\right)^6 P(x,y,z)\,b$$

$$- 2\cos\left(\frac{\pi}{n}\right)^5 \sin\left(\frac{\pi}{n}\right) Q(x,y,z)\,a + 2\cos\left(\frac{\pi}{n}\right)^7 \sin\left(\frac{\pi}{n}\right) Q(x,y,z)\,a$$

$$- 2\cos\left(\frac{\pi}{n}\right)^8 P(x,y,z)\,b + 2\cos\left(\frac{\pi}{n}\right)^5 \sin\left(\frac{\pi}{n}\right)^3 Q(x,y,z)\,a$$

$$+ \cos\left(\frac{\pi}{n}\right)^4 \sin\left(\frac{\pi}{n}\right)^2 b\,P(x,y,z) - 2\cos\left(\frac{\pi}{n}\right)^6 \sin\left(\frac{\pi}{n}\right)^2 P(x,y,z)\,b \Bigg) \Bigg/ \Bigg($$

$$\left(-1 + \cos\left(\frac{\pi}{n}\right)^2\right)^2 n^2 \Bigg)$$

```
> delta:=value(%);
```

$$\delta := 0$$

```
> x:='x':y:='y':z:='z':
```

```
> [r*CS[1],r*CS[2],r*CS[3]];
```

Entirely multiply radius vector 'r' (suppose r>0) by each item of surface 'CS' parametrized expression, convert "Parametrized expression of surface 'CS'" to "Parametrized expression of surface 'CS' coordinates"

$$[\,r\,a\sin(u)\cos(v),\, r\,b\sin(u)\sin(v),\, r\,c\cos(u)\,] \qquad \textbf{(8)}$$

```
> x:=r*CS[1]:y:=r*CS[2]:z:=r*CS[3]:
```

```
> matrix(3,3,[Diff(r*CS[1],r),Diff(r*CS[1],u),Diff(r*CS[1],v),
Diff(r*CS[2],r),Diff(r*CS[2],u),Diff(r*CS[2],v),Diff(r*CS[3],r),
Diff(r*CS[3],u),Diff(r*CS[3],v)])=
matrix(3,3,[diff(r*CS[1],r),diff(r*CS[1],u),diff(r*CS[1],v),
diff(r*CS[2],r),diff(r*CS[2],u),diff(r*CS[2],v),diff(r*CS[3],r),
diff(r*CS[3],u),diff(r*CS[3],v)]);m2:=rhs(%);
```

Define and calculate matrix of partial derivatives 'M2', obtain general expression of surface 'CS' coordinates volume element coefficient

$$\begin{bmatrix} \dfrac{\partial}{\partial r}(r\,a\sin(u)\cos(v)) & \dfrac{\partial}{\partial u}(r\,a\sin(u)\cos(v)) & \dfrac{\partial}{\partial v}(r\,a\sin(u)\cos(v)) \\[2mm] \dfrac{\partial}{\partial r}(r\,b\sin(u)\sin(v)) & \dfrac{\partial}{\partial u}(r\,b\sin(u)\sin(v)) & \dfrac{\partial}{\partial v}(r\,b\sin(u)\sin(v)) \\[2mm] \dfrac{\partial}{\partial r}(r\,c\cos(u)) & \dfrac{\partial}{\partial u}(r\,c\cos(u)) & \dfrac{\partial}{\partial v}(r\,c\cos(u)) \end{bmatrix} =$$

$$\begin{bmatrix} a\,\sin(u)\,\cos(v) & r\,a\,\cos(u)\,\cos(v) & -r\,a\,\sin(u)\,\sin(v) \\ b\,\sin(u)\,\sin(v) & r\,b\,\cos(u)\,\sin(v) & r\,b\,\sin(u)\,\cos(v) \\ c\,\cos(u) & -r\,c\,\sin(u) & 0 \end{bmatrix}$$

$$m2 := \begin{bmatrix} a\,\sin(u)\,\cos(v) & r\,a\,\cos(u)\,\cos(v) & -r\,a\,\sin(u)\,\sin(v) \\ b\,\sin(u)\,\sin(v) & r\,b\,\cos(u)\,\sin(v) & r\,b\,\sin(u)\,\cos(v) \\ c\,\cos(u) & -r\,c\,\sin(u) & 0 \end{bmatrix}$$

```
> det(m2);
```

$$a\,\sin(u)^3\,\cos(v)^2\,r^2\,b\,c + b\,\sin(u)^3\,\sin(v)^2\,r^2\,a\,c + c\,\cos(u)^2\,r^2\,a\,\cos(v)^2\,b\,\sin(u)$$
$$+ c\,\cos(u)^2\,r^2\,a\,\sin(u)\,\sin(v)^2\,b$$

```
> J:=simplify(%);   (9)
```

\# General expression of surface 'CS' coordinates volume element coefficient

$$J := a\,\sin(u)\,r^2\,b\,c$$

```
> x:='x':y:='y':z:='z':
```

// Transform variables, prevent that "Import 'x = r*a*sin(u)*cos(v),
y = r*b*sin(u)*sin(v),z = r*c*cos(u)' to abstract vector field
'[P(x,y,z),Q(x,y,z),R(x,y,z)]'"

```
> Diff(V[1],x)*Diff(r*CS[1],r)*Diff(r*CS[1],u)*Diff(r*CS[1],v)
+Diff(V[2],y)*Diff(r*CS[2],r)*Diff(r*CS[2],u)*Diff(r*CS[2],v)
+Diff(V[3],z)*Diff(r*CS[3],r)*Diff(r*CS[3],u)*Diff(r*CS[3],v);
diV2:=value(%);   (10)
```

\# Convert 'diV1' from Cartesian Coordinates expression to Surface 'CS' Coordinates expression

$$\left(\frac{\partial}{\partial x}P(x,y,z)\right)\left(\frac{\partial}{\partial r}(r\,a\,\sin(u)\,\cos(v))\right)\left(\frac{\partial}{\partial u}(r\,a\,\sin(u)\,\cos(v))\right) ,$$

$$\left(\frac{\partial}{\partial v}(r\,a\,\sin(u)\,\cos(v))\right) + \left(\frac{\partial}{\partial y}Q(x,y,z)\right)\left(\frac{\partial}{\partial r}(r\,b\,\sin(u)\,\sin(v))\right)$$

$$\left(\frac{\partial}{\partial u}(r\,b\,\sin(u)\,\sin(v))\right)\left(\frac{\partial}{\partial v}(r\,b\,\sin(u)\,\sin(v))\right)$$

$$+ \left(\frac{\partial}{\partial z}R(x,y,z)\right)\left(\frac{\partial}{\partial r}(r\,c\,\cos(u))\right)\left(\frac{\partial}{\partial u}(r\,c\,\cos(u))\right)\left(\frac{\partial}{\partial v}(r\,c\,\cos(u))\right)$$

$$diV2 := -\left(\frac{\partial}{\partial x}P(x,y,z)\right)a^3\,\sin(u)^2\,\cos(v)^2\,r^2\,\cos(u)\,\sin(v)$$

$$+ \left(\frac{\partial}{\partial y}Q(x,y,z)\right)b^3\,\sin(u)^2\,\sin(v)^2\,r^2\,\cos(u)\,\cos(v)$$

```
// Segment range `[0,1],[0,Pi],[0,2*Pi]' of `r,u,v':
> dus:=20;
```
Set amount of spacial bounded closed region `Ω's (Surface `CS' surrounds) segmental cells, this amount can be discretional natural number

$$dus := 20$$

```
> rgr:=[0,1];
```
// Be different to ordinary triple integrals, in this position, first integral range is `r[0,1]' always, and not `r[0,n]' or `r[0,∞]'; Because of `1', it is possible to hold a correct ratio between `a,b,c,r,u,v'

$$rgr := [0, 1]$$

```
> dr:=(rgr[2]-rgr[1])/dus; # Segment range of `r'
```

$$dr := \frac{1}{20}$$

```
> du:=(rgu[2]-rgu[1])/dus; # Segment range of `u'
```

$$du := \frac{\pi}{20}$$

```
> dv:=(rgv[2]-rgv[1])/dus; # Segment range of `v'
```

$$dv := \frac{\pi}{10}$$

Microcosmic Triple integrals course of spacial bounded closed region `Ω's first segmental cell (20 segmental cells):

// Segment volume element coefficient `J':
```
> dJ:=subs(r=rgr[1]+dr,subs(v=rgv[1]+dv,subs(u=rgu[1]+du,J)));
```
Volume element coefficient `J's average value that corresponds to spacial bounded closed region `Ω's first segmental cell

$$dJ := \frac{1}{400}\, a \sin\!\left(\frac{\pi}{20}\right) b\, c$$

// Segment abstract divergence `diV2':
```
> ddiV2:=subs(r=rgr[1]+dr,subs(v=rgv[1]+dv,subs(u=rgu[1]+du,
DiV2)));
```
Divergence `diV2's average value that corresponds to spacial bounded closed region `Ω's first segmental cell;

$$ddiV2 := -\frac{1}{400}\left(\frac{\partial}{\partial x}\,\mathrm{P}(x,y,z)\right) a^3 \sin\!\left(\frac{\pi}{20}\right)^2 \cos\!\left(\frac{\pi}{10}\right)^2 \cos\!\left(\frac{\pi}{20}\right) \sin\!\left(\frac{\pi}{10}\right)$$

$$+\frac{1}{400}\left(\frac{\partial}{\partial y}\,\mathrm{Q}(x,y,z)\right) b^3 \sin\!\left(\frac{\pi}{20}\right)^2 \sin\!\left(\frac{\pi}{10}\right)^2 \cos\!\left(\frac{\pi}{20}\right) \cos\!\left(\frac{\pi}{10}\right)$$

```
//Calculate microcosmic triple integrals value spacial bounded
closed region 'Ω's first segmental cell:
> ddiV2*dJ*dr*du*dv;
```

\# Base on integral median theorem, integral value of product (Divergence 'diV2' and volume element) in first segmental cell

$$\frac{1}{1600000}\left(-\frac{1}{400}\left(\frac{\partial}{\partial x}P(x,y,z)\right)a^3\sin\left(\frac{\pi}{20}\right)^2\cos\left(\frac{\pi}{10}\right)^2\cos\left(\frac{\pi}{20}\right)\sin\left(\frac{\pi}{10}\right)\right.$$

$$\left.+\frac{1}{400}\left(\frac{\partial}{\partial y}Q(x,y,z)\right)b^3\sin\left(\frac{\pi}{20}\right)^2\sin\left(\frac{\pi}{10}\right)^2\cos\left(\frac{\pi}{20}\right)\cos\left(\frac{\pi}{10}\right)\right)a\sin\left(\frac{\pi}{20}\right)b\,c\,\pi^2$$

\# Microcosmic triple integrals course of spacial closed region 'Ω's all segmental cells (20 segmental cells):

```
// Segment volume element coefficient 'J':
> ijkdJ:=subs(r=rgr[1]+k*dr,subs(v=rgv[1]+j*dv,subs(u=rgu[1]
+i*du,J)));
```

\# Volume element coefficient 'J's average values that correspond to spacial bounded closed region 'Ω's all segmental cells

$$ijkdJ:=\frac{1}{400}a\sin\left(\frac{i\,\pi}{20}\right)k^2\,b\,c$$

```
// In actual expressions, 'i,j,k' stand for natural number 1~20
//'ijkdJ' isn't that 'single value', but that 'set of finite
quantitative (viz.20) values'
```

```
// Segment abstract divergence 'diV2':
> ijkddiV2:=subs(r=rgr[1]+k*dr,subs(v=rgv[1]+j*dv,subs(u=rgu[1]
+i*du,diV2)));
```

\# Divergence 'diV2's average values that correspond to spacial bounded closed region 'Ω's all segmental cells

$$ijkddiV2:=-\frac{1}{400}\left(\frac{\partial}{\partial x}P(x,y,z)\right)a^3\sin\left(\frac{i\,\pi}{20}\right)^2\cos\left(\frac{j\,\pi}{10}\right)^2k^2\cos\left(\frac{i\,\pi}{20}\right)\sin\left(\frac{j\,\pi}{10}\right)$$

$$+\frac{1}{400}\left(\frac{\partial}{\partial y}Q(x,y,z)\right)b^3\sin\left(\frac{i\,\pi}{20}\right)^2\sin\left(\frac{j\,\pi}{10}\right)^2k^2\cos\left(\frac{i\,\pi}{20}\right)\cos\left(\frac{j\,\pi}{10}\right)$$

```
// In actual expressions, 'i,j,k' stand for natural number 1~20
//'ijkddiV2' isn't that 'single value', but that 'set of finite
quantitative (viz.20) values'
```

```
// Calculate microcosmic triple integrals values of spacial bounded
closed region 'Ω's all segmental cells:
> ijkddiV2*ijkdJ*dr*du*dv;
```

\# Base on integral median theorem, integral value of product
(Divergence 'diV2' and volume element) in all segmental cells

$$\frac{1}{1600000}\left(-\frac{1}{400}\left(\frac{\partial}{\partial x}P(x,y,z)\right)a^3\sin\left(\frac{i\,\pi}{20}\right)^2\cos\left(\frac{j\,\pi}{10}\right)^2 k^2\cos\left(\frac{i\,\pi}{20}\right)\sin\left(\frac{j\,\pi}{10}\right)\right.$$

$$\left.+\frac{1}{400}\left(\frac{\partial}{\partial y}Q(x,y,z)\right)b^3\sin\left(\frac{i\,\pi}{20}\right)^2\sin\left(\frac{j\,\pi}{10}\right)^2 k^2\cos\left(\frac{i\,\pi}{20}\right)\cos\left(\frac{j\,\pi}{10}\right)\right)a\sin\left(\frac{i\,\pi}{20}\right)k^2 b$$

$$c\,\pi^2$$

```
// In actual expressions, 'i,j,k 'stand for natural number 1~20
//  Expression above-mentioned isn't that 'single value', but that
'set of finite quantitative (viz.20) values'


// List of finite quantitative(viz.20) microcosmic triple integrals
values:
> seq(seq(seq([i*j*k,ijkddiV2*ijkdJ*dr*du*dv,
evalf(ijkddiV2*ijkdJ*dr*du*dv)],i=1..dus),j=1..dus),k=1..dus):
```

\# List of integral values of product (Divergence 'diV2' and volume
element)in all segmental cells, be elided

```
// Structure sequence that is composed of finite quantitative
microcomosic triple integrals values:
> sqn:=seq(seq(seq(ijkddiV2*ijkdJ*dr*du*dv,i=1..dus),j=1..dus),
k=1..dus):
// Accumulational sum of sequence, obtain triple integrals value at
manifold (Entire spacial bounded closed region 'Ω'):
> add(k,k=sqn):omega:=evalf(%):
```
 (11)

\# Integral values' sum of product(Divergence 'diV2' and volume element)
in all segmental cells; Verbose analytic and float expression are
elided; Amount of spacial bounded closed region 'Ω's segmental cells
increases infinitely, sum of integral values above-mentioned tends
to '0' infinitely

```
// Renewedly Segment the range '[0,1],[0,Pi],[0,2*Pi]' of 'r,u,v':
> dus:=n;
```

\# Set amount of spacial bounded closed region `Ω`'s (Surface `CS` surrounds) segmental cells as natural number `n`

$$dus := n$$

> dr:=(rgr[2]-rgr[1])/dus; # Segment range of `r`

$$dr := \frac{1}{n}$$

> du:=(rgu[2]-rgu[1])/dus; # Segment range of `u`

$$du := \frac{\pi}{n}$$

> dv:=(rgv[2]-rgv[1])/dus; # Segment range of `v`

$$dv := \frac{2\,\pi}{n}$$

\# Microcosmic Triple integrals course of spacial bounded closed region `Ω`'s first segmental cell (n segmental cells):

// Segment volume element coefficient `J`:

> dJ:=subs(r=rgr[1]+dr,subs(v=rgv[1]+dv,subs(u=rgu[1]+du,J)));

\# Volume element coefficient `J`'s average value that corresponds to spacial bounded closed region `Ω`'s first segmental cell

$$dJ := \frac{a \sin\!\left(\dfrac{\pi}{n}\right) b\,c}{n^2}$$

// Segment abstract divergence `diV2`:

> ddiV2:=subs(r=rgr[1]+dr,subs(v=rgv[1]+dv,subs(u=rgu[1]+du, diV2)));

\# Divergence `diV2`'s average value that corresponds to spacial bounded closed region `Ω`'s first segmental cell

$$ddiV2 := -\frac{\left(\dfrac{\partial}{\partial x}\mathrm{P}(x,y,z)\right) a^3 \sin\!\left(\dfrac{\pi}{n}\right)^2 \cos\!\left(\dfrac{2\,\pi}{n}\right)^2 \cos\!\left(\dfrac{\pi}{n}\right) \sin\!\left(\dfrac{2\,\pi}{n}\right)}{n^2}$$

$$+\frac{\left(\dfrac{\partial}{\partial y}\mathrm{Q}(x,y,z)\right) b^3 \sin\!\left(\dfrac{\pi}{n}\right)^2 \sin\!\left(\dfrac{2\,\pi}{n}\right)^2 \cos\!\left(\dfrac{\pi}{n}\right) \cos\!\left(\dfrac{2\,\pi}{n}\right)}{n^2}$$

// Calculate microcosmic triple integrals value spacial bouned closed region `Ω`'s first segmental cell:

> ddiV2*dJ*dr*du*dv;

\# Base on integral median theorem, integral value of product (Divergence `diV2` and volume element) in first segmental cell

$$2\left(-\frac{\left(\frac{\partial}{\partial x}P(x,y,z)\right)a^3\sin\left(\frac{\pi}{n}\right)^2\cos\left(\frac{2\pi}{n}\right)^2\cos\left(\frac{\pi}{n}\right)\sin\left(\frac{2\pi}{n}\right)}{n^2}\right.$$

$$\left.+\frac{\left(\frac{\partial}{\partial y}Q(x,y,z)\right)b^3\sin\left(\frac{\pi}{n}\right)^2\sin\left(\frac{2\pi}{n}\right)^2\cos\left(\frac{\pi}{n}\right)\cos\left(\frac{2\pi}{n}\right)}{n^2}\right)a\sin\left(\frac{\pi}{n}\right)b\,c\,\pi^2\,/\,n^5$$

Microcosmic triple integrals course of spacial bounded closed region 'Ω's all segmental cells (n segmental cells):

// Segment volume element coefficient 'J':

```
> ijkdJ:=subs(r=rgr[1]+k*dr,subs(v=rgv[1]+j*dv,subs(u=rgu[1]
+i*du,J)));
```

Volume element coefficient 'J's average values that correspond to spacial bounded closed region 'Ω's all segmental cells

$$ijkdJ:=\frac{a\sin\left(\frac{i\pi}{n}\right)k^2\,b\,c}{n^2}$$

// In actual expressions, 'i,j,k' stand for natural number 1~n

//'ijkdJ' isn't that 'single value', but that 'set of finite quantitative (viz.n) values'

// Segment abstract divergence 'diV2':

```
> ijkddiV2:=subs(r=rgr[1]+k*dr,subs(v=rgv[1]+j*dv,
subs(u=rgu[1]+i*du,diV2)));
```

Divergence 'diV2's average values that correspond to spacial bounded closed region 'Ω's all segmental cells

$$ijkddiV2:=-\frac{\left(\frac{\partial}{\partial x}P(x,y,z)\right)a^3\sin\left(\frac{i\pi}{n}\right)^2\cos\left(\frac{2j\pi}{n}\right)^2k^2\cos\left(\frac{i\pi}{n}\right)\sin\left(\frac{2j\pi}{n}\right)}{n^2}$$

$$+\frac{\left(\frac{\partial}{\partial y}Q(x,y,z)\right)b^3\sin\left(\frac{i\pi}{n}\right)^2\sin\left(\frac{2j\pi}{n}\right)^2k^2\cos\left(\frac{i\pi}{n}\right)\cos\left(\frac{2j\pi}{n}\right)}{n^2}$$

// In actual expressions, 'i,j,k' stand for natural number 1~n

//'ijkddiV2' isn't that 'single value', but that 'set of finite quantitative (viz.n) values'

978-1-62265-930-2 (online)   978-1-62265-931-9 (paper)   Yang Ke

```
// Calculate microcosmic triple integrals values of spacial bounded
closed region 'Ω's all segmental cells:
> ijkddiV2*ijkdJ*dr*du*dv;
# Base on integral median theorem, integral value of product
(Divergence 'diV2' and volume element) in all segmental cells
```

$$2\left(-\frac{\left(\frac{\partial}{\partial x}P(x,y,z)\right)a^3\sin\left(\frac{i\,\pi}{n}\right)^2\cos\left(\frac{2j\,\pi}{n}\right)^2 k^2\cos\left(\frac{i\,\pi}{n}\right)\sin\left(\frac{2j\,\pi}{n}\right)}{n^2}\right.$$

$$\left.+\frac{\left(\frac{\partial}{\partial y}Q(x,y,z)\right)b^3\sin\left(\frac{i\,\pi}{n}\right)^2\sin\left(\frac{2j\,\pi}{n}\right)^2 k^2\cos\left(\frac{i\,\pi}{n}\right)\cos\left(\frac{2j\,\pi}{n}\right)}{n^2}\right)a\sin\left(\frac{i\,\pi}{n}\right)k^2 b$$

$$c\,\pi^2 / n^5$$

```
// In actual expressions, 'i,j,k' stand for natural number 1~n
// Expression above-mentioned isn't that 'Single value', but that
'Set of finite quantitative (viz.n) values'


// Structure finite sums:
> Sum(Sum(Sum(ijkddiV2*ijkdJ*dr*du*dv,i=1..dus),j=1..dus),
k=1..dus);
# In the case of "Amount 'n' of spacial bounded closed region 'Ω's
segmental cells is pending", sum of integral values (Product about
Divergence 'diV2' and Volume element) in all segmental cells (12)
```

$$\sum_{k=1}^{n}\left(\sum_{j=1}^{n}\left(\sum_{i=1}^{n}\left(2\left(-\frac{\left(\frac{\partial}{\partial x}P(x,y,z)\right)a^3\sin\left(\frac{i\,\pi}{n}\right)^2\cos\left(\frac{2j\,\pi}{n}\right)^2 k^2\cos\left(\frac{i\,\pi}{n}\right)\sin\left(\frac{2j\,\pi}{n}\right)}{n^2}\right.\right.\right.\right.$$

$$\left.+\frac{\left(\frac{\partial}{\partial y}Q(x,y,z)\right)b^3\sin\left(\frac{i\,\pi}{n}\right)^2\sin\left(\frac{2j\,\pi}{n}\right)^2 k^2\cos\left(\frac{i\,\pi}{n}\right)\cos\left(\frac{2j\,\pi}{n}\right)}{n^2}\right)a\sin\left(\frac{i\,\pi}{n}\right)k^2 b$$

$$\left.\left.\left.c\,\pi^2 / n^5\right)\right)\right)$$

```
> vs:=value(%);
```

$$vs := 0$$

```
> Limit(vs,n=infinity);
// Infinitize finite sums, its limit operational value is 'Triple
```

Integrals Value at Manifold'

In the case of "Amount 'n' of spacial bounded closed region 'Ω's segmental cells tends to infinity", limit value of integral values' sums (Product about Divergence 'diV2' and Volume element) in all segmental cells

$$\lim_{n \to \infty} 0$$

> epsilon:=value(%);

$$\varepsilon := 0$$

Thereinto, $\displaystyle\lim_{n \to \infty} \sum_{k=1}^{n}\left(\sum_{j=1}^{n}\left(\sum_{i=1}^{n}\left(\frac{2\,a\,\sin\left(\dfrac{i\,\pi}{n}\right)k^2\,b\,c\,\pi^2}{n^5}\right)\right)\right) \neq 0$

Viz. Suppose sum of triple integrals values (volume element itself) in all segmental cells can't be '0'; Explain visually ,it can be regarded as "Suppose spacial bounded closed region 'Ω' can't be '0 volume'"

Viz. In the case of 'n →∞', (7)=(12):

$$\sum_{t=1}^{n}\left(\sum_{s=1}^{n}\left(2\left(P(x,y,z)\,c\,\sin\left(\frac{s\,\pi}{n}\right)^2 b\,\cos\left(\frac{2\,t\,\pi}{n}\right) + Q(x,y,z)\,\sin\left(\frac{s\,\pi}{n}\right)^2 a\,\sin\left(\frac{2\,t\,\pi}{n}\right)c\right.\right.\right.$$

$$\left.\left.\left. + R(x,y,z)\,\sin\left(\frac{s\,\pi}{n}\right)a\,b\,\cos\left(\frac{s\,\pi}{n}\right)\right)\pi^2 \,/\, n^2\right)\right)$$

=

$$\sum_{k=1}^{n}\left(\sum_{j=1}^{n}\left(\sum_{i=1}^{n}\left(2\left(-\frac{\left(\dfrac{\partial}{\partial x}P(x,y,z)\right)a^3\,\sin\left(\dfrac{i\,\pi}{n}\right)^2\cos\left(\dfrac{2\,j\,\pi}{n}\right)^2 k^2\,\cos\left(\dfrac{i\,\pi}{n}\right)\sin\left(\dfrac{2\,j\,\pi}{n}\right)}{n^2}\right.\right.\right.\right.$$

$$\left.\left.\left.\left. + \frac{\left(\dfrac{\partial}{\partial y}Q(x,y,z)\right)b^3\,\sin\left(\dfrac{i\,\pi}{n}\right)^2\sin\left(\dfrac{2\,j\,\pi}{n}\right)^2 k^2\,\cos\left(\dfrac{i\,\pi}{n}\right)\cos\left(\dfrac{2\,j\,\pi}{n}\right)}{n^2}\right)a\,\sin\left(\dfrac{i\,\pi}{n}\right)k^2\,b\right.\right.\right.$$

$$\left.\left.\left. c\,\pi^2 \,/\, n^5\right)\right)\right)$$

Above-mentioned equation can be described as:

$$\iint\limits_{S} A \cdot n\, dS = \iiint\limits_{\Omega} div A\, d\, \omega$$ (1), Complete Proof.

4.3 Discuss About Abstract Vector Field、 Finite Sums Limit、 Integral and Riemann Sums (with friend in UESTC)

December 2014, May 2016, Chengdu, China

In finite sums limits proof procedure of Divergence theorem at Manifold, it is ideal state that calculate individual corresponding value of vector field at every segmental cell of abstract simply connected orientable closed surface admittedly; but its prerequisite is that vector field must be given beforehand (Viz. vector field must be functions with idiographic expression, shaped like [x, x*z, y^2] and so on).

In the case of vector field has not been given (Viz. vector field is abstract functions, shaped like [P(x,y,z), Q(x,y,z), R(x,y,z)]), individual corresponding value of vector field at every segmental cell of abstract simply connected orientable closed surface is absence in numerical meaning.

Admittedly, in the case of vector field has not been given, it is possible to obtain individual corresponding value of vector field at every segmental cell of abstract simply connected orientable closed surface by method of setting variable subscript. (Viz. Set -

$[P(x_1,y_1,z_1), Q(x_1,y_1,z_1), R(x_1,y_1,z_1)]$,

$[P(x_2,y_2,z_2), Q(x_2,y_2,z_2), R(x_2,y_2,z_2)]$,

 . . .

$[P(x_n,y_n,z_n), Q(x_n,y_n,z_n), R(x_n,y_n,z_n)]$)

 --But objects above-mentioned are artificial setting scantly (Even they can be understood as setting out of thin air).

 --Because of, known conditions of proof, only expression of integral surface:

 [a*sin(u)cos(v),b*sin(u)sin(v),c*cos(u)],u∈[0,π],v∈[0,2π]

and expression of vector field: [P(x,y,z),Q(x,y,z),R(x,y,z)]

In other words, no segmented vectors (of vector field) with variable subscript in known conditions of proof (viz. Start point of logical deduction and calculation); Segmented vectors (of vector field) with variable subscript (under-mentioned) have not been derived out in logical deduction and calculation of proof.

$$[\mathrm{P}(\mathrm{x}_1, y_1, z_1), Q(\mathrm{x}_1, y_1, z_1), R(\mathrm{x}_1, y_1, z_1)] \,,$$

$$[P(x_2, y_2, z_2), Q(x_2, y_2, z_2), R(x_2, y_2, z_2)] \,,$$

$$\ldots$$

$$[P(x_n, y_n, z_n), Q(x_n, y_n, z_n), R(x_n, y_n, z_n)]$$

--In other words, expressions above-mentioned are artificial data format, instead of real calculational values.

If take these artificial data format as basis of continual logical deduction and calculation, then whole procedure of logical deduction and calculation will fall into chaos consequentially, cause incomputable situation or no result (For example, Verified by Program Template of Waterloo Maplesoft in 4.2 Section this Chapter).

--In other words, artificial data format above-mentioned can not form chain of causality with logical context of proof procedure, thus be independent of whole formular proof's logical deductional and calculational procedure.

--As above-mentioned, in the case of vector field has not been given, individual corresponding value of vector field at every segmental cell is absence in numerical meaning.

Incidentally, integral vector field is not given, it does not influence proof of integral theorem. For example, in proof of traditional Остроградский-Gauss theorem in Cartesian coordinates, integral vector field is not given similarly.

--In other words, at every segmental cell, abstract vector field [P(x,y,z),Q(x,y,z),R(x,y,z)] possesses formal logical meaning scantly, possesses meaning of connecting logical context scantly, and no numerical meaning.

And in formal logical meaning, about abstract vector field [P(x,y,z),Q(x,y,z),R(x,y,z)], in its definition domain of abstract simply connected orientable closed surface, local expression (at a segmental cell of definition domain) and overall expression are consistent.

Vivid metaphor, there is a piece of land (definition domain) in North American continent, an abstract vector field named 'Canada' is defined on this land; Be differentiated (Derivative) by certain rule, there is a local expression named 'Quebec' of abstract vector field above-mentioned;
 --In formal logic meaning, this local expression can only be called 'Quebec' and cannot be called 'Canada'?

Another vivid metaphor, fancy historical scene, in World War II, an allied transport aircraft (C46) flew over the hump and entered China; Due to various reasons such as injuries and lost course, Pilot abandoned aircraft and parachuted, landed in Shangri-La Valley in western mountains area of China's Yunnan province; Be unaware of administrative divisions about China, Pilot said, 'I landed in territory of China' (Rather than claiming that 'I landed in territory of Yunnan province');
 --In formal logic meaning, statement of Pilot is wrong or not?

Mathematical-calculus question is replaced by logic questions;

Recognize from aspect of integral, about abstract vector field [P(x,y,z), Q(x,y,z),R(x,y,z)], at its definition domain of abstract simply connected orientable closed surface, local expression at a segmental cell (of definition domain) is [P(x,y,z),Q(x,y,z),R(x,y,z)], local expression at another segmental cell (of definition domain) is [P(x,y,z),Q(x,y,z),R(x,y,z)], local expressions at umpty segmental cells (of definition domain) are [P(x,y,z),Q(x,y,z),R(x,y,z)], then local expressions at umpty segmental cells (of definition domain, about abstract vector field) are integrated (integral) by certain rule, expression of generated new object is [P(x,y,z),Q(x,y,z),R(x,y,z)] similarly -- it is also understood that 'Abstract vector field [P(x,y,z),Q(x,y,z),R(x,y,z)] can preserve primary form after integral'.

Vivid metaphor, differentiate abstract vector field 'Canada' by certain rule, there is a local expression 'Quebec' of this abstract vector field, another local expression 'Ontario', another local expression 'Newfoundland-Labrador'; Integrate (integral) three local expressions by certain rule, resulting in expression of new object 'Eastern 3 Provinces';
 --In formal logic meaning, this expression of new object can only be called 'Eastern 3 Provinces' and cannot be called 'Canada'?

Another vivid metaphor, another two crew members of allied transport aircraft above-mentioned, 'Andre' parachuted and landed in Tibet of China, 'Bob' parachuted and landed in Sichuan province of China; Be unaware of administrative divisions about China similarly, Rescuer said, 'I obtained three survivors in territory of China' (Rather than claiming that 'I obtained three survivors in territory of Yunnan、Tibet And Sichuan');
 --In formal logic meaning, statement of Rescuer is wrong or not?

Mathematical-calculus question is replaced by logic questions similarly.

Base on same logic, in proof of divergence theorem finite sums limits at manifold, about divergence

'$\frac{\partial}{\partial x}P(x,y,z)+\frac{\partial}{\partial y}Q(x,y,z)+\frac{\partial}{\partial z}R(x,y,z)$' of abstract vector field

'[P(x,y,z),Q(x,y,z),R(x,y,z)]' (Belongs to abstract scalar field, viz. abstract 3-variables funtion), in its definition domain of spacial bounded closed region 'Ω', local expression (at a segmental cell of definition domain) and overall expression are consistent; local expressions (at umpty segmental cells of definition domain) are integrated (integral) by certain rule, expression of generated new object and overall expression are consistent similarly;

Bases on same logic, in proof of Green theorem finite sums limits at manifold, about abstract vector field `[P(x,y),Q(x,y)]', at its definition domain of abstract plane simply connected closed curve `L', local expression (at a segmental cell of definition domain) and overall expression are consistent; local expressions (at umpty segmental cells of definition domain) are integrated (integral) by certain rule, expression of generated new object and overall expression are consistent similarly; About Differential function

$$` \frac{\partial}{\partial x}Q(x,y) - \frac{\partial}{\partial y}P(x,y) \ '$$ of abstract vector field `[P(x,y),Q(x,y)]'

(Belongs to abstract scalar field, viz. abstract 2-Variables funtion), in its definition domain of plane bounded closed region`S', local expression (in a segmental cell of definition domain) and overall expression are consistent; local expressions (in umpty segmental cells of definition domain) are integrated (integral) by certain rule, expression of generated new object and overall expression are consistent similarly;

Bases on same logic, in proof of Curl theorem finite sums limits at manifold, about abstract vector field `[P(x,y,z),Q(x,y,z), R(x,y,z)]', at its definition domain of abstract spacial closed curve `L', local expression (at a segmental cell of definition domain) and overall expression are consistent; local expressions (at umpty segmental cells of definition domain) are integrated (integral) by certain rule, expression of generated new object and overall expression are consistent similarly; About Curl

$$\left[\left(\frac{\partial}{\partial y}R(x,y,z) \right) - \left(\frac{\partial}{\partial z}Q(x,y,z) \right), \left(\frac{\partial}{\partial z}P(x,y,z) \right) - \left(\frac{\partial}{\partial x}R(x,y,z) \right), \right.$$
$$\left. \left(\frac{\partial}{\partial x}Q(x,y,z) \right) - \left(\frac{\partial}{\partial y}P(x,y,z) \right) \right]$$

of abstract vector field `[P(x,y,z),Q(x,y,z),R(x,y,z)]' (Abstract curl can be regarded as differential metamorphose of abstract vector field `[P(x,y,z),Q(x,y,z),R(x,y,z)]`, it belongs to another abstract vector field), at its definition domain of abstract simply or multiply connected orientable unclosed surface `S`, local expression (at a segmental cell of definition domain) and overall expression are consistent; local expressions (at umpty segmental cells of definition domain) are integrated (integral) by certain rule, expression of generated new object and overall expression are consistent similarly.

Actually, final results of proofs--
Identical equation of surface integral and triple integrals--

$$\sum_{t=1}^{\infty}\left(\sum_{s=1}^{\infty}\left(2\left(c\sin\left(\frac{\pi s}{n}\right)^2 b\cos\left(\frac{2\pi t}{n}\right)P(x,y,z)+\sin\left(\frac{\pi s}{n}\right)^2\sin\left(\frac{2\pi t}{n}\right)a\,c\,Q(x,y,z)\right.\right.$$
$$\left.\left.+\sin\left(\frac{\pi s}{n}\right)a\cos\left(\frac{\pi s}{n}\right)b\,R(x,y,z)\right)\pi^2/n^2\right)\right)$$

$$=$$

$$\sum_{k=1}^{\infty}\left(\sum_{j=1}^{\infty}\left(\sum_{i=1}^{\infty}\left(2\left(-\frac{\left(\frac{\partial}{\partial x}P(x,y,z)\right)a^3\sin\left(\frac{\pi i}{n}\right)^2\cos\left(\frac{2\pi j}{n}\right)^2 k^2\cos\left(\frac{\pi i}{n}\right)\sin\left(\frac{2\pi j}{n}\right)}{n^2}\right.\right.\right.$$

$$\left.\left.\left.+\frac{\left(\frac{\partial}{\partial y}Q(x,y,z)\right)b^3\sin\left(\frac{\pi i}{n}\right)^2\sin\left(\frac{2\pi j}{n}\right)^2 k^2\cos\left(\frac{\pi i}{n}\right)\cos\left(\frac{2\pi j}{n}\right)}{n^2}\right)a\sin\left(\frac{\pi i}{n}\right)k^2 b\,c\right.\right.$$

$$\left.\left.\pi^2/n^5\right)\right)\right)$$

And equivalent expression—

$$\int_0^{2\pi}\int_0^{\pi}P(x,y,z)\,c\sin(u)^2 b\cos(v)+Q(x,y,z)\sin(u)^2 a\sin(v)\,c$$
$$+R(x,y,z)\sin(u)\,a\,b\cos(u)\,du\,dv$$

$$=$$

978-1-62265-930-2 (online)   978-1-62265-931-9 (paper)   Yang Ke

$$\int_0^{2\pi}\int_0^{\pi}\int_0^1 \left(-\left(\frac{\partial}{\partial x}P(x,y,z)\right)a^3\sin(u)^2\cos(v)^2 r^2\cos(u)\sin(v)\right.$$

$$\left.+\left(\frac{\partial}{\partial y}Q(x,y,z)\right)b^3\sin(u)^2\sin(v)^2 r^2\cos(u)\cos(v)\right)a\sin(u)r^2 b\,c\,dr\,du\,dv$$

Identical Equation of Plane Closed Curve Integral and Double Integrals—

$$\sum_{i=1}^{\infty}\left(\frac{2\pi\left(-a\sin\left(\frac{2\pi i}{n}\right)P(x,y)+b\cos\left(\frac{2\pi i}{n}\right)Q(x,y)\right)}{n}\right)$$

$$=$$

$$\sum_{j=1}^{\infty}\left(\sum_{i=1}^{\infty}\left(2\right.\right.$$

$$-\frac{\left(\frac{\partial}{\partial x}Q(x,y)\right)a^2\cos\left(\frac{2\pi i}{n}\right)j\sin\left(\frac{2\pi i}{n}\right)}{n}-\frac{\left(\frac{\partial}{\partial y}P(x,y)\right)b^2\sin\left(\frac{2\pi i}{n}\right)j\cos\left(\frac{2\pi i}{n}\right)}{n}$$

$$\left.\left.\right)a\,b\,j\,\pi/n^3\right)\right)$$

And equivalent expression—

$$\int_0^{2\pi}P(x,y)\left(\frac{\partial}{\partial t}(a\cos(t))\right)+Q(x,y)\left(\frac{\partial}{\partial t}(b\sin(t))\right)dt=$$

$$\int_0^{2\pi}\int_0^1\left(-\left(\frac{\partial}{\partial x}Q(x,y)\right)a^2\cos(u)r\sin(u)-\left(\frac{\partial}{\partial y}P(x,y)\right)b^2\sin(u)r\cos(u)\right)a\,r\,b\,dr\,du$$

Identical Equation of Spacial Closed Curve Integral and Surface Integral—

978-1-62265-930-2 (online)   978-1-62265-931-9 (paper)

Yang Ke

$$\sum_{i=1}^{\infty} \left(\frac{2\pi\left(-P(x,y,z)\,\alpha\,\sin\left(\frac{2\pi i}{w}\right)+Q(x,y,z)\,\beta\,\cos\left(\frac{2\pi i}{w}\right)\right)}{w} \right)$$

$$=$$

$$\sum_{t=1}^{\infty}\left(\sum_{s=1}^{\infty}\left(2\left(\vphantom{\sum}\right.\right.\right.$$

$$c\,\sin\!\left(\frac{\left(\frac{\pi}{n}-\theta\right)s}{w}\right)^{3} b^{3}\cos\!\left(\frac{2\pi t}{w}\right)^{2}\left(\frac{\partial}{\partial y}R(x,y,z)\right)\cos\!\left(\frac{\left(\frac{\pi}{n}-\theta\right)s}{w}\right)\sin\!\left(\frac{2\pi t}{w}\right)$$

$$+\sin\!\left(\frac{\left(\frac{\pi}{n}-\theta\right)s}{w}\right)^{3}\sin\!\left(\frac{2\pi t}{w}\right)^{2} a^{3}c\left(\frac{\partial}{\partial x}R(x,y,z)\right)\cos\!\left(\frac{\left(\frac{\pi}{n}-\theta\right)s}{w}\right)\cos\!\left(\frac{2\pi t}{w}\right)+$$

$$\sin\!\left(\frac{\left(\frac{\pi}{n}-\theta\right)s}{w}\right) a\,\cos\!\left(\frac{\left(\frac{\pi}{n}-\theta\right)s}{w}\right) b\left(\vphantom{\sum}\right.$$

$$-\left(\frac{\partial}{\partial x}Q(x,y,z)\right) a^{2}\cos\!\left(\frac{\left(\frac{\pi}{n}-\theta\right)s}{w}\right)\cos\!\left(\frac{2\pi t}{w}\right)\sin\!\left(\frac{\left(\frac{\pi}{n}-\theta\right)s}{w}\right)\sin\!\left(\frac{2\pi t}{w}\right)$$

$$-\left(\frac{\partial}{\partial y}P(x,y,z)\right) b^{2}\cos\!\left(\frac{\left(\frac{\pi}{n}-\theta\right)s}{w}\right)\sin\!\left(\frac{2\pi t}{w}\right)\sin\!\left(\frac{\left(\frac{\pi}{n}-\theta\right)s}{w}\right)\cos\!\left(\frac{2\pi t}{w}\right)\left.\right)\left.\right)$$

$$\left(\frac{\pi}{n}-\theta\right)\pi\,/\,w^{2}\left.\right)\left.\right)$$

And equivalent expression—

$$\int_{0}^{2\pi} P(x,y,z)\left(\frac{\partial}{\partial v}(\alpha\cos(v))\right)+Q(x,y,z)\left(\frac{\partial}{\partial v}(\beta\sin(v))\right)+R(x,y,z)\left(\frac{\partial}{\partial v}\gamma\right)dv=$$

$$\int_{0}^{2\pi}\int_{0}^{\frac{\pi}{n}-\theta}\left(\frac{\partial}{\partial y}R(x,y,z)\right) b^{3}\cos(u)\sin(v)\sin(u)^{3}\cos(v)^{2}c$$

$$+\left(\frac{\partial}{\partial x}R(x,y,z)\right)a^3\cos(u)\cos(v)\sin(u)^3\sin(v)^2\,c+\Bigg($$

$$-\left(\frac{\partial}{\partial x}Q(x,y,z)\right)a^2\cos(u)\cos(v)\sin(u)\sin(v)$$

$$-\left(\frac{\partial}{\partial y}P(x,y,z)\right)b^2\cos(u)\sin(v)\sin(u)\cos(v)\Bigg)\sin(u)\,a\,b\,\cos(u)\,du\,dv$$

--Be equal in formal logic meaning simply, not to be equal in numerical meaning.

--Only in idiographic numerical models, in the case of vector fields and integral curves、surfaces have been given, six identical equations above-mentioned are equal in numerical meaning.

As friend' inequality:
$$\sum_{i=1}^{n}P(x_i)\Delta x_i \neq \sum_{i=1}^{n}P(x)\Delta x_i$$

or
$$\sum_{i=1}^{n}P(x_i,y_i,z_i)\Delta v_i \neq \sum_{i=1}^{n}P(x,y,z)\Delta v_i$$

Author thinks, left of inequality, depicts conventional numerical calculation by algebraic symbols [Viz. 'Riemann sums', its geometrical intuition is 'Trapezoid with Curved Edge (Corresponding to 1-variable function)' or 'Column with Surface (Corresponding to 2-variables function)', 3-variables function and more variables function without geometrical intuitions]. In traditional mathematic system, 'Riemann Sums' is pure、independent numerical calculational method, be unprecedented that 'Riemann Sums' is used in certain formular proof; 'Riemann Sums' didn't become a tache of a formal logical reasoning, without any logical context constraints.

Right of inequality, depicts certain a tache of a formal logical reasoning under specified conditions, with certain logical context constraints. Actually, summed function 'P(x)'or 'P(x,y,z)'on right of inequality is limited to abstract functions (viz. functions without idiographic expression) by certain prerequisite (For example, prerequisite of formular proof in this works), science explorers can't directly obtain individual corresponding value of summed abstract function `P(x)`or`P(x,y,z)` at every segmental cell in numerical calculational meaning, and artificial algebraic expression `P(x_i)`or`P(x_i,y_i,z_i)` aren't real calculational values, can't form a chain of causality with logical context of formular proof; therefore right of inequality belongs to metamorphosis of `Riemann sums` in especial logical environment, belongs to broader concept of finite sums limits.

Unequal two sides can't prove that latter is false.

Chapter 5 Numerical Model of Divergence Theorem
Finite Sums Limits at Manifold

5.1 Numerical Models of Divergence Theorem Finite Sums Limits at Manifold

Known: Expression of Simply connected、Orientable、Closed Surface (Irregular、Asymmetrical)

$$[\sin(u)(1-\cos(v)), \sin(u)\sin(v)+2u, u(1-\cos(u))]$$ (1)

thereinto, u∈[0,π],v∈[0,2π];

and Integral Vector Field $\left[\dfrac{x^2}{2}, \dfrac{yz}{2}, \dfrac{z^2}{3}\right]$ (2)

Calculate and Validate Divergence Theorem at Manifold (Finite Sums Limits).

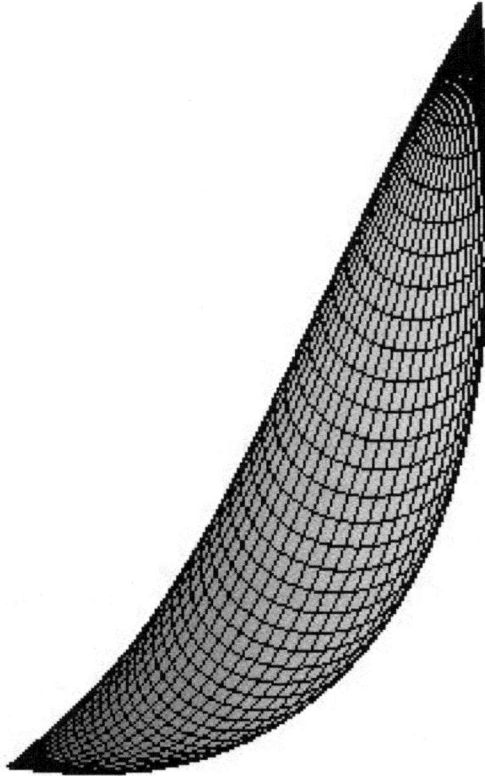

Figure(I)5.1 Simply Connected Orientable Closed Surface 'CS'

[Asymmetrical、Irregular]

Solution:

First, Free Surface Integral (Finite Sums Limits):

According to surface parametric expression (1), define and calculate First Matrix of Partial Derivatives, obtain tangent plane normal vector of surface:

$$\begin{bmatrix} i & j & k \\ \dfrac{\partial}{\partial u}(\sin(u)(1-\cos(v))) & \dfrac{\partial}{\partial u}(\sin(u)\sin(v)+2\,u) & \dfrac{d}{du}(u(1-\cos(u))) \\ \dfrac{\partial}{\partial v}(\sin(u)(1-\cos(v))) & \dfrac{\partial}{\partial v}(\sin(u)\sin(v)+2\,u) & \dfrac{\partial}{\partial v}(u(1-\cos(u))) \end{bmatrix}$$

=

$$\sin(u)\,(-\cos(v)\sin(u)\,i\,u + \sin(u)\sin(v)\,j\,u + \cos(v)\cos(u)\,i + \cos(v)\cos(u)\,k$$
$$- \cos(u)\sin(v)\,j - \cos(v)\,i - \cos(u)\,k + \sin(v)\,j - 2\sin(v)\,k)$$
$$(3)$$

From expression(3), respectively pick up coefficient of item 'i,j,k', obtain tangent plane normal vector of surface (4):

$$[\sin(u)\,(-\cos(v)\sin(u)\,u + \cos(v)\cos(u) - \cos(v)),$$
$$\sin(u)\,(\sin(u)\sin(v)\,u - \cos(u)\sin(v) + \sin(v)),$$
$$\sin(u)\,(\cos(v)\cos(u) - \cos(u) - 2\sin(v))]$$
$$(4)$$

Set Amount of surface 'S's segmental cells as 50 (5)

(This amount can be discretional natural number)

1.Microcosmic surface integral course at surface 'S's first segmental cell:

Segment range $[0,\pi]$ of 'u': du = $\dfrac{\pi}{50}$

Segment range $[0,2\pi]$ of 'v': dv = $\dfrac{2\pi}{50} = \dfrac{\pi}{25}$ (6)

Segment 'tangent plane normal vector'(7): [viz. Input (6) to (4)]

$$\left[\sin\!\left(\frac{\pi}{50}\right)\!\left(-\frac{1}{50}\cos\!\left(\frac{\pi}{25}\right)\sin\!\left(\frac{\pi}{50}\right)\pi + \cos\!\left(\frac{\pi}{25}\right)\cos\!\left(\frac{\pi}{50}\right) - \cos\!\left(\frac{\pi}{25}\right)\right),\right.$$
$$\sin\!\left(\frac{\pi}{50}\right)\!\left(\frac{1}{50}\sin\!\left(\frac{\pi}{50}\right)\sin\!\left(\frac{\pi}{25}\right)\pi - \cos\!\left(\frac{\pi}{50}\right)\sin\!\left(\frac{\pi}{25}\right) + \sin\!\left(\frac{\pi}{25}\right)\right),$$
$$\left.\sin\!\left(\frac{\pi}{50}\right)\!\left(\cos\!\left(\frac{\pi}{25}\right)\cos\!\left(\frac{\pi}{50}\right) - \cos\!\left(\frac{\pi}{50}\right) - 2\sin\!\left(\frac{\pi}{25}\right)\right)\right]$$
$$(7)$$

Segment idiographic vector field $\left[\dfrac{x^2}{2}, \dfrac{yz}{2}, \dfrac{z^2}{3}\right]$ (8):

[Viz. Input(1) to (2), then input (6)]

$$\left[\frac{1}{2}\sin\!\left(\frac{\pi}{50}\right)^2\!\left(1 - \cos\!\left(\frac{\pi}{25}\right)\right)^2, \frac{1}{100}\!\left(\sin\!\left(\frac{\pi}{50}\right)\sin\!\left(\frac{\pi}{25}\right) + \frac{\pi}{25}\right)\pi\!\left(1 - \cos\!\left(\frac{\pi}{50}\right)\right),\right.$$
$$\left.\frac{1}{7500}\pi^2\!\left(1 - \cos\!\left(\frac{\pi}{50}\right)\right)^2\right]$$
$$(8)$$

Calculate microcosmic surface integral value at surface'S's first segmental cell: (9)

Base on integral median theorem, dot product of abstract vector field (8) and tangent plane normal vector (7), times segment unit of 'u,v'(6), that is microcosmic surface integral value at first segmental cell (9)

$$\frac{1}{1250}\left(\frac{1}{2}\sin\left(\frac{\pi}{50}\right)\right)^3\left(-\frac{1}{50}\cos\left(\frac{\pi}{25}\right)\sin\left(\frac{\pi}{50}\right)\pi+\cos\left(\frac{\pi}{25}\right)\cos\left(\frac{\pi}{50}\right)-\cos\left(\frac{\pi}{25}\right)\right)$$

$$\left(1-\cos\left(\frac{\pi}{25}\right)\right)^2+\frac{1}{100}\sin\left(\frac{\pi}{50}\right)$$

$$\left(\frac{1}{50}\sin\left(\frac{\pi}{50}\right)\sin\left(\frac{\pi}{25}\right)\pi-\cos\left(\frac{\pi}{50}\right)\sin\left(\frac{\pi}{25}\right)+\sin\left(\frac{\pi}{25}\right)\right)\left(\sin\left(\frac{\pi}{50}\right)\sin\left(\frac{\pi}{25}\right)+\frac{\pi}{25}\right)\pi$$

$$\left(1-\cos\left(\frac{\pi}{50}\right)\right)$$

$$+\frac{1}{7500}\sin\left(\frac{\pi}{50}\right)\left(\cos\left(\frac{\pi}{25}\right)\cos\left(\frac{\pi}{50}\right)-\cos\left(\frac{\pi}{50}\right)-2\sin\left(\frac{\pi}{25}\right)\right)\pi^2\left(1-\cos\left(\frac{\pi}{50}\right)\right)^2\right)$$

$$\pi^2$$

2.Microcosmic surface integral course at surface 'S's all (viz.50) segmental cells:

Segment range [0,π] of 'u': du = $\dfrac{s\pi}{50}$

Segment range [0,2π] of 'v': dv = $\dfrac{2t\pi}{50}=\dfrac{t\pi}{25}$ (10)

(thereinto, 's' and 't' stand for natural number 1~50)

Segment 'tangent plane normal vector'(11): [viz. Input (10) to (4)]

$$\left[\sin\left(\frac{\pi s}{50}\right)\left(-\frac{1}{50}\cos\left(\frac{\pi t}{25}\right)\sin\left(\frac{\pi s}{50}\right)\pi s+\cos\left(\frac{\pi t}{25}\right)\cos\left(\frac{\pi s}{50}\right)-\cos\left(\frac{\pi t}{25}\right)\right),\right.$$

$$\sin\left(\frac{\pi s}{50}\right)\left(\frac{1}{50}\sin\left(\frac{\pi s}{50}\right)\sin\left(\frac{\pi t}{25}\right)\pi s-\cos\left(\frac{\pi s}{50}\right)\sin\left(\frac{\pi t}{25}\right)+\sin\left(\frac{\pi t}{25}\right)\right),$$

$$\left.\sin\left(\frac{\pi s}{50}\right)\left(\cos\left(\frac{\pi t}{25}\right)\cos\left(\frac{\pi s}{50}\right)-\cos\left(\frac{\pi s}{50}\right)-2\sin\left(\frac{\pi t}{25}\right)\right)\right]$$

(11)

// Expression (11) isn't that 'Single vector value',but that 'Set of finite quantitative (viz. 50) vector values'

Segment idiographic vector field $\left[\dfrac{x^2}{2},\dfrac{yz}{2},\dfrac{z^2}{3}\right]$ (12):

[Viz. Input(1) to (2), then input (10)]

$$\left[\frac{1}{2}\sin\left(\frac{\pi s}{50}\right)^2\left(1-\cos\left(\frac{\pi t}{25}\right)\right)^2, \frac{1}{100}\left(\sin\left(\frac{\pi s}{50}\right)\sin\left(\frac{\pi t}{25}\right)+\frac{\pi s}{25}\right)\pi s\left(1-\cos\left(\frac{\pi s}{50}\right)\right),\right.$$
$$\left.\frac{1}{7500}\pi^2 s^2\left(1-\cos\left(\frac{\pi s}{50}\right)\right)^2\right]$$

// Expression (12) isn't that 'Single vector value', but that 'Set of finite quantitative (viz. 50) vector values'

Calculate microcosmic surface integral value at surface 'S's all segmental cells (13):

Base on integral median theorem, dot product of idiographic vector field (12) and tangent plane normal vector (11), times segment unit of 'u,v' (10), that is microcosmic surface integral values at all segmental cells (13)

$$\frac{1}{1250}\left(\frac{1}{2}\sin\left(\frac{\pi s}{50}\right)^3\left(-\frac{1}{50}\cos\left(\frac{\pi t}{25}\right)\sin\left(\frac{\pi s}{50}\right)\pi s+\cos\left(\frac{\pi t}{25}\right)\cos\left(\frac{\pi s}{50}\right)-\cos\left(\frac{\pi t}{25}\right)\right)\right.$$
$$\left(1-\cos\left(\frac{\pi t}{25}\right)\right)^2+\frac{1}{100}\sin\left(\frac{\pi s}{50}\right)$$
$$\left(\frac{1}{50}\sin\left(\frac{\pi s}{50}\right)\sin\left(\frac{\pi t}{25}\right)\pi s-\cos\left(\frac{\pi s}{50}\right)\sin\left(\frac{\pi t}{25}\right)+\sin\left(\frac{\pi t}{25}\right)\right)$$
$$\left(\sin\left(\frac{\pi s}{50}\right)\sin\left(\frac{\pi t}{25}\right)+\frac{\pi s}{25}\right)\pi s\left(1-\cos\left(\frac{\pi s}{50}\right)\right)+$$
$$\left.\frac{1}{7500}\sin\left(\frac{\pi s}{50}\right)\left(\cos\left(\frac{\pi t}{25}\right)\cos\left(\frac{\pi s}{50}\right)-\cos\left(\frac{\pi s}{50}\right)-2\sin\left(\frac{\pi t}{25}\right)\right)\pi^2 s^2\left(1-\cos\left(\frac{\pi s}{50}\right)\right)^2\right)$$
$$\pi^2$$

//Expression (13) isn't that 'single value', but that 'set of finite quantitative (viz.50) values'

Structure sequence that is composed of finite quantitative (viz. 50) microcomosic surface integral values (14):

(Lengthy expression of sequence is elided;

In program template of Waterloo Maplesoft (See also Section 5.2):

```
sqn:=seq(seq((stdV[1]*stdA+stdV[2]*stdB+stdV[3]*stdC)*du*dv,
s=1..dus),t=1..dus):
```

change ':'(last) to ';', then obtain expression of sequence)

Accumulational sum of sequence (viz. Sum of integral values of dot product [idiographic vector field (12) and tangent plane normal vector (11) at all segmental cells]), obtain surface integral value at manifold (15):

[Lengthy expression of accumulational sum result is elided;

In program template of Waterloo Maplesoft (See also Section 5.2):

```
add(k,k=sqn):xi:=evalf(%);
```

change ':'(middle) to ';', then obtain expression of result]

Change expression of accumulational sum result to float value:

$$41.18505570$$

Set Amount of surface 'S's segmental cells as n

(viz. uncertain natural number)   (16)

3.Microcosmic surface integral course at surface 'S's first segmental cell:

Segment range $[0,\pi]$ of 'u': $du = \dfrac{\pi}{n}$

Segment range $[0,2\pi]$ of 'v': $dv = \dfrac{2\pi}{n}$ (17)

Segment 'tangent plane normal vector'(18): [viz. Input (17) to (4)]

$$\left[\sin\left(\frac{\pi}{n}\right)\left(-\frac{\cos\left(\frac{2\pi}{n}\right)\sin\left(\frac{\pi}{n}\right)\pi}{n} + \cos\left(\frac{2\pi}{n}\right)\cos\left(\frac{\pi}{n}\right) - \cos\left(\frac{2\pi}{n}\right) \right), \right.$$

$$\sin\left(\frac{\pi}{n}\right)\left(\frac{\sin\left(\frac{\pi}{n}\right)\sin\left(\frac{2\pi}{n}\right)\pi}{n} - \cos\left(\frac{\pi}{n}\right)\sin\left(\frac{2\pi}{n}\right) + \sin\left(\frac{2\pi}{n}\right) \right),$$

$$\left. \sin\left(\frac{\pi}{n}\right)\left(\cos\left(\frac{2\pi}{n}\right)\cos\left(\frac{\pi}{n}\right) - \cos\left(\frac{\pi}{n}\right) - 2\sin\left(\frac{2\pi}{n}\right) \right) \right]$$ (18)

Segment idiographic vector field $\left[\dfrac{x^2}{2}, \dfrac{yz}{2}, \dfrac{z^2}{3} \right]$ (19):

[Viz. Input(1) to (2), then input (17)]

$$\left[\frac{1}{2}\sin\left(\frac{\pi}{n}\right)^2\left(1 - \cos\left(\frac{2\pi}{n}\right)\right)^2, \frac{1}{2}\frac{\left(\sin\left(\frac{\pi}{n}\right)\sin\left(\frac{2\pi}{n}\right) + \frac{2\pi}{n}\right)\pi\left(1 - \cos\left(\frac{\pi}{n}\right)\right)}{n}, \right.$$

$$\left. \frac{1}{3} \frac{\pi^2 \left(1 - \cos\left(\frac{\pi}{n}\right)\right)^2}{n^2} \right]$$

(19)

Calculate microcosmic surface integral value at surface 'S's first segmental cell: (20)

Base on integral median theorem, dot product of idiographic vector field (19) and tangent plane normal vector (18), times segment unit of 'u,v' (17), that is microcosmic surface integral value at first segmental cell (20)

$$2\left(\frac{1}{2}\sin\left(\frac{\pi}{n}\right)^3 \left(-\frac{\cos\left(\frac{2\pi}{n}\right)\sin\left(\frac{\pi}{n}\right)\pi}{n} + \cos\left(\frac{2\pi}{n}\right)\cos\left(\frac{\pi}{n}\right) - \cos\left(\frac{2\pi}{n}\right) \right)\left(1 - \cos\left(\frac{2\pi}{n}\right)\right)^2 \right.$$

$$+ \frac{1}{2}\sin\left(\frac{\pi}{n}\right)\left(\frac{\sin\left(\frac{\pi}{n}\right)\sin\left(\frac{2\pi}{n}\right)\pi}{n} - \cos\left(\frac{\pi}{n}\right)\sin\left(\frac{2\pi}{n}\right) + \sin\left(\frac{2\pi}{n}\right) \right)$$

$$\left(\sin\left(\frac{\pi}{n}\right)\sin\left(\frac{2\pi}{n}\right) + \frac{2\pi}{n} \right)\pi\left(1 - \cos\left(\frac{\pi}{n}\right)\right)\Big/n$$

$$\left. + \frac{1}{3} \frac{\sin\left(\frac{\pi}{n}\right)\left(\cos\left(\frac{2\pi}{n}\right)\cos\left(\frac{\pi}{n}\right) - \cos\left(\frac{\pi}{n}\right) - 2\sin\left(\frac{2\pi}{n}\right)\right)\pi^2\left(1 - \cos\left(\frac{\pi}{n}\right)\right)^2}{n^2} \right)\pi^2 \Big/ n^2$$

4. Microcosmic surface integral course at surface 'S's all (viz. n cells) segmental cells:

Segment range [0,π] of 'u': du = $\frac{s\pi}{n}$

Segment range [0,2π] of 'v': dv = $\frac{2t\pi}{n}$ (21)

(thereinto, 's' and 't' stand for natural number 1~n)

Segment 'tangent plane normal vector' (22): [viz. Input (21) to (4)]

$$\left[\sin\left(\frac{\pi s}{n}\right)\left(-\frac{\cos\left(\frac{2\pi t}{n}\right)\sin\left(\frac{\pi s}{n}\right)\pi s}{n} + \cos\left(\frac{2\pi t}{n}\right)\cos\left(\frac{\pi s}{n}\right) - \cos\left(\frac{2\pi t}{n}\right) \right), \right.$$

$$\sin\left(\frac{\pi s}{n}\right)\left(\frac{\sin\left(\frac{\pi s}{n}\right)\sin\left(\frac{2\pi t}{n}\right)\pi s}{n} - \cos\left(\frac{\pi s}{n}\right)\sin\left(\frac{2\pi t}{n}\right) + \sin\left(\frac{2\pi t}{n}\right) \right),$$

$$\sin\left(\frac{\pi s}{n}\right)\left(\cos\left(\frac{2\pi t}{n}\right)\cos\left(\frac{\pi s}{n}\right)-\cos\left(\frac{\pi s}{n}\right)-2\sin\left(\frac{2\pi t}{n}\right)\right)\right] \quad (22)$$

//Expression (22) isn't that 'single vector value', but that 'set of finite quantitative (viz. n) vector values'

Segment idiographic vector field $\left[\dfrac{x^2}{2},\dfrac{yz}{2},\dfrac{z^2}{3}\right]$ (23):

[Viz. Input(1) to (2), then input (21)]

$$\left[\frac{1}{2}\sin\left(\frac{\pi s}{n}\right)^2\left(1-\cos\left(\frac{2\pi t}{n}\right)\right)^2,\frac{1}{2}\frac{\left(\sin\left(\frac{\pi s}{n}\right)\sin\left(\frac{2\pi t}{n}\right)+\frac{2\pi s}{n}\right)\pi s\left(1-\cos\left(\frac{\pi s}{n}\right)\right)}{n},\right.$$

$$\left.\frac{1}{3}\frac{\pi^2 s^2\left(1-\cos\left(\frac{\pi s}{n}\right)\right)^2}{n^2}\right]$$

$$(23)$$

//Expression (23) isn't that 'single vector value', but that 'set of finite quantitative (viz. n) vector values'

Calculate microcosmic surface integral value at surface 'S's all segmental cells (24):

Base on integral median theorem, dot product of idiographic vector field (23) and tangent plane normal vector (22), times segment unit of 'u,v'(21), that is microcosmic surface integral values at all segmental cells(24)

$$2\left(\frac{1}{2}\sin\left(\frac{\pi s}{n}\right)^3\left(-\frac{\cos\left(\frac{2\pi t}{n}\right)\sin\left(\frac{\pi s}{n}\right)\pi s}{n}+\cos\left(\frac{2\pi t}{n}\right)\cos\left(\frac{\pi s}{n}\right)-\cos\left(\frac{2\pi t}{n}\right)\right)\right.$$

$$\left(1-\cos\left(\frac{2\pi t}{n}\right)\right)^2+\frac{1}{2}\sin\left(\frac{\pi s}{n}\right)$$

$$\left(\frac{\sin\left(\frac{\pi s}{n}\right)\sin\left(\frac{2\pi t}{n}\right)\pi s}{n}-\cos\left(\frac{\pi s}{n}\right)\sin\left(\frac{2\pi t}{n}\right)+\sin\left(\frac{2\pi t}{n}\right)\right)$$

$$\left(\sin\left(\frac{\pi s}{n}\right)\sin\left(\frac{2\pi t}{n}\right)+\frac{2\pi s}{n}\right)\pi s\left(1-\cos\left(\frac{\pi s}{n}\right)\right)\Big/n+$$

$$\frac{1}{3}\frac{\sin\left(\frac{\pi s}{n}\right)\left(\cos\left(\frac{2\pi t}{n}\right)\cos\left(\frac{\pi s}{n}\right)-\cos\left(\frac{\pi s}{n}\right)-2\sin\left(\frac{2\pi t}{n}\right)\right)\pi^2 s^2\left(1-\cos\left(\frac{\pi s}{n}\right)\right)^2}{n^2}$$

$$\Bigg) \pi^2 / n^2$$

$$(24)$$

//Expression (24) isn't that 'single value', but that 'set of finite quantitative (viz.n) values'

Structure finite sums (25):

(In the case of "Amount 'n' of surface 'S's segmental cells is uncertain", sum of integral values [dot product about absract vector field (23) and tangent plane normal vector (22)] at all segmental cells)

$$\sum_{t=1}^{n}\Bigg(\sum_{s=1}^{n}\Bigg(2\Bigg(\frac{1}{2}\sin\left(\frac{\pi s}{n}\right)\Bigg)^{3}$$

$$\left(-\frac{\cos\left(\frac{2\pi t}{n}\right)\sin\left(\frac{\pi s}{n}\right)\pi s}{n}+\cos\left(\frac{2\pi t}{n}\right)\cos\left(\frac{\pi s}{n}\right)-\cos\left(\frac{2\pi t}{n}\right)\right)\left(1-\cos\left(\frac{2\pi t}{n}\right)\right)^{2}$$

$$+\frac{1}{2}\sin\left(\frac{\pi s}{n}\right)\left(\frac{\sin\left(\frac{\pi s}{n}\right)\sin\left(\frac{2\pi t}{n}\right)\pi s}{n}-\cos\left(\frac{\pi s}{n}\right)\sin\left(\frac{2\pi t}{n}\right)+\sin\left(\frac{2\pi t}{n}\right)\right)$$

$$\left(\sin\left(\frac{\pi s}{n}\right)\sin\left(\frac{2\pi t}{n}\right)+\frac{2\pi s}{n}\right)\pi s\left(1-\cos\left(\frac{\pi s}{n}\right)\right)\Big)/n+$$

$$\frac{1}{3}\frac{\sin\left(\frac{\pi s}{n}\right)\left(\cos\left(\frac{2\pi t}{n}\right)\cos\left(\frac{\pi s}{n}\right)-\cos\left(\frac{\pi s}{n}\right)-2\sin\left(\frac{2\pi t}{n}\right)\right)\pi^{2}s^{2}\left(1-\cos\left(\frac{\pi s}{n}\right)\right)^{2}}{n^{2}}$$

$$\Bigg)\pi^2 / n^2\Bigg)\Bigg)$$

$$(25)$$

Infinitize finite sums, its limit operational value is 'Surface Integral Value at Manifold' (26):

(In the case of "Amount 'n' of surface 'S's segmental cells tends to infinity", limit value of integral values' sums [dot product about idiographic vector field (23) and tangent plane normal vector (22)] at all segmental cells)

$$\lim_{n\to\infty}\sum_{t=1}^{n}\Bigg(\sum_{s=1}^{n}\Bigg(2\Bigg(\frac{1}{2}\sin\left(\frac{\pi s}{n}\right)\Bigg)^{3}$$

$$\Bigg(\Bigg(-\frac{\cos\left(\frac{2\pi t}{n}\right)\sin\left(\frac{\pi s}{n}\right)\pi s}{n}+\cos\left(\frac{2\pi t}{n}\right)\cos\left(\frac{\pi s}{n}\right)-\cos\left(\frac{2\pi t}{n}\right)\Bigg)\left(1-\cos\left(\frac{2\pi t}{n}\right)\right)^2$$

$$+\frac{1}{2}\sin\left(\frac{\pi s}{n}\right)\Bigg(\frac{\sin\left(\frac{\pi s}{n}\right)\sin\left(\frac{2\pi t}{n}\right)\pi s}{n}-\cos\left(\frac{\pi s}{n}\right)\sin\left(\frac{2\pi t}{n}\right)+\sin\left(\frac{2\pi t}{n}\right)\Bigg)$$

$$\left(\sin\left(\frac{\pi s}{n}\right)\sin\left(\frac{2\pi t}{n}\right)+\frac{2\pi s}{n}\right)\pi s\left(1-\cos\left(\frac{\pi s}{n}\right)\right)\Bigg/n+$$

$$\frac{1}{3}\frac{\sin\left(\frac{\pi s}{n}\right)\left(\cos\left(\frac{2\pi t}{n}\right)\cos\left(\frac{\pi s}{n}\right)-\cos\left(\frac{\pi s}{n}\right)-2\sin\left(\frac{2\pi t}{n}\right)\right)\pi^2 s^2\left(1-\cos\left(\frac{\pi s}{n}\right)\right)^2}{n^2}$$

$$\Bigg)\pi^2/n^2\Bigg)\Bigg)$$

$$=\ -\frac{88}{81}\pi+\frac{829}{576}\pi^3 \tag{26}$$

Second, Free Triple Integrals (Finite Sums Limits):

Entirely multiply radius vector 'r' (suppose r > 0) by each item of surface 'S' parametrized expression (1), convert "Parametrized expression of surface 'S'" to "Parametrized expression of surface 'S' coordinates":

$$[r\sin(u)(1-\cos(v)),\, r(\sin(u)\sin(v)+2u),\, ru(1-\cos(u))] \tag{27}$$

Base on parametrized expression of surface 'S' coordinates (27), define and calculate second matrix of partial derivatives, obtain general expression of surface 'S' coordinates volume element coefficient (28):

$$\begin{bmatrix}\frac{\partial}{\partial r}(r\sin(u)(1-\cos(v))), & \frac{\partial}{\partial u}(r\sin(u)(1-\cos(v))), & \frac{\partial}{\partial v}(r\sin(u)(1-\cos(v)))\\ \frac{\partial}{\partial r}(r(\sin(u)\sin(v)+2u)), & \frac{\partial}{\partial u}(r(\sin(u)\sin(v)+2u)), & \frac{\partial}{\partial v}(r(\sin(u)\sin(v)+2u))\\ \frac{\partial}{\partial r}(ru(1-\cos(u))), & \frac{\partial}{\partial u}(ru(1-\cos(u))), & \frac{\partial}{\partial v}(ru(1-\cos(u)))\end{bmatrix}$$

$$=$$

$$\sin(u)(2\sin(u)\sin(v)u^2+\cos(u)\cos(v)\sin(u)+\cos(u)\cos(v)u-\cos(u)\sin(u)$$
$$-\cos(u)u-\sin(u)\cos(v)-\cos(v)u+\sin(u)+u)r^2$$

Calculate divergence of idiographic vector field (2):

$$\left(\frac{d}{dx}\left(\frac{x^2}{2}\right)\right)+\left(\frac{\partial}{\partial y}\left(\frac{yz}{2}\right)\right)+\left(\frac{d}{dz}\left(\frac{z^2}{3}\right)\right)=x+\frac{7z}{6} \tag{29}$$

Set Amount of spacial bounded closed region 'Ω's (Surface 'S' surrounds) segmental cells as 20

(This amount can be discretional natural number)   (30)

5.Microcosmic Triple Integrals Course in spacial bounded closed region 'Ω's first segmental cell:

Segment range [0,1] of 'r': dr = $\dfrac{1}{20}$

Segment range [0,π] of 'u': du = $\dfrac{\pi}{20}$

Segment range [0,2π] of 'v': dv = $\dfrac{2\pi}{20}=\dfrac{\pi}{10}$ (31)

Segment volume element coefficient 'J'(32):

[viz. Input (31) to (28)]

(Volume element coefficient (28), average value that corresponds to spacial bounded closed region 'Ω's first segmental cell)

$$\frac{1}{400}\sin\left(\frac{\pi}{20}\right)\left(\frac{1}{200}\sin\left(\frac{\pi}{20}\right)\sin\left(\frac{\pi}{10}\right)\pi^2+\cos\left(\frac{\pi}{20}\right)\cos\left(\frac{\pi}{10}\right)\sin\left(\frac{\pi}{20}\right)\right.$$
$$+\frac{1}{20}\cos\left(\frac{\pi}{20}\right)\cos\left(\frac{\pi}{10}\right)\pi-\cos\left(\frac{\pi}{20}\right)\sin\left(\frac{\pi}{20}\right)-\frac{1}{20}\cos\left(\frac{\pi}{20}\right)\pi-\sin\left(\frac{\pi}{20}\right)\cos\left(\frac{\pi}{10}\right)$$
$$\left.-\frac{1}{20}\cos\left(\frac{\pi}{10}\right)\pi+\sin\left(\frac{\pi}{20}\right)+\frac{\pi}{20}\right) \tag{32}$$

Segment idiographic divergence (33):

[viz. Input (27) to (29),then input (31)]

(Idiographic Divergence (29), average value that corresponds to spacial bounded closed region 'Ω's first segmental cell)

$$\frac{1}{20}\sin\left(\frac{\pi}{20}\right)\left(1-\cos\left(\frac{\pi}{10}\right)\right)+\frac{7}{2400}\pi\left(1-\cos\left(\frac{\pi}{20}\right)\right) \tag{33}$$

Calculate microcosmic triple integrals value in spacial bounded closed region 'Ω's first segmental cell (34):

Base on integral median theorem, product of Idiographic Divergence

(33) and Volume Element Coefficient (32), times segment unit of 'r,u,v'(31), that is microcosmic triple integrals value in first segmental cell (34)

$$
\frac{1}{1600000}\left(\frac{1}{20}\sin\left(\frac{\pi}{20}\right)\left(1-\cos\left(\frac{\pi}{10}\right)\right)+\frac{7}{2400}\pi\left(1-\cos\left(\frac{\pi}{20}\right)\right)\right)\sin\left(\frac{\pi}{20}\right)\Bigg(
$$
$$
\frac{1}{200}\sin\left(\frac{\pi}{20}\right)\sin\left(\frac{\pi}{10}\right)\pi^2+\cos\left(\frac{\pi}{20}\right)\cos\left(\frac{\pi}{10}\right)\sin\left(\frac{\pi}{20}\right)+\frac{1}{20}\cos\left(\frac{\pi}{20}\right)\cos\left(\frac{\pi}{10}\right)\pi
$$
$$
-\cos\left(\frac{\pi}{20}\right)\sin\left(\frac{\pi}{20}\right)-\frac{1}{20}\cos\left(\frac{\pi}{20}\right)\pi-\sin\left(\frac{\pi}{20}\right)\cos\left(\frac{\pi}{10}\right)-\frac{1}{20}\cos\left(\frac{\pi}{10}\right)\pi
$$
$$
+\sin\left(\frac{\pi}{20}\right)+\frac{\pi}{20}\bigg)\pi^2
$$
(34)

6.Microcosmic triple integrals course in spacial bounded closed region 'Ω's all (viz. 20)segmental cells:

Segment range $[0,\pi]$ of 'u': du $=\dfrac{i\pi}{20}$

Segment range $[0,2\pi]$ of 'v': dv $=\dfrac{2j\pi}{20}=\dfrac{j\pi}{10}$

Segment range $[0,1]$ of 'r': dr $=\dfrac{k}{20}$ (35)

(thereinto, 'i,j,k' stand for natural number 1~20)

Segment volume element coefficient (36): [viz. Input (35) to (28)]

(Volume element coefficient (28),average values that correspond to spacial bounded closed region 'Ω's all (viz. 20) segmental cells)

$$
\frac{1}{400}\sin\left(\frac{\pi i}{20}\right)\left(\frac{1}{200}\sin\left(\frac{\pi i}{20}\right)\sin\left(\frac{\pi j}{10}\right)\pi^2 i^2+\cos\left(\frac{\pi i}{20}\right)\cos\left(\frac{\pi j}{10}\right)\sin\left(\frac{\pi i}{20}\right)\right.
$$
$$
+\frac{1}{20}\cos\left(\frac{\pi i}{20}\right)\cos\left(\frac{\pi j}{10}\right)\pi i-\cos\left(\frac{\pi i}{20}\right)\sin\left(\frac{\pi i}{20}\right)-\frac{1}{20}\cos\left(\frac{\pi i}{20}\right)\pi i
$$
$$
-\sin\left(\frac{\pi i}{20}\right)\cos\left(\frac{\pi j}{10}\right)-\frac{1}{20}\cos\left(\frac{\pi j}{10}\right)\pi i+\sin\left(\frac{\pi i}{20}\right)+\frac{\pi i}{20}\bigg)k^2
$$
(36)

// Expression (36) isn't that 'single value', but that 'Set of finite quantitative (viz. 20) values'

Segment idiographic divergence (37):
[viz. Input (27) to (29),then input (35)]
(Idiographic Divergence (29), average values that correspond to spacial bounded closed region 'Ω's all (viz. 20) segmental cells)

$$\frac{1}{20}\,k\,\sin\!\left(\frac{\pi\,i}{20}\right)\!\left(1-\cos\!\left(\frac{\pi\,j}{10}\right)\right)+\frac{7}{2400}\,k\,\pi\,i\left(1-\cos\!\left(\frac{\pi\,i}{20}\right)\right)\text{,} \tag{37}$$

// Expression (37) isn't that 'single value', but that 'Set of finite
quantitative (viz. 20) values'

Calculate microcosmic triple integrals values in spacial bounded
closed region 'Ω's all segmental cells (38):

Base on integral median theorem, Product of Idiographic Divergence
(37) and Volume Element Coefficient (36), times segment unit of
'r,u,v' (35), that is microcosmic triple integrals values in all (viz.
20) segmental cells (38)

$$\frac{1}{1600000}\left(\frac{1}{20}\,k\,\sin\!\left(\frac{\pi\,i}{20}\right)\!\left(1-\cos\!\left(\frac{\pi\,j}{10}\right)\right)+\frac{7}{2400}\,k\,\pi\,i\left(1-\cos\!\left(\frac{\pi\,i}{20}\right)\right)\right)\sin\!\left(\frac{\pi\,i}{20}\right)\Bigg($$

$$\frac{1}{200}\sin\!\left(\frac{\pi\,i}{20}\right)\sin\!\left(\frac{\pi\,j}{10}\right)\pi^2\,i^2+\cos\!\left(\frac{\pi\,i}{20}\right)\cos\!\left(\frac{\pi\,j}{10}\right)\sin\!\left(\frac{\pi\,i}{20}\right)$$

$$+\frac{1}{20}\cos\!\left(\frac{\pi\,i}{20}\right)\cos\!\left(\frac{\pi\,j}{10}\right)\pi\,i-\cos\!\left(\frac{\pi\,i}{20}\right)\sin\!\left(\frac{\pi\,i}{20}\right)-\frac{1}{20}\cos\!\left(\frac{\pi\,i}{20}\right)\pi\,i$$

$$-\sin\!\left(\frac{\pi\,i}{20}\right)\cos\!\left(\frac{\pi\,j}{10}\right)-\frac{1}{20}\cos\!\left(\frac{\pi\,j}{10}\right)\pi\,i+\sin\!\left(\frac{\pi\,i}{20}\right)+\frac{\pi\,i}{20}\Bigg)k^2\,\pi^2 \tag{38}$$

// Expression (38) isn't that 'Single value', but that 'Set of finite
quantitative (viz. 20) values'

Structure sequence that is composed of finite quantitative (viz.
20) microcomosic triple integrals values (39):

  [Lengthy expression of sequence is elided;

  In program template of Waterloo Maplesoft (see also Section 5.2):

  sqn:=seq(seq(seq(ijkddiV*ijkdJ*dr*du*dv,i=1..dus),j=1..dus),

  k=1..dus):

  change ':' (last) to ';', then obtain expression of sequence]

Accumulational sum of sequence (viz. Sum of integral values of
Product [Idiographic Divergence (37) and Volume Element at all (viz.
20) segmental cells], obtain triple integrals value at manifold (40):

  [Lengthy expression of accumulational sum result is elided;

  In program template of Waterloo Maplesoft (See also Section 5.2):

  add(k,k=sqn):omega:=evalf(%):

change ':'(middle and last) to ';', then obtain expression and float value of result]

Change expression of accumulational sum result to float value:

$$4527224382$$

Set Amount of spacial bounded closed region 'Ω's (Surface 'S' surrounds) segmental cells as n (viz.uncertain natural number) (41)

7.Microcosmic Triple Integrals Course in spacial bounded closed region 'Ω's first segmental cell:

Segment range [0,1] of 'r': dr = $\dfrac{1}{n}$

Segment range [0,π] of 'u': du = $\dfrac{\pi}{n}$

Segment range [0,2π] of 'v': dv = $\dfrac{2\pi}{n}$ (42)

Segment volume element coefficient 'J' (43):

(viz. Input (42) to (28))

(Volume element coefficient (28),average value that corresponds to spacial bounded closed region 'Ω's first segmental cell)

$$\sin\left(\frac{\pi}{n}\right)\left(\frac{2\sin\left(\frac{\pi}{n}\right)\sin\left(\frac{2\pi}{n}\right)\pi^2}{n^2} + \cos\left(\frac{\pi}{n}\right)\cos\left(\frac{2\pi}{n}\right)\sin\left(\frac{\pi}{n}\right) + \frac{\cos\left(\frac{\pi}{n}\right)\cos\left(\frac{2\pi}{n}\right)\pi}{n} \right.$$

$$\left. - \cos\left(\frac{\pi}{n}\right)\sin\left(\frac{\pi}{n}\right) - \frac{\cos\left(\frac{\pi}{n}\right)\pi}{n} - \sin\left(\frac{\pi}{n}\right)\cos\left(\frac{2\pi}{n}\right) - \frac{\cos\left(\frac{2\pi}{n}\right)\pi}{n} + \sin\left(\frac{\pi}{n}\right) + \frac{\pi}{n} \right) / n^2$$

Segment idiographic divergence (44):

[viz. Input (27) to (29), then input (42)]

(Idiographic divergence (29), average value that corresponds to spacial bounded closed region 'Ω's first segmental cell)

$$\frac{\sin\left(\frac{\pi}{n}\right)\left(1 - \cos\left(\frac{2\pi}{n}\right)\right)}{n} + \frac{7}{6}\frac{\pi\left(1 - \cos\left(\frac{\pi}{n}\right)\right)}{n^2}$$ (44)

Calculate microcosmic triple integrals value in spacial bounded closed region 'Ω's first segmental cell (45):

Base on integral median theorem, product of Idiographic Divergence (44) and Volume Element Coefficient (43), times segment unit of 'r,u,v'(42), that is microcosmic triple integrals value in first segmental cell (45)

$$2\left(\frac{\sin\left(\frac{\pi}{n}\right)\left(1-\cos\left(\frac{2\pi}{n}\right)\right)}{n}+\frac{7}{6}\frac{\pi\left(1-\cos\left(\frac{\pi}{n}\right)\right)}{n^2}\right)\sin\left(\frac{\pi}{n}\right)\left(\frac{2\sin\left(\frac{\pi}{n}\right)\sin\left(\frac{2\pi}{n}\right)\pi^2}{n^2}\right.$$

$$+\cos\left(\frac{\pi}{n}\right)\cos\left(\frac{2\pi}{n}\right)\sin\left(\frac{\pi}{n}\right)+\frac{\cos\left(\frac{\pi}{n}\right)\cos\left(\frac{2\pi}{n}\right)\pi}{n}-\cos\left(\frac{\pi}{n}\right)\sin\left(\frac{\pi}{n}\right)-\frac{\cos\left(\frac{\pi}{n}\right)\pi}{n}$$

$$\left.-\sin\left(\frac{\pi}{n}\right)\cos\left(\frac{2\pi}{n}\right)-\frac{\cos\left(\frac{2\pi}{n}\right)\pi}{n}+\sin\left(\frac{\pi}{n}\right)+\frac{\pi}{n}\right)\pi^2/n^5$$

(45)

8.Microcosmic triple integrals course in spacial bounded closed region 'Ω's all (viz. n)segmental cells:

Segment range $[0,\pi]$ of 'u': du $= \dfrac{i\pi}{n}$

Segment range $[0,2\pi]$ of 'v': dv $= \dfrac{2j\pi}{n}$

Segment range $[0,1]$ of 'r': dr $= \dfrac{k}{n}$

(thereinto, 'i,j,k' stand for natural number 1~n)   (46)

Segment volume element coefficient (47): [viz. Input (46) to (28)]

(Volume element coefficient (28),average values that correspond to spacial bounded closed region 'Ω's all (viz. n) segmental cells)

$$\sin\left(\frac{\pi i}{n}\right)\left(\frac{2\sin\left(\frac{\pi i}{n}\right)\sin\left(\frac{2\pi j}{n}\right)\pi^2 i^2}{n^2}+\cos\left(\frac{\pi i}{n}\right)\cos\left(\frac{2\pi j}{n}\right)\sin\left(\frac{\pi i}{n}\right)\right.$$

$$+\frac{\cos\left(\frac{\pi i}{n}\right)\cos\left(\frac{2\pi j}{n}\right)\pi i}{n}-\cos\left(\frac{\pi i}{n}\right)\sin\left(\frac{\pi i}{n}\right)-\frac{\cos\left(\frac{\pi i}{n}\right)\pi i}{n}$$

$$\left.-\sin\left(\frac{\pi i}{n}\right)\cos\left(\frac{2\pi j}{n}\right)-\frac{\cos\left(\frac{2\pi j}{n}\right)\pi i}{n}+\sin\left(\frac{\pi i}{n}\right)+\frac{\pi i}{n}\right)k^2/n^2$$

(47)

//Expression (47) isn't that 'Single value', but that 'Set of finite quantitative (viz. n) values'

Segment idiographic divergence (48):

[viz. Input (27) to (29),then input (46)]

(Idiographic Divergence (29), average values that correspond to spacial bounded closed region 'Ω's all (viz. n) segmental cells)

$$\frac{k \sin\left(\frac{\pi i}{n}\right)\left(1 - \cos\left(\frac{2 \pi j}{n}\right)\right)}{n} + \frac{7}{6} \frac{k \pi i \left(1 - \cos\left(\frac{\pi i}{n}\right)\right)}{n^2} \tag{48}$$

// Expression (48) isn't that 'Single value', but that 'Set of finite quantitative (viz. n) values'

Calculate microcosmic triple integrals values in spacial bounded closed region 'Ω's all segmental cells (49):

Base on integral median theorem, product of Idiographic Divergence (48) and Volume Element Coefficient (47), times segment unit of 'r,u,v' (46), that is microcosmic triple integrals values in all (viz. n) segmental cells (49)

$$2 \left(\frac{k \sin\left(\frac{\pi i}{n}\right)\left(1 - \cos\left(\frac{2 \pi j}{n}\right)\right)}{n} + \frac{7}{6} \frac{k \pi i \left(1 - \cos\left(\frac{\pi i}{n}\right)\right)}{n^2}\right) \sin\left(\frac{\pi i}{n}\right) \Bigg($$

$$\frac{2 \sin\left(\frac{\pi i}{n}\right) \sin\left(\frac{2 \pi j}{n}\right) \pi^2 i^2}{n^2} + \cos\left(\frac{\pi i}{n}\right) \cos\left(\frac{2 \pi j}{n}\right) \sin\left(\frac{\pi i}{n}\right)$$

$$+ \frac{\cos\left(\frac{\pi i}{n}\right) \cos\left(\frac{2 \pi j}{n}\right) \pi i}{n} - \cos\left(\frac{\pi i}{n}\right) \sin\left(\frac{\pi i}{n}\right) - \frac{\cos\left(\frac{\pi i}{n}\right) \pi i}{n}$$

$$- \sin\left(\frac{\pi i}{n}\right) \cos\left(\frac{2 \pi j}{n}\right) - \frac{\cos\left(\frac{2 \pi j}{n}\right) \pi i}{n} + \sin\left(\frac{\pi i}{n}\right) + \frac{\pi i}{n}\Bigg) k^2 \pi^2 / n^5 \tag{49}$$

// Expression (49) isn't that 'Single value', but that 'Set of finite quantitative (viz. n) values'

Structure finite sums (50):

$$\sum_{k=1}^{n}\left(\sum_{j=1}^{n}\left(\sum_{i=1}^{n}\left(2\left(\frac{k \sin\left(\frac{\pi i}{n}\right)\left(1 - \cos\left(\frac{2 \pi j}{n}\right)\right)}{n} + \frac{7}{6} \frac{k \pi i \left(1 - \cos\left(\frac{\pi i}{n}\right)\right)}{n^2}\right) \sin\left(\frac{\pi i}{n}\right) \Bigg(\right.\right.\right.$$

$$
\frac{2 \sin\left(\dfrac{\pi i}{n}\right) \sin\left(\dfrac{2 \pi j}{n}\right) \pi^2 i^2}{n^2} + \cos\left(\frac{\pi i}{n}\right) \cos\left(\frac{2 \pi j}{n}\right) \sin\left(\frac{\pi i}{n}\right)
$$

$$
+ \frac{\cos\left(\dfrac{\pi i}{n}\right) \cos\left(\dfrac{2 \pi j}{n}\right) \pi i}{n} - \cos\left(\frac{\pi i}{n}\right) \sin\left(\frac{\pi i}{n}\right) - \frac{\cos\left(\dfrac{\pi i}{n}\right) \pi i}{n}
$$

$$
\left. \left. \left. - \sin\left(\frac{\pi i}{n}\right) \cos\left(\frac{2 \pi j}{n}\right) - \frac{\cos\left(\dfrac{2 \pi j}{n}\right) \pi i}{n} + \sin\left(\frac{\pi i}{n}\right) + \frac{\pi i}{n} \right) k^2 \pi^2 \Big/ n^5 \right) \right) \right)
$$

(50)

Infinitize finite sums, its limit operational value is 'Triple Integrals Value at Manifold' (51):

(In the case of "Amount 'n' of spacial bounded closed region 'Ω's (Surface 'S' surrounds) segmental cells tends to infinity", limit value of integral values' sums [Product about Idiographic Divergence (48) and Volume Element] in all segmental cells)

$$
\lim_{n \to \infty} \sum_{k=1}^{n} \left(\sum_{j=1}^{n} \left(\sum_{i=1}^{n} \left(2 \left(\frac{k \sin\left(\dfrac{\pi i}{n}\right)\left(1 - \cos\left(\dfrac{2 \pi j}{n}\right)\right)}{n} + \frac{7}{6} \frac{k \pi i \left(1 - \cos\left(\dfrac{\pi i}{n}\right)\right)}{n^2} \right) \sin\left(\frac{\pi i}{n}\right) \right. \right. \right.
$$

$$
\left(\frac{2 \sin\left(\dfrac{\pi i}{n}\right) \sin\left(\dfrac{2 \pi j}{n}\right) \pi^2 i^2}{n^2} + \cos\left(\frac{\pi i}{n}\right) \cos\left(\frac{2 \pi j}{n}\right) \sin\left(\frac{\pi i}{n}\right) \right.
$$

$$
+ \frac{\cos\left(\dfrac{\pi i}{n}\right) \cos\left(\dfrac{2 \pi j}{n}\right) \pi i}{n} - \cos\left(\frac{\pi i}{n}\right) \sin\left(\frac{\pi i}{n}\right) - \frac{\cos\left(\dfrac{\pi i}{n}\right) \pi i}{n}
$$

$$
\left. \left. \left. \left. - \sin\left(\frac{\pi i}{n}\right) \cos\left(\frac{2 \pi j}{n}\right) - \frac{\cos\left(\dfrac{2 \pi j}{n}\right) \pi i}{n} + \sin\left(\frac{\pi i}{n}\right) + \frac{\pi i}{n} \right) k^2 \pi^2 \Big/ n^5 \right) \right) \right)
$$

$$
= -\frac{88}{81} \pi + \frac{829}{576} \pi^3
$$

(51)

Surface Integral Precision Value (26) [Vector Field (2) on Target Surface (1)], is equal to Volume Element Integral Precision Value (51) [Divergence (29) in Spacial Bounded Closed Region 'Ω' that Target Surface (1) surrounds], Complete Calculation and Validation of Divergence Theorem at Manifold (Finite Sums Limits)

5.2 Numerical Model of Divergence Theorem Finite Sums Limits at Manifold
[Program Template of Waterloo Maplesoft, Optional]

```
> restart;
> with(plots):with(linalg):
> CS:=[sin(u)*(1-cos(v)),sin(u)*sin(v)+2*u,u*(1-cos(u))];  (1)
# Define parametrized expression of discretional simply connected
orientable closed surface 'CS'
```

$$CS := [\sin(u)(1-\cos(v)),\ \sin(u)\sin(v)+2u,\ u(1-\cos(u))]$$

```
> rgu:=[0,Pi];
```

$$rgu := [0, \pi]$$

```
> rgv:=[0,2*Pi]; # Define range of 'u,v', make surface 'CS' closed
```

$$rgv := [0, 2\pi]$$

```
> plot3d(CS,u=rgu[1]..rgu[2],v=rgv[1]..rgv[2],scaling=constrained,
projection=0.9,numpoints=2000):g1:=%: # Draw surface 'CS'
> V:=[x^2/2,y*z/2,z^2/3];   (2)
# Define discretional spacial vector field 'V'
 (Suppose Vector Field 'V' possesses 1th order continuous partial
derivatives in spacial bounded closed region 'Ω' that Surface 'CS'
surrounds)
```

$$V := \left[\frac{x^2}{2}, \frac{yz}{2}, \frac{z^2}{3} \right]$$

```
> Diff(V[1],x)+Diff(V[2],y)+Diff(V[3],z)=diff(V[1],x)+diff(V[2],y)
+diff(V[3],z);diV:=rhs(%);  (3)
# Calculate divergence 'diV' of vector field 'V'
```

$$\left(\frac{d}{dx}\left(\frac{x^2}{2} \right) \right) + \left(\frac{\partial}{\partial y}\left(\frac{yz}{2} \right) \right) + \left(\frac{d}{dz}\left(\frac{z^2}{3} \right) \right) = x + \frac{7z}{6}$$

$$diV := x + \frac{7z}{6}$$

```
> rgx:=[-2*Pi/3,4*Pi/3];
```

$$rgx := \left[-\frac{2\pi}{3}, \frac{4\pi}{3} \right]$$

```
> rgy:=[0,2*Pi];
```

$$rgy := [0, 2\pi]$$

```
> rgz:=[0,2*Pi];
```

$$rgz := [0, 2\pi]$$

```
> fieldplot3d(V,x=rgx[1]..rgx[2],y=rgy[1]..rgy[2],z=rgz[1]..
rgz[2],arrows=SLIM,color=blue):g2:=%:
# Draw spacial vector field 'V'
> implicitplot3d(diV,x=rgx[1]..rgx[2],y=rgy[1]..rgy[2],z=rgz[1]..
rgz[2],style=wireframe,color=cyan,numpoints=2000):g3:=%:
# Draw level surface of divergence 'diV'
> display(g1,g2,g3): # Synthesize figures
```

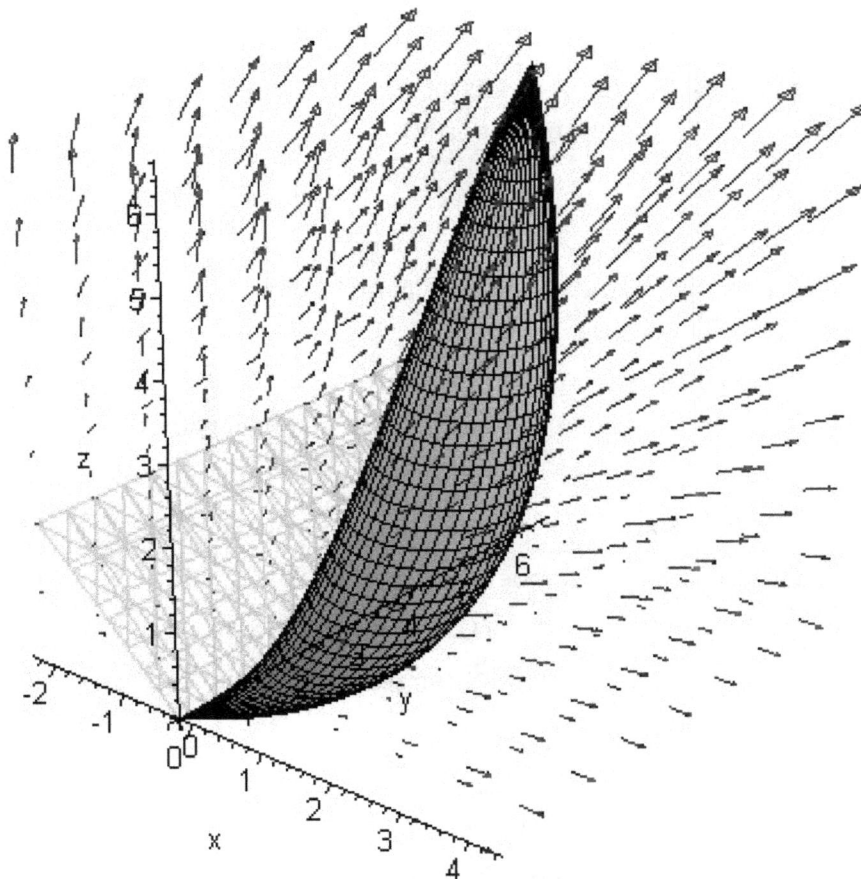

Figure(I)5.2 Surface 'CS'、Vector Field 'V' and
Level Surface of Divergence 'diV'

```
> x:=CS[1]:y:=CS[2]:z:=CS[3]:
> matrix(3,3,[i,j,k,Diff(x,u),Diff(y,u),Diff(z,u),Diff(x,v),
Diff(y,v),Diff(z,v)])=
matrix(3,3,[i,j,k,diff(x,u),diff(y,u),diff(z,u),diff(x,v),
```

```
diff(y,v),diff(z,v)]);m1:=rhs(%);
```

Define and calculate matrix of partial derivatives 'm1'; obtain 'tangent plane normal vector' of surface 'CS'

$$
\begin{bmatrix}
i & j & k \\
\dfrac{\partial}{\partial u}(\sin(u)(1-\cos(v))) & \dfrac{\partial}{\partial u}(\sin(u)\sin(v)+2\,u) & \dfrac{d}{du}(u(1-\cos(u))) \\
\dfrac{\partial}{\partial v}(\sin(u)(1-\cos(v))) & \dfrac{\partial}{\partial v}(\sin(u)\sin(v)+2\,u) & \dfrac{\partial}{\partial v}(u(1-\cos(u)))
\end{bmatrix} =
$$

$$
\begin{bmatrix}
i & j & k \\
\cos(u)(1-\cos(v)) & \cos(u)\sin(v)+2 & 1-\cos(u)+u\sin(u) \\
\sin(u)\sin(v) & \sin(u)\cos(v) & 0
\end{bmatrix}
$$

$$
m1 := \begin{bmatrix}
i & j & k \\
\cos(u)(1-\cos(v)) & \cos(u)\sin(v)+2 & 1-\cos(u)+u\sin(u) \\
\sin(u)\sin(v) & \sin(u)\cos(v) & 0
\end{bmatrix}
$$

```
> det(m1):
> mn:=simplify(%);
```

$$
mn := -\sin(u)(i\cos(v)-i\cos(v)\cos(u)+i\sin(u)\cos(v)u-\cos(u)k\cos(v) \\
-\sin(v)j+\sin(v)j\cos(u)-\sin(u)\sin(v)ju+k\cos(u)+2\sin(v)k)
$$

```
> A:=coeff(mn,i); # Obtain coefficient of 'i'
```

$$
A := -\sin(u)(\cos(v)-\cos(v)\cos(u)+\sin(u)\cos(v)u)
$$

```
> B:=coeff(mn,j); # Obtain coefficient of 'j'
```

$$
B := -\sin(u)(-\sin(v)+\cos(u)\sin(v)-\sin(u)\sin(v)u)
$$

```
> C:=coeff(mn,k); # Obtain coefficient of 'k'
```

$$
C := -\sin(u)(-\cos(v)\cos(u)+\cos(u)+2\sin(v))
$$

```
> [A,B,C]: # [A,B,C] structure 'tangent plane normal vector' (4)


// Segment range '[0,Pi],[0,2*Pi]' of 'u,v':
> dus:=50;
```

Set amount of surface 'CS's segmental cells, this amount can be discretional natural number

$$
dus := 50
$$

```
> du:=(rgu[2]-rgu[1])/dus; # Segment range of 'u'
```

$$
du := \frac{\pi}{50}
$$

```
> dv:=(rgv[2]-rgv[1])/dus; # Segment range of 'v'
```

$$
dv := \frac{\pi}{25}
$$

Microcosmic surface integral course of surface 'CS's first segmental cell (50 segmental cells):

// Segment 'tangent plane normal vector'

> dA:=subs(v=rgv[1]+dv,subs(u=rgu[1]+du,A));

$$dA := -\sin\left(\frac{\pi}{50}\right)\left(\cos\left(\frac{\pi}{25}\right) - \cos\left(\frac{\pi}{25}\right)\cos\left(\frac{\pi}{50}\right) + \frac{1}{50}\sin\left(\frac{\pi}{50}\right)\cos\left(\frac{\pi}{25}\right)\pi\right)$$

> dB:=subs(v=rgv[1]+dv,subs(u=rgu[1]+du,B));

$$dB := -\sin\left(\frac{\pi}{50}\right)\left(-\sin\left(\frac{\pi}{25}\right) + \cos\left(\frac{\pi}{50}\right)\sin\left(\frac{\pi}{25}\right) - \frac{1}{50}\sin\left(\frac{\pi}{50}\right)\sin\left(\frac{\pi}{25}\right)\pi\right)$$

> dC:=subs(v=rgv[1]+dv,subs(u=rgu[1]+du,C));

$$dC := -\sin\left(\frac{\pi}{50}\right)\left(-\cos\left(\frac{\pi}{25}\right)\cos\left(\frac{\pi}{50}\right) + \cos\left(\frac{\pi}{50}\right) + 2\sin\left(\frac{\pi}{25}\right)\right)$$

> [dA,dB,dC]: #'[dA,dB,dC]' is 'tangent plane normal vector's average value that corresponds to surface 'CS's first segmental cell

// Segment spacial vector field 'V':

(This step possesses formal meaning in 'proof' only, and possesses essential meaning in 'Numerical Model')

> dV:=subs(v=rgv[1]+dv,subs(u=rgu[1]+du,V));

Spacial vector field 'V's average value that corresponds to surface 'CS's first segmental cell

$$dV := \left[\frac{1}{2}\sin\left(\frac{\pi}{50}\right)^2\left(1 - \cos\left(\frac{\pi}{25}\right)\right)^2, \frac{1}{100}\left(\sin\left(\frac{\pi}{50}\right)\sin\left(\frac{\pi}{25}\right) + \frac{\pi}{25}\right)\pi\left(1 - \cos\left(\frac{\pi}{50}\right)\right),\right.$$
$$\left. \frac{1}{7500}\pi^2\left(1 - \cos\left(\frac{\pi}{50}\right)\right)^2\right]$$

// Calculate microcosmic surface integral value of surface 'CS's first segmental cell:

> (dV[1]*dA+dV[2]*dB+dV[3]*dC)*du*dv;

Base on integral median theorem, integral expression of dot product (Spacial vector field 'V' and '[dA,dB,dC]') at first segmental cell

$$\frac{1}{1250}\left(-\frac{1}{2}\right.$$
$$\left. \sin\left(\frac{\pi}{50}\right)^3\left(1 - \cos\left(\frac{\pi}{25}\right)\right)^2\left(\cos\left(\frac{\pi}{25}\right) - \cos\left(\frac{\pi}{25}\right)\cos\left(\frac{\pi}{50}\right) + \frac{1}{50}\sin\left(\frac{\pi}{50}\right)\cos\left(\frac{\pi}{25}\right)\pi\right)\right.$$

$$-\frac{1}{100}\left(\sin\left(\frac{\pi}{50}\right)\sin\left(\frac{\pi}{25}\right)+\frac{\pi}{25}\right)\pi\left(1-\cos\left(\frac{\pi}{50}\right)\right)\sin\left(\frac{\pi}{50}\right)$$

$$\left(-\sin\left(\frac{\pi}{25}\right)+\cos\left(\frac{\pi}{50}\right)\sin\left(\frac{\pi}{25}\right)-\frac{1}{50}\sin\left(\frac{\pi}{50}\right)\sin\left(\frac{\pi}{25}\right)\pi\right)$$

$$-\frac{1}{7500}\pi^2\left(1-\cos\left(\frac{\pi}{50}\right)\right)^2\sin\left(\frac{\pi}{50}\right)\left(-\cos\left(\frac{\pi}{25}\right)\cos\left(\frac{\pi}{50}\right)+\cos\left(\frac{\pi}{50}\right)+2\sin\left(\frac{\pi}{25}\right)\right)\right)\right)$$

$$\pi^2$$

Microcosmic surface integral courses of surface 'CS's all segmental cells (50 segmental cells):

```
// Segment 'tangent plane normal vector'
> stdA:=subs(v=rgv[1]+t*dv,subs(u=rgu[1]+s*du,A));
```

$$stdA:=-\sin\left(\frac{s\,\pi}{50}\right)\left(\cos\left(\frac{t\,\pi}{25}\right)-\cos\left(\frac{t\,\pi}{25}\right)\cos\left(\frac{s\,\pi}{50}\right)+\frac{1}{50}\sin\left(\frac{s\,\pi}{50}\right)\cos\left(\frac{t\,\pi}{25}\right)s\,\pi\right)$$

```
> stdB:=subs(v=rgv[1]+t*dv,subs(u=rgu[1]+s*du,B));
```

$$stdB:=-\sin\left(\frac{s\,\pi}{50}\right)\left(-\sin\left(\frac{t\,\pi}{25}\right)+\cos\left(\frac{s\,\pi}{50}\right)\sin\left(\frac{t\,\pi}{25}\right)-\frac{1}{50}\sin\left(\frac{s\,\pi}{50}\right)\sin\left(\frac{t\,\pi}{25}\right)s\,\pi\right)$$

```
> stdC:=subs(v=rgv[1]+t*dv,subs(u=rgu[1]+s*du,C));
```

$$stdC:=-\sin\left(\frac{s\,\pi}{50}\right)\left(-\cos\left(\frac{t\,\pi}{25}\right)\cos\left(\frac{s\,\pi}{50}\right)+\cos\left(\frac{s\,\pi}{50}\right)+2\sin\left(\frac{t\,\pi}{25}\right)\right)$$

```
> [stdA,stdB,stdC]: #'[stdA,stdB,stdC]' are 'tangent plane normal
```
vector's average values that correspond to surface 'CS's all segmental cells
//'[stdA,stdB,stdC]' isn't that 'Single vector value', but that 'Set of finite quantitative (Viz.50) vector values'


```
// Segment spacial vector field 'V'
```
(This step possesses formal meaning in 'Proof' only, and possesses essential meaning in 'Numerical Model')
```
> stdV:=subs(v=rgv[1]+t*dv,subs(u=rgu[1]+s*du,V));
```
Spacial vector field 'V's average values that correspond to surface'CS's all segmental cells

$$stdV:=\left[\frac{1}{2}\sin\left(\frac{s\,\pi}{50}\right)^2\left(1-\cos\left(\frac{t\,\pi}{25}\right)\right)^2,\right.$$

$$\left.\frac{1}{100}\left(\sin\left(\frac{s\,\pi}{50}\right)\sin\left(\frac{t\,\pi}{25}\right)+\frac{s\,\pi}{25}\right)s\,\pi\left(1-\cos\left(\frac{s\,\pi}{50}\right)\right),\frac{1}{7500}s^2\,\pi^2\left(1-\cos\left(\frac{s\,\pi}{50}\right)\right)^2\right]$$

//'stdV' isn't that 'Single vector value', but that 'Set of finite quantitative (Viz.50) vector values'

```
// Calculate  microcosmosic surface integral value (All segmental
cells of surface 'CS')
> (stdV[1]*stdA+stdV[2]*stdB+stdV[3]*stdC)*du*dv;
# Base on integral median theorem, integral expression of dot product
(Spacial vector field 'V' and '[stdA,stdB,stdC]') at all segmental
cells
```

$$\frac{1}{1250}\left(-\frac{1}{2}\sin\left(\frac{s\,\pi}{50}\right)^3\left(1-\cos\left(\frac{t\,\pi}{25}\right)\right)^2\right.$$

$$\left(\cos\left(\frac{t\,\pi}{25}\right)-\cos\left(\frac{t\,\pi}{25}\right)\cos\left(\frac{s\,\pi}{50}\right)+\frac{1}{50}\sin\left(\frac{s\,\pi}{50}\right)\cos\left(\frac{t\,\pi}{25}\right)s\,\pi\right)-\frac{1}{100}$$

$$\left(\sin\left(\frac{s\,\pi}{50}\right)\sin\left(\frac{t\,\pi}{25}\right)+\frac{s\,\pi}{25}\,s\,\pi\left(1-\cos\left(\frac{s\,\pi}{50}\right)\right)\sin\left(\frac{s\,\pi}{50}\right)\right)$$

$$\left(-\sin\left(\frac{t\,\pi}{25}\right)+\cos\left(\frac{s\,\pi}{50}\right)\sin\left(\frac{t\,\pi}{25}\right)-\frac{1}{50}\sin\left(\frac{s\,\pi}{50}\right)\sin\left(\frac{t\,\pi}{25}\right)s\,\pi\right)-\frac{1}{7500}$$

$$\left.s^2\,\pi^2\left(1-\cos\left(\frac{s\,\pi}{50}\right)\right)^2\sin\left(\frac{s\,\pi}{50}\right)\left(-\cos\left(\frac{t\,\pi}{25}\right)\cos\left(\frac{s\,\pi}{50}\right)+\cos\left(\frac{s\,\pi}{50}\right)+2\sin\left(\frac{t\,\pi}{25}\right)\right)\right)\pi^2$$

```
//Expression above-mentioned isn't that 'Single value', but that 'Set
of finite quantitative (Viz.50) values'


// List of finite quantitative microcosmic surface integral values:
> seq(seq([s*t,(stdV[1]*stdA+stdV[2]*stdB+stdV[3]*stdC)*du*dv],
s=1..dus),t=1..dus):
# List of integral values of dot product(Spacial vector field 'V'
and '[stdA,stdB,stdC]')at all segmental cells, be elided
// Structure sequence that is composed of finite quantitative
microcomosic surface integral values:
> sqn:=seq(seq((stdV[1]*stdA+stdV[2]*stdB+stdV[3]*stdC)*du*dv,
s=1..dus),t=1..dus):
// Accumulational sum of sequence, obtain surface integral value at
manifold:
> add(k,k=sqn):xi:=evalf(%);   (5)
# Integral values' sum of dot product (Spacial vector field 'V' and
'[stdA,stdB,stdC]') at all segmental cells (Verbose analytical
expression is elided)
```

$$\xi := 41.18505568$$

```
// Renewedly segment range '[0,Pi],[0,2*Pi]' of 'u,v':
> dus:=n;
# Set amount of surface 'CS's segmental cells as natural number 'n'
```

$$dus := n$$

```
> du:=(rgu[2]-rgu[1])/dus; # Segment range of 'u'
```

$$du := \frac{\pi}{n}$$

```
> dv:=(rgv[2]-rgv[1])/dus; # Segment range of 'v'
```

$$dv := \frac{2\,\pi}{n}$$

Microcosmic surface integral course of surface 'CS's first segmental cell (n segmental cells):

```
// Segment 'tangent plane normal vector'
> dA:=subs(v=rgv[1]+dv,subs(u=rgu[1]+du,A));
```

$$dA := -\sin\!\left(\frac{\pi}{n}\right)\!\left(\cos\!\left(\frac{2\,\pi}{n}\right) - \cos\!\left(\frac{2\,\pi}{n}\right)\cos\!\left(\frac{\pi}{n}\right) + \frac{\sin\!\left(\frac{\pi}{n}\right)\cos\!\left(\frac{2\,\pi}{n}\right)\pi}{n}\right)$$

```
> dB:=subs(v=rgv[1]+dv,subs(u=rgu[1]+du,B));
```

$$dB := -\sin\!\left(\frac{\pi}{n}\right)\!\left(-\sin\!\left(\frac{2\,\pi}{n}\right) + \cos\!\left(\frac{\pi}{n}\right)\sin\!\left(\frac{2\,\pi}{n}\right) - \frac{\sin\!\left(\frac{\pi}{n}\right)\sin\!\left(\frac{2\,\pi}{n}\right)\pi}{n}\right)$$

```
> dC:=subs(v=rgv[1]+dv,subs(u=rgu[1]+du,C));
```

$$dC := -\sin\!\left(\frac{\pi}{n}\right)\!\left(-\cos\!\left(\frac{2\,\pi}{n}\right)\cos\!\left(\frac{\pi}{n}\right) + \cos\!\left(\frac{\pi}{n}\right) + 2\,\sin\!\left(\frac{2\,\pi}{n}\right)\right)$$

```
> [dA,dB,dC]: #'[dA,dB,dC]' is 'tangent plane normal vector's average
value that corresponds to surface 'CS's first segmental cell
```

```
// Segment spacial vector field 'V':
(This step possesses formal meaning in 'proof' only, and possesses
essential meaning in 'Numerical Model')
> dV:=subs(v=rgv[1]+dv,subs(u=rgu[1]+du,V));
# Spacial vector field 'V's average value that corresponds to
surface 'CS's first segmental cell
```

$$dV := \left[\frac{1}{2}\sin\!\left(\frac{\pi}{n}\right)^{2}\!\left(1 - \cos\!\left(\frac{2\,\pi}{n}\right)\right)^{2}, \frac{1}{2}\frac{\left(\sin\!\left(\frac{\pi}{n}\right)\sin\!\left(\frac{2\,\pi}{n}\right) + \frac{2\,\pi}{n}\right)\pi\!\left(1 - \cos\!\left(\frac{\pi}{n}\right)\right)}{n},\right.$$

$$\left. \frac{1}{3} \frac{\pi^2 \left(1 - \cos\left(\frac{\pi}{n}\right)\right)^2}{n^2} \right]$$

```
// Calculate microcosmic surface integral value of surface 'CS's
first segmental cell:
> (dV[1]*dA+dV[2]*dB+dV[3]*dC)*du*dv;
# Base on integral median theorem, integral value of dot product
(Spacial vector field 'V' and '[dA,dB,dC]') at first segmental cell
```

$$2\left(-\frac{1}{2}\sin\left(\frac{\pi}{n}\right)^3 \left(1-\cos\left(\frac{2\pi}{n}\right)\right)^2 \left(\cos\left(\frac{2\pi}{n}\right) - \cos\left(\frac{2\pi}{n}\right)\cos\left(\frac{\pi}{n}\right) + \frac{\sin\left(\frac{\pi}{n}\right)\cos\left(\frac{2\pi}{n}\right)\pi}{n} \right) \right.$$

$$-\frac{1}{2}\left(\sin\left(\frac{\pi}{n}\right)\sin\left(\frac{2\pi}{n}\right) + \frac{2\pi}{n} \right)\pi\left(1 - \cos\left(\frac{\pi}{n}\right)\right)\sin\left(\frac{\pi}{n}\right)$$

$$\left(-\sin\left(\frac{2\pi}{n}\right) + \cos\left(\frac{\pi}{n}\right)\sin\left(\frac{2\pi}{n}\right) - \frac{\sin\left(\frac{\pi}{n}\right)\sin\left(\frac{2\pi}{n}\right)\pi}{n} \right) /n$$

$$\left. -\frac{1}{3}\frac{\pi^2\left(1-\cos\left(\frac{\pi}{n}\right)\right)^2 \sin\left(\frac{\pi}{n}\right)\left(-\cos\left(\frac{2\pi}{n}\right)\cos\left(\frac{\pi}{n}\right) + \cos\left(\frac{\pi}{n}\right) + 2\sin\left(\frac{2\pi}{n}\right)\right)}{n^2}\right)\pi^2 /$$

$$n^2$$

Microcosmic surface integral course of surface 'CS's all segmental cells (n segmental cells):

```
// Segment 'tangent plane normal vector'
> stdA:=subs(v=rgv[1]+t*dv,subs(u=rgu[1]+s*du,A));
```

$$stdA := -\sin\left(\frac{s\pi}{n}\right)\left(\cos\left(\frac{2t\pi}{n}\right) - \cos\left(\frac{2t\pi}{n}\right)\cos\left(\frac{s\pi}{n}\right) + \frac{\sin\left(\frac{s\pi}{n}\right)\cos\left(\frac{2t\pi}{n}\right)s\pi}{n} \right)$$

```
> stdB:=subs(v=rgv[1]+t*dv,subs(u=rgu[1]+s*du,B));
```

$$stdB := -\sin\left(\frac{s\pi}{n}\right)\left(-\sin\left(\frac{2t\pi}{n}\right) + \cos\left(\frac{s\pi}{n}\right)\sin\left(\frac{2t\pi}{n}\right) - \frac{\sin\left(\frac{s\pi}{n}\right)\sin\left(\frac{2t\pi}{n}\right)s\pi}{n} \right)$$

```
> stdC:=subs(v=rgv[1]+t*dv,subs(u=rgu[1]+s*du,C));
```

$$stdC := -\sin\left(\frac{s\pi}{n}\right)\left(-\cos\left(\frac{2t\pi}{n}\right)\cos\left(\frac{s\pi}{n}\right) + \cos\left(\frac{s\pi}{n}\right) + 2\sin\left(\frac{2t\pi}{n}\right) \right)$$

```
> [stdA,stdB,stdC]: #'[stdA,stdB,stdC]' are 'tangent plane normal
vector's average values that correspond to surface 'CS's all segmental
cells
```

//`[stdA,stdB,stdC]' isn't that `Single vector value', but that `Set
of finite quantitative (Viz.n) vector values'

// Segment spacial vector field `V'
(This step possesses formal meaning in `Proof' only, and possesses
essential meaning in `Numerical Model')
> stdV:=subs(v=rgv[1]+t*dv,subs(u=rgu[1]+s*du,V));
Spacial vector field `V's average values that correspond to
surface`CS's all segmental cells

$$stdV := \left[\frac{1}{2} \sin\left(\frac{s\,\pi}{n}\right)^2 \left(1 - \cos\left(\frac{2\,t\,\pi}{n}\right)\right)^2, \right.$$

$$\frac{1}{2} \frac{\left(\sin\left(\frac{s\,\pi}{n}\right)\sin\left(\frac{2\,t\,\pi}{n}\right) + \frac{2\,s\,\pi}{n}\right) s\,\pi \left(1 - \cos\left(\frac{s\,\pi}{n}\right)\right)}{n}, \left. \frac{1}{3} \frac{s^2\,\pi^2 \left(1 - \cos\left(\frac{s\,\pi}{n}\right)\right)^2}{n^2} \right]$$

//`stdV' isn't that `Single vector value', but that `Set of finite
quantitative (Viz.n) vector values'

// Calculate microcosmosic surface integral values (All segmental
cells of surface `CS'):
> (stdV[1]*stdA+stdV[2]*stdB+stdV[3]*stdC)*du*dv;
Base on integral median theorem, integral values of dot product
(Spacial vector field `V' and `[stdA,stdB,stdC]') at all segmental
cells

$$2\left(-\frac{1}{2}\sin\left(\frac{s\,\pi}{n}\right)^3 \left(1 - \cos\left(\frac{2\,t\,\pi}{n}\right)\right)^2 \right.$$

$$\left(\cos\left(\frac{2\,t\,\pi}{n}\right) - \cos\left(\frac{2\,t\,\pi}{n}\right)\cos\left(\frac{s\,\pi}{n}\right) + \frac{\sin\left(\frac{s\,\pi}{n}\right)\cos\left(\frac{2\,t\,\pi}{n}\right)s\,\pi}{n}\right) - \frac{1}{2}$$

$$\left(\sin\left(\frac{s\,\pi}{n}\right)\sin\left(\frac{2\,t\,\pi}{n}\right) + \frac{2\,s\,\pi}{n}\right) s\,\pi \left(1 - \cos\left(\frac{s\,\pi}{n}\right)\right)\sin\left(\frac{s\,\pi}{n}\right)$$

$$\left. \left(-\sin\left(\frac{2\,t\,\pi}{n}\right) + \cos\left(\frac{s\,\pi}{n}\right)\sin\left(\frac{2\,t\,\pi}{n}\right) - \frac{\sin\left(\frac{s\,\pi}{n}\right)\sin\left(\frac{2\,t\,\pi}{n}\right)s\,\pi}{n}\right)\right/n - \frac{1}{3}$$

$$\frac{s^2 \pi^2 \left(1 - \cos\left(\frac{s\pi}{n}\right)\right)^2 \sin\left(\frac{s\pi}{n}\right)\left(-\cos\left(\frac{2t\pi}{n}\right)\cos\left(\frac{s\pi}{n}\right) + \cos\left(\frac{s\pi}{n}\right) + 2\sin\left(\frac{2t\pi}{n}\right)\right)}{n^2}$$

$$\Bigg) \; \pi^2 \, / \, n^2$$

//Expression above-mentioned isn't that 'Single value', but that 'Set of finite quantitative (Viz. n) values'

// Structure finite sums:

```
> Sum(Sum((stdV[1]*stdA+stdV[2]*stdB+stdV[3]*stdC)*du*dv,
s=1..dus),t=1..dus);
```

\# In the case of "Amount 'n' of surface 'CS's segmental cells is pending", sum of integral values (dot product about Spacial vector field 'V' and '[stdA,stdB,stdC]'at all segmental cells

$$\sum_{t=1}^{n}\left(\sum_{s=1}^{n}\left(2\left(-\frac{1}{2}\sin\left(\frac{s\pi}{n}\right)^3\left(1-\cos\left(\frac{2t\pi}{n}\right)\right)^2\right.\right.\right.$$

$$\left(\cos\left(\frac{2t\pi}{n}\right)-\cos\left(\frac{2t\pi}{n}\right)\cos\left(\frac{s\pi}{n}\right)+\frac{\sin\left(\frac{s\pi}{n}\right)\cos\left(\frac{2t\pi}{n}\right)s\pi}{n}\right)-\frac{1}{2}$$

$$\left(\sin\left(\frac{s\pi}{n}\right)\sin\left(\frac{2t\pi}{n}\right)+\frac{2s\pi}{n}\right)s\pi\left(1-\cos\left(\frac{s\pi}{n}\right)\right)\sin\left(\frac{s\pi}{n}\right)$$

$$\left(-\sin\left(\frac{2t\pi}{n}\right)+\cos\left(\frac{s\pi}{n}\right)\sin\left(\frac{2t\pi}{n}\right)-\frac{\sin\left(\frac{s\pi}{n}\right)\sin\left(\frac{2t\pi}{n}\right)s\pi}{n}\right)/n-\frac{1}{3}$$

$$\frac{s^2\pi^2\left(1-\cos\left(\frac{s\pi}{n}\right)\right)^2\sin\left(\frac{s\pi}{n}\right)\left(-\cos\left(\frac{2t\pi}{n}\right)\cos\left(\frac{s\pi}{n}\right)+\cos\left(\frac{s\pi}{n}\right)+2\sin\left(\frac{2t\pi}{n}\right)\right)}{n^2}$$

$$\Bigg)\;\pi^2\,/\,n^2\Bigg)\Bigg)$$

```
> vs:=value(%):
> Limit(vs,n=infinity):   (6)
```

// Infinitize finite sums, its limit operational value is 'Surface Integral Value at Manifold'

```
# In the case of "Amount 'n' of surface 'CS's segmental cells tends
to infinity", limit value of integral values' sums (dot product about
spacial vector field 'V' and '[stdA,stdB,stdC]') at all segmental cells
> alpha:=value(%);delta:=evalf(alpha);
```

$$\alpha := -\frac{88}{81}\pi + \frac{829}{576}\pi^3$$
$$\delta := 41.21226476$$

```
> xi;delta;
# There is tiny difference between 'Integral values' sum of finite
quantitative segmental cells' and 'Integral values' sum of infinite
quantitative segmental cells'
```

$$41.18505568$$
$$41.21226476$$

```
> x:='x':y:='y':z:='z':
> [r*CS[1],r*CS[2],r*CS[3]];
# Entirely multiply radius vector 'r' (suppose r>0) by each item
of surface 'CS' parametrized expression, convert "Parametrized
expression of surface 'CS'" to "Parametrized expression of surface
'CS' coordinates"
```

$$[r\sin(u)(1-\cos(v)), r(\sin(u)\sin(v)+2u), ru(1-\cos(u))] \quad (7)$$

```
> x:=r*CS[1]:y:=r*CS[2]:z:=r*CS[3]:
> matrix(3,3,[Diff(x,r),Diff(x,u),Diff(x,v),Diff(y,r),Diff(y,u),
Diff(y,v),Diff(z,r),Diff(z,u),Diff(z,v)])=
matrix(3,3,[diff(x,r),diff(x,u),diff(x,v),diff(y,r),diff(y,u),
diff(y,v),diff(z,r),diff(z,u),diff(z,v)]);m2:=rhs(%);
# Define and calculate matrix of partial derivatives 'M2', obtain
general expression of surface 'CS' coordinates volume element
coefficient
```

$$\left[\frac{\partial}{\partial r}(r\sin(u)(1-\cos(v))), \frac{\partial}{\partial u}(r\sin(u)(1-\cos(v))),\right.$$
$$\left.\frac{\partial}{\partial v}(r\sin(u)(1-\cos(v)))\right]$$
$$\left[\frac{\partial}{\partial r}(r(\sin(u)\sin(v)+2u)), \frac{\partial}{\partial u}(r(\sin(u)\sin(v)+2u)),\right.$$
$$\left.\frac{\partial}{\partial v}(r(\sin(u)\sin(v)+2u))\right]$$

$$\left[\frac{\partial}{\partial r}\left(r\,u\,(1-\cos(u))\right),\frac{\partial}{\partial u}\left(r\,u\,(1-\cos(u))\right),\frac{\partial}{\partial v}\left(r\,u\,(1-\cos(u))\right)\right]=$$

$$\begin{bmatrix} \sin(u)\,(1-\cos(v)) & r\cos(u)\,(1-\cos(v)) & r\sin(u)\sin(v) \\ \sin(u)\sin(v)+2\,u & r\,(\cos(u)\sin(v)+2) & r\sin(u)\cos(v) \\ u\,(1-\cos(u)) & r\,(1-\cos(u))+r\,u\sin(u) & 0 \end{bmatrix}$$

$$m2:=\begin{bmatrix} \sin(u)\,(1-\cos(v)) & r\cos(u)\,(1-\cos(v)) & r\sin(u)\sin(v) \\ \sin(u)\sin(v)+2\,u & r\,(\cos(u)\sin(v)+2) & r\sin(u)\cos(v) \\ u\,(1-\cos(u)) & r\,(1-\cos(u))+r\,u\sin(u) & 0 \end{bmatrix}$$

```
> det(m2):

> J:=simplify(%);   (8)
```

General expression of surface 'CS' coordinates volume element coefficient

$$J:=\sin(u)\,(u+\sin(u)\cos(v)\cos(u)+2\,\sin(u)\sin(v)\,u^2+u\cos(u)\cos(v)$$
$$-\sin(u)\cos(u)-\cos(v)\,u-u\cos(u)-\sin(u)\cos(v)+\sin(u))\,r^2$$

```
// Segment range '[0,1],[0,Pi],[0,2*Pi]' of 'r,u,v':

> dus:=20;
```

Set amount of spacial bounded closed region 'Ω's (Surface 'CS' surrounds) segmental cells, this amount can be discretional natural number

$$dus:=20$$

```
> rgr:=[0,1];
```

// Be different to ordinary triple integrals, in this position, first integral range is 'r[0,1]' always, and not 'r[0,n] or r[0, ∞]'; Because of '1', it is possible to hold a correct ratio between 'r,u,v'

$$rgr:=[0,1]$$

```
> dr:=(rgr[2]-rgr[1])/dus; # Segment range of 'r'
```

$$dr:=\frac{1}{20}$$

```
> du:=(rgu[2]-rgu[1])/dus; # Segment range of 'u'
```

$$du:=\frac{\pi}{20}$$

```
> dv:=(rgv[2]-rgv[1])/dus; # Segment range of 'v'
```

$$dv:=\frac{\pi}{10}$$

Microcosmic triple integrals course of spacial bounded closed

region 'Ω's first segmental cell (20 segmental cells):

// Segment volume element coefficient 'J':

```
> dJ:=subs(r=rgr[1]+dr,subs(v=rgv[1]+dv,subs(u=rgu[1]+du,J)));
```

Volume element coefficient 'J's average value that corresponds to spacial bounded closed region 'Ω's first segmental cell

$$dJ := \frac{1}{400} \sin\left(\frac{\pi}{20}\right)\left(\frac{\pi}{20} + \sin\left(\frac{\pi}{20}\right)\cos\left(\frac{\pi}{10}\right)\cos\left(\frac{\pi}{20}\right) + \frac{1}{200} \sin\left(\frac{\pi}{20}\right)\sin\left(\frac{\pi}{10}\right)\pi^2$$
$$+ \frac{1}{20}\pi\cos\left(\frac{\pi}{20}\right)\cos\left(\frac{\pi}{10}\right) - \sin\left(\frac{\pi}{20}\right)\cos\left(\frac{\pi}{20}\right) - \frac{1}{20}\cos\left(\frac{\pi}{10}\right)\pi - \frac{1}{20}\pi\cos\left(\frac{\pi}{20}\right)$$
$$- \sin\left(\frac{\pi}{20}\right)\cos\left(\frac{\pi}{10}\right) + \sin\left(\frac{\pi}{20}\right)\right)$$

// Segment idiographic divergence 'diV':

// Be different to abstract divergence 'diV1,diV2' that 'Proof' refers to, in 'Numerical Model, be able to directly input idiographic segmented values of 'r,u,v' to idiographic divergence 'diV', then segment 'diV' in accordance with amount 'dus' of parameter segmental cells

```
> ddiV:=subs(r=rgr[1]+dr,subs(v=rgv[1]+dv,subs(u=rgu[1]+du,diV)));
```

Divergence 'diV's average value that corresponds to spacial bounded closed region 'Ω's first segmental cell

$$ddiV := \frac{1}{20}\sin\left(\frac{\pi}{20}\right)\left(1 - \cos\left(\frac{\pi}{10}\right)\right) + \frac{7}{2400}\pi\left(1 - \cos\left(\frac{\pi}{20}\right)\right)$$

// Calculate microcosmic triple integrals value of spacial bounded closed region 'Ω's first segmental cell:

```
> ddiV*dJ*dr*du*dv;
```

Base on integral median theorem, integral value of product (Divergence 'diV' and volume element) in first segmental cell

$$\frac{1}{1600000}\left(\frac{1}{20}\sin\left(\frac{\pi}{20}\right)\left(1 - \cos\left(\frac{\pi}{10}\right)\right) + \frac{7}{2400}\pi\left(1 - \cos\left(\frac{\pi}{20}\right)\right)\right)\sin\left(\frac{\pi}{20}\right)\left(\frac{\pi}{20}\right.$$
$$+ \sin\left(\frac{\pi}{20}\right)\cos\left(\frac{\pi}{10}\right)\cos\left(\frac{\pi}{20}\right) + \frac{1}{200}\sin\left(\frac{\pi}{20}\right)\sin\left(\frac{\pi}{10}\right)\pi^2 + \frac{1}{20}\pi\cos\left(\frac{\pi}{20}\right)\cos\left(\frac{\pi}{10}\right)$$
$$- \sin\left(\frac{\pi}{20}\right)\cos\left(\frac{\pi}{20}\right) - \frac{1}{20}\cos\left(\frac{\pi}{10}\right)\pi - \frac{1}{20}\pi\cos\left(\frac{\pi}{20}\right) - \sin\left(\frac{\pi}{20}\right)\cos\left(\frac{\pi}{10}\right)$$
$$+ \sin\left(\frac{\pi}{20}\right)\right)\pi^2$$

Microcosmic triple integrals course of spacial bounded closed region'Ω's all segmental cells (20 segmental cells):

// Segment volume element coefficient 'J':

```
> ijkdJ:=subs(r=rgr[1]+k*dr,subs(v=rgv[1]+j*dv,subs(u=rgu[1]
+i*du,J)));
```

Volume element coefficient 'J's average values that correspond to spacial bounded closed region 'Ω's all segmental cells

$$ijkdJ := \frac{1}{400}\sin\left(\frac{i\,\pi}{20}\right)\left(\frac{i\,\pi}{20} + \sin\left(\frac{i\,\pi}{20}\right)\cos\left(\frac{j\,\pi}{10}\right)\cos\left(\frac{i\,\pi}{20}\right) + \frac{1}{200}\sin\left(\frac{i\,\pi}{20}\right)\sin\left(\frac{j\,\pi}{10}\right)i^2\,\pi^2\right.$$

$$+ \frac{1}{20}i\,\pi\cos\left(\frac{i\,\pi}{20}\right)\cos\left(\frac{j\,\pi}{10}\right) - \sin\left(\frac{i\,\pi}{20}\right)\cos\left(\frac{i\,\pi}{20}\right) - \frac{1}{20}\cos\left(\frac{j\,\pi}{10}\right)i\,\pi$$

$$- \frac{1}{20}i\,\pi\cos\left(\frac{i\,\pi}{20}\right) - \sin\left(\frac{i\,\pi}{20}\right)\cos\left(\frac{j\,\pi}{10}\right) + \left.\sin\left(\frac{i\,\pi}{20}\right)\right)k^2$$

//'ijkdJ' isn't that 'Single value', but that 'Set of finite quantitative (Viz.20) values'

// Segment idiographic divergence 'diV':

// Be different to abstract divergence 'diV1,diV2' that 'Proof' refers to, in 'Numerical Model', be able to directly input idiographic segmented values of 'r,u,v' to idiographic divergence 'diV', then segment 'diV' in accordance with amount 'dus' of parameter segmental cells

```
> ijkddiV:=subs(r=rgr[1]+k*dr,subs(v=rgv[1]+j*dv,subs(u=rgu[1]
+i*du,diV)));
```

Divergence 'diV's average values that correspond to spacial bounded closed region 'Ω's all segmental cells

$$ijkddiV := \frac{1}{20}k\sin\left(\frac{i\,\pi}{20}\right)\left(1 - \cos\left(\frac{j\,\pi}{10}\right)\right) + \frac{7}{2400}k\,i\,\pi\left(1 - \cos\left(\frac{i\,\pi}{20}\right)\right)$$

//'ijkddiV' isn't that 'Single value', but that 'Set of finite quantitative (Viz.20) values'

// Calculate microcosmic triple integrals values of spacial bounded closed region 'Ω's all segmental cells:

```
> ijkddiV*ijkdJ*dr*du*dv;
```

Base on integral median theorem, integral value of product (Divergence 'diV' and volume element) in all segmental cells

$$\frac{1}{1600000}\left(\frac{1}{20}k\sin\left(\frac{i\,\pi}{20}\right)\left(1 - \cos\left(\frac{j\,\pi}{10}\right)\right) + \frac{7}{2400}k\,i\,\pi\left(1 - \cos\left(\frac{i\,\pi}{20}\right)\right)\right)\sin\left(\frac{i\,\pi}{20}\right)\left(\frac{i\,\pi}{20}\right.$$

$$+ \sin\left(\frac{i\,\pi}{20}\right)\cos\left(\frac{j\,\pi}{10}\right)\cos\left(\frac{i\,\pi}{20}\right) + \frac{1}{200}\sin\left(\frac{i\,\pi}{20}\right)\sin\left(\frac{j\,\pi}{10}\right)i^2\,\pi^2$$

$$+ \frac{1}{20} i\,\pi \cos\left(\frac{i\,\pi}{20}\right) \cos\left(\frac{j\,\pi}{10}\right) - \sin\left(\frac{i\,\pi}{20}\right) \cos\left(\frac{i\,\pi}{20}\right) - \frac{1}{20} \cos\left(\frac{j\,\pi}{10}\right) i\,\pi$$

$$- \frac{1}{20} i\,\pi \cos\left(\frac{i\,\pi}{20}\right) - \sin\left(\frac{i\,\pi}{20}\right) \cos\left(\frac{j\,\pi}{10}\right) + \sin\left(\frac{i\,\pi}{20}\right)\Bigg) k^2\,\pi^2$$

```
//Expression above-mentioned isn't that 'Single value', but that 'Set
of finite quantitative (Viz.20) values
```

```
// List of finite quantitative microcosmic triple integrals values:
> seq(seq(seq([i*j*k,ijkddiV*ijkdJ*dr*du*dv],i=1..dus),j=1..dus),
k=1..dus):
# List of integral values of product (Divergence 'diV' and volume
element) in all segmental cells, be elided
// Structure sequence that is composed of finite quantitative
microcomosic triple integrals values:
> sqn:=seq(seq(seq(ijkddiV*ijkdJ*dr*du*dv,i=1..dus),j=1..dus),
k=1..dus):
// Accumulational sum of sequence, obtain triple integrals value at
manifold (Entire spacial bounded closed region 'Ω'):
> add(k,k=sqn):omega:=evalf(%);  (9)
# Integral values' sum of product (Divergence 'diV' and volume element)
in all segmental cells; Elide verbose analytic expression
```

$$\omega := 45.27224393$$

```
// Renewedly Segment range '[0,1],[0,Pi],[0,2*Pi]' of 'r,u,v':
> dus:=n;
# Set amount of spacial bounded closed region 'Ω's (Surface 'CS'
surrounds) segmental cells as natural number 'n'
```

$$dus := n$$

```
> dr:=(rgr[2]-rgr[1])/dus; # Segment range of 'r'
```

$$dr := \frac{1}{n}$$

```
> du:=(rgu[2]-rgu[1])/dus; # Segment range of 'u'
```

$$du := \frac{\pi}{n}$$

```
> dv:=(rgv[2]-rgv[1])/dus; # Segment range of 'v'
```

$$dv := \frac{2\,\pi}{n}$$

Microcosmic Triple integrals course of spacial bounded closed region 'Ω's first segmental cell (n segmental cell):

// Segment volume element coefficient 'J':

> dJ:=subs(r=rgr[1]+dr,subs(v=rgv[1]+dv,subs(u=rgu[1]+du,J)));

Volume element coefficient 'J's average value that corresponds to spacial bounded closed region 'Ω's first segmental cell

$$
dJ := \sin\left(\frac{\pi}{n}\right)\left(\frac{\pi}{n} + \sin\left(\frac{\pi}{n}\right)\cos\left(\frac{2\pi}{n}\right)\cos\left(\frac{\pi}{n}\right) + \frac{2\sin\left(\frac{\pi}{n}\right)\sin\left(\frac{2\pi}{n}\right)\pi^2}{n^2} \right.
$$

$$
+ \frac{\pi\cos\left(\frac{\pi}{n}\right)\cos\left(\frac{2\pi}{n}\right)}{n} - \sin\left(\frac{\pi}{n}\right)\cos\left(\frac{\pi}{n}\right) - \frac{\cos\left(\frac{2\pi}{n}\right)\pi}{n} - \frac{\pi\cos\left(\frac{\pi}{n}\right)}{n}
$$

$$
\left. - \sin\left(\frac{\pi}{n}\right)\cos\left(\frac{2\pi}{n}\right) + \sin\left(\frac{\pi}{n}\right) \right) / n^2
$$

// Segment idiographic divergence 'diV':

// Be different to abstract divergence 'diV1,diV2' that 'Proof' refers to, in 'Numerical Model', be able to directly input idiographic segmented values of 'r,u,v' to idiographic divergence 'diV', then segment 'diV' in accordance with amount 'dus' of parameter segmental cells

> ddiV:=subs(r=rgr[1]+dr,subs(v=rgv[1]+dv,subs(u=rgu[1]+du,diV)));

Divergence 'diV's average value that corresponds to spacial bounded closed region 'Ω's first segmental cell

$$
ddiV := \frac{\sin\left(\frac{\pi}{n}\right)\left(1 - \cos\left(\frac{2\pi}{n}\right)\right)}{n} + \frac{7}{6}\frac{\pi\left(1 - \cos\left(\frac{\pi}{n}\right)\right)}{n^2}
$$

// Calculate microcosmic triple integrals value of spacial bounded closed region 'Ω's first segmental cell:

> ddiV*dJ*dr*du*dv;

Base on integral median theorem, integral value of product (Divergence 'diV' and volume element) in first segmental cell

$$
2\left(\frac{\sin\left(\frac{\pi}{n}\right)\left(1 - \cos\left(\frac{2\pi}{n}\right)\right)}{n} + \frac{7}{6}\frac{\pi\left(1 - \cos\left(\frac{\pi}{n}\right)\right)}{n^2} \right)\sin\left(\frac{\pi}{n}\right)\left(\frac{\pi}{n} \right.
$$

$$
+ \sin\left(\frac{\pi}{n}\right)\cos\left(\frac{2\pi}{n}\right)\cos\left(\frac{\pi}{n}\right) + \frac{2\sin\left(\frac{\pi}{n}\right)\sin\left(\frac{2\pi}{n}\right)\pi^2}{n^2} + \frac{\pi\cos\left(\frac{\pi}{n}\right)\cos\left(\frac{2\pi}{n}\right)}{n}
$$

$$-\sin\left(\frac{\pi}{n}\right)\cos\left(\frac{\pi}{n}\right)-\frac{\cos\left(\frac{2\pi}{n}\right)\pi}{n}-\frac{\pi\cos\left(\frac{\pi}{n}\right)}{n}-\sin\left(\frac{\pi}{n}\right)\cos\left(\frac{2\pi}{n}\right)+\sin\left(\frac{\pi}{n}\right)\Bigg)\pi^2\Bigg/$$

$$n^5$$

Microcosmic triple integrals course of spacial bounded closed region 'Ω's all segmental cells (n segmental cells):

// Segment volume element coefficient 'J':

```
> ijkdJ:=subs(r=rgr[1]+k*dr,subs(v=rgv[1]+j*dv,subs(u=rgu[1]
+i*du,J)));
```

Volume element coefficient 'J's average values that correspond to spacial bounded closed region 'Ω's all segmental cells

$$ijkdJ:=\sin\left(\frac{i\pi}{n}\right)\Bigg(\frac{i\pi}{n}+\sin\left(\frac{i\pi}{n}\right)\cos\left(\frac{2j\pi}{n}\right)\cos\left(\frac{i\pi}{n}\right)+\frac{2\sin\left(\frac{i\pi}{n}\right)\sin\left(\frac{2j\pi}{n}\right)i^2\pi^2}{n^2}$$

$$+\frac{i\pi\cos\left(\frac{i\pi}{n}\right)\cos\left(\frac{2j\pi}{n}\right)}{n}-\sin\left(\frac{i\pi}{n}\right)\cos\left(\frac{i\pi}{n}\right)-\frac{\cos\left(\frac{2j\pi}{n}\right)i\pi}{n}-\frac{i\pi\cos\left(\frac{i\pi}{n}\right)}{n}$$

$$-\sin\left(\frac{i\pi}{n}\right)\cos\left(\frac{2j\pi}{n}\right)+\sin\left(\frac{i\pi}{n}\right)\Bigg)k^2\Big/n^2$$

//'ijkdJ' isn't that 'Single value', but that 'Set of finite quantitative (Viz.n) values'

// Segment idiographic divergence 'diV':

// Be different to abstract divergence 'diV1,diV2' that 'Proof' refers to, in 'Numerical Model', be able to directly input idiographic segmented values of 'r,u,v' to idiographic divergence 'diV', then segment 'diV' in accordance with amount 'dus' of parameter segmental cells

```
> ijkddiV:=subs(r=rgr[1]+k*dr,subs(v=rgv[1]+j*dv,subs(u=rgu[1]
+i*du,diV)));
```

Divergence 'diV's average values that correspond to spacial bounded closed region 'Ω's all segmental cells

$$ijkddiV:=\frac{k\sin\left(\frac{i\pi}{n}\right)\left(1-\cos\left(\frac{2j\pi}{n}\right)\right)}{n}+\frac{7}{6}\frac{ki\pi\left(1-\cos\left(\frac{i\pi}{n}\right)\right)}{n^2}$$

//'ijkddiV' isn't that 'Single value', but that 'Set of finite quantitative (Viz.n) values'

// Calculate microcosmic triple integrals values of spacial bounded closed region 'Ω's all segmental cells:

> ijkddiV*ijkdJ*dr*du*dv;

Base on integral median theorem, integral value of product (Divergence 'diV' and volume element) in all segmental cells

$$2\left(\frac{k\sin\left(\frac{i\pi}{n}\right)\left(1-\cos\left(\frac{2j\pi}{n}\right)\right)}{n}+\frac{7}{6}\frac{ki\pi\left(1-\cos\left(\frac{i\pi}{n}\right)\right)}{n^2}\right)\sin\left(\frac{i\pi}{n}\right)\left(\frac{i\pi}{n}\right.$$

$$+\sin\left(\frac{i\pi}{n}\right)\cos\left(\frac{2j\pi}{n}\right)\cos\left(\frac{i\pi}{n}\right)+\frac{2\sin\left(\frac{i\pi}{n}\right)\sin\left(\frac{2j\pi}{n}\right)i^2\pi^2}{n^2}$$

$$+\frac{i\pi\cos\left(\frac{i\pi}{n}\right)\cos\left(\frac{2j\pi}{n}\right)}{n}-\sin\left(\frac{i\pi}{n}\right)\cos\left(\frac{i\pi}{n}\right)-\frac{\cos\left(\frac{2j\pi}{n}\right)i\pi}{n}-\frac{i\pi\cos\left(\frac{i\pi}{n}\right)}{n}$$

$$\left.-\sin\left(\frac{i\pi}{n}\right)\cos\left(\frac{2j\pi}{n}\right)+\sin\left(\frac{i\pi}{n}\right)\right)k^2\pi^2/n^5$$

//Expression above-mentioned isn't that 'Single value', but that 'Set of finite quantitative (Viz.n) values'

// Structure finite sums:

> Sum(Sum(Sum(ijkddiV*ijkdJ*dr*du*dv,i=1..dus),j=1..dus), k=1..dus);

In the case of "Amount 'n' of spacial closed region 'Ω's segmental cells is pending", sum of integral values (product about divergence 'diV' and volume element) in all segmental cells

$$\sum_{k=1}^{n}\left(\sum_{j=1}^{n}\left(\sum_{i=1}^{n}\left(2\left(\frac{k\sin\left(\frac{i\pi}{n}\right)\left(1-\cos\left(\frac{2j\pi}{n}\right)\right)}{n}+\frac{7}{6}\frac{ki\pi\left(1-\cos\left(\frac{i\pi}{n}\right)\right)}{n^2}\right)\sin\left(\frac{i\pi}{n}\right)\left(\frac{i\pi}{n}\right.\right.\right.\right.$$

$$+\sin\left(\frac{i\pi}{n}\right)\cos\left(\frac{2j\pi}{n}\right)\cos\left(\frac{i\pi}{n}\right)+\frac{2\sin\left(\frac{i\pi}{n}\right)\sin\left(\frac{2j\pi}{n}\right)i^2\pi^2}{n^2}$$

$$+\frac{i\pi\cos\left(\frac{i\pi}{n}\right)\cos\left(\frac{2j\pi}{n}\right)}{n}-\sin\left(\frac{i\pi}{n}\right)\cos\left(\frac{i\pi}{n}\right)-\frac{\cos\left(\frac{2j\pi}{n}\right)i\pi}{n}-\frac{i\pi\cos\left(\frac{i\pi}{n}\right)}{n}$$

$$- \sin\left(\frac{i\,\pi}{n}\right) \cos\left(\frac{2\,j\,\pi}{n}\right) + \sin\left(\frac{i\,\pi}{n}\right)\right) k^2\,\pi^2\,/\,n^5 \Bigg)\Bigg)\Bigg)$$

```
> vs:=value(%):
> Limit(vs,n=infinity):  (10)
```

// Infinitize finite sums, its limit operational value is 'Triple Integrals Value at Manifold'

\# In the case of "Amount 'n' of spacial bounded closed region 'Ω's segmental cells tends to infinity", limit value of integral values' sums (Product about Divergence 'diV' and volume element) in all segmental cells

```
> beta:=value(%);epsilon:=evalf(beta);
```

$$\beta := -\frac{88}{81}\,\pi + \frac{829}{576}\,\pi^3$$

$$\varepsilon := 41.21226476$$

```
> omega;epsilon;
```

\# There is tiny difference between 'Integral values' sum of finite quantitative segmental cells' and 'Integral values' sum of infinite quantitative segmental cells'

$$45.27224393$$

$$41.21226476$$

```
> alpha,beta;  # Two analytic values are equal
```

$$-\frac{88}{81}\,\pi + \frac{829}{576}\,\pi^3, -\frac{88}{81}\,\pi + \frac{829}{576}\,\pi^3$$

```
> delta,epsilon;  # Two float values are equal
```

$$41.21226476 \quad 41.21226476$$

Chapter 6 Counterexample of Divergence Theorem at Manifold
--Surface Integral and Triple Integrals about Klein Bottle

6.1 Divergence Theorem at Manifold and Klein Bottle

Confessedly, Klein bottle is representative unorientable closed surface; If try to conclude and deduce 'Klein bottle' by logic method of 'Prove Divergence Theorem at Manifold', what instance will appear possibly ?

Divergence Theorem

Suppose spacial bounded closed region 'Ω' is surrounded by smooth or piecewise smooth closed surface 'S'. Functions 'P(x,y,z),Q(x,y,z), R(x,y,z)' [Structure vector field 'A'] and its partial derivatives are continuous at spacial bounded closed region 'Ω', then:

$$\iint\limits_{S} A \cdot n \, dS = \iiint\limits_{\Omega} divA \cdot d\omega \qquad\qquad (1)$$

thereinto, surface 'S' is outer side of spacial bounded closed region 'Ω's entire boundary surface, 'n' is unit normal vector of surface 'S's outer side, and 'divA' is divergence of vector field 'A'.

Proof (Counterexample):

Define parameterized expression of Klein Bottle:

[cos(u)*(cos(u/2)*sin(v)-sin(u/2)*sin(2*v)+3),

 sin(u)*(cos(u/2)*sin(v)-sin(u/2)*sin(2*v)+3),

 (sin(u/2)*sin(v)+cos(u/2)*sin(2*v))]   (2)

Set range of u∈[0,4π] and v∈[0,π], make Klein Bottle close.

Figure(I)6.1 Klein Bottle(unorientable, closed)

According to expression of Klein Bottle (2), define and calculate first matrix of partial derivatives, obtain tangent plane normal vector of Klein Bottle (3):

$$
\begin{bmatrix}
i & j & k \\
\frac{\partial}{\partial u}\cos(u)(\cos(u/2)\sin(v)-\sin(u/2)\sin(2v)+3) & \frac{\partial}{\partial u}\sin(u)(\cos(u/2)\sin(v)-\sin(u/2)\sin(2v)+3) & \frac{\partial}{\partial u}\sin(u/2)\sin(v)+\cos(u/2)\sin(2v) \\
\frac{\partial}{\partial v}\cos(u)(\cos(u/2)\sin(v)-\sin(u/2)\sin(2v)+3) & \frac{\partial}{\partial v}\sin(u)(\cos(u/2)\sin(v)-\sin(u/2)\sin(2v)+3) & \frac{\partial}{\partial v}\sin(u/2)\sin(v)+\cos(u/2)\sin(2v)
\end{bmatrix}
$$

$$
=
$$

$$
i\cos(u)\cos\left(\frac{u}{2}\right)\sin(v)\sin\left(\frac{u}{2}\right)\cos(v) - 2\,i\cos(u)\cos\left(\frac{u}{2}\right)\sin\left(\frac{u}{2}\right)\sin(2v)\cos(2v)
$$

$$
+ 2\,i\cos(u)\cos\left(\frac{u}{2}\right)^2\sin(v)\cos(2v) + 2\cos\left(\frac{u}{2}\right)^2\sin(v)\sin(u)\cos(2v)\,j
$$

$$
+ 2\cos\left(\frac{u}{2}\right)\sin(v)\sin\left(\frac{u}{2}\right)\cos(2v)\,k + 6\,i\cos(u)\cos\left(\frac{u}{2}\right)\cos(2v)
$$

$$
+ 3\,i\cos(u)\sin\left(\frac{u}{2}\right)\cos(v) + 6\cos\left(\frac{u}{2}\right)\sin(u)\cos(2v)\,j + 3\sin\left(\frac{u}{2}\right)\sin(u)\cos(v)\,j
$$

$$
- i\cos(u)\sin(2v)\cos(v) - \frac{1}{2}\,i\sin(v)\sin(u)\cos(v) - i\sin(2v)\sin(u)\cos(2v)
$$

$$
- \sin(2v)\sin(u)\cos(v)\,j + \frac{1}{2}\cos(u)\sin(v)\cos(v)\,j + \cos(u)\sin(2v)\cos(2v)\,j
$$

$$
+ 2\cos\left(\frac{u}{2}\right)^2\sin(2v)\cos(2v)\,k - \cos\left(\frac{u}{2}\right)^2\sin(v)\cos(v)\,k
$$

$$
+ \cos(u)\cos\left(\frac{u}{2}\right)^2\sin(2v)\cos(v)\,i + \cos\left(\frac{u}{2}\right)^2\sin(2v)\sin(u)\cos(v)\,j
$$

$$
+ \cos\left(\frac{u}{2}\right)\sin\left(\frac{u}{2}\right)\sin(2v)\cos(v)\,k - 3\cos\left(\frac{u}{2}\right)\cos(v)\,k + 6\sin\left(\frac{u}{2}\right)\cos(2v)\,k
$$

$$
- 2\sin(2v)\cos(2v)\,k + \cos\left(\frac{u}{2}\right)\sin(v)\sin\left(\frac{u}{2}\right)\sin(u)\cos(v)\,j
$$

$$
- 2\cos\left(\frac{u}{2}\right)\sin\left(\frac{u}{2}\right)\sin(2v)\sin(u)\cos(2v)\,j
$$

$$
\tag{3}
$$

From expression(3), respectively pick up coefficients of item 'i,j,k', obtain tangent plane normal vector of Klein Bottle (4):

$$
\left[\; \cos(u)\cos\left(\frac{u}{2}\right)\sin(v)\sin\left(\frac{u}{2}\right)\cos(v) - 2\cos(u)\cos\left(\frac{u}{2}\right)\sin\left(\frac{u}{2}\right)\sin(2v)\cos(2v) \right.
$$

$$
+ 2\cos(u)\cos\left(\frac{u}{2}\right)^2\sin(v)\cos(2v) + 6\cos(u)\cos\left(\frac{u}{2}\right)\cos(2v)
$$

$$+ 3\cos(u)\sin\left(\frac{u}{2}\right)\cos(v) - \cos(u)\sin(2v)\cos(v) - \frac{1}{2}\sin(v)\sin(u)\cos(v)$$

$$- \sin(2v)\sin(u)\cos(2v) + \cos(u)\cos\left(\frac{u}{2}\right)^2\sin(2v)\cos(v),$$

$$2\cos\left(\frac{u}{2}\right)^2\sin(v)\sin(u)\cos(2v) + 6\cos\left(\frac{u}{2}\right)\sin(u)\cos(2v)$$

$$+ 3\sin\left(\frac{u}{2}\right)\sin(u)\cos(v) - \sin(2v)\sin(u)\cos(v) + \frac{1}{2}\cos(u)\sin(v)\cos(v)$$

$$+ \cos(u)\sin(2v)\cos(2v) + \cos\left(\frac{u}{2}\right)^2\sin(2v)\sin(u)\cos(v)$$

$$+ \cos\left(\frac{u}{2}\right)\sin(v)\sin\left(\frac{u}{2}\right)\sin(u)\cos(v) - 2\cos\left(\frac{u}{2}\right)\sin\left(\frac{u}{2}\right)\sin(2v)\sin(u)\cos(2v),$$

$$2\cos\left(\frac{u}{2}\right)\sin(v)\sin\left(\frac{u}{2}\right)\cos(2v) + 2\cos\left(\frac{u}{2}\right)^2\sin(2v)\cos(2v)$$

$$- \cos\left(\frac{u}{2}\right)^2\sin(v)\cos(v) + \cos\left(\frac{u}{2}\right)\sin\left(\frac{u}{2}\right)\sin(2v)\cos(v) - 3\cos\left(\frac{u}{2}\right)\cos(v)$$

$$+ 6\sin\left(\frac{u}{2}\right)\cos(2v) - 2\sin(2v)\cos(2v)\Bigg] \tag{4}$$

Surface Integral of Dot Product [Vector Field 'A' and Tangent Plane Normal Vector(4)] in range of 'u,v' (5):

$$\int_0^\pi\int_0^{4\pi}\Bigg(\cos(u)\cos\left(\frac{u}{2}\right)\sin(v)\sin\left(\frac{u}{2}\right)\cos(v) - 2\cos(u)\cos\left(\frac{u}{2}\right)\sin\left(\frac{u}{2}\right)\sin(2v)\cos(2v)$$

$$+ 2\cos(u)\cos\left(\frac{u}{2}\right)^2\sin(v)\cos(2v) + 6\cos(u)\cos\left(\frac{u}{2}\right)\cos(2v)$$

$$+ 3\cos(u)\sin\left(\frac{u}{2}\right)\cos(v) - \cos(u)\sin(2v)\cos(v) - \frac{1}{2}\sin(v)\sin(u)\cos(v)$$

$$- \sin(2v)\sin(u)\cos(2v) + \cos(u)\cos\left(\frac{u}{2}\right)^2\sin(2v)\cos(v)\Bigg)P(x,y,z) + \Bigg($$

$$2\cos\left(\frac{u}{2}\right)^2\sin(v)\sin(u)\cos(2v) + 6\cos\left(\frac{u}{2}\right)\sin(u)\cos(2v)$$

$$+ 3\sin\left(\frac{u}{2}\right)\sin(u)\cos(v) - \sin(2v)\sin(u)\cos(v) + \frac{1}{2}\cos(u)\sin(v)\cos(v)$$

$$+ \cos(u)\sin(2v)\cos(2v) + \cos\left(\frac{u}{2}\right)^2\sin(2v)\sin(u)\cos(v)$$

$$+ \cos\left(\frac{u}{2}\right)\sin(v)\sin\left(\frac{u}{2}\right)\sin(u)\cos(v) - 2\cos\left(\frac{u}{2}\right)\sin\left(\frac{u}{2}\right)\sin(2v)\sin(u)\cos(2v) \Bigg)$$

$$Q(x,y,z) + \left(2\cos\left(\frac{u}{2}\right)\sin(v)\sin\left(\frac{u}{2}\right)\cos(2v) + 2\cos\left(\frac{u}{2}\right)^2\sin(2v)\cos(2v)\right.$$

$$- \cos\left(\frac{u}{2}\right)^2\sin(v)\cos(v) + \cos\left(\frac{u}{2}\right)\sin\left(\frac{u}{2}\right)\sin(2v)\cos(v) - 3\cos\left(\frac{u}{2}\right)\cos(v)$$

$$+ 6\sin\left(\frac{u}{2}\right)\cos(2v) - 2\sin(2v)\cos(2v)\bigg) R(x,y,z) \, du \, dv$$

$$= 0 \qquad (5)$$

Entirely multiply radius vector 'r' (suppose r>0) by each item of Klein Bottle surface parametrized expression (2), convert 'Parametrized expression of Klein Bottle surface' to 'Parametrized expression of Klein Bottle surface coordinates'(6):

$$\left[r\cos(u)\left(\cos\left(\frac{u}{2}\right)\sin(v) - \sin\left(\frac{u}{2}\right)\sin(2v) + 3\right),\right.$$

$$r\sin(u)\left(\cos\left(\frac{u}{2}\right)\sin(v) - \sin\left(\frac{u}{2}\right)\sin(2v) + 3\right),$$

$$\left. r\left(\sin\left(\frac{u}{2}\right)\sin(v) + \cos\left(\frac{u}{2}\right)\sin(2v)\right)\right] \qquad (6)$$

According to parametrized expression of Klein Bottle surface coordinates(6), define and calculate Second Matrix of Partial Derivatives, obtain general expression of Klein Bottle surface coordinates volume element coefficient (7):

$$\begin{bmatrix}
\frac{\partial}{\partial r}r\cos(u)(\cos(\frac{u}{2})\sin(v)-\sin(\frac{u}{2})\sin(2v)+3) & \frac{\partial}{\partial u}r\cos(u)(\cos(\frac{u}{2})\sin(v)-\sin(\frac{u}{2})\sin(2v)+3) & \frac{\partial}{\partial v}r\cos(u)(\cos(\frac{u}{2})\sin(v)-\sin(\frac{u}{2})\sin(2v)+3) \\
\frac{\partial}{\partial r}r\sin(u)(\cos(\frac{u}{2})\sin(v)-\sin(\frac{u}{2})\sin(2v)+3) & \frac{\partial}{\partial u}r\sin(u)(\cos(\frac{u}{2})\sin(v)-\sin(\frac{u}{2})\sin(2v)+3) & \frac{\partial}{\partial v}r\sin(u)(\cos(\frac{u}{2})\sin(v)-\sin(\frac{u}{2})\sin(2v)+3) \\
\frac{\partial}{\partial r}r(\sin(\frac{u}{2})\sin(v)+\cos(\frac{u}{2})\sin(2v)) & \frac{\partial}{\partial u}r(\sin(\frac{u}{2})\sin(v)+\cos(\frac{u}{2})\sin(2v)) & \frac{\partial}{\partial v}r(\sin(\frac{u}{2})\sin(v)+\cos(\frac{u}{2})\sin(2v))
\end{bmatrix}$$

$$=$$

$$r^2\left(6\cos\left(\frac{u}{2}\right)^2\sin(v)\cos(2v) + 3\cos\left(\frac{u}{2}\right)^2\sin(2v)\cos(v)\right.$$

$$+ 3\cos\left(\frac{u}{2}\right)\sin(v)\sin\left(\frac{u}{2}\right)\cos(v) - \cos\left(\frac{u}{2}\right)\sin(v)\sin(2v)\cos(v)$$

$$- 6\cos\left(\frac{u}{2}\right)\sin\left(\frac{u}{2}\right)\sin(2v)\cos(2v) - 2\cos\left(\frac{u}{2}\right)\cos(v)^2\cos(2v)$$

$$- 2\sin(v)\sin\left(\frac{u}{2}\right)\sin(2v)\cos(2v) - \sin\left(\frac{u}{2}\right)\cos(v)\cos(2v)^2 + 20\cos\left(\frac{u}{2}\right)\cos(2v)$$

$$+ 6 \sin(v) \cos(2\,v) + 10 \sin\left(\frac{u}{2}\right) \cos(v) - 6 \sin(2\,v) \cos(v) \Bigg) \qquad (7)$$

Calculate Divergence of Vector Field 'A', and convert Divergence from Cartesian coordinates expression(8) to Klein Bottle surface coordinates expression(9):

$$\frac{\partial}{\partial x} P(x,y,z) + \frac{\partial}{\partial y} Q(x,y,z) + \frac{\partial}{\partial z} R(x,y,z) \qquad (8)$$

$$\frac{\partial}{\partial x} P(x,y,z)(\frac{\partial}{\partial r} r \cos(u)(\cos(\frac{u}{2})\sin(v) - \sin(\frac{u}{2})\sin(2v) + 3))(\frac{\partial}{\partial u} r \cos(u)(\cos(\frac{u}{2})\sin(v) - \sin(\frac{u}{2})\sin(2v) + 3))(\frac{\partial}{\partial v} r \cos(u)(\cos(\frac{u}{2})\sin(v) - \sin(\frac{u}{2})\sin(2v) + 3))$$

$$+ \frac{\partial}{\partial y} Q(x,y,z)(\frac{\partial}{\partial r} r \sin(u)(\cos(\frac{u}{2})\sin(v) - \sin(\frac{u}{2})\sin(2v) + 3))(\frac{\partial}{\partial u} r \sin(u)(\cos(\frac{u}{2})\sin(v) - \sin(\frac{u}{2})\sin(2v) + 3))(\frac{\partial}{\partial v} r \sin(u)(\cos(\frac{u}{2})\sin(v) - \sin(\frac{u}{2})\sin(2v) + 3))$$

$$+ \frac{\partial}{\partial z} R(x,y,z)(\frac{\partial}{\partial r} r(\sin(\frac{u}{2})\sin(v) + \cos(\frac{u}{2})\sin(2v))(\frac{\partial}{\partial u} r(\sin(\frac{u}{2})\sin(v) + \cos(\frac{u}{2})\sin(2v))(\frac{\partial}{\partial v} r(\sin(\frac{u}{2})\sin(v) + \cos(\frac{u}{2})\sin(2v))$$

$$=$$

$$\left(\frac{\partial}{\partial x} P(x,y,z)\right) \cos(u)^2 \left(\cos\left(\frac{u}{2}\right)\sin(v) - \sin\left(\frac{u}{2}\right)\sin(2\,v) + 3\right)\Bigg($$

$$-r \sin(u)\left(\cos\left(\frac{u}{2}\right)\sin(v) - \sin\left(\frac{u}{2}\right)\sin(2\,v) + 3\right)$$

$$+ r \cos(u)\left(-\frac{1}{2}\sin\left(\frac{u}{2}\right)\sin(v) - \frac{1}{2}\cos\left(\frac{u}{2}\right)\sin(2\,v)\right)\Bigg) r$$

$$\left(\cos\left(\frac{u}{2}\right)\cos(v) - 2\sin\left(\frac{u}{2}\right)\cos(2\,v)\right) + \left(\frac{\partial}{\partial y} Q(x,y,z)\right)\sin(u)^2$$

$$\left(\cos\left(\frac{u}{2}\right)\sin(v) - \sin\left(\frac{u}{2}\right)\sin(2\,v) + 3\right)\Bigg($$

$$r \cos(u)\left(\cos\left(\frac{u}{2}\right)\sin(v) - \sin\left(\frac{u}{2}\right)\sin(2\,v) + 3\right)$$

$$+ r \sin(u)\left(-\frac{1}{2}\sin\left(\frac{u}{2}\right)\sin(v) - \frac{1}{2}\cos\left(\frac{u}{2}\right)\sin(2\,v)\right)\Bigg) r$$

$$\left(\cos\left(\frac{u}{2}\right)\cos(v) - 2\sin\left(\frac{u}{2}\right)\cos(2\,v)\right) + \left(\frac{\partial}{\partial z} R(x,y,z)\right)$$

$$\left(\sin\left(\frac{u}{2}\right)\sin(v) + \cos\left(\frac{u}{2}\right)\sin(2\,v)\right) r^2 \left(\frac{1}{2}\cos\left(\frac{u}{2}\right)\sin(v) - \frac{1}{2}\sin\left(\frac{u}{2}\right)\sin(2\,v)\right)$$

$$\left(\sin\left(\frac{u}{2}\right)\cos(v) + 2\cos\left(\frac{u}{2}\right)\cos(2\,v)\right) \qquad (9)$$

Triple Integrals of Product [Divergence(9) and Klein Bottle Surface Coordinates Volume Element] in range of 'r,u,v' (10):

$$\int_0^\pi \int_0^{4\pi} \int_0^1 \left(\left(\frac{\partial}{\partial x} P(x,y,z)\right) \cos(u)^2 \left(\cos\left(\frac{u}{2}\right)\sin(v) - \sin\left(\frac{u}{2}\right)\sin(2\,v) + 3\right)\Bigg($$

$$-r\sin(u)\left(\cos\left(\frac{u}{2}\right)\sin(v)-\sin\left(\frac{u}{2}\right)\sin(2v)+3\right)$$

$$+r\cos(u)\left(-\frac{1}{2}\sin\left(\frac{u}{2}\right)\sin(v)-\frac{1}{2}\cos\left(\frac{u}{2}\right)\sin(2v)\right)\Bigg)r$$

$$\left(\cos\left(\frac{u}{2}\right)\cos(v)-2\sin\left(\frac{u}{2}\right)\cos(2v)\right)+\left(\frac{\partial}{\partial y}Q(x,y,z)\right)\sin(u)^2$$

$$\left(\cos\left(\frac{u}{2}\right)\sin(v)-\sin\left(\frac{u}{2}\right)\sin(2v)+3\right)\Bigg($$

$$r\cos(u)\left(\cos\left(\frac{u}{2}\right)\sin(v)-\sin\left(\frac{u}{2}\right)\sin(2v)+3\right)$$

$$+r\sin(u)\left(-\frac{1}{2}\sin\left(\frac{u}{2}\right)\sin(v)-\frac{1}{2}\cos\left(\frac{u}{2}\right)\sin(2v)\right)\Bigg)r$$

$$\left(\cos\left(\frac{u}{2}\right)\cos(v)-2\sin\left(\frac{u}{2}\right)\cos(2v)\right)+\left(\frac{\partial}{\partial z}R(x,y,z)\right)$$

$$\left(\sin\left(\frac{u}{2}\right)\sin(v)+\cos\left(\frac{u}{2}\right)\sin(2v)\right)r^2\left(\frac{1}{2}\cos\left(\frac{u}{2}\right)\sin(v)-\frac{1}{2}\sin\left(\frac{u}{2}\right)\sin(2v)\right)$$

$$\left(\sin\left(\frac{u}{2}\right)\cos(v)+2\cos\left(\frac{u}{2}\right)\cos(2v)\right)\Bigg)r^2\left(6\cos\left(\frac{u}{2}\right)^2\sin(v)\cos(2v)\right.$$

$$+3\cos\left(\frac{u}{2}\right)^2\sin(2v)\cos(v)+3\cos\left(\frac{u}{2}\right)\sin(v)\sin\left(\frac{u}{2}\right)\cos(v)$$

$$-\cos\left(\frac{u}{2}\right)\sin(v)\sin(2v)\cos(v)-6\cos\left(\frac{u}{2}\right)\sin\left(\frac{u}{2}\right)\sin(2v)\cos(2v)$$

$$-2\cos\left(\frac{u}{2}\right)\cos(v)^2\cos(2v)-2\sin(v)\sin\left(\frac{u}{2}\right)\sin(2v)\cos(2v)$$

$$-\sin\left(\frac{u}{2}\right)\cos(v)\cos(2v)^2+20\cos\left(\frac{u}{2}\right)\cos(2v)+6\sin(v)\cos(2v)$$

$$\left.+10\sin\left(\frac{u}{2}\right)\cos(v)-6\sin(2v)\cos(v)\right)dr\,du\,dv$$

$$=\quad\frac{471}{64}\left(\frac{\partial}{\partial x}P(x,y,z)\right)\pi^2 \tag{10}$$

Viz. Expression(5) ≠ Expression(10)

Conclude and deduce 'Klein Bottle' by logic method of 'Prove Divergence Theorem at Manifold', Surface Integral at 'Klein Bottle' and Triple Integrals in Spacial Bounded Closed Region (Klein Bottle surface surrounds) are unequal in logic.

Complete Proof (Counterexample).

6.2 Divergence Theorem at Manifold and Klein Bottle
[Program Template of Waterloo Maplesoft, Optional]

Divergence Theorem

Suppose spacial bounded closed region 'Ω' is surrounded by smooth or piecewise smooth closed surface 'S'. Functions 'P(x,y,z),Q(x,y,z), R(x,y,z)'[Structure vector field 'A'] and its partial derivatives are continuous at spacial bounded closed region 'Ω', then:

$$\iint\limits_{S} A \cdot n \, dS = \iiint\limits_{\Omega} divA \cdot d\omega \qquad (1)$$

thereinto, surface 'S' is outer side of spacial bounded closed region 'Ω's entire boundary surface, 'n' is unit normal vector of surface 'S's outer side, and 'divA' is divergence of vector field 'A'.

Symbol System:

Spacial Vector Field 'V',

Divergence of Vector Field 'V' as 'diV1,diV2',

Unorientable closed surface 'CS' (Viz. Klein bottle),

'Tangent plane normal vector [A,B,C]' of surface 'CS',

Spacial bounded closed region 'Ω' surrounded by Torus surface 'CS',

General expression of Klein Bottle surface coordinates volume element coefficient 'J'

```
> restart;
> with(plots):with(linalg):
> CS:=[cos(u)*(cos(u/2)*sin(v)-sin(u/2)*sin(2*v)+3),
sin(u)*(cos(u/2)*sin(v)-sin(u/2)*sin(2*v)+3),
(sin(u/2)*sin(v)+cos(u/2)*sin(2*v))];
# Define unorientable closed parameterized surface 'CS' (Viz. Klein
Bottle)
```

$$CS := \left[\cos(u) \left(\cos\left(\frac{u}{2}\right) \sin(v) - \sin\left(\frac{u}{2}\right) \sin(2\,v) + 3 \right),\right.$$
$$\left. \sin(u) \left(\cos\left(\frac{u}{2}\right) \sin(v) - \sin\left(\frac{u}{2}\right) \sin(2\,v) + 3 \right), \sin\left(\frac{u}{2}\right) \sin(v) + \cos\left(\frac{u}{2}\right) \sin(2\,v) \right]$$

```
> rgu:=[0,4*Pi];
```

$$rgu := [0, 4\,\pi]$$

```
> rgv:=[0,Pi]; # Define range of 'u,v'
```

$$rgv := [0, \pi]$$

```
> plot3d(CS,u=rgu[1]..rgu[2],v=rgv[1]..rgv[2],scaling=constrained,
projection=0.9,numpoints=6000): # Draw Klein bottle
```

 [See also Figure(I)6.1]

```
> V:=[(P)(x,y,z),(Q)(x,y,z),(R)(x,y,z)];
```

\# Define abstract spacial vector field 'V' (Suppose spacial vector field 'V' possesses 1th order continuous partial derivatives in spacial bounded closed region that 'Klein bottle' surrounds)

$$V := [\mathrm{P}(x, y, z), \mathrm{Q}(x, y, z), \mathrm{R}(x, y, z)]$$

```
> Diff(V[1],x)+Diff(V[2],y)+Diff(V[3],z)
= diff(V[1],x)+diff(V[2],y)+diff(V[3],z);diV1:=rhs(%);
```

\# Calculate divergence 'diV1' of spacial vector field 'V'

$$\left(\frac{\partial}{\partial x} \mathrm{P}(x, y, z)\right) + \left(\frac{\partial}{\partial y} \mathrm{Q}(x, y, z)\right) + \left(\frac{\partial}{\partial z} \mathrm{R}(x, y, z)\right) =$$

$$\left(\frac{\partial}{\partial x} \mathrm{P}(x, y, z)\right) + \left(\frac{\partial}{\partial y} \mathrm{Q}(x, y, z)\right) + \left(\frac{\partial}{\partial z} \mathrm{R}(x, y, z)\right)$$

$$diV1 := \left(\frac{\partial}{\partial x} \mathrm{P}(x, y, z)\right) + \left(\frac{\partial}{\partial y} \mathrm{Q}(x, y, z)\right) + \left(\frac{\partial}{\partial z} \mathrm{R}(x, y, z)\right)$$

```
> x:=CS[1]:y:=CS[2]:z:=CS[3]:
> matrix(3,3,[i,j,k,Diff(CS[1],u),Diff(CS[2],u),Diff(CS[3],u),
Diff(CS[1],v),Diff(CS[2],v),Diff(CS[3],v)])=
matrix(3,3,[i,j,k,diff(CS[1],u),diff(CS[2],u),diff(CS[3],u),
diff(CS[1],v),diff(CS[2],v),diff(CS[3],v)]);m1:=rhs(%);
```

\# Define and calculate matrix of partial derivatives 'm1', obtain 'tangent plane normal vector' of 'Klein bottle' surface

$$[i, j, k]$$

$$\left[\frac{\partial}{\partial u}\left(\cos(u)\left(\cos\left(\frac{u}{2}\right)\sin(v) - \sin\left(\frac{u}{2}\right)\sin(2v) + 3\right)\right),\right.$$

$$\frac{\partial}{\partial u}\left(\sin(u)\left(\cos\left(\frac{u}{2}\right)\sin(v) - \sin\left(\frac{u}{2}\right)\sin(2v) + 3\right)\right),$$

$$\left.\frac{\partial}{\partial u}\left(\sin\left(\frac{u}{2}\right)\sin(v) + \cos\left(\frac{u}{2}\right)\sin(2v)\right)\right]$$

$$\left[\frac{\partial}{\partial v}\left(\cos(u)\left(\cos\left(\frac{u}{2}\right)\sin(v) - \sin\left(\frac{u}{2}\right)\sin(2v) + 3\right)\right),\right.$$

$$\frac{\partial}{\partial v}\left(\sin(u)\left(\cos\left(\frac{u}{2}\right)\sin(v) - \sin\left(\frac{u}{2}\right)\sin(2v) + 3\right)\right),$$

$$\left.\frac{\partial}{\partial v}\left(\sin\left(\frac{u}{2}\right)\sin(v) + \cos\left(\frac{u}{2}\right)\sin(2v)\right)\right] =$$

$$[i, j, k]$$

$$\Bigg[-\sin(u)\left(\cos\!\left(\frac{u}{2}\right)\sin(v)-\sin\!\left(\frac{u}{2}\right)\sin(2\,v)+3\right)$$

$$+\cos(u)\left(-\frac{1}{2}\sin\!\left(\frac{u}{2}\right)\sin(v)-\frac{1}{2}\cos\!\left(\frac{u}{2}\right)\sin(2\,v)\right),$$

$$\cos(u)\left(\cos\!\left(\frac{u}{2}\right)\sin(v)-\sin\!\left(\frac{u}{2}\right)\sin(2\,v)+3\right)$$

$$+\sin(u)\left(-\frac{1}{2}\sin\!\left(\frac{u}{2}\right)\sin(v)-\frac{1}{2}\cos\!\left(\frac{u}{2}\right)\sin(2\,v)\right),$$

$$\frac{1}{2}\cos\!\left(\frac{u}{2}\right)\sin(v)-\frac{1}{2}\sin\!\left(\frac{u}{2}\right)\sin(2\,v)\Bigg]$$

$$\Bigg[\cos(u)\left(\cos\!\left(\frac{u}{2}\right)\cos(v)-2\sin\!\left(\frac{u}{2}\right)\cos(2\,v)\right),$$

$$\sin(u)\left(\cos\!\left(\frac{u}{2}\right)\cos(v)-2\sin\!\left(\frac{u}{2}\right)\cos(2\,v)\right),\ \sin\!\left(\frac{u}{2}\right)\cos(v)+2\cos\!\left(\frac{u}{2}\right)\cos(2\,v)\Bigg]$$

m1 :=

$$[i\,,j\,,k]$$

$$\Bigg[-\sin(u)\left(\cos\!\left(\frac{u}{2}\right)\sin(v)-\sin\!\left(\frac{u}{2}\right)\sin(2\,v)+3\right)$$

$$+\cos(u)\left(-\frac{1}{2}\sin\!\left(\frac{u}{2}\right)\sin(v)-\frac{1}{2}\cos\!\left(\frac{u}{2}\right)\sin(2\,v)\right),$$

$$\cos(u)\left(\cos\!\left(\frac{u}{2}\right)\sin(v)-\sin\!\left(\frac{u}{2}\right)\sin(2\,v)+3\right)$$

$$+\sin(u)\left(-\frac{1}{2}\sin\!\left(\frac{u}{2}\right)\sin(v)-\frac{1}{2}\cos\!\left(\frac{u}{2}\right)\sin(2\,v)\right),$$

$$\frac{1}{2}\cos\!\left(\frac{u}{2}\right)\sin(v)-\frac{1}{2}\sin\!\left(\frac{u}{2}\right)\sin(2\,v)\Bigg]$$

$$\Bigg[\cos(u)\left(\cos\!\left(\frac{u}{2}\right)\cos(v)-2\sin\!\left(\frac{u}{2}\right)\cos(2\,v)\right),$$

$$\sin(u)\left(\cos\!\left(\frac{u}{2}\right)\cos(v)-2\sin\!\left(\frac{u}{2}\right)\cos(2\,v)\right),\ \sin\!\left(\frac{u}{2}\right)\cos(v)+2\cos\!\left(\frac{u}{2}\right)\cos(2\,v)\Bigg]$$

```
> det(m1);
```

$$i\cos(u)\cos\!\left(\frac{u}{2}\right)\sin(v)\sin\!\left(\frac{u}{2}\right)\cos(v)+2\sin(u)^2\cos\!\left(\frac{u}{2}\right)\sin(v)\,k\sin\!\left(\frac{u}{2}\right)\cos(2\,v)$$

$$-2\sin(u)\sin\!\left(\frac{u}{2}\right)\sin(2\,v)\,j\cos\!\left(\frac{u}{2}\right)\cos(2\,v)$$

$$-2\,i\cos(u)\sin\!\left(\frac{u}{2}\right)\sin(2\,v)\cos\!\left(\frac{u}{2}\right)\cos(2\,v)$$

$$+\cos\!\left(\frac{u}{2}\right)\cos(v)\,k\cos(u)^2\sin\!\left(\frac{u}{2}\right)\sin(2\,v)$$

$$+2\sin\!\left(\frac{u}{2}\right)\cos(2\,v)\,k\cos(u)^2\cos\!\left(\frac{u}{2}\right)\sin(v)$$

$$+\sin(u)^2\sin\!\left(\frac{u}{2}\right)\sin(2\,v)\,k\cos\!\left(\frac{u}{2}\right)\cos(v)+\sin(u)\cos\!\left(\frac{u}{2}\right)\sin(v)\,j\sin\!\left(\frac{u}{2}\right)\cos(v)$$

$$-\frac{1}{2}\,i\,\sin(u)\,\sin\!\left(\frac{u}{2}\right)^{2}\sin(v)\cos(v)-i\,\sin(u)\,\cos\!\left(\frac{u}{2}\right)^{2}\sin(2\,v)\cos(2\,v)$$

$$-\frac{1}{2}\,i\,\sin(u)\,\cos\!\left(\frac{u}{2}\right)^{2}\sin(v)\cos(v)-i\,\sin(u)\,\sin\!\left(\frac{u}{2}\right)^{2}\sin(2\,v)\cos(2\,v)$$

$$+2\,\sin(u)\,\cos\!\left(\frac{u}{2}\right)^{2}\sin(v)\,j\,\cos(2\,v)-\sin(u)^{2}\cos\!\left(\frac{u}{2}\right)^{2}\sin(v)\,k\,\cos(v)$$

$$-\sin(u)\,\sin\!\left(\frac{u}{2}\right)^{2}\sin(2\,v)\,j\,\cos(v)-2\,\sin(u)^{2}\sin\!\left(\frac{u}{2}\right)^{2}\sin(2\,v)\,k\,\cos(2\,v)$$

$$+\frac{1}{2}\,\cos(u)\,\sin\!\left(\frac{u}{2}\right)^{2}\sin(v)\,j\,\cos(v)+2\,i\,\cos(u)\,\cos\!\left(\frac{u}{2}\right)^{2}\sin(v)\cos(2\,v)$$

$$-i\,\cos(u)\,\sin\!\left(\frac{u}{2}\right)^{2}\sin(2\,v)\cos(v)+\cos(u)\,\sin\!\left(\frac{u}{2}\right)^{2}\cos(2\,v)\,j\,\sin(2\,v)$$

$$+\cos(u)\,\cos\!\left(\frac{u}{2}\right)^{2}\sin(2\,v)\,j\,\cos(2\,v)-\cos\!\left(\frac{u}{2}\right)^{2}\cos(v)\,k\,\cos(u)^{2}\sin(v)$$

$$-2\,\sin\!\left(\frac{u}{2}\right)^{2}\cos(2\,v)\,k\,\cos(u)^{2}\sin(2\,v)+\frac{1}{2}\,\cos(u)\,\cos\!\left(\frac{u}{2}\right)^{2}\cos(v)\,j\,\sin(v)$$

$$+3\,i\,\cos(u)\,\sin\!\left(\frac{u}{2}\right)\cos(v)+6\,i\,\cos(u)\,\cos\!\left(\frac{u}{2}\right)\cos(2\,v)$$

$$-3\,k\,\sin(u)^{2}\cos\!\left(\frac{u}{2}\right)\cos(v)+6\,k\,\sin(u)^{2}\sin\!\left(\frac{u}{2}\right)\cos(2\,v)$$

$$+3\,\sin(u)\,j\,\sin\!\left(\frac{u}{2}\right)\cos(v)+6\,\sin(u)\,j\,\cos\!\left(\frac{u}{2}\right)\cos(2\,v)$$

$$-3\,\cos\!\left(\frac{u}{2}\right)\cos(v)\,k\,\cos(u)^{2}+6\,\sin\!\left(\frac{u}{2}\right)\cos(2\,v)\,k\,\cos(u)^{2}$$

```
> mn:=simplify(%);
```

$$mn:=-3\,k\,\cos\!\left(\frac{u}{2}\right)\cos(v)-2\,\sin(2\,v)\,k\,\cos(2\,v)+3\,i\,\cos(u)\,\sin\!\left(\frac{u}{2}\right)\cos(v)$$

$$+6\,i\,\cos(u)\,\cos\!\left(\frac{u}{2}\right)\cos(2\,v)+3\,\sin(u)\,j\,\sin\!\left(\frac{u}{2}\right)\cos(v)$$

$$+6\,\sin(u)\,j\,\cos\!\left(\frac{u}{2}\right)\cos(2\,v)+i\,\cos(u)\,\cos\!\left(\frac{u}{2}\right)\sin(v)\,\sin\!\left(\frac{u}{2}\right)\cos(v)$$

$$+2\,i\,\cos(u)\,\cos\!\left(\frac{u}{2}\right)^{2}\sin(v)\cos(2\,v)-2\,i\,\cos(u)\,\sin\!\left(\frac{u}{2}\right)\sin(2\,v)\cos\!\left(\frac{u}{2}\right)\cos(2\,v)$$

$$+\sin(u)\,\cos\!\left(\frac{u}{2}\right)\sin(v)\,j\,\sin\!\left(\frac{u}{2}\right)\cos(v)+2\,\sin(u)\,\cos\!\left(\frac{u}{2}\right)^{2}\sin(v)\,j\,\cos(2\,v)$$

$$-2\,\sin(u)\,\sin\!\left(\frac{u}{2}\right)\sin(2\,v)\,j\,\cos\!\left(\frac{u}{2}\right)\cos(2\,v)-i\,\cos(u)\,\sin(2\,v)\cos(v)$$

$$+ 2 \sin(2v) \, k \cos(2v) \cos\left(\frac{u}{2}\right)^2 + \cos(u) \cos(2v) \, j \sin(2v)$$

$$- \frac{1}{2} i \sin(u) \sin(v) \cos(v) - i \sin(u) \sin(2v) \cos(2v) - \cos\left(\frac{u}{2}\right)^2 \sin(v) \, k \cos(v)$$

$$- \sin(u) \sin(2v) \, j \cos(v) + \frac{1}{2} \cos(u) \sin(v) \, j \cos(v) + 6 \, k \sin\left(\frac{u}{2}\right) \cos(2v)$$

$$+ \sin\left(\frac{u}{2}\right) \sin(2v) \, k \cos\left(\frac{u}{2}\right) \cos(v) + i \cos(u) \sin(2v) \cos(v) \cos\left(\frac{u}{2}\right)^2$$

$$+ 2 \cos\left(\frac{u}{2}\right) \sin(v) \, k \sin\left(\frac{u}{2}\right) \cos(2v) + \sin(u) \sin(2v) \, j \cos(v) \cos\left(\frac{u}{2}\right)^2$$

```
> A:=coeff(mn,i); # Obtain coefficient of 'i'
```

$$A := 3 \cos(u) \sin\left(\frac{u}{2}\right) \cos(v) + 6 \cos(u) \cos\left(\frac{u}{2}\right) \cos(2v)$$

$$+ \cos(u) \cos\left(\frac{u}{2}\right) \sin(v) \sin\left(\frac{u}{2}\right) \cos(v) + 2 \cos(u) \cos\left(\frac{u}{2}\right)^2 \sin(v) \cos(2v)$$

$$- 2 \cos(u) \sin\left(\frac{u}{2}\right) \sin(2v) \cos\left(\frac{u}{2}\right) \cos(2v) - \cos(u) \sin(2v) \cos(v)$$

$$- \frac{1}{2} \sin(u) \sin(v) \cos(v) - \sin(u) \sin(2v) \cos(2v)$$

$$+ \cos(u) \sin(2v) \cos(v) \cos\left(\frac{u}{2}\right)^2$$

```
> B:=coeff(mn,j); # Obtain coefficient of 'j'
```

$$B := 3 \sin(u) \sin\left(\frac{u}{2}\right) \cos(v) + 6 \sin(u) \cos\left(\frac{u}{2}\right) \cos(2v)$$

$$+ \sin(u) \cos\left(\frac{u}{2}\right) \sin(v) \sin\left(\frac{u}{2}\right) \cos(v) + 2 \sin(u) \cos\left(\frac{u}{2}\right)^2 \sin(v) \cos(2v)$$

$$- 2 \sin(u) \sin\left(\frac{u}{2}\right) \sin(2v) \cos\left(\frac{u}{2}\right) \cos(2v) + \cos(u) \cos(2v) \sin(2v)$$

$$- \sin(u) \sin(2v) \cos(v) + \frac{1}{2} \cos(u) \sin(v) \cos(v) + \sin(u) \sin(2v) \cos(v) \cos\left(\frac{u}{2}\right)^2$$

```
> C:=coeff(mn,k); # Obtain coefficient of 'k'
```

$$C := -3 \cos\left(\frac{u}{2}\right) \cos(v) - 2 \sin(2v) \cos(2v) + 2 \sin(2v) \cos(2v) \cos\left(\frac{u}{2}\right)^2$$

$$- \cos\left(\frac{u}{2}\right)^2 \sin(v) \cos(v) + 6 \sin\left(\frac{u}{2}\right) \cos(2v) + \sin\left(\frac{u}{2}\right) \sin(2v) \cos\left(\frac{u}{2}\right) \cos(v)$$

$$+ 2 \cos\left(\frac{u}{2}\right) \sin(v) \sin\left(\frac{u}{2}\right) \cos(2 v)$$

```
> [A,B,C]; # [A,B,C] structure 'tangent plane normal vector'
```

$$\left[\cos(u) \cos\left(\frac{u}{2}\right) \sin(v) \sin\left(\frac{u}{2}\right) \cos(v) - 2 \cos(u) \cos\left(\frac{u}{2}\right) \sin\left(\frac{u}{2}\right) \sin(2 v) \cos(2 v) \right.$$

$$+ 2 \cos(u) \cos\left(\frac{u}{2}\right)^2 \sin(v) \cos(2 v) + 6 \cos(u) \cos\left(\frac{u}{2}\right) \cos(2 v)$$

$$+ 3 \cos(u) \sin\left(\frac{u}{2}\right) \cos(v) - \cos(u) \sin(2 v) \cos(v) - \frac{1}{2} \sin(v) \sin(u) \cos(v)$$

$$- \sin(2 v) \sin(u) \cos(2 v) + \cos(u) \cos\left(\frac{u}{2}\right)^2 \sin(2 v) \cos(v),$$

$$2 \cos\left(\frac{u}{2}\right)^2 \sin(v) \sin(u) \cos(2 v) + 6 \cos\left(\frac{u}{2}\right) \sin(u) \cos(2 v)$$

$$+ 3 \sin\left(\frac{u}{2}\right) \sin(u) \cos(v) - \sin(2 v) \sin(u) \cos(v) + \frac{1}{2} \cos(u) \sin(v) \cos(v)$$

$$+ \cos(u) \sin(2 v) \cos(2 v) + \cos\left(\frac{u}{2}\right)^2 \sin(2 v) \sin(u) \cos(v)$$

$$+ \cos\left(\frac{u}{2}\right) \sin(v) \sin\left(\frac{u}{2}\right) \sin(u) \cos(v) - 2 \cos\left(\frac{u}{2}\right) \sin\left(\frac{u}{2}\right) \sin(2 v) \sin(u) \cos(2 v),$$

$$2 \cos\left(\frac{u}{2}\right) \sin(v) \sin\left(\frac{u}{2}\right) \cos(2 v) + 2 \cos\left(\frac{u}{2}\right)^2 \sin(2 v) \cos(2 v)$$

$$- \cos\left(\frac{u}{2}\right)^2 \sin(v) \cos(v) + \cos\left(\frac{u}{2}\right) \sin\left(\frac{u}{2}\right) \sin(2 v) \cos(v) - 3 \cos\left(\frac{u}{2}\right) \cos(v)$$

$$\left. + 6 \sin\left(\frac{u}{2}\right) \cos(2 v) - 2 \sin(2 v) \cos(2 v) \right]$$

```
> x:='x':y:='y':z:='z':
> Int(Int(V[1]*A+V[2]*B+V[3]*C,u=rgu[1]..rgu[2]),v=rgv[1]..rgv[2])
=Int(int(V[1]*A+V[2]*B+V[3]*C,u=rgu[1]..rgu[2]),v=rgv[1]..rgv[2]);
# Integrals of Dot Product (Vector 'V' and Tangent plane normal vector
[A,B,C] of 'Klein bottle') in range of 'u,v'
```

$$\int_0^\pi \int_0^{4\pi} P(x, y, z) \left(3 \cos(u) \sin\left(\frac{u}{2}\right) \cos(v) + 6 \cos(u) \cos\left(\frac{u}{2}\right) \cos(2 v) \right.$$

$$+ \cos(u)\cos\left(\frac{u}{2}\right)\sin(v)\sin\left(\frac{u}{2}\right)\cos(v) + 2\cos(u)\cos\left(\frac{u}{2}\right)^2\sin(v)\cos(2v)$$

$$- 2\cos(u)\sin\left(\frac{u}{2}\right)\sin(2v)\cos\left(\frac{u}{2}\right)\cos(2v) - \cos(u)\sin(2v)\cos(v)$$

$$- \frac{1}{2}\sin(u)\sin(v)\cos(v) - \sin(u)\sin(2v)\cos(2v)$$

$$+ \cos(u)\sin(2v)\cos(v)\cos\left(\frac{u}{2}\right)^2\Bigg) + Q(x,y,z)\Bigg(3\sin(u)\sin\left(\frac{u}{2}\right)\cos(v)$$

$$+ 6\sin(u)\cos\left(\frac{u}{2}\right)\cos(2v) + \sin(u)\cos\left(\frac{u}{2}\right)\sin(v)\sin\left(\frac{u}{2}\right)\cos(v)$$

$$+ 2\sin(u)\cos\left(\frac{u}{2}\right)^2\sin(v)\cos(2v) - 2\sin(u)\sin\left(\frac{u}{2}\right)\sin(2v)\cos\left(\frac{u}{2}\right)\cos(2v)$$

$$+ \cos(u)\cos(2v)\sin(2v) - \sin(u)\sin(2v)\cos(v) + \frac{1}{2}\cos(u)\sin(v)\cos(v)$$

$$+ \sin(u)\sin(2v)\cos(v)\cos\left(\frac{u}{2}\right)^2\Bigg) + R(x,y,z)\Bigg(-3\cos\left(\frac{u}{2}\right)\cos(v)$$

$$- 2\sin(2v)\cos(2v) + 2\sin(2v)\cos(2v)\cos\left(\frac{u}{2}\right)^2 - \cos\left(\frac{u}{2}\right)^2\sin(v)\cos(v)$$

$$+ 6\sin\left(\frac{u}{2}\right)\cos(2v) + \sin\left(\frac{u}{2}\right)\sin(2v)\cos\left(\frac{u}{2}\right)\cos(v)$$

$$+ 2\cos\left(\frac{u}{2}\right)\sin(v)\sin\left(\frac{u}{2}\right)\cos(2v)\Bigg) du\, dv = \int_0^\pi -16\,R(x,y,z)\,\pi\sin(v)\cos(v)^3$$

$$+ 6\,R(x,y,z)\sin(v)\cos(v)\,\pi + 5\,Q(x,y,z)\sin(v)\cos(v)\,\pi$$
$$+ 6\,P(x,y,z)\cos(v)^2\,\pi\sin(v) - 8\,Q(x,y,z)\,\pi\sin(v)\cos(v)^3$$
$$- 2\,P(x,y,z)\sin(v)\,\pi\, dv$$

```
> alpha:=rhs(value(%)); # Calculate value of surface integral
```

$$\alpha := 0$$

```
> x:='x':y:='y':z:='z':
> [r*CS[1],r*CS[2],r*CS[3]];
```

Entirely multiply radius vector 'r' (suppose r>0) by each item of Klein Bottle surface 'CS' parametrized expression, Convert 'Parametrized expression of Klein Bottle surface' to 'Parametrized expression of Klein Bottle surface coordinates'

$$\left[r\cos(u)\left(\cos\!\left(\frac{u}{2}\right)\sin(v)-\sin\!\left(\frac{u}{2}\right)\sin(2\,v)+3\right),\right.$$

$$r\sin(u)\left(\cos\!\left(\frac{u}{2}\right)\sin(v)-\sin\!\left(\frac{u}{2}\right)\sin(2\,v)+3\right),$$

$$\left. r\left(\sin\!\left(\frac{u}{2}\right)\sin(v)+\cos\!\left(\frac{u}{2}\right)\sin(2\,v)\right)\right]$$

```
> x:=r*CS[1]:y:=r*CS[2]:z:=r*CS[3]:
> matrix(3,3,[Diff(r*CS[1],r),Diff(r*CS[1],u),Diff(r*CS[1],v),
Diff(r*CS[2],r),Diff(r*CS[2],u),Diff(r*CS[2],v),Diff(r*CS[3],r),
Diff(r*CS[3],u),Diff(r*CS[3],v)])=
matrix(3,3,[diff(r*CS[1],r),diff(r*CS[1],u),diff(r*CS[1],v),
diff(r*CS[2],r),diff(r*CS[2],u),diff(r*CS[2],v),diff(r*CS[3],r),
diff(r*CS[3],u),diff(r*CS[3],v)]);m2:=rhs(%);
```

Define and calculate matrix of partial derivatives 'm2', obtain general expression of Klein Bottle surface coordinates volume element coefficient

$$\left[\frac{\partial}{\partial r}\left(r\cos(u)\left(\cos\!\left(\frac{u}{2}\right)\sin(v)-\sin\!\left(\frac{u}{2}\right)\sin(2\,v)+3\right)\right),\right.$$

$$\frac{\partial}{\partial u}\left(r\cos(u)\left(\cos\!\left(\frac{u}{2}\right)\sin(v)-\sin\!\left(\frac{u}{2}\right)\sin(2\,v)+3\right)\right),$$

$$\left.\frac{\partial}{\partial v}\left(r\cos(u)\left(\cos\!\left(\frac{u}{2}\right)\sin(v)-\sin\!\left(\frac{u}{2}\right)\sin(2\,v)+3\right)\right)\right]$$

$$\left[\frac{\partial}{\partial r}\left(r\sin(u)\left(\cos\!\left(\frac{u}{2}\right)\sin(v)-\sin\!\left(\frac{u}{2}\right)\sin(2\,v)+3\right)\right),\right.$$

$$\frac{\partial}{\partial u}\left(r\sin(u)\left(\cos\!\left(\frac{u}{2}\right)\sin(v)-\sin\!\left(\frac{u}{2}\right)\sin(2\,v)+3\right)\right),$$

$$\left.\frac{\partial}{\partial v}\left(r\sin(u)\left(\cos\!\left(\frac{u}{2}\right)\sin(v)-\sin\!\left(\frac{u}{2}\right)\sin(2\,v)+3\right)\right)\right]$$

$$\left[\frac{\partial}{\partial r}\left(r\left(\sin\!\left(\frac{u}{2}\right)\sin(v)+\cos\!\left(\frac{u}{2}\right)\sin(2\,v)\right)\right),\right.$$

$$\frac{\partial}{\partial u}\left(r\left(\sin\!\left(\frac{u}{2}\right)\sin(v)+\cos\!\left(\frac{u}{2}\right)\sin(2\,v)\right)\right),$$

$$\left.\frac{\partial}{\partial v}\left(r\left(\sin\!\left(\frac{u}{2}\right)\sin(v)+\cos\!\left(\frac{u}{2}\right)\sin(2\,v)\right)\right)\right]=$$

$$\left[\cos(u)\left(\cos\!\left(\frac{u}{2}\right)\sin(v)-\sin\!\left(\frac{u}{2}\right)\sin(2\,v)+3\right),\right.$$

$$-r\sin(u)\left(\cos\!\left(\frac{u}{2}\right)\sin(v)-\sin\!\left(\frac{u}{2}\right)\sin(2\,v)+3\right)$$

$$+r\cos(u)\left(-\frac{1}{2}\sin\!\left(\frac{u}{2}\right)\sin(v)-\frac{1}{2}\cos\!\left(\frac{u}{2}\right)\sin(2\,v)\right),$$

$$r \cos(u) \left(\cos\left(\frac{u}{2}\right) \cos(v) - 2 \sin\left(\frac{u}{2}\right) \cos(2\,v) \right) \Big]$$

$$\Big[\sin(u) \left(\cos\left(\frac{u}{2}\right) \sin(v) - \sin\left(\frac{u}{2}\right) \sin(2\,v) + 3 \right),$$

$$r \cos(u) \left(\cos\left(\frac{u}{2}\right) \sin(v) - \sin\left(\frac{u}{2}\right) \sin(2\,v) + 3 \right)$$

$$+ r \sin(u) \left(-\frac{1}{2} \sin\left(\frac{u}{2}\right) \sin(v) - \frac{1}{2} \cos\left(\frac{u}{2}\right) \sin(2\,v) \right),$$

$$r \sin(u) \left(\cos\left(\frac{u}{2}\right) \cos(v) - 2 \sin\left(\frac{u}{2}\right) \cos(2\,v) \right) \Big]$$

$$\Big[\sin\left(\frac{u}{2}\right) \sin(v) + \cos\left(\frac{u}{2}\right) \sin(2\,v), r \left(\frac{1}{2} \cos\left(\frac{u}{2}\right) \sin(v) - \frac{1}{2} \sin\left(\frac{u}{2}\right) \sin(2\,v) \right),$$

$$r \left(\sin\left(\frac{u}{2}\right) \cos(v) + 2 \cos\left(\frac{u}{2}\right) \cos(2\,v) \right) \Big]$$

$m2 :=$

$$\Big[\cos(u) \left(\cos\left(\frac{u}{2}\right) \sin(v) - \sin\left(\frac{u}{2}\right) \sin(2\,v) + 3 \right),$$

$$-r \sin(u) \left(\cos\left(\frac{u}{2}\right) \sin(v) - \sin\left(\frac{u}{2}\right) \sin(2\,v) + 3 \right)$$

$$+ r \cos(u) \left(-\frac{1}{2} \sin\left(\frac{u}{2}\right) \sin(v) - \frac{1}{2} \cos\left(\frac{u}{2}\right) \sin(2\,v) \right),$$

$$r \cos(u) \left(\cos\left(\frac{u}{2}\right) \cos(v) - 2 \sin\left(\frac{u}{2}\right) \cos(2\,v) \right) \Big]$$

$$\Big[\sin(u) \left(\cos\left(\frac{u}{2}\right) \sin(v) - \sin\left(\frac{u}{2}\right) \sin(2\,v) + 3 \right),$$

$$r \cos(u) \left(\cos\left(\frac{u}{2}\right) \sin(v) - \sin\left(\frac{u}{2}\right) \sin(2\,v) + 3 \right)$$

$$+ r \sin(u) \left(-\frac{1}{2} \sin\left(\frac{u}{2}\right) \sin(v) - \frac{1}{2} \cos\left(\frac{u}{2}\right) \sin(2\,v) \right),$$

$$r \sin(u) \left(\cos\left(\frac{u}{2}\right) \cos(v) - 2 \sin\left(\frac{u}{2}\right) \cos(2\,v) \right) \Big]$$

$$\Big[\sin\left(\frac{u}{2}\right) \sin(v) + \cos\left(\frac{u}{2}\right) \sin(2\,v), r \left(\frac{1}{2} \cos\left(\frac{u}{2}\right) \sin(v) - \frac{1}{2} \sin\left(\frac{u}{2}\right) \sin(2\,v) \right),$$

$$r \left(\sin\left(\frac{u}{2}\right) \cos(v) + 2 \cos\left(\frac{u}{2}\right) \cos(2\,v) \right) \Big]$$

> det(m2);

$$-2 \sin\left(\frac{u}{2}\right)^3 \sin(v)\, r^2 \cos(u)^2 \cos(2\,v) \sin(2\,v) + 2 \cos\left(\frac{u}{2}\right)^3 \sin(v)^2\, r^2 \cos(u)^2 \cos(2\,v)$$

$$- \cos\left(\frac{u}{2}\right) \sin(v)\, r^2 \cos(u)^2 \sin\left(\frac{u}{2}\right)^2 \sin(2\,v) \cos(v)$$

$$-2\cos\left(\frac{u}{2}\right)^2 \sin(v)\, r^2 \cos(u)^2 \sin\left(\frac{u}{2}\right) \sin(2v) \cos(2v)$$

$$+\sin\left(\frac{u}{2}\right)^3 \sin(2v)^2 r^2 \cos(u)^2 \cos(v) + 3\, r^2 \cos(u)^2 \cos\left(\frac{u}{2}\right) \sin(v) \sin\left(\frac{u}{2}\right) \cos(v)$$

$$+12\, r^2 \cos(u)^2 \cos\left(\frac{u}{2}\right)^2 \sin(v) \cos(2v) - 6\, r^2 \cos(u)^2 \sin\left(\frac{u}{2}\right)^2 \sin(2v) \cos(v)$$

$$-6\, r^2 \cos(u)^2 \sin\left(\frac{u}{2}\right) \sin(2v) \cos\left(\frac{u}{2}\right) \cos(2v)$$

$$+12\, r^2 \sin(u)^2 \cos\left(\frac{u}{2}\right)^2 \sin(v) \cos(2v)$$

$$+3\, r^2 \sin(u)^2 \cos\left(\frac{u}{2}\right) \sin(v) \sin\left(\frac{u}{2}\right) \cos(v) - 6\, r^2 \sin(u)^2 \sin\left(\frac{u}{2}\right)^2 \sin(2v) \cos(v)$$

$$-6\, r^2 \sin(u)^2 \sin\left(\frac{u}{2}\right) \sin(2v) \cos\left(\frac{u}{2}\right) \cos(2v)$$

$$+2\cos\left(\frac{u}{2}\right)^3 \sin(v)^2 r^2 \sin(u)^2 \cos(2v)$$

$$-\cos\left(\frac{u}{2}\right) \sin(v)\, r^2 \sin(u)^2 \sin\left(\frac{u}{2}\right)^2 \sin(2v) \cos(v)$$

$$-2\cos\left(\frac{u}{2}\right)^2 \sin(v)\, r^2 \sin(u)^2 \sin\left(\frac{u}{2}\right) \sin(2v) \cos(2v)$$

$$+\sin\left(\frac{u}{2}\right)^3 \sin(2v)^2 r^2 \sin(u)^2 \cos(v) + 6\sin\left(\frac{u}{2}\right)^2 \sin(v)\, r^2 \cos(u)^2 \cos(2v)$$

$$-\cos\left(\frac{u}{2}\right)^3 \sin(2v)\, r^2 \sin(u)^2 \sin(v) \cos(v)$$

$$+\cos\left(\frac{u}{2}\right)^2 \sin(2v)^2 r^2 \sin(u)^2 \sin\left(\frac{u}{2}\right) \cos(v)$$

$$-3\cos\left(\frac{u}{2}\right)^2 \sin(2v)\, r^2 \sin(u)^2 \cos(v) - \cos\left(\frac{u}{2}\right)^3 \sin(2v)\, r^2 \cos(u)^2 \cos(v) \sin(v)$$

$$+\cos\left(\frac{u}{2}\right)^2 \sin(2v)^2 r^2 \cos(u)^2 \sin\left(\frac{u}{2}\right) \cos(v)$$

$$-3\cos\left(\frac{u}{2}\right)^2 \sin(2v)\, r^2 \cos(u)^2 \cos(v)$$

$$+ 2 \sin\left(\frac{u}{2}\right)^2 \sin(v)^2 \, r^2 \sin(u)^2 \cos\left(\frac{u}{2}\right) \cos(2\,v)$$

$$- 2 \sin\left(\frac{u}{2}\right)^3 \sin(v) \, r^2 \sin(u)^2 \sin(2\,v) \cos(2\,v)$$

$$+ 6 \sin\left(\frac{u}{2}\right)^2 \sin(v) \, r^2 \sin(u)^2 \cos(2\,v)$$

$$+ 2 \sin\left(\frac{u}{2}\right)^2 \sin(v)^2 \, r^2 \cos(u)^2 \cos\left(\frac{u}{2}\right) \cos(2\,v) + 9\, r^2 \sin(u)^2 \sin\left(\frac{u}{2}\right) \cos(v)$$

$$+ 9\, r^2 \cos(u)^2 \sin\left(\frac{u}{2}\right) \cos(v) + 18\, r^2 \cos(u)^2 \cos\left(\frac{u}{2}\right) \cos(2\,v)$$

$$+ 18\, r^2 \sin(u)^2 \cos\left(\frac{u}{2}\right) \cos(2\,v)$$

```
> J:=simplify(%);
```

$$J := r^2 \left(-2 \sin\left(\frac{u}{2}\right) \sin(v) \sin(2\,v) \cos(2\,v) - \sin\left(\frac{u}{2}\right) \cos(v) \cos(2\,v)^2 \right.$$

$$+ 3 \cos\left(\frac{u}{2}\right) \sin(v) \sin\left(\frac{u}{2}\right) \cos(v) - 6 \sin\left(\frac{u}{2}\right) \sin(2\,v) \cos\left(\frac{u}{2}\right) \cos(2\,v)$$

$$- \cos\left(\frac{u}{2}\right) \sin(v) \sin(2\,v) \cos(v) + 3 \cos\left(\frac{u}{2}\right)^2 \sin(2\,v) \cos(v) + 6 \sin(v) \cos(2\,v)$$

$$- 6 \sin(2\,v) \cos(v) + 10 \sin\left(\frac{u}{2}\right) \cos(v) - 2 \cos\left(\frac{u}{2}\right) \cos(2\,v) \cos(v)^2$$

$$\left. + 20 \cos\left(\frac{u}{2}\right) \cos(2\,v) + 6 \sin(v) \cos(2\,v) \cos\left(\frac{u}{2}\right)^2 \right)$$

```
> x:='x':y:='y':z:='z':
> Diff(V[1],x)*Diff(r*CS[1],r)*Diff(r*CS[1],u)*Diff(r*CS[1],v)
+Diff(V[2],y)*Diff(r*CS[2],r)*Diff(r*CS[2],u)*Diff(r*CS[2],v)
+Diff(V[3],z)*Diff(r*CS[3],r)*Diff(r*CS[3],u)*Diff(r*CS[3],v);
diV2:=value(%);
# Convert 'diV1' from Cartesian coordinates expression to Klein bottle
surface Coordinates expression
```

$$\left(\frac{\partial}{\partial x} P(x,y,z)\right)\left(\frac{\partial}{\partial r}\left(r \cos(u)\left(\cos\left(\frac{u}{2}\right)\sin(v) - \sin\left(\frac{u}{2}\right)\sin(2\,v) + 3\right)\right)\right)$$

$$\left(\frac{\partial}{\partial u}\left(r \cos(u)\left(\cos\left(\frac{u}{2}\right)\sin(v) - \sin\left(\frac{u}{2}\right)\sin(2\,v) + 3\right)\right)\right)$$

$$\left(\frac{\partial}{\partial v}\left(r\cos(u)\left(\cos\left(\frac{u}{2}\right)\sin(v)-\sin\left(\frac{u}{2}\right)\sin(2\,v)+3\right)\right)\right)+\left(\frac{\partial}{\partial y}\,Q(x,y,z)\right)$$

$$\left(\frac{\partial}{\partial r}\left(r\sin(u)\left(\cos\left(\frac{u}{2}\right)\sin(v)-\sin\left(\frac{u}{2}\right)\sin(2\,v)+3\right)\right)\right)$$

$$\left(\frac{\partial}{\partial u}\left(r\sin(u)\left(\cos\left(\frac{u}{2}\right)\sin(v)-\sin\left(\frac{u}{2}\right)\sin(2\,v)+3\right)\right)\right)$$

$$\left(\frac{\partial}{\partial v}\left(r\sin(u)\left(\cos\left(\frac{u}{2}\right)\sin(v)-\sin\left(\frac{u}{2}\right)\sin(2\,v)+3\right)\right)\right)+\left(\frac{\partial}{\partial z}\,R(x,y,z)\right)$$

$$\left(\frac{\partial}{\partial r}\left(r\left(\sin\left(\frac{u}{2}\right)\sin(v)+\cos\left(\frac{u}{2}\right)\sin(2\,v)\right)\right)\right)$$

$$\left(\frac{\partial}{\partial u}\left(r\left(\sin\left(\frac{u}{2}\right)\sin(v)+\cos\left(\frac{u}{2}\right)\sin(2\,v)\right)\right)\right)$$

$$\left(\frac{\partial}{\partial v}\left(r\left(\sin\left(\frac{u}{2}\right)\sin(v)+\cos\left(\frac{u}{2}\right)\sin(2\,v)\right)\right)\right)$$

$$diV2:=\left(\frac{\partial}{\partial x}\,P(x,y,z)\right)\cos(u)^2\left(\cos\left(\frac{u}{2}\right)\sin(v)-\sin\left(\frac{u}{2}\right)\sin(2\,v)+3\right)\left(\right.$$

$$-r\sin(u)\left(\cos\left(\frac{u}{2}\right)\sin(v)-\sin\left(\frac{u}{2}\right)\sin(2\,v)+3\right)$$

$$+r\cos(u)\left(-\frac{1}{2}\sin\left(\frac{u}{2}\right)\sin(v)-\frac{1}{2}\cos\left(\frac{u}{2}\right)\sin(2\,v)\right)\right)r$$

$$\left(\cos\left(\frac{u}{2}\right)\cos(v)-2\sin\left(\frac{u}{2}\right)\cos(2\,v)\right)+\left(\frac{\partial}{\partial y}\,Q(x,y,z)\right)\sin(u)^2$$

$$\left(\cos\left(\frac{u}{2}\right)\sin(v)-\sin\left(\frac{u}{2}\right)\sin(2\,v)+3\right)\left(\right.$$

$$r\cos(u)\left(\cos\left(\frac{u}{2}\right)\sin(v)-\sin\left(\frac{u}{2}\right)\sin(2\,v)+3\right)$$

$$+r\sin(u)\left(-\frac{1}{2}\sin\left(\frac{u}{2}\right)\sin(v)-\frac{1}{2}\cos\left(\frac{u}{2}\right)\sin(2\,v)\right)\right)r$$

$$\left(\cos\left(\frac{u}{2}\right)\cos(v)-2\sin\left(\frac{u}{2}\right)\cos(2\,v)\right)+\left(\frac{\partial}{\partial z}\,R(x,y,z)\right)$$

$$\left(\sin\left(\frac{u}{2}\right)\sin(v)+\cos\left(\frac{u}{2}\right)\sin(2\,v)\right)r^2\left(\frac{1}{2}\cos\left(\frac{u}{2}\right)\sin(v)-\frac{1}{2}\sin\left(\frac{u}{2}\right)\sin(2\,v)\right)$$

$$\left(\sin\left(\frac{u}{2}\right)\cos(v)+2\cos\left(\frac{u}{2}\right)\cos(2\,v)\right)$$

```
> Int(Int(Int(diV2*J,r=0..1),u=rgu[1]..rgu[2]),v=rgv[1]..rgv[2]);
```

\# Triple Integrals of Product (Divergence 'diV2' and Klein Bottle Surface Coordinates Volume Element) in range of 'r,u,v'

$$\int_0^\pi\int_0^{4\pi}\int_0^1\left(\left(\frac{\partial}{\partial x}\,P(x,y,z)\right)\cos(u)^2\left(\cos\left(\frac{u}{2}\right)\sin(v)-\sin\left(\frac{u}{2}\right)\sin(2\,v)+3\right)\left(\right.\right.$$

$$-r\sin(u)\left(\cos\left(\frac{u}{2}\right)\sin(v)-\sin\left(\frac{u}{2}\right)\sin(2v)+3\right)$$

$$+r\cos(u)\left(-\frac{1}{2}\sin\left(\frac{u}{2}\right)\sin(v)-\frac{1}{2}\cos\left(\frac{u}{2}\right)\sin(2v)\right)\right)r$$

$$\left(\cos\left(\frac{u}{2}\right)\cos(v)-2\sin\left(\frac{u}{2}\right)\cos(2v)\right)+\left(\frac{\partial}{\partial y}Q(x,y,z)\right)\sin(u)^2$$

$$\left(\cos\left(\frac{u}{2}\right)\sin(v)-\sin\left(\frac{u}{2}\right)\sin(2v)+3\right)\bigg($$

$$r\cos(u)\left(\cos\left(\frac{u}{2}\right)\sin(v)-\sin\left(\frac{u}{2}\right)\sin(2v)+3\right)$$

$$+r\sin(u)\left(-\frac{1}{2}\sin\left(\frac{u}{2}\right)\sin(v)-\frac{1}{2}\cos\left(\frac{u}{2}\right)\sin(2v)\right)\right)r$$

$$\left(\cos\left(\frac{u}{2}\right)\cos(v)-2\sin\left(\frac{u}{2}\right)\cos(2v)\right)+\left(\frac{\partial}{\partial z}R(x,y,z)\right)$$

$$\left(\sin\left(\frac{u}{2}\right)\sin(v)+\cos\left(\frac{u}{2}\right)\sin(2v)\right)r^2\left(\frac{1}{2}\cos\left(\frac{u}{2}\right)\sin(v)-\frac{1}{2}\sin\left(\frac{u}{2}\right)\sin(2v)\right)$$

$$\left(\sin\left(\frac{u}{2}\right)\cos(v)+2\cos\left(\frac{u}{2}\right)\cos(2v)\right)\right)r^2\left(-2\sin\left(\frac{u}{2}\right)\sin(v)\sin(2v)\cos(2v)\right.$$

$$-\sin\left(\frac{u}{2}\right)\cos(v)\cos(2v)^2+3\cos\left(\frac{u}{2}\right)\sin(v)\sin\left(\frac{u}{2}\right)\cos(v)$$

$$-6\sin\left(\frac{u}{2}\right)\sin(2v)\cos\left(\frac{u}{2}\right)\cos(2v)-\cos\left(\frac{u}{2}\right)\sin(v)\sin(2v)\cos(v)$$

$$+3\cos\left(\frac{u}{2}\right)^2\sin(2v)\cos(v)+6\sin(v)\cos(2v)-6\sin(2v)\cos(v)$$

$$+10\sin\left(\frac{u}{2}\right)\cos(v)-2\cos\left(\frac{u}{2}\right)\cos(2v)\cos(v)^2+20\cos\left(\frac{u}{2}\right)\cos(2v)$$

$$+6\sin(v)\cos(2v)\cos\left(\frac{u}{2}\right)^2\right)dr\,du\,dv$$

```
> beta:=value(%); # Calculate value of volume element integral
```

$$\beta:=\frac{471}{64}\left(\frac{\partial}{\partial x}P(x,y,z)\right)\pi^2$$

```
> alpha;beta;
```

$$0$$

$$\frac{471}{64}\left(\frac{\partial}{\partial x}P(x,y,z)\right)\pi^2$$

// Conclude and deduce 'Klein bottle' by logic method of 'Prove Divergence Theorem at Manifold', surface integral at Klein bottle surface and triple integrals in spacial bounded closed region of 'Klein bottle' are unequal in logic.

6.3 Surface Integral and Triple Integrals about Klein Bottle, Numerical Model

Known: Parametric Expression of Klein Bottle:

(Unorientable、 Closed Surface)

[cos(u)*(cos(u/2)*sin(v)-sin(u/2)*sin(2*v)+3),

  sin(u)*(cos(u/2)*sin(v)-sin(u/2)*sin(2*v)+3),

  (sin(u/2)*sin(v)+cos(u/2)*sin(2*v))]   (1)

thereinto, u∈[0,4π], v∈[0,π]; and Integral Vector Field:

$$\left[\frac{\left(\frac{x}{3}+\frac{y}{6}+\frac{z}{5}\right)^2}{3} - \frac{y^3}{5} - \frac{xz^2}{7}, \frac{1}{7}xyz - \frac{1}{9}yz^2, \frac{1}{6}x^2y - \frac{1}{12}xyz - \frac{1}{9}z^3 \right]$$ (2)

Calculate and Validate Divergence Theorem at Manifold (Counterexample)

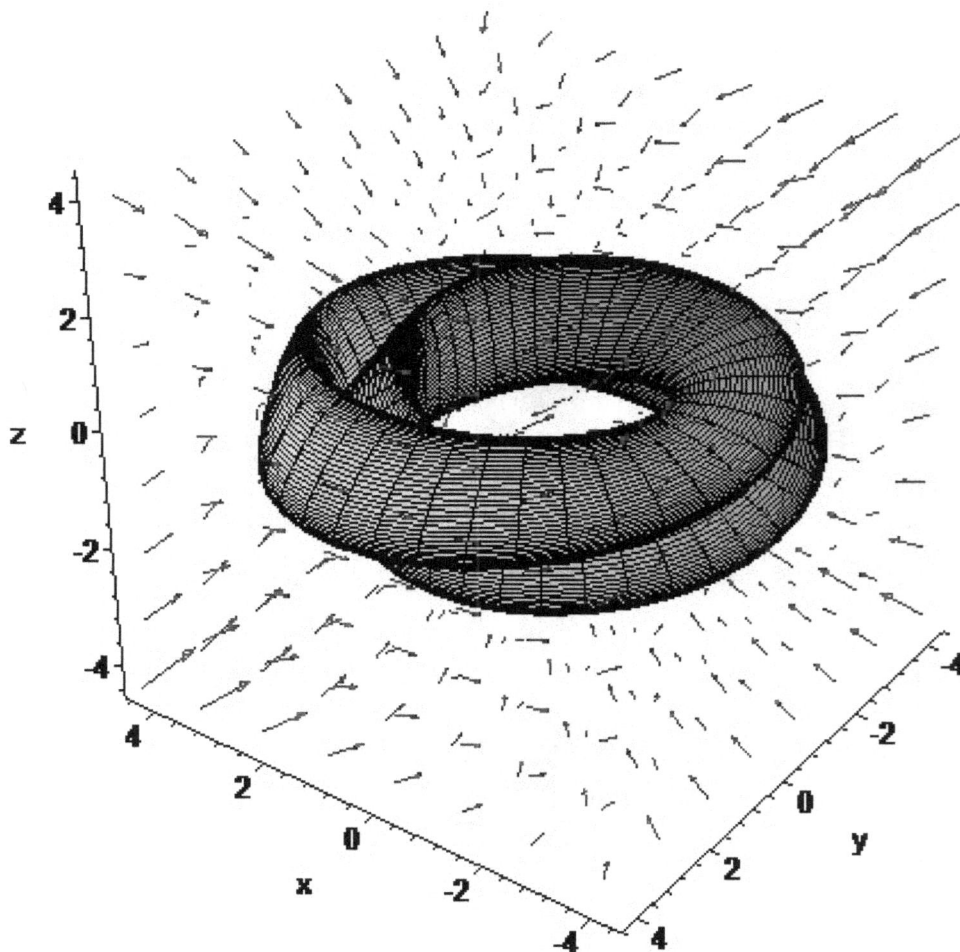

Figure(I)6.3 Klein bottle,Vector Field

Solution:

First, Surface Integral about Klein Bottle:

According to parametric expression of Klein Bottle (1), define and calculate First Matrix of Partial Derivatives, obtain tangent plane normal vector of Klein Bottle surface (3):

$$
\begin{bmatrix}
i & j & k \\
\frac{\partial}{\partial u}\cos(u)(\cos(u/2)\sin(v)-\sin(u/2)\sin(2v)+3) & \frac{\partial}{\partial u}\sin(u)(\cos(u/2)\sin(v)-\sin(u/2)\sin(2v)+3) & \frac{\partial}{\partial u}\sin(u/2)\sin(v)+\cos(u/2)\sin(2v) \\
\frac{\partial}{\partial v}\cos(u)(\cos(u/2)\sin(v)-\sin(u/2)\sin(2v)+3) & \frac{\partial}{\partial v}\sin(u)(\cos(u/2)\sin(v)-\sin(u/2)\sin(2v)+3) & \frac{\partial}{\partial v}\sin(u/2)\sin(v)+\cos(u/2)\sin(2v)
\end{bmatrix}
$$

$$
=
$$

$$
i\cos(u)\cos\!\left(\frac{u}{2}\right)\sin(v)\sin\!\left(\frac{u}{2}\right)\cos(v)-2\,i\cos(u)\cos\!\left(\frac{u}{2}\right)\sin\!\left(\frac{u}{2}\right)\sin(2\,v)\cos(2\,v)
$$

$$
+2\,i\cos(u)\cos\!\left(\frac{u}{2}\right)^2\sin(v)\cos(2\,v)+2\cos\!\left(\frac{u}{2}\right)^2\sin(v)\sin(u)\cos(2\,v)\,j
$$

$$
+2\cos\!\left(\frac{u}{2}\right)\sin(v)\sin\!\left(\frac{u}{2}\right)\cos(2\,v)\,k+6\,i\cos(u)\cos\!\left(\frac{u}{2}\right)\cos(2\,v)
$$

$$
+3\,i\cos(u)\sin\!\left(\frac{u}{2}\right)\cos(v)+6\cos\!\left(\frac{u}{2}\right)\sin(u)\cos(2\,v)\,j+3\sin\!\left(\frac{u}{2}\right)\sin(u)\cos(v)\,j
$$

$$
-i\cos(u)\sin(2\,v)\cos(v)-\frac{1}{2}i\sin(v)\sin(u)\cos(v)-i\sin(2\,v)\sin(u)\cos(2\,v)
$$

$$
-\sin(2\,v)\sin(u)\cos(v)\,j+\frac{1}{2}\cos(u)\sin(v)\cos(v)\,j+\cos(u)\sin(2\,v)\cos(2\,v)\,j
$$

$$
+2\cos\!\left(\frac{u}{2}\right)^2\sin(2\,v)\cos(2\,v)\,k-\cos\!\left(\frac{u}{2}\right)^2\sin(v)\cos(v)\,k
$$

$$
+\cos(u)\cos\!\left(\frac{u}{2}\right)^2\sin(2\,v)\cos(v)\,i+\cos\!\left(\frac{u}{2}\right)^2\sin(2\,v)\sin(u)\cos(v)\,j
$$

$$
+\cos\!\left(\frac{u}{2}\right)\sin\!\left(\frac{u}{2}\right)\sin(2\,v)\cos(v)\,k-3\cos\!\left(\frac{u}{2}\right)\cos(v)\,k+6\sin\!\left(\frac{u}{2}\right)\cos(2\,v)\,k
$$

$$
-2\sin(2\,v)\cos(2\,v)\,k+\cos\!\left(\frac{u}{2}\right)\sin(v)\sin\!\left(\frac{u}{2}\right)\sin(u)\cos(v)\,j
$$

$$
-2\cos\!\left(\frac{u}{2}\right)\sin\!\left(\frac{u}{2}\right)\sin(2\,v)\sin(u)\cos(2\,v)\,j
$$

$$
\tag{3}
$$

From expression (3), respectively pick up coefficients of item 'i,j,k', obtain tangent plane normal vector of Klein Bottle surface (4):

$$
\begin{aligned}
\Bigg[& \cos(u)\cos\!\left(\frac{u}{2}\right)\sin(v)\sin\!\left(\frac{u}{2}\right)\cos(v) - 2\cos(u)\cos\!\left(\frac{u}{2}\right)\sin\!\left(\frac{u}{2}\right)\sin(2v)\cos(2v) \\
& + 2\cos(u)\cos\!\left(\frac{u}{2}\right)^2\sin(v)\cos(2v) + 6\cos(u)\cos\!\left(\frac{u}{2}\right)\cos(2v) \\
& + 3\cos(u)\sin\!\left(\frac{u}{2}\right)\cos(v) - \cos(u)\sin(2v)\cos(v) - \frac{1}{2}\sin(v)\sin(u)\cos(v) \\
& - \sin(2v)\sin(u)\cos(2v) + \cos(u)\cos\!\left(\frac{u}{2}\right)^2\sin(2v)\cos(v),
\end{aligned}
$$

$$
\begin{aligned}
& 2\cos\!\left(\frac{u}{2}\right)^2\sin(v)\sin(u)\cos(2v) + 6\cos\!\left(\frac{u}{2}\right)\sin(u)\cos(2v) \\
& + 3\sin\!\left(\frac{u}{2}\right)\sin(u)\cos(v) - \sin(2v)\sin(u)\cos(v) + \frac{1}{2}\cos(u)\sin(v)\cos(v) \\[4pt]
& + \cos(u)\sin(2v)\cos(2v) + \cos\!\left(\frac{u}{2}\right)^2\sin(2v)\sin(u)\cos(v) \\
& + \cos\!\left(\frac{u}{2}\right)\sin(v)\sin\!\left(\frac{u}{2}\right)\sin(u)\cos(v) - 2\cos\!\left(\frac{u}{2}\right)\sin\!\left(\frac{u}{2}\right)\sin(2v)\sin(u)\cos(2v),
\end{aligned}
$$

$$
\begin{aligned}
& 2\cos\!\left(\frac{u}{2}\right)\sin(v)\sin\!\left(\frac{u}{2}\right)\cos(2v) + 2\cos\!\left(\frac{u}{2}\right)^2\sin(2v)\cos(2v) \\
& - \cos\!\left(\frac{u}{2}\right)^2\sin(v)\cos(v) + \cos\!\left(\frac{u}{2}\right)\sin\!\left(\frac{u}{2}\right)\sin(2v)\cos(v) - 3\cos\!\left(\frac{u}{2}\right)\cos(v) \\
& + 6\sin\!\left(\frac{u}{2}\right)\cos(2v) - 2\sin(2v)\cos(2v) \Bigg]
\end{aligned}
$$

(4)

Input expression (1) to Vector Field (2); and Calculate Integral of Dot Product [Vetor Field (2) and tangent plane normal vector (4)] in range of 'u,v' (5):

$$
\begin{aligned}
\int_0^{\pi}\int_0^{4\pi} & \left(\frac{1}{3}\left(\frac{1}{3}\cos(u)\left(\cos\!\left(\frac{u}{2}\right)\sin(v) - \sin\!\left(\frac{u}{2}\right)\sin(2v) + 3 \right) \right. \right. \\
& + \frac{1}{6}\sin(u)\left(\cos\!\left(\frac{u}{2}\right)\sin(v) - \sin\!\left(\frac{u}{2}\right)\sin(2v) + 3 \right) + \frac{1}{5}\sin\!\left(\frac{u}{2}\right)\sin(v) \\
& + \frac{1}{5}\cos\!\left(\frac{u}{2}\right)\sin(2v) \Big)^2 - \frac{1}{5}\sin(u)^3\left(\cos\!\left(\frac{u}{2}\right)\sin(v) - \sin\!\left(\frac{u}{2}\right)\sin(2v) + 3 \right)^3 - \frac{1}{7}
\end{aligned}
$$

$$\cos(u)\left(\cos\left(\frac{u}{2}\right)\sin(v)-\sin\left(\frac{u}{2}\right)\sin(2v)+3\right)\left(\sin\left(\frac{u}{2}\right)\sin(v)+\cos\left(\frac{u}{2}\right)\sin(2v)\right)^2\right)$$

$$\left(\cos(u)\cos\left(\frac{u}{2}\right)\sin(v)\sin\left(\frac{u}{2}\right)\cos(v)-2\cos(u)\sin\left(\frac{u}{2}\right)\sin(2v)\cos\left(\frac{u}{2}\right)\cos(2v)\right.$$

$$+2\cos(u)\cos\left(\frac{u}{2}\right)^2\sin(v)\cos(2v)+\cos(u)\sin(2v)\cos(v)\cos\left(\frac{u}{2}\right)^2$$

$$-\sin(u)\sin(2v)\cos(2v)-\cos(u)\sin(2v)\cos(v)-\frac{1}{2}\sin(u)\sin(v)\cos(v)$$

$$+3\cos(u)\sin\left(\frac{u}{2}\right)\cos(v)+6\cos(u)\cos\left(\frac{u}{2}\right)\cos(2v)\right)+\left(\frac{1}{7}\cos(u)\right.$$

$$\left(\cos\left(\frac{u}{2}\right)\sin(v)-\sin\left(\frac{u}{2}\right)\sin(2v)+3\right)^2\sin(u)\left(\sin\left(\frac{u}{2}\right)\sin(v)+\cos\left(\frac{u}{2}\right)\sin(2v)\right)$$

$$-\frac{1}{9}$$

$$\sin(u)\left(\cos\left(\frac{u}{2}\right)\sin(v)-\sin\left(\frac{u}{2}\right)\sin(2v)+3\right)\left(\sin\left(\frac{u}{2}\right)\sin(v)+\cos\left(\frac{u}{2}\right)\sin(2v)\right)^2\right)$$

$$\left(-2\sin(u)\sin\left(\frac{u}{2}\right)\sin(2v)\cos\left(\frac{u}{2}\right)\cos(2v)+\sin(u)\cos\left(\frac{u}{2}\right)\sin(v)\sin\left(\frac{u}{2}\right)\cos(v)\right.$$

$$+2\sin(u)\cos\left(\frac{u}{2}\right)^2\sin(v)\cos(2v)+\frac{1}{2}\cos(u)\sin(v)\cos(v)$$

$$+\sin(u)\sin(2v)\cos(v)\cos\left(\frac{u}{2}\right)^2+\cos(u)\cos(2v)\sin(2v)$$

$$-\sin(u)\sin(2v)\cos(v)+3\sin(u)\sin\left(\frac{u}{2}\right)\cos(v)+6\sin(u)\cos\left(\frac{u}{2}\right)\cos(2v)\right)+\left(\rule{0pt}{2.5em}\right.$$

$$\frac{1}{6}\cos(u)^2\left(\cos\left(\frac{u}{2}\right)\sin(v)-\sin\left(\frac{u}{2}\right)\sin(2v)+3\right)^3\sin(u)-\frac{1}{12}\cos(u)$$

$$\left(\cos\left(\frac{u}{2}\right)\sin(v)-\sin\left(\frac{u}{2}\right)\sin(2v)+3\right)^2\sin(u)\left(\sin\left(\frac{u}{2}\right)\sin(v)+\cos\left(\frac{u}{2}\right)\sin(2v)\right)$$

$$-\frac{1}{9}\left(\sin\left(\frac{u}{2}\right)\sin(v)+\cos\left(\frac{u}{2}\right)\sin(2v)\right)^3\right)\left(-2\sin(2v)\cos(2v)\right.$$

$$+6\sin\left(\frac{u}{2}\right)\cos(2v)-3\cos\left(\frac{u}{2}\right)\cos(v)-\cos\left(\frac{u}{2}\right)^2\sin(v)\cos(v)$$

$$+ 2\sin(2v)\cos(2v)\cos\left(\frac{u}{2}\right)^2 + 2\cos\left(\frac{u}{2}\right)\sin(v)\sin\left(\frac{u}{2}\right)\cos(2v)$$

$$+ \sin\left(\frac{u}{2}\right)\sin(2v)\cos\left(\frac{u}{2}\right)\cos(v)\Bigg)\, du\, dv$$

$$= \frac{19424\,\pi}{6615}$$

(5)

Second, Volume Element Integral about Klein Bottle:

Entirely multiply radius vector 'r' (suppose r>0) by each item of Klein Bottle surface parametrized expression (1), convert 'Parametrized expression of Klein Bottle surface' to 'Parametrized expression of Klein Bottle surface coordinates' (6):

$$\left[r\cos(u)\left(\cos\left(\frac{u}{2}\right)\sin(v) - \sin\left(\frac{u}{2}\right)\sin(2v) + 3 \right), \right.$$

$$r\sin(u)\left(\cos\left(\frac{u}{2}\right)\sin(v) - \sin\left(\frac{u}{2}\right)\sin(2v) + 3 \right),$$

$$\left. r\left(\sin\left(\frac{u}{2}\right)\sin(v) + \cos\left(\frac{u}{2}\right)\sin(2v) \right) \right]$$

(6)

According to parametrized expression of Klein Bottle surface coordinates (6), define and calculate Second Matrix of Partial Derivatives, obtain general expression of Klein Bottle surface coordinates volume element coefficient (7):

$$\begin{bmatrix} \frac{\partial}{\partial r} r\cos(u)(\cos(\frac{u}{2})\sin(v) - \sin(\frac{u}{2})\sin(2v) + 3) & \frac{\partial}{\partial u} r\cos(u)(\cos(\frac{u}{2})\sin(v) - \sin(\frac{u}{2})\sin(2v) + 3) & \frac{\partial}{\partial v} r\cos(u)(\cos(\frac{u}{2})\sin(v) - \sin(\frac{u}{2})\sin(2v) + 3) \\ \frac{\partial}{\partial r} r\sin(u)(\cos(\frac{u}{2})\sin(v) - \sin(\frac{u}{2})\sin(2v) + 3) & \frac{\partial}{\partial u} r\sin(u)(\cos(\frac{u}{2})\sin(v) - \sin(\frac{u}{2})\sin(2v) + 3) & \frac{\partial}{\partial v} r\sin(u)(\cos(\frac{u}{2})\sin(v) - \sin(\frac{u}{2})\sin(2v) + 3) \\ \frac{\partial}{\partial r} r(\sin(\frac{u}{2})\sin(v) + \cos(\frac{u}{2})\sin(2v)) & \frac{\partial}{\partial u} r(\sin(\frac{u}{2})\sin(v) + \cos(\frac{u}{2})\sin(2v)) & \frac{\partial}{\partial v} r(\sin(\frac{u}{2})\sin(v) + \cos(\frac{u}{2})\sin(2v)) \end{bmatrix}$$

$$=$$

$$r^2 \left(6\cos\left(\frac{u}{2}\right)^2 \sin(v)\cos(2v) + 3\cos\left(\frac{u}{2}\right)^2 \sin(2v)\cos(v) \right.$$

$$+ 3\cos\left(\frac{u}{2}\right)\sin(v)\sin\left(\frac{u}{2}\right)\cos(v) - \cos\left(\frac{u}{2}\right)\sin(v)\sin(2v)\cos(v)$$

$$- 6\cos\left(\frac{u}{2}\right)\sin\left(\frac{u}{2}\right)\sin(2v)\cos(2v) - 2\cos\left(\frac{u}{2}\right)\cos(v)^2\cos(2v)$$

$$- 2\sin(v)\sin\left(\frac{u}{2}\right)\sin(2v)\cos(2v) - \sin\left(\frac{u}{2}\right)\cos(v)\cos(2v)^2 + 20\cos\left(\frac{u}{2}\right)\cos(2v)$$

$$+\ 6\sin(v)\cos(2\,v)+10\sin\!\left(\frac{u}{2}\right)\cos(v)-6\sin(2\,v)\cos(v)\Bigg) \tag{7}$$

Calculate Divergence of Vector Field (8):

$$\left(\frac{\partial}{\partial x}\left(\frac{\left(\frac{x}{3}+\frac{y}{6}+\frac{z}{5}\right)^2}{3}-\frac{y^3}{5}-\frac{x\,z^2}{7}\right)\right)+\left(\frac{\partial}{\partial y}\left(\frac{1}{7}x\,y\,z-\frac{1}{9}y\,z^2\right)\right)$$
$$+\left(\frac{\partial}{\partial z}\left(\frac{1}{6}x^2\,y-\frac{1}{12}x\,y\,z-\frac{1}{9}z^3\right)\right)$$

$$=$$

$$\frac{2}{27}x+\frac{1}{27}y+\frac{2}{45}z-\frac{37}{63}z^2+\frac{1}{7}x\,z-\frac{1}{12}x\,y \tag{8}$$

Input parametrized expression of Klein Bottle Surface Coordinates (6) to Divergence (8); and Calculate Triple Integrals of Product [Divergence (8) and Klein Bottle Surface Coordinates Volume Element] in range of 'r,u,v' (9):

$$\int_0^\pi\int_0^{4\pi}\Bigg(2\cos(u)\cos\!\left(\frac{u}{2}\right)^2\sin(v)\cos(2\,v)+6\cos(u)\cos\!\left(\frac{u}{2}\right)\cos(2\,v)$$

$$+\ 3\cos(u)\sin\!\left(\frac{u}{2}\right)\cos(v)-\cos(u)\sin(2\,v)\cos(v)-\frac{1}{2}\sin(v)\sin(u)\cos(v)$$

$$-\sin(2\,v)\sin(u)\cos(2\,v)+\cos(u)\cos\!\left(\frac{u}{2}\right)^2\sin(2\,v)\cos(v)$$

$$+\cos(u)\cos\!\left(\frac{u}{2}\right)\sin(v)\sin\!\left(\frac{u}{2}\right)\cos(v)-2\cos(u)\cos\!\left(\frac{u}{2}\right)\sin\!\left(\frac{u}{2}\right)\sin(2\,v)\cos(2\,v)$$

$$\Bigg)\Bigg(\frac{1}{3}\left(\frac{1}{3}\cos(u)\left(\cos\!\left(\frac{u}{2}\right)\sin(v)-\sin\!\left(\frac{u}{2}\right)\sin(2\,v)+3\right)\right.$$

$$+\frac{1}{6}\sin(u)\left(\cos\!\left(\frac{u}{2}\right)\sin(v)-\sin\!\left(\frac{u}{2}\right)\sin(2\,v)+3\right)+\frac{1}{5}\sin\!\left(\frac{u}{2}\right)\sin(v)$$

$$+\frac{1}{5}\cos\!\left(\frac{u}{2}\right)\sin(2\,v)\Bigg)^2-\frac{1}{5}\sin(u)^3-\left(\cos\!\left(\frac{u}{2}\right)\sin(v)-\sin\!\left(\frac{u}{2}\right)\sin(2\,v)+3\right)^3-$$

$$\frac{1}{7}\cos(u)\left(\cos\!\left(\frac{u}{2}\right)\sin(v)-\sin\!\left(\frac{u}{2}\right)\sin(2\,v)+3\right)\left(\sin\!\left(\frac{u}{2}\right)\sin(v)+\cos\!\left(\frac{u}{2}\right)\sin(2\,v)\right)^2$$

$$\Bigg)+\left(2\cos\!\left(\frac{u}{2}\right)^2\sin(v)\sin(u)\cos(2\,v)+6\cos\!\left(\frac{u}{2}\right)\sin(u)\cos(2\,v)\right.$$

$$+ 3 \sin\left(\frac{u}{2}\right) \sin(u) \cos(v) - \sin(2v) \sin(u) \cos(v) + \frac{1}{2} \sin(v) \cos(v) \cos(u)$$

$$+ \sin(2v) \cos(2v) \cos(u) + \cos\left(\frac{u}{2}\right)^2 \sin(2v) \sin(u) \cos(v)$$

$$+ \cos\left(\frac{u}{2}\right) \sin(v) \sin\left(\frac{u}{2}\right) \sin(u) \cos(v) - 2 \cos\left(\frac{u}{2}\right) \sin\left(\frac{u}{2}\right) \sin(2v) \sin(u) \cos(2v) \Bigg)$$

$$\left(\frac{1}{7} \cos(u) \left(\cos\left(\frac{u}{2}\right) \sin(v) - \sin\left(\frac{u}{2}\right) \sin(2v) + 3\right)^2 \sin(u)\right)$$

$$\left(\sin\left(\frac{u}{2}\right) \sin(v) + \cos\left(\frac{u}{2}\right) \sin(2v)\right) -$$

$$\frac{1}{9} \sin(u) \left(\cos\left(\frac{u}{2}\right) \sin(v) - \sin\left(\frac{u}{2}\right) \sin(2v) + 3\right) \left(\sin\left(\frac{u}{2}\right) \sin(v) + \cos\left(\frac{u}{2}\right) \sin(2v)\right)^2$$

$$\Bigg) + \left(-\cos\left(\frac{u}{2}\right)^2 \sin(v) \cos(v) + 2 \cos\left(\frac{u}{2}\right)^2 \sin(2v) \cos(2v)\right)$$

$$+ 2 \cos\left(\frac{u}{2}\right) \sin(v) \sin\left(\frac{u}{2}\right) \cos(2v) + \cos\left(\frac{u}{2}\right) \sin\left(\frac{u}{2}\right) \sin(2v) \cos(v)$$

$$- 3 \cos\left(\frac{u}{2}\right) \cos(v) + 6 \sin\left(\frac{u}{2}\right) \cos(2v) - 2 \sin(2v) \cos(2v) \Bigg)\Bigg($$

$$\frac{1}{6} \cos(u)^2 \left(\cos\left(\frac{u}{2}\right) \sin(v) - \sin\left(\frac{u}{2}\right) \sin(2v) + 3\right)^3 \sin(u) - \frac{1}{12} \cos(u)$$

$$\left(\cos\left(\frac{u}{2}\right) \sin(v) - \sin\left(\frac{u}{2}\right) \sin(2v) + 3\right)^2 \sin(u) \left(\sin\left(\frac{u}{2}\right) \sin(v) + \cos\left(\frac{u}{2}\right) \sin(2v)\right)$$

$$- \frac{1}{9} \left(\sin\left(\frac{u}{2}\right) \sin(v) + \cos\left(\frac{u}{2}\right) \sin(2v)\right)^3 \Bigg) du\, dv$$

$$= \frac{19424\,\pi}{6615} \qquad\qquad\qquad (9)$$

Surface Integral Precision Value (5) [Vector Field (2) at Klein Bottle surface (1)], is equal to Volume Element Integral Precision Value (9) [Divergence (8) in Spacial Bounded Closed Region 'Ω' (Klein Bottle Surrounds)], Complete Calculation and Validation of Divergence Theorem at Manifold (Counterexample).

In the case of Abstract Vector Field '[P(x,y,z),Q(x,y,z), R(x,y,z)]', Surface Integral about Klein Bottle surface is unequal to triple intergrals of spacial bounded closed region (Klein Bottle Surrounds) in logic;

Owning to various values of idiographic spacial vector field, Surface Integral about Klein Bottle surface is equal to Triple Intergrals about spacial bounded closed region (Klein Bottle surface surrounds) possibly.

See also Chapter 6 (Part III), 'Counterexample of Curl Theorem at Manifold - Spacial Closed Curve Integral and Surface Integral about Mobius Strip'

6.4 Surface Integral and Triple Integrals about Klein Bottle, Numerical Model [Program Template of Waterloo Maplesoft, Optional]

```
> restart;
> with(plots):with(linalg):
> CS:=[cos(u)*(cos(u/2)*sin(v)-sin(u/2)*sin(2*v)+3),
sin(u)*(cos(u/2)*sin(v)-sin(u/2)*sin(2*v)+3),
sin(u/2)*sin(v)+cos(u/2)*sin(2*v)];
# Define unorientable closed parameterized surface 'CS' (Viz. Klein
Bottle)
```

$$CS := \left[\cos(u) \left(\cos\left(\frac{u}{2}\right) \sin(v) - \sin\left(\frac{u}{2}\right) \sin(2\,v) + 3 \right), \right.$$
$$\left. \sin(u) \left(\cos\left(\frac{u}{2}\right) \sin(v) - \sin\left(\frac{u}{2}\right) \sin(2\,v) + 3 \right), \sin\left(\frac{u}{2}\right) \sin(v) + \cos\left(\frac{u}{2}\right) \sin(2\,v) \right]$$

```
> rgu:=[0,4*Pi];
```
$$rgu := [\,0, 4\,\pi\,]$$

```
> rgv:=[0,Pi]; # Define range of 'u,v'
```
$$rgv := [\,0, \pi\,]$$

```
> plot3d(CS,u=rgu[1]..rgu[2],v=rgv[1]..rgv[2],scaling=constrained,
projection=0.9,numpoints=6000);g1:=%: # Draw Klein bottle, be elided
> V:=[(x/3+y/6+z/5)^2/3-y^3/5-x*z^2/7,x*y*z/7-y*z^2/9,x^2*y/6
-x*y*z/12-z^3/9];
# Define discretional spacial vector field 'V' (Suppose spacial vector
field 'V' possesses 1th order continuous  partial derivatives in
spacial bounded closed region that 'Klein bottle' surrounds)
```

$$V := \left[\frac{\left(\dfrac{x}{3}+\dfrac{y}{6}+\dfrac{z}{5}\right)^2}{3} - \frac{y^3}{5} - \frac{x\,z^2}{7}, \frac{1}{7}x\,y\,z - \frac{1}{9}y\,z^2, \frac{1}{6}x^2\,y - \frac{1}{12}x\,y\,z - \frac{1}{9}z^3 \right]$$

```
> Diff(V[1],x)+Diff(V[2],y)+Diff(V[3],z)
= diff(V[1],x)+diff(V[2],y)+diff(V[3],z);diV:=rhs(%);
# Calculate divergence 'diV' of spacial vector field 'V'
```

$$\left(\frac{\partial}{\partial x}\left(\frac{\left(\frac{x}{3}+\frac{y}{6}+\frac{z}{5}\right)^2}{3}-\frac{y^3}{5}-\frac{x\,z^2}{7}\right)\right)+\left(\frac{\partial}{\partial y}\left(\frac{1}{7}x\,y\,z-\frac{1}{9}y\,z^2\right)\right)$$

$$+\left(\frac{\partial}{\partial z}\left(\frac{1}{6}x^2\,y-\frac{1}{12}x\,y\,z-\frac{1}{9}z^3\right)\right)=\frac{2}{27}x+\frac{1}{27}y+\frac{2}{45}z-\frac{37}{63}z^2+\frac{1}{7}x\,z-\frac{1}{12}x\,y$$

$$diV:=\frac{2}{27}x+\frac{1}{27}y+\frac{2}{45}z-\frac{37}{63}z^2+\frac{1}{7}x\,z-\frac{1}{12}x\,y$$

```
> rgx:=[-4,4];
```

$$rgx:=[-4,4]$$

```
> rgy:=[-4,4];
```

$$rgy:=[-4,4]$$

```
> rgz:=[-4,4];
```

$$rgz:=[-4,4]$$

```
> fieldplot3d(V,x=rgx[1]..rgx[2],y=rgy[1]..rgy[2],z=rgz[1]..
rgz[2],arrows=SLIM,color=blue):g2:=%: # Draw spacial vector field 'V'
> implicitplot3d(diV,x=rgx[1]..rgx[2],y=rgy[1]..rgy[2],z=rgz[1]..
rgz[2],style=wireframe,color=cyan,numpoints=3000):g3:=%:
# Draw level surface of divergence 'diV'
> display(g1,g2,g3): # Synthesize figures
```

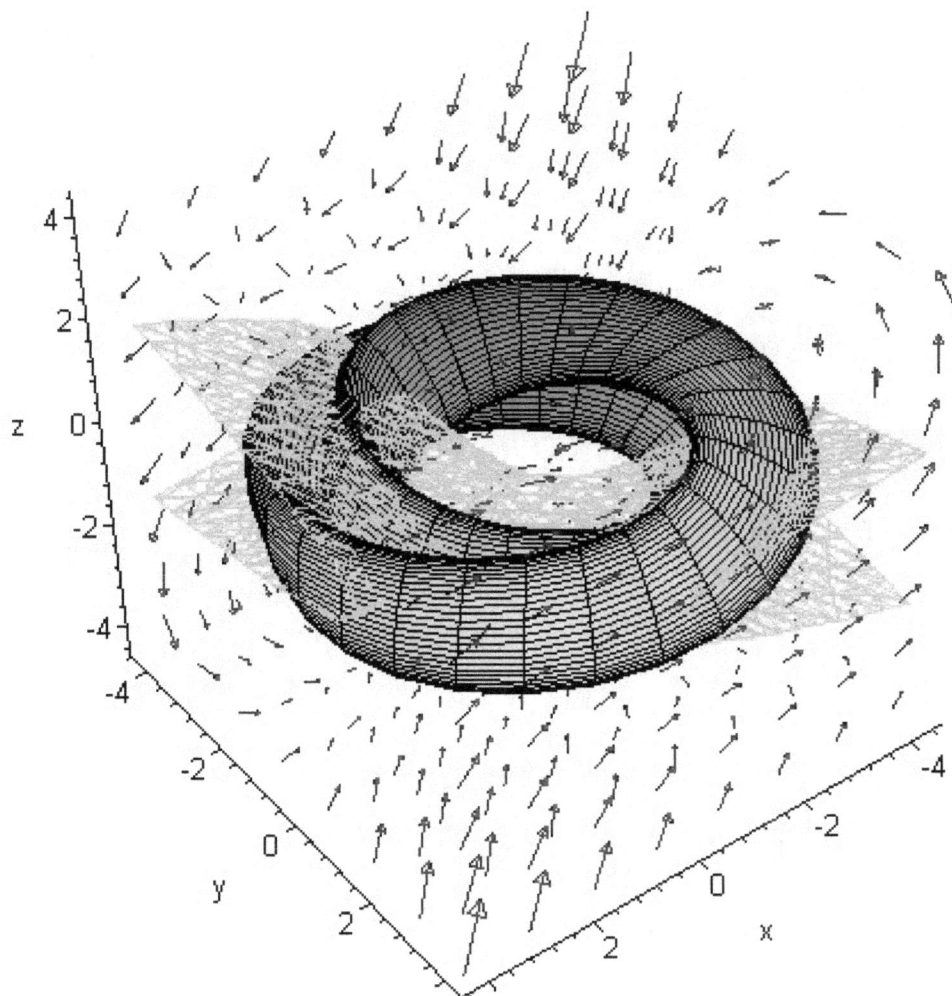

Figure(I)6.4 Klein bottle,Vector Field
and Level Surface of Divergence

```
> x:=CS[1]:y:=CS[2]:z:=CS[3]:
> matrix(3,3,[i,j,k,Diff(CS[1],u),Diff(CS[2],u),Diff(CS[3],u),
Diff(CS[1],v),Diff(CS[2],v),Diff(CS[3],v)])=
matrix(3,3,[i,j,k,diff(CS[1],u),diff(CS[2],u),diff(CS[3],u),
diff(CS[1],v),diff(CS[2],v),diff(CS[3],v)]);m1:=rhs(%);
# Define and calculate matrix of partial derivatives 'm1'; obtain
'tangent plane normal vector' of 'Klein bottle' surface
```

$$[i,j,k]$$

$$\left[\frac{\partial}{\partial u}\left(\cos(u)\left(\cos\left(\frac{u}{2}\right)\sin(v) - \sin\left(\frac{u}{2}\right)\sin(2\,v) + 3 \right) \right), \right.$$

$$\frac{\partial}{\partial u}\left(\sin(u)\left(\cos\left(\frac{u}{2}\right)\sin(v) - \sin\left(\frac{u}{2}\right)\sin(2\,v) + 3 \right) \right),$$

$$\frac{\partial}{\partial u}\left(\sin\left(\frac{u}{2}\right)\sin(v)+\cos\left(\frac{u}{2}\right)\sin(2\,v)\right)\Bigg]$$

$$\left[\frac{\partial}{\partial v}\left(\cos(u)\left(\cos\left(\frac{u}{2}\right)\sin(v)-\sin\left(\frac{u}{2}\right)\sin(2\,v)+3\right)\right),\right.$$

$$\frac{\partial}{\partial v}\left(\sin(u)\left(\cos\left(\frac{u}{2}\right)\sin(v)-\sin\left(\frac{u}{2}\right)\sin(2\,v)+3\right)\right),$$

$$\frac{\partial}{\partial v}\left(\sin\left(\frac{u}{2}\right)\sin(v)+\cos\left(\frac{u}{2}\right)\sin(2\,v)\right)\Bigg]=$$

$$[i,j,k]$$

$$\left[-\sin(u)\left(\cos\left(\frac{u}{2}\right)\sin(v)-\sin\left(\frac{u}{2}\right)\sin(2\,v)+3\right)\right.$$

$$+\cos(u)\left(-\frac{1}{2}\sin\left(\frac{u}{2}\right)\sin(v)-\frac{1}{2}\cos\left(\frac{u}{2}\right)\sin(2\,v)\right),$$

$$\cos(u)\left(\cos\left(\frac{u}{2}\right)\sin(v)-\sin\left(\frac{u}{2}\right)\sin(2\,v)+3\right)$$

$$+\sin(u)\left(-\frac{1}{2}\sin\left(\frac{u}{2}\right)\sin(v)-\frac{1}{2}\cos\left(\frac{u}{2}\right)\sin(2\,v)\right),$$

$$\frac{1}{2}\cos\left(\frac{u}{2}\right)\sin(v)-\frac{1}{2}\sin\left(\frac{u}{2}\right)\sin(2\,v)\Bigg]$$

$$\left[\cos(u)\left(\cos\left(\frac{u}{2}\right)\cos(v)-2\sin\left(\frac{u}{2}\right)\cos(2\,v)\right),\right.$$

$$\sin(u)\left(\cos\left(\frac{u}{2}\right)\cos(v)-2\sin\left(\frac{u}{2}\right)\cos(2\,v)\right),\sin\left(\frac{u}{2}\right)\cos(v)+2\cos\left(\frac{u}{2}\right)\cos(2\,v)\Bigg]$$

$$m1 :=$$

$$[i,j,k]$$

$$\left[-\sin(u)\left(\cos\left(\frac{u}{2}\right)\sin(v)-\sin\left(\frac{u}{2}\right)\sin(2\,v)+3\right)\right.$$

$$+\cos(u)\left(-\frac{1}{2}\sin\left(\frac{u}{2}\right)\sin(v)-\frac{1}{2}\cos\left(\frac{u}{2}\right)\sin(2\,v)\right),$$

$$\cos(u)\left(\cos\left(\frac{u}{2}\right)\sin(v)-\sin\left(\frac{u}{2}\right)\sin(2\,v)+3\right)$$

$$+\sin(u)\left(-\frac{1}{2}\sin\left(\frac{u}{2}\right)\sin(v)-\frac{1}{2}\cos\left(\frac{u}{2}\right)\sin(2\,v)\right),$$

$$\frac{1}{2}\cos\left(\frac{u}{2}\right)\sin(v)-\frac{1}{2}\sin\left(\frac{u}{2}\right)\sin(2\,v)\Bigg]$$

$$\left[\cos(u)\left(\cos\left(\frac{u}{2}\right)\cos(v)-2\sin\left(\frac{u}{2}\right)\cos(2\,v)\right),\right.$$

$$\sin(u)\left(\cos\left(\frac{u}{2}\right)\cos(v)-2\sin\left(\frac{u}{2}\right)\cos(2\,v)\right),\sin\left(\frac{u}{2}\right)\cos(v)+2\cos\left(\frac{u}{2}\right)\cos(2\,v)\Bigg]$$

```
> det(m1):
> mn:=simplify(%);
```

$$mn := i \cos(u) \cos\left(\frac{u}{2}\right) \sin(v) \sin\left(\frac{u}{2}\right) \cos(v)$$

$$- 2 \sin(u) \sin\left(\frac{u}{2}\right) \sin(2v) j \cos\left(\frac{u}{2}\right) \cos(2v)$$

$$- 2 i \cos(u) \sin\left(\frac{u}{2}\right) \sin(2v) \cos\left(\frac{u}{2}\right) \cos(2v)$$

$$+ \sin(u) \cos\left(\frac{u}{2}\right) \sin(v) j \sin\left(\frac{u}{2}\right) \cos(v) - 2 \sin(2v) k \cos(2v)$$

$$+ 6 k \sin\left(\frac{u}{2}\right) \cos(2v) - 3 k \cos\left(\frac{u}{2}\right) \cos(v) + 2 \sin(u) \cos\left(\frac{u}{2}\right)^2 \sin(v) j \cos(2v)$$

$$+ 2 i \cos(u) \cos\left(\frac{u}{2}\right)^2 \sin(v) \cos(2v) - \cos\left(\frac{u}{2}\right)^2 \sin(v) k \cos(v)$$

$$+ \frac{1}{2} \cos(u) \sin(v) j \cos(v) + 2 \sin(2v) k \cos(2v) \cos\left(\frac{u}{2}\right)^2$$

$$+ 2 \cos\left(\frac{u}{2}\right) \sin(v) k \sin\left(\frac{u}{2}\right) \cos(2v) + i \cos(u) \sin(2v) \cos(v) \cos\left(\frac{u}{2}\right)^2$$

$$- i \sin(u) \sin(2v) \cos(2v) - i \cos(u) \sin(2v) \cos(v)$$

$$+ \sin(u) \sin(2v) j \cos(v) \cos\left(\frac{u}{2}\right)^2 + \sin\left(\frac{u}{2}\right) \sin(2v) k \cos\left(\frac{u}{2}\right) \cos(v)$$

$$+ \cos(u) \cos(2v) j \sin(2v) - \sin(u) \sin(2v) j \cos(v) - \frac{1}{2} i \sin(u) \sin(v) \cos(v)$$

$$+ 3 i \cos(u) \sin\left(\frac{u}{2}\right) \cos(v) + 6 i \cos(u) \cos\left(\frac{u}{2}\right) \cos(2v)$$

$$+ 3 \sin(u) j \sin\left(\frac{u}{2}\right) \cos(v) + 6 \sin(u) j \cos\left(\frac{u}{2}\right) \cos(2v)$$

```
> A:=coeff(mn,i); # Obtain coefficient of 'i'
```

$$A := \cos(u) \cos\left(\frac{u}{2}\right) \sin(v) \sin\left(\frac{u}{2}\right) \cos(v) - 2 \cos(u) \sin\left(\frac{u}{2}\right) \sin(2v) \cos\left(\frac{u}{2}\right) \cos(2v)$$

$$+ 2 \cos(u) \cos\left(\frac{u}{2}\right)^2 \sin(v) \cos(2v) + \cos(u) \sin(2v) \cos(v) \cos\left(\frac{u}{2}\right)^2$$

$$- \sin(u) \sin(2v) \cos(2v) - \cos(u) \sin(2v) \cos(v) - \frac{1}{2} \sin(u) \sin(v) \cos(v)$$

$$+ 3 \cos(u) \sin\left(\frac{u}{2}\right) \cos(v) + 6 \cos(u) \cos\left(\frac{u}{2}\right) \cos(2v)$$

```
> B:=coeff(mn,j); # Obtain coefficient of 'j'
```

$$B := -2 \sin(u) \sin\left(\frac{u}{2}\right) \sin(2v) \cos\left(\frac{u}{2}\right) \cos(2v) + \sin(u) \cos\left(\frac{u}{2}\right) \sin(v) \sin\left(\frac{u}{2}\right) \cos(v)$$

$$+ 2 \sin(u) \cos\left(\frac{u}{2}\right)^2 \sin(v) \cos(2v) + \frac{1}{2} \cos(u) \sin(v) \cos(v)$$

$$+ \sin(u)\sin(2v)\cos(v)\cos\left(\frac{u}{2}\right)^2 + \cos(u)\cos(2v)\sin(2v)$$

$$- \sin(u)\sin(2v)\cos(v) + 3\sin(u)\sin\left(\frac{u}{2}\right)\cos(v) + 6\sin(u)\cos\left(\frac{u}{2}\right)\cos(2v)$$

> C:=coeff(mn,k); # Obtain coefficient of 'k'

$$C := -2\sin(2v)\cos(2v) + 6\sin\left(\frac{u}{2}\right)\cos(2v) - 3\cos\left(\frac{u}{2}\right)\cos(v)$$

$$- \cos\left(\frac{u}{2}\right)^2 \sin(v)\cos(v) + 2\sin(2v)\cos(2v)\cos\left(\frac{u}{2}\right)^2$$

$$+ 2\cos\left(\frac{u}{2}\right)\sin(v)\sin\left(\frac{u}{2}\right)\cos(2v) + \sin\left(\frac{u}{2}\right)\sin(2v)\cos\left(\frac{u}{2}\right)\cos(v)$$

>[A,B,C]: # [A,B,C] structure 'tangent plane normal vector'

> Int(Int(V[1]*A+V[2]*B+V[3]*C,u=rgu[1]..rgu[2]),v=rgv[1]..

rgv[2]);

Integrals of Dot Product (Vector 'V' and Tangent plane normal vector

[A,B,C] of 'Klein bottle' surface) in range of 'u,v'

$$\int_0^\pi \int_0^{4\pi} \left(\frac{1}{3}\left(\frac{1}{3}\cos(u)\left(\cos\left(\frac{u}{2}\right)\sin(v) - \sin\left(\frac{u}{2}\right)\sin(2v) + 3 \right) \right. \right.$$

$$+ \frac{1}{6}\sin(u)\left(\cos\left(\frac{u}{2}\right)\sin(v) - \sin\left(\frac{u}{2}\right)\sin(2v) + 3 \right) + \frac{1}{5}\sin\left(\frac{u}{2}\right)\sin(v)$$

$$+ \frac{1}{5}\cos\left(\frac{u}{2}\right)\sin(2v)\right)^2 - \frac{1}{5}\sin(u)^3\left(\cos\left(\frac{u}{2}\right)\sin(v) - \sin\left(\frac{u}{2}\right)\sin(2v) + 3 \right)^3 - \frac{1}{7}$$

$$\cos(u)\left(\cos\left(\frac{u}{2}\right)\sin(v) - \sin\left(\frac{u}{2}\right)\sin(2v) + 3 \right)\left(\sin\left(\frac{u}{2}\right)\sin(v) + \cos\left(\frac{u}{2}\right)\sin(2v) \right)^2 \right)$$

$$\left(\cos(u)\cos\left(\frac{u}{2}\right)\sin(v)\sin\left(\frac{u}{2}\right)\cos(v) - 2\cos(u)\sin\left(\frac{u}{2}\right)\sin(2v)\cos\left(\frac{u}{2}\right)\cos(2v) \right.$$

$$+ 2\cos(u)\cos\left(\frac{u}{2}\right)^2 \sin(v)\cos(2v) + \cos(u)\sin(2v)\cos(v)\cos\left(\frac{u}{2}\right)^2$$

$$- \sin(u)\sin(2v)\cos(2v) - \cos(u)\sin(2v)\cos(v) - \frac{1}{2}\sin(u)\sin(v)\cos(v)$$

$$+ 3\cos(u)\sin\left(\frac{u}{2}\right)\cos(v) + 6\cos(u)\cos\left(\frac{u}{2}\right)\cos(2v) \right) + \left(\frac{1}{7}\cos(u) \right)$$

$$\left(\cos\left(\frac{u}{2}\right)\sin(v) - \sin\left(\frac{u}{2}\right)\sin(2v) + 3 \right)^2 \sin(u)\left(\sin\left(\frac{u}{2}\right)\sin(v) + \cos\left(\frac{u}{2}\right)\sin(2v) \right)$$

$$-\frac{1}{9}$$

$$\sin(u)\left(\cos\left(\frac{u}{2}\right)\sin(v)-\sin\left(\frac{u}{2}\right)\sin(2v)+3\right)\left(\sin\left(\frac{u}{2}\right)\sin(v)+\cos\left(\frac{u}{2}\right)\sin(2v)\right)^2\right)$$

$$\left(-2\sin(u)\sin\left(\frac{u}{2}\right)\sin(2v)\cos\left(\frac{u}{2}\right)\cos(2v)+\sin(u)\cos\left(\frac{u}{2}\right)\sin(v)\sin\left(\frac{u}{2}\right)\cos(v)\right.$$

$$+2\sin(u)\cos\left(\frac{u}{2}\right)^2\sin(v)\cos(2v)+\frac{1}{2}\cos(u)\sin(v)\cos(v)$$

$$+\sin(u)\sin(2v)\cos(v)\cos\left(\frac{u}{2}\right)^2+\cos(u)\cos(2v)\sin(2v)$$

$$-\sin(u)\sin(2v)\cos(v)+3\sin(u)\sin\left(\frac{u}{2}\right)\cos(v)+6\sin(u)\cos\left(\frac{u}{2}\right)\cos(2v)\right)+\left(\vphantom{\frac12}\right.$$

$$\frac{1}{6}\cos(u)^2\left(\cos\left(\frac{u}{2}\right)\sin(v)-\sin\left(\frac{u}{2}\right)\sin(2v)+3\right)^3\sin(u)-\frac{1}{12}\cos(u)$$

$$\left(\cos\left(\frac{u}{2}\right)\sin(v)-\sin\left(\frac{u}{2}\right)\sin(2v)+3\right)^2\sin(u)\left(\sin\left(\frac{u}{2}\right)\sin(v)+\cos\left(\frac{u}{2}\right)\sin(2v)\right)$$

$$-\frac{1}{9}\left(\sin\left(\frac{u}{2}\right)\sin(v)+\cos\left(\frac{u}{2}\right)\sin(2v)\right)^3\right)\left(-2\sin(2v)\cos(2v)\right.$$

$$+6\sin\left(\frac{u}{2}\right)\cos(2v)-3\cos\left(\frac{u}{2}\right)\cos(v)-\cos\left(\frac{u}{2}\right)^2\sin(v)\cos(v)$$

$$+2\sin(2v)\cos(2v)\cos\left(\frac{u}{2}\right)^2+2\cos\left(\frac{u}{2}\right)\sin(v)\sin\left(\frac{u}{2}\right)\cos(2v)$$

$$+\sin\left(\frac{u}{2}\right)\sin(2v)\cos\left(\frac{u}{2}\right)\cos(v)\right)du\,dv$$

```
> alpha:=value(%);delta:=evalf(alpha);
# Calculate value of surface integral
```

$$\alpha:=\frac{19424\pi}{6615}$$

$$\delta:=9.224836842$$

```
> x:='x':y:='y':z:='z': # Reset variable 'x', 'y','z'
> [r*CS[1],r*CS[2],r*CS[3]];
 # Entirely multiply radius vector 'r'(Suppose r>0) by each item of
Klein  Bottle  surface  'CS'  parametrized  expression,  convert
'Parametrized expression of Klein Bottle surface' to 'Parametrized
expression of Klein Bottle surface coordinates'
```

$$\left[\, r\cos(u)\left(\cos\!\left(\frac{u}{2}\right)\sin(v) - \sin\!\left(\frac{u}{2}\right)\sin(2\,v) + 3\right),\right.$$

$$r\sin(u)\left(\cos\!\left(\frac{u}{2}\right)\sin(v) - \sin\!\left(\frac{u}{2}\right)\sin(2\,v) + 3\right),$$

$$\left. r\left(\sin\!\left(\frac{u}{2}\right)\sin(v) + \cos\!\left(\frac{u}{2}\right)\sin(2\,v)\right)\right]$$

```
> x:=r*CS[1]:y:=r*CS[2]:z:=r*CS[3]:
> matrix(3,3,[Diff(r*CS[1],r),Diff(r*CS[1],u),Diff(r*CS[1],v),
Diff(r*CS[2],r),Diff(r*CS[2],u),Diff(r*CS[2],v),Diff(r*CS[3],r),
Diff(r*CS[3],u),Diff(r*CS[3],v)])=
matrix(3,3,[diff(r*CS[1],r),diff(r*CS[1],u),diff(r*CS[1],v),
diff(r*CS[2],r),diff(r*CS[2],u),diff(r*CS[2],v),diff(r*CS[3],r),
diff(r*CS[3],u),diff(r*CS[3],v)]);m2:=rhs(%);
```

Define and calculate matrix of partial derivatives 'm2', obtain general expression of Klein bottle surface coordinates volume element coefficient

$$\left[\frac{\partial}{\partial r}\left(r\cos(u)\left(\cos\!\left(\frac{u}{2}\right)\sin(v) - \sin\!\left(\frac{u}{2}\right)\sin(2\,v) + 3\right)\right),\right.$$

$$\frac{\partial}{\partial u}\left(r\cos(u)\left(\cos\!\left(\frac{u}{2}\right)\sin(v) - \sin\!\left(\frac{u}{2}\right)\sin(2\,v) + 3\right)\right),$$

$$\left.\frac{\partial}{\partial v}\left(r\cos(u)\left(\cos\!\left(\frac{u}{2}\right)\sin(v) - \sin\!\left(\frac{u}{2}\right)\sin(2\,v) + 3\right)\right)\right]$$

$$\left[\frac{\partial}{\partial r}\left(r\sin(u)\left(\cos\!\left(\frac{u}{2}\right)\sin(v) - \sin\!\left(\frac{u}{2}\right)\sin(2\,v) + 3\right)\right),\right.$$

$$\frac{\partial}{\partial u}\left(r\sin(u)\left(\cos\!\left(\frac{u}{2}\right)\sin(v) - \sin\!\left(\frac{u}{2}\right)\sin(2\,v) + 3\right)\right),$$

$$\left.\frac{\partial}{\partial v}\left(r\sin(u)\left(\cos\!\left(\frac{u}{2}\right)\sin(v) - \sin\!\left(\frac{u}{2}\right)\sin(2\,v) + 3\right)\right)\right]$$

$$\left[\frac{\partial}{\partial r}\left(r\left(\sin\!\left(\frac{u}{2}\right)\sin(v) + \cos\!\left(\frac{u}{2}\right)\sin(2\,v)\right)\right),\right.$$

$$\frac{\partial}{\partial u}\left(r\left(\sin\!\left(\frac{u}{2}\right)\sin(v) + \cos\!\left(\frac{u}{2}\right)\sin(2\,v)\right)\right),$$

$$\left.\frac{\partial}{\partial v}\left(r\left(\sin\!\left(\frac{u}{2}\right)\sin(v) + \cos\!\left(\frac{u}{2}\right)\sin(2\,v)\right)\right)\right] =$$

$$\left[\cos(u)\left(\cos\!\left(\frac{u}{2}\right)\sin(v) - \sin\!\left(\frac{u}{2}\right)\sin(2\,v) + 3\right),\right.$$

$$-r\sin(u)\left(\cos\!\left(\frac{u}{2}\right)\sin(v) - \sin\!\left(\frac{u}{2}\right)\sin(2\,v) + 3\right)$$

$$+ r\cos(u)\left(-\frac{1}{2}\sin\!\left(\frac{u}{2}\right)\sin(v) - \frac{1}{2}\cos\!\left(\frac{u}{2}\right)\sin(2\,v)\right),$$

$$r \cos(u) \left(\cos\left(\frac{u}{2}\right) \cos(v) - 2 \sin\left(\frac{u}{2}\right) \cos(2\,v) \right) \Big]$$

$$\left[\sin(u) \left(\cos\left(\frac{u}{2}\right) \sin(v) - \sin\left(\frac{u}{2}\right) \sin(2\,v) + 3 \right), \right.$$

$$r \cos(u) \left(\cos\left(\frac{u}{2}\right) \sin(v) - \sin\left(\frac{u}{2}\right) \sin(2\,v) + 3 \right)$$

$$+ r \sin(u) \left(-\frac{1}{2} \sin\left(\frac{u}{2}\right) \sin(v) - \frac{1}{2} \cos\left(\frac{u}{2}\right) \sin(2\,v) \right),$$

$$r \sin(u) \left(\cos\left(\frac{u}{2}\right) \cos(v) - 2 \sin\left(\frac{u}{2}\right) \cos(2\,v) \right) \Big]$$

$$\left[\sin\left(\frac{u}{2}\right) \sin(v) + \cos\left(\frac{u}{2}\right) \sin(2\,v), r \left(\frac{1}{2} \cos\left(\frac{u}{2}\right) \sin(v) - \frac{1}{2} \sin\left(\frac{u}{2}\right) \sin(2\,v) \right), \right.$$

$$r \left(\sin\left(\frac{u}{2}\right) \cos(v) + 2 \cos\left(\frac{u}{2}\right) \cos(2\,v) \right) \Big]$$

$m2 :=$

$$\left[\cos(u) \left(\cos\left(\frac{u}{2}\right) \sin(v) - \sin\left(\frac{u}{2}\right) \sin(2\,v) + 3 \right), \right.$$

$$-r \sin(u) \left(\cos\left(\frac{u}{2}\right) \sin(v) - \sin\left(\frac{u}{2}\right) \sin(2\,v) + 3 \right)$$

$$+ r \cos(u) \left(-\frac{1}{2} \sin\left(\frac{u}{2}\right) \sin(v) - \frac{1}{2} \cos\left(\frac{u}{2}\right) \sin(2\,v) \right),$$

$$r \cos(u) \left(\cos\left(\frac{u}{2}\right) \cos(v) - 2 \sin\left(\frac{u}{2}\right) \cos(2\,v) \right) \Big]$$

$$\left[\sin(u) \left(\cos\left(\frac{u}{2}\right) \sin(v) - \sin\left(\frac{u}{2}\right) \sin(2\,v) + 3 \right), \right.$$

$$r \cos(u) \left(\cos\left(\frac{u}{2}\right) \sin(v) - \sin\left(\frac{u}{2}\right) \sin(2\,v) + 3 \right)$$

$$+ r \sin(u) \left(-\frac{1}{2} \sin\left(\frac{u}{2}\right) \sin(v) - \frac{1}{2} \cos\left(\frac{u}{2}\right) \sin(2\,v) \right),$$

$$r \sin(u) \left(\cos\left(\frac{u}{2}\right) \cos(v) - 2 \sin\left(\frac{u}{2}\right) \cos(2\,v) \right) \Big]$$

$$\left[\sin\left(\frac{u}{2}\right) \sin(v) + \cos\left(\frac{u}{2}\right) \sin(2\,v), r \left(\frac{1}{2} \cos\left(\frac{u}{2}\right) \sin(v) - \frac{1}{2} \sin\left(\frac{u}{2}\right) \sin(2\,v) \right), \right.$$

$$r \left(\sin\left(\frac{u}{2}\right) \cos(v) + 2 \cos\left(\frac{u}{2}\right) \cos(2\,v) \right) \Big]$$

```
> det(m2):

> J:=simplify(%);
```

$$J := r^2 \left(-2 \cos\left(\frac{u}{2}\right) \cos(2\,v) \cos(v)^2 + 3 \sin(2\,v) \cos(v) \cos\left(\frac{u}{2}\right)^2 \right.$$

$$- \sin\left(\frac{u}{2}\right) \cos(v) \cos(2\,v)^2 + 10 \sin\left(\frac{u}{2}\right) \cos(v) + 20 \cos\left(\frac{u}{2}\right) \cos(2\,v)$$

$$- 2 \sin\left(\frac{u}{2}\right) \sin(v) \sin(2v) \cos(2v) + 3 \cos\left(\frac{u}{2}\right) \sin(v) \sin\left(\frac{u}{2}\right) \cos(v)$$

$$- 6 \sin\left(\frac{u}{2}\right) \sin(2v) \cos\left(\frac{u}{2}\right) \cos(2v) - \cos\left(\frac{u}{2}\right) \sin(v) \sin(2v) \cos(v)$$

$$+ 6 \cos\left(\frac{u}{2}\right)^2 \sin(v) \cos(2v) - 6 \sin(2v) \cos(v) + 6 \sin(v) \cos(2v) \Bigg)$$

```
> Int(Int(Int(diV*J,r=0..1),u=rgu[1]..rgu[2]),v=rgv[1]..rgv[2]);
```

```
# Triple Integrals of Product (Divergence 'diV' and Klein Bottle
  Surface Coordinates Volume Element) in range of 'r,u,v'
```

$$\int_0^\pi \int_0^{4\pi} \int_0^1 \left(\frac{2}{27} r \cos(u) \left(\cos\left(\frac{u}{2}\right) \sin(v) - \sin\left(\frac{u}{2}\right) \sin(2v) + 3 \right) \right.$$

$$+ \frac{1}{27} r \sin(u) \left(\cos\left(\frac{u}{2}\right) \sin(v) - \sin\left(\frac{u}{2}\right) \sin(2v) + 3 \right)$$

$$+ \frac{2}{45} r \left(\sin\left(\frac{u}{2}\right) \sin(v) + \cos\left(\frac{u}{2}\right) \sin(2v) \right)$$

$$- \frac{37}{63} r^2 \left(\sin\left(\frac{u}{2}\right) \sin(v) + \cos\left(\frac{u}{2}\right) \sin(2v) \right)^2 + \frac{1}{7} r^2 \cos(u)$$

$$\left(\cos\left(\frac{u}{2}\right) \sin(v) - \sin\left(\frac{u}{2}\right) \sin(2v) + 3 \right) \left(\sin\left(\frac{u}{2}\right) \sin(v) + \cos\left(\frac{u}{2}\right) \sin(2v) \right)$$

$$- \frac{1}{12} r^2 \cos(u) \left(\cos\left(\frac{u}{2}\right) \sin(v) - \sin\left(\frac{u}{2}\right) \sin(2v) + 3 \right)^2 \sin(u) \Bigg) r^2 \Bigg($$

$$-2 \cos\left(\frac{u}{2}\right) \cos(2v) \cos(v)^2 + 3 \sin(2v) \cos(v) \cos\left(\frac{u}{2}\right)^2 - \sin\left(\frac{u}{2}\right) \cos(v) \cos(2v)^2$$

$$+ 10 \sin\left(\frac{u}{2}\right) \cos(v) + 20 \cos\left(\frac{u}{2}\right) \cos(2v) - 2 \sin\left(\frac{u}{2}\right) \sin(v) \sin(2v) \cos(2v)$$

$$+ 3 \cos\left(\frac{u}{2}\right) \sin(v) \sin\left(\frac{u}{2}\right) \cos(v) - 6 \sin\left(\frac{u}{2}\right) \sin(2v) \cos\left(\frac{u}{2}\right) \cos(2v)$$

$$- \cos\left(\frac{u}{2}\right) \sin(v) \sin(2v) \cos(v) + 6 \cos\left(\frac{u}{2}\right)^2 \sin(v) \cos(2v) - 6 \sin(2v) \cos(v)$$

$$+ 6 \sin(v) \cos(2v) \Bigg) dr\, du\, dv$$

```
> beta:=value(%);epsilon:=evalf(beta);
```

```
# Calculate value of volume element integral
```

$$\beta := \frac{19424\,\pi}{6615}$$

$$\varepsilon := 9.224836842$$

```
> alpha,beta; # Two analytic values are equal
```

$$\frac{19424\,\pi}{6615}\;,\;\frac{19424\,\pi}{6615}$$

> delta,epsilon; # Two float values are equal

9.224836842, 9.224836842

// In the case of idiographic spacial vector field, surface integral
about Klein Bottom surface is equal to triple intergrals about space
bounded closed region (Klein Bottom surface surrounds)

// In the case of abstract spacial vector field '[P(x,y,z),Q(x,y,z),
R(x,y,z)]', surface integral about Klein bottle surface is unequal
to triple intergrals about space bounded closed region (Klein Bottom
surface surrounds) in logic;

   owning to various values of idiographic spacial vector field,
surface integral about Klein bottle surface is equal to triple
intergrals about space bounded closed region (Klein Bottom surface
surrounds) possibly.

Conclusion A in Part I

   Constitute individualized geometric object coordinates that
matches with idiographic geometric object [Manifold] (Viz. What
idiographic surface, what coordinates of idiographic surface, what
volume element coefficient of idiographic surface coordinates; no
longer rely on a few existent coordinates: Cartesian coordinates、
Spherical coordinates、Cylindrical coordinates、Generalized Spherical
coordinates 、 Generalized Cylindrical coordinates and their
correlative volume element coefficients etc.), by two methods
(Integral、 Finite Sums Limits), prove presence of Divergence Theorem
in unlimited quantitative Individualized Surface coordinates
[Summarized and described by abstract simply connected orientable
closed surface coordinates (Bases on Poincare Conjecture) and multiple
connected orientable closed surface (Torus) coordinates], enable
Divergence Theorem surpass traditional architecture of 3-Dimensional
Cartesian coordinates, in Individualized Surface coordinates,
establish new formular association between surface integral (Bases
on parameterized dot product method) and triple integrals (Bases on

idiographic volume element coefficient method), and realize mutual validation between two typical integrals in unlimited quantitative individualized and gorgeous formular numerical model operations, radicate theoretical logic basis and numerical models of two new typical integral methods.

'Prove Divergence Theorem at Manifold' itself is not sole purpose, 'Establish new formular association between surface integral (Bases on parameterized dot product method) and triple integrals (Bases on idiographic volume element coefficient method) in Individualized Surface coordinates, radicate theoretical logic basis and numerical models of two new typical integral methods' is prime purpose.

Correlative numerical models of this part have indicated, by surface integral (Bases on parameterized dot product method) and triple integrals (Bases on idiographic volume element coefficient method) in Individualized Surface coordinates, science explorers can obtain analytic integral value or float integral value in discretional precision about complicated geometric objects (Manifold, Irregular、 Asymmetrical surface and its embedded spacial bounded closed region especially). Realize 'free surface integral' and 'free triple integrals in discretional space bounded closed region', indeed realize artistic integral interval, realize exact integral calculation of vector field [Electric field、 magnetic field、 hydromechanical field、gravitational field etc.] and scalar field [Electric potential field、temperature field、density field etc.] in discretional free surface and its embedded spacial bounded closed region, radicate logic associated relationship of two typical integrals, seek direct joint point between calculus、topology and physical engineering calculation, realize Divergence Theorem at Manifold and manifold integral in physical、engineering meaning, realize much more vast and free physical、mathematical explorations and engineering practices.

978-1-62265-930-2 (online)   978-1-62265-931-9 (paper)

Yang Ke

Conclusion B Comparison of Different Methods in Part I

Method	Normalization	Application Range	Calculation Precision	Result Verification
Traditional Surface Integral bases on Projective Method & Triple Integrals in 3-Dimensional Cartesian Coordinates	Tedious and Divergent Calculational Method	Symmetric & Regular Geometric Objects, such as Sphere, Ellipsoid, Cone, Cylinder, Hexahedron & their simple combinations	Analytic Value	No Verification
Traditional Partial Differential Equations, such as Finite Difference Method、 Finite Element Method、 Boundary Element Method etc.	Tedious & Divergent Calculational Method; Method & Geometric Object, One-to-One; 'Algorithm' Design is a huge project & tedious process	Slightly Complex Geometric objects; Mainstream & Research Direction of Numerical Analysis in Contemporary Physical & Engineering Fields	Approximate Value	Convergence & Stability Verification
New Surface Integral bases on Parameterized Dot Product Method & Triple Integrals bases on Idiographic Volume Element Coefficient Method in Individualized Surface Coordinates	Standardized & Matrixing Calculational Method, Simple & Convergent; Method & Geometric Object, One-to-Many	Discretional Complex Geometric Objects	Analytic Value	One to One Verification by Numerical Models of New Integral Theorem

Part II Green Theorem at Manifold

Chapter 1 Precondition of Proofs
--Constitute Abstract Simply Connected Closed Parametrized Curve Coordinates (Plane)

1.1 Introspection between Green Theorem and Poincare Conjecture

Review object of Proof - Green Theorem:

Green Theorem Suppose boundary curve 'L' of plane bounded closed region 'S' is composed by numbered smooth or piecewise smooth curves. If exist 1th order continuous partial derivatives about functions 'P(x,y),Q(x,y)' [Structure plane vector field 'A'] at plane bounded closed region 'S', then:

$$\int_{L+} A \cdot dL = \iint_S \left(\frac{\partial}{\partial x} Q(x,y) - \frac{\partial}{\partial y} P(x,y) \right) dS .$$

In definiens of theorem, it emphasizes that boundary curve 'L' of plane bounded closed region 'S' must be 'Closed' curve

In proof of traditional Green theorem, 'Abstract closed curve L' is described as that:

Abstract plane closed curve that constructed by 'a,b,y=ϕ1(x), y=ϕ2(x)' or 'x=ψ1(y),x=ψ2(y),c,d'.

(See also <Calculus [6th Edition]> Tongji University, Department of Mathematics; Higher Education Press, Beijing; ed.1, 1978.10; ed.6, 2007.6; 9th Printing, 2009.8; pp.142-145)

In other words, objectively, Green Theorem requests that no matter in Cartesian coordinates or in other coordinates, correlative curve must possess two attributes: (1)Simply connected; (2)Closed.

Depart from traditional Cartesian coordinates, how to depict abstract universal 'Simply connected closed curve', and constitute 'Simply connected closed curve Coordinates'? Ready-made solution is absence.

Poincare Conjecture [19] concludes that 'Every closed n-Manifold which is homotopy equivalent to n-Sphere is homeomorphic to n-Sphere'.

In 2-Dimensional Euclidean space that Green theorem refers, corresponding opinion is that: 'Every simply connected closed 1-Manifold is homeomorphic to 1-Sphere (viz. Circle)'.

In other words, according as Poincare Conjecture, in 2-Dimensional Euclidean space that Green theorem refers, every simply connected closed curve, no matter its Protean geometrical shape, 'Be homeomorphic to 1-Sphere(Circle)' is universal attribute.

Naturally, next question is 'In 2-Dimensional Euclidean space, base on Poincare Conjecture, it is possible to define an abstract universal expression of simply connected closed curve ?' -- It is kernel content of this Chapter.

In plane analytic geometry, parametrized expression of '1-Sphere' (Circle) above-mentioned is '[cos(t),sin(t)]', thereinto, t[0,2*Pi] (In proper meaning, this parametrized expression is the transform expression of '1-Sphere' between '2-Dimensional Cartesian coordinates' and 'Polar coordinates'; expression of '1-Sphere' in Polar coordinates is constant 1).

In topology field, define 'Homeomorphism' as 'Two manifolds, if it is possible to change one to another by the operations of bending、extending 、 cutting etc., then recognise that two manifolds above-mentioned are homeomorphous'.

Reconsider Poincare Conjecture by the visual angle of topology and analytic geometry, since the parametrized expression of '1-Sphere' is '[cos(t),sin(t)],t[0,2*Pi]', then its transfiguration '[a*cos(t), b*sin(t)],t[0,2*Pi]' (Thereinto, 'a,b' are discretional nonzero constants) is the parametrized expression of discretional ellipse. In 2-Dimensional Euclidean space, discretional ellipse is homeomorphic to 1-Sphere, this is the general knowledge of topology, dispense with discussion;

If 'a,b' are discretional '1th order derivable continuous functions', what instance appears possibly ?

See also below figures:

Figure(II)1.1.1 Non Simply Connected Closed Curve, Deduced from parametrized expression '[a*cos(t), b*sin(t)], t∈[0,2π]'

Suppose discretional pending coefficients

a = cos(2*t)+2*sin(t)/3, b = sin(3*t)/3

(viz. a, b are discretional "1th order derivable continuous functions"), then target parametrized curve '[a*cos(t), b*sin(t)], t∈[0,2π]' equals '[(cos(2*t)+2*sin(t)/3)*cos(t), sin(3*t)/3*sin(t)], t∈[0,2π]'.

```
// Corresponding drawing instructions of Waterloo Maplesoft 17:
> restart;
> with(plots):with(linalg):
> a:=cos(2*t)+2*sin(t)/3;
# Input discretional "1th order derivable continuous functions" by
  pending coefficients a,b
```

$$a := \cos(2\,t) + \frac{2}{3}\sin(t)$$

```
> b:=sin(3*t)/3;
```

$$b := \frac{1}{3}\sin(3\,t)$$

```
> CO:=[a*cos(t),b*sin(t)]; # Target parametrized expression
```

$$CO := \left[\left(\cos(2\,t) + \frac{2}{3}\sin(t)\right)\cos(t),\, \frac{1}{3}\sin(3\,t)\sin(t)\right]$$

```
> rgt:=[0,2*Pi]; # Set range of "u,v"
```

$$rgt := [0, 2\,\pi]$$

```
> plot([CO[1],CO[2],t=rgt[1]..rgt[2]],scaling=constrained,
color=red,numpoints=1000);
# Drawing instruction of Waterloo Maplesoft 17
```

Actual shape of target parametrized curve:

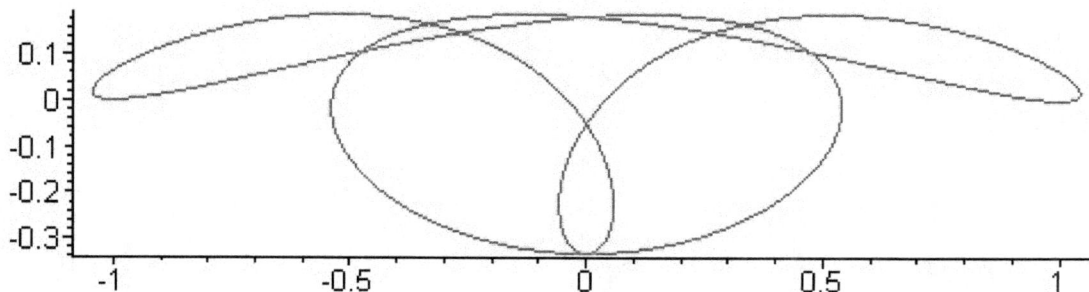

Figure(II)1.1.1 Non Simply Connected Closed Curve, Deduced from
   parametrized expression '[a*cos(t),b*sin(t)],t∈[0,2π]'

   Observe shape of Figure(II)1.1.1 by intuitionistic vision, parts
Of target parametrized curve above-mentioned show intersected、
overlapped states; therefore target parametrized surface belongs to
non simply connected closed curve, be independent of "Poincare
Conjecture" and "Green Theorem at Manifold".

Figure(II)1.1.2 Simply Connected Closed Curve, Deduced from
   parametrized expression '[a*cos(t),b*sin(t)],t∈[0,2π]'

   Suppose discretional pending coefficients
   a = cos(t-1)+sin(9*t-2)/12, b = sin(t-1)-cos(t)
(viz. a, b are discretional "1th order derivable continuous functions"),
then target parametrized curve '[a*cos(t), b*sin(t)], t∈[0,2π]'
equals '[(cos(t-1)+sin(9*t-2)/12)*cos(t), (sin(t-1)-cos(t))*sin(t)],
t∈[0,2π]'.

   Actual shape of target parametrized curve:

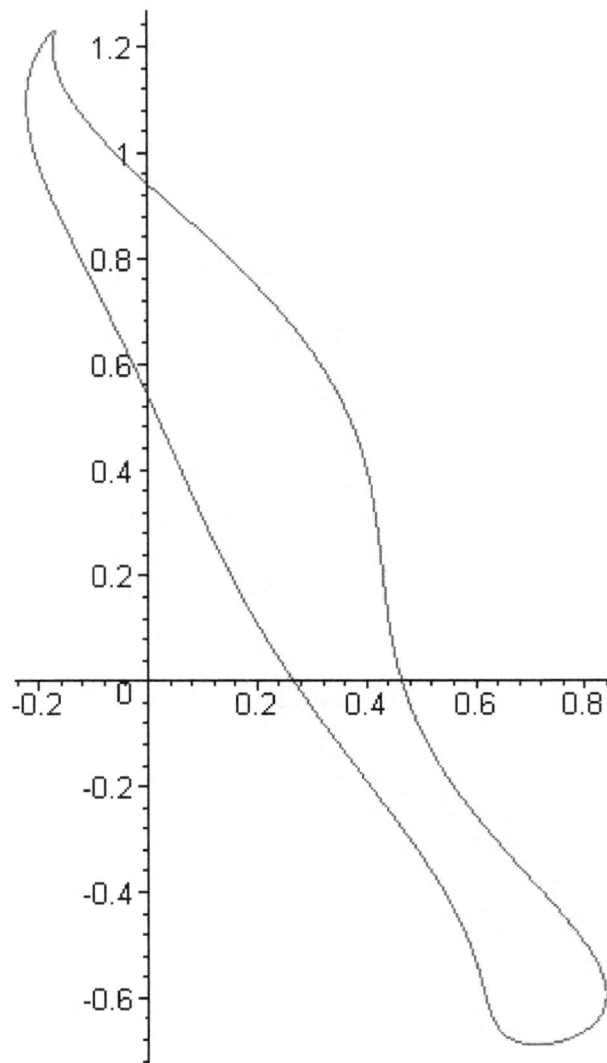

Figure(II)1.1.2 Simply Connected Closed Curve, Deduced from parametrized expression'[a*cos(t),b*sin(t)],t∈[0,2π]'

Observe shape of Figure(II)1.1.2 by intuitionistic vision, there aren't intersected、overlapped states about target parametrized curve above-mentioned, therefore target parametrized curve belongs to simply connected closed curve, be relational to "Poincare Conjecture" and "Green Theorem at Manifold".

In lay of actual operation, draw certain a parametrized curve by 'Plot'(Command of Waterloo Maplesoft), must judge that 'This parametrized curve is simply connected closed curve or not' by intuitionistic vision at first, then decide that 'This parametrized curve is the same with numerical models of Green theorem at manifold or not' at second. It is impossible to judge that 'Certain a parametrized curve is simply connected closed curve or not' by parametrized expression itself.

Determinant diathesis of 'Parametrized curve (plane) is simply connected closed curve or not' is in topology field and not in analytical geometry field; Only method of parametrized expression in analytical geometry is unable to conclude and deduce certain a curve as a simply connected closed curve.

By original phenomenons, experimental data above-meitioned indicate that, Similarly belong to parametrized curve'[a*cos(t), b*sin(t)],t[0,2*Pi]', because of different values of pending coefficients 'a,b', a part of curves belong to simply connected closed curve, and another part of curves are exceptional.

In other words, there are two instances about parametrized curve '[a*cos(t),b*sin(t)],t[0,2*Pi]':

(1) In the case of pending coefficients 'a,b' are discretional nonzero constants, parametrized curve is ellipse (Be homeomorphic to 1-Sphere naturally);

(2) In the case of pending coefficients 'a,b' are discretional 1th order derivable continuous functions, parametrized curve is simply connected closed curve possibly (Be homeomorphic to 1-Sphere), and parametrized curve isn't simply connected closed curve possibly (Not to be homeomorphic to 1-Sphere).

Naturally, next question is "In mode of parametrized curve [a*cos(t),b*sin(t)],t[0,2*Pi], it is possible to exclude instance of 'Parametrized curve is not simply connected closed curve' by certain logical defining ?"

1.2 Base on Poincare Conjecture,
Define Abstract Simply Connected Closed Curve (Plane)

Suppose 'Discretional Curves' as a set, then 'Discretional Simply Connected Closed Curves' is subset of former, Poincare Conjecture is attribute of this subset, current thesis 'Proof of Green Theorem at Manifold' and its 'Proof form **of** Finite Sums Limits' discuss that Green Theorem is the same with this subset or not.

Poincare Conjecture supplies the realizing route for describing 'discretional simply connected closed curve's certain attribute (Viz. 'Be homeomorphic to 1-Sphere' this attribute) by parametrized expression.

Base on situation above-mentioned, abstract unlimited quantitative idiographic simply connected closed curves as an uniform expression:

[a*cos(t),b*sin(t)],t[0,2*Pi]

(Thereinto, pending coefficients 'a,b' can't be designated discretionarily, 'a,b' must obey topology attribute of simply connected closed curve)

In other words, if pending coefficients 'a,b' can be designated discretionarily, then target curve '[a*cos(t),b*sin(t)],t[0,2*Pi]' may be simply connected closed curve possibly, or may be not.

If preestablish that 'Target curve [a*cos(t),b*sin(t)],t[0,2*Pi] itself is simply connected closed curve', then pending coefficients 'a,b' can't be designated discretionarily.

Explain phenomenons above-mentioned on geometrical meaning, in 2- Dimensional Cartesian coordinates, circle (Viz. [cos(t), sin(t)], t[0,2*Pi]) itself extends continuously discretionarily along two directions of axis 'x,y' (Viz. [a*cos(t),b*sin(t)],t[0,2*Pi], thereinto, 'a,b' are discretional 1th order derivable continuous functions, not always comes into being simply connected closed curve;

Contrarily, in 2-Dimensional Cartesian coordinates, discretional simply connected closed curve, consequentially, based on circle (Viz. [cos(t),sin(t)],t[0,2*Pi]), extended continuously along two directions of axis 'x,y', came into being (Likewise consequentially, discretional simply connected closed curve extends continuously along two directions of axis 'x,y', comes into being circle)-- Base on Poincare Conjecture.

For example, side lines of triangle or square can be regarded as simply connected closed curve -- But side lines of triangle or square is difficult or can't be described by parametrized expression -- But it is incontestable: Base on Poincare Conjecture, side lines of triangle or square is homeomorphic to 1-Sphere consequentially, side lines of triangle or square based on circle (Viz. [cos(t),sin(t)], t[0,2*Pi]), extended continuously along two directions of axis 'x,y', came into being (Likewise consequentially, side lines of triangle or square extends continuously along two directions of axis 'x,y', comes into being circle); Base on Poincare Conjecture, similarly, side lines of triangle or square can be described by '[a*cos(t),b*sin(t)], t[0,2*Pi]' mode.

Describe abstract universal simply connected closed curve by '[a*cos(t),b*sin(t)],t[0,2*Pi]' Mode -- Actually, describe certain inner configuration and attribute (Viz.'Be homeomorphic to 1-Sphere' this inner configuration and attribute) of simply connected closed curve by Poincare Conjecture, define a felicitous precondition for proof.

1.3 Constitute Abstract Simply Connected Closed Curve Coordinates (Plane)

Base on Poincare Conjecture, '[a*cos(t),b*sin(t)],t[0,2*Pi]' is just abstract universal expression of simply connected closed curve, but it isn't coordinates; Abstract simply connected closed curve coordinates is [r*a*cos(t), r*b*sin(t)], r[0, ∞], t[0,2*Pi], thereinto, 'r' is radius, 'a,b' is pending coefficients (Because of 'a,b' maybe nonzero constants, maybe 1th order derivable continuous functions), be uncertain.

Actually, expression of abstract simply connected closed curve '[a*cos(t),b*sin(t)],t[0,2*Pi]' is same as expression of ellipse '[a*cos(t),b*sin(t)],t[0,2*Pi]' formally, only explanations of pending coefficients 'a,b' are different: Former explains 'a,b' as 'Discretional nonzero constants or 1th order derivable continuous functions (Non discretional, be restricted by attribute of simply connected closed curve)'; and latter explains 'a,b' as 'Discretional nonzero constants' -- So that corresponding relation of abstract simply connected closed curve coordinates and Cartesian coordinates is 'x=r*a*cos(t),y=r*b*sin(t)' (Be same as transform expression of ellipse coordinates and Cartesian coordinates).

Chapter 2 Prove Green Theorem at Manifold

2.1 Prove Green Theorem at Manifold

Green Theorem   Suppose boundary curve 'L' of plane bounded closed region 'S' is composed by numbered smooth or piecewise smooth curves. If exist 1th order continuous partial derivatives about functions 'P(x,y),Q(x,y)'[Structure plane vector field 'A'] at plane bounded closed region 'S', then:

$$\int_{L+} A \cdot dL = \iint_{S} \left(\frac{\partial}{\partial x} Q(x,y) - \frac{\partial}{\partial y} P(x,y) \right) dS \qquad (1)$$

Proof:

Define parametrized expression of abstract (plane) simply connected、closed curve 'L':

  [a*cos(t),b*sin(t)]   (2)

thereinto, 'a,b' are nonzero constants or 1th order derivable continuous functions, simply connected closed curve 'L' determines value of 'a,b'; set range of parameters 't' in [0,2π], make boundary curve 'L' closed. (See also Poincare Conjecture: 'Every closed n-Manifold which is homotopy equivalent to n-Sphere is homeomorphic to n-Sphere' [19], and Chapter 1 this Part 'Precondition of Proofs —Constitute Abstract Simply Connected Closed Parametrized Curve Coordinates [Plane]')

Calculate Closed Curve Integral (Plane Vector Field 'A' at Boundary Curve 'L'):

$$\int_{0}^{2\pi} P(x,y) \left(\frac{\partial}{\partial t} (a\cos(t)) \right) + Q(x,y) \left(\frac{\partial}{\partial t} (b\sin(t)) \right) dt =$$

$$\int_{0}^{2\pi} -P(x,y) a \sin(t) + Q(x,y) b \cos(t) dt = 0$$

$$(3)$$

// Relative to Protean given 2-Dimensional plane vector fields (Composed of idiographic 2-variable functions),abstract 2-Dimensional plane vector field '[P(x,y),Q(x,y)]' is a sort of balanced、symmetrical data structure;

In proof of Green Theorem at Manifold, objectively require a sort of balanced、symmetrical abstract simply connected、closed curve's expression, match with abstract 2-Dimensional plane vector field `[P(x,y),Q(x,y)]';

In meaning of data structure, abstract simply connected closed curve's expression `[a*cos(t),b*sin(t)],t[0,2*Pi]' also possesses balanced、symmetrical attribute.

//Owing to universality and homogeneity of abstract vector field [P(x,y),Q(x,y)], Integral by Embedded Sub-Variable 't' of Variable 'x,y', Result of Integral can be described as `P(x,y)[or Q(x,y)]'.

In other words, Integral by Embedded Sub-Variable 't' of Variable 'x,y', Can't change structure of abstract function P(x,y)[or Q(x,y)], so that abstract function structure P(x,y) [or Q(x,y)] can preserve primary form after integral.

(See also Chapter 4 (This Part) Section 4.1 'Prove Green Theorem Finite Sums Limits at Manifold' and Chapter 4 (Part I) Section 4.3 'Discuss About Abstract Vector Field、Finite Sums Limit、Integral and Riemann Sums')

//'Integral value equals 0' possesses specific mathematical and physical meaning:

In mathematical meaning, 'Integral value equals 0' is the inevitable result of logic illation, incarnates the logic balanced state between various integral elements;

In physical meaning, 'Integral value equals 0' implies that the circumfluent value of 'Abstract Plane Vector Field at Abstract plane simply connected closed curve' is constant for '0' (immobile and pending) identically; If integral value equals certain a positive/negative number or an expression, it implies that always there is positive/ reverse directional flux or unknown flux about 'Abstract Plane Vector Field at Abstract simply connected closed curve', this instance will be unaccountable.

Change character 't' to 'u' in parametrized expression boundary curve 'L', and entirely multiply radius vector 'r' (suppose r>0) by each item of boundary curve 'L' parametrized expression, convert "Parametrized expression of boundary curve 'L'" to "Parametrized expression of curve 'L' coordinates":

[r*a*cos(u),r*b*sin(u)]   (4)

According to parametrized expression of curve 'L' coordinates (4), define and calculate matrix of partial derivatives, obtain general expression of curve 'L' coordinates area element coefficient:

$$\begin{bmatrix} \dfrac{\partial}{\partial r}ra\cos(u) & \dfrac{\partial}{\partial u}ra\cos(u) \\ \dfrac{\partial}{\partial r}rb\sin(u) & \dfrac{\partial}{\partial u}rb\sin(u) \end{bmatrix} = \begin{bmatrix} a\cos(u) & -ra\sin(u) \\ b\sin(u) & rb\cos(u) \end{bmatrix} = \text{abr} \quad (5)$$

// In proper meaning, curve 'L' coordinates area element coefficient is ratio of "Curve 'L' coordinates area element" and 'Plane Cartesian coordinates area element'

Convert differential function '$\dfrac{\partial}{\partial x}Q(x,y)-\dfrac{\partial}{\partial y}P(x,y)$' from Cartesian coordinates to curve 'L' coordinates:

$$\left(\frac{\partial}{\partial x}Q(x,y)\right)\left(\frac{\partial}{\partial r}(r\,a\,\cos(u))\right)\left(\frac{\partial}{\partial u}(r\,a\,\cos(u))\right)$$
$$-\left(\frac{\partial}{\partial y}P(x,y)\right)\left(\frac{\partial}{\partial r}(r\,b\,\sin(u))\right)\left(\frac{\partial}{\partial u}(r\,b\,\sin(u))\right)=$$
$$-\left(\frac{\partial}{\partial x}Q(x,y)\right)a^2\cos(u)\,r\sin(u)-\left(\frac{\partial}{\partial y}P(x,y)\right)b^2\sin(u)\,r\cos(u)$$
$$(6)$$

//In Cartesian Coordinates, Differential Function of abstract vector field [P(x,y),Q(x,y)] is $\dfrac{\partial}{\partial x}Q(x,y)-\dfrac{\partial}{\partial y}P(x,y)$.

In logic deduction of this Proof, it is necessary to import Differential Function of abstract vector field above-mentioned to abstract simply connected closed curve coordinates.

Two elements of abstract vector field's Differential Function:

$\dfrac{\partial}{\partial x}Q(x,y)$, $\dfrac{\partial}{\partial y}P(x,y)$ are abstract differential function, and their variables 'x,y' contain sub-variables 'r,u'.

978-1-62265-930-2 (online)   978-1-62265-931-9 (paper)

Transform Expression between 'Cartesian coordinates' and 'abstract simply connected closed curve coordinates' is:

x = r*a*cos(u),y = r*b*sin(v)

--"Coordinates Transform Differential Functions" (Correspond to "Two

Differential Variables '$\frac{\partial}{\partial x}$, $\frac{\partial}{\partial y}$' of Differential Functions '$\frac{\partial}{\partial x}Q(x,y)$,

$\frac{\partial}{\partial y}P(x,y)$'") are: $\frac{\partial}{\partial r}ra\cos(u)\frac{\partial}{\partial u}ra\cos(u)$ and $\frac{\partial}{\partial r}rb\sin(u)\frac{\partial}{\partial u}rb\sin(u)$.

Product of "Differential Functions '$\frac{\partial}{\partial x}Q(x,y)$, $\frac{\partial}{\partial y}P(x,y)$'" and

"Coordinates Transform Differential Functions" (Viz. Product of two different functions) constitute "Differential Function in Abstract Simply Connected Closed Curve Coordinates"

"Chain differentiation" or "Coordinates conversion" ?

If it is "Chain differentiation", according to principle of "multiply in identical chain, plus in subdivision chain", it ought to be:

$$\left(\frac{\partial}{\partial x}Q(x,y)\right)\left(\left(\frac{\partial}{\partial r}(r\,a\,\cos(u))\right)+\left(\frac{\partial}{\partial u}(r\,a\,\cos(u))\right)\right)$$
$$-\left(\frac{\partial}{\partial y}P(x,y)\right)\left(\left(\frac{\partial}{\partial r}(r\,b\,\sin(u))\right)+\left(\frac{\partial}{\partial u}(r\,b\,\sin(u))\right)\right)$$

// No matter "Chain differentiation" or "Solve differential function

'$\frac{\partial}{\partial x}Q(x,y)-\frac{\partial}{\partial y}P(x,y)$'", question for discussion is "How to solve

differential coefficient" or "Method of differentiation" about abstract vector field "[P(x,y),Q(x,y)]"; and this step, question for discussion is "How to convert result of solving differential function

'$\frac{\partial}{\partial x}Q(x,y)-\frac{\partial}{\partial y}P(x,y)$' from a coordinates to another"; character and

hiberarchy of "Two questions" above-mentioned are disparate, it is "multiply", not "plus" - Be determined by 2-Dimensional spacial attribute of coordinates.

Double Integrals of Product [Differential Function (6) and

Curve 'L' Coordinates Area Element] in range of 'r,u' (7):

$$\int_0^{2\pi} \int_0^1 \left(-\left(\frac{\partial}{\partial x} Q(x,y) \right) a^2 \cos(u)\, r\sin(u) - \left(\frac{\partial}{\partial y} P(x,y) \right) b^2 \sin(u)\, r\cos(u) \right) a\, r\, b\, dr\, du$$

$$=$$

$$\int_0^{2\pi} \frac{1}{3} \left(-\left(\frac{\partial}{\partial x} Q(x,y) \right) a^2 \cos(u)\sin(u) - \left(\frac{\partial}{\partial y} P(x,y) \right) b^2 \sin(u)\cos(u) \right) a\, b\, du = 0$$

// Owing to universality and homogeneity of abstract vector field's

Differential Function ' $\frac{Q(x,y)}{\partial x} - \frac{P(x,y)}{\partial y}$ ', (or their elements ' $\frac{\partial}{\partial x} Q(x,y)$,

$\frac{\partial}{\partial y} P(x,y)$ '), Integral by Embedded Sub-Variable 'r,u' of Variable 'x,y',

character of integral can be regarded as 'Integral about Differential

Function Elements of Differential Function (Viz. $\frac{\partial}{\partial x} Q(x,y)$, $\frac{\partial}{\partial y} P(x,y)$)、

Coordinates Transform Differential Functions、Curve 'L' Coordinates

Area Element'. Integral by Embedded Sub-Variable 'r,u' of Variable

'x,y', Can't change structure of abstract Differential Function (Viz.

' $\frac{Q(x,y)}{\partial x} - \frac{P(x,y)}{\partial y}$ ') or their Differential Function Elements (Viz.

$\frac{\partial}{\partial x} Q(x,y)$, or $\frac{\partial}{\partial y} P(x,y)$), they can preserve primary form after integral.

Two Coordinates Transform Differential Functions (Correspond to

Two Differential Variables ' $\frac{\partial}{\partial x}$, $\frac{\partial}{\partial y}$ '), Viz: $\frac{\partial}{\partial r} ra\cos(u) \frac{\partial}{\partial u} ra\cos(u)$ and

$\frac{\partial}{\partial r} rb\sin(u) \frac{\partial}{\partial u} rb\sin(u)$ will be changed after integral.

(See also Chapter 4 (This Part) Section 4.1 'Prove Green Theorem Finite Sums Limits at Manifold' and Chapter 4 (Part I) Section 4.3 'Discuss About Abstract Vector Field、Finite Sums Limit、Integral and Riemann Sums')

//Be different to ordinary double integrals, in this position, first

integral range is 'r∈[0,1]' always, and not 'r∈[0,n] or r∈[0,∞]';

Because of '1', it is possible to hold a correct ratio between

'a,b,r,u'.

　　thereinto:

$$\int_0^{2\pi} \int_0^1 a\, b\, r\, dr\, du \neq 0$$

　　Viz. Suppose double integrals (Curve 'L' coordinates area element
in range of 'r,u') can't be '0';

　　Explain visually, it can be regarded as "Suppose plane bounded
closed region 'S' can't be '0 area'"

　　Viz. expression(3) = expression(7):

$$\int_0^{2\pi} P(x,y)\left[\frac{\partial}{\partial t}(a\cos(t))\right] + Q(x,y)\left[\frac{\partial}{\partial t}(b\sin(t))\right] dt =$$

$$\int_0^{2\pi}\int_0^1 \left(-\left(\frac{\partial}{\partial x}Q(x,y)\right)a^2\cos(u)\,r\sin(u) - \left(\frac{\partial}{\partial y}P(x,y)\right)b^2\sin(u)\,r\cos(u)\right) a\,r\,b\,dr\,du$$

　　Above-mentioned equation can be described as:

$$\int_{L+} A\cdot dL = \iint_S \left(\frac{\partial}{\partial x}Q(x,y) - \frac{\partial}{\partial y}P(x,y)\right)dS \quad (1),\ \text{Complete Proof.}$$

2.2 Prove Green Theorem at Manifold

[Program Template of Waterloo Maplesoft, Optional]

Green Theorem Suppose boundary curve 'L' of plane bounded
region 'S' is composed by numbered smooth or piecewise smooth curves.
If exist 1th order continuous partial derivatives about functions

'P(x,y),Q(x,y)' [Structure plane vector field 'A'] at plane bounded closed region 'S', then:

$$\int_{L+} A \cdot dL = \iint_S \left(\frac{\partial}{\partial x} Q(x,y) - \frac{\partial}{\partial y} P(x,y) \right) dS$$

Symbol System:

Abstract (plane) simply connected closed parametrized curve 'CO' (closed),

Plane vector field 'PV',

Differential function 'dPV1,dPV2' of plane vector field 'PV',

Plane bounded closed region 'S' that curve 'CO' surrounds,

General expression of curve 'CO' coordinates area element coefficient 'J'

> restart; # Reset computer algebraic system

> with(linalg): # Load 'linear algebra' package

> CO:=[a*cos(t),b*sin(t)];

 # Define parametrized expression of abstract (plane) simply connected closed curve 'CO'; 'a,b' are nonzero constants or 1th order derivable continuous functions, simply connected closed curve 'CO' determines value of 'a,b'. (See also Poincare Conjecture: 'Every closed n-Manifold which is homotopy equivalent to n-Sphere is homeomorphic to n-Sphere' [19], and Chapter 1 this Part 'Constitute Abstract Simply Connected Closed Parametrized Curve Coordinates [Plane]')

$$CO := [a \cos(t), b \sin(t)]$$

> rgt:=[0,2*Pi]; # Set range of parameter 't', make curve 'CO' closed

$$rgt := [0, 2\pi]$$

> PV:=[(P)(x,y),(Q)(x,y)];

Define abstract plane vector field 'PV' (Suppose Vector Field 'PV' possesses 1th order continuous partial derivatives in plane bounded closed region 'S' that curve 'CO' surrounds)

$$PV := [P(x,y), Q(x,y)]$$

> Diff(PV[2],x)-Diff(PV[1],y);dPV:=%;

Calculate differential function of abstract plane vector field 'PV'

$$\left(\frac{\partial}{\partial x} Q(x,y) \right) - \left(\frac{\partial}{\partial y} P(x,y) \right)$$

$$dPVl := \left(\frac{\partial}{\partial x} Q(x,y)\right) - \left(\frac{\partial}{\partial y} P(x,y)\right)$$

```
> x:=CO[1];y:=CO[2];
```

$$x := a\cos(t)$$
$$y := b\sin(t)$$

```
> Int(PV[1]*Diff(CO[1],t)+PV[2]*Diff(CO[2],t),t=rgt[1]..rgt[2]);
# Closed Curve Integral (Plane Vector Field 'PV' at Curve 'CO')
```

$$\int_0^{2\pi} P(a\cos(t),b\sin(t))\left(\frac{\partial}{\partial t}(a\cos(t))\right) + Q(a\cos(t),b\sin(t))\left(\frac{\partial}{\partial t}(b\sin(t))\right)dt$$

```
> value(%);
```

$$\int_0^{2\pi} -P(a\cos(t),b\sin(t))\,a\sin(t) + Q(a\cos(t),b\sin(t))\,b\cos(t)\,dt$$

```
# Import 'x = a*cos(t), y = b*sin(t)' to abstract vector field '[P(x,y),
Q(x,y)]', obtain insoluble result, abnegate it
> x:='x':y:='y':
> Int(PV[1]*Diff(CO[1],t)+PV[2]*Diff(CO[2],t),t=rgt[1]..rgt[2])
=Int(PV[1]*diff(CO[1],t)+PV[2]*diff(CO[2],t),t=rgt[1]..rgt[2]);
```

$$\int_0^{2\pi} P(x,y)\left(\frac{\partial}{\partial t}(a\cos(t))\right) + Q(x,y)\left(\frac{\partial}{\partial t}(b\sin(t))\right)dt =$$

$$\int_0^{2\pi} -P(x,y)\,a\sin(t) + Q(x,y)\,b\cos(t)\,dt$$

```
# Closed Curve Integral about Vector Field 'PV' at Curve 'CO'[Reserve
original form of Abstract Vector Field 'P(x,y),Q(x,y)']
> alpha:=rhs(value(%));
# Calculate value of plane closed curve integral
```

$$\alpha := 0$$ # Obtain a constant '0'

```
# Owing to universality and homogeneity of abstract vector field
[P(x,y),Q(x,y)], Integral by Embedded Sub-Variable 't' of Variable
'x,y', Result of Integral can be described as 'P(x,y)[or Q(x,y)]'.
```

   In other words, Integral by Embedded Sub-Variable 't' of Variable 'x,y', Can't change structure of abstract function P(x,y)[or Q(x,y)], so that abstract function structure P(x,y) [or Q(x,y)] can preserve primary form after integral.

  [See also Chapter 4 (This Part) Section 4.1 'Prove Green Theorem Finite Sums Limits at Manifold' and Chapter 4 (Part I) Section 4.3 'Discuss About Abstract Vector Field、Finite Sums Limit、Integral and Riemann Sums']

```
> x:='x':y:='y':
> COc:=subs(t=u,CO);
```

Change character 't' to'u' in parametrized expression of curve 'CO'

$$COc := [a\cos(u), b\sin(u)]$$

```
> [r*COc[1],r*COc[2]];
```

Entirely multiply radius vector 'r' (suppose r>0) by each item of curve 'CO' parametrized expression, convert "Parametrized expression of curve 'CO'" to "Parametrized expression of curve 'CO' coordinates"

$$[r\,a\cos(u), r\,b\sin(u)]$$

```
> x:=r*COc[1]:y:=r*COc[2]:
> matrix(2,2,[Diff(r*COc[1],r),Diff(r*COc[1],u),Diff(r*COc[2],r),
Diff(r*COc[2],u)])=
matrix(2,2,[diff(r*COc[1],r),diff(r*COc[1],u),diff(r*COc[2],r),
diff(r*COc[2],u)]);m:=rhs(%);
```

Define and calculate matrix of partial derivatives 'm', obtain general expression of curve 'CO' coordinates area element coefficient

$$\begin{bmatrix} \dfrac{\partial}{\partial r}(r\,a\cos(u)) & \dfrac{\partial}{\partial u}(r\,a\cos(u)) \\ \dfrac{\partial}{\partial r}(r\,b\sin(u)) & \dfrac{\partial}{\partial u}(r\,b\sin(u)) \end{bmatrix} = \begin{bmatrix} a\cos(u) & -r\,a\sin(u) \\ b\sin(u) & r\,b\cos(u) \end{bmatrix}$$

$$m := \begin{bmatrix} a\cos(u) & -r\,a\sin(u) \\ b\sin(u) & r\,b\cos(u) \end{bmatrix}$$

```
> det(m);
```

$$a\cos(u)^2\,r\,b + r\,a\sin(u)^2\,b$$

```
> J:=simplify(%);
```

General expression of curve 'CO' coordinates area element coefficient

$$J := a\,r\,b$$

```
> x:='x':y:='y':
> Diff(PV[2],x)*Diff(r*COc[1],r)*Diff(r*COc[1],u)
- Diff(PV[1],y)*Diff(r*COc[2],r)*Diff(r*COc[2],u);
```

\# Convert 'dPV1' from Cartesian coordinates expression to Curve 'CO' coordinates expression

\# Reserve original form of abstract vector field '[P(x,y), Q(x,y)]' and its partial derivatives (insoluble), calculate partial derivatives of 'r*a*cos(u),r*b*sin(u)'(computable), obtain a new differential function 'dPV2'.

\> dPV2:=value(%);

$$\left(\frac{\partial}{\partial x} Q(x,y)\right)\left(\frac{\partial}{\partial r}(r\,a\cos(u))\right)\left(\frac{\partial}{\partial u}(r\,a\cos(u))\right)$$

$$-\left(\frac{\partial}{\partial y} P(x,y)\right)\left(\frac{\partial}{\partial r}(r\,b\sin(u))\right)\left(\frac{\partial}{\partial u}(r\,b\sin(u))\right)$$

$$dPV2 := -\left(\frac{\partial}{\partial x} Q(x,y)\right)a^2\cos(u)\,r\sin(u)-\left(\frac{\partial}{\partial y} P(x,y)\right)b^2\sin(u)\,r\cos(u)$$

\> Int(Int(dPV2*J,r=0..1),u=rgt[1]..rgt[2])=Int(int(dPV2*J,

r=0..1),u=rgt[1]..rgt[2]);

\# Double Integrals of Product (Differential Function 'dPV2' and Curve 'CO' Coordinates Area Element) in range of 'r,u'

$$\int_0^{2\pi}\int_0^1\left(\left(-\left(\frac{\partial}{\partial x} Q(x,y)\right)a^2\cos(u)\,r\sin(u)-\left(\frac{\partial}{\partial y} P(x,y)\right)b^2\sin(u)\,r\cos(u)\right)a\,r\,b\,\boldsymbol{dr}\,\boldsymbol{du}\right.$$

$$=\int_0^{2\pi}\frac{1}{3}\left(-\left(\frac{\partial}{\partial x} Q(x,y)\right)a^2\cos(u)\sin(u)-\left(\frac{\partial}{\partial y} P(x,y)\right)b^2\sin(u)\cos(u)\right)a\,b\,\boldsymbol{du}$$

\> beta:=rhs(value(%));

\# Calculate value of plane area element integral

$$\beta := 0$$ \# Obtain a constant '0'

\# Owing to universality and homogeneity of abstract vector field's

Differential Function '$\frac{Q(x,y)}{\partial x}-\frac{P(x,y)}{\partial y}$', (or their elements '$\frac{\partial}{\partial x}Q(x,y)$,

$\frac{\partial}{\partial y}P(x,y)$'), Integral by Embedded Sub-Variable 'r,u' of Variable 'x,y',

character of integral can be regarded as "Integral about Differential

Function Elements of Differential Function (Viz. $\frac{\partial}{\partial x}Q(x,y)$, $\frac{\partial}{\partial y}P(x,y)$) 、

Coordinates Transform Differential Functions、Curve 'L' Coordinates Area Element". Integral by Embedded Sub-Variable 'r,u' of Variable 'x,y', Can't change structure of abstract Differential Function (Viz.

、$\dfrac{Q(x,y)}{\partial x} - \dfrac{P(x,y)}{\partial y}$ ') or their Differential Function Elements (Viz.

$\dfrac{\partial}{\partial \text{x}} Q(x,y)$, or $\dfrac{\partial}{\partial y} P(x,y)$) , they can preserve primary form after integral.

Two Coordinates Transform Differential Functions (Correspond to

Two Differential Variables '$\dfrac{\partial}{\partial x}$, $\dfrac{\partial}{\partial y}$'), Viz:

$$\frac{\partial}{\partial \text{r}} ra\cos(u) \frac{\partial}{\partial u} ra\cos(u) \text{ and } \frac{\partial}{\partial r} rb\sin(u) \frac{\partial}{\partial u} rb\sin(u)$$

will be changed after integral.

[See also Chapter 4 (This Part) Section 4.1 'Prove Green Theorem Finite Sums Limits at Manifold' and Chapter 4 (Part I) Section 4.3 'Discuss About Abstract Vector Field、Finite Sums Limit、Integral and Riemann Sums']

Viz:

$$\int_0^{2\pi} P(x,y)\left(\frac{\partial}{\partial t}(a\cos(t))\right) + Q(x,y)\left(\frac{\partial}{\partial t}(b\sin(t))\right) dt =$$

$$\int_0^{2\pi}\int_0^1 \left(-\left(\frac{\partial}{\partial x}Q(x,y)\right)a^2\cos(u)\,r\sin(u) - \left(\frac{\partial}{\partial y}P(x,y)\right)b^2\sin(u)\,r\cos(u)\right)a\,r\,b\,dr\,du$$

Value of 'Closed Curve Integral' is equal to 'Plane Area Element Integral', Complete Proof

Chapter 3 Numerical Models of Green Theorem at Manifold

3.1 Numerical Model of Green Theorem at Manifold (I)

Known: Parametrized Expression of Simply Connected Closed Curve (Irregular、Asymmetrical)

$$\left[\cos(t-1)+\frac{1}{5}\sin(5t-1),-t\sin(t)\right] \tag{1}$$

thereinto, $t\in[0,2\pi]$; and Integral Vector Field (Plane)

$$\left[\left(\frac{1}{3}x+\frac{1}{5}y+\frac{1}{15}xy\right)^2-\frac{xy^2}{3}-\frac{y^3}{6},\frac{\left(\frac{x}{4}-\frac{y}{5}-\frac{1}{9}\right)^2}{5}-\frac{x^2}{5}+\frac{xy^2}{3}\right] \tag{2}$$

Calculate and Validate Green Theorem at Manifold

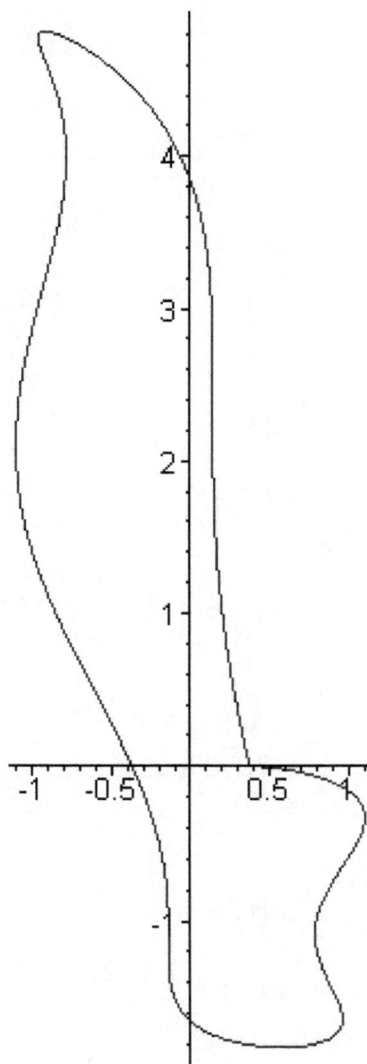

Figure(II)3.1 Simply connected、Closed Curve (1)

[Irregular、Asymmetrical]

Solution:

First, Free Closed Curve Integral:

According to curve parametrized expression (1), Calculate Tangent

Vector of Curve:

$$\left[\frac{d}{dt}\left(\cos(t-1)+\frac{1}{5}\sin(5\,t-1)\right),\frac{d}{dt}(-t\sin(t))\right]=$$
$$[-\sin(t-1)+\cos(5\,t-1),-\sin(t)-t\cos(t)] \tag{3}$$

Input curve parametric expression (1) to Vector Field (2); and calculate Integral of dot product [Vetor Field (2) and Tangent Vector (3)] in range of 't'(4):

$$\int_{0}^{2\pi}\bigg($$

$$\left(\frac{1}{3}\cos(t-1)+\frac{1}{15}\sin(5\,t-1)-\frac{1}{5}t\sin(t)-\frac{1}{15}\left(\cos(t-1)+\frac{1}{5}\sin(5\,t-1)\right)t\sin(t)\right)^{\!\!\prime}$$
$$-\frac{1}{3}\left(\cos(t-1)+\frac{1}{5}\sin(5\,t-1)\right)t^2\sin(t)^2+\frac{1}{6}t^3\sin(t)^3\bigg)$$

$$\left(\frac{d}{dt}\left(\cos(t-1)+\frac{1}{5}\sin(5\,t-1)\right)\right)+\bigg($$

$$\frac{1}{5}\left(\frac{1}{4}\cos(t-1)+\frac{1}{20}\sin(5\,t-1)+\frac{1}{5}t\sin(t)-\frac{1}{9}\right)^2-\frac{1}{5}\left(\cos(t-1)+\frac{1}{5}\sin(5\,t-1)\right)^2$$

$$+\frac{1}{3}\left(\cos(t-1)+\frac{1}{5}\sin(5\,t-1)\right)t^2\sin(t)^2\bigg)\left(\frac{d}{dt}(-t\sin(t))\right)dt$$

$$=$$

$$-\frac{113\,\pi}{13500}-\frac{6105379}{1694013750}\sin(3)\,\pi-\frac{364513}{7200000}\sin(1)\,\pi^2-\frac{388}{658125}\cos(3)\,\pi^2$$

$$+\frac{25752307}{864000000}\cos(1)\,\pi-\frac{6691}{504000}\cos(2)\,\pi-\frac{10373}{90000}\cos(2)\,\pi^2-\frac{23}{3000}\sin(2)\,\pi^2$$

$$-\frac{5}{144}\cos(1)\,\pi^3+\frac{2204}{55940625}\sin(3)\,\pi^2-\frac{82508547313}{124617811500000}\cos(3)\,\pi$$

$$-\frac{27986861}{118800000}\sin(2)\,\pi-\frac{5}{12}\cos(1)\,\pi^4+\frac{15253597}{27720000}\cos(1)\,\pi^2$$

$$+\frac{11896690513}{76839840000}\sin(1)\,\pi-\frac{5}{24}\sin(1)\,\pi^3+\frac{23}{225}\sin(2)\,\pi^3 \tag{4}$$

Second, Free Area Element Integral:

Change character 't' to 'u' in parametrized expression (1),entirely multiply radius vector 'r'(suppose r>0) by each item of curve parametrized expression (1), convert 'Parametrized expression of curve' to 'Parametrized expression of curve coordinates':

$$\left[r\left(\cos(u-1) + \frac{1}{5}\sin(5\,u-1) \right), -r\,u\,\sin(u) \right] \tag{5}$$

According to parametrized expression of curve coordinates (5), define and calculate matrix of partial derivatives, obtain general expression of curve coordinates area element coefficient (6):

$$\left[\begin{array}{cc} \dfrac{\partial}{\partial r}\left(r\left(\cos(u-1) + \frac{1}{5}\sin(5\,u-1) \right) \right) & \dfrac{\partial}{\partial u}\left(r\left(\cos(u-1) + \frac{1}{5}\sin(5\,u-1) \right) \right) \\[2ex] \dfrac{\partial}{\partial r}(-r\,u\,\sin(u)) & \dfrac{\partial}{\partial u}(-r\,u\,\sin(u)) \end{array} \right]$$

$$=$$

$$-\frac{1}{5}r\,(5\cos(u-1)\sin(u) + 5\cos(u-1)\,u\cos(u) + \sin(5\,u-1)\sin(u)$$

$$+ \sin(5\,u-1)\,u\cos(u) + 5\,u\sin(u)\sin(u-1) - 5\,u\sin(u)\cos(5\,u-1)) \tag{6}$$

Calculate Differential Function '$\frac{\partial}{\partial x}Q(x,y) - \frac{\partial}{\partial y}P(x,y)$' of Vector Field (2):

$$\left(\frac{\partial}{\partial x}\left(\frac{\left(\frac{x}{4} - \frac{y}{5} - \frac{1}{9} \right)^2}{5} - \frac{x^2}{5} + \frac{x\,y^2}{3} \right) \right) - \left(\frac{\partial}{\partial y}\left(\left(\frac{1}{3}x + \frac{1}{5}y + \frac{1}{15}x\,y \right)^2 - \frac{x\,y^2}{3} - \frac{y^3}{6} \right) \right) =$$

$$-\frac{3\,x}{8} - \frac{y}{50} - \frac{1}{90} + \frac{5\,y^2}{6} - 2\left(\frac{1}{3}x + \frac{1}{5}y + \frac{1}{15}x\,y \right)\left(\frac{1}{5} + \frac{x}{15} \right) + \frac{2\,x\,y}{3} \tag{7}$$

Input Parametrized Expression of Curve Coordinates (5) to Differential Function (7); and Calculate Double Integrals of Product [Differential Function (7) and Curve Coordinates Area Element] in range of 'r,u' (8):

$$\int_0^{2\pi}\int_0^1 -\frac{1}{5}\left(-\frac{3}{8}r\left(\cos(u-1) + \frac{1}{5}\sin(5\,u-1) \right) + \frac{1}{50}r\,u\,\sin(u) - \frac{1}{90} + \frac{5}{6}r^2\,u^2\,\sin(u)^2 - 2 \right.$$

$$\left(\frac{1}{3}r\left(\cos(u-1) + \frac{1}{5}\sin(5\,u-1) \right) - \frac{1}{5}r\,u\,\sin(u) \right.$$

$$\left. -\frac{1}{15}r^2\left(\cos(u-1) + \frac{1}{5}\sin(5\,u-1) \right)u\,\sin(u) \right)$$

$$\left. \left(\frac{1}{5} + \frac{1}{15}r\left(\cos(u-1) + \frac{1}{5}\sin(5\,u-1) \right) \right) \right)$$

978-1-62265-930-2 (online)　　978-1-62265-931-9 (paper)　　　　　　Yang Ke

$$-\frac{2}{3}r^2\left(\cos(u-1)+\frac{1}{5}\sin(5\,u-1)\right)u\sin(u)\Big)r\left(5\cos(u-1)\sin(u)\right.$$

$$+5\cos(u-1)\,u\cos(u)+\sin(5\,u-1)\sin(u)+\sin(5\,u-1)\,u\cos(u)$$

$$+5\,u\sin(u)\sin(u-1)-5\,u\sin(u)\cos(5\,u-1))\,dr\,du$$

$$=$$

$$-\frac{113\,\pi}{13500}-\frac{6105379}{1694013750}\sin(3)\,\pi-\frac{364513}{7200000}\sin(1)\,\pi^2-\frac{388}{658125}\cos(3)\,\pi^2$$

$$+\frac{25752307}{864000000}\cos(1)\,\pi-\frac{6691}{504000}\cos(2)\,\pi-\frac{10373}{90000}\cos(2)\,\pi^2-\frac{23}{3000}\sin(2)\,\pi^2$$

$$-\frac{5}{144}\cos(1)\,\pi^3+\frac{2204}{55940625}\sin(3)\,\pi^2-\frac{82508547313}{124617811500000}\cos(3)\,\pi$$

$$-\frac{27986861}{118800000}\sin(2)\,\pi-\frac{5}{12}\cos(1)\,\pi^4+\frac{15253597}{27720000}\cos(1)\,\pi^2$$

$$+\frac{11896690513}{76839840000}\sin(1)\,\pi-\frac{5}{24}\sin(1)\,\pi^3+\frac{23}{225}\sin(2)\,\pi^3$$

$$(8)$$

// Be different to ordinary double integrals, in this position, first integral range is 'r∈[0,1]' always, and not 'r∈[0,n]' or 'r∈[0,∞]'; Because of '1', it is possible to hold a correct ratio between 'r,u';

// Be different to abstract Differential Function in 'Proof', and in 'Numerical Models', be able to input idiographic parametrized expression of curve coordinates (5) to idiographic Differential Function(7) directly, then calculate product of Differential Function(7) and curve coordinates area element, complete double integrals

Integral Precision Value (4) [Vector Field (2) on Target Curve (1)], is equal to Area Element Integral Precision Value (8) [Differential Function (7) in Plane Bounded Closed Region that Target Curve (1) surrounds], Complete Calculation and Validation of Green Theorem at Manifold.

Double Integrals of 'Curve Coordinates Area Element' in range of 'r,u', absolute value of result is area value of plane bounded closed region surrounded by closed curve (9):

$$\int_0^{2\pi}\int_0^1 -\frac{1}{5}r\left(\sin(5\,u-1)\cos(u)\,u+5\,u\sin(u)\sin(u-1)-5\,u\sin(u)\cos(5\,u-1)\right.$$

$$+5\cos(u)\cos(u-1)\,u+\sin(5\,u-1)\sin(u)+5\sin(u)\cos(u-1))\,dr\,du$$

$$= -\cos(1)\,\pi^2 + \frac{1}{12}\cos(1)\,\pi - \frac{1}{2}\sin(1)\,\pi$$

(9)

3.2 Numerical Model of Green Theorem at Manifold (I)
[Program Template of Waterloo Maplesoft, Optional]

```
> restart;
> with(plots):with(linalg):
> CO:=[ cos(t-1/5)+sin(5*t)*cos(9*t-3/2)/14-cos(4*t-2/3)/7,
sin(t-2/3)-cos(5*t-2/3)*sin(3*t-1/3)/7+cos(6*t-1)*sin(4*t)/9
-cos(7*t-1)/9];
# Define parametrized expression of discretional (plane) simply
connected closed curve 'CO'
```

$$CO := \left[\cos\left(t - \frac{1}{5}\right) + \frac{1}{14}\sin(5\,t)\cos\left(9\,t - \frac{3}{2}\right) - \frac{1}{7}\cos\left(4\,t - \frac{2}{3}\right),\right.$$
$$\left.\sin\left(t - \frac{2}{3}\right) - \frac{1}{7}\cos\left(5\,t - \frac{2}{3}\right)\sin\left(3\,t - \frac{1}{3}\right) + \frac{1}{9}\cos(6\,t - 1)\sin(4\,t) - \frac{1}{9}\cos(7\,t - 1)\right]$$

```
> rgt:=[0,2*Pi];
# Set range of parameter 't', make plane curve 'CO' closed
```

$$rgt := [0, 2\,\pi]$$

```
> plot([CO[1],CO[2],t=rgt[1]..rgt[2]],scaling=constrained,
color=blue,numpoints=1000);g1:=%:
```

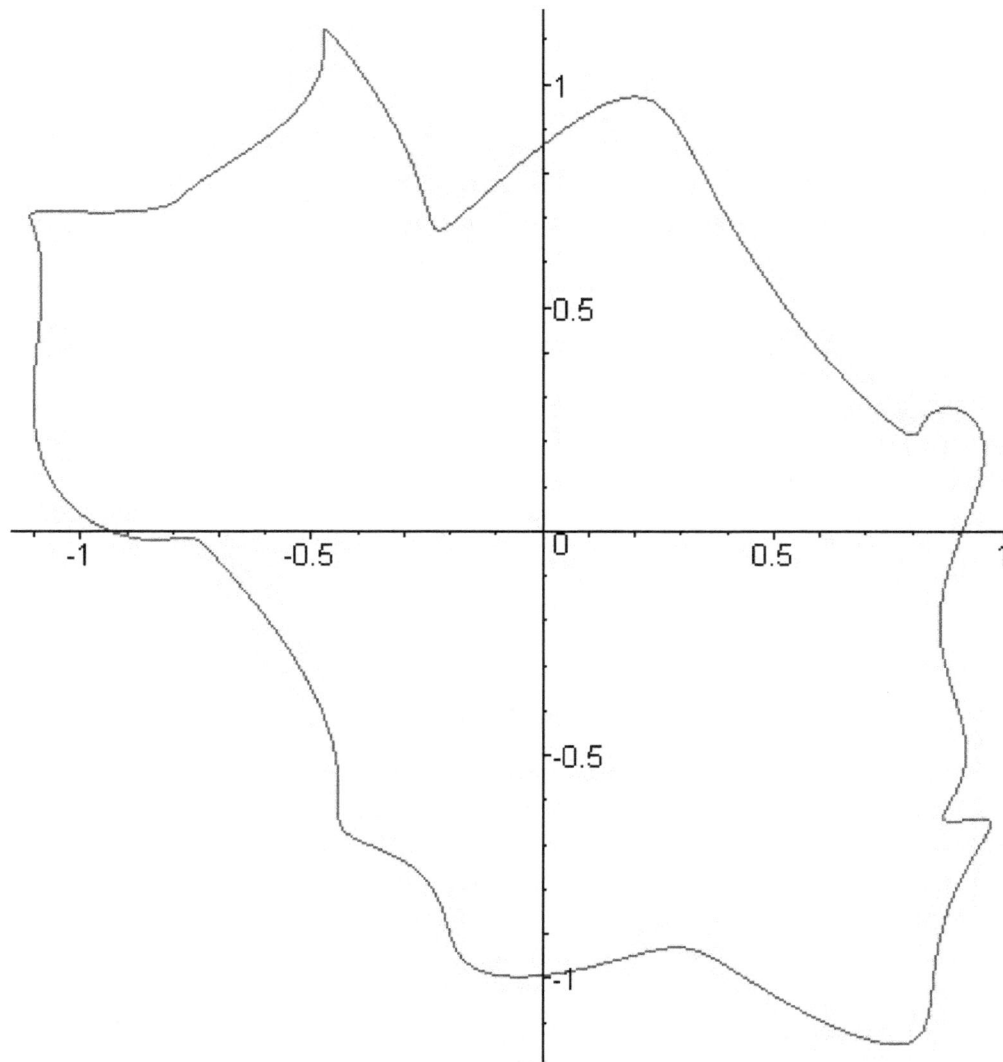

Figure(II)3.2.1 Asymmetrical、Irregular Discretional
Simply Connected Closed Curve 'CO'

> PV:=[(x/2-y/5-1)^2/3-x^2/7-x*y/3,(x/4-y/5-1/21)^2/2-x^3/5-y^3/3];
Define discretional plane vector field 'PV' (Suppose Vector Field
'PV' possesses 1th order continuous partial derivatives in plane
bounded closed region 'S' that curve 'CO' surrounds)

$$PV := \left[\frac{\left(\dfrac{x}{2} - \dfrac{y}{5} - 1 \right)^2}{3} - \frac{x^2}{7} - \frac{xy}{3}, \frac{\left(\dfrac{x}{4} - \dfrac{y}{5} - \dfrac{1}{21} \right)^2}{2} - \frac{x^3}{5} - \frac{y^3}{3} \right]$$

> Diff(PV[2],x)-Diff(PV[1],y);
Calculate differential function of idiographic plane vector field
'PV', result is an idiographic differential function 'dPV'

$$\left(\frac{\partial}{\partial x}\left(\frac{\left(\frac{x}{4}-\frac{y}{5}-\frac{1}{21}\right)^2}{2}-\frac{x^3}{5}-\frac{y^3}{3}\right)\right)-\left(\frac{\partial}{\partial y}\left(\frac{\left(\frac{x}{2}-\frac{y}{5}-1\right)^2}{3}-\frac{x^2}{7}-\frac{xy}{3}\right)\right)$$

```
> dPV:=value(%);
```

$$dPV:=\frac{37}{80}x-\frac{23}{300}y-\frac{61}{420}-\frac{3}{5}x^2$$

```
> rgx:=[-1.2,1.1];
```

$$rgx:=[-1.2, 1.1]$$

```
> rgy:=[-1.15,1.15]; # Set drawing range of plane vector field 'PV'
```

$$rgy:=[-1.15, 1.15]$$

```
> plot3d(dPV,x=rgx[1]..rgx[2],y=rgy[1]..rgy[2],projection=0.9,
axes=normal); # Drawing differential function 'dPV' in 3-Dimensional
Euclidean space
```

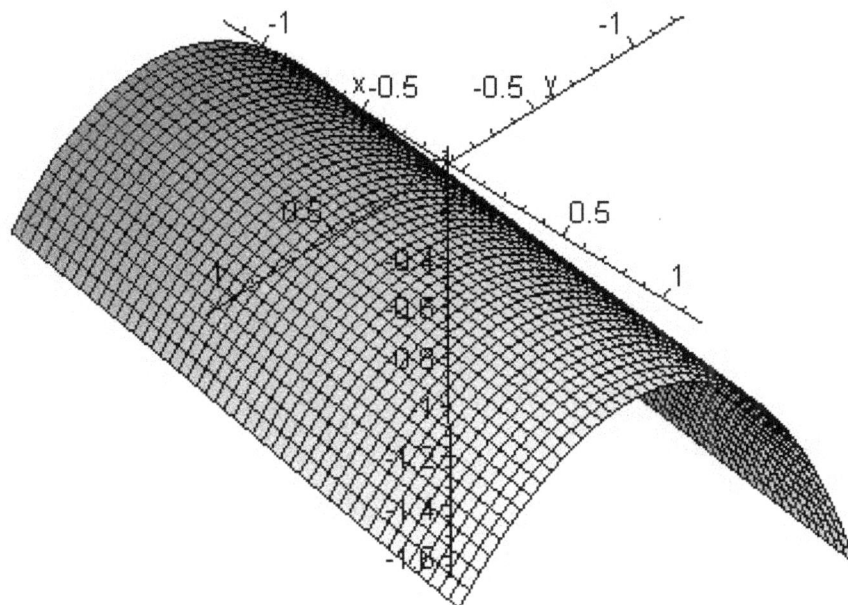

Figure(II)3.2.2 Space View of Differential Function 'dPV'

```
> fieldplot(PV,x=rgx[1]..rgx[2],y=rgy[1]..rgy[2],arrows=SLIM,
color=blue):g2:=%: # Draw plane vector field 'PV'
> implicitplot(dPV,x=rgx[1]..rgx[2],y=rgy[1]..rgy[2],color=red,
thickness=1,numpoints=5000):g3:=%:
# Draw contour curve of differential function 'dPV'
```

```
> display(g1,g2,g3);
```

Plane closed curve 'CO', plane vector field 'PV', contour curve of 'dPV', three figures are synthesized

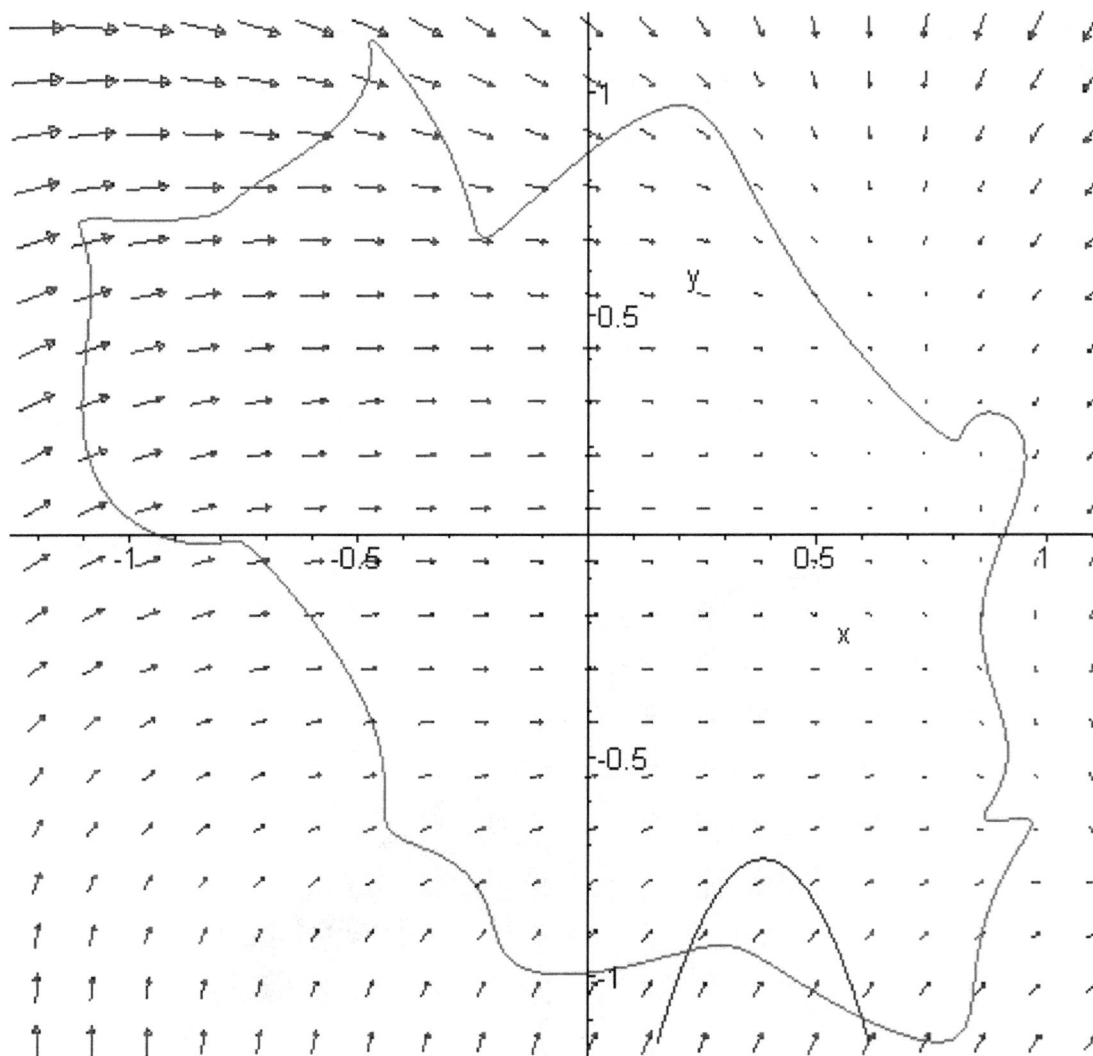

Figure(II)3.2.3 Plane Closed Curve 'CO', Vector Field 'PV' and Contour Curve of 'dPV'

```
> x:=CO[1];y:=CO[2];
```

$$x := \cos\left(t - \frac{1}{5}\right) + \frac{1}{14}\sin(5\,t)\cos\left(9\,t - \frac{3}{2}\right) - \frac{1}{7}\cos\left(4\,t - \frac{2}{3}\right)$$

$$y := \sin\left(t - \frac{2}{3}\right) - \frac{1}{7}\cos\left(5\,t - \frac{2}{3}\right)\sin\left(3\,t - \frac{1}{3}\right) + \frac{1}{9}\cos(6\,t - 1)\sin(4\,t) - \frac{1}{9}\cos(7\,t - 1)$$

```
> Int(PV[1]*Diff(CO[1],t)+PV[2]*Diff(CO[2],t),t=rgt[1]..rgt[2]);
```

Closed Curve Integral (Plane Vector Field 'PV' at Plane Closed Curve 'CO')

$$\int_0^{2\pi} \left(\frac{1}{3}\left(\frac{1}{2}\cos\left(t-\frac{1}{5}\right)+\frac{1}{28}\sin(5t)\cos\left(9t-\frac{3}{2}\right)-\frac{1}{14}\cos\left(4t-\frac{2}{3}\right)-\frac{1}{5}\sin\left(t-\frac{2}{3}\right)\right.\right.$$

$$\left.+\frac{1}{35}\cos\left(5t-\frac{2}{3}\right)\sin\left(3t-\frac{1}{3}\right)-\frac{1}{45}\cos(6t-1)\sin(4t)+\frac{1}{45}\cos(7t-1)-1\right)^2$$

$$-\frac{1}{7}\left(\cos\left(t-\frac{1}{5}\right)+\frac{1}{14}\sin(5t)\cos\left(9t-\frac{3}{2}\right)-\frac{1}{7}\cos\left(4t-\frac{2}{3}\right)\right)^2-\frac{1}{3}$$

$$\left(\cos\left(t-\frac{1}{5}\right)+\frac{1}{14}\sin(5t)\cos\left(9t-\frac{3}{2}\right)-\frac{1}{7}\cos\left(4t-\frac{2}{3}\right)\right)$$

$$\left(\sin\left(t-\frac{2}{3}\right)-\frac{1}{7}\cos\left(5t-\frac{2}{3}\right)\sin\left(3t-\frac{1}{3}\right)+\frac{1}{9}\cos(6t-1)\sin(4t)-\frac{1}{9}\cos(7t-1)\right)\right)$$

$$\left(\frac{d}{dt}\left(\cos\left(t-\frac{1}{5}\right)+\frac{1}{14}\sin(5t)\cos\left(9t-\frac{3}{2}\right)-\frac{1}{7}\cos\left(4t-\frac{2}{3}\right)\right)\right)+\left(\frac{1}{2}\left(\frac{1}{4}\cos\left(t-\frac{1}{5}\right)\right.\right.$$

$$\left.+\frac{1}{56}\sin(5t)\cos\left(9t-\frac{3}{2}\right)-\frac{1}{28}\cos\left(4t-\frac{2}{3}\right)-\frac{1}{5}\sin\left(t-\frac{2}{3}\right)\right.$$

$$\left.+\frac{1}{35}\cos\left(5t-\frac{2}{3}\right)\sin\left(3t-\frac{1}{3}\right)-\frac{1}{45}\cos(6t-1)\sin(4t)+\frac{1}{45}\cos(7t-1)-\frac{1}{21}\right)^2$$

$$-\frac{1}{5}\left(\cos\left(t-\frac{1}{5}\right)+\frac{1}{14}\sin(5t)\cos\left(9t-\frac{3}{2}\right)-\frac{1}{7}\cos\left(4t-\frac{2}{3}\right)\right)^3-$$

$$\frac{1}{3}\left(\sin\left(t-\frac{2}{3}\right)-\frac{1}{7}\cos\left(5t-\frac{2}{3}\right)\sin\left(3t-\frac{1}{3}\right)+\frac{1}{9}\cos(6t-1)\sin(4t)-\frac{1}{9}\cos(7t-1)\right)^3$$

$$\left.\right)\left(\right.$$

$$\frac{d}{dt}\left(\sin\left(t-\frac{2}{3}\right)-\frac{1}{7}\cos\left(5t-\frac{2}{3}\right)\sin\left(3t-\frac{1}{3}\right)+\frac{1}{9}\cos(6t-1)\sin(4t)-\frac{1}{9}\cos(7t-1)\right)$$

$$\left.\right)dt$$

```
> alpha:=value(%);delta:=evalf(alpha);
# Calculate value of plane closed curve integral
```

$$\alpha:=-\frac{37}{225792}\cos(1)\pi-\frac{37}{27440}\cos\left(\frac{1}{3}\right)\pi+\frac{1}{6720}\sin\left(\frac{9}{5}\right)\pi-\frac{1}{840}\cos\left(\frac{29}{30}\right)\pi$$

$$+\frac{3}{1960}\sin\left(\frac{13}{10}\right)\pi+\frac{3}{1960}\sin\left(\frac{11}{30}\right)\pi-\frac{3}{3920}\sin\left(\frac{23}{30}\right)\pi+\frac{1}{1680}\sin\left(\frac{1}{10}\right)\pi$$

$$-\frac{1}{420}\cos\left(\frac{11}{15}\right)\pi+\frac{23}{181440}\cos\left(\frac{1}{2}\right)\pi-\frac{23}{680400}\sin\left(\frac{4}{3}\right)\pi+\frac{307}{15680}\cos\left(\frac{1}{15}\right)\pi$$

$$-\frac{37}{2880}\cos\left(\frac{3}{5}\right)\pi+\frac{23}{8400}\sin\left(\frac{8}{15}\right)\pi-\frac{19}{75600}\sin\left(\frac{2}{15}\right)\pi-\frac{23}{75600}\cos\left(\frac{1}{5}\right)\pi$$

$$+ \frac{23}{264600} \sin\left(\frac{2}{3}\right)\pi - \frac{23}{1058400} \cos\left(\frac{1}{6}\right)\pi - \frac{37}{56448} \sin\left(\frac{1}{6}\right)\pi + \frac{37}{54880} \sin\left(\frac{7}{6}\right)\pi$$

$$+ \frac{37}{439040} \cos(2)\pi + \frac{23}{1646400} \cos\left(\frac{5}{6}\right)\pi - \frac{3553}{11760} \cos\left(\frac{7}{15}\right)\pi$$

$$\delta := -0.8210311510$$

```
> x:='x':y:='y':
> COc:=subs(t=u,CO);
```

\# Convert character 't' to 'u' in parametrized expression of curve 'CO'

$$COc := \left[\cos\left(u - \frac{1}{5}\right) + \frac{1}{14}\sin(5\,u)\cos\left(9\,u - \frac{3}{2}\right) - \frac{1}{7}\cos\left(4\,u - \frac{2}{3}\right),\right.$$
$$\left. \sin\left(u - \frac{2}{3}\right) - \frac{1}{7}\cos\left(5\,u - \frac{2}{3}\right)\sin\left(3\,u - \frac{1}{3}\right) + \frac{1}{9}\cos(6\,u - 1)\sin(4\,u) - \frac{1}{9}\cos(7\,u - 1)\right]$$

```
> [r*COc[1],r*COc[2]];
```

\# Entirely multiply radius vector 'r' (suppose r>0) by each item of curve 'CO' parametrized expression, convert "Parametrized expression of curve 'CO'" to "Parametrized expression of curve 'CO' coordinates"

$$\left[r\left(\cos\left(u - \frac{1}{5}\right) + \frac{1}{14}\sin(5\,u)\cos\left(9\,u - \frac{3}{2}\right) - \frac{1}{7}\cos\left(4\,u - \frac{2}{3}\right)\right), r\left(\right.\right.$$
$$\left. \sin\left(u - \frac{2}{3}\right) - \frac{1}{7}\cos\left(5\,u - \frac{2}{3}\right)\sin\left(3\,u - \frac{1}{3}\right) + \frac{1}{9}\cos(6\,u - 1)\sin(4\,u) - \frac{1}{9}\cos(7\,u - 1)\right)$$
$$\left.\vphantom{\frac{1}{1}}\right]$$

```
> x:=r*COc[1]:y:=r*COc[2]:
> matrix(2,2,[Diff(x,r),Diff(x,u),Diff(y,r),Diff(y,u)])=
matrix(2,2,[diff(x,r),diff(x,u),diff(y,r),diff(y,u)]);m:=rhs(%);
```

\# Define and calculate matrix of partial derivatives 'm', obtain general expression of curve 'CO' coordinates area element coefficient

$$\left[\frac{\partial}{\partial r}\left(r\left(\cos\left(u - \frac{1}{5}\right) + \frac{1}{14}\sin(5\,u)\cos\left(9\,u - \frac{3}{2}\right) - \frac{1}{7}\cos\left(4\,u - \frac{2}{3}\right)\right)\right),\right.$$
$$\left.\frac{\partial}{\partial u}\left(r\left(\cos\left(u - \frac{1}{5}\right) + \frac{1}{14}\sin(5\,u)\cos\left(9\,u - \frac{3}{2}\right) - \frac{1}{7}\cos\left(4\,u - \frac{2}{3}\right)\right)\right)\right]$$
$$\left[\frac{\partial}{\partial r}\left(r\left(\vphantom{\frac{1}{1}}\right.\right.\right.$$
$$\left.\sin\left(u - \frac{2}{3}\right) - \frac{1}{7}\cos\left(5\,u - \frac{2}{3}\right)\sin\left(3\,u - \frac{1}{3}\right) + \frac{1}{9}\cos(6\,u - 1)\sin(4\,u) - \frac{1}{9}\cos(7\,u - 1)\right)$$
$$\left.\right), \frac{\partial}{\partial u}\left(r\left(\vphantom{\frac{1}{1}}\right.\right.$$
$$\left.\sin\left(u - \frac{2}{3}\right) - \frac{1}{7}\cos\left(5\,u - \frac{2}{3}\right)\sin\left(3\,u - \frac{1}{3}\right) + \frac{1}{9}\cos(6\,u - 1)\sin(4\,u) - \frac{1}{9}\cos(7\,u - 1)\right)$$

$$\Big)\Big]=$$

$$\left[\cos\!\left(u-\frac{1}{5}\right)+\frac{1}{14}\sin(5\,u)\cos\!\left(9\,u-\frac{3}{2}\right)-\frac{1}{7}\cos\!\left(4\,u-\frac{2}{3}\right),\,r\Big(\right.$$

$$\left.-\sin\!\left(u-\frac{1}{5}\right)+\frac{5}{14}\cos(5\,u)\cos\!\left(9\,u-\frac{3}{2}\right)-\frac{9}{14}\sin(5\,u)\sin\!\left(9\,u-\frac{3}{2}\right)+\frac{4}{7}\sin\!\left(4\,u-\frac{2}{3}\right)\Big)\right]$$

$$\left[\sin\!\left(u-\frac{2}{3}\right)-\frac{1}{7}\cos\!\left(5\,u-\frac{2}{3}\right)\sin\!\left(3\,u-\frac{1}{3}\right)+\frac{1}{9}\cos(6\,u-1)\sin(4\,u)-\frac{1}{9}\cos(7\,u-1)\right.$$

$$,\,r\left(\cos\!\left(u-\frac{2}{3}\right)+\frac{5}{7}\sin\!\left(5\,u-\frac{2}{3}\right)\sin\!\left(3\,u-\frac{1}{3}\right)-\frac{3}{7}\cos\!\left(5\,u-\frac{2}{3}\right)\cos\!\left(3\,u-\frac{1}{3}\right)\right.$$

$$\left.\left.-\frac{2}{3}\sin(6\,u-1)\sin(4\,u)+\frac{4}{9}\cos(6\,u-1)\cos(4\,u)+\frac{7}{9}\sin(7\,u-1)\right)\right]$$

$$m:=$$

$$\left[\cos\!\left(u-\frac{1}{5}\right)+\frac{1}{14}\sin(5\,u)\cos\!\left(9\,u-\frac{3}{2}\right)-\frac{1}{7}\cos\!\left(4\,u-\frac{2}{3}\right),\,r\Big(\right.$$

$$\left.-\sin\!\left(u-\frac{1}{5}\right)+\frac{5}{14}\cos(5\,u)\cos\!\left(9\,u-\frac{3}{2}\right)-\frac{9}{14}\sin(5\,u)\sin\!\left(9\,u-\frac{3}{2}\right)+\frac{4}{7}\sin\!\left(4\,u-\frac{2}{3}\right)\Big)\right]$$

$$\left[\sin\!\left(u-\frac{2}{3}\right)-\frac{1}{7}\cos\!\left(5\,u-\frac{2}{3}\right)\sin\!\left(3\,u-\frac{1}{3}\right)+\frac{1}{9}\cos(6\,u-1)\sin(4\,u)-\frac{1}{9}\cos(7\,u-1)\right.$$

$$,\,r\left(\cos\!\left(u-\frac{2}{3}\right)+\frac{5}{7}\sin\!\left(5\,u-\frac{2}{3}\right)\sin\!\left(3\,u-\frac{1}{3}\right)-\frac{3}{7}\cos\!\left(5\,u-\frac{2}{3}\right)\cos\!\left(3\,u-\frac{1}{3}\right)\right.$$

$$\left.\left.-\frac{2}{3}\sin(6\,u-1)\sin(4\,u)+\frac{4}{9}\cos(6\,u-1)\cos(4\,u)+\frac{7}{9}\sin(7\,u-1)\right)\right]$$

```
> det(m):
> J:=simplify(%);
```

General expression of curve 'CO' coordinates area element coefficient

$$J:=\frac{1}{882}\,r\left(882\cos\!\left(u-\frac{1}{5}\right)\cos\!\left(u-\frac{2}{3}\right)+686\cos\!\left(u-\frac{1}{5}\right)\sin(7\,u-1)\right.$$

$$-126\cos\!\left(4\,u-\frac{2}{3}\right)\cos\!\left(u-\frac{2}{3}\right)-98\cos\!\left(4\,u-\frac{2}{3}\right)\sin(7\,u-1)$$

$$+882\sin\!\left(u-\frac{1}{5}\right)\sin\!\left(u-\frac{2}{3}\right)-98\sin\!\left(u-\frac{1}{5}\right)\cos(7\,u-1)$$

$$-504\sin\!\left(4\,u-\frac{2}{3}\right)\sin\!\left(u-\frac{2}{3}\right)+56\sin\!\left(4\,u-\frac{2}{3}\right)\cos(7\,u-1)$$

$$+45\sin(5\,u)\cos\!\left(9\,u-\frac{3}{2}\right)\sin\!\left(5\,u-\frac{2}{3}\right)\sin\!\left(3\,u-\frac{1}{3}\right)$$

$$-27\sin(5\,u)\cos\!\left(9\,u-\frac{3}{2}\right)\cos\!\left(5\,u-\frac{2}{3}\right)\cos\!\left(3\,u-\frac{1}{3}\right)$$

$$- 42 \sin(5u) \cos\left(9u - \frac{3}{2}\right) \sin(6u - 1) \sin(4u)$$

$$+ 28 \sin(5u) \cos\left(9u - \frac{3}{2}\right) \cos(6u - 1) \cos(4u)$$

$$+ 45 \cos(5u) \cos\left(9u - \frac{3}{2}\right) \cos\left(5u - \frac{2}{3}\right) \sin\left(3u - \frac{1}{3}\right)$$

$$- 35 \cos(5u) \cos\left(9u - \frac{3}{2}\right) \cos(6u - 1) \sin(4u)$$

$$- 81 \sin(5u) \sin\left(9u - \frac{3}{2}\right) \cos\left(5u - \frac{2}{3}\right) \sin\left(3u - \frac{1}{3}\right)$$

$$+ 63 \sin(5u) \sin\left(9u - \frac{3}{2}\right) \cos(6u - 1) \sin(4u)$$

$$+ 35 \cos(5u) \cos\left(9u - \frac{3}{2}\right) \cos(7u - 1) - 315 \cos(5u) \cos\left(9u - \frac{3}{2}\right) \sin\left(u - \frac{2}{3}\right)$$

$$+ 49 \sin(5u) \cos\left(9u - \frac{3}{2}\right) \sin(7u - 1) + 567 \sin(5u) \sin\left(9u - \frac{3}{2}\right) \sin\left(u - \frac{2}{3}\right)$$

$$+ 72 \sin\left(4u - \frac{2}{3}\right) \cos\left(5u - \frac{2}{3}\right) \sin\left(3u - \frac{1}{3}\right)$$

$$- 126 \sin\left(u - \frac{1}{5}\right) \cos\left(5u - \frac{2}{3}\right) \sin\left(3u - \frac{1}{3}\right) - 56 \cos\left(4u - \frac{2}{3}\right) \cos(6u - 1) \cos(4u)$$

$$- 63 \sin(5u) \sin\left(9u - \frac{3}{2}\right) \cos(7u - 1) + 84 \cos\left(4u - \frac{2}{3}\right) \sin(6u - 1) \sin(4u)$$

$$- 378 \cos\left(u - \frac{1}{5}\right) \cos\left(5u - \frac{2}{3}\right) \cos\left(3u - \frac{1}{3}\right) - 588 \cos\left(u - \frac{1}{5}\right) \sin(6u - 1) \sin(4u)$$

$$+ 630 \cos\left(u - \frac{1}{5}\right) \sin\left(5u - \frac{2}{3}\right) \sin\left(3u - \frac{1}{3}\right) + 98 \sin\left(u - \frac{1}{5}\right) \cos(6u - 1) \sin(4u)$$

$$+ 392 \cos\left(u - \frac{1}{5}\right) \cos(6u - 1) \cos(4u)$$

$$+ 54 \cos\left(4u - \frac{2}{3}\right) \cos\left(5u - \frac{2}{3}\right) \cos\left(3u - \frac{1}{3}\right)$$

$$- 90 \cos\left(4u - \frac{2}{3}\right) \sin\left(5u - \frac{2}{3}\right) \sin\left(3u - \frac{1}{3}\right) + 63 \sin(5u) \cos\left(9u - \frac{3}{2}\right) \cos\left(u - \frac{2}{3}\right)$$

$$- 56 \sin\left(4u - \frac{2}{3}\right) \cos(6u - 1) \sin(4u) \Bigg)$$

```
> Int(Int(dPV*J,r=0..1),u=rgt[1]..rgt[2]);
```

Double Integrals of Product (Differential Function 'dPV' and "Curve 'CO' Coordinates Area Element") in range of 'r,u'

// Be different to abstract differential function 'dPV1,dPV2' that 'Proof' refers to, in 'Numerical Models', be able to input idiographic values 'x=r*COc[1], y=r*COc[2]' to idiographic differential function 'dPV', then calculate product of idiographic differential function 'dPV' and area element, complete double integrals

$$\int_0^{2\pi}\int_0^1 \frac{1}{882}\left(\frac{37}{80}r\left(\cos\left(u-\frac{1}{5}\right)+\frac{1}{14}\sin(5\,u)\cos\left(9\,u-\frac{3}{2}\right)-\frac{1}{7}\cos\left(4\,u-\frac{2}{3}\right)\right)-\frac{23}{300}r\,\Bigg(\right.$$

$$\sin\left(u-\frac{2}{3}\right)-\frac{1}{7}\cos\left(5\,u-\frac{2}{3}\right)\sin\left(3\,u-\frac{1}{3}\right)+\frac{1}{9}\cos(6\,u-1)\sin(4\,u)-\frac{1}{9}\cos(7\,u-1)\Bigg)$$

$$\left.-\frac{61}{420}-\frac{3}{5}r^2\left(\cos\left(u-\frac{1}{5}\right)+\frac{1}{14}\sin(5\,u)\cos\left(9\,u-\frac{3}{2}\right)-\frac{1}{7}\cos\left(4\,u-\frac{2}{3}\right)\right)^2\right)r\,\Bigg($$

$$882\cos\left(u-\frac{1}{5}\right)\cos\left(u-\frac{2}{3}\right)+686\cos\left(u-\frac{1}{5}\right)\sin(7\,u-1)$$

$$-126\cos\left(4\,u-\frac{2}{3}\right)\cos\left(u-\frac{2}{3}\right)-98\cos\left(4\,u-\frac{2}{3}\right)\sin(7\,u-1)$$

$$+882\sin\left(u-\frac{1}{5}\right)\sin\left(u-\frac{2}{3}\right)-98\sin\left(u-\frac{1}{5}\right)\cos(7\,u-1)$$

$$-504\sin\left(4\,u-\frac{2}{3}\right)\sin\left(u-\frac{2}{3}\right)+56\sin\left(4\,u-\frac{2}{3}\right)\cos(7\,u-1)$$

$$+45\sin(5\,u)\cos\left(9\,u-\frac{3}{2}\right)\sin\left(5\,u-\frac{2}{3}\right)\sin\left(3\,u-\frac{1}{3}\right)$$

$$-27\sin(5\,u)\cos\left(9\,u-\frac{3}{2}\right)\cos\left(5\,u-\frac{2}{3}\right)\cos\left(3\,u-\frac{1}{3}\right)$$

$$-42\sin(5\,u)\cos\left(9\,u-\frac{3}{2}\right)\sin(6\,u-1)\sin(4\,u)$$

$$+28\sin(5\,u)\cos\left(9\,u-\frac{3}{2}\right)\cos(6\,u-1)\cos(4\,u)$$

$$+45\cos(5\,u)\cos\left(9\,u-\frac{3}{2}\right)\cos\left(5\,u-\frac{2}{3}\right)\sin\left(3\,u-\frac{1}{3}\right)$$

$$-35\cos(5\,u)\cos\left(9\,u-\frac{3}{2}\right)\cos(6\,u-1)\sin(4\,u)$$

$$-81\sin(5\,u)\sin\left(9\,u-\frac{3}{2}\right)\cos\left(5\,u-\frac{2}{3}\right)\sin\left(3\,u-\frac{1}{3}\right)$$

$$+63\sin(5\,u)\sin\left(9\,u-\frac{3}{2}\right)\cos(6\,u-1)\sin(4\,u)$$

$$+35\cos(5\,u)\cos\left(9\,u-\frac{3}{2}\right)\cos(7\,u-1)-315\cos(5\,u)\cos\left(9\,u-\frac{3}{2}\right)\sin\left(u-\frac{2}{3}\right)$$

$$+49\sin(5\,u)\cos\left(9\,u-\frac{3}{2}\right)\sin(7\,u-1)+567\sin(5\,u)\sin\left(9\,u-\frac{3}{2}\right)\sin\left(u-\frac{2}{3}\right)$$

$$+72\sin\left(4\,u-\frac{2}{3}\right)\cos\left(5\,u-\frac{2}{3}\right)\sin\left(3\,u-\frac{1}{3}\right)$$

$$-126\sin\left(u-\frac{1}{5}\right)\cos\left(5\,u-\frac{2}{3}\right)\sin\left(3\,u-\frac{1}{3}\right)-56\cos\left(4\,u-\frac{2}{3}\right)\cos(6\,u-1)\cos(4\,u)$$

$$- 63 \sin(5\,u) \sin\left(9\,u - \frac{3}{2}\right) \cos(7\,u - 1) + 84 \cos\left(4\,u - \frac{2}{3}\right) \sin(6\,u - 1) \sin(4\,u)$$

$$- 378 \cos\left(u - \frac{1}{5}\right) \cos\left(5\,u - \frac{2}{3}\right) \cos\left(3\,u - \frac{1}{3}\right) - 588 \cos\left(u - \frac{1}{5}\right) \sin(6\,u - 1) \sin(4\,u)$$

$$+ 630 \cos\left(u - \frac{1}{5}\right) \sin\left(5\,u - \frac{2}{3}\right) \sin\left(3\,u - \frac{1}{3}\right) + 98 \sin\left(u - \frac{1}{5}\right) \cos(6\,u - 1) \sin(4\,u)$$

$$+ 392 \cos\left(u - \frac{1}{5}\right) \cos(6\,u - 1) \cos(4\,u)$$

$$+ 54 \cos\left(4\,u - \frac{2}{3}\right) \cos\left(5\,u - \frac{2}{3}\right) \cos\left(3\,u - \frac{1}{3}\right)$$

$$- 90 \cos\left(4\,u - \frac{2}{3}\right) \sin\left(5\,u - \frac{2}{3}\right) \sin\left(3\,u - \frac{1}{3}\right) + 63 \sin(5\,u) \cos\left(9\,u - \frac{3}{2}\right) \cos\left(u - \frac{2}{3}\right)$$

$$- 56 \sin\left(4\,u - \frac{2}{3}\right) \cos(6\,u - 1) \sin(4\,u) \Bigg) dr\, du$$

```
> beta:=value(%);epsilon:=evalf(beta);
```

Calculate value of plane area element integral

$$\beta := -\frac{37}{225792} \cos(1)\,\pi - \frac{37}{27440} \cos\left(\frac{1}{3}\right) \pi + \frac{1}{6720} \sin\left(\frac{9}{5}\right) \pi - \frac{1}{840} \cos\left(\frac{29}{30}\right) \pi$$

$$+ \frac{3}{1960} \sin\left(\frac{13}{10}\right) \pi + \frac{3}{1960} \sin\left(\frac{11}{30}\right) \pi - \frac{3}{3920} \sin\left(\frac{23}{30}\right) \pi + \frac{1}{1680} \sin\left(\frac{1}{10}\right) \pi$$

$$- \frac{1}{420} \cos\left(\frac{11}{15}\right) \pi + \frac{23}{181440} \cos\left(\frac{1}{2}\right) \pi - \frac{23}{680400} \sin\left(\frac{4}{3}\right) \pi + \frac{307}{15680} \cos\left(\frac{1}{15}\right) \pi$$

$$- \frac{37}{2880} \cos\left(\frac{3}{5}\right) \pi + \frac{23}{8400} \sin\left(\frac{8}{15}\right) \pi - \frac{19}{75600} \sin\left(\frac{2}{15}\right) \pi - \frac{23}{75600} \cos\left(\frac{1}{5}\right) \pi$$

$$+ \frac{23}{264600} \sin\left(\frac{2}{3}\right) \pi - \frac{23}{1058400} \cos\left(\frac{1}{6}\right) \pi - \frac{37}{56448} \sin\left(\frac{1}{6}\right) \pi + \frac{37}{54880} \sin\left(\frac{7}{6}\right) \pi$$

$$+ \frac{37}{439040} \cos(2)\,\pi + \frac{23}{1646400} \cos\left(\frac{5}{6}\right) \pi - \frac{3553}{11760} \cos\left(\frac{7}{15}\right) \pi$$

$$\varepsilon := -0.8210311510$$

```
> alpha;beta; # Two analytic values are equal
```

$$-\frac{37}{225792} \cos(1)\,\pi - \frac{37}{27440} \cos\left(\frac{1}{3}\right) \pi + \frac{1}{6720} \sin\left(\frac{9}{5}\right) \pi - \frac{1}{840} \cos\left(\frac{29}{30}\right) \pi$$

$$+ \frac{3}{1960} \sin\left(\frac{13}{10}\right) \pi + \frac{3}{1960} \sin\left(\frac{11}{30}\right) \pi - \frac{3}{3920} \sin\left(\frac{23}{30}\right) \pi + \frac{1}{1680} \sin\left(\frac{1}{10}\right) \pi$$

$$- \frac{1}{420} \cos\left(\frac{11}{15}\right) \pi + \frac{23}{181440} \cos\left(\frac{1}{2}\right) \pi - \frac{23}{680400} \sin\left(\frac{4}{3}\right) \pi + \frac{307}{15680} \cos\left(\frac{1}{15}\right) \pi$$

$$- \frac{37}{2880} \cos\left(\frac{3}{5}\right) \pi + \frac{23}{8400} \sin\left(\frac{8}{15}\right) \pi - \frac{19}{75600} \sin\left(\frac{2}{15}\right) \pi - \frac{23}{75600} \cos\left(\frac{1}{5}\right) \pi$$

$$+ \frac{23}{264600} \sin\left(\frac{2}{3}\right) \pi - \frac{23}{1058400} \cos\left(\frac{1}{6}\right) \pi - \frac{37}{56448} \sin\left(\frac{1}{6}\right) \pi + \frac{37}{54880} \sin\left(\frac{7}{6}\right) \pi$$

$$+ \frac{37}{439040} \cos(2)\,\pi + \frac{23}{1646400} \cos\left(\frac{5}{6}\right) \pi - \frac{3553}{11760} \cos\left(\frac{7}{15}\right) \pi$$

$$-\frac{37}{225792}\cos(1)\,\pi-\frac{37}{27440}\cos\!\left(\frac{1}{3}\right)\pi+\frac{1}{6720}\sin\!\left(\frac{9}{5}\right)\pi-\frac{1}{840}\cos\!\left(\frac{29}{30}\right)\pi$$

$$+\frac{3}{1960}\sin\!\left(\frac{13}{10}\right)\pi+\frac{3}{1960}\sin\!\left(\frac{11}{30}\right)\pi-\frac{3}{3920}\sin\!\left(\frac{23}{30}\right)\pi+\frac{1}{1680}\sin\!\left(\frac{1}{10}\right)\pi$$

$$-\frac{1}{420}\cos\!\left(\frac{11}{15}\right)\pi+\frac{23}{181440}\cos\!\left(\frac{1}{2}\right)\pi-\frac{23}{680400}\sin\!\left(\frac{4}{3}\right)\pi+\frac{307}{15680}\cos\!\left(\frac{1}{15}\right)\pi$$

$$-\frac{37}{2880}\cos\!\left(\frac{3}{5}\right)\pi+\frac{23}{8400}\sin\!\left(\frac{8}{15}\right)\pi-\frac{19}{75600}\sin\!\left(\frac{2}{15}\right)\pi-\frac{23}{75600}\cos\!\left(\frac{1}{5}\right)\pi$$

$$+\frac{23}{264600}\sin\!\left(\frac{2}{3}\right)\pi-\frac{23}{1058400}\cos\!\left(\frac{1}{6}\right)\pi-\frac{37}{56448}\sin\!\left(\frac{1}{6}\right)\pi+\frac{37}{54880}\sin\!\left(\frac{7}{6}\right)\pi$$

$$+\frac{37}{439040}\cos(2)\,\pi+\frac{23}{1646400}\cos\!\left(\frac{5}{6}\right)\pi-\frac{3553}{11760}\cos\!\left(\frac{7}{15}\right)\pi$$

> delta;epsilon; # Two float values are equal

$$-0.8210311510$$

$$-0.8210311510$$

> Int(Int(J,r=0..1),u=rgt[1]..rgt[2]);

Double Integrals of 'Curve Coordinates Area Element' in range of 'r,u', absolute value of result is area value of plane bounded closed region surrounded by closed curve

$$\int_0^{2\pi}\int_0^1 \frac{1}{882}r\left(882\cos\!\left(u-\frac{1}{5}\right)\cos\!\left(u-\frac{2}{3}\right)+686\cos\!\left(u-\frac{1}{5}\right)\sin(7u-1)\right.$$

$$-126\cos\!\left(4u-\frac{2}{3}\right)\cos\!\left(u-\frac{2}{3}\right)-98\cos\!\left(4u-\frac{2}{3}\right)\sin(7u-1)$$

$$+882\sin\!\left(u-\frac{1}{5}\right)\sin\!\left(u-\frac{2}{3}\right)-98\sin\!\left(u-\frac{1}{5}\right)\cos(7u-1)$$

$$-504\sin\!\left(4u-\frac{2}{3}\right)\sin\!\left(u-\frac{2}{3}\right)+56\sin\!\left(4u-\frac{2}{3}\right)\cos(7u-1)$$

$$+45\sin(5u)\cos\!\left(9u-\frac{3}{2}\right)\sin\!\left(5u-\frac{2}{3}\right)\sin\!\left(3u-\frac{1}{3}\right)$$

$$-27\sin(5u)\cos\!\left(9u-\frac{3}{2}\right)\cos\!\left(5u-\frac{2}{3}\right)\cos\!\left(3u-\frac{1}{3}\right)$$

$$-42\sin(5u)\cos\!\left(9u-\frac{3}{2}\right)\sin(6u-1)\sin(4u)$$

$$+28\sin(5u)\cos\!\left(9u-\frac{3}{2}\right)\cos(6u-1)\cos(4u)$$

$$+45\cos(5u)\cos\!\left(9u-\frac{3}{2}\right)\cos\!\left(5u-\frac{2}{3}\right)\sin\!\left(3u-\frac{1}{3}\right)$$

$$-35\cos(5u)\cos\!\left(9u-\frac{3}{2}\right)\cos(6u-1)\sin(4u)$$

$$- 81 \sin(5u) \sin\left(9u - \frac{3}{2}\right) \cos\left(5u - \frac{2}{3}\right) \sin\left(3u - \frac{1}{3}\right)$$

$$+ 63 \sin(5u) \sin\left(9u - \frac{3}{2}\right) \cos(6u - 1) \sin(4u)$$

$$+ 35 \cos(5u) \cos\left(9u - \frac{3}{2}\right) \cos(7u - 1) - 315 \cos(5u) \cos\left(9u - \frac{3}{2}\right) \sin\left(u - \frac{2}{3}\right)$$

$$+ 49 \sin(5u) \cos\left(9u - \frac{3}{2}\right) \sin(7u - 1) + 567 \sin(5u) \sin\left(9u - \frac{3}{2}\right) \sin\left(u - \frac{2}{3}\right)$$

$$+ 72 \sin\left(4u - \frac{2}{3}\right) \cos\left(5u - \frac{2}{3}\right) \sin\left(3u - \frac{1}{3}\right)$$

$$- 126 \sin\left(u - \frac{1}{5}\right) \cos\left(5u - \frac{2}{3}\right) \sin\left(3u - \frac{1}{3}\right) - 56 \cos\left(4u - \frac{2}{3}\right) \cos(6u - 1) \cos(4u)$$

$$- 63 \sin(5u) \sin\left(9u - \frac{3}{2}\right) \cos(7u - 1) + 84 \cos\left(4u - \frac{2}{3}\right) \sin(6u - 1) \sin(4u)$$

$$- 378 \cos\left(u - \frac{1}{5}\right) \cos\left(5u - \frac{2}{3}\right) \cos\left(3u - \frac{1}{3}\right) - 588 \cos\left(u - \frac{1}{5}\right) \sin(6u - 1) \sin(4u)$$

$$+ 630 \cos\left(u - \frac{1}{5}\right) \sin\left(5u - \frac{2}{3}\right) \sin\left(3u - \frac{1}{3}\right) + 98 \sin\left(u - \frac{1}{5}\right) \cos(6u - 1) \sin(4u)$$

$$+ 392 \cos\left(u - \frac{1}{5}\right) \cos(6u - 1) \cos(4u)$$

$$+ 54 \cos\left(4u - \frac{2}{3}\right) \cos\left(5u - \frac{2}{3}\right) \cos\left(3u - \frac{1}{3}\right)$$

$$- 90 \cos\left(4u - \frac{2}{3}\right) \sin\left(5u - \frac{2}{3}\right) \sin\left(3u - \frac{1}{3}\right) + 63 \sin(5u) \cos\left(9u - \frac{3}{2}\right) \cos\left(u - \frac{2}{3}\right)$$

$$\left. - 56 \sin\left(4u - \frac{2}{3}\right) \cos(6u - 1) \sin(4u) \right) dr\, du$$

```
> value(%);evalf(%);
```

```
# Area value of plane bounded closed region surrounded by closed curve
```

$$\cos\left(\frac{7}{15}\right) \pi$$

$$2.805671429$$

3.3 Numerical Model of Green Theorem at Manifold (II)

```
Known: Parametrized Expression of Simply connected、Closed Curve
(Irregular、Asymmetrical)
```

$$\left[\left(1 + \frac{1}{2}\cos(t)\right) \cos\left(\frac{t}{7} - \frac{1}{16}\sin(2t) - \frac{1}{3}\right) - \frac{1}{15}\cos(t), \right.$$
$$\left. \left(1 + \frac{1}{2}\cos(t)\right) \sin\left(\frac{t}{7} - \frac{1}{16}\sin(2t)\right) + \frac{1}{7}\cos(t)\right] \tag{1}$$

```
thereinto, t∈[0,14π];
```

and Integral Vector Field (Plane) $\left[y\cos(x-y), x\sin(y)-\dfrac{1}{2} \right]$ (2)

Calculate and Validate Green Theorem at Manifold

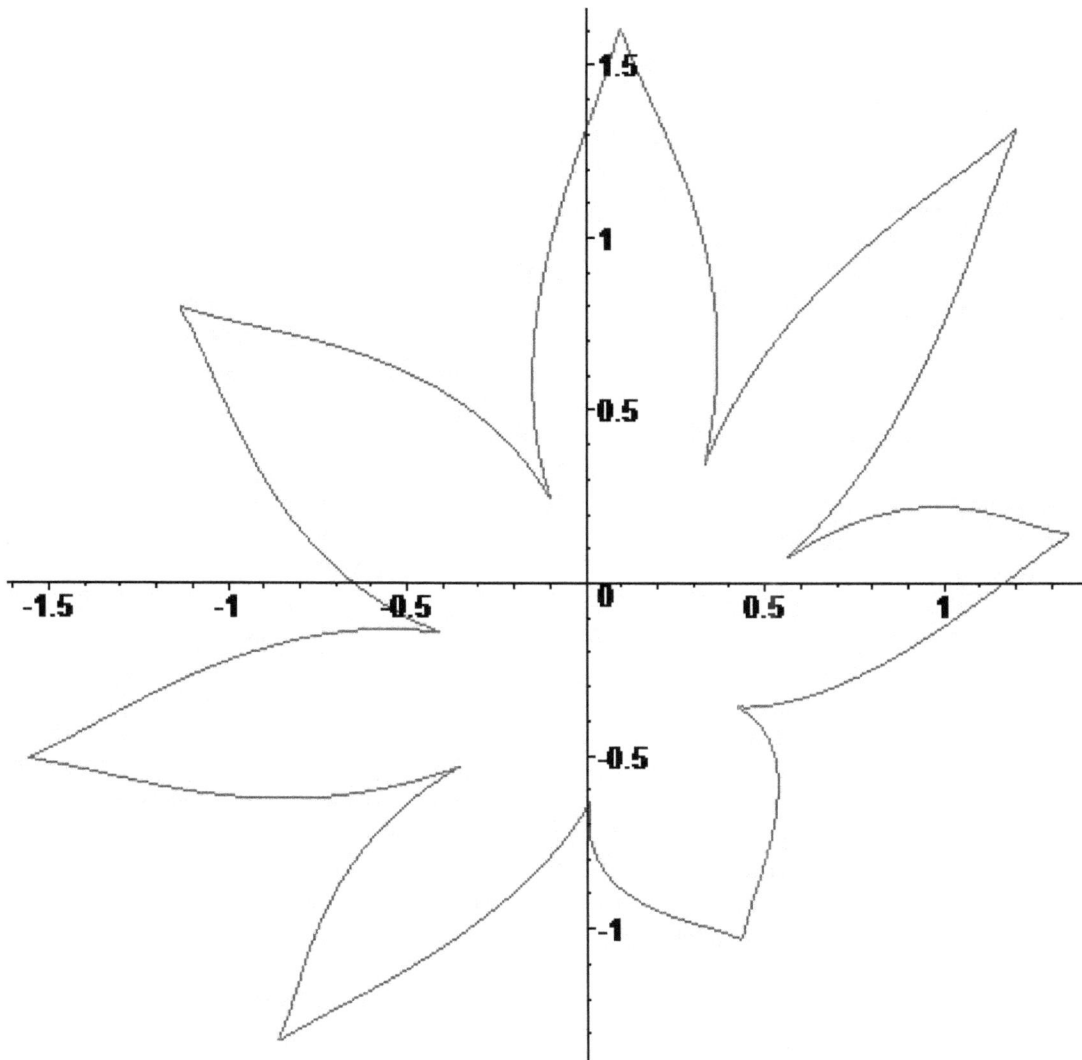

Figure(II)3.3 Simply connected、Closed Curve (1)

[Irregular、Asymmetrical]

Solution:

First, Free Closed Curve Integral:

According to parametric expression (1), Calculate Tangent Vector of Curve:

$$\left[\frac{d}{dt}\left(\left(1+\frac{1}{2}\cos(t)\right)\cos\left(\frac{t}{7}-\frac{1}{16}\sin(2t)-\frac{1}{3}\right)-\frac{1}{15}\cos(t) \right), \right.$$
$$\left. \frac{d}{dt}\left(\left(1+\frac{1}{2}\cos(t)\right)\sin\left(\frac{t}{7}-\frac{1}{16}\sin(2t)\right)+\frac{1}{7}\cos(t) \right) \right]=\left[\rule{0pt}{2.2em}\right.$$

$$-\frac{1}{2}\sin(t)\cos\left(\frac{t}{7}-\frac{1}{16}\sin(2\,t)-\frac{1}{3}\right)$$

$$-\left(1+\frac{1}{2}\cos(t)\right)\sin\left(\frac{t}{7}-\frac{1}{16}\sin(2\,t)-\frac{1}{3}\right)\left(\frac{1}{7}-\frac{1}{8}\cos(2\,t)\right)+\frac{1}{15}\sin(t),$$

$$-\frac{1}{2}\sin(t)\sin\left(\frac{t}{7}-\frac{1}{16}\sin(2\,t)\right)$$

$$+\left(1+\frac{1}{2}\cos(t)\right)\cos\left(\frac{t}{7}-\frac{1}{16}\sin(2\,t)\right)\left(\frac{1}{7}-\frac{1}{8}\cos(2\,t)\right)-\frac{1}{7}\sin(t)\bigg] \tag{3}$$

Input Curve parametrized expression (1) to Vector Field (2); and calculate Integral of Dot Product [Vetor Field (2) and Tangent Vector (3)] in range of 't' (4):

$$\int_0^{14\pi}\left(\left(\left(1+\frac{1}{2}\cos(t)\right)\sin\left(\frac{t}{7}-\frac{1}{16}\sin(2\,t)\right)+\frac{1}{7}\cos(t)\right)\cos\left(\right.\right.$$

$$\left(1+\frac{1}{2}\cos(t)\right)\cos\left(\frac{t}{7}-\frac{1}{16}\sin(2\,t)-\frac{1}{3}\right)-\frac{22}{105}\cos(t)$$

$$\left.-\left(1+\frac{1}{2}\cos(t)\right)\sin\left(\frac{t}{7}-\frac{1}{16}\sin(2\,t)\right)\right)$$

$$\left(\frac{d}{dt}\left(\left(1+\frac{1}{2}\cos(t)\right)\cos\left(\frac{t}{7}-\frac{1}{16}\sin(2\,t)-\frac{1}{3}\right)-\frac{1}{15}\cos(t)\right)\right)+\left(\right.$$

$$\left(\left(1+\frac{1}{2}\cos(t)\right)\cos\left(\frac{t}{7}-\frac{1}{16}\sin(2\,t)-\frac{1}{3}\right)-\frac{1}{15}\cos(t)\right)$$

$$\sin\left(\left(1+\frac{1}{2}\cos(t)\right)\sin\left(\frac{t}{7}-\frac{1}{16}\sin(2\,t)\right)+\frac{1}{7}\cos(t)\right)-\frac{1}{2}\right)$$

$$\left.\left(\frac{d}{dt}\left(\left(1+\frac{1}{2}\cos(t)\right)\sin\left(\frac{t}{7}-\frac{1}{16}\sin(2\,t)\right)+\frac{1}{7}\cos(t)\right)\right)dt$$

$$=$$

-1.801423627 (4)

// Because of similar structures such as 'cos(cos(...))', 'sin(sin(...))', there is no analytical value(Expressed by Elementary Function) about integral result, but only arbitrary precision floating value

Second, Free Area Element Integral:

Change character 't' to 'u' in expression (1), entirely multiply radius vector 'r' (suppose r>0) by each item of curve parametrized expression (1), convert 'Parametrized expression of curve' to 'Parametrized expression of curve coordinates':

$$\left[r\left(\left(1+\frac{1}{2}\cos(u)\right)\cos\left(\frac{u}{7}-\frac{1}{16}\sin(2\,u)-\frac{1}{3}\right)-\frac{1}{15}\cos(u)\right),\right.$$
$$\left. r\left(\left(1+\frac{1}{2}\cos(u)\right)\sin\left(\frac{u}{7}-\frac{1}{16}\sin(2\,u)\right)+\frac{1}{7}\cos(u)\right)\right] \tag{5}$$

According to parametrized expression of curve coordinates (5), define and calculate matrix of partial derivatives, obtain general expression of curve coordinates area element coefficient (6):

$$\left[\begin{array}{cc} \frac{\partial}{\partial r}(r((1+\frac{1}{2}\cos(u))\cos(\frac{u}{7}-\frac{1}{16}\sin(2u)-\frac{1}{3})-\frac{1}{15}\cos(u))) & \frac{\partial}{\partial u}(r((1+\frac{1}{2}\cos(u))\cos(\frac{u}{7}-\frac{1}{16}\sin(2u)-\frac{1}{3})-\frac{1}{15}\cos(u))) \\ \frac{\partial}{\partial r}(r((1+\frac{1}{2}\cos(u))\sin(\frac{u}{7}-\frac{1}{16}\sin(2u))+\frac{1}{7}\cos(u))) & \frac{\partial}{\partial u}(r((1+\frac{1}{2}\cos(u))\sin(\frac{u}{7}-\frac{1}{16}\sin(2u))+\frac{1}{7}\cos(u))) \end{array}\right]$$

$$=$$

$$-\frac{1}{23520}r\left(-240\sin\left(\frac{u}{7}-\frac{1}{16}\sin(2\,u)-\frac{1}{3}\right)\cos(u)^2\right.$$
$$-3360\cos\left(\frac{u}{7}-\frac{1}{16}\sin(2\,u)-\frac{1}{3}\right)\cos\left(\frac{u}{7}-\frac{1}{16}\sin(2\,u)\right)$$
$$+224\cos\left(\frac{u}{7}-\frac{1}{16}\sin(2\,u)\right)\cos(u)+3360\sin(u)\cos\left(\frac{u}{7}-\frac{1}{16}\sin(2\,u)-\frac{1}{3}\right)$$
$$+1568\sin(u)\sin\left(\frac{u}{7}-\frac{1}{16}\sin(2\,u)\right)$$
$$-3360\sin\left(\frac{u}{7}-\frac{1}{16}\sin(2\,u)-\frac{1}{3}\right)\sin\left(\frac{u}{7}-\frac{1}{16}\sin(2\,u)\right)$$
$$+112\cos\left(\frac{u}{7}-\frac{1}{16}\sin(2\,u)\right)\cos(u)^2-480\sin\left(\frac{u}{7}-\frac{1}{16}\sin(2\,u)-\frac{1}{3}\right)\cos(u)$$
$$+2940\cos\left(\frac{u}{7}-\frac{1}{16}\sin(2\,u)-\frac{1}{3}\right)\cos\left(\frac{u}{7}-\frac{1}{16}\sin(2\,u)\right)\cos(2\,u)$$
$$-3360\cos\left(\frac{u}{7}-\frac{1}{16}\sin(2\,u)-\frac{1}{3}\right)\cos\left(\frac{u}{7}-\frac{1}{16}\sin(2\,u)\right)\cos(u)$$
$$-840\cos\left(\frac{u}{7}-\frac{1}{16}\sin(2\,u)-\frac{1}{3}\right)\cos(u)^2\cos\left(\frac{u}{7}-\frac{1}{16}\sin(2\,u)\right).$$
$$-196\cos\left(\frac{u}{7}-\frac{1}{16}\sin(2\,u)\right)\cos(u)\cos(2\,u)$$
$$-98\cos\left(\frac{u}{7}-\frac{1}{16}\sin(2\,u)\right)\cos(u)^2\cos(2\,u)$$
$$-3360\sin\left(\frac{u}{7}-\frac{1}{16}\sin(2\,u)-\frac{1}{3}\right)\sin\left(\frac{u}{7}-\frac{1}{16}\sin(2\,u)\right)\cos(u)$$
$$+2940\sin\left(\frac{u}{7}-\frac{1}{16}\sin(2\,u)-\frac{1}{3}\right)\cos(2\,u)\sin\left(\frac{u}{7}-\frac{1}{16}\sin(2\,u)\right)$$
$$+420\sin\left(\frac{u}{7}-\frac{1}{16}\sin(2\,u)-\frac{1}{3}\right)\cos(u)\cos(2\,u)$$

$$- 840 \sin\left(\frac{u}{7} - \frac{1}{16}\sin(2\,u) - \frac{1}{3}\right)\cos(u)^2 \sin\left(\frac{u}{7} - \frac{1}{16}\sin(2\,u)\right)\Bigg)$$

$$+ 210 \sin\left(\frac{u}{7} - \frac{1}{16}\sin(2\,u) - \frac{1}{3}\right)\cos(u)^2 \cos(2\,u)$$

$$+ 2940 \cos\left(\frac{u}{7} - \frac{1}{16}\sin(2\,u) - \frac{1}{3}\right)\cos\left(\frac{u}{7} - \frac{1}{16}\sin(2\,u)\right)\cos(u)\cos(2\,u)$$

$$+ 735 \cos\left(\frac{u}{7} - \frac{1}{16}\sin(2\,u) - \frac{1}{3}\right)\cos(u)^2 \cos\left(\frac{u}{7} - \frac{1}{16}\sin(2\,u)\right)\cos(2\,u)$$

$$+ 2940 \sin\left(\frac{u}{7} - \frac{1}{16}\sin(2\,u) - \frac{1}{3}\right)\cos(2\,u)\sin\left(\frac{u}{7} - \frac{1}{16}\sin(2\,u)\right)\cos(u)$$

$$+ 735 \sin\left(\frac{u}{7} - \frac{1}{16}\sin(2\,u) - \frac{1}{3}\right)\cos(u)^2 \cos(2\,u)\sin\left(\frac{u}{7} - \frac{1}{16}\sin(2\,u)\right)\Bigg)\Bigg) \quad (6)$$

Calculate Differential Function '$\frac{\partial}{\partial x}Q(x,y) - \frac{\partial}{\partial y}P(x,y)$' of Vector

Field (2):

$$\left(\frac{\partial}{\partial x}\left(x\sin(y) - \frac{1}{2}\right)\right) - \left(\frac{\partial}{\partial y}(y\cos(x-y))\right) = \sin(y) - \cos(x-y) - y\sin(x-y) \quad (7)$$

Input parametrized expression of curve coordinates (5) to Differential Function (7); and Calculate Double Integrals of Product [Differential Function (7) and Curve Coordinates Area Element] in range of 'r,u' (8):

$$\int_0^{14\pi}\int_0^1 -\frac{1}{23520}\Bigg(\sin\left(r\left(\left(1+\frac{1}{2}\cos(u)\right)\sin\left(\frac{u}{7} - \frac{1}{16}\sin(2\,u)\right) + \frac{1}{7}\cos(u)\right)\right) - \cos\Bigg($$

$$r\left(\left(1+\frac{1}{2}\cos(u)\right)\cos\left(\frac{u}{7} - \frac{1}{16}\sin(2\,u) - \frac{1}{3}\right) - \frac{1}{15}\cos(u)\right)$$

$$- r\left(\left(1+\frac{1}{2}\cos(u)\right)\sin\left(\frac{u}{7} - \frac{1}{16}\sin(2\,u)\right) + \frac{1}{7}\cos(u)\right)\Bigg) - r$$

$$\left(\left(1+\frac{1}{2}\cos(u)\right)\sin\left(\frac{u}{7} - \frac{1}{16}\sin(2\,u)\right) + \frac{1}{7}\cos(u)\right)\sin\Bigg($$

$$r\left(\left(1+\frac{1}{2}\cos(u)\right)\cos\left(\frac{u}{7} - \frac{1}{16}\sin(2\,u) - \frac{1}{3}\right) - \frac{1}{15}\cos(u)\right)$$

$$- r\left(\left(1+\frac{1}{2}\cos(u)\right)\sin\left(\frac{u}{7} - \frac{1}{16}\sin(2\,u)\right) + \frac{1}{7}\cos(u)\right)\Bigg)\Bigg)r\Bigg($$

$$-240 \sin\left(\frac{u}{7} - \frac{1}{16}\sin(2\,u) - \frac{1}{3}\right)\cos(u)^2$$

$$- 3360 \cos\left(\frac{u}{7} - \frac{1}{16}\sin(2\,u) - \frac{1}{3}\right)\cos\left(\frac{u}{7} - \frac{1}{16}\sin(2\,u)\right)$$

$$+ 224 \cos\left(\frac{u}{7} - \frac{1}{16}\sin(2u)\right)\cos(u) + 3360\sin(u)\cos\left(\frac{u}{7} - \frac{1}{16}\sin(2u) - \frac{1}{3}\right)$$

$$+ 1568\sin(u)\sin\left(\frac{u}{7} - \frac{1}{16}\sin(2u)\right)$$

$$- 3360\sin\left(\frac{u}{7} - \frac{1}{16}\sin(2u) - \frac{1}{3}\right)\sin\left(\frac{u}{7} - \frac{1}{16}\sin(2u)\right)$$

$$+ 112\cos\left(\frac{u}{7} - \frac{1}{16}\sin(2u)\right)\cos(u)^2 - 480\sin\left(\frac{u}{7} - \frac{1}{16}\sin(2u) - \frac{1}{3}\right)\cos(u)$$

$$+ 2940\cos\left(\frac{u}{7} - \frac{1}{16}\sin(2u) - \frac{1}{3}\right)\cos\left(\frac{u}{7} - \frac{1}{16}\sin(2u)\right)\cos(2u)$$

$$- 3360\cos\left(\frac{u}{7} - \frac{1}{16}\sin(2u) - \frac{1}{3}\right)\cos\left(\frac{u}{7} - \frac{1}{16}\sin(2u)\right)\cos(u)$$

$$- 840\cos\left(\frac{u}{7} - \frac{1}{16}\sin(2u) - \frac{1}{3}\right)\cos(u)^2\cos\left(\frac{u}{7} - \frac{1}{16}\sin(2u)\right)$$

$$- 196\cos\left(\frac{u}{7} - \frac{1}{16}\sin(2u)\right)\cos(u)\cos(2u)$$

$$- 98\cos\left(\frac{u}{7} - \frac{1}{16}\sin(2u)\right)\cos(u)^2\cos(2u)$$

$$- 3360\sin\left(\frac{u}{7} - \frac{1}{16}\sin(2u) - \frac{1}{3}\right)\sin\left(\frac{u}{7} - \frac{1}{16}\sin(2u)\right)\cos(u)$$

$$+ 2940\sin\left(\frac{u}{7} - \frac{1}{16}\sin(2u) - \frac{1}{3}\right)\cos(2u)\sin\left(\frac{u}{7} - \frac{1}{16}\sin(2u)\right)$$

$$+ 420\sin\left(\frac{u}{7} - \frac{1}{16}\sin(2u) - \frac{1}{3}\right)\cos(u)\cos(2u)$$

$$- 840\sin\left(\frac{u}{7} - \frac{1}{16}\sin(2u) - \frac{1}{3}\right)\cos(u)^2\sin\left(\frac{u}{7} - \frac{1}{16}\sin(2u)\right)$$

$$+ 210\sin\left(\frac{u}{7} - \frac{1}{16}\sin(2u) - \frac{1}{3}\right)\cos(u)^2\cos(2u)$$

$$+ 2940\cos\left(\frac{u}{7} - \frac{1}{16}\sin(2u) - \frac{1}{3}\right)\cos\left(\frac{u}{7} - \frac{1}{16}\sin(2u)\right)\cos(u)\cos(2u)$$

$$+ 735\cos\left(\frac{u}{7} - \frac{1}{16}\sin(2u) - \frac{1}{3}\right)\cos(u)^2\cos\left(\frac{u}{7} - \frac{1}{16}\sin(2u)\right)\cos(2u)$$

$$+ 2940\sin\left(\frac{u}{7} - \frac{1}{16}\sin(2u) - \frac{1}{3}\right)\cos(2u)\sin\left(\frac{u}{7} - \frac{1}{16}\sin(2u)\right)\cos(u)$$

$$+ 735\sin\left(\frac{u}{7} - \frac{1}{16}\sin(2u) - \frac{1}{3}\right)\cos(u)^2\cos(2u)\sin\left(\frac{u}{7} - \frac{1}{16}\sin(2u)\right)\right) dr\, du$$

$$=$$

$$-1.801423627 \qquad\qquad (8)$$

Because of similar structures such as 'cos(cos(...))', 'sin(sin (...))', there is no analytical value(Expressed by Elementary Function) about integral result, but only arbitrary precision floating value

Integral Value(Arbitrary Precision Floating Value)(4) [Vector Field (2) on Target Curve (1)], is equal to Area Element Integral Value (Arbitrary Precision Floating Value)(8) [Differential Function (7) in Plane Bounded Closed Region that Target Curve (1) surrounds], Complete Calculation and Validation of Green Theorem at Manifold.

3.4 Numerical Model of Green Theorem at Manifold (II) [Program Template of Waterloo Maplesoft, Optional]

```
> restart;
> with(plots):with(linalg):
> CO:=[(1+cos(t)/2)*cos(t/7-sin(2*t)/16-1/3)-cos(t)/15,
(1+cos(t)/2)*sin(t/7-sin(2*t)/16)+cos(t)/7];
# Define parametrized expression of discretional (plane) simply
connected closed curve 'CO'
```

$$CO := \left[\left(1 + \frac{1}{2}\cos(t) \right) \cos\left(-\frac{t}{7} + \frac{1}{16}\sin(2\,t) + \frac{1}{3} \right) - \frac{1}{15}\cos(t), \right.$$
$$\left. -\left(1 + \frac{1}{2}\cos(t) \right) \sin\left(-\frac{t}{7} + \frac{1}{16}\sin(2\,t) \right) + \frac{1}{7}\cos(t) \right]$$

```
> rgt:=[0,14*Pi];
# Relative to parameter range '[0,2*Pi]' of abstract simply connected
closed curve, aberrant range of current curve is '[0,14*Pi]'
```

$$rgt := [\, 0,\, 14\,\pi \,]$$

```
> plot([CO[1],CO[2],t=rgt[1]..rgt[2]],scaling=constrained,
color=red,numpoints=1000):g1:=%:
> PV:=[y*cos(x-y),x*sin(y)-1/2];
# Define discretional plane vector field 'PV' (Suppose Vector Field
'PV' possesses 1th order continuous partial derivatives in plane
bounded closed region 'S' that curve 'CO' surrounds)
```

$$PV := \left[y\cos(x - y),\, x\sin(y) - \frac{1}{2} \right]$$

```
> Diff(PV[2],x)-Diff(PV[1],y);
# Calculate differential function of idiographic  plane vector field
'PV', result is an idiographic differential function 'dPV'
```

$$\left(\frac{\partial}{\partial x}\left(x\sin(y) - \frac{1}{2} \right) \right) - \left(\frac{\partial}{\partial y}\left(y\cos(x - y) \right) \right)$$

```
> dPV:=value(%);
```

$$dPV := \sin(y) - \cos(x - y) - y \sin(x - y)$$

```
> rgx:=[-1.5,1.5];
```

$$rgx := [-1.5, 1.5]$$

```
> rgy:=[-1.35,1.65];
```

$$rgy := [-1.35, 1.65]$$

```
> plot3d(dPV,x=rgx[1]..rgx[2],y=rgy[1]..rgy[2],projection=0.9,
axes=normal); # Drawing differential function 'dPV' in 3-Dimensional
Euclidean space
```

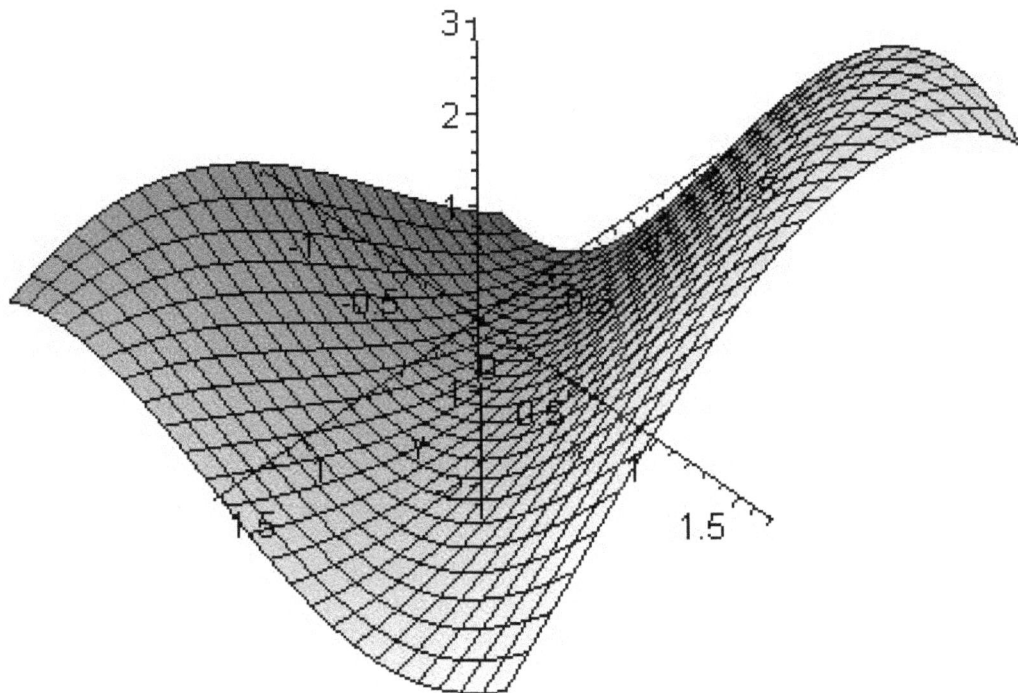

Figure(II)3.4.1 Space View of Differential Function 'dPV'

```
> fieldplot(PV,x=rgx[1]..rgx[2],y=rgy[1]..rgy[2],arrows=SLIM,
color=blue):g2:=%: # Draw plane vector field 'PV'
> implicitplot(dPV,x=rgx[1]..rgx[2],y=rgy[1]..rgy[2],color=navy,
thickness=1,numpoints=5000):g3:=%:
 # Draw contour curve of 2-variable function 'dPV'
> display(g1,g2,g3);
# Plane closed curve 'CO', plane vector field 'PV', contour curve of
```

`dPV`, three figures are synthesized

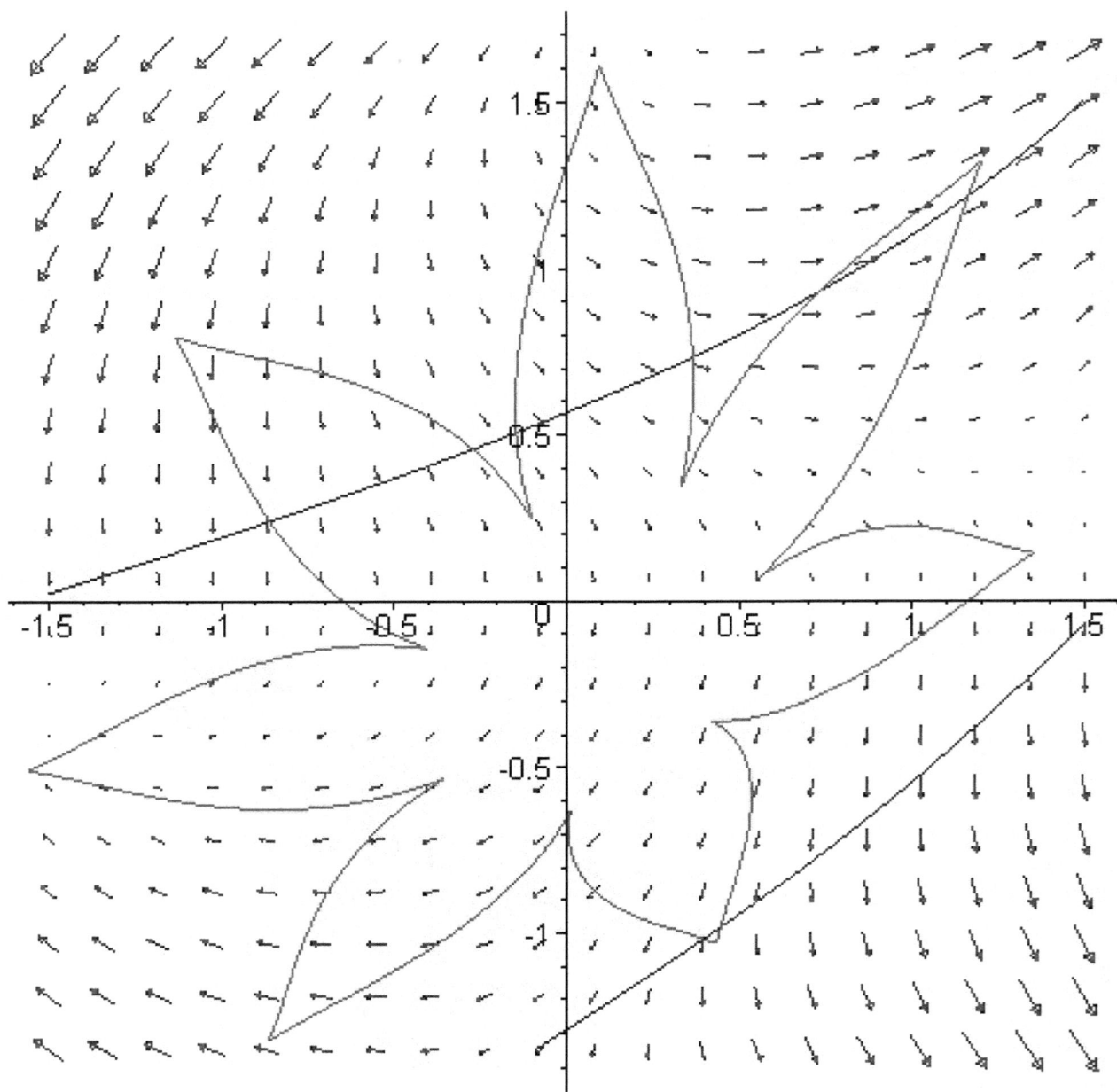

Figure(II)3.4.2 Plane Closed Curve `CO`

Vector Field `PV` and Contour Curve of `dPV`

```
> x:=CO[1];y:=CO[2];
```

$$x := \left(1 + \frac{1}{2}\cos(t)\right)\cos\left(-\frac{t}{7} + \frac{1}{16}\sin(2\,t) + \frac{1}{3}\right) - \frac{1}{15}\cos(t)$$

$$y := -\left(1 + \frac{1}{2}\cos(t)\right)\sin\left(-\frac{t}{7} + \frac{1}{16}\sin(2\,t)\right) + \frac{1}{7}\cos(t)$$

```
> Int(PV[1]*diff(CO[1],t)+PV[2]*diff(CO[2],t),t=rgt[1]..rgt[2]);
# Closed Curve Integral (Plane Vector Field `PV` at Plane Closed
```

Curve 'CO')

$$
\int_0^{14\pi} \left(-\left(1+\frac{1}{2}\cos(t)\right)\sin\left(-\frac{t}{7}+\frac{1}{16}\sin(2\,t)\right)+\frac{1}{7}\cos(t)\right)\cos\Bigg(
$$

$$
\left(1+\frac{1}{2}\cos(t)\right)\cos\left(-\frac{t}{7}+\frac{1}{16}\sin(2\,t)+\frac{1}{3}\right)-\frac{22}{105}\cos(t)
$$

$$
+\left(1+\frac{1}{2}\cos(t)\right)\sin\left(-\frac{t}{7}+\frac{1}{16}\sin(2\,t)\right)\Bigg)\left(-\frac{1}{2}\sin(t)\cos\left(-\frac{t}{7}+\frac{1}{16}\sin(2\,t)+\frac{1}{3}\right)\right.
$$

$$
\left.-\left(1+\frac{1}{2}\cos(t)\right)\sin\left(-\frac{t}{7}+\frac{1}{16}\sin(2\,t)+\frac{1}{3}\right)\left(-\frac{1}{7}+\frac{1}{8}\cos(2\,t)\right)+\frac{1}{15}\sin(t)\right)+\Bigg(-
$$

$$
\left(\left(1+\frac{1}{2}\cos(t)\right)\cos\left(-\frac{t}{7}+\frac{1}{16}\sin(2\,t)+\frac{1}{3}\right)-\frac{1}{15}\cos(t)\right)
$$

$$
\sin\left(\left(1+\frac{1}{2}\cos(t)\right)\sin\left(-\frac{t}{7}+\frac{1}{16}\sin(2\,t)\right)-\frac{1}{7}\cos(t)\right)-\frac{1}{2}\Bigg)\Bigg(
$$

$$
\frac{1}{2}\sin(t)\sin\left(-\frac{t}{7}+\frac{1}{16}\sin(2\,t)\right)
$$

$$
-\left(1+\frac{1}{2}\cos(t)\right)\cos\left(-\frac{t}{7}+\frac{1}{16}\sin(2\,t)\right)\left(-\frac{1}{7}+\frac{1}{8}\cos(2\,t)\right)-\frac{1}{7}\sin(t)\Bigg)\,dt
$$

```
> alpha:=evalf(%); # Calculate value of plane closed curve integral
```
$$
\alpha := -1.801423627
$$

```
> x:='x':y:='y':
> COc:=subs(t=u,CO);
```
Convert character 't' to 'u' in expression of curve 'CO'

$$
COc := \left[\left(1+\frac{1}{2}\cos(u)\right)\cos\left(-\frac{u}{7}+\frac{1}{16}\sin(2\,u)+\frac{1}{3}\right)-\frac{1}{15}\cos(u),\right.
$$

$$
\left.-\left(1+\frac{1}{2}\cos(u)\right)\sin\left(-\frac{u}{7}+\frac{1}{16}\sin(2\,u)\right)+\frac{1}{7}\cos(u)\right]
$$

```
> [r*COc[1],r*COc[2]];
```
Entirely multiply radius vector 'r' (suppose r>0) by each item of
curve 'CO' parametrized expression, convert 'Parametrized expression
of curve' to 'Parametrized expression of curve coordinates'

$$
\left[r\left(\left(1+\frac{1}{2}\cos(u)\right)\cos\left(\frac{u}{7}-\frac{1}{16}\sin(2\,u)-\frac{1}{3}\right)-\frac{1}{15}\cos(u)\right),\right.
$$

$$
\left.r\left(\left(1+\frac{1}{2}\cos(u)\right)\sin\left(\frac{u}{7}-\frac{1}{16}\sin(2\,u)\right)+\frac{1}{7}\cos(u)\right)\right]
$$

```
> x:=r*COc[1]:y:=r*COc[2]:
> matrix(2,2,[Diff(r*COc[1],r),Diff(r*COc[1],u),Diff(r*COc[2],r),
Diff(r*COc[2],u)])=
matrix(2,2,[diff(r*COc[1],r),diff(r*COc[1],u),diff(r*COc[2],r),
diff(r*COc[2],u)]);m:=rhs(%);
```

Define and calculate matrix of partial derivatives 'm', obtain
general expression of curve coordinates area element coefficient

$$\left[\frac{\partial}{\partial r}\left(r\left(\left(1+\frac{1}{2}\cos(u)\right)\cos\left(\frac{u}{7}-\frac{1}{16}\sin(2\,u)-\frac{1}{3}\right)-\frac{1}{15}\cos(u)\right)\right),\right.$$

$$\left.\frac{\partial}{\partial u}\left(r\left(\left(1+\frac{1}{2}\cos(u)\right)\cos\left(\frac{u}{7}-\frac{1}{16}\sin(2\,u)-\frac{1}{3}\right)-\frac{1}{15}\cos(u)\right)\right)\right]$$

$$\left[\frac{\partial}{\partial r}\left(r\left(\left(1+\frac{1}{2}\cos(u)\right)\sin\left(\frac{u}{7}-\frac{1}{16}\sin(2\,u)\right)+\frac{1}{7}\cos(u)\right)\right),\right.$$

$$\left.\frac{\partial}{\partial u}\left(r\left(\left(1+\frac{1}{2}\cos(u)\right)\sin\left(\frac{u}{7}-\frac{1}{16}\sin(2\,u)\right)+\frac{1}{7}\cos(u)\right)\right)\right]=$$

$$\left[\left(1+\frac{1}{2}\cos(u)\right)\cos\left(\frac{u}{7}-\frac{1}{16}\sin(2\,u)-\frac{1}{3}\right)-\frac{1}{15}\cos(u),r\left(\vphantom{\frac{1}{2}}\right.\right.$$

$$-\frac{1}{2}\sin(u)\cos\left(\frac{u}{7}-\frac{1}{16}\sin(2\,u)-\frac{1}{3}\right)$$

$$\left.\left.-\left(1+\frac{1}{2}\cos(u)\right)\sin\left(\frac{u}{7}-\frac{1}{16}\sin(2\,u)-\frac{1}{3}\right)\left(\frac{1}{7}-\frac{1}{8}\cos(2\,u)\right)+\frac{1}{15}\sin(u)\right)\right]$$

$$\left[\left(1+\frac{1}{2}\cos(u)\right)\sin\left(\frac{u}{7}-\frac{1}{16}\sin(2\,u)\right)+\frac{1}{7}\cos(u),r\left(-\frac{1}{2}\sin(u)\sin\left(\frac{u}{7}-\frac{1}{16}\sin(2\,u)\right)\right.\right.$$

$$\left.\left.+\left(1+\frac{1}{2}\cos(u)\right)\cos\left(\frac{u}{7}-\frac{1}{16}\sin(2\,u)\right)\left(\frac{1}{7}-\frac{1}{8}\cos(2\,u)\right)-\frac{1}{7}\sin(u)\right)\right]$$

$m :=$

$$\left[\left(1+\frac{1}{2}\cos(u)\right)\cos\left(\frac{u}{7}-\frac{1}{16}\sin(2\,u)-\frac{1}{3}\right)-\frac{1}{15}\cos(u),r\left(\vphantom{\frac{1}{2}}\right.\right.$$

$$-\frac{1}{2}\sin(u)\cos\left(\frac{u}{7}-\frac{1}{16}\sin(2\,u)-\frac{1}{3}\right)$$

$$\left.\left.-\left(1+\frac{1}{2}\cos(u)\right)\sin\left(\frac{u}{7}-\frac{1}{16}\sin(2\,u)-\frac{1}{3}\right)\left(\frac{1}{7}-\frac{1}{8}\cos(2\,u)\right)+\frac{1}{15}\sin(u)\right)\right]$$

$$\left[\left(1+\frac{1}{2}\cos(u)\right)\sin\left(\frac{u}{7}-\frac{1}{16}\sin(2\,u)\right)+\frac{1}{7}\cos(u),r\left(-\frac{1}{2}\sin(u)\sin\left(\frac{u}{7}-\frac{1}{16}\sin(2\,u)\right)\right.\right.$$

$$\left.\left.+\left(1+\frac{1}{2}\cos(u)\right)\cos\left(\frac{u}{7}-\frac{1}{16}\sin(2\,u)\right)\left(\frac{1}{7}-\frac{1}{8}\cos(2\,u)\right)-\frac{1}{7}\sin(u)\right)\right]$$

```
> det(m):
> J:=simplify(%);
```

General expression of curve coordinates area element coefficient

$$J:=-\frac{1}{23520}r\left(210\sin\left(\frac{u}{7}-\frac{1}{16}\sin(2\,u)-\frac{1}{3}\right)\cos(u)^2\cos(2\,u)\right.$$

$$-840\sin\left(\frac{u}{7}-\frac{1}{16}\sin(2\,u)-\frac{1}{3}\right)\cos(u)^2\sin\left(\frac{u}{7}-\frac{1}{16}\sin(2\,u)\right)$$

$$+420\sin\left(\frac{u}{7}-\frac{1}{16}\sin(2\,u)-\frac{1}{3}\right)\cos(u)\cos(2\,u)$$

$$\left.+2940\sin\left(\frac{u}{7}-\frac{1}{16}\sin(2\,u)-\frac{1}{3}\right)\cos(2\,u)\sin\left(\frac{u}{7}-\frac{1}{16}\sin(2\,u)\right)\right)$$

$$- 3360 \sin\left(\frac{u}{7} - \frac{1}{16}\sin(2\,u) - \frac{1}{3}\right)\sin\left(\frac{u}{7} - \frac{1}{16}\sin(2\,u)\right)\cos(u)$$

$$- 98 \cos\left(\frac{u}{7} - \frac{1}{16}\sin(2\,u)\right)\cos(u)^2 \cos(2\,u)$$

$$- 196 \cos\left(\frac{u}{7} - \frac{1}{16}\sin(2\,u)\right)\cos(u)\cos(2\,u)$$

$$- 840 \cos\left(\frac{u}{7} - \frac{1}{16}\sin(2\,u) - \frac{1}{3}\right)\cos(u)^2 \cos\left(\frac{u}{7} - \frac{1}{16}\sin(2\,u)\right)$$

$$- 3360 \cos\left(\frac{u}{7} - \frac{1}{16}\sin(2\,u) - \frac{1}{3}\right)\cos\left(\frac{u}{7} - \frac{1}{16}\sin(2\,u)\right)\cos(u)$$

$$+ 2940 \cos\left(\frac{u}{7} - \frac{1}{16}\sin(2\,u) - \frac{1}{3}\right)\cos\left(\frac{u}{7} - \frac{1}{16}\sin(2\,u)\right)\cos(2\,u)$$

$$+ 1568 \sin(u)\sin\left(\frac{u}{7} - \frac{1}{16}\sin(2\,u)\right)$$

$$- 3360 \sin\left(\frac{u}{7} - \frac{1}{16}\sin(2\,u) - \frac{1}{3}\right)\sin\left(\frac{u}{7} - \frac{1}{16}\sin(2\,u)\right)$$

$$- 240 \sin\left(\frac{u}{7} - \frac{1}{16}\sin(2\,u) - \frac{1}{3}\right)\cos(u)^2 - 480 \sin\left(\frac{u}{7} - \frac{1}{16}\sin(2\,u) - \frac{1}{3}\right)\cos(u)$$

$$+ 112 \cos\left(\frac{u}{7} - \frac{1}{16}\sin(2\,u)\right)\cos(u)^2$$

$$- 3360 \cos\left(\frac{u}{7} - \frac{1}{16}\sin(2\,u) - \frac{1}{3}\right)\cos\left(\frac{u}{7} - \frac{1}{16}\sin(2\,u)\right)$$

$$+ 224 \cos\left(\frac{u}{7} - \frac{1}{16}\sin(2\,u)\right)\cos(u) + 3360 \sin(u)\cos\left(\frac{u}{7} - \frac{1}{16}\sin(2\,u) - \frac{1}{3}\right)$$

$$+ 2940 \cos\left(\frac{u}{7} - \frac{1}{16}\sin(2\,u) - \frac{1}{3}\right)\cos\left(\frac{u}{7} - \frac{1}{16}\sin(2\,u)\right)\cos(u)\cos(2\,u)$$

$$+ 735 \cos\left(\frac{u}{7} - \frac{1}{16}\sin(2\,u) - \frac{1}{3}\right)\cos(u)^2 \cos\left(\frac{u}{7} - \frac{1}{16}\sin(2\,u)\right)\cos(2\,u)$$

$$+ 2940 \sin\left(\frac{u}{7} - \frac{1}{16}\sin(2\,u) - \frac{1}{3}\right)\cos(2\,u)\sin\left(\frac{u}{7} - \frac{1}{16}\sin(2\,u)\right)\cos(u)$$

$$+ 735 \sin\left(\frac{u}{7} - \frac{1}{16}\sin(2\,u) - \frac{1}{3}\right)\cos(u)^2 \cos(2\,u)\sin\left(\frac{u}{7} - \frac{1}{16}\sin(2\,u)\right)\bigg)\bigg)$$

```
> Int(Int(dPV*J,r=0..1),u=rgt[1]..rgt[2]);
```

\# Double Integrals of Product (Differential Function 'dPV' and Curve
  Coordinates Area Element) in range of 'r,u'

$$\int_0^{14\pi}\int_0^1 -\frac{1}{23520}\Bigg(\sin\left(r\left(\left(1 + \frac{1}{2}\cos(u)\right)\sin\left(\frac{u}{7} - \frac{1}{16}\sin(2\,u)\right) + \frac{1}{7}\cos(u)\right)\right) - \cos\Bigg($$

$$-r\left(\left(1 + \frac{1}{2}\cos(u)\right)\cos\left(\frac{u}{7} - \frac{1}{16}\sin(2\,u) - \frac{1}{3}\right) - \frac{1}{15}\cos(u)\right)\Bigg)$$

$$+ r\left(\left(1 + \frac{1}{2}\cos(u)\right)\sin\left(\frac{u}{7} - \frac{1}{16}\sin(2\,u)\right) + \frac{1}{7}\cos(u)\right)\Bigg) + r$$

$$\left(\left(\left(1+\frac{1}{2}\cos(u)\right)\sin\left(\frac{u}{7}-\frac{1}{16}\sin(2\,u)\right)+\frac{1}{7}\cos(u)\right)\sin\left(\right.\right.$$

$$-r\left(\left(1+\frac{1}{2}\cos(u)\right)\cos\left(\frac{u}{7}-\frac{1}{16}\sin(2\,u)-\frac{1}{3}\right)-\frac{1}{15}\cos(u)\right)$$

$$+r\left(\left(1+\frac{1}{2}\cos(u)\right)\sin\left(\frac{u}{7}-\frac{1}{16}\sin(2\,u)\right)+\frac{1}{7}\cos(u)\right)\right)\right)r\left(\right.$$

$$210\sin\left(\frac{u}{7}-\frac{1}{16}\sin(2\,u)-\frac{1}{3}\right)\cos(u)^2\cos(2\,u)$$

$$-840\sin\left(\frac{u}{7}-\frac{1}{16}\sin(2\,u)-\frac{1}{3}\right)\cos(u)^2\sin\left(\frac{u}{7}-\frac{1}{16}\sin(2\,u)\right)$$

$$+420\sin\left(\frac{u}{7}-\frac{1}{16}\sin(2\,u)-\frac{1}{3}\right)\cos(u)\cos(2\,u)$$

$$+2940\sin\left(\frac{u}{7}-\frac{1}{16}\sin(2\,u)-\frac{1}{3}\right)\cos(2\,u)\sin\left(\frac{u}{7}-\frac{1}{16}\sin(2\,u)\right)$$

$$-3360\sin\left(\frac{u}{7}-\frac{1}{16}\sin(2\,u)-\frac{1}{3}\right)\sin\left(\frac{u}{7}-\frac{1}{16}\sin(2\,u)\right)\cos(u)$$

$$-98\cos\left(\frac{u}{7}-\frac{1}{16}\sin(2\,u)\right)\cos(u)^2\cos(2\,u)$$

$$-196\cos\left(\frac{u}{7}-\frac{1}{16}\sin(2\,u)\right)\cos(u)\cos(2\,u)$$

$$-840\cos\left(\frac{u}{7}-\frac{1}{16}\sin(2\,u)-\frac{1}{3}\right)\cos(u)^2\cos\left(\frac{u}{7}-\frac{1}{16}\sin(2\,u)\right)$$

$$-3360\cos\left(\frac{u}{7}-\frac{1}{16}\sin(2\,u)-\frac{1}{3}\right)\cos\left(\frac{u}{7}-\frac{1}{16}\sin(2\,u)\right)\cos(u)$$

$$+2940\cos\left(\frac{u}{7}-\frac{1}{16}\sin(2\,u)-\frac{1}{3}\right)\cos\left(\frac{u}{7}-\frac{1}{16}\sin(2\,u)\right)\cos(2\,u)$$

$$+1568\sin(u)\sin\left(\frac{u}{7}-\frac{1}{16}\sin(2\,u)\right)$$

$$-3360\sin\left(\frac{u}{7}-\frac{1}{16}\sin(2\,u)-\frac{1}{3}\right)\sin\left(\frac{u}{7}-\frac{1}{16}\sin(2\,u)\right)$$

$$-240\sin\left(\frac{u}{7}-\frac{1}{16}\sin(2\,u)-\frac{1}{3}\right)\cos(u)^2-480\sin\left(\frac{u}{7}-\frac{1}{16}\sin(2\,u)-\frac{1}{3}\right)\cos(u)$$

$$+112\cos\left(\frac{u}{7}-\frac{1}{16}\sin(2\,u)\right)\cos(u)^2$$

$$-3360\cos\left(\frac{u}{7}-\frac{1}{16}\sin(2\,u)-\frac{1}{3}\right)\cos\left(\frac{u}{7}-\frac{1}{16}\sin(2\,u)\right)$$

$$+224\cos\left(\frac{u}{7}-\frac{1}{16}\sin(2\,u)\right)\cos(u)+3360\sin(u)\cos\left(\frac{u}{7}-\frac{1}{16}\sin(2\,u)-\frac{1}{3}\right)$$

$$+2940\cos\left(\frac{u}{7}-\frac{1}{16}\sin(2\,u)-\frac{1}{3}\right)\cos\left(\frac{u}{7}-\frac{1}{16}\sin(2\,u)\right)\cos(u)\cos(2\,u)$$

$$+735\cos\left(\frac{u}{7}-\frac{1}{16}\sin(2\,u)-\frac{1}{3}\right)\cos(u)^2\cos\left(\frac{u}{7}-\frac{1}{16}\sin(2\,u)\right)\cos(2\,u)$$

$$+2940\sin\left(\frac{u}{7}-\frac{1}{16}\sin(2\,u)-\frac{1}{3}\right)\cos(2\,u)\sin\left(\frac{u}{7}-\frac{1}{16}\sin(2\,u)\right)\cos(u)$$

$$+ 735 \sin\left(\frac{u}{7} - \frac{1}{16} \sin(2\,u) - \frac{1}{3}\right) \cos(u)^2 \cos(2\,u) \sin\left(\frac{u}{7} - \frac{1}{16} \sin(2\,u)\right)\Bigg)\Bigg) dr\, du$$

```
> beta:=evalf(%); # Calculate value of plane area element integral
```
$$\beta := -1.801423627$$

```
> alpha;beta; # Two float values are equal
```
$$-1.801423627, -1.801423627$$

3.5 Numerical Model of Green Theorem at Manifold (III)

```
Known:
A.Parametrized Expression of Unclosed Curve 1
```

$$\left[2\sin(t), t + \frac{1}{5}\sin(12\,t) + \cos(3\,t) \right] \quad \text{thereinto}, t \in [0, \pi]; \qquad (1)$$

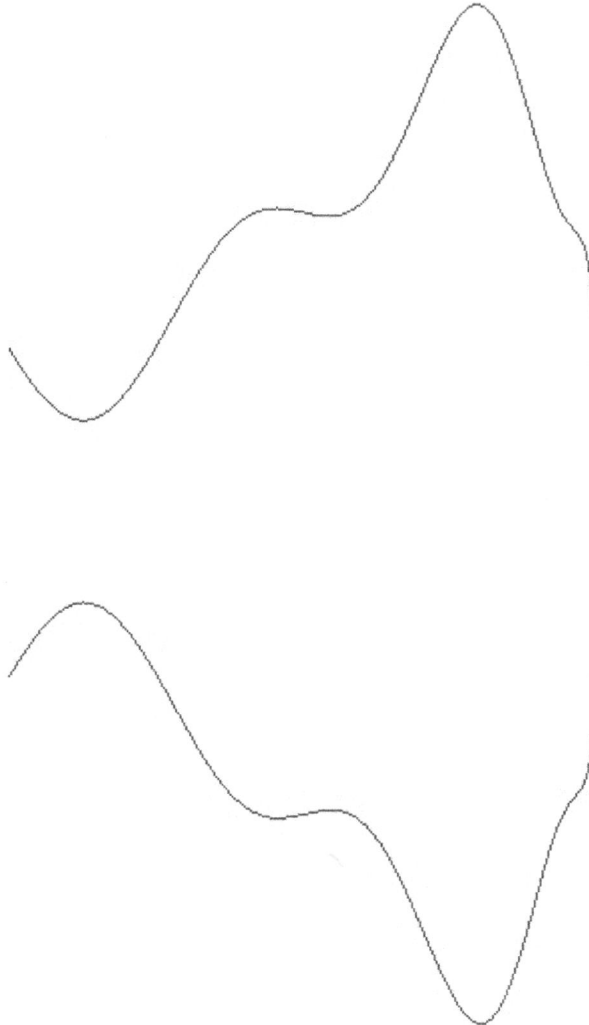

Figure(II)3.5.1 Unclosed Curve 1

B.Parametrized Expression of Unclosed Curve 2

$$\left[-2\sin(t), t - \frac{1}{5}\sin(9\,t) + \cos(3\,t) + \sin(t) \right]$$ (2)

thereinto,t∈ [0, π];

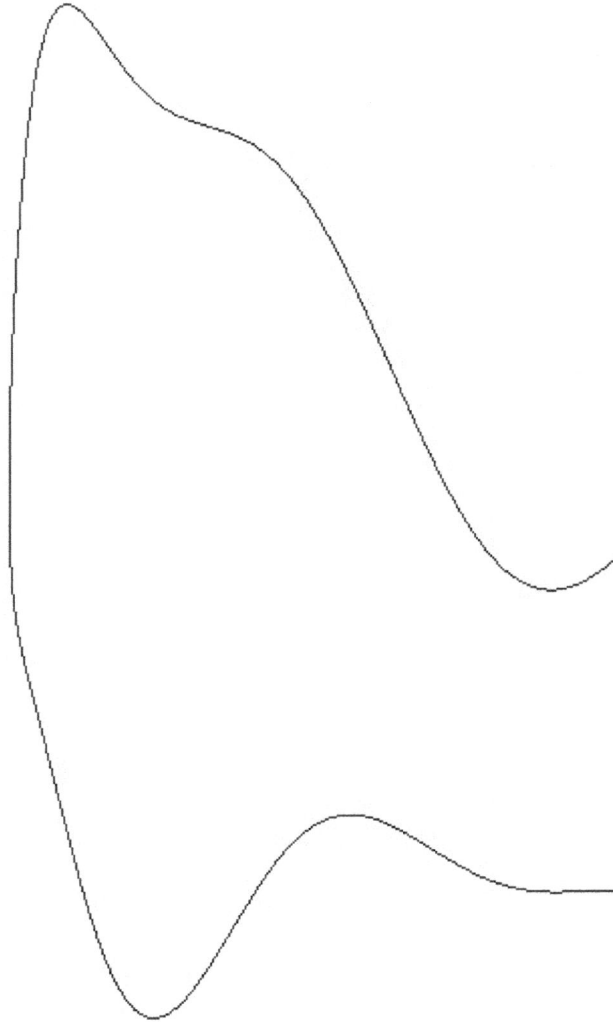

Figure(II)3.5.2 Unclosed Curve 2

C. Patchwork of Unclosed Curve 1 & 2, Structures (Composite) Simply
Connected Closed Curve 3:

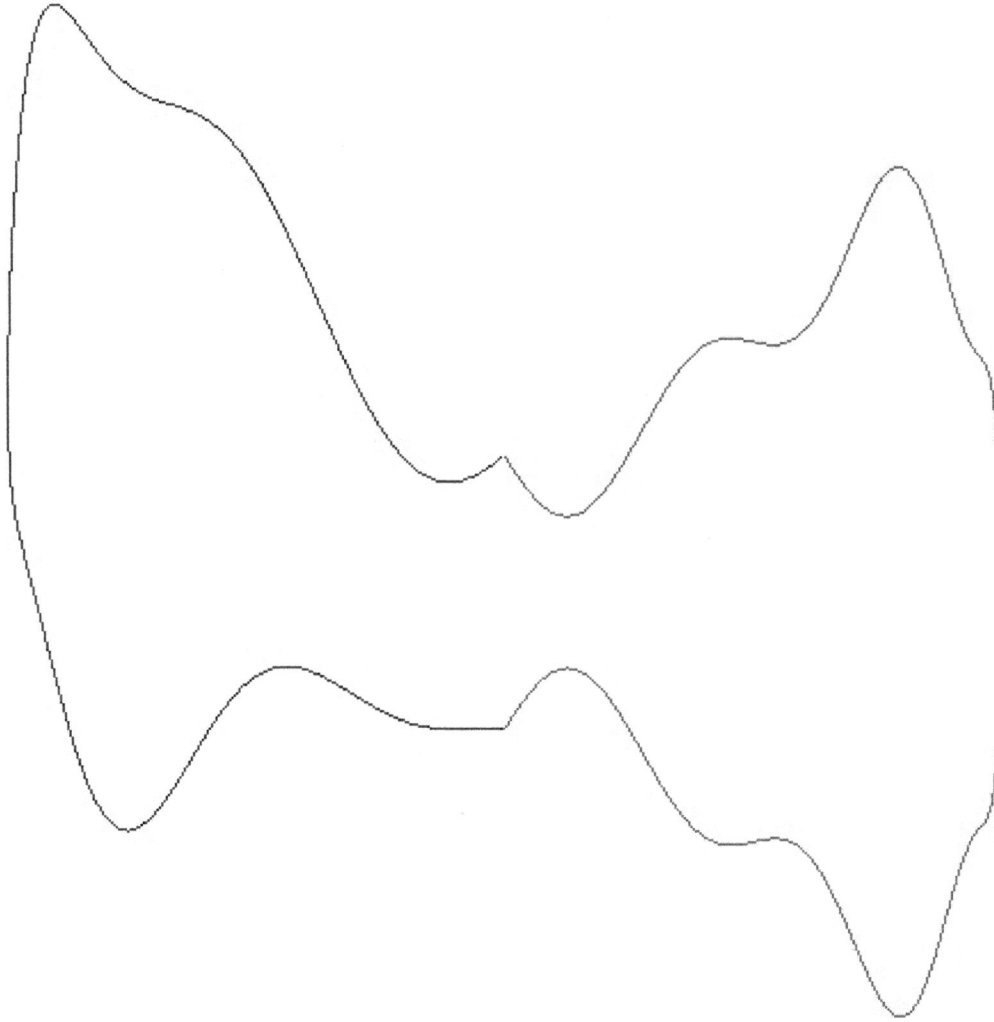

Figure(II)3.5.3 Composite Simply Connected Closed Curve 3;

Patchwork of Unclosed Curve 1 & 2; [Irregular、Asymmetrical]

and Integral Vector Field

$$\left[\left(\frac{1}{3}x+\frac{1}{5}y+\frac{1}{15}xy\right)^2-\frac{xy^2}{3}-\frac{y^3}{6},\frac{\left(\frac{x}{4}-\frac{y}{5}-\frac{1}{9}\right)^3}{5}-\frac{x^2}{5}+\frac{xy^2}{3}\right]$$ (3)

Calculate and Validate Green Theorem at Manifold.

Solution:

  First, Free Closed Curve Integral:

  According to parametrized expression (1), Calculate Tangent Vector

of Curve 1:

978-1-62265-930-2 (online)   978-1-62265-931-9 (paper) Yang Ke

$$\left[\frac{d}{dt}(2\sin(t)), \frac{d}{dt}\left(t+\frac{1}{5}\sin(12\,t)+\cos(3\,t)\right)\right] = \left[2\cos(t), 1+\frac{12}{5}\cos(12\,t)-3\sin(3\,t)\right] \quad (4)$$

Input parametric expression (1) to Vector Field (3); and calculate Integral of Dot Product [Vetor Field (3) and Tangent Vector (4)] in range of 't' (5):

$$\int_0^\pi 2\left(\left(\left(\frac{2}{3}\sin(t)+\frac{t}{5}+\frac{1}{25}\sin(12\,t)+\frac{1}{5}\cos(3\,t)+\frac{2}{15}\sin(t)\left(t+\frac{1}{5}\sin(12\,t)+\cos(3\,t)\right)\right)^2\right.\right.$$

$$-\frac{2}{3}\sin(t)\left(t+\frac{1}{5}\sin(12\,t)+\cos(3\,t)\right)^2 - \frac{1}{6}\left(t+\frac{1}{5}\sin(12\,t)+\cos(3\,t)\right)^3\right)\cos(t)+\left(\vphantom{\frac{1}{5}}\right.$$

$$\frac{1}{5}\left(\frac{1}{2}\sin(t)-\frac{t}{5}-\frac{1}{25}\sin(12\,t)-\frac{1}{5}\cos(3\,t)-\frac{1}{9}\right)^3 - \frac{4}{5}\sin(t)^2$$

$$\left.\left.+\frac{2}{3}\sin(t)\left(t+\frac{1}{5}\sin(12\,t)+\cos(3\,t)\right)^2\right)\left(1+\frac{12}{5}\cos(12\,t)-3\sin(3\,t)\right)\right)dt$$

$$=$$

$$-\frac{153902621568172874331135807}{2579595875428888722656250} - \frac{13176815923}{18243225000}\pi - \frac{1}{2500}\pi^4 + \frac{15017293}{7722000}\pi^2$$

$$+\frac{4}{5625}\pi^3$$

$$(5)$$

According to parametrized expression (2), Calculate Tangent Vector of Curve 2:

$$\left[\frac{d}{dt}(-2\sin(t)), \frac{d}{dt}\left(t-\frac{1}{5}\sin(9\,t)+\cos(3\,t)+\sin(t)\right)\right] =$$

$$\left[-2\cos(t), 1-\frac{9}{5}\cos(9\,t)-3\sin(3\,t)+\cos(t)\right] \quad (6)$$

Input parametric expression (2) to Vector Field (3); and calculate Integral of Dot Product [Vetor Field (3) and Tangent Vector (6)] in range of 't' (7):

$$\int_0^\pi -2\left(\left(-\frac{7}{15}\sin(t)+\frac{t}{5}-\frac{1}{25}\sin(9\,t)+\frac{1}{5}\cos(3\,t)\right.\right.$$

$$\left.-\frac{2}{15}\sin(t)\left(t-\frac{1}{5}\sin(9\,t)+\cos(3\,t)+\sin(t)\right)\right)^2$$

$$+\frac{2}{3}\sin(t)\left(t-\frac{1}{5}\sin(9\,t)+\cos(3\,t)+\sin(t)\right)^2$$

$$-\frac{1}{6}\left(t-\frac{1}{5}\sin(9\,t)+\cos(3\,t)+\sin(t)\right)^{3}\Bigg)\cos(t)+\Bigg($$

$$\frac{1}{5}\left(-\frac{7}{10}\sin(t)-\frac{t}{5}+\frac{1}{25}\sin(9\,t)-\frac{1}{5}\cos(3\,t)-\frac{1}{9}\right)^{3}-\frac{4}{5}\sin(t)^{2}$$

$$-\frac{2}{3}\sin(t)\left(t-\frac{1}{5}\sin(9\,t)+\cos(3\,t)+\sin(t)\right)^{2}\Bigg)\left(1-\frac{9}{5}\cos(9\,t)-3\sin(3\,t)+\cos(t)\right)$$

$$dt$$

$$=$$

$$\frac{23974351850937133}{5234067907818750}+\frac{4594244203}{145945800000}\pi-\frac{1}{2500}\pi^{4}-\frac{4745051}{2700000}\pi^{2}+\frac{4}{5625}\pi^{3} \qquad (7)$$

Difference between (5) and (7) [viz.(5)-(7)] is curve integral value of composite simply connected closed curve (8):

$$-\frac{162731916511763840309216459}{1542980495858390 8470703125}-\frac{85476901}{113400000}\pi+\frac{476468981}{128700000}\pi^{2} \qquad (8)$$

Second, Free Area Element Integral:

Change 't' to 'u' in parametrized expression (1), and entirely multiply radius vector 'r' (suppose r>0) by each item of parametrized expression(1), convert 'Parametrized expression of curve 1' to 'Parametrized expression of curve 1 coordinates' (9):

$$\left[2\,r\sin(u),\,r\left(u+\frac{1}{5}\sin(12\,u)+\cos(3\,u)\right)\right] \qquad (9)$$

According to parametrized expression of curve 1 coordinates (9), define and calculate matrix of partial derivatives, obtain general expression of curve 1 coordinates area element coefficient (10):

$$\left[\begin{array}{cc} \dfrac{\partial}{\partial r}(2\,r\sin(u)) & \dfrac{\partial}{\partial u}(2\,r\sin(u)) \\[2ex] \dfrac{\partial}{\partial r}\left(r\left(u+\dfrac{1}{5}\sin(12\,u)+\cos(3\,u)\right)\right) & \dfrac{\partial}{\partial u}\left(r\left(u+\dfrac{1}{5}\sin(12\,u)+\cos(3\,u)\right)\right) \end{array}\right]$$

$$=$$

$$2\,r\sin(u)+\frac{24}{5}\sin(u)\,r\cos(12\,u)-6\sin(u)\,r\sin(3\,u)-2\,r\cos(u)\,u$$

$$-\frac{2}{5}r\cos(u)\sin(12\,u)-2\,r\cos(u)\cos(3\,u) \qquad (10)$$

Calculate Differential Function `$\dfrac{\partial}{\partial x}Q(x,y)-\dfrac{\partial}{\partial y}P(x,y)$` of Vector

Field (3):

$$
\left(\frac{\partial}{\partial x}\left(\frac{\left(\frac{x}{4}-\frac{y}{5}-\frac{1}{9}\right)^{3}}{5}-\frac{x^{2}}{5}+\frac{x\,y^{2}}{3}\right)\right)-\left(\frac{\partial}{\partial y}\left(\left(\frac{1}{3}x+\frac{1}{5}y+\frac{1}{15}x\,y\right)^{2}-\frac{x\,y^{2}}{3}-\frac{y^{3}}{6}\right)\right)=
$$

$$
\frac{3\left(\frac{x}{4}-\frac{y}{5}-\frac{1}{9}\right)^{2}}{20}-\frac{2\,x}{5}+\frac{5\,y^{2}}{6}-2\left(\frac{1}{3}x+\frac{1}{5}y+\frac{1}{15}x\,y\right)\left(\frac{1}{5}+\frac{x}{15}\right)+\frac{2\,x\,y}{3}
$$

(11)

Input Parametrized Expression of Curve 1 Coordinates (9) to Differential Function (11); and Calculate Double Integrals of Product [Differential Function (11) and Curve 1 Coordinates Area Element] in range of 'r,u' (12):

$$
\int_{0}^{\pi}\int_{0}^{1}\left(\frac{3}{20}\left(\frac{1}{2}r\sin(u)-\frac{1}{5}r\left(u+\frac{1}{5}\sin(12\,u)+\cos(3\,u)\right)-\frac{1}{9}\right)^{2}-\frac{4}{5}r\sin(u)\right.
$$

$$
+\frac{5}{6}r^{2}\left(u+\frac{1}{5}\sin(12\,u)+\cos(3\,u)\right)^{2}-2\left(\frac{2}{3}r\sin(u)\right.
$$

$$
+\frac{1}{5}r\left(u+\frac{1}{5}\sin(12\,u)+\cos(3\,u)\right)+\frac{2}{15}r^{2}\sin(u)\left(u+\frac{1}{5}\sin(12\,u)+\cos(3\,u)\right)\right)
$$

$$
\left(\frac{1}{5}+\frac{2}{15}r\sin(u)\right)+\frac{4}{3}r^{2}\sin(u)\left(u+\frac{1}{5}\sin(12\,u)+\cos(3\,u)\right)\right)\left(2\,r\sin(u)\right)
$$

$$
+\frac{24}{5}\sin(u)\,r\cos(12\,u)-6\sin(u)\,r\sin(3\,u)-2\,r\cos(u)\,u-\frac{2}{5}r\cos(u)\sin(12\,u)
$$

$$
\left.-2\,r\cos(u)\cos(3\,u)\right)dr\,du
$$

$$
=
$$

$$
-\frac{633591091405698653332349}{1061562088653863671875}-\frac{488125129}{675675000}\pi+\frac{25034923}{12870000}\pi^{2}
$$

(12)

Change 't' to 'u' in parametrized expression (2), and entirely multiply radius vector 'r' (suppose r>0) by each item of parametrized expression(2), convert 'Parametrized expression of curve 2' to 'Parametrized expression of curve 2 coordinates' (13):

978-1-62265-930-2 (online)   978-1-62265-931-9 (paper)

$$\left[-2\,r\sin(u), r\left(u - \frac{1}{5}\sin(9\,u) + \cos(3\,u) + \sin(u) \right) \right]$$ (13)

According to parametrized expression of curve 2 coordinates (13), define and calculate matrix of partial derivatives, obtain general expression of curve 2 coordinates area element coefficient (14):

$$\left[\frac{\partial}{\partial r}(-2\,r\sin(u)), \frac{\partial}{\partial u}(-2\,r\sin(u)) \right]$$

$$\left[\frac{\partial}{\partial r}\left(r\left(u - \frac{1}{5}\sin(9\,u) + \cos(3\,u) + \sin(u) \right) \right), \right.$$

$$\left. \frac{\partial}{\partial u}\left(r\left(u - \frac{1}{5}\sin(9\,u) + \cos(3\,u) + \sin(u) \right) \right) \right]$$

$$=$$

$$-2\,r\sin(u) + \frac{18}{5}\sin(u)\,r\cos(9\,u) + 6\sin(u)\,r\sin(3\,u) + 2\,r\cos(u)\,u$$

$$-\frac{2}{5}r\cos(u)\sin(9\,u) + 2\,r\cos(u)\cos(3\,u)$$

(14)

Input Parametrized Expression of Curve 2 Coordinates (13) to Differential Function (11); and Calculate Double Integrals of Product [Differential Function (11) and Curve 2 Coordinates Area Element] in range of 'r,u' (15):

$$\int_0^\pi \int_0^1 \left(\frac{3}{20}\left(-\frac{1}{2}r\sin(u) - \frac{1}{5}r\left(u - \frac{1}{5}\sin(9\,u) + \cos(3\,u) + \sin(u) \right) - \frac{1}{9} \right)^2 + \frac{4}{5}r\sin(u) \right.$$

$$+ \frac{5}{6}r^2\left(u - \frac{1}{5}\sin(9\,u) + \cos(3\,u) + \sin(u) \right)^2 - 2\left(-\frac{2}{3}r\sin(u) \right.$$

$$+ \frac{1}{5}r\left(u - \frac{1}{5}\sin(9\,u) + \cos(3\,u) + \sin(u) \right)$$

$$- \frac{2}{15}r^2\sin(u)\left(u - \frac{1}{5}\sin(9\,u) + \cos(3\,u) + \sin(u) \right) \right)\left(\frac{1}{5} - \frac{2}{15}r\sin(u) \right)$$

$$- \frac{4}{3}r^2\sin(u)\left(u - \frac{1}{5}\sin(9\,u) + \cos(3\,u) + \sin(u) \right) \right)\left(-2\,r\sin(u) \right.$$

$$+ \frac{18}{5}\sin(u)\,r\cos(9\,u) + 6\sin(u)\,r\sin(3\,u) + 2\,r\cos(u)\,u - \frac{2}{5}r\cos(u)\sin(9\,u)$$

$$\left. + 2\,r\cos(u)\cos(3\,u) \right)dr\,du$$

$$=$$

$$\frac{887487960176879}{193854366956250} + \frac{508193747}{16216200000}\,\pi - \frac{1581257}{900000}\,\pi^2$$

(15)

Difference between (12) and (15) [viz.(12)-(15)] is area element integral value of plane bounded closed region that composite simply connected closed curve 3 surrounds (16):

$$-\frac{162731916511763840309216459}{1542980495858390847070703125} - \frac{85476901}{113400000}\,\pi + \frac{476468981}{128700000}\,\pi^2$$

(16)

Curve Integral Precision Value (8) [Vector Field (3) at Composite Simply Connected Closed Curve 3], is equal to Area Element Integral Precision Value (16) [Differential Function (11) in Plane Bounded Closed Region that Target Curve 3 surrounds], Complete Calculation and Validation of Green Theorem at Manifold

3.6 Numerical Model of Green Theorem at Manifold (III)
   ### [Program Template of Waterloo Maplesoft, Optional]

```
> restart;
> with(plots):with(linalg):
> CL1:=[2*sin(t),t+1/5*sin(12*t)+cos(3*t)];
# Define parametrized expression of discretional curve 'CL1'
```

$$CL1 := \left[2\sin(t), t + \frac{1}{5}\sin(12\,t) + \cos(3\,t) \right]$$

```
> CL2:=[-2*sin(t),t-1/5*sin(9*t)+cos(3*t)+sin(t)];
# Define parametrized expression of discretional curve 'CL2'
```

$$CL2 := \left[-2\sin(t), t - \frac{1}{5}\sin(9\,t) + \cos(3\,t) + \sin(t) \right]$$

```
> rgt:=[0,Pi]; # Set range of parameter 't'
```

$$rgt := [\,0, \pi\,]$$

```
> plot([CL1[1],CL1[2],t=rgt[1]..rgt[2]],color=red,
scaling=constrained,axes=none):g1:=%:
# Draw unclosed curve 'CL1'
> plot([CL2[1],CL2[2],t=rgt[1]..rgt[2]],color=blue,
scaling=constrained,axes=none):g2:=%:
# Draw unclosed curve 'CL2'
> display(g1,g2): # Synthesize figures
```

```
> PV:=[(x/3+y/5+x*y/15)^2-x*y^2/3-y^3/6,(x/4-y/5-1/9)^3/5-x^2/5
  +x*y^2/3];
```

Define discretional vector field 'PV' (Suppose Vector Field 'PV' possesses 1th order continuous partial derivatives in plane bounded closed region that composite simply connected closed curve 'CL1/CL2' surrounds)

$$PV := \left[\left(\frac{1}{3}x + \frac{1}{5}y + \frac{1}{15}xy\right)^2 - \frac{xy^2}{3} - \frac{y^3}{6}, \frac{\left(\frac{x}{4} - \frac{y}{5} - \frac{1}{9}\right)^3}{5} - \frac{x^2}{5} + \frac{xy^2}{3} \right]$$

```
> Diff(PV[2],x)-Diff(PV[1],y)=diff(PV[2],x)-diff(PV[1],y);
 dPV:=rhs(%):
```

Calculate differential function 'dPV' of vector field 'PV'

$$\left(\frac{\partial}{\partial x}\left(\frac{\left(\frac{x}{4} - \frac{y}{5} - \frac{1}{9}\right)^3}{5} - \frac{x^2}{5} + \frac{xy^2}{3}\right)\right) - \left(\frac{\partial}{\partial y}\left(\left(\frac{1}{3}x + \frac{1}{5}y + \frac{1}{15}xy\right)^2 - \frac{xy^2}{3} - \frac{y^3}{6}\right)\right) =$$

$$\frac{3\left(\frac{x}{4} - \frac{y}{5} - \frac{1}{9}\right)^2}{20} - \frac{2x}{5} + \frac{5y^2}{6} - 2\left(\frac{1}{3}x + \frac{1}{5}y + \frac{1}{15}xy\right)\left(\frac{1}{5} + \frac{x}{15}\right) + \frac{2xy}{3}$$

```
> rgx:=[-2.2,2.2]; # Set drawing range of vector field 'PV'
```
$$rgx := [-2.2, 2.2]$$

```
> rgy:=[-0.3,4.1];
```
$$rgy := [-0.3, 4.1]$$

```
> fieldplot(PV,x=rgx[1]..rgx[2],y=rgy[1]..rgy[2],arrows=SLIM,
  color=blue):g3:=%: # Draw vector field 'PV'
> implicitplot(dPV,x=rgx[1]..rgx[2],y=rgy[1]..rgy[2],color=black,
  thickness=1,numpoints=2000):g4:=%:
 # Draw contour curve of differential function 'dPV'
> display(g1,g2,g3,g4,axes=normal);
```

Composite simply connected closed curve 'CL1/CL2', vector field 'PV', contour curve of 'dPV', four figures are synthesized

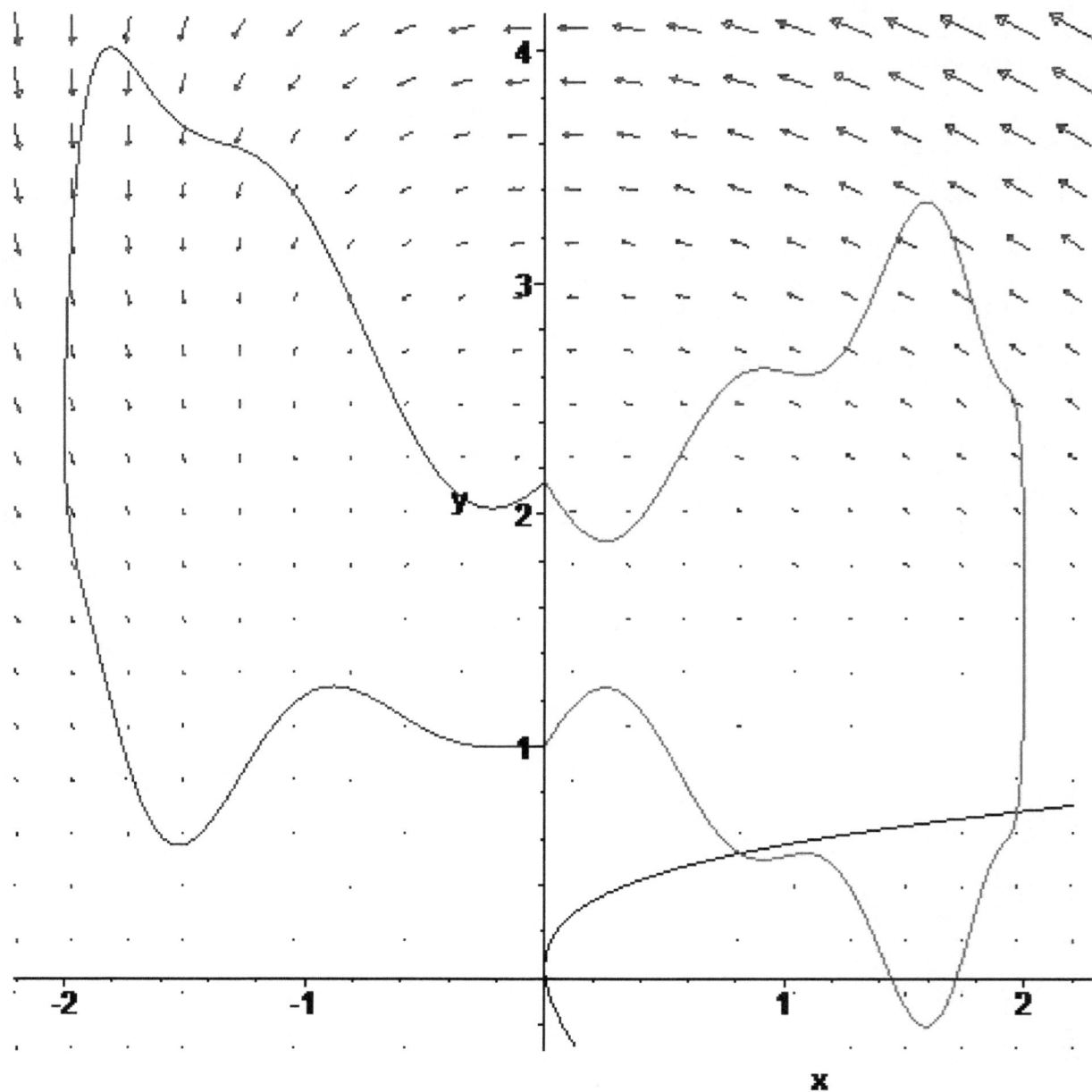

Figure(II)3.6.1 Composite Simply Connected Closed Curve 'CL1/CL2' Vector Field 'PV' and Contour Curve of Differential Function 'dPV'

```
> plot3d(dPV,x=rgx[1]..rgx[2],y=rgy[1]..rgy[2],axes=normal);
# Drawing differential function 'dPV' in 3-Dimensional Euclidean space
```

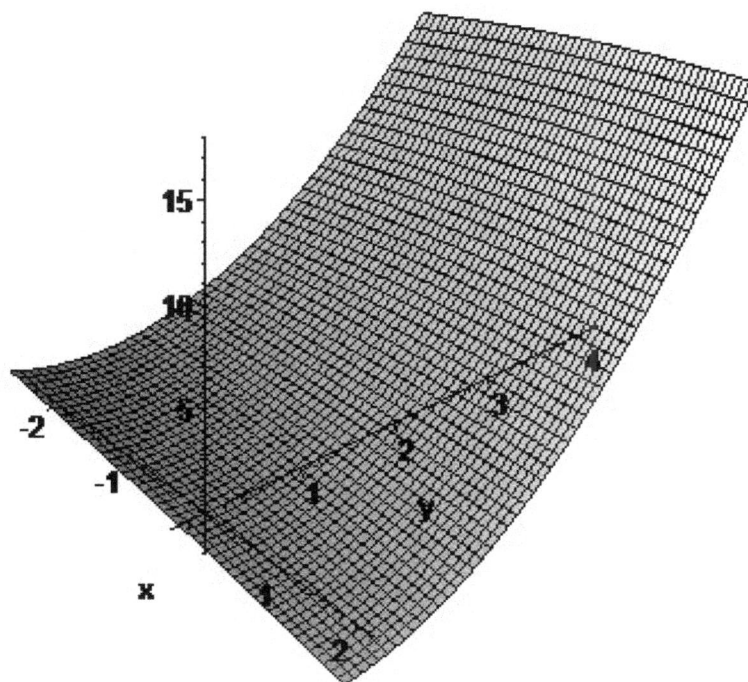

Figure(II)3.6.2 Space View of Differential Function 'dPV

```
> x:=CL1[1];y:=CL1[2];  # Variables assignment
```

$$x := 2 \sin(t)$$

$$y := t + \frac{1}{5} \sin(12\,t) + \cos(3\,t)$$

```
> [Diff(x,t),Diff(y,t)]=[diff(x,t),diff(y,t)];
# Calculate Tangent Vector of Curve 'CL1':
```

$$\left[\frac{d}{dt}(2\sin(t)), \frac{d}{dt}\left(t + \frac{1}{5}\sin(12\,t) + \cos(3\,t)\right)\right] = \left[2\cos(t), 1 + \frac{12}{5}\cos(12\,t) - 3\sin(3\,t)\right]$$

```
> Int(PV[1]*diff(x,t)+PV[2]*diff(y,t),t=rgt[1]..rgt[2]);
# Curve Integral (Vector Field 'PV' at Unclosed Curve 'CL1')
```

$$\int_0^\pi 2\left(\left(\frac{2}{3}\sin(t) + \frac{t}{5} + \frac{1}{25}\sin(12\,t) + \frac{1}{5}\cos(3\,t) + \frac{2}{15}\sin(t)\left(t + \frac{1}{5}\sin(12\,t) + \cos(3\,t)\right)\right)^2\right.$$

$$\left. - \frac{2}{3}\sin(t)\left(t + \frac{1}{5}\sin(12\,t) + \cos(3\,t)\right)^2 - \frac{1}{6}\left(t + \frac{1}{5}\sin(12\,t) + \cos(3\,t)\right)^3\right)\cos(t) + \left(\vphantom{\frac{1}{5}}\right.$$

$$\frac{1}{5}\left(\frac{1}{2}\sin(t) - \frac{t}{5} - \frac{1}{25}\sin(12\,t) - \frac{1}{5}\cos(3\,t) - \frac{1}{9}\right)^3 - \frac{4}{5}\sin(t)^2$$

$$+\frac{2}{3}\sin(t)\left(t+\frac{1}{5}\sin(12\,t)+\cos(3\,t)\right)^2\right)\left(1+\frac{12}{5}\cos(12\,t)-3\sin(3\,t)\right)dt$$

> v1:=value(%); # Calculate curve integral value of 'CL1'

$$v1:=-\frac{153902621568172874331335807}{257959587542888872656250}-\frac{13176815923}{18243225000}\pi-\frac{1}{2500}\pi^4+\frac{15017293}{7722000}\pi^2$$
$$+\frac{4}{5625}\pi^3$$

> x:=CL2[1];y:=CL2[2]; # Variables assignment

$$x:=-2\sin(t)$$

$$y:=t-\frac{1}{5}\sin(9\,t)+\cos(3\,t)+\sin(t)$$

> [Diff(x,t),Diff(y,t)]=[diff(x,t),diff(y,t)];

\# Calculate Tangent Vector of Curve 'CL2':

$$\left[\frac{d}{dt}(-2\sin(t)),\frac{d}{dt}\left(t-\frac{1}{5}\sin(9\,t)+\cos(3\,t)+\sin(t)\right)\right]=$$
$$\left[-2\cos(t),1-\frac{9}{5}\cos(9\,t)-3\sin(3\,t)+\cos(t)\right]$$

> Int(PV[1]*diff(x,t)+PV[2]*diff(y,t),t=rgt[1]..rgt[2]);

\# Curve Integral (Vector Field 'PV' at Unclosed Curve 'CL2')

$$\int_0^{\pi}-2\left(\left(-\frac{7}{15}\sin(t)+\frac{t}{5}-\frac{1}{25}\sin(9\,t)+\frac{1}{5}\cos(3\,t)\right.\right.$$

$$-\frac{2}{15}\sin(t)\left(t-\frac{1}{5}\sin(9\,t)+\cos(3\,t)+\sin(t)\right)\right)^2$$

$$+\frac{2}{3}\sin(t)\left(t-\frac{1}{5}\sin(9\,t)+\cos(3\,t)+\sin(t)\right)^2$$

$$-\frac{1}{6}\left(t-\frac{1}{5}\sin(9\,t)+\cos(3\,t)+\sin(t)\right)^3\right)\cos(t)+\left(\vphantom{\frac{1}{5}}\right.$$

$$\frac{1}{5}\left(-\frac{7}{10}\sin(t)-\frac{t}{5}+\frac{1}{25}\sin(9\,t)-\frac{1}{5}\cos(3\,t)-\frac{1}{9}\right)^3-\frac{4}{5}\sin(t)^2$$

$$-\frac{2}{3}\sin(t)\left(t-\frac{1}{5}\sin(9\,t)+\cos(3\,t)+\sin(t)\right)^2\right)\left(1-\frac{9}{5}\cos(9\,t)-3\sin(3\,t)+\cos(t)\right)$$

$$dt$$

> v2:=value(%); # Calculate curve integral value of 'CL2'

$$v2:=\frac{23974351850937133}{5234067907818750}+\frac{4594244203}{145945800000}\pi-\frac{1}{2500}\pi^4-\frac{4745051}{2700000}\pi^2+\frac{4}{5625}\pi^3$$

```
> alpha:=(v1-v2);delta:=evalf(alpha);
```

$$\alpha := -\frac{162731916511763840309216459}{1542980495858390847070703125} - \frac{85476901}{113400000}\pi + \frac{476468981}{128700000}\pi^2$$

$$\delta := 23.62431347$$

Closed curve integral value of composite simply connected closed curve 'CL1/CL2'


```
> x:='x':y:='y':
> CL1c:=subs(t=u,CL1);
```

Change character 't' to 'u' in parametrized expression of curve 'CL1'

$$CL1c := \left[\, 2\,\sin(u),\, u + \frac{1}{5}\sin(12\,u) + \cos(3\,u)\,\right]$$

```
> [r*CL1c[1],r*CL1c[2]];
```

Entirely multiply radius vector 'r' (suppose r>0) by each item of parametrized expression 'CL1c', Convert "Parametrized expression of curve 'CL1'" to "Parametrized expression of curve 'CL1' coordinates"

$$\left[\, 2\,r\,\sin(u),\, r\left(u + \frac{1}{5}\sin(12\,u) + \cos(3\,u)\right)\,\right]$$

```
> x:=r*CL1c[1]:y:=r*CL1c[2]:
> matrix(2,2,[Diff(x,r),Diff(x,u),Diff(y,r),Diff(y,u)])=
matrix(2,2,[diff(x,r),diff(x,u),diff(y,r),diff(y,u)]);m:=rhs(%);
```

Define and calculate matrix of partial derivatives 'm', obtain general expression of curve 'CL1' coordinates area element coefficient

$$\begin{bmatrix} \dfrac{\partial}{\partial r}(2\,r\,\sin(u)) & \dfrac{\partial}{\partial u}(2\,r\,\sin(u)) \\[2mm] \dfrac{\partial}{\partial r}\left(r\left(u + \frac{1}{5}\sin(12\,u) + \cos(3\,u)\right)\right) & \dfrac{\partial}{\partial u}\left(r\left(u + \frac{1}{5}\sin(12\,u) + \cos(3\,u)\right)\right) \end{bmatrix} =$$

$$\begin{bmatrix} 2\,\sin(u) & 2\,r\,\cos(u) \\[2mm] u + \frac{1}{5}\sin(12\,u) + \cos(3\,u) & r\left(1 + \frac{12}{5}\cos(12\,u) - 3\,\sin(3\,u)\right) \end{bmatrix}$$

$$m := \begin{bmatrix} 2\,\sin(u) & 2\,r\,\cos(u) \\[2mm] u + \frac{1}{5}\sin(12\,u) + \cos(3\,u) & r\left(1 + \frac{12}{5}\cos(12\,u) - 3\,\sin(3\,u)\right) \end{bmatrix}$$

```
> J:=det(m);
```

General expression of curve 'CL1' coordinates area element coefficient

$$J := 2\,r\sin(u) + \frac{24}{5}\sin(u)\,r\cos(12\,u) - 6\sin(u)\,r\sin(3\,u) - 2\,r\cos(u)\,u$$
$$- \frac{2}{5}\,r\cos(u)\sin(12\,u) - 2\,r\cos(u)\cos(3\,u)$$

```
> #J:=simplify(%);
> Int(Int(dPV*J,r=0..1),u=rgt[1]..rgt[2]);
```

Double Integrals of Product (Differential Function 'dPV' and "Curve

'CL1' Coordinates Area Element") in range of 'r,u'

$$\int_0^\pi \int_0^1 \left(\frac{3}{20}\left(\frac{1}{2}r\sin(u) - \frac{1}{5}r\left(u + \frac{1}{5}\sin(12\,u) + \cos(3\,u)\right) - \frac{1}{9} \right)^2 - \frac{4}{5}r\sin(u) \right.$$

$$+ \frac{5}{6}r^2\left(u + \frac{1}{5}\sin(12\,u) + \cos(3\,u)\right)^2 - 2\left(\frac{2}{3}r\sin(u)\right.$$
$$+ \frac{1}{5}r\left(u + \frac{1}{5}\sin(12\,u) + \cos(3\,u)\right) + \frac{2}{15}r^2\sin(u)\left(u + \frac{1}{5}\sin(12\,u) + \cos(3\,u)\right) \Big) $$

$$\left(\frac{1}{5} + \frac{2}{15}r\sin(u) \right) + \frac{4}{3}r^2\sin(u)\left(u + \frac{1}{5}\sin(12\,u) + \cos(3\,u)\right) \Big) \Big)(2\,r\sin(u)$$
$$+ \frac{24}{5}\sin(u)\,r\cos(12\,u) - 6\sin(u)\,r\sin(3\,u) - 2\,r\cos(u)\,u - \frac{2}{5}r\cos(u)\sin(12\,u)$$

$$\left. - 2\,r\cos(u)\cos(3\,u)\right)dr\,du$$

```
> v3:=value(%); # Calculate value of plane area element integral
```

$$v3 := -\frac{63359109140569865332349}{1061562088653863718750} - \frac{488125129}{675675000}\pi + \frac{25034923}{12870000}\pi^2$$

```
> x:='x':y:='y':
> CL2c:=subs(t=u,CL2);
```

Change character 't' to 'u' in parametrized expression of curve 'CL2'

$$CL2c := \left[-2\sin(u),\, u - \frac{1}{5}\sin(9\,u) + \cos(3\,u) + \sin(u) \right]$$

```
> [r*CL2c[1],r*CL2c[2]];
```

Entirely multiply radius vector 'r' (suppose r>0) by each item of

parametrized expression 'CL2c', Convert "Parametrized expression of

curve 'CL2'" to "Parametrized expression of curve 'CL2' coordinates"

$$\left[-2\,r\sin(u),\, r\left(u - \frac{1}{5}\sin(9\,u) + \cos(3\,u) + \sin(u)\right) \right]$$

```
> x:=r*CL2c[1]:y:=r*CL2c[2]:
> matrix(2,2,[Diff(x,r),Diff(x,u),Diff(y,r),Diff(y,u)])=
```

- 281 -

```
matrix(2,2,[diff(x,r),diff(x,u),diff(y,r),diff(y,u)]);m:=rhs(%);
```

Define and calculate matrix of partial derivatives 'm', obtain
general expression of curve 'CL2' coordinates area element coefficient

$$\left[\frac{\partial}{\partial r}(-2\,r\sin(u)), \frac{\partial}{\partial u}(-2\,r\sin(u)) \right]$$

$$\left[\frac{\partial}{\partial r}\left(r\left(u - \frac{1}{5}\sin(9\,u) + \cos(3\,u) + \sin(u) \right) \right), \right.$$

$$\left. \frac{\partial}{\partial u}\left(r\left(u - \frac{1}{5}\sin(9\,u) + \cos(3\,u) + \sin(u) \right) \right) \right] =$$

$$\left[\begin{array}{cc} -2\sin(u) & -2\,r\cos(u) \\ u - \frac{1}{5}\sin(9\,u) + \cos(3\,u) + \sin(u) & r\left(1 - \frac{9}{5}\cos(9\,u) - 3\sin(3\,u) + \cos(u) \right) \end{array} \right]$$

$$m := \left[\begin{array}{cc} -2\sin(u) & -2\,r\cos(u) \\ u - \frac{1}{5}\sin(9\,u) + \cos(3\,u) + \sin(u) & r\left(1 - \frac{9}{5}\cos(9\,u) - 3\sin(3\,u) + \cos(u) \right) \end{array} \right]$$

```
> J:=det(m);
```

General expression of curve 'CL2' coordinates area element
coefficient

$$J := -2\,r\sin(u) + \frac{18}{5}\sin(u)\,r\cos(9\,u) + 6\sin(u)\,r\sin(3\,u) + 2\,r\cos(u)\,u$$

$$- \frac{2}{5}r\cos(u)\sin(9\,u) + 2\,r\cos(u)\cos(3\,u)$$

```
> #J:=simplify(%);
> Int(Int(dPV*J,r=0..1),u=rgt[1]..rgt[2]);
```

Double Integrals of Product (Differential Function 'dPV' and "Curve
'CL2' Coordinates Area Element") in range of 'r,u'

$$\int_0^\pi \int_0^1 \left(\frac{3}{20}\left(-\frac{1}{2}r\sin(u) - \frac{1}{5}r\left(u - \frac{1}{5}\sin(9\,u) + \cos(3\,u) + \sin(u) \right) - \frac{1}{9} \right)^2 + \frac{4}{5}r\sin(u) \right.$$

$$+ \frac{5}{6}r^2\left(u - \frac{1}{5}\sin(9\,u) + \cos(3\,u) + \sin(u) \right)^2 - 2\left(-\frac{2}{3}r\sin(u) \right.$$

$$+ \frac{1}{5}r\left(u - \frac{1}{5}\sin(9\,u) + \cos(3\,u) + \sin(u) \right)$$

$$\left. - \frac{2}{15}r^2\sin(u)\left(u - \frac{1}{5}\sin(9\,u) + \cos(3\,u) + \sin(u) \right) \right)\left(\frac{1}{5} - \frac{2}{15}r\sin(u) \right)$$

$$\left. - \frac{4}{3}r^2\sin(u)\left(u - \frac{1}{5}\sin(9\,u) + \cos(3\,u) + \sin(u) \right) \right)\left(-2\,r\sin(u) \right)$$

$$+ \frac{18}{5} \sin(u) \, r \cos(9\,u) + 6 \sin(u) \, r \sin(3\,u) + 2\,r\cos(u)\,u - \frac{2}{5} r \cos(u) \sin(9\,u)$$

$$+ 2\,r\cos(u)\cos(3\,u) \bigg) dr\, du$$

```
> v4:=value(%); # Calculate value of plane area element integral
```

$$v4 := \frac{887487960176879}{193854366956250} + \frac{508193747}{16216200000} \pi - \frac{1581257}{900000} \pi^2$$

```
> beta:=v3-v4;epsilon:=evalf(beta);
```

```
# Area element integral value of plane bounded closed region that
composite simply connected closed surface 'CL1/CL2' surrounds
```

$$\beta := - \frac{16273191651176384030921 6459}{1542980495858390847070312 5} - \frac{85476901}{113400000} \pi + \frac{476468981}{128700000} \pi^2$$

$$\varepsilon := 23.62431347$$

```
> alpha;beta; # Two analytic values are equal
```

$$- \frac{16273191651176384030921 6459}{1542980495858390847070312 5} - \frac{85476901}{113400000} \pi + \frac{476468981}{128700000} \pi^2$$

$$- \frac{16273191651176384030921 6459}{1542980495858390847070312 5} - \frac{85476901}{113400000} \pi + \frac{476468981}{128700000} \pi^2$$

```
> delta;epsilon; # Two float values are equal
```

$$23.62431347$$

$$23.62431347$$

Chapter 4 Prove Green Theorem Finite Sums Limits at Manifold

4.1 Prove Green Theorem Finite Sums Limits at Manifold

Green Theorem Suppose boundary curve 'L' of plane bounded closed region 'S' is composed by numbered smooth or piecewise smooth curves. If exist 1th order continuous partial derivatives about functions 'P(x,y),Q(x,y)'[Structure plane vector field 'A'] at plane bounded closed region 'S', then:

$$\int_{L+} A \cdot dL = \iint_S \left(\frac{\partial}{\partial x} Q(x,y) - \frac{\partial}{\partial y} P(x,y) \right) dS \qquad (1)$$

Proof:

Define parametrized expression of abstract (plane) simply connected closed curve 'L':

```
    [a*cos(t),b*sin(t)]                                     (2)
```

thereinto, 'a,b' are nonzero constants or 1th order derivable continuous functions, simply connected closed curve 'L' determines value of 'a,b'; set range of parameters 't' in $[0,2\pi]$, make boundary curve 'L' closed. (See also Poincare Conjecture: 'Every closed n-Manifold which is homotopy equivalent to n-Sphere is homeomorphic to n-Sphere' [19], and Chapter 1 this Part 'Constitute Abstract Simply Connected Closed Parametrized Curve Coordinates [Plane]')

Calculate tangent vector of plane closed curve 'L' (3):

$$\left[\frac{\partial}{\partial t}(a\cos(t)),\frac{\partial}{\partial t}(b\sin(t))\right]=[-a\sin(t),b\cos(t)] \tag{3}$$

Set Amount of boundary curve 'L's parametrized segmental cells as 50 (It can be discretional natural number)   (4)

1.Microcosmic Curve Integral Course at Boundary Curve 'L's first segmental cell:

Segment range $[0,2\pi]$ of 't': dt $=\dfrac{2\pi}{50}=\dfrac{\pi}{25}$ (5)

Segment Abstract Vector Field [P(x,y),Q(x,y)]:

[P(x,y),Q(x,y)]   (6)

// Owing to universality and homogeneity of abstract vector field [P(x,y),Q(x,y)], its value at first segment cell is [P(x,y),Q(x,y)]; See also Chapter 4 (Part I) Section 4.3 'Discuss About Abstract Vector Field、Finite Sums Limit、Integral and Riemann Sums'

Segment Tangent Vector (7): [Viz. Input (5) to (3)]

$$\left[-a\sin\left(\frac{\pi}{25}\right),b\cos\left(\frac{\pi}{25}\right)\right] \tag{7}$$

Calculate Microcosmic Curve Integral Value at Closed Boundary Curve 'L's first segmental cell:

Base on integral median theorem, dot product of abstract vector field (6) and tangent vector (7), times segment unit of 't' (5), that is Microcosmic Curve Integral Value at first segmental cell (8)

$$\frac{1}{25}\pi\left(-a\sin\left(\frac{\pi}{25}\right)P(x,y)+b\cos\left(\frac{\pi}{25}\right)Q(x,y)\right) \tag{8}$$

2.Microcosmic Curve Integral Course at Closed Boundary Curve 'L's all (Viz. 50) segmental cells:

Segment range [0,2π] of 't': dt $= \dfrac{i2\pi}{50} = \dfrac{i\pi}{25}$

(thereinto, 'i' stand for natural number 1~50)   (9)

Segment Abstract Vector Field [P(x,y),Q(x,y)]:

    [P(x,y),Q(x,y)]   (10)

// Owing to universality and homogeneity of abstract vector field [P(x,y),Q(x,y)], its values at umpty segment cells are [P(x,y),Q(x,y)]; See also Chapter 4 (Part I) Section 4.3 'Discuss About Abstract Vector Field、Finite Sums Limit、Integral and Riemann Sums'

Segment Tangent Vector (11): [Viz. Input (9) to (3)]

$$\left[-a \sin\left(\frac{\pi i}{25}\right), b \cos\left(\frac{\pi i}{25}\right) \right]$$ (11)

// Expression (11) isn't that 'single vector value', but that 'Set of finite quantitative (viz.50) vector values'

Calculate Microcosmic Curve Integral Values at Closed Boundary Curve 'L's all segmental cells (12):

Base on Integral Median Theorem, dot product of abstract vector field (10) and tangent vector (11), times segment unit of 't' (9), these are Microcosmic Curve Integral Values at all segmental cells (13)

$$\frac{1}{25}\pi \left(-a \sin\left(\frac{\pi i}{25}\right) P(x,y) + b \cos\left(\frac{\pi i}{25}\right) Q(x,y) \right)$$ (12)

// Expression (12) isn't that 'single value', but that 'Set of finite quantitative (viz.50) values'

Structure Sequence that is composed of finite quantitative (viz. 50) Microcosmic Curve Integral Values (13):

[Lengthy expression of sequence is elided;

In program template of Waterloo Maplesoft (see also Section 4.2):

sqn:=seq(dt*(idPV[1]*idCO[1]+idPV[2]*idCO[2]),i=1..dus):

change ':' (last) to ';', then obtain expression of sequence]

Accumulational sum of sequence (viz. Sum of integral values of dot product [abstract vector field (10) and tangent vector (11) at all (Viz. 50) segmental cells], obtain Curve Integral Value at manifold (14):

[Lengthy expression of accumulation sum result is elided;

In program template of Waterloo Maplesoft (see also Section 4.2):

add(k,k=sqn):xi:=evalf(%);

change ':'(middle) to ';', then obtain expression of result]

Change expression of accumulational sum result to float value:

$$-0.60 \ 10^{-9} \ b \ Q(x,y)$$

// Amount of Closed Boundary Curve 'L's segmental cells increases infinitely, sum of integral values above-mentioned tends to '0' infinitely

Set Amount of closed boundary curve 'L's parametrized segmental cells as uncertain natural number 'n'   (15)

3.Microcosmic Curve Integral Course at Closed Boundary Curve 'L's first segmental cell:

Segment range $[0,2\pi]$ of 't': dt = $\dfrac{2\pi}{n}$ (16)

Segment Abstract Vector Field [P(x,y),Q(x,y)]:

[P(x,y),Q(x,y)]   (17)

// Owing to universality and homogeneity of abstract vector field [P(x,y),Q(x,y)], its value at first segment cell is [P(x,y),Q(x,y)] ; See also Chapter 4 (Part I) Section 4.3 'Discuss About Abstract Vector Field、Finite Sums Limit、Integral and Riemann Sums'

Segment Tangent Vector (18): [Viz. Input (16) to (3)]

$$\left[-a \sin\left(\frac{2\pi}{n}\right), b \cos\left(\frac{2\pi}{n}\right) \right]$$

(18)

Calculate Microcosmic Curve Integral Value at Closed Boundary Curve 'L's first segmental cell (19):

Base on Integral Median Theorem, dot product of abstract vector field (17) and tangent vector (18), times segment unit of 't' (16), that is Microcosmic Curve Integral Value at first segmental cell (19)

$$\frac{2\pi\left(-a\sin\left(\frac{2\pi}{n}\right)P(x,y)+b\cos\left(\frac{2\pi}{n}\right)Q(x,y)\right)}{n}$$

(19)

4.Microcosmic Curve Integral Course at Closed Boundary Curve 'L's all (Viz. 50) segmental cells:

Segment range [0,2π] of 't': dt = $\dfrac{2\pi i}{n}$

(thereinto, 'i' stand for natural number 1~n)   (20)

Segment Abstract Vector Field [P(x,y),Q(x,y)]:

[P(x,y),Q(x,y)]   (21)

// Owing to universality and homogeneity of abstract vector field [P(x,y),Q(x,y)],its values at umpty segment cells are [P(x,y),Q(x,y)] ; See also Chapter 4 (Part I) Section 4.3 'Discuss About Abstract Vector Field、Finite Sums Limit、Integral and Riemann Sums'

Segment Tangent Vector (22): [Viz. Input (20) to (3)]

$$\left[-a\sin\left(\frac{2\pi i}{n}\right), b\cos\left(\frac{2\pi i}{n}\right)\right]$$

(22)

// Expression (22) isn't that 'single vector value', but that 'Set of finite quantitative (viz.n) vector values'

Calculate Microcosmic Curve Integral Values at Closed Boundary Curve 'L's all segmental cells (23):

Base on Integral Median Theorem, dot product of abstract vector field (21) and tangent vector (22), times segment unit of 't'(20), these are Microcosmic Curve Integral Values at all segmental cells (23)

$$\frac{2\pi\left(-a\sin\left(\frac{2\pi i}{n}\right)P(x,y)+b\cos\left(\frac{2\pi i}{n}\right)Q(x,y)\right)}{n}$$

(23)

// Expression (23) isn't that 'single value', but that 'Set of finite quantitative (viz.n) values'

Structure finite sums (24):

(In the case of "Amount 'n' of closed boundary curve 'L's segmental cells is uncertain", sum of integral values [dot product about absract vector field (21) and tangent vector (22)] at all segmental cells)

$$\sum_{i=1}^{n}\left(\frac{2\pi\left(-a\sin\left(\frac{2\pi i}{n}\right)P(x,y)+b\cos\left(\frac{2\pi i}{n}\right)Q(x,y)\right)}{n}\right) \quad (24)$$

Infinitize finite sums, its limit operational value is 'Curl Integral Value at Manifold' (25):

(In the case of "Amount 'n' of closed boundary curve 'L's segmental cells tends to infinity", limit value of integral values' sums [dot product about abstract vector field (21) and tangent vector (22)] at all segmental cells)

$$\lim_{n\to\infty}\sum_{i=1}^{n}\left(\frac{2\pi\left(-a\sin\left(\frac{2\pi i}{n}\right)P(x,y)+b\cos\left(\frac{2\pi i}{n}\right)Q(x,y)\right)}{n}\right)$$

$$=$$

$$\lim_{n\to\infty}-2\pi\cos\left(\frac{\pi}{n}\right)^2\left(P(x,y)\,a\sin\left(\frac{\pi}{n}\right)\cos\left(\frac{\pi}{n}\right)^3-P(x,y)\,a\sin\left(\frac{\pi}{n}\right)\cos\left(\frac{\pi}{n}\right)\right.$$

$$+Q(x,y)\,b\cos\left(\frac{\pi}{n}\right)^4-2\,Q(x,y)\,b\cos\left(\frac{\pi}{n}\right)^2+Q(x,y)\,b$$

$$+\sin\left(\frac{\pi}{n}\right)^3\cos\left(\frac{\pi}{n}\right)P(x,y)\,a+\sin\left(\frac{\pi}{n}\right)^2\cos\left(\frac{\pi}{n}\right)^2Q(x,y)\,b-\sin\left(\frac{\pi}{n}\right)^2Q(x,y)\,b\right)$$

$$(n+1)\ \Big/\ \left(\left(\left(\cos\left(\frac{\pi}{n}\right)^2-1\right)^2n\right)-2\pi\left(\sin\left(\frac{\pi}{n}\right)^2\cos\left(\frac{\pi}{n}\right)^2P(x,y)\,a\right.\right.$$

$$+\sin\left(\frac{\pi}{n}\right)\cos\left(\frac{\pi}{n}\right)^3Q(x,y)\,b-\sin\left(\frac{\pi}{n}\right)\cos\left(\frac{\pi}{n}\right)Q(x,y)\,b+P(x,y)\,a\cos\left(\frac{\pi}{n}\right)^2$$

$$\left.-P(x,y)\,a\right)\sin\left(\frac{(n+1)\pi}{n}\right)\cos\left(\frac{(n+1)\pi}{n}\right)\ \Big/\ \left(\left(\cos\left(\frac{\pi}{n}\right)^2-1\right)^2n\right)-$$

$$\frac{2\pi\left(P(x,y)\,a\sin\left(\frac{\pi}{n}\right)\cos\left(\frac{\pi}{n}\right)+Q(x,y)\,b\cos\left(\frac{\pi}{n}\right)^2-Q(x,y)\,b\right)\cos\left(\frac{(n+1)\pi}{n}\right)^2}{n\left(\cos\left(\frac{\pi}{n}\right)^2-1\right)}$$

$$+ 2\,\pi \cos\!\left(\frac{\pi}{n}\right)^{2}\left(\mathrm{P}(x,y)\,a\,\sin\!\left(\frac{\pi}{n}\right)\cos\!\left(\frac{\pi}{n}\right)^{3} - \mathrm{P}(x,y)\,a\,\sin\!\left(\frac{\pi}{n}\right)\cos\!\left(\frac{\pi}{n}\right)\right.$$

$$+\, \mathrm{Q}(x,y)\,b\,\cos\!\left(\frac{\pi}{n}\right)^{4} - 2\,\mathrm{Q}(x,y)\,b\,\cos\!\left(\frac{\pi}{n}\right)^{2} + \mathrm{Q}(x,y)\,b$$

$$+\, \sin\!\left(\frac{\pi}{n}\right)^{3}\cos\!\left(\frac{\pi}{n}\right)\mathrm{P}(x,y)\,a + \sin\!\left(\frac{\pi}{n}\right)^{2}\cos\!\left(\frac{\pi}{n}\right)^{2}\mathrm{Q}(x,y)\,b - \sin\!\left(\frac{\pi}{n}\right)^{2}\mathrm{Q}(x,y)\,b\right)$$

$$\Bigg/ \left(\left(\cos\!\left(\frac{\pi}{n}\right)^{2}-1\right)^{2}n\right) + 2\,\pi\left(\sin\!\left(\frac{\pi}{n}\right)^{2}\cos\!\left(\frac{\pi}{n}\right)^{2}\mathrm{P}(x,y)\,a\right.$$

$$+\, \sin\!\left(\frac{\pi}{n}\right)\cos\!\left(\frac{\pi}{n}\right)^{3}\mathrm{Q}(x,y)\,b - \sin\!\left(\frac{\pi}{n}\right)\cos\!\left(\frac{\pi}{n}\right)\mathrm{Q}(x,y)\,b + \mathrm{P}(x,y)\,a\,\cos\!\left(\frac{\pi}{n}\right)^{2}$$

$$-\, \mathrm{P}(x,y)\,a\Bigg)\sin\!\left(\frac{\pi}{n}\right)\cos\!\left(\frac{\pi}{n}\right) \Bigg/ \left(\left(\cos\!\left(\frac{\pi}{n}\right)^{2}-1\right)^{2}n\right)$$

$$+\, \frac{2\,\pi\left(\mathrm{P}(x,y)\,a\,\sin\!\left(\frac{\pi}{n}\right)\cos\!\left(\frac{\pi}{n}\right) + \mathrm{Q}(x,y)\,b\,\cos\!\left(\frac{\pi}{n}\right)^{2} - \mathrm{Q}(x,y)\,b\right)\cos\!\left(\frac{\pi}{n}\right)^{2}}{n\left(\cos\!\left(\frac{\pi}{n}\right)^{2}-1\right)}$$

$$= 0 \tag{25}$$

//'Integral value equals 0' possesses specific mathematical and physical meaning:

In mathematical meaning, 'Integral value equals 0' is inevitable result of logic illation, incarnates logic balanced state between various integral elements;

In physical meaning, 'Integral value equals 0' implies that circumfluent value of 'Abstract Plane Vector Field at Abstract plane simply connected closed curve' is constant for '0' (immobile and pending) identically; If integral value equals certain a positive/negative number or an expression, it implies that always there is positive/ reverse directional flux or unknown flux about 'Abstract Plane Vector Field at Abstract simply connected closed curve', this instance will be unaccountable.

Entirely multiply radius vector 'r' (suppose r > 0) by each item of closed boundary curve 'L' parametrized expression (2), convert "Parametrized expression of curve 'L'" to "Parametrized expression of curve 'L' coordinates"; and change 't' to 'u' in expression:

$$[r\,a\,\cos(u), r\,b\,\sin(u)]\qquad(26)$$

Base on parametrized expression of curve 'L' coordinates (26), define and calculate matrix of partial derivatives, obtain general expression of curve 'L' coordinates area element coefficient(27):

$$\begin{bmatrix} \dfrac{\partial}{\partial r}(r\,a\,\cos(u)) & \dfrac{\partial}{\partial u}(r\,a\,\cos(u)) \\[2mm] \dfrac{\partial}{\partial r}(r\,b\,\sin(u)) & \dfrac{\partial}{\partial u}(r\,b\,\sin(u)) \end{bmatrix} = \text{abr} \qquad(27)$$

Convert differential function '$\dfrac{\partial Q(x,y)}{\partial x} - \dfrac{\partial P(x,y)}{\partial y}$', from Cartesian coordinates expression to curve 'L' coordinates expression (28):

$$\left(\frac{\partial}{\partial x}Q(x,y)\right)\left(\frac{\partial}{\partial r}(r\,a\,\cos(u))\right)\left(\frac{\partial}{\partial u}(r\,a\,\cos(u))\right)$$
$$-\left(\frac{\partial}{\partial y}P(x,y)\right)\left(\frac{\partial}{\partial r}(r\,b\,\sin(u))\right)\left(\frac{\partial}{\partial u}(r\,b\,\sin(u))\right)$$
$$=$$
$$-\left(\frac{\partial}{\partial x}Q(x,y)\right)a^2\cos(u)\,r\,\sin(u) - \left(\frac{\partial}{\partial y}P(x,y)\right)b^2\sin(u)\,r\,\cos(u)$$
$$\qquad(28)$$

//In Cartesian Coordinates, Differential Function of abstract vector field [P(x,y),Q(x,y)] is $\dfrac{\partial}{\partial x}Q(x,y) - \dfrac{\partial}{\partial y}P(x,y)$

In logic deduction of this Proof, it is necessary to import Differential Function of abstract vector field above-mentioned to abstract simply connected closed curve coordinates.

Two elements of abstract vector field's Differential Function:

$\dfrac{\partial}{\partial x}Q(x,y)$, $\dfrac{\partial}{\partial y}P(x,y)$ are abstract differential function, and their variables 'x,y' contain sub-variables 'r,u'.

Transform Expression between 'Cartesian coordinates' and 'abstract simply connected closed curve coordinates' is:

x = r*a*cos(u), y = r*b*sin(v)

--"Coordinates Transform Differential Functions" (Correspond to "Two Differential Variables '$\dfrac{\partial}{\partial x}$, $\dfrac{\partial}{\partial y}$' of Differential Functions '$\dfrac{\partial}{\partial x}Q(x,y)$,

$\frac{\partial}{\partial y}P(x,y)\ '\ ''$) are $\frac{\partial}{\partial r}ra\cos(u)\frac{\partial}{\partial u}ra\cos(u)$ and $\frac{\partial}{\partial r}rb\sin(u)\frac{\partial}{\partial u}rb\sin(u)$.

Product of "Differential Functions '$\frac{\partial}{\partial x}Q(x,y)$, $\frac{\partial}{\partial y}P(x,y)$ '" and "Coordinates Transform Differential Functions" (Viz. Product of two different functions) constitute "Differential Function in Abstract Simply Connected Closed Curve Coordinates"

// 'Chain differentiation' or 'Coordinates conversion' ?

If it is 'Chain differentiation', according to the principle of 'multiply in identical chain, plus in subdivision chain', it ought to be:

$$\left(\frac{\partial}{\partial x}Q(x,y)\right)\left(\left(\frac{\partial}{\partial r}(r\,a\cos(u))\right)+\left(\frac{\partial}{\partial u}(r\,a\cos(u))\right)\right)$$

$$-\left(\frac{\partial}{\partial y}P(x,y)\right)\left(\left(\frac{\partial}{\partial r}(r\,b\sin(u))\right)+\left(\frac{\partial}{\partial u}(r\,b\sin(u))\right)\right)$$

// No matter "Chain differentiation" or "Solve differential function '$\frac{\partial}{\partial x}Q(x,y)-\frac{\partial}{\partial y}P(x,y)$ '", question for discussion is 'How to solve differential coefficient' or 'method of differentiation' about abstract vector field '[P(x,y),Q(x,y)]'; and this step, question for discussion is "How to convert result of solving differential function '$\frac{\partial}{\partial x}Q(x,y)-\frac{\partial}{\partial y}P(x,y)$' from a coordinates to another"; character and hiberarchy of 'two questions' above-mentioned are disparate, it is 'multiply', not 'plus' - be determined by 2- Dimensional Euclidean spacial attribute of coordinates.

Set Amount of plane bounded closed region 'S's (Curve 'L' surrounds) segmental cells as 50

(This amount can be discretional natural number)   (29)

5.Microcosmic Double Integrals Course in plane bounded closed region 'S's first segmental cell:

Segment range [0,1] of 'r': dr = $\dfrac{1}{50}$

Segment range [0,2π] of 'u': du = $\dfrac{2\pi}{50}=\dfrac{\pi}{25}$ (30)

Segment area element coefficient (31): [viz. Input (30) to (27)]

(Area element coefficient(27), average value that corresponds to plane bounded closed region 'S's first segmental cell)

$$\frac{a\,b}{50}$$
                    (31)

Segment abstract differential function (32):

[viz. Input (30) to (28)]

(Abstract differential function (28), average value that corresponds to area bounded closed region 'S's first segmental cell)

$$-\frac{1}{50}\left(\frac{\partial}{\partial x}Q(x,y)\right)a^2\cos\left(\frac{\pi}{25}\right)\sin\left(\frac{\pi}{25}\right)-\frac{1}{50}\left(\frac{\partial}{\partial y}P(x,y)\right)b^2\sin\left(\frac{\pi}{25}\right)\cos\left(\frac{\pi}{25}\right)$$ (32)

// Owing to universality and homogeneity of abstract differential

function $\dfrac{\partial Q(x,y)}{\partial x}-\dfrac{\partial P(x,y)}{\partial y}$, if this abstract differential function is

continuous in plane bounded closed region 'S', then this abstract

differential function still can be described as $\dfrac{\partial Q(x,y)}{\partial x}-\dfrac{\partial P(x,y)}{\partial y}$ in one

or more segment cells of plane bounded closed region 'S'.

See also Chapter 4 (Part I) Section 4.3 'Discuss About Abstract Vector Field、Finite Sums Limit、Integral and Riemann Sums'

Calculate Microcosmic Double Integrals Value in plane bounded closed region 'S's first segmental cell (33):

Base on integral median theorem, product of Abstract differential function (32) and Area Element Coefficient (31), times segment unit of 'r,u'(30), that is Microcosmic Double Integrals Value in first segmental cell (33)

$$\frac{1}{62500}\left(-\frac{1}{50}\left(\frac{\partial}{\partial x}Q(x,y)\right)a^2\cos\left(\frac{\pi}{25}\right)\sin\left(\frac{\pi}{25}\right)-\frac{1}{50}\left(\frac{\partial}{\partial y}P(x,y)\right)b^2\sin\left(\frac{\pi}{25}\right)\cos\left(\frac{\pi}{25}\right)\right)$$
$$a\,b\,\pi$$

6.Microcosmic double integrals course in plane bounded closed region 'S's all (viz. 50) segmental cells:

Segment range [0,1] of 'r': dr = $\dfrac{j}{50}$

Segment range [0,2π] of 'u': du = $\dfrac{2i\pi}{50}=\dfrac{i\pi}{25}$ (34)

(thereinto, 'i,j' stand for natural number 1~50)

Segment area element coefficient (35): [viz. Input (34) to (27)]

(Area element coefficient (27), average values that correspond to plane bounded closed region 'S's all segmental cells)

$$\frac{a\,b\,j}{50}$$

(35)

// Expression (35) isn't that 'single value', but that 'Set of finite quantitative (viz. 50) values'

Segment abstract differential function (36):

[viz. Input (34) to (28)]

(Abstract differential function (28), average values that correspond to plane bounded closed region 'S's all(viz. 20)segmental cells)

$$-\frac{1}{50}\left(\frac{\partial}{\partial x}Q(x,y)\right)a^2\cos\left(\frac{\pi i}{25}\right)j\sin\left(\frac{\pi i}{25}\right)-\frac{1}{50}\left(\frac{\partial}{\partial y}P(x,y)\right)b^2\sin\left(\frac{\pi i}{25}\right)j\cos\left(\frac{\pi i}{25}\right)$$

// Expression (36) isn't that 'single value', but that 'Set of finite quantitative (viz. 50) values'

// Owing to universality and homogeneity of abstract differential function $\dfrac{\partial Q(x,y)}{\partial x}-\dfrac{\partial P(x,y)}{\partial y}$, if this abstract differential function is continuous in plane bounded closed region 'S', then this abstract differential function still can be described as $\dfrac{\partial Q(x,y)}{\partial x}-\dfrac{\partial P(x,y)}{\partial y}$ in one or more segment cells of plane bounded closed region 'S'.

See also Chapter 4 (Part I) Section 4.3 'Discuss About Abstract Vector Field、Finite Sums Limit、Integral and Riemann Sums'

Calculate Microcosmic Double Integrals Values in Plane Bounded Closed Region 'S's all segmental cells (37):

Base on integral median theorem, product of Abstract differential function (36) and Area Element Coefficient (35), times segment unit of 'r,u' (34), these are microcosmic double integrals values in all (viz. 50) segmental cells (37)

$$\frac{1}{62500}$$

$$\left(-\frac{1}{50}\left(\frac{\partial}{\partial x} Q(x,y) \right) a^2 \cos\left(\frac{\pi\, i}{25} \right) j \sin\left(\frac{\pi\, i}{25} \right) - \frac{1}{50}\left(\frac{\partial}{\partial y} P(x,y) \right) b^2 \sin\left(\frac{\pi\, i}{25} \right) j \cos\left(\frac{\pi\, i}{25} \right) \right)$$

$$a\, b\, j\, \pi$$

// 'Integral Value' (37) isn't that 'single value', but that 'Set of finite quantitative (viz. 50) values'

Structure sequence that is composed of finite quantitative (viz. 50) microcomosic double integrals values (38):

[Lengthy expression of sequence is elided;

In program template of Waterloo Maplesoft (see also Section 4.2):

sqn:=seq(seq(ijddPV2*ijdJ*du*dr,i=1..dus),j=1..dus):

change ':' (last) to ';', then obtain expression of sequence]

Accumulational sum of sequence (viz. Sum of integral values of product [Abstract differential function (36) and Volume Element at all (viz. 50) segmental cells], obtain double integrals value at manifold (39):

[Lengthy expression of accumulational sum result is elided;

In program template of Waterloo Maplesoft (see also Section 4.2):

add(k,k=sqn):omega:=evalf(%):

change ':' (middle and last) to ';', then obtain expression and float value of result]

 // Amount of plane bounded closed region 'S's (Curve 'L' surrounds) segmental cells increases infinitely, sum of integral values above-mentioned tends to '0' infinitely

Set Amount of plane bounded closed region 'S's (Curve 'L' surrounds) segmental cells as uncertain natural number 'n'   (40)

7.Microcosmic Double Integrals Course in plane bounded closed region 'S's first segmental cell:

Segment range [0,1] of 'r': dr = $\dfrac{1}{n}$

Segment range [0,2π] of 'u': du = $\dfrac{2\pi}{n}$ (41)

Segment area element coefficient 'J' (42): [viz. Input (41) to (27)]

(Area element coefficient (27),average value that corresponds to plane bounded closed region 'S's first segmental cell)

$$\dfrac{a\,b}{n}$$ (42)

Segment abstract differential function (43):

[viz. Input (41) to (28)]

(Abstract differential function (28), average value that corresponds to plane bounded closed region 'S's first segmental cell)

$$-\dfrac{\left(\dfrac{\partial}{\partial x}Q(x,y)\right)a^2\cos\left(\dfrac{2\pi}{n}\right)\sin\left(\dfrac{2\pi}{n}\right)}{n}-\dfrac{\left(\dfrac{\partial}{\partial y}P(x,y)\right)b^2\sin\left(\dfrac{2\pi}{n}\right)\cos\left(\dfrac{2\pi}{n}\right)}{n}$$ (43)

// Owing to universality and homogeneity of abstract differential function $\dfrac{\partial Q(x,y)}{\partial x}-\dfrac{\partial P(x,y)}{\partial y}$, if this abstract differential function is continuous in plane bounded closed region 'S', then this abstract differential function still can be described as $\dfrac{\partial Q(x,y)}{\partial x}-\dfrac{\partial P(x,y)}{\partial y}$ in one or more segment cells of plane bounded closed region 'S'.

See also Chapter 4 (Part I) Section 4.3 'Discuss About Abstract Vector Field、Finite Sums Limit、Integral and Riemann Sums'

Calculate microcosmic double integrals value in plane bounded closed region 'S's first segmental cell (44):

Base on integral median theorem, product of Abstract differential function (43) and Area Element Coefficient (42), times segment unit of 'r,u'(41), that is microcosmic double integrals value in first segmental cell (44)

$$2\frac{\left(-\dfrac{\left(\dfrac{\partial}{\partial x}Q(x,y)\right)a^2\cos\left(\dfrac{2\pi}{n}\right)\sin\left(\dfrac{2\pi}{n}\right)}{n}-\dfrac{\left(\dfrac{\partial}{\partial y}P(x,y)\right)b^2\sin\left(\dfrac{2\pi}{n}\right)\cos\left(\dfrac{2\pi}{n}\right)}{n}\right)ab\pi}{n^3}$$

8.Microcosmic double integrals course in plane bounded closed region 'S's all (viz. n)segmental cells:

　　Segment range [0,1] of 'r':　dr = $\dfrac{j}{n}$

　　Segment range [0,2π] of 'u':　du = $\dfrac{2i\pi}{n}$

　　(thereinto, 'i,j' stand for natural number 1~n)　　　(45)

　　Segment area element coefficient (46): [viz. Input (45) to (27)]

　　(Area element coefficient (27), average values that correspond to plane bounded closed region 'S's all (viz. n) segmental cells)

$$\frac{abj}{n}$$

(46)

　//Expression (46) isn't that 'single value', but that 'Set of finite quantitative (viz. n) values'

　　Segment abstract differential function (47):

　　[viz. Input (45) to (28)]

　　(Abstract differential function (28), average values that correspond to plane bounded closed region 'S's all (viz. n) segmental cells)

$$-\frac{\left(\dfrac{\partial}{\partial x}Q(x,y)\right)a^2\cos\left(\dfrac{2\pi i}{n}\right)j\sin\left(\dfrac{2\pi i}{n}\right)}{n}-\frac{\left(\dfrac{\partial}{\partial y}P(x,y)\right)b^2\sin\left(\dfrac{2\pi i}{n}\right)j\cos\left(\dfrac{2\pi i}{n}\right)}{n}$$

　// Expression (47) isn't that 'single value', but that 'Set of finite quantitative (viz. n) values'

// Owing to universality and homogeneity of abstract differential

function $\dfrac{\partial Q(x,y)}{\partial x}-\dfrac{\partial P(x,y)}{\partial y}$, if this abstract differential function is

continuous in plane bounded closed region 'S', then this abstract

differential function still can be described as $\dfrac{\partial Q(x,y)}{\partial x}-\dfrac{\partial P(x,y)}{\partial y}$ in one or more segment cells of plane bounded closed region 'S'.

See also Chapter 4 (Part I) Section 4.3 'Discuss About Abstract Vector Field、Finite Sums Limit、Integral and Riemann Sums'

Calculate microcosmic double integrals values in plane bounded closed region 'S's all segmental cells (48):

Base on integral median theorem, product of Abstract differential function (47) and Area Element Coefficient (46), times segment unit of 'r,u'(45), these are microcosmic double integrals values in all (viz. n) segmental cells (48)

$$2\left(-\frac{\left(\frac{\partial}{\partial x}Q(x,y)\right)a^2\cos\left(\frac{2\pi i}{n}\right)j\sin\left(\frac{2\pi i}{n}\right)}{n}-\frac{\left(\frac{\partial}{\partial y}P(x,y)\right)b^2\sin\left(\frac{2\pi i}{n}\right)j\cos\left(\frac{2\pi i}{n}\right)}{n}\right)$$
$$a\,b\,j\,\pi\,/\,n^3$$

// Expression (48) isn't that 'single integral value', but that 'Set of finite quantitative (viz. n) integral values'

Structure finite sums (49):

$$\sum_{j=1}^{n}\left(\sum_{i=1}^{n}\left(2\left(-\frac{\left(\frac{\partial}{\partial x}Q(x,y)\right)a^2\cos\left(\frac{2\,i\,\pi}{n}\right)j\sin\left(\frac{2\,i\,\pi}{n}\right)}{n}\right.\right.\right.$$
$$\left.\left.\left.-\frac{\left(\frac{\partial}{\partial y}P(x,y)\right)b^2\sin\left(\frac{2\,i\,\pi}{n}\right)j\cos\left(\frac{2\,i\,\pi}{n}\right)}{n}\right)a\,j\,b\,\pi\,/\,n^3\right)\right)$$

(49)

Infinitize finite sums, its limit operational value is 'Double Integrals Value in Manifold'(50):

(In the case of "Amount 'n' of plane bounded closed region 'S's (Curve 'L' surrounds) segmental cells tends to infinity", limit value of integral values' sums [product about Abstract differential function (47) and Volume Element] at all segmental cells)

$$\lim_{n\to\infty}\sum_{j=1}^{n}\left(\sum_{i=1}^{n}\left(2\left(\right.\right.\right.$$
$$-\frac{\left(\frac{\partial}{\partial x}Q(x,y)\right)a^2\cos\left(\frac{2\pi i}{n}\right)j\sin\left(\frac{2\pi i}{n}\right)}{n}-\frac{\left(\frac{\partial}{\partial y}P(x,y)\right)b^2\sin\left(\frac{2\pi i}{n}\right)j\cos\left(\frac{2\pi i}{n}\right)}{n}$$

$$\Bigg) a\,b\,j\,\pi\,/\,n^3\Bigg)\Bigg)$$

$$= 0 \hspace{8cm} (50)$$

Thereinto, Suppose: $\displaystyle\lim_{n\to\infty}\sum_{j=1}^{n}\left(\sum_{i=1}^{n}\left(\frac{2\,a\,j\,b\,\pi}{n^3}\right)\right)\neq 0$

Viz. Suppose sum of double integrals values (Curve 'L' coordinates area element itself) at all segmental cells can't be '0'; Explain visually, it can be regarded as 'Suppose plane bounded closed region 'S' can't be '0 area''

Viz. In the case of 'n →∞', **(24)=(49)**,

$$\sum_{i=1}^{n}\left(\frac{2\,\pi\left(-\mathrm{P}(x,y)\,a\,\sin\!\left(\dfrac{2\,i\,\pi}{n}\right)+\mathrm{Q}(x,y)\,b\,\cos\!\left(\dfrac{2\,i\,\pi}{n}\right)\right)}{n}\right)$$

$$=$$

$$\sum_{j=1}^{n}\left(\sum_{i=1}^{n}\left(2\left(-\frac{\left(\dfrac{\partial}{\partial x}Q(x,y)\right)a^2\cos\!\left(\dfrac{2\,i\,\pi}{n}\right)j\sin\!\left(\dfrac{2\,i\,\pi}{n}\right)}{n}\right.\right.\right.$$

$$\left.\left.\left.-\frac{\left(\dfrac{\partial}{\partial y}\mathrm{P}(x,y)\right)b^2\sin\!\left(\dfrac{2\,i\,\pi}{n}\right)j\cos\!\left(\dfrac{2\,i\,\pi}{n}\right)}{n}\right)a\,j\,b\,\pi\,/\,n^3\right)\right)$$

Equation above-mentioned can be described as:

$$\int_{L+} A\cdot dL = \iint_{S}\left(\frac{\partial}{\partial x}Q(x,y)-\frac{\partial}{\partial y}P(x,y)\right)dS \quad \textbf{(1)}, \text{ Complete Proof.}$$

4.2 Prove Green Theorem Finite Sums Limits at Manifold [Program Template of Waterloo Maplesoft, Optional]

Green Theorem Suppose boundary curve 'L' of plane bounded closed region 'S' is composed by numbered smooth or piecewise smooth curves.

If exist 1th order continuous partial derivatives about functions 'P(x,y),Q(x,y)' [Structure plane vector field 'A'] at plane bounded closed region 'S', then:

$$\int_{L+} A \cdot dL = \iint_S \left(\frac{\partial}{\partial x} Q(x,y) - \frac{\partial}{\partial y} P(x,y) \right) dS \qquad (1)$$

Symbol System:
Abstract simply connected closed curve 'CO'(Plane),

Plane Vector Field 'PV',

Differential function 'dPV1,dPV2' of Plane Vector Field 'PV',

Plane bounded closed region 'S' that closed curve 'CO' surrounds,

General expression of curve 'CO' coordinates area element coefficient 'J';

Amount 'dus' of plane closed curve 'CO's parameter segmental cells (It can be discretional natural number),

Segmental range 'dt' of parameter 't',

Tangent vector 'dCO' of plane closed curve 'CO';

Plane vector field 'PV's average value 'dPVm' that corresponds to plane closed curve 'CO's first segmental cell,

Tangent vector 'dCO's average value 'dCOm' that corresponds to plane closed curve 'CO's first segmental cell;

Plane vector field 'PV's average values 'idPV' that correspond to plane closed curve 'CO's all segmental cells,

Tangent vector 'dCO's average values 'idCO' that correspond to plane closed curve 'CO's all segmental cells;

(In actual expressions, 'i' stands for natural number)

Amount 'dus' of plane bounded closed region 'S's parameter segmental cells (It can be discretional natural number),

Segmental range 'dr' of parameter 'r',

Segmental range 'du' of parameter 'u';

Area element coefficient 'J's average value 'dJ' that corresponds to plane bounded closed region 'S's first segmental cell,

Differential function 'dPV2's average value 'ddPV2' that corresponds to plane bounded closed region 'S's first segmental cell;

Area element coefficient 'J's average values 'ijdJ' that correspond to plane bounded closed region 'S's all segmental cells,

Differential function 'dPV2's average values 'ijddPV2' that correspond to plane bounded closed region 'S's all segmental cells

(In actual expressions, 'i,j' stand for natural number)

```
> restart;
> with(linalg):
```
Define parametrized expression of abstract (plane) simply connected closed curve 'CO':
```
> CO:=[a*cos(t),b*sin(t)];        (2)
```
$$CO := [a\cos(t), b\sin(t)]$$

Thereinto, 'a,b' are nonzero constants or 1th order derivable continuous functions, simply connected closed curve 'CO' determines value of 'a,b'
```
> rgt:=[0,2*Pi];
```
Set range of parameters 't' in $[0, 2\pi]$, make boundary curve 'CO' closed. (See also Poincare Conjecture: "Every closed n-Manifold which is homotopy equivalent to n-Sphere is homeomorphic to n-Sphere"[19], and Chapter 1 this Part 'Constitute Abstract Simply Connected Closed Parametrized Curve Coordinates [Plane]')
$$rgt := [0, 2\pi]$$
```
>[Diff(CO[1],t),Diff(CO[2],t)]=[diff(CO[1],t),diff(CO[2],t)];
dCO:=rhs(%);
```
Calculate tangent vector 'dCO' of plane closed curve 'CO'
$$\left[\frac{\partial}{\partial t}(a\cos(t)), \frac{\partial}{\partial t}(b\sin(t))\right] = [-a\sin(t), b\cos(t)]$$
$$dCO := [-a\sin(t), b\cos(t)]$$

```
> PV:=[(P)(x,y),(Q)(x,y)];        (3)
```
Define abstract plane vector field 'PV' (Suppose Vector Field 'PV' possesses 1th order continuous partial derivatives in plane bounded closed region 'S' that curve 'CO' surrounds)
$$PV := [P(x,y), Q(x,y)]$$

```
> Diff(PV[2],x)-Diff(PV[1],y)=diff(PV[2],x)-diff(PV[1],y);

dPV1:=rhs(%);        (4)
```

\# Calculate differential function 'dPV1' of plane vector field 'PV'

$$\left(\frac{\partial}{\partial x}Q(x,y)\right)-\left(\frac{\partial}{\partial y}P(x,y)\right)=\left(\frac{\partial}{\partial x}Q(x,y)\right)-\left(\frac{\partial}{\partial y}P(x,y)\right)$$

$$dPV1:=\left(\frac{\partial}{\partial x}Q(x,y)\right)-\left(\frac{\partial}{\partial y}P(x,y)\right)$$

```
> x:=CO[1];y:=CO[2];
```

$$x:=a\cos(t)$$
$$y:=b\sin(t)$$

// Segment range '[0,2*Pi]' of 't':

```
> dus:=50;
```

\# Define amount 'dus' of plane closed curve 'CO's parameter segmental cells (It can be discretional natural number)

$$dus:=50$$

```
> dt:=(rgt[2]-rgt[1])/dus; # Segment range '[0,2*Pi]' of 't'
```

$$dt:=\frac{\pi}{25}$$

```
> x:='x':y:='y':
```

// Transform variables, prevent that 'Import 'x=a*cos(t),y=b*sin(t)' to abstract vector field '[P(x,y),Q(x,y)]''

Microcosmic curve integral course of plane closed curve 'CO's first segmental cell (50 segmental cells):

// Segment plane vector field 'PV':

(This step possesses formal meaning in 'Proof' only, and possesses essential meaning in 'Numerical Models')

```
> dPVm:=subs(t=rgt[1]+dt,PV);
```

\# Plane vector field 'PV's average value that corresponds to plane closed curve 'CO's first segmental cell

$$dPVm:=[P(x,y),Q(x,y)]$$

// Segment tangent vector 'dCO':

```
> dCOm:=subs(t=rgt[1]+dt,dCO);
```

\# Tangent vector 'dCO's average value that corresponds to plane closed curve 'CO's first segmental cell

$$dCOm := \left[-a\sin\left(\frac{\pi}{25}\right), b\cos\left(\frac{\pi}{25}\right) \right]$$

// Calculate microcosmic curve integral value of plane closed curve 'CO's first segmental cell:

> dt*(dPVm[1]*dCOm[1]+dPVm[2]*dCOm[2]);

Base on integral median theorem, integral value of dot product(plane vector field 'V' and tangent vector 'dCO') at first segmental cell

$$\frac{1}{25}\pi\left(-\mathrm{P}(x,y)\,a\sin\left(\frac{\pi}{25}\right) + \mathrm{Q}(x,y)\,b\cos\left(\frac{\pi}{25}\right) \right)$$

Microcosmic curve integral course of plane closed curve 'CO's all segmental cells (50 segmental cells):

// Segment plane vector field 'PV':

(This step possesses formal meaning in 'Proof' only, and possesses essential meaning in 'Numerical Models')

> idPV:=subs(t=rgt[1]+i*dt,PV);

Plane vector field 'PV's average values that correspond to plane closed curve 'CO's all segmental cells,

$$idPV := [\mathrm{P}(x,y), \mathrm{Q}(x,y)]$$

> seq(([i,idPV,evalf(idPV)]),i=1..dus):

List of plane vector field 'PV's average values that correspond to plane closed curve 'CO's all segmental cells, be elided

// Segment tangent vector 'dCO':

> idCO:=subs(t=rgt[1]+i*dt,dCO);

Tangent vector 'dCO's average values that correspond to plane closed curve 'CO's all segmental cells

$$idCO := \left[-a\sin\left(\frac{i\pi}{25}\right), b\cos\left(\frac{i\pi}{25}\right) \right]$$

> seq(([i,idCO,evalf(idCO)]),i=1..dus):

List of tangent vector 'dCO's average values that correspond to plane closed curve 'CO's all segmental cells

// In actual expressions,'i'stands for natural number 1~50

//'idCO' isn't that 'single vector value', but that 'set of finite quantitative (viz.50) vector values'

// Calculate microcosmic curve integral values of plane closed curve

- 302 -

```
'CO's all segmental cells:
> dt*(idPV[1]*idCO[1]+idPV[2]*idCO[2]);
# Base on integral median theorem, integral values of dot product (Plane
vector field 'PV' and tangent vector 'dCO') at all segmental cells
```

$$\frac{1}{25}\pi\left(-\mathrm{P}(x,y)\,a\sin\left(\frac{i\,\pi}{25}\right)+\mathrm{Q}(x,y)\,b\cos\left(\frac{i\,\pi}{25}\right)\right)$$

```
//Expression above-mentioned isn't that 'single value', but that
'set of finite quantitative (viz.50) values'


// List of finite quantitative microcosmic curl integral values:
>seq(([i,dt*(idPV[1]*idCO[1]+idPV[2]*idCO[2]),
evalf(dt*(idPV[1]*idCO[1]+idPV[2]*idCO[2]))]),i=1..dus):
# List of integral values of dot product(Plane vector field 'PV' and
tangent vector 'dCO') at all segmental cells, be elided
// Structure sequence that is composed of finite quantitative
microcomosic curl integral values:
> sqn:=seq(dt*(idPV[1]*idCO[1]+idPV[2]*idCO[2]),i=1..dus):
// Accumulational sum of sequence, obtain plane closed curve integral
value at manifold:
> add(k,k=sqn):xi:=evalf(%);          (5)
# Sum of integral values of dot product(Plane vector field 'PV' and
tangent vector 'dCO') at all segmental cells; Amount 'dus' of plane
closed curve 'CO's segmental cells increases infinitely, sum of
integral values above-mentioned tends to '0' infinitely (Verbose
analytical expression is elided)
```

$$\xi := -0.20\,10^{-9}\,\mathrm{P}(x,y)\,a$$

```
// Renewedly segment range '[0,2*Pi]' of 't':
> dus:=n;
# Set amount 'dus' of plane closed curve 'CO's parameter segmental
cells as natural number 'n'
```

$$dus := n$$

```
> dt:=(rgt[2]-rgt[1])/dus; # Segment range '[0,2*Pi]' of 't'
```

$$dt := \frac{2\,\pi}{n}$$

Microcosmic curve integral course of plane closed curve 'CO's first segmental cell (n segmental cells):

// Segment plane vector field 'PV':

(This step possesses formal meaning in 'Proof' only, and possesses essential meaning in 'Numerical Models')

```
> dPVm:=subs(t=rgt[1]+dt,PV);
```

Plane vector field 'PV's average value that corresponds to plane closed curve 'CO's first segmental cell

$$dPVm := [\mathrm{P}(x,y), \mathrm{Q}(x,y)]$$

// Segment tangent vector 'dCO':

```
> dCOm:=subs(t=rgt[1]+dt,dCO);
```

Tangent vector 'dCO's average value that corresponds to plane closed curve 'CO's first segmental cell

$$dCOm := \left[-a\sin\left(\frac{2\pi}{n}\right), b\cos\left(\frac{2\pi}{n}\right)\right]$$

// Calculate microcosmic curve integral value of plane closed curve 'CO's first segmental cell:

```
> dt*(dPVm[1]*dCOm[1]+dPVm[2]*dCOm[2]);
```

Base on integral median theorem, integral value of dot product (plane vector field 'V' and tangent vector 'dCO') at first segmental cell

$$\frac{2\pi\left(-\mathrm{P}(x,y)\,a\sin\left(\frac{2\pi}{n}\right) + \mathrm{Q}(x,y)\,b\cos\left(\frac{2\pi}{n}\right)\right)}{n}$$

Microcosmic curve integral course of plane closed curve 'CO's all segmental cells (n segmental cells):

// Segment plane vector field 'PV':

(This step possesses formal meaning in 'Proof' only, and possesses essential meaning in 'Numerical Models')

```
> idPV:=subs(t=rgt[1]+i*dt,PV);
```

Plane vector field 'PV's average values that correspond to plane closed curve 'CO's all segmental cells

$$idPV := [\mathrm{P}(x,y), \mathrm{Q}(x,y)]$$

// Segment tangent vector 'dCO':

```
> idCO:=subs(t=rgt[1]+i*dt,dCO);
```

\# Tangent vector 'dCO's average values that correspond to plane closed curve 'CO's all segmental cells

$$idCO := \left[-a \sin\left(\frac{2\,i\,\pi}{n} \right),\, b \cos\left(\frac{2\,i\,\pi}{n} \right) \right]$$

// In actual expressions, 'i' stands for natural number 1~n

// 'idCO' isn't that 'single vector value', but that 'set of finite quantitative (viz. n) vector values'

// Calculate microcosmic curve integral values of plane closed curve 'CO's all segmental cells:

```
> dt*(idPV[1]*idCO[1]+idPV[2]*idCO[2]);
```

\# Base on integral median theorem, integral values of dot product(Plane vector field 'PV' and tangent vector 'dCO') at all segmental cells

$$\frac{2\,\pi\left(-P(x,y)\,a\sin\left(\frac{2\,i\,\pi}{n}\right) + Q(x,y)\,b\cos\left(\frac{2\,i\,\pi}{n}\right) \right)}{n}$$

//Expression above-mentioned isn't that 'single value', but that 'set of finite quantitative(viz.n)values'

// Structure finite sums:

```
> Sum(dt*(idPV[1]*idCO[1]+idPV[2]*idCO[2]),i=1..dus);        (6)
```

\# In the case of "Amount 'dus' of plane closed curve 'CO's parameter segmental cells is pending", sum of integral values (dot product about plane vector field 'PV' and tangent vector 'dCO') at all segmental cells

$$\sum_{i=1}^{n} \left(\frac{2\,\pi\left(-P(x,y)\,a\sin\left(\frac{2\,i\,\pi}{n}\right) + Q(x,y)\,b\cos\left(\frac{2\,i\,\pi}{n}\right) \right)}{n} \right)$$

```
> vs:=value(%):
```

```
> Limit(vs,n=infinity);
```

// Infinitize finite sums, its limit operational value is 'Closed Curve Integral Value at Manifold'

\# In the case of "Amount 'n' of plane closed curve 'CO's segmental cells tends to infinity", limit value of integral values' sums (Dot product about plane vector field 'PV' and tangent vector 'dCO') at all segmental cells

$$\lim_{n \to \infty} -2\pi \cos\left(\frac{\pi}{n}\right)^2 \left(P(x,y)\, a \sin\left(\frac{\pi}{n}\right) \cos\left(\frac{\pi}{n}\right)^3 - P(x,y)\, a \sin\left(\frac{\pi}{n}\right) \cos\left(\frac{\pi}{n}\right) \right.$$

$$+ Q(x,y)\, b \cos\left(\frac{\pi}{n}\right)^4 - 2\, Q(x,y)\, b \cos\left(\frac{\pi}{n}\right)^2 + Q(x,y)\, b$$

$$+ \sin\left(\frac{\pi}{n}\right)^3 \cos\left(\frac{\pi}{n}\right) P(x,y)\, a + \sin\left(\frac{\pi}{n}\right)^2 \cos\left(\frac{\pi}{n}\right)^2 Q(x,y)\, b - \sin\left(\frac{\pi}{n}\right)^2 Q(x,y)\, b \Bigg)$$

$$(n+1) \Bigg/ \left(\left(\cos\left(\frac{\pi}{n}\right)^2 - 1 \right)^2 n \right) - 2\pi \left(\sin\left(\frac{\pi}{n}\right)^2 \cos\left(\frac{\pi}{n}\right)^2 P(x,y)\, a \right.$$

$$+ \sin\left(\frac{\pi}{n}\right) \cos\left(\frac{\pi}{n}\right)^3 Q(x,y)\, b - \sin\left(\frac{\pi}{n}\right) \cos\left(\frac{\pi}{n}\right) Q(x,y)\, b + P(x,y)\, a \cos\left(\frac{\pi}{n}\right)^2$$

$$\left. - P(x,y)\, a \right) \sin\left(\frac{(n+1)\pi}{n}\right) \cos\left(\frac{(n+1)\pi}{n}\right) \Bigg/ \left(\left(\cos\left(\frac{\pi}{n}\right)^2 - 1 \right)^2 n \right) -$$

$$\frac{2\pi \left(P(x,y)\, a \sin\left(\frac{\pi}{n}\right) \cos\left(\frac{\pi}{n}\right) + Q(x,y)\, b \cos\left(\frac{\pi}{n}\right)^2 - Q(x,y)\, b \right) \cos\left(\frac{(n+1)\pi}{n}\right)^2}{n \left(\cos\left(\frac{\pi}{n}\right)^2 - 1 \right)}$$

$$+ 2\pi \cos\left(\frac{\pi}{n}\right)^2 \left(P(x,y)\, a \sin\left(\frac{\pi}{n}\right) \cos\left(\frac{\pi}{n}\right)^3 - P(x,y)\, a \sin\left(\frac{\pi}{n}\right) \cos\left(\frac{\pi}{n}\right) \right.$$

$$+ Q(x,y)\, b \cos\left(\frac{\pi}{n}\right)^4 - 2\, Q(x,y)\, b \cos\left(\frac{\pi}{n}\right)^2 + Q(x,y)\, b$$

$$+ \sin\left(\frac{\pi}{n}\right)^3 \cos\left(\frac{\pi}{n}\right) P(x,y)\, a + \sin\left(\frac{\pi}{n}\right)^2 \cos\left(\frac{\pi}{n}\right)^2 Q(x,y)\, b - \sin\left(\frac{\pi}{n}\right)^2 Q(x,y)\, b \Bigg)$$

$$\Bigg/ \left(\left(\cos\left(\frac{\pi}{n}\right)^2 - 1 \right)^2 n \right) + 2\pi \left(\sin\left(\frac{\pi}{n}\right)^2 \cos\left(\frac{\pi}{n}\right)^2 P(x,y)\, a \right.$$

$$+ \sin\left(\frac{\pi}{n}\right) \cos\left(\frac{\pi}{n}\right)^3 Q(x,y)\, b - \sin\left(\frac{\pi}{n}\right) \cos\left(\frac{\pi}{n}\right) Q(x,y)\, b + P(x,y)\, a \cos\left(\frac{\pi}{n}\right)^2$$

$$\left. - P(x,y)\, a \right) \sin\left(\frac{\pi}{n}\right) \cos\left(\frac{\pi}{n}\right) \Bigg/ \left(\left(\cos\left(\frac{\pi}{n}\right)^2 - 1 \right)^2 n \right)$$

$$+ \frac{2\pi \left(P(x,y)\, a \sin\left(\frac{\pi}{n}\right) \cos\left(\frac{\pi}{n}\right) + Q(x,y)\, b \cos\left(\frac{\pi}{n}\right)^2 - Q(x,y)\, b \right) \cos\left(\frac{\pi}{n}\right)^2}{n \left(\cos\left(\frac{\pi}{n}\right)^2 - 1 \right)}$$

```
> delta:=value(%);
```
$$\delta := 0$$

```
> x:='x':y:='y':
> COc:=subs(t=u,CO);
# Change character 't' to 'u' in expression of curve 'CO'
```
$$COc := [a\cos(u), b\sin(u)]$$

```
> [r*COc[1],r*COc[2]];        (7)
```
Entirely multiply radius vector 'r' (suppose r>0) by each item of curve 'CO' parametrized expression, convert 'Parametrized expression of curve' to 'Parametrized expression of curve coordinates'
$$[r\,a\cos(u), r\,b\sin(u)]$$

```
> x:=r*COc[1]:y:=r*COc[2]:
> matrix(2,2,[Diff(r*COc[1],r),Diff(r*COc[1],u),Diff(r*COc[2],r),
Diff(r*COc[2],u)])=
matrix(2,2,[diff(r*COc[1],r),diff(r*COc[1],u),diff(r*COc[2],r),
diff(r*COc[2],u)]);m:=rhs(%);
```
Define and calculate matrix of partial derivatives 'm', obtain general expression of curve coordinates area element coefficient

$$\begin{bmatrix} \dfrac{\partial}{\partial r}(r\,a\cos(u)) & \dfrac{\partial}{\partial u}(r\,a\cos(u)) \\ \dfrac{\partial}{\partial r}(r\,b\sin(u)) & \dfrac{\partial}{\partial u}(r\,b\sin(u)) \end{bmatrix} = \begin{bmatrix} a\cos(u) & -r\,a\sin(u) \\ b\sin(u) & r\,b\cos(u) \end{bmatrix}$$

$$m := \begin{bmatrix} a\cos(u) & -r\,a\sin(u) \\ b\sin(u) & r\,b\cos(u) \end{bmatrix}$$

```
> det(m);
```

$$a\cos(u)^2\,r\,b + r\,a\sin(u)^2\,b$$

```
> J:=simplify(%);        (8)
```
General expression of curve coordinates area element coefficient
$$J := a\,r\,b$$

```
> x:='x':y:='y':
// Transform variables, prevent that "Import 'x=r*a*cos(u),
y=r*b*sin(u)' to abstract vector field '[P(x,y),Q(x,y)]'"
>Diff(PV[2],x)*Diff(r*COc[1],r)*Diff(r*COc[1],u)
-Diff(PV[1],y)*Diff(r*COc[2],r)*Diff(r*COc[2],u);dPV2:=value(%);
```
Convert 'dPV1' from Cartesian coordinates expression to Curve

coordinates expression   **(9)**

$$\left(\frac{\partial}{\partial x}Q(x,y)\right)\left(\frac{\partial}{\partial r}(r\,a\cos(u))\right)\left(\frac{\partial}{\partial u}(r\,a\cos(u))\right)$$

$$-\left(\frac{\partial}{\partial y}P(x,y)\right)\left(\frac{\partial}{\partial r}(r\,b\sin(u))\right)\left(\frac{\partial}{\partial u}(r\,b\sin(u))\right)$$

$$dPV2:=-\left(\frac{\partial}{\partial x}Q(x,y)\right)a^2\cos(u)\,r\sin(u)-\left(\frac{\partial}{\partial y}P(x,y)\right)b^2\sin(u)\,r\cos(u)$$

// Segment range `[0,1],[0,2*Pi]` of `r,u`:

> dus:=50;

\# Set amount of plane bounded closed region `S`s (Plane closed curve `CO` surrounds) parameter segmental cells, this amount can be discretional natural number

$$dus:=50$$

> rgr:=[0,1];

// Be different to ordinary double integrals, in this position, first integral range is `r[0,1]` always, and not `r[0,n]` or r[0,∞]`; Because of `1`,it is possible to hold a correct ratio between `a,b,r,u`

$$rgr:=[0,1]$$

> dr:=(rgr[2]-rgr[1])/dus; \# Segment range of `r`

$$dr:=\frac{1}{50}$$

> du:=(rgt[2]-rgt[1])/dus; \# Segment range of `u`

$$du:=\frac{\pi}{25}$$

\# **Microcosmic double integrals course of plane bounded closed region`S`s first segmental cell (50 segmental cells):**

// Segment area element coefficient `J`:

> dJ:=subs(r=rgr[1]+dr,subs(u=rgt[1]+du,J));

\# Plane area element coefficient `J`s average value `dJ` that corresponds to plane bounded closed region `S`s first segmental cell

$$dJ:=\frac{a\,b}{50}$$

// Segment abstract differential function `dPV2`:

> ddPV2:=subs(r=rgr[1]+dr,subs(u=rgt[1]+du,dPV2));

\# Differential function `dPV2`s average value that corresponds to plane bounded closed region `S`s first segmental cell

$$ddPV2 := -\frac{1}{50}\left(\frac{\partial}{\partial x}Q(x,y)\right)a^2\cos\left(\frac{\pi}{25}\right)\sin\left(\frac{\pi}{25}\right) - \frac{1}{50}\left(\frac{\partial}{\partial y}P(x,y)\right)b^2\sin\left(\frac{\pi}{25}\right)\cos\left(\frac{\pi}{25}\right)$$

```
// Calculate microcosmic double integrals value of plane bounded
closed region'S's first segmental cell:
> ddPV2*dJ*du*dr;
# Base on integral median theorem, integral value of product
(Differential function 'dPV2' and Curve coordinates area element) in
first segmental cell
```

$$\frac{1}{62500}\left(-\frac{1}{50}\left(\frac{\partial}{\partial x}Q(x,y)\right)a^2\cos\left(\frac{\pi}{25}\right)\sin\left(\frac{\pi}{25}\right) - \frac{1}{50}\left(\frac{\partial}{\partial y}P(x,y)\right)b^2\sin\left(\frac{\pi}{25}\right)\cos\left(\frac{\pi}{25}\right)\right)$$
$$a\,b\,\pi$$

Microcosmic double integrals course of plane bounded closed region'S's all segmental cells (50 segmental cells):

```
// Segment area element coefficient 'J':
> ijdJ:=subs(r=rgr[1]+j*dr,subs(u=rgt[1]+i*du,J));
# Area element coefficient 'J's average values that correspond to plane
bounded closed region 'S's all segmental cells
```

$$ijdJ := \frac{a\,j\,b}{50}$$

```
// In actual expressions,'i,j'stand for natural number 1~50
//'ijdJ' isn't that 'single value', but that 'set of finite
quantitative (viz. 50) values'


// Segment abstract differential function 'dPV2':
> ijddPV2:=subs(r=rgr[1]+j*dr,subs(u=rgt[1]+i*du,dPV2));
# Differential function 'dPV2's average values  that correspond to
plane bounded closed region 'S's all segmental cells
```

$$ijddPV2 :=$$
$$-\frac{1}{50}\left(\frac{\partial}{\partial x}Q(x,y)\right)a^2\cos\left(\frac{i\,\pi}{25}\right)j\sin\left(\frac{i\,\pi}{25}\right) - \frac{1}{50}\left(\frac{\partial}{\partial y}P(x,y)\right)b^2\sin\left(\frac{i\,\pi}{25}\right)j\cos\left(\frac{i\,\pi}{25}\right)$$

```
// In actual expressions,'i,j'stand for natural number 1~50
//'ijddPV2' isn't that 'single value', but that 'set of finite
quantitative (viz. 50) values'


// Calculate microcosmic double integrals values of plane bounded
closed region'S's all segmental cells:
```

```
> ijddPV2*ijdJ*du*dr;
```

Base on integral median theorem, integral value of product (Differential function 'dPV2' and curve coordinates area element) in all segmental cells

$$\frac{1}{62500}$$

$$\left(-\frac{1}{50}\left(\frac{\partial}{\partial x}Q(x,y)\right)a^2\cos\left(\frac{i\,\pi}{25}\right)j\sin\left(\frac{i\,\pi}{25}\right)-\frac{1}{50}\left(\frac{\partial}{\partial y}P(x,y)\right)b^2\sin\left(\frac{i\,\pi}{25}\right)j\cos\left(\frac{i\,\pi}{25}\right)\right)$$

$$a\,j\,b\,\pi$$

// Expression above-mentioned isn't that 'single value', but that 'set of finite quantitative(viz.50)values'

// List of finite quantitative microcosmic double integrals values:
```
> seq(seq([i*j,ijddPV2*ijdJ*du*dr,evalf(ijddPV2*ijdJ*du*dr)],
i=1..dus),j=1..dus):
```
List of integral value of product(Differential function 'dPV2' and curve coordinates area element) in all segmental cells, be elided
// Structure sequence that is composed of finite quantitative microcomosic double integrals values:
```
> sqn:=seq(seq(ijddPV2*ijdJ*du*dr,i=1..dus),j=1..dus):
```
// Accumulational sum of sequence, obtain double integrals value at manifold (Entire plane bounded closed region 'S'):
```
> add(k,k=sqn):omega:=evalf(%):        (10)
```
Elide verbose analytical and float expressions
Integral values' sum of product (Differential function 'dPV2' and curve coordinates area element)in all segmental cells; Amount of plane bounded closed region 'S's segmental cells increases infinitely, sum of integral values above-mentioned tends to '0' infinitely

// Renewedly segment range '[0,1],[0,2*Pi]' of 'r,u':
```
> dus:=n;
```
Set amount of plane bounded closed region 'S's (plane closed curve 'CO' surrounds) parameter segmental cells as natural number 'n'

$$dus := n$$

```
> dr:=(rgr[2]-rgr[1])/dus; # Segment range of 'r'
```

$$dr := \frac{1}{n}$$

```
> du:=(rgt[2]-rgt[1])/dus; # Segment range of 'u'
```

$$du := \frac{2\pi}{n}$$

Microcosmic double integrals course of plane bounded closed region 'S's first segmental cell (n segmental cells):

// Segment area element coefficient 'J':

```
> dJ:=subs(r=rgr[1]+dr,subs(u=rgt[1]+du,J));
```

Plane area element coefficient 'J's average value that corresponds to plane bounded closed region 'S's first segmental cell

$$dJ := \frac{a\,b}{n}$$

// Segment abstract differential function 'dPV2':

```
> ddPV2:=subs(r=rgr[1]+dr,subs(u=rgt[1]+du,dPV2));
```

Differential function 'dPV2's average value that corresponds to plane bounded closed region 'S's first segmental cell

$$ddPV2 := -\frac{\left(\frac{\partial}{\partial x}Q(x,y)\right)a^2\cos\left(\frac{2\pi}{n}\right)\sin\left(\frac{2\pi}{n}\right)}{n} - \frac{\left(\frac{\partial}{\partial y}P(x,y)\right)b^2\sin\left(\frac{2\pi}{n}\right)\cos\left(\frac{2\pi}{n}\right)}{n}$$

// Calculate microcosmic double integrals value of plane bounded closed region 'S's first segmental cell:

```
> ddPV2*dJ*du*dr;
```

Base on integral median theorem, integral value of product (Differential function 'dPV2' and Curve coordinates area element) in first segmental cell

$$\frac{2\left(-\frac{\left(\frac{\partial}{\partial x}Q(x,y)\right)a^2\cos\left(\frac{2\pi}{n}\right)\sin\left(\frac{2\pi}{n}\right)}{n} - \frac{\left(\frac{\partial}{\partial y}P(x,y)\right)b^2\sin\left(\frac{2\pi}{n}\right)\cos\left(\frac{2\pi}{n}\right)}{n}\right)a\,b\,\pi}{n^3}$$

Microcosmic double integrals course of plane bounded closed region 'S's all segmental cells (n segmental cells):

// Segment area element coefficient 'J':

```
> ijdJ:=subs(r=rgr[1]+j*dr,subs(u=rgt[1]+i*du,J));
```

Area element coefficient 'J's average values that correspond to plane bounded closed region 'S's all segmental cells

$$ijdJ := \frac{a\,j\,b}{n}$$

```
// In actual expressions, 'i,j' stand for natural number 1~n
//'ijdJ' isn't that 'single value', but that 'set of finite
quantitative (viz. n) values'


// Segment abstract differential function 'dPV2':
> ijddPV2:=subs(r=rgr[1]+j*dr,subs(u=rgt[1]+i*du,dPV2));
# Differential function 'dPV2's average values that correspond to
plane bounded closed region 'S's all segmental cells
```

$$ijddPV2 := -\frac{\left(\dfrac{\partial}{\partial x} Q(x, y)\right) a^2 \cos\left(\dfrac{2\,i\,\pi}{n}\right) j \sin\left(\dfrac{2\,i\,\pi}{n}\right)}{n}$$

$$-\frac{\left(\dfrac{\partial}{\partial y} P(x, y)\right) b^2 \sin\left(\dfrac{2\,i\,\pi}{n}\right) j \cos\left(\dfrac{2\,i\,\pi}{n}\right)}{n}$$

```
// In actual expressions, 'i,j' stand for natural number 1~n
//'ijddPV2' isn't that 'single value', but that 'set of finite
quantitative (viz. n) values'


// Calculate microcosmic double integrals values of plane bounded
closed region 'S's all segmental cells:
> ijddPV2*ijdJ*du*dr;
# Base on integral median theorem, integral value of product
(Differential function 'dPV2' and Curve coordinates area element) in
all segmental cells
```

$$2\left(-\frac{\left(\dfrac{\partial}{\partial x} Q(x, y)\right) a^2 \cos\left(\dfrac{2\,i\,\pi}{n}\right) j \sin\left(\dfrac{2\,i\,\pi}{n}\right)}{n} - \frac{\left(\dfrac{\partial}{\partial y} P(x, y)\right) b^2 \sin\left(\dfrac{2\,i\,\pi}{n}\right) j \cos\left(\dfrac{2\,i\,\pi}{n}\right)}{n}\right) a\,j\,b\,\pi / n^3$$

```
//Expression above-mentioned isn't that 'single value', but that 'set
of finite quantitative(viz.n)values'


// Structure finite sums:
> Sum(Sum(ijddPV2*ijdJ*du*dr,i=1..dus),j=1..dus);        (11)
# In the case of "Amount 'n' of plane bounded closed region 'S's
segmental cells is pending', sum of integral values (Product about
Differential function 'dPV2' and Curve coordinates area element) in
```

all segmental cells

$$\sum_{j=1}^{n}\left(\sum_{i=1}^{n}\left(2\left(-\frac{\left(\frac{\partial}{\partial x}Q(x,y)\right)a^2\cos\left(\frac{2\,i\,\pi}{n}\right)j\sin\left(\frac{2\,i\,\pi}{n}\right)}{n}\right.\right.\right.$$

$$\left.\left.\left.-\frac{\left(\frac{\partial}{\partial y}P(x,y)\right)b^2\sin\left(\frac{2\,i\,\pi}{n}\right)j\cos\left(\frac{2\,i\,\pi}{n}\right)}{n}\right)aj\,b\,\pi\,/\,n^3\right)\right)$$

> vs:=value(%);

$$vs := 0$$

> Limit(vs,n=infinity):

// Infinitize finite sums, its limit operational value is 'Double Integrals Value at Manifold'

\# In the case of "Amount 'n' of plane bounded closed region 'S's segmental cells tends to infinity", limit value of integral values' sums (Product about Differential function 'dPV2' and Curve coordinates area element)in all segmental cells

> epsilon:=value(%);

$$\varepsilon := 0$$

Thereinto, $\displaystyle\lim_{n\to\infty}\sum_{j=1}^{n}\left(\sum_{i=1}^{n}\left(\frac{2\,a\,j\,b\,\pi}{n^3}\right)\right)\neq 0$

Viz. Suppose sum of double integrals values (Curve coordinates area element itself) in all segmental cells can't be '0'; Explain visually ,it can be regarded as "Suppose plane bounded closed region 'S' can't be '0 area'"

Viz. In the case of 'n → ∞', **(6) = (11)**

$$\sum_{i=1}^{n}\left(\frac{2\,\pi\left(-P(x,y)\,a\sin\left(\frac{2\,i\,\pi}{n}\right)+Q(x,y)\,b\cos\left(\frac{2\,i\,\pi}{n}\right)\right)}{n}\right)$$

$$=$$

$$\sum_{j=1}^{n}\left(\sum_{i=1}^{n}\left(2\left(-\frac{\left(\frac{\partial}{\partial x}Q(x,y)\right)a^2\cos\left(\frac{2\,i\,\pi}{n}\right)j\sin\left(\frac{2\,i\,\pi}{n}\right)}{n}\right.\right.\right.$$

$$-\frac{\left(\frac{\partial}{\partial y}\mathrm{P}(x,y)\right)b^2\sin\left(\frac{2\,i\,\pi}{n}\right)j\cos\left(\frac{2\,i\,\pi}{n}\right)}{n}\Bigg)a\,j\,b\,\pi\,/\,n^3\Bigg)\Bigg)$$

Above-mentioned equation can be described as:

$$\int_{L+} A\cdot dL = \iint_{S}\left(\frac{\partial}{\partial x}Q(x,y)-\frac{\partial}{\partial y}P(x,y)\right)dS \quad \textbf{(1)}, \;\text{Complete Proof.}$$

Chapter 5 Numerical Models of Green Theorem Finite Sums Limits at Manifold

5.1 Numerical Model of Green Theorem Finite Sums Limits at Manifold

Known: Parametrized Expression of Simply connected、Closed Curve (Irregular、Asymmetrical)

$$\left[2\cos(t)-\sin(2\,t),4\sin\left(t-\frac{1}{5}\right)\right] \qquad (1)$$

thereinto, t∈[0,2π];

and Integral Vector Field $\left[\left(\frac{x}{2}-\frac{y}{3}\right)^2,x\,y\right]$ (2)

Calculate and Validate Green Theorem at Manifold
(Finite Sums Limits)

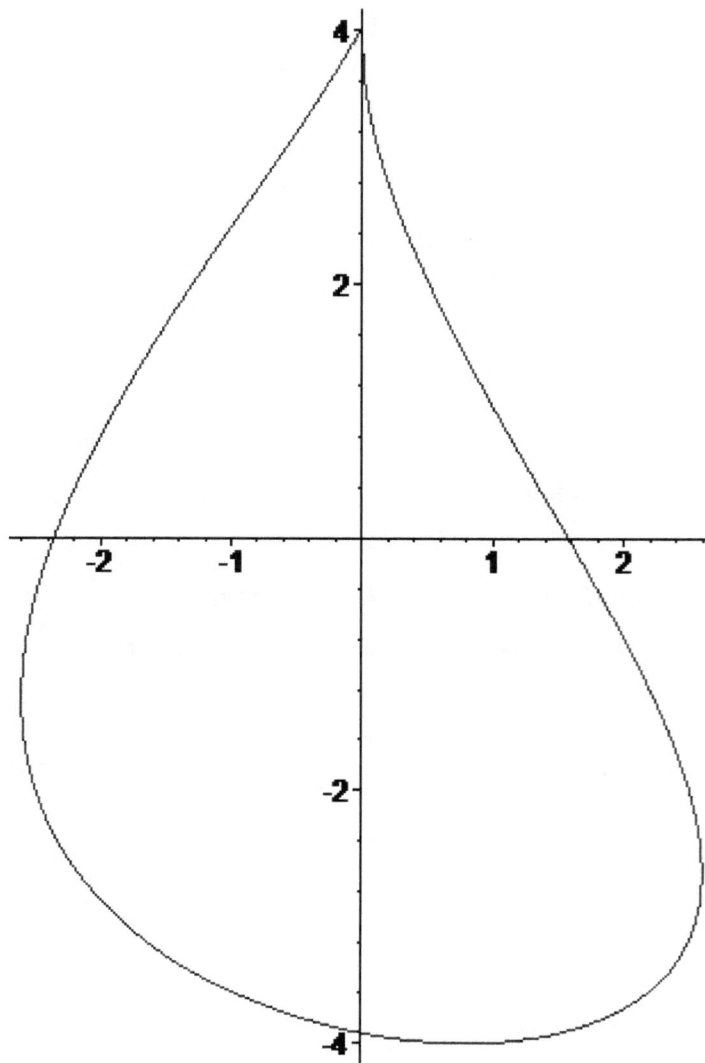

Figure(II)5.1 Simply Connected、Closed Curve (1)

[Irregular、Asymmetrical]

Solution:

  First, Free Closed Curve Integral:

  Calculate tangent vector of plane closed curve 'L'(3):

$$\left[\frac{d}{dt}(2\cos(t)-\sin(2\,t)), \frac{d}{dt}\left(4\sin\left(t-\frac{1}{5}\right)\right)\right] = \left[-2\sin(t)-2\cos(2\,t), 4\cos\left(t-\frac{1}{5}\right)\right]$$

  Set Amount of boundary curve 'L's parametrized segmental cells as 50 (It can be discretional natural number)   (4)

  1.Microcosmic Curve Integral Course at Boundary Curve 'L's first segmental cell:

Segment range $[0,2\pi]$ of 't': dt $= \dfrac{2\pi}{50} = \dfrac{\pi}{25}$ (5)

Segment Vector Field (6): [Viz. Input (1) to (2), then input (5)]

$$\left[\left(\cos\left(\frac{\pi}{25}\right)-\frac{1}{2}\sin\left(\frac{2\pi}{25}\right)-\frac{4}{3}\cos\left(\frac{23\pi}{50}+\frac{1}{5}\right)\right)^2,\right.$$
$$\left.4\left(2\cos\left(\frac{\pi}{25}\right)-\sin\left(\frac{2\pi}{25}\right)\right)\cos\left(\frac{23\pi}{50}+\frac{1}{5}\right)\right]$$

       (6)

Segment Tangent Vector (7): [Viz. Input (5) to (3)]

$$\left[-2\sin\left(\frac{\pi}{25}\right)-2\cos\left(\frac{2\pi}{25}\right),4\sin\left(\frac{23\pi}{50}+\frac{1}{5}\right)\right]$$

       (7)

Calculate Microcosmic Curve Integral Value at Closed Boundary Curve 'L's first segmental cell:

Base on integral median theorem, dot product of abstract vector field (6) and tangent vector (7), times segment unit of 't'(5), that is Microcosmic Curve Integral Value at first segmental cell (8)

$$\frac{1}{25}\pi\left(\left(-2\sin\left(\frac{\pi}{25}\right)-2\cos\left(\frac{2\pi}{25}\right)\right)\left(\cos\left(\frac{\pi}{25}\right)-\frac{1}{2}\sin\left(\frac{2\pi}{25}\right)-\frac{4}{3}\cos\left(\frac{23\pi}{50}+\frac{1}{5}\right)\right)^2\right.$$
$$\left.+16\sin\left(\frac{23\pi}{50}+\frac{1}{5}\right)\left(2\cos\left(\frac{\pi}{25}\right)-\sin\left(\frac{2\pi}{25}\right)\right)\cos\left(\frac{23\pi}{50}+\frac{1}{5}\right)\right)$$

       (8)

2.Microcosmic Curve Integral Course at Closed Boundary Curve 'L's all (Viz. 50) segmental cells:

Segment range $[0,2\pi]$ of 't': dt $= \dfrac{i2\pi}{50} = \dfrac{i\pi}{25}$ (9)

(thereinto, 'i' stand for natural number 1~50)

Segment Vector Field (10): [Viz. Input(1) to (2),then input (9)]

$$\left[\left(\cos\left(\frac{\pi i}{25}\right)-\frac{1}{2}\sin\left(\frac{2\pi i}{25}\right)-\frac{4}{3}\sin\left(\frac{\pi i}{25}-\frac{1}{5}\right)\right)^2,4\left(2\cos\left(\frac{\pi i}{25}\right)-\sin\left(\frac{2\pi i}{25}\right)\right)\sin\left(\frac{\pi i}{25}-\frac{1}{5}\right)\right]$$

// Expression (10) isn't that 'Single vector value', but that 'Set of finite quantitative (Viz.50) vector values'

Segment Tangent Vector (11): [Viz. Input (9) to (3)]

$$\left[-2\sin\left(\frac{\pi i}{25}\right)-2\cos\left(\frac{2\pi i}{25}\right),4\cos\left(\frac{\pi i}{25}-\frac{1}{5}\right)\right]$$

       (11)

// Expression (11) isn't that 'Single vector value', but that 'Set of finite quantitative (Viz.50) vector values'

Calculate Microcosmic Curve Integral Values at Closed Boundary Curve 'L's all segmental cells:

Base on Integral Median Theorem, dot product of abstract vector field (10) and tangent vector (11), times segment unit of 't' (9), these are Microcosmic Curve Integral Values at all segmental cells (12)

$$\frac{1}{25}\pi\left(\left(-2\sin\left(\frac{\pi i}{25}\right)-2\cos\left(\frac{2\pi i}{25}\right)\right)\left(\cos\left(\frac{\pi i}{25}\right)-\frac{1}{2}\sin\left(\frac{2\pi i}{25}\right)-\frac{4}{3}\sin\left(\frac{\pi i}{25}-\frac{1}{5}\right)\right)^2\right.$$
$$\left.+16\cos\left(\frac{\pi i}{25}-\frac{1}{5}\right)\left(2\cos\left(\frac{\pi i}{25}\right)-\sin\left(\frac{2\pi i}{25}\right)\right)\sin\left(\frac{\pi i}{25}-\frac{1}{5}\right)\right)$$

(12)

// Expression (12) isn't that 'Single value', but that 'Set of finite quantitative (Viz.50) values'

Structure Sequence that is composed of finite quantitative (viz. 50) Microcosmic Curve Integral Values (13):

  [Lengthy expression of sequence is elided;

  In program template of Waterloo Maplesoft (see also Section 5.2):

  sqn:=seq(dt*(idPV[1]*idCO[1]+idPV[2]*idCO[2]),i=1..dus):

change ':' (last) to ';', then obtain expression of sequence]

Accumulational sum of sequence (viz. Sum of integral values of dot product [abstract vector field (10) and tangent vector (11) at all (Viz. 50) segmental cells], obtain Curve Integral Value at manifold (14):

  [Lengthy expression of accumulational sum result is elided;

  In program template of Waterloo Maplesoft (see also Section 5.2):

  add(k,k=sqn):xi:=evalf(%);

change ':' (middle) to ';', then obtain expression of result]

  Change expression of accumulational sum result to float value:
                     -18.83679674

Set Amount of closed boundary curve 'L's parametrized segmental cells as uncertain natural number 'n'   (15)

3.Microcosmic Curve Integral Course at Closed Boundary Curve 'L's first segmental cell:

Segment range [0,2π] of 't': dt = $\dfrac{2\pi}{n}$ (16)

Segment Vector Field (17): [Viz. Input (1) to(2),then input (16)]

$$\left[\left(\cos\left(\frac{2\pi}{n}\right) - \frac{1}{2}\sin\left(\frac{4\pi}{n}\right) + \frac{4}{3}\sin\left(-\frac{2\pi}{n}+\frac{1}{5}\right) \right)^2 , \right.$$
$$\left. -4\left(2\cos\left(\frac{2\pi}{n}\right) - \sin\left(\frac{4\pi}{n}\right) \right)\sin\left(-\frac{2\pi}{n}+\frac{1}{5}\right) \right]$$ (17)

Segment Tangent Vector (18): [Viz. Input (16) to (3)]

$$\left[-2\sin\left(\frac{2\pi}{n}\right) - 2\cos\left(\frac{4\pi}{n}\right), 4\cos\left(-\frac{2\pi}{n}+\frac{1}{5}\right) \right]$$ (18)

Calculate Microcosmic Curve Integral Value at Closed Boundary Curve 'L's first segmental cell:

Base on Integral Median Theorem, dot product of abstract vector field (17) and tangent vector (18), times segment unit of 't' (16), that is Microcosmic Curve Integral Value at first segmental cell (19)

$$2\pi\left(\left(-2\sin\left(\frac{2\pi}{n}\right) - 2\cos\left(\frac{4\pi}{n}\right)\right)\left(\cos\left(\frac{2\pi}{n}\right) - \frac{1}{2}\sin\left(\frac{4\pi}{n}\right) + \frac{4}{3}\sin\left(-\frac{2\pi}{n}+\frac{1}{5}\right)\right)^2 \right.$$
$$\left. -16\cos\left(-\frac{2\pi}{n}+\frac{1}{5}\right)\left(2\cos\left(\frac{2\pi}{n}\right) - \sin\left(\frac{4\pi}{n}\right)\right)\sin\left(-\frac{2\pi}{n}+\frac{1}{5}\right) \right)/n$$ (19)

4.Microcosmic Curve Integral Course at Closed Boundary Curve 'L's all (Viz. 50) segmental cells:

Segment range [0,2π] of 't': dt = $\dfrac{2\pi i}{n}$ (20)

(thereinto, 'i' stand for natural number 1~n)

Segment Vector Field (21): [Viz. Input (1) to (2),then input (20)]

$$\left[\left(\cos\left(\frac{2\pi i}{n}\right) - \frac{1}{2}\sin\left(\frac{4\pi i}{n}\right) - \frac{4}{3}\sin\left(\frac{2\pi i}{n}-\frac{1}{5}\right) \right)^2 , \right.$$
$$\left. 4\left(2\cos\left(\frac{2\pi i}{n}\right) - \sin\left(\frac{4\pi i}{n}\right) \right)\sin\left(\frac{2\pi i}{n}-\frac{1}{5}\right) \right]$$ (21)

// Expression (21) isn't that 'Single vector value', but that 'Set of finite quantitative (Viz.n) vector values'

Segment Tangent Vector (22): [Viz. Input (20) to (3)]

$$\left[-2\sin\left(\frac{2\pi i}{n}\right)-2\cos\left(\frac{4\pi i}{n}\right),\ 4\cos\left(\frac{2\pi i}{n}-\frac{1}{5}\right)\right] \tag{22}$$

// Expression (21) isn't that 'Single vector value', but that 'Set of finite quantitative (Viz.n) vector values'

Calculate Microcosmic Curve Integral Values at Closed Boundary Curve 'L's all segmental cells:

Base on Integral Median Theorem, dot product of vector field (21) and tangent vector (22), times segment unit of 't'(20), these are Microcosmic Curve Integral Values at all segmental cells (23)

$$2\pi\left(\left(\left(-2\sin\left(\frac{2\pi i}{n}\right)-2\cos\left(\frac{4\pi i}{n}\right)\right)\left(\cos\left(\frac{2\pi i}{n}\right)-\frac{1}{2}\sin\left(\frac{4\pi i}{n}\right)-\frac{4}{3}\sin\left(\frac{2\pi i}{n}-\frac{1}{5}\right)\right)\right)^{2}\right.$$
$$\left.+16\cos\left(\frac{2\pi i}{n}-\frac{1}{5}\right)\left(2\cos\left(\frac{2\pi i}{n}\right)-\sin\left(\frac{4\pi i}{n}\right)\right)\sin\left(\frac{2\pi i}{n}-\frac{1}{5}\right)\right)\Big/n$$

// Expression (23) isn't that 'Single value', but that 'Set of finite quantitative (Viz.n) values'

Structure finite sums (24):

(In the case of "Amount 'n' of closed boundary curve 'L's segmental cells is uncertain", sum of integral values [dot product about absract vector field (21) and tangent vector (22)] at all segmental cells)

$$\sum_{i=1}^{n}\left(2\pi\left(\vphantom{\frac{1}{1}}\right.\right.$$
$$\left(\left(-2\sin\left(\frac{2\pi i}{n}\right)-2\cos\left(\frac{4\pi i}{n}\right)\right)\left(\cos\left(\frac{2\pi i}{n}\right)-\frac{1}{2}\sin\left(\frac{4\pi i}{n}\right)-\frac{4}{3}\sin\left(\frac{2\pi i}{n}-\frac{1}{5}\right)\right)\right)^{2}$$
$$\left.\left.+16\cos\left(\frac{2\pi i}{n}-\frac{1}{5}\right)\left(2\cos\left(\frac{2\pi i}{n}\right)-\sin\left(\frac{4\pi i}{n}\right)\right)\sin\left(\frac{2\pi i}{n}-\frac{1}{5}\right)\right)\Big/n\right) \tag{24}$$

Infinitize finite sums, its limit operational value is 'Curve Integral Value at Manifold' (25):

(In the case of "Amount 'n' of closed boundary curve 'L's segmental cells tends to infinity", limit value of integral values' sums [dot product about abstract vector field (21) and tangent vector (22)] at all segmental cells)

$$\lim_{n \to \infty} \sum_{i=1}^{n} \Bigg(2\,\pi \Bigg($$

$$\left(-2\sin\left(\frac{2\pi i}{n}\right) - 2\cos\left(\frac{4\pi i}{n}\right) \right)\left(\cos\left(\frac{2\pi i}{n}\right) - \frac{1}{2}\sin\left(\frac{4\pi i}{n}\right) - \frac{4}{3}\sin\left(\frac{2\pi i}{n} - \frac{1}{5}\right) \right)^2$$

$$+ 16\cos\left(\frac{2\pi i}{n} - \frac{1}{5}\right)\left(2\cos\left(\frac{2\pi i}{n}\right) - \sin\left(\frac{4\pi i}{n}\right) \right)\sin\left(\frac{2\pi i}{n} - \frac{1}{5}\right) \Bigg)\Big/n \Bigg)$$

$$= \frac{56\,\pi}{9} - \frac{112}{9}\,\pi\cos\left(\frac{1}{5}\right)^2 - \frac{4}{3}\,\pi\sin\left(\frac{1}{5}\right) \tag{25}$$

Second, Free Area Element Integral (Finite Sums Limits):

Entirely multiply radius vector 'r' (suppose r > 0) by each item of closed boundary curve 'L' parametrized expression (1), convert 'Parametrized expression of curve' to 'Parametrized expression of curve coordinates'; and change 't' to 'u' in expression:

$$\left[r\,(2\cos(u) - \sin(2\,u)),\, 4\,r\sin\left(u - \frac{1}{5}\right) \right] \tag{26}$$

Base on parametrized expression of curve coordinates (26), define and calculate matrix of partial derivatives, obtain general expression of curve coordinates area element coefficient (27):

$$\left[\begin{array}{cc} \frac{\partial}{\partial r}(r\,(2\cos(u) - \sin(2\,u))) & \frac{\partial}{\partial u}(r\,(2\cos(u) - \sin(2\,u))) \\ \frac{\partial}{\partial r}\left(4\,r\sin\left(u - \frac{1}{5}\right)\right) & \frac{\partial}{\partial u}\left(4\,r\sin\left(u - \frac{1}{5}\right)\right) \end{array} \right]$$

$$=$$

$$4\,r\left(2\cos(u)\cos\left(u - \frac{1}{5}\right) - \sin(2\,u)\cos\left(u - \frac{1}{5}\right) + 2\sin(u)\sin\left(u - \frac{1}{5}\right) \right.$$

$$\left. + 2\cos(2\,u)\sin\left(u - \frac{1}{5}\right) \right) \tag{27}$$

Calculate differential function '$\frac{\partial Q(x,y)}{\partial x} - \frac{\partial P(x,y)}{\partial y}$' of vector field (2):

$$\left(\frac{\partial}{\partial x}(x\,y) \right) - \left(\frac{\partial}{\partial y}\left(\left(\frac{x}{2} - \frac{y}{3} \right)^2 \right) \right) = \frac{7\,y}{9} + \frac{x}{3} \tag{28}$$

Set Amount of plane bounded closed region'S's (Curve 'L' surrounds) segmental cells as 50

(This amount can be discretional natural number)   (29)

5.Microcosmic Double Integrals Course in plane bounded closed region 'S's first segmental cell:

Segment range [0,1] of 'r': dr = $\dfrac{1}{50}$

Segment range [0,2π] of 'u': du = $\dfrac{2\pi}{50} = \dfrac{\pi}{25}$ (30)

Segment area element coefficient (31): [viz. Input (30) to (27)]

(Area element coefficient(27), average value that corresponds to plane bounded closed region 'S's first segmental cell)

$$\frac{4}{25}\cos\left(\frac{\pi}{25}\right)\sin\left(\frac{23\pi}{50}+\frac{1}{5}\right) - \frac{2}{25}\sin\left(\frac{2\pi}{25}\right)\sin\left(\frac{23\pi}{50}+\frac{1}{5}\right) + \frac{4}{25}\sin\left(\frac{\pi}{25}\right)\cos\left(\frac{23\pi}{50}+\frac{1}{5}\right)$$
$$+\frac{4}{25}\cos\left(\frac{2\pi}{25}\right)\cos\left(\frac{23\pi}{50}+\frac{1}{5}\right)$$

Segment differential function (32):

[viz. Input (26) to (28), then input (30)]

(Differential function (28), average value that corresponds to plane bounded closed region 'S's first segmental cell)

$$\frac{14}{225}\cos\left(\frac{23\pi}{50}+\frac{1}{5}\right) + \frac{1}{75}\cos\left(\frac{\pi}{25}\right) - \frac{1}{150}\sin\left(\frac{2\pi}{25}\right)$$ (32)

Calculate Microcosmic Double Integrals Value in plane bounded closed region 'S's first segmental cell:

Base on integral median theorem, product of Abstract differential function (32) and Area Element Coefficient (31), times segment unit of 'r,u'(30), that is Microcosmic Double Integrals Value in first segmental cell (33)

$$\frac{1}{1250}\left(\frac{14}{225}\cos\left(\frac{23\pi}{50}+\frac{1}{5}\right) + \frac{1}{75}\cos\left(\frac{\pi}{25}\right) - \frac{1}{150}\sin\left(\frac{2\pi}{25}\right)\right)\left(\frac{4}{25}\cos\left(\frac{\pi}{25}\right)\sin\left(\frac{23\pi}{50}+\frac{1}{5}\right)\right.$$
$$-\frac{2}{25}\sin\left(\frac{2\pi}{25}\right)\sin\left(\frac{23\pi}{50}+\frac{1}{5}\right) + \frac{4}{25}\sin\left(\frac{\pi}{25}\right)\cos\left(\frac{23\pi}{50}+\frac{1}{5}\right)$$
$$+\left.\frac{4}{25}\cos\left(\frac{2\pi}{25}\right)\cos\left(\frac{23\pi}{50}+\frac{1}{5}\right)\right)\pi$$ (33)

6.Microcosmic double integrals course in plane bounded closed region

'S's all (viz. 50) segmental cells:

Segment range [0,1] of 'r': dr = $\dfrac{j}{50}$

Segment range [0,2π] of 'u': du = $\dfrac{2i\pi}{50} = \dfrac{i\pi}{25}$ (34)

(thereinto, 'i,j' stand for natural number 1~50)

Segment area element coefficient (35): [viz. Input (34) to (27)]

(Area element coefficient (27), average values that correspond to plane bounded closed region 'S's all segmental cells)

$$\frac{2}{25}j\left(2\cos\left(\frac{\pi i}{25}\right)\cos\left(\frac{\pi i}{25}-\frac{1}{5}\right)-\sin\left(\frac{2\pi i}{25}\right)\cos\left(\frac{\pi i}{25}-\frac{1}{5}\right)+2\sin\left(\frac{\pi i}{25}\right)\sin\left(\frac{\pi i}{25}-\frac{1}{5}\right)\right.$$
$$\left.+2\cos\left(\frac{2\pi i}{25}\right)\sin\left(\frac{\pi i}{25}-\frac{1}{5}\right)\right)$$

// Expression (35) isn't that 'Single value', but that 'Set of finite quantitative (Viz.50) values'

Segment differential function (36)

[Viz. Input (26) to (28),then Input (34)]:

(Differential function (28), average values that correspond to plane bounded closed region 'S's all(viz. 50)segmental cells)

$$\frac{14}{225}j\sin\left(\frac{\pi i}{25}-\frac{1}{5}\right)+\frac{1}{150}j\left(2\cos\left(\frac{\pi i}{25}\right)-\sin\left(\frac{2\pi i}{25}\right)\right)$$ (36)

// Expression (36) isn't that 'Single value', but that 'Set of finite quantitative (Viz.50) values'

Calculate Microcosmic Double Integrals Values in Plane Bounded Closed Region 'S's all segmental cells:

Base on integral median theorem, product of Abstract differential function (36) and Area Element Coefficient (35), times segment unit of 'r,u'(34), these are microcosmic double integrals values in all (viz. 50) segmental cells (37)

$$\frac{1}{15625}\left(\frac{14}{225}j\sin\left(\frac{\pi i}{25}-\frac{1}{5}\right)+\frac{1}{150}j\left(2\cos\left(\frac{\pi i}{25}\right)-\sin\left(\frac{2\pi i}{25}\right)\right)\right)j\left(\right.$$
$$2\cos\left(\frac{\pi i}{25}\right)\cos\left(\frac{\pi i}{25}-\frac{1}{5}\right)-\sin\left(\frac{2\pi i}{25}\right)\cos\left(\frac{\pi i}{25}-\frac{1}{5}\right)+2\sin\left(\frac{\pi i}{25}\right)\sin\left(\frac{\pi i}{25}-\frac{1}{5}\right)$$

$$+ 2\cos\left(\frac{2\pi i}{25}\right)\sin\left(\frac{\pi i}{25} - \frac{1}{5}\right)\right)\pi$$

(37)

// Expression (37) isn't that 'Single value', but that 'Set of finite
quantitative (Viz.50) values'

Structure sequence that is composed of finite quantitative (viz.
50) microcomosic double integrals values (38):

   [Lengthy expression of sequence is elided;

   In program template of Waterloo Maplesoft (see also Section 5.2):

   sqn:=seq(seq(ijddPV2*ijdJ*du*dr,i=1..dus),j=1..dus):

change ':'(last) to ';', then obtain expression of sequence]

   Accumulational sum of sequence (viz. Sum of integral values of
product [Abstract differential function (36) and Volume Element at
all (viz. 50) segmental cells], obtain double integrals value at
manifold (39):

   [Lengthy expression of accumulational sum result is elided;

   In program template of Waterloo Maplesoft (see also Section 5.2):

   add(k,k=sqn):omega:=evalf(%):

   change ':'(middle and last) to ';', then obtain expression and
float value of result]

   Change expression of accumulational sum result to float value:
                          -19.40566799

   Set Amount of plane bounded closed region 'S's (Curve 'L' surrounds)
segmental cells as uncertain natural number 'n'   (40)

   7.Microcosmic Double Integrals Course in plane bounded closed region
'S's first segmental cell:

   Segment range [0,1] of 'r': dr = $\dfrac{1}{n}$

   Segment range [0,2π] of 'u': du = $\dfrac{2\pi}{n}$ (41)

   Segment area element coefficient (42): [viz. Input (41) to (27)]

   (Area element coefficient (27), average value that corresponds to
plane bounded closed region 'S's first segmental cell)

$$4\left(2\cos\left(\frac{2\pi}{n}\right)\cos\left(-\frac{2\pi}{n}+\frac{1}{5}\right)-\sin\left(\frac{4\pi}{n}\right)\cos\left(-\frac{2\pi}{n}+\frac{1}{5}\right)-2\sin\left(\frac{2\pi}{n}\right)\sin\left(-\frac{2\pi}{n}+\frac{1}{5}\right)\right.$$
$$\left.-2\cos\left(\frac{4\pi}{n}\right)\sin\left(-\frac{2\pi}{n}+\frac{1}{5}\right)\right)/n$$

Segment differential function (43):

[Viz. Input (26) to (28),then Input (41)]

(Differential function (28), average value that corresponds to plane bounded closed region 'S's first segmental cell)

$$-\frac{28}{9}\frac{\sin\left(-\frac{2\pi}{n}+\frac{1}{5}\right)}{n}+\frac{1}{3}\frac{2\cos\left(\frac{2\pi}{n}\right)-\sin\left(\frac{4\pi}{n}\right)}{n} \qquad (43)$$

Calculate microcosmic double integrals value in plane bounded closed region 'S's first segmental cell (44):

Base on integral median theorem, product of Abstract differential function (43) and Area Element Coefficient (42), times segment unit of 'r,u' (41), that is microcosmic double integrals value in first segmental cell (44)

$$8\left(-\frac{28}{9}\frac{\sin\left(-\frac{2\pi}{n}+\frac{1}{5}\right)}{n}+\frac{1}{3}\frac{2\cos\left(\frac{2\pi}{n}\right)-\sin\left(\frac{4\pi}{n}\right)}{n}\right)\left(2\cos\left(\frac{2\pi}{n}\right)\cos\left(-\frac{2\pi}{n}+\frac{1}{5}\right)\right.$$
$$\left.-\sin\left(\frac{4\pi}{n}\right)\cos\left(-\frac{2\pi}{n}+\frac{1}{5}\right)-2\sin\left(\frac{2\pi}{n}\right)\sin\left(-\frac{2\pi}{n}+\frac{1}{5}\right)\right.$$
$$\left.-2\cos\left(\frac{4\pi}{n}\right)\sin\left(-\frac{2\pi}{n}+\frac{1}{5}\right)\right)\pi/n^3 \qquad (44)$$

8.Microcosmic double integrals course in plane bounded closed region 'S's all (viz. n) segmental cells:

Segment range [0,1] of 'r': dr = $\frac{j}{n}$

Segment range [0,2π] of 'u': du = $\frac{2i\pi}{n}$

(thereinto, 'i,j' stand for natural number 1~n)   (45)

Segment area element coefficient (46): [viz. Input (45) to (27)]

(Area element coefficient (27),average values that correspond to plane bounded closed region 'S's all (viz. n) segmental cells)

$$4j\left(2\cos\left(\frac{2\pi i}{n}\right)\cos\left(\frac{2\pi i}{n}-\frac{1}{5}\right)-\sin\left(\frac{4\pi i}{n}\right)\cos\left(\frac{2\pi i}{n}-\frac{1}{5}\right)\right.$$
$$\left.+2\sin\left(\frac{2\pi i}{n}\right)\sin\left(\frac{2\pi i}{n}-\frac{1}{5}\right)+2\cos\left(\frac{4\pi i}{n}\right)\sin\left(\frac{2\pi i}{n}-\frac{1}{5}\right)\right)/n \tag{46}$$

// Expression (46) isn't that 'Single value', but that 'Set of finite quantitative (Viz.n) values'

Segment differential function (47):

[viz. Input (26) to (28),then input (45)]

(Differential function (28), average values that correspond to plane bounded closed region 'S's all (viz. n) segmental cells)

$$\frac{28}{9}\frac{j\sin\left(\frac{2\pi i}{n}-\frac{1}{5}\right)}{n}+\frac{1}{3}\frac{j\left(2\cos\left(\frac{2\pi i}{n}\right)-\sin\left(\frac{4\pi i}{n}\right)\right)}{n} \tag{47}$$

// Expression (47) isn't that 'Single value', but that 'Set of finite quantitative (Viz.n) values'

Calculate microcosmic double integrals values in plane bounded closed region 'S's all segmental cells (48):

Base on integral median theorem, Product of Abstract differential function (47) and Area Element Coefficient (46), times segment unit of 'r,u' (45), these are microcosmic double integrals values in all (viz. n) segmental cells (48)

$$8\left(\frac{28}{9}\frac{j\sin\left(\frac{2\pi i}{n}-\frac{1}{5}\right)}{n}+\frac{1}{3}\frac{j\left(2\cos\left(\frac{2\pi i}{n}\right)-\sin\left(\frac{4\pi i}{n}\right)\right)}{n}\right)j\left(2\cos\left(\frac{2\pi i}{n}\right)\cos\left(\frac{2\pi i}{n}-\frac{1}{5}\right)\right.$$
$$\left.-\sin\left(\frac{4\pi i}{n}\right)\cos\left(\frac{2\pi i}{n}-\frac{1}{5}\right)+2\sin\left(\frac{2\pi i}{n}\right)\sin\left(\frac{2\pi i}{n}-\frac{1}{5}\right)\right.$$
$$\left.+2\cos\left(\frac{4\pi i}{n}\right)\sin\left(\frac{2\pi i}{n}-\frac{1}{5}\right)\right)\pi/n^3 \tag{48}$$

// Expression (48) isn't that 'Single value', but that 'Set of finite quantitative (Viz.n) values'

Structure finite sums (49):

$$\sum_{j=1}^{n}\left(\sum_{i=1}^{n}\left(8\left(\frac{28}{9}\frac{j\sin\left(\frac{2\pi i}{n}-\frac{1}{5}\right)}{n}+\frac{1}{3}\frac{j\left(2\cos\left(\frac{2\pi i}{n}\right)-\sin\left(\frac{4\pi i}{n}\right)\right)}{n}\right)j\left(\right.\right.\right.$$

$$2\cos\left(\frac{2\pi i}{n}\right)\cos\left(\frac{2\pi i}{n}-\frac{1}{5}\right)-\sin\left(\frac{4\pi i}{n}\right)\cos\left(\frac{2\pi i}{n}-\frac{1}{5}\right)$$

$$+2\sin\left(\frac{2\pi i}{n}\right)\sin\left(\frac{2\pi i}{n}-\frac{1}{5}\right)+2\cos\left(\frac{4\pi i}{n}\right)\sin\left(\frac{2\pi i}{n}-\frac{1}{5}\right)\right)\pi/n^3\right)\right) \qquad (49)$$

Infinitize finite sums, its limit operational value is 'Double Integrals Value in Manifold' (50):

(In the case of "Amount 'n' of plane bounded closed region 'S's (Curve 'L' surrounds) segmental cells tends to infinity", limit value of integral values' sums [Product about Abstract differential function (47) and Volume Element] at all segmental cells)

$$\lim_{n\to\infty}\sum_{j=1}^{n}\left(\sum_{i=1}^{n}\left(8\left(\frac{28}{9}\frac{j\sin\left(\frac{2\pi i}{n}-\frac{1}{5}\right)}{n}+\frac{1}{3}\frac{j\left(2\cos\left(\frac{2\pi i}{n}\right)-\sin\left(\frac{4\pi i}{n}\right)\right)}{n}\right)j\left(\right.\right.\right.$$

$$2\cos\left(\frac{2\pi i}{n}\right)\cos\left(\frac{2\pi i}{n}-\frac{1}{5}\right)-\sin\left(\frac{4\pi i}{n}\right)\cos\left(\frac{2\pi i}{n}-\frac{1}{5}\right)$$

$$+2\sin\left(\frac{2\pi i}{n}\right)\sin\left(\frac{2\pi i}{n}-\frac{1}{5}\right)+2\cos\left(\frac{4\pi i}{n}\right)\sin\left(\frac{2\pi i}{n}-\frac{1}{5}\right)\right)\pi/n^3\right)\right)$$

$$=\quad\frac{56\pi}{9}-\frac{112}{9}\pi\cos\left(\frac{1}{5}\right)^2-\frac{4}{3}\pi\sin\left(\frac{1}{5}\right) \qquad (50)$$

Viz. In the case of 'n → ∞', **(25)=(50)**
Complete Calculate and Validate Green Theorem at Manifold (Finite Sums Limits)

5.2 Numerical Models of Green Theorem Finite Sums Limits at Manifold [Program Template of Waterloo Maplesoft, Optional]

```
> restart;
> with(plots):with(linalg):
```

```
> CO:=[2*cos(t)-sin(2*t),4*sin(t-1/5)];    (1)
```
Define parametrized expression of discretional (plane) simply connected closed curve 'CO':

$$CO := \left[2\cos(t) - \sin(2t), 4\sin\left(t - \frac{1}{5}\right) \right]$$

```
> rgt:=[0,2*Pi];
```
Set range of 't' in [0,2π], make boundary curve 'CO' closed

$$rgt := [0, 2\pi]$$

```
> [Diff(CO[1],t),Diff(CO[2],t)]=[diff(CO[1],t),diff(CO[2],t)];
dCO:=rhs(%);
```
Calculate tangent vector 'dCO' of plane closed curve 'CO'

$$\left[\frac{d}{dt}(2\cos(t) - \sin(2t)), \frac{d}{dt}\left(4\sin\left(t - \frac{1}{5}\right)\right) \right] = \left[-2\sin(t) - 2\cos(2t), 4\cos\left(t - \frac{1}{5}\right) \right]$$

$$dCO := \left[-2\sin(t) - 2\cos(2t), 4\cos\left(t - \frac{1}{5}\right) \right]$$

```
> plot([CO[1],CO[2],t=rgt[1]..rgt[2]],scaling=constrained,
  color=blue,numpoints=1000):g1:=%:
> PV:=[(x/2-y/3)^2,x*y];    (2)
```
Define discretional plane vector field 'PV' (Suppose Vector Field 'PV' possesses 1th order continuous partial derivatives in plane bounded closed region 'S' that curve 'CO' surrounds)

$$PV := \left[\left(\frac{x}{2} - \frac{y}{3}\right)^2, xy \right]$$

```
> Diff(PV[2],x)-Diff(PV[1],y)=diff(PV[2],x)-diff(PV[1],y);
  dPV:=rhs(%);    (3)
```
Calculate differential function 'dPV' of plane vector field 'PV'

$$\left(\frac{\partial}{\partial x}(xy) \right) - \left(\frac{\partial}{\partial y}\left(\left(\frac{x}{2} - \frac{y}{3}\right)^2\right) \right) = \frac{7y}{9} + \frac{x}{3}$$

$$dPV := \frac{7y}{9} + \frac{x}{3}$$

```
> rgx:=[-4,4];
```

$$rgx := [-4, 4]$$

```
> rgy:=[-4,4];
```

$$rgy := [-4, 4]$$

```
> fieldplot(PV,x=rgx[1]..rgx[2],y=rgy[1]..rgy[2],arrows=SLIM,
color=black):g2:=%: # Draw plane vector field 'PV'
```

```
> implicitplot(dPV,x=rgx[1]..rgx[2],y=rgy[1]..rgy[2],
color=red,thickness=1,numpoints=2000):g3:=%:
# Draw the contour curve of 2-variable function 'dPV'
> display(g1,g2,g3);
# Synthesize three figures: Plane closed curve 'CO', plane vector field
'PV', contour curve of 'dPV'
```

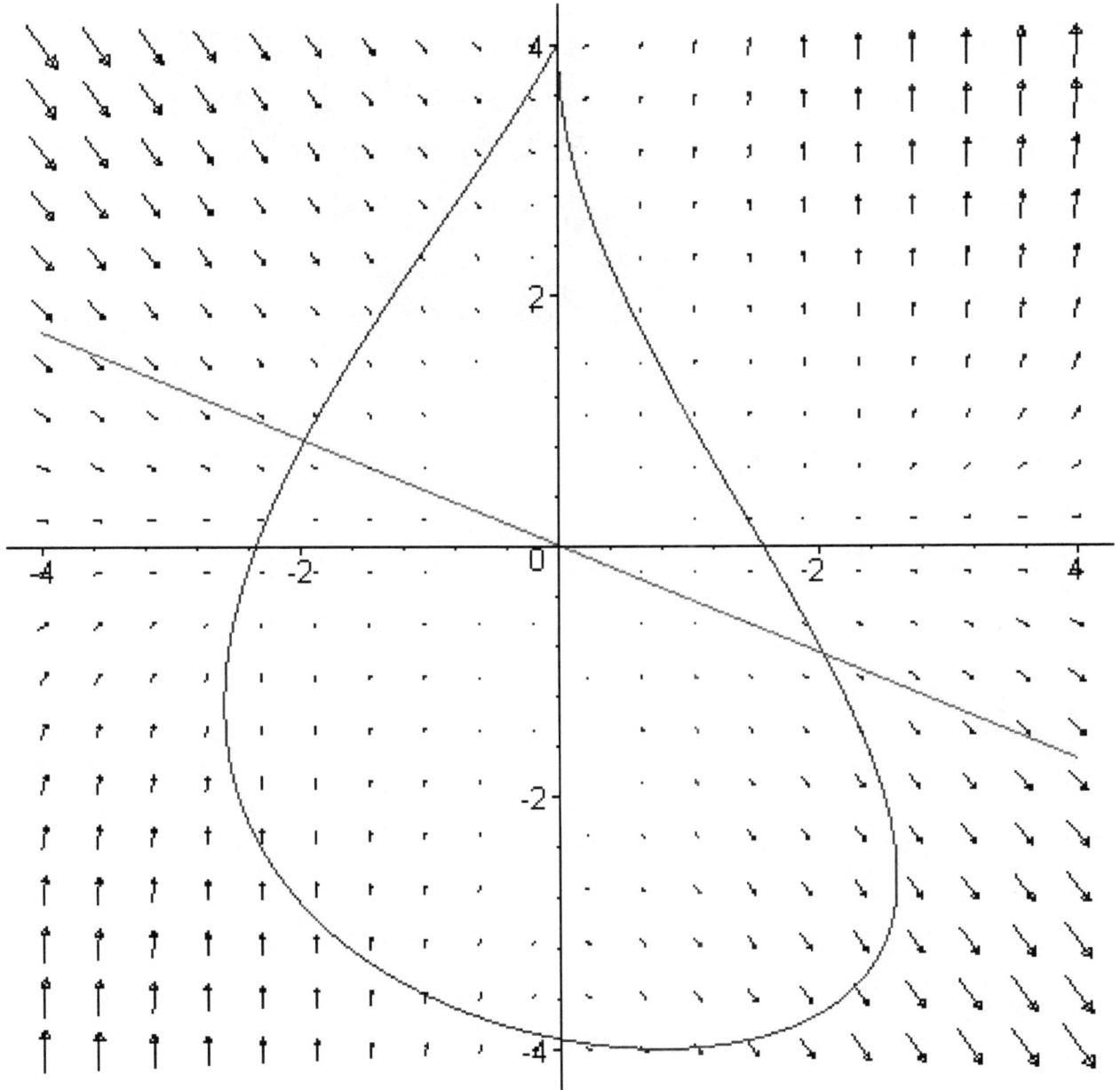

Figure(II)5.2.1 Plane Closed Curve 'CO',Vector Field 'PV'
and Contour Curve of 'dPV'

```
> plot3d(dPV,x=rgx[1]..rgx[2],y=rgy[1]..rgy[2]);
# Drawing differential function 'dPV' in 3-Dimensional Euclidean space
```

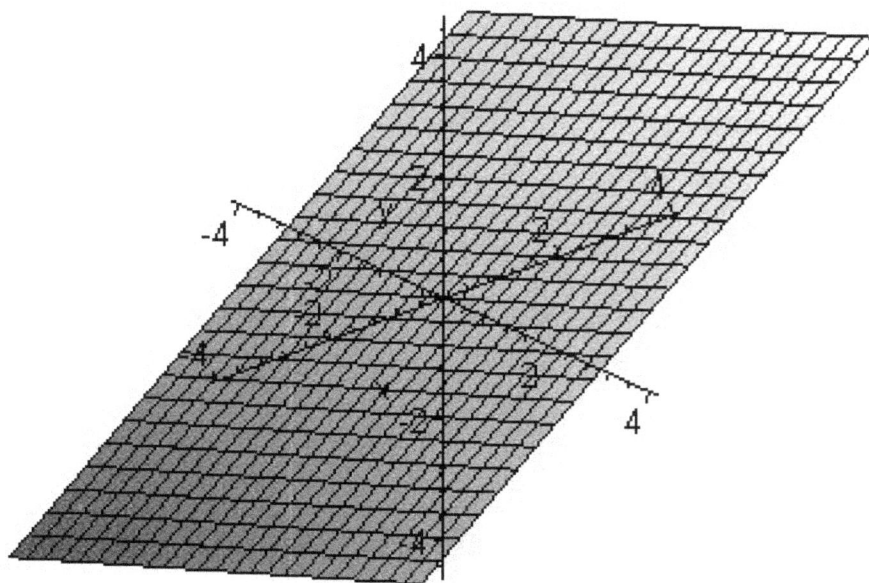

Figure(II)5.2.2 Space View of Differential Function 'dPV'

```
> x:=CO[1];y:=CO[2];
```

$$x := 2\cos(t) - \sin(2t)$$

$$y := 4\sin\left(t - \frac{1}{5}\right)$$

// Segment range '[0,2*Pi]' of 't':

```
> dus:=50;
```

Define amount 'dus' of plane closed curve 'CO's parameter segmental cells (It can be discretional natural number)

$$dus := 50$$

```
> dt:=(rgt[2]-rgt[1])/dus; # Segment range '[0,2*Pi]' of 't'
```

$$dt := \frac{\pi}{25}$$

Microcosmic curve integral course of plane bounded closed curve 'CO's first segmental cell (50 segmental cells):

// Segment plane vector field 'PV':

(This step possesses formal meaning in 'Proof' only, and possesses essential meaning in 'Numerical Models')

```
> dPVm:=subs(t=rgt[1]+dt,PV);
```

Plane vector field 'PV's average value that corresponds to plane closed curve 'CO's first segmental cell

$$dPVm := \left[\left(\cos\left(\frac{\pi}{25}\right) - \frac{1}{2}\sin\left(\frac{2\pi}{25}\right) - \frac{4}{3}\sin\left(\frac{\pi}{25} - \frac{1}{5}\right)\right)^2, \ 4\left(2\cos\left(\frac{\pi}{25}\right) - \sin\left(\frac{2\pi}{25}\right)\right)\sin\left(\frac{\pi}{25} - \frac{1}{5}\right)\right]$$

// Segment tangent vector 'dCO':

```
> dCOm:=subs(t=rgt[1]+dt,dCO);
```

Tangent vector 'dCO's average value that corresponds to plane closed curve 'CO's first segmental cell

$$dCOm := \left[-2\sin\left(\frac{\pi}{25}\right) - 2\cos\left(\frac{2\pi}{25}\right), \ 4\cos\left(\frac{\pi}{25} - \frac{1}{5}\right)\right]$$

// Calculate microcosmic curve integral value of plane closed curve 'CO's first segmental cell:

```
> dt*(dPVm[1]*dCOm[1]+dPVm[2]*dCOm[2]);
```

Base on integral median theorem, integral value of dot product (plane vector field 'V' and tangent vector 'dCO') at first segmental cell

$$\frac{1}{25}\pi\left(\left(\cos\left(\frac{\pi}{25}\right) - \frac{1}{2}\sin\left(\frac{2\pi}{25}\right) - \frac{4}{3}\sin\left(\frac{\pi}{25} - \frac{1}{5}\right)\right)^2\left(-2\sin\left(\frac{\pi}{25}\right) - 2\cos\left(\frac{2\pi}{25}\right)\right)\right.$$
$$\left. + 16\left(2\cos\left(\frac{\pi}{25}\right) - \sin\left(\frac{2\pi}{25}\right)\right)\sin\left(\frac{\pi}{25} - \frac{1}{5}\right)\sin\left(\frac{23\pi}{50} + \frac{1}{5}\right)\right)$$

Microcosmic curve integral course of plane closed curve 'CO's all segmental cells (50 segmental cells):

// Segment plane vector field 'PV':

(This step possesses formal meaning in 'Proof' only, and possesses essential meaning in 'Numerical Models')

```
> idPV:=subs(t=rgt[1]+i*dt,PV);
```

Plane vector field 'PV's average values that correspond to plane closed curve 'CO's all segmental cells

$$idPV := \left[\left(\cos\left(\frac{i\pi}{25}\right) - \frac{1}{2}\sin\left(\frac{2i\pi}{25}\right) - \frac{4}{3}\sin\left(\frac{i\pi}{25} - \frac{1}{5}\right)\right)^2, \right.$$
$$\left. 4\left(2\cos\left(\frac{i\pi}{25}\right) - \sin\left(\frac{2i\pi}{25}\right)\right)\sin\left(\frac{i\pi}{25} - \frac{1}{5}\right)\right]$$

```
> seq(([i,idPV]),i=1..dus):
```

```
# List of plane vector field 'PV's average values that correspond to
plane closed curve 'CO's all segmental cells, be elided
//'idPV' isn't that 'Single vector value', but that 'Set of finite
quantitative (Viz.50) vector values'
```

```
// Segment tangent vector 'dCO':
> idCO:=subs(t=rgt[1]+i*dt,dCO);
# Tangent vector 'dCO's average values that correspond to plane closed
curve 'CO's all segmental cells
```

$$idCO := \left[-2\sin\left(\frac{i\,\pi}{25}\right) - 2\cos\left(\frac{2\,i\,\pi}{25}\right),\, 4\cos\left(\frac{i\,\pi}{25} - \frac{1}{5}\right) \right]$$

```
> seq(([i,idCO]),i=1..dus):
# List of tangent vector 'dCO's average values that correspond to plane
closed curve 'CO's all segmental cells, be elided
//'idCO' isn't that 'single vector value', but that 'set of finite
quantitative (Viz. 50) vector values'
```

```
// Calculate microcosmic curve integral values of plane closed curve
'CO's all segmental cells:
> dt*(idPV[1]*idCO[1]+idPV[2]*idCO[2]);
# Base on integral median theorem, integral values of dot product (Plane
vector field 'PV' and tangent vector 'dCO') at all segmental cells
```

$$\frac{1}{25}\pi\left(\left(\cos\left(\frac{i\,\pi}{25}\right) - \frac{1}{2}\sin\left(\frac{2\,i\,\pi}{25}\right) - \frac{4}{3}\sin\left(\frac{i\,\pi}{25} - \frac{1}{5}\right)\right)^2\left(-2\sin\left(\frac{i\,\pi}{25}\right) - 2\cos\left(\frac{2\,i\,\pi}{25}\right)\right)\right.$$
$$\left. + 16\left(2\cos\left(\frac{i\,\pi}{25}\right) - \sin\left(\frac{2\,i\,\pi}{25}\right)\right)\sin\left(\frac{i\,\pi}{25} - \frac{1}{5}\right)\cos\left(\frac{i\,\pi}{25} - \frac{1}{5}\right)\right)$$

```
// Expression above-mentioned isn't that 'single value', but that 'set
of finite quantitative (viz.50) values'
```

```
// List of finite quantitative microcosmic curve integral values:
> seq(([i,dt*(idPV[1]*idCO[1]+idPV[2]*idCO[2])]),i=1..dus):
# List of integral values of dot product(Plane vector field 'PV' and
tangent vector 'dCO') at all segmental cells, be elided
// Structure sequence that is composed of finite quantitative
microcomosic curl integral values:
> sqn:=seq(dt*(idPV[1]*idCO[1]+idPV[2]*idCO[2]),i=1..dus):
```

```
// Accumulational sum of sequence, obtain plane closed curve integral
value at manifold:
> add(k,k=sqn):xi:=evalf(%);     (4)
```
Sum of integral values of dot product(Plane vector field 'PV' and
tangent vector 'dCO')at all segmental cells

$$\xi := -18.83679675$$


```
// Renewedly segment range '[0,2*Pi]' of 't':
> dus:=n;
```
Set amount 'dus' of plane closed curve 'CO's parameter segmental
cells as natural number 'n'

$$dus := n$$

```
> dt:=(rgt[2]-rgt[1])/dus; # Segment range '[0,2*Pi]' of 't'
```

$$dt := \frac{2\pi}{n}$$

Microcosmic curve integral course of plane closed curve 'CO's first
segmental cell (n segmental cells):

```
// Segment plane vector field 'PV':
```
(This step possesses formal meaning in 'Proof' only, and possesses
essential meaning in 'Numerical Models')
```
> dPVm:=subs(t=rgt[1]+dt,PV);
```
Plane vector field 'PV's average value that corresponds to plane
closed curve 'CO's first segmental cell

$$dPVm := \left[\left(\cos\left(\frac{2\pi}{n}\right) - \frac{1}{2}\sin\left(\frac{4\pi}{n}\right) - \frac{4}{3}\sin\left(\frac{2\pi}{n} - \frac{1}{5}\right) \right)^2, \right.$$
$$\left. 4\left(2\cos\left(\frac{2\pi}{n}\right) - \sin\left(\frac{4\pi}{n}\right) \right)\sin\left(\frac{2\pi}{n} - \frac{1}{5}\right) \right]$$

```
// Segment tangent vector 'dCO':
> dCOm:=subs(t=rgt[1]+dt,dCO);
```
Tangent vector 'dCO's average value that corresponds to plane closed
curve 'CO's first segmental cell

$$dCOm := \left[-2\sin\left(\frac{2\pi}{n}\right) - 2\cos\left(\frac{4\pi}{n}\right), 4\cos\left(\frac{2\pi}{n} - \frac{1}{5}\right) \right]$$

```
// Calculate microcosmic curve integral value of plane closed
curve 'CO's first segmental cell:
```

```
> dt*(dPVm[1]*dCOm[1]+dPVm[2]*dCOm[2]);
```
Base on integral median theorem, integral value of dot product (plane vector field 'V' and tangent vector 'dCO') at first segmental cell

$$2\pi\left(\left(\cos\left(\frac{2\pi}{n}\right)-\frac{1}{2}\sin\left(\frac{4\pi}{n}\right)+\frac{4}{3}\sin\left(-\frac{2\pi}{n}+\frac{1}{5}\right)\right)^2\left(-2\sin\left(\frac{2\pi}{n}\right)-2\cos\left(\frac{4\pi}{n}\right)\right)\right.$$
$$\left.-16\left(2\cos\left(\frac{2\pi}{n}\right)-\sin\left(\frac{4\pi}{n}\right)\right)\sin\left(-\frac{2\pi}{n}+\frac{1}{5}\right)\cos\left(-\frac{2\pi}{n}+\frac{1}{5}\right)\right)/n$$

Microcosmic curve integral course of plane closed curve 'CO's all segmental cells (n segmental cells):

// Segment plane vector field 'PV':

(This step possesses formal meaning in 'Proof' only, and possesses essential meaning in 'Numerical Models')

```
> idPV:=subs(t=rgt[1]+i*dt,PV);
```
Plane vector field 'PV's average values that correspond to plane closed curve 'CO's all segmental cells

$$idPV:=\left[\left(\cos\left(\frac{2i\pi}{n}\right)-\frac{1}{2}\sin\left(\frac{4i\pi}{n}\right)-\frac{4}{3}\sin\left(\frac{2i\pi}{n}-\frac{1}{5}\right)\right)^2,\right.$$
$$\left.4\left(2\cos\left(\frac{2i\pi}{n}\right)-\sin\left(\frac{4i\pi}{n}\right)\right)\sin\left(\frac{2i\pi}{n}-\frac{1}{5}\right)\right]$$

// 'idPV' isn't that 'Single vector value', but that 'Set of finite quantitative (viz.n) vector values'

// Segment tangent vector 'dCO':
```
> idCO:=subs(t=rgt[1]+i*dt,dCO);
```
Tangent vector 'dCO's average values that correspond to plane closed curve 'CO's all segmental cells

$$idCO:=\left[-2\sin\left(\frac{2i\pi}{n}\right)-2\cos\left(\frac{4i\pi}{n}\right),4\cos\left(\frac{2i\pi}{n}-\frac{1}{5}\right)\right]$$

// 'idCO' isn't that 'Single vector value', but that 'Set of finite quantitative (viz.n) vector values'

// Calculate microcosmic curve integral values of plane closed curve 'CO's all segmental cells:
```
> dt*(idPV[1]*idCO[1]+idPV[2]*idCO[2]);
```
Base on integral median theorem, integral values of dot product (Plane

vector field 'PV' and tangent vector 'dCO') at all segmental cells

$$2\pi\left(\left(\cos\left(\frac{2\,i\,\pi}{n}\right)-\frac{1}{2}\sin\left(\frac{4\,i\,\pi}{n}\right)-\frac{4}{3}\sin\left(\frac{2\,i\,\pi}{n}-\frac{1}{5}\right)\right)^2\left(-2\sin\left(\frac{2\,i\,\pi}{n}\right)-2\cos\left(\frac{4\,i\,\pi}{n}\right)\right)\right.$$
$$\left.+16\left(2\cos\left(\frac{2\,i\,\pi}{n}\right)-\sin\left(\frac{4\,i\,\pi}{n}\right)\right)\sin\left(\frac{2\,i\,\pi}{n}-\frac{1}{5}\right)\cos\left(\frac{2\,i\,\pi}{n}-\frac{1}{5}\right)\right)/n$$

// Expression above-mentioned isn't that 'Single value', but that 'Set of finite quantitative (viz.n) values'

// Structure finite sums:

> Sum(dt*(idPV[1]*idCO[1]+idPV[2]*idCO[2]),i=1..dus);

In the case of "Amount 'dus' of plane closed curve 'CO's parameter segmental cells is pending", sum of integral values (dot product about plane vector field 'PV' and tangent vector 'dCO') at all segmental cells

$$\sum_{i=1}^{n}\left(2\pi\left(\vphantom{\frac{2\,i\,\pi}{n}}\right.\right.$$
$$\left(\cos\left(\frac{2\,i\,\pi}{n}\right)-\frac{1}{2}\sin\left(\frac{4\,i\,\pi}{n}\right)-\frac{4}{3}\sin\left(\frac{2\,i\,\pi}{n}-\frac{1}{5}\right)\right)^2\left(-2\sin\left(\frac{2\,i\,\pi}{n}\right)-2\cos\left(\frac{4\,i\,\pi}{n}\right)\right)$$
$$\left.\left.+16\left(2\cos\left(\frac{2\,i\,\pi}{n}\right)-\sin\left(\frac{4\,i\,\pi}{n}\right)\right)\sin\left(\frac{2\,i\,\pi}{n}-\frac{1}{5}\right)\cos\left(\frac{2\,i\,\pi}{n}-\frac{1}{5}\right)\right)/n\right)$$

> vs:=value(%):

> Limit(vs,n=infinity):   **(5)**

// Infinitize finite sums, its limit operational value is 'Closed Curve Integral Value at Manifold'

In the case of "Amount 'n' of plane closed curve 'CO's segmental cells tends to infinity", limit value of integral values' sums (dot product about plane vector field 'PV' and tangent vector 'dCO') at all segmental cells

> alpha:=value(%);delta:=evalf(alpha);

$$\alpha := -\frac{112}{9}\pi\cos\left(\frac{1}{5}\right)^2+\frac{56\pi}{9}-\frac{4}{3}\pi\sin\left(\frac{1}{5}\right)$$

$$\delta := -18.83679673$$

> xi;delta;

There is tiny difference between 'Integral values' sum of finite quantitative segmental cells' and 'Integral values' sum of infinite quantitative segmental cells'

$$-18.83679675$$

$$-18.83679673$$

```
> x:='x':y:='y':
> COc:=subs(t=u,CO);
```

\# Change character 't' to 'u' in expression of curve 'CO'

$$COc := \left[2\cos(u) - \sin(2u),\, 4\sin\!\left(u - \frac{1}{5}\right) \right]$$

```
> [r*COc[1],r*COc[2]];    (6)
```

\# Entirely multiply radius vector 'r' (Suppose r>0) by each item of curve 'CO' parametrized expression, Convert 'Parametrized expression of curve' to 'Parametrized expression of curve coordinates'

$$\left[r\,(2\cos(u) - \sin(2u)),\, 4\,r\sin\!\left(u - \frac{1}{5}\right) \right]$$

```
> x:=r*COc[1]:y:=r*COc[2]:
> matrix(2,2,[Diff(r*COc[1],r),Diff(r*COc[1],u),Diff(r*COc[2],r),
Diff(r*COc[2],u)])=
matrix(2,2,[diff(r*COc[1],r),diff(r*COc[1],u),diff(r*COc[2],r),
diff(r*COc[2],u)]);m:=rhs(%);
```

\# Define and calculate matrix of partial derivatives 'm', obtain general expression of curve coordinates area element coefficient

$$\left[\begin{array}{cc} \dfrac{\partial}{\partial r}\,(r\,(2\cos(u) - \sin(2u))) & \dfrac{\partial}{\partial u}\,(r\,(2\cos(u) - \sin(2u))) \\[2ex] \dfrac{\partial}{\partial r}\!\left(4\,r\sin\!\left(u - \dfrac{1}{5}\right)\right) & \dfrac{\partial}{\partial u}\!\left(4\,r\sin\!\left(u - \dfrac{1}{5}\right)\right) \end{array} \right] =$$

$$\left[\begin{array}{cc} 2\cos(u) - \sin(2u) & r\,(-2\sin(u) - 2\cos(2u)) \\[2ex] 4\sin\!\left(u - \dfrac{1}{5}\right) & 4\,r\cos\!\left(u - \dfrac{1}{5}\right) \end{array} \right]$$

$$m := \left[\begin{array}{cc} 2\cos(u) - \sin(2u) & r\,(-2\sin(u) - 2\cos(2u)) \\[2ex] 4\sin\!\left(u - \dfrac{1}{5}\right) & 4\,r\cos\!\left(u - \dfrac{1}{5}\right) \end{array} \right]$$

```
> det(m):
> J:=simplify(%);    (7)
```

\# General expression of curve coordinates area element coefficient

$$J := -4\,r\left(-2\cos\!\left(u - \frac{1}{5}\right)\cos(u) + \cos\!\left(u - \frac{1}{5}\right)\sin(2u) - 2\sin\!\left(u - \frac{1}{5}\right)\sin(u)\right.$$
$$\left. - 2\sin\!\left(u - \frac{1}{5}\right)\cos(2u)\right)$$

```
// Segment range '[0,1],[0,2*Pi]' of 'r,u':
```

```
> dus:=50;
```
Set amount of plane bounded closed region 'S's (Plane closed curve
'CO' surrounds) parameter segmental cells, this amount can be
discretional natural number

$$dus := 50$$

```
> rgr:=[0,1];
```
// Be different to ordinary double integrals, in this position, first
integral range is 'r[0,1]' always, and not 'r[0,n] or r[0,∞]'; Because
of '1', it is possible to hold a correct ratio between 'a,b,r,u'

$$rgr := [0, 1]$$

```
> dr:=(rgr[2]-rgr[1])/dus; # Segment range of 'r'
```

$$dr := \frac{1}{50}$$

```
> du:=(rgt[2]-rgt[1])/dus; # Segment range of 'u'
```

$$du := \frac{\pi}{25}$$

Microcosmic double integrals course of plane bounded closed region 'S's first segmental cell (50 segmental cells):
// Segment area element coefficient 'J':
```
> dJ:=subs(r=rgr[1]+dr,subs(u=rgt[1]+du,J));
```
Plane area element coefficient 'J's average value that corresponds
to plane bounded closed region 'S's first segmental cell

$$dJ := \frac{4}{25}\cos\left(\frac{\pi}{25}-\frac{1}{5}\right)\cos\left(\frac{\pi}{25}\right) - \frac{2}{25}\cos\left(\frac{\pi}{25}-\frac{1}{5}\right)\sin\left(\frac{2\pi}{25}\right) + \frac{4}{25}\sin\left(\frac{\pi}{25}-\frac{1}{5}\right)\sin\left(\frac{\pi}{25}\right)$$
$$+ \frac{4}{25}\sin\left(\frac{\pi}{25}-\frac{1}{5}\right)\cos\left(\frac{2\pi}{25}\right)$$

// Segment idiographic differential function 'dPV':
// Be different to abstract differential function 'dPV1,dPV2' that
'Proof' refers to, in 'Numerical Models', be able to directly input
idiographic segmented values of 'r,u' to idiographic differential
function 'dPV', then segment 'dPV' in accordance with amount 'dus'
of parameter segmental cells
```
> ddPV:=subs(r=rgr[1]+dr,subs(u=rgt[1]+du,dPV));
```
Differential function 'dPV's average value that corresponds to
plane bounded closed region 'S's first segmental cell

$$ddPV := \frac{14}{225}\sin\left(\frac{\pi}{25}-\frac{1}{5}\right)+\frac{1}{75}\cos\left(\frac{\pi}{25}\right)-\frac{1}{150}\sin\left(\frac{2\pi}{25}\right)$$

// Calculate microcosmic double integrals value of plane bounded closed region 'S's first segmental cell:

> ddPV*dJ*du*dr;

Base on integral median theorem, integral value of product (differential function 'dPV' and curve coordinates area element) in first segmental cell

$$\frac{1}{1250}\left(\frac{14}{225}\sin\left(\frac{\pi}{25}-\frac{1}{5}\right)+\frac{1}{75}\cos\left(\frac{\pi}{25}\right)-\frac{1}{150}\sin\left(\frac{2\pi}{25}\right)\right)\left(\frac{4}{25}\sin\left(\frac{23\pi}{50}+\frac{1}{5}\right)\cos\left(\frac{\pi}{25}\right)\right.$$

$$-\frac{2}{25}\sin\left(\frac{23\pi}{50}+\frac{1}{5}\right)\sin\left(\frac{2\pi}{25}\right)+\frac{4}{25}\sin\left(\frac{\pi}{25}-\frac{1}{5}\right)\sin\left(\frac{\pi}{25}\right)+\frac{4}{25}\sin\left(\frac{\pi}{25}-\frac{1}{5}\right)\cos\left(\frac{2\pi}{25}\right)$$

$$\left.\right)\pi$$

Microcosmic double integrals course of plane bounded closed region 'S's all segmental cells (50 segmental cells):

// Segment area element coefficient 'J':

> ijdJ:=subs(r=rgr[1]+j*dr,subs(u=rgt[1]+i*du,J));

Area element coefficient 'J's average values that correspond to plane bounded closed region 'S's all segmental cells

$$ijdJ := -\frac{2}{25}j\left(-2\cos\left(\frac{i\pi}{25}-\frac{1}{5}\right)\cos\left(\frac{i\pi}{25}\right)+\cos\left(\frac{i\pi}{25}-\frac{1}{5}\right)\sin\left(\frac{2i\pi}{25}\right)\right.$$

$$-2\sin\left(\frac{i\pi}{25}-\frac{1}{5}\right)\sin\left(\frac{i\pi}{25}\right)-2\sin\left(\frac{i\pi}{25}-\frac{1}{5}\right)\cos\left(\frac{2i\pi}{25}\right)\left.\right)$$

//'ijdJ' isn't that 'Single value', but that 'Set of finite quantitative (viz.50) values'

// Segment idiographic differential function 'dPV':

// Be different to abstract differential function 'dPV1,dPV2' that 'Proof' refers to, in 'Numerical Models', be able to directly input idiographic segmented values of 'r,u' to idiographic differential function 'dPV', then segment 'dPV' in accordance with amount 'dus' of parameter segmental cells

> ijddPV:=subs(r=rgr[1]+j*dr,subs(u=rgt[1]+i*du,dPV));

Differential function 'dPV's average values that correspond to plane bounded closed region 'S's all segmental cells

$$ijddPV := \frac{14}{225}j\sin\left(\frac{i\pi}{25}-\frac{1}{5}\right) + \frac{1}{150}j\left(2\cos\left(\frac{i\pi}{25}\right)-\sin\left(\frac{2i\pi}{25}\right)\right)$$

//'ijddPV' isn't that 'Single value', but that 'Set of finite
quantitative (viz.50) values'

// Calculate microcosmic double integrals values of plane closed
region 'S's all segmental cells:

> ijddPV*ijdJ*du*dr;

Base on integral median theorem, integral value of product
(Differential function 'dPV' and curve coordinates area element) in
all segmental cells

$$-\frac{1}{15625}\left(\frac{14}{225}j\sin\left(\frac{i\pi}{25}-\frac{1}{5}\right)+\frac{1}{150}j\left(2\cos\left(\frac{i\pi}{25}\right)-\sin\left(\frac{2i\pi}{25}\right)\right)\right)j\left(\right.$$
$$\left. -2\cos\left(\frac{i\pi}{25}-\frac{1}{5}\right)\cos\left(\frac{i\pi}{25}\right)+\cos\left(\frac{i\pi}{25}-\frac{1}{5}\right)\sin\left(\frac{2i\pi}{25}\right)-2\sin\left(\frac{i\pi}{25}-\frac{1}{5}\right)\sin\left(\frac{i\pi}{25}\right)\right.$$
$$\left. -2\sin\left(\frac{i\pi}{25}-\frac{1}{5}\right)\cos\left(\frac{2i\pi}{25}\right)\right)\pi$$

// Expression above-mentioned isn't that 'Single value', but that
'Set of finite quantitative (viz.50) values'

// List of finite quantitative microcosmic double integrals values:
> seq(seq([i*j,ijddPV*ijdJ*du*dr],i=1..dus),j=1..dus):
List of integral value of product (Differential function 'dPV' and
curve 'L' coordinates area element) in all segmental cells, be elided
// Structure sequence that is composed of finite quantitative
microcomosic double integrals values:
> sqn:=seq(seq(ijddPV*ijdJ*du*dr,i=1..dus),j=1..dus):
// Accumulational sum of sequence, obtain double integrals value at
manifold (Entire plane bounded closed region 'S'):
> add(k,k=sqn):omega:=evalf(%);   **(8)**
Integral values' sum of product (Differential function 'dPV' and
curve 'L' coordinates area element)in all segmental cells
$$\omega := -19.40566801$$

// Renewedly segment range '[0,1],[0,2*Pi]' of 'r,u':

> dus:=n;

Set amount of plane bounded closed region 'S's (plane closed curve

'CO' surrounds) parameter segmental cells as natural number 'n'

$$dus := n$$

```
> dr:=(rgr[2]-rgr[1])/dus; # Segment range of 'r'
```

$$dr := \frac{1}{n}$$

```
> du:=(rgt[2]-rgt[1])/dus; # Segment range of 'u'
```

$$du := \frac{2\pi}{n}$$

Microcosmic double integrals course of plane bounded closed region 'S's first segmental cell (n segmental cells):

// Segment area element coefficient 'J':

```
> dJ:=subs(r=rgr[1]+dr,subs(u=rgt[1]+du,J));
```

Plane area element coefficient 'J's average value that corresponds to plane bounded closed region 'S's first segmental cell

$$dJ := -4\left(-2\cos\left(\frac{2\pi}{n}-\frac{1}{5}\right)\cos\left(\frac{2\pi}{n}\right)+\cos\left(\frac{2\pi}{n}-\frac{1}{5}\right)\sin\left(\frac{4\pi}{n}\right)\right.$$
$$\left.-2\sin\left(\frac{2\pi}{n}-\frac{1}{5}\right)\sin\left(\frac{2\pi}{n}\right)-2\sin\left(\frac{2\pi}{n}-\frac{1}{5}\right)\cos\left(\frac{4\pi}{n}\right)\right)/n$$

// Segment idiographic differential function 'dPV':

// Be different to abstract differential function 'dPV1,dPV2' that 'Proof' refers to, in 'Numerical Models', be able to directly input idiographic segmented values of 'r,u' to idiographic differential function 'dPV', then segment 'dPV' in accordance with amount 'dus' of parameter segmental cells

```
> ddPV:=subs(r=rgr[1]+dr,subs(u=rgt[1]+du,dPV));
```

Differential function 'dPV's average value that corresponds to plane bounded closed region 'S's first segmental cell

$$ddPV := \frac{28}{9}\frac{\sin\left(\frac{2\pi}{n}-\frac{1}{5}\right)}{n}+\frac{1}{3}\frac{2\cos\left(\frac{2\pi}{n}\right)-\sin\left(\frac{4\pi}{n}\right)}{n}$$

// Calculate microcosmic double integrals value of plane closed region 'S's first segmental cell:

```
> ddPV*dJ*du*dr;
```

Base on integral median theorem, integral value of product (Differential function 'dPV' and Curve coordinates area element) in

first segmental cell

$$-8\left(-\frac{28}{9}\frac{\sin\left(-\frac{2\pi}{n}+\frac{1}{5}\right)}{n}+\frac{1}{3}\frac{2\cos\left(\frac{2\pi}{n}\right)-\sin\left(\frac{4\pi}{n}\right)}{n}\right)\left(-2\cos\left(-\frac{2\pi}{n}+\frac{1}{5}\right)\cos\left(\frac{2\pi}{n}\right)\right.$$

$$+\cos\left(-\frac{2\pi}{n}+\frac{1}{5}\right)\sin\left(\frac{4\pi}{n}\right)+2\sin\left(-\frac{2\pi}{n}+\frac{1}{5}\right)\sin\left(\frac{2\pi}{n}\right)$$

$$\left.+2\sin\left(-\frac{2\pi}{n}+\frac{1}{5}\right)\cos\left(\frac{4\pi}{n}\right)\right)\pi\,/\,n^3$$

Microcosmic double integrals course of plane bounded closed region 'S's all segmental cells (n segmental cells):

// Segment area element coefficient 'J':

> ijdJ:=subs(r=rgr[1]+j*dr,subs(u=rgt[1]+i*du,J));

Area element coefficient 'J's average values that correspond to plane bounded closed region 'S's all segmental cells

$$ijdJ:=-4j\left(-2\cos\left(\frac{2i\pi}{n}-\frac{1}{5}\right)\cos\left(\frac{2i\pi}{n}\right)+\cos\left(\frac{2i\pi}{n}-\frac{1}{5}\right)\sin\left(\frac{4i\pi}{n}\right)\right.$$

$$\left.-2\sin\left(\frac{2i\pi}{n}-\frac{1}{5}\right)\sin\left(\frac{2i\pi}{n}\right)-2\sin\left(\frac{2i\pi}{n}-\frac{1}{5}\right)\cos\left(\frac{4i\pi}{n}\right)\right)/n$$

//'ijdJ' isn't that 'Single value', but that 'Set of finite quantitative (viz.n) values'

// Segment idiographic differential function 'dPV':

// Be different to abstract differential function 'dPV1,dPV2' that 'Proof' refers to, in'Numerical Models', be able to directly input idiographic segmented values of 'r,u' to idiographic differential function 'dPV', then segment 'dPV' in accordance with amount 'dus' of parameter segmental cells

> ijddPV:=subs(r=rgr[1]+j*dr,subs(u=rgt[1]+i*du,dPV));

Differential function 'dPV's average values that correspond to plane bounded closed region 'S's all segmental cells

$$ijddPV:=\frac{28}{9}\frac{j\sin\left(\frac{2i\pi}{n}-\frac{1}{5}\right)}{n}+\frac{1}{3}\frac{j\left(2\cos\left(\frac{2i\pi}{n}\right)-\sin\left(\frac{4i\pi}{n}\right)\right)}{n}$$

//'ijddPV' isn't that 'Single value', but that 'Set of finite quantitative (viz.n) values'

// Calculate microcosmic double integrals values of plane bounded closed region 'S's all segmental cells:

```
> ijddPV*ijdJ*du*dr;
```

\# Base on integral median theorem, integral value of product (Differential function 'dPV' and Curve 'L' coordinates area element) in all segmental cells

$$-8\left(\frac{28}{9}\frac{j\sin\left(\frac{2\,i\,\pi}{n}-\frac{1}{5}\right)}{n}+\frac{1}{3}\frac{j\left(2\cos\left(\frac{2\,i\,\pi}{n}\right)-\sin\left(\frac{4\,i\,\pi}{n}\right)\right)}{n}\right)j\Bigg($$

$$-2\cos\left(\frac{2\,i\,\pi}{n}-\frac{1}{5}\right)\cos\left(\frac{2\,i\,\pi}{n}\right)+\cos\left(\frac{2\,i\,\pi}{n}-\frac{1}{5}\right)\sin\left(\frac{4\,i\,\pi}{n}\right)$$

$$-2\sin\left(\frac{2\,i\,\pi}{n}-\frac{1}{5}\right)\sin\left(\frac{2\,i\,\pi}{n}\right)-2\sin\left(\frac{2\,i\,\pi}{n}-\frac{1}{5}\right)\cos\left(\frac{4\,i\,\pi}{n}\right)\Bigg)\pi\,/\,n^3$$

// Expression above-mentioned isn't that 'Single value', but that 'Set of finite quantitative (viz.n) values'

// Structure finite sums:

```
> Sum(Sum(ijddPV*ijdJ*du*dr,i=1..dus),j=1..dus);
```

\# In the case of "Amount 'n' of plane bounded closed region 'S's segmental cells is pending", sum of integral values (product about differential function 'dPV' and curve 'L' coordinates area element) in all segmental cells

$$\sum_{j=1}^{n}\left(\sum_{i=1}^{n}\left(-8\left(\frac{28}{9}\frac{j\sin\left(\frac{2\,i\,\pi}{n}-\frac{1}{5}\right)}{n}+\frac{1}{3}\frac{j\left(2\cos\left(\frac{2\,i\,\pi}{n}\right)-\sin\left(\frac{4\,i\,\pi}{n}\right)\right)}{n}\right)j\Bigg(\right.\right.$$

$$-2\cos\left(\frac{2\,i\,\pi}{n}-\frac{1}{5}\right)\cos\left(\frac{2\,i\,\pi}{n}\right)+\cos\left(\frac{2\,i\,\pi}{n}-\frac{1}{5}\right)\sin\left(\frac{4\,i\,\pi}{n}\right)$$

$$\left.\left.-2\sin\left(\frac{2\,i\,\pi}{n}-\frac{1}{5}\right)\sin\left(\frac{2\,i\,\pi}{n}\right)-2\sin\left(\frac{2\,i\,\pi}{n}-\frac{1}{5}\right)\cos\left(\frac{4\,i\,\pi}{n}\right)\Bigg)\pi\,/\,n^3\right)\right)$$

```
> vs:=value(%):
> Limit(vs,n=infinity):     (9)
```

// Infinitize finite sums, its limit operational value is 'Double Integrals Value at Manifold'

\# In the case of "Amount 'n' of plane bounded closed region 'S's segmental cells tends to infinity", limit value of integral values' sums (product about differential function 'dPV' and curve 'L' coordinates area element) in all segmental cells

```
> beta:=value(%);epsilon:=evalf(beta);
```

$$\beta := -\frac{112}{9}\,\pi\cos\left(\frac{1}{5}\right)^2 + \frac{56\,\pi}{9} - \frac{4}{3}\,\pi\sin\left(\frac{1}{5}\right)$$

$$\epsilon := -18.83679673$$

```
> omega;epsilon;
```

There is tiny difference between "Integral values' sum of finite quantitative segmental cells" and "Integral values' sum of infinite quantitative segmental cells"

$$-19.40566801$$

$$-18.83679673$$

```
> alpha,beta; # Two analytic values are equal
```

$$-\frac{112}{9}\,\pi\cos\left(\frac{1}{5}\right)^2 + \frac{56\,\pi}{9} - \frac{4}{3}\,\pi\sin\left(\frac{1}{5}\right),\; -\frac{112}{9}\,\pi\cos\left(\frac{1}{5}\right)^2 + \frac{56\,\pi}{9} - \frac{4}{3}\,\pi\sin\left(\frac{1}{5}\right)$$

```
> delta,epsilon; # Two float values are equal
```

$$-18.83679673\; -18.83679673$$

Conclusion A in Part II

Constitute individualized geometric object coordinates that matches with idiographic geometric object (Manifold) (Viz. What idiographic curve, what coordinates of idiographic curve, what area element coefficient of idiographic curve coordinates; no longer rely on a few existent coordinates: Cartesian coordinates、Polar coordinates、Generalized Polar coordinates and their correlative area element coefficients etc.), by two methods (Integral、Finite Sums Limits), prove presence of Green Theorem in unlimited quantitative Individualized Curve coordinates [Summarized and described by abstract simply connected orientable closed surface coordinates (Bases on Poincare Conjecture)], enable Green Theorem surpass traditional architecture of 2-Dimensional Cartesian coordinates, in Individualized Curve coordinates, establish new formular association between double integrals (Bases on idiographic area element coefficient method) and plane closed curve integral, and realize mutual validation between two typical integrals in in unlimited quantitative individualized and gorgeous formular numerical model operations,

978-1-62265-930-2 (online)   978-1-62265-931-9 (paper)

radicate theoretical logic basis and numerical models of new typical double integrals (Bases on idiographic area element coefficient method in Individualized Curve coordinates).

'Prove Green Theorem at Manifold' itself is not sole purpose, 'Establish new formular association between double integrals (Bases on idiographic area element coefficient) and plane closed curve integral in Individualized Curve coordinates, radicate theoretical logic basis and numerical models of new typical double integrals (Bases on idiographic area element coefficient method in Individualized Curve coordinates)' is prime purpose.

Correlative numerical models of this part have indicated, by double integrals (Bases on idiographic area element coefficient method in Individualized Curve coordinates), science explorers can obtain analytic integral value and float integral value in discretional precision about complicated geometric objects (Manifold, Irregular、 Asymmetrical plane bounded closed region especially). Realize free double integrals in discretional plane bounded closed region, indeed realize artistic integral interval; realize exact integral calculation of vector field [Plane electric field、plane magnetic field、plane hydromechanical field etc.] and scalar field [Plane electric potential field、plane temperature field etc.] in discretional free plane bounded closed region and its bounded closed route, radicate logic associated relationship of two typical integrals, seek direct joint point between calculus 、topology and physical engineering calculation, realize Green Theorem at Manifold and manifold integral in physical 、 engineering meaning, realize much more vast and free physical 、 mathematical explorations and engineering practices.

Conclusion B Comparison of Different Methods in Part II

Method	Normalization	Application Range	Calculation Precision	Result Verification
Traditional Double Integrals in 2-Dimensional Cartesian Coordinates	Tedious & Divergent Calculational Method	Symmetric & Regular Geometric Objects, such as Circle, Ellipse, Fan, Rectangle, Triangle & their simple combinations	Analytic Value	No Verification
Traditional Partial Differential Equations, such as Finite Difference Method、Finite Element Method、Boundary Element Method etc.	Tedious & Divergent Calculational Method; Method & Geometric Object, One-to-One; 'Algorithm' Design is a huge project and tedious process	Slightly Complex Geometric objects; Mainstream & Research Direction of Numerical Analysis in Contemporary Physical & Engineering Fields	Approximate Value	Convergence & Stability Verification
New Double Integrals bases on Idiographic Area Element Coefficient Method in Individualized Curve Coordinates	Standardized & Matrixing Calculational Method, Simple & Convergent; Method & Geometric Object, One-to-Many	Discretional Complex Geometric Objects	Analytic Value	One to One Verification by Numerical Models of New Integral Theorem

Part III Curl Theorem at Manifold

Chapter 1 Precondition of Proof

1.1 Introspection between Curl Theorem and Poincare Conjecture

Review object of Proof - Curl Theorem:

Curl Theorem Suppose positive directional boundary 'L+' of smooth or piecewise smooth orientational surface 'S' is smooth or piecewise smooth closed curve, Right Hand Rule is associated basis between positive direction of boundary curve 'L' and outer side of orientational surface 'S'. If there are 1th order continuous partial derivatives about functions 'P(x,y,z),Q(x,y,z),R(x,y,z)' [Structure vector field 'A'] at orientational surface 'S', then:

$$\int_{L+} A \cdot dL = \iint_{S} rotA \cdot n \, dS \qquad (1)$$

thereinto, 'rotA' is curl of vector field 'A', 'n' is unit normal vector of orientational surface 'S's outer side.

In definiens of theorem, it emphasizes that unclosed surface must be orientable surface (Observer can discriminate its outer side or inner side)

In proof of traditional Stokes theorem in Cartesian coordinates, 'Abstract orientable unclosed surface Σ and its boundary curve Γ' is described as that: $\Sigma: z = f(x,y)$ and Γ; thereinto, surface $\Sigma: z = f(x,y)$ is abstract 2-variable function.

(See also <Calculus [6th Edition]> Tongji University, Department of Mathematics; Higher Education Press, Beijing; ed.1, 1978.10; ed.6, 2007.6; 9th Printing, 2009.8; pp.237-239)

In other words, objectively, Curl Theorem requests that no matter in Cartesian coordinates or in other coordinates, correlative surface Σ must possess three attributes: (1) Unclosed; (2) Orientable; (3) Its boundary curve Γ must be closed.

Depart from traditional Cartesian coordinates, How to depict abstract universal 'Orientable Closed Surface' and 'Orientable Unclosed Surface'? How to depict abstract universal 'Space Closed

Curve'? Ready-made solutions are absence.

Poincare Conjecture [19] concludes that 'Every closed n-Manifold which is homotopy equivalent to n-Sphere is homeomorphic to n-Sphere'. In 3-Dimensional Euclidean space that Divergence or Curl theorem refers, corresponding opinion is that: 'Every simply connected、 orientable closed 2-Manifold is homeomorphic to 2-Sphere'.

In other words, according as Poincare Conjecture, in 3-Dimensional Euclidean space that Divergence or Curl theorem refers, every simply connected、orientable closed surface(although it is 'simply connected' scantly),no matter its Protean geometrical shape, 'Be homeomorphic to 2-Sphere' is universal attribute.

Naturally, next question is 'In 3-Dimensional Euclidean space, base on Poincare Conjecture, it is possible to define an abstract universal expression of simply connected、orientable closed surface ?' -- It is one of kernel content in this Chapter.

In spacial analytic geometry, the parametrized expression of '2-Sphere' above-mentioned is '[sin(u)cos(v),sin(u)sin(v), cos(u)]', thereinto, u[0,Pi], v[0,2*Pi] (In proper meaning, this parametrized expression is transform expression of '2-Sphere' between '3-Dimensional Cartesian coordinates' and 'spherical coordinates'; expression of '2-Sphere' in spherical coordinates is constant 1);

In topology field, define 'Homeomorphism' as 'Two manifolds, if it is possible to change one to another by operations of bending、 extending、cutting etc., then recognise that two manifolds above-mentioned are homeomorphous'.

Reconsider Poincare Conjecture by visual angle of topology and analytic geometry, since parametrized expression of '2-Sphere' is '[sin(u)cos(v), sin(u)sin(v), cos(u)], u[0,Pi], v[0,2*Pi]', then its transfiguration '[a*sin(u)cos(v),b*sin(u)sin(v),c*cos(u)],u[0,Pi], v[0,2*Pi]'(thereinto, 'a,b,c' are discretional nonzero constants) is parametrized expression of discretional ellipsoid. In 3-Dimensional Euclidean space, discretional ellipsoid is homeomorphic to 2-Sphere, this is general knowledge of topology, dispense with discussion;

If 'a,b,c' are discretional '1th order derivable continuous functions', what instance appears possibly ? See also below figures:

Figure(III)1.1.1 Non Simply Connected Orientable Closed Surface, Deduced from parametrized expression of `[a*sin(u)cos(v), b*sin(u)sin(v), c*cos(u)], u[0,Pi], v[0, 2*Pi]`

Suppose discretional pending coefficients

a = sin(u)+cos(v), b = cos(u), c = cos(v/2)

(viz. `a,b,c` are discretional `1th order derivable continuous functions`), then target parametrized surface [a*sin(u)*cos(v), b*sin(u)*sin(v), c*cos(u)], u∈[0,π], v∈[0,2π] equals

[(sin(u)+cos(v))*sin(u)*cos(v), cos(u)*sin(u)*sin(v), cos(v/2)*cos(u)], u∈[0,π], v∈[0,2π].

Actual shape of target parametrized surface:

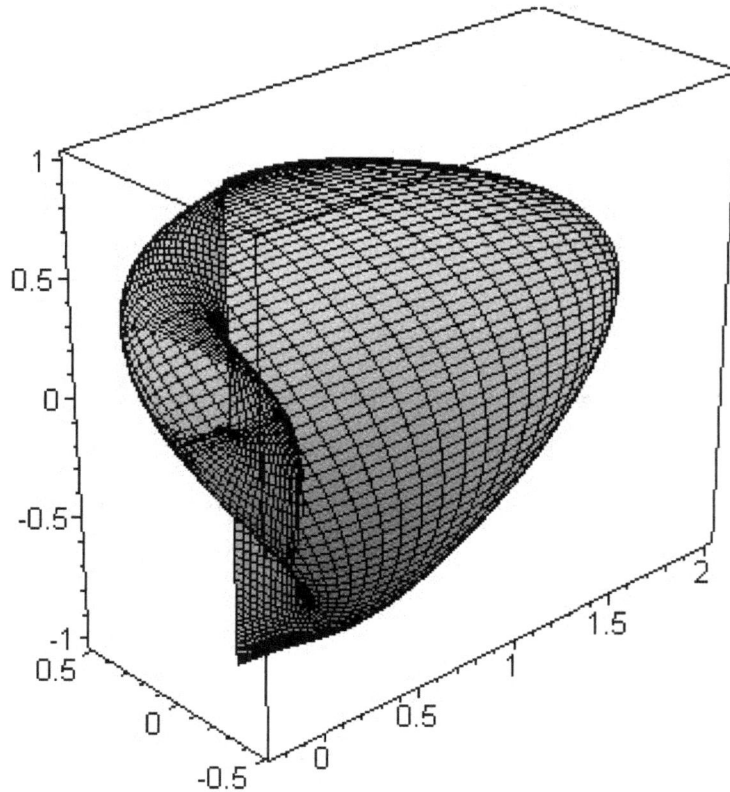

Figure(III)1.1.1 Non Simply Connected Orientable Closed Surface, Deduced from parametrized expression `[a*sin(u)cos(v), b*sin(u)sin(v), c*cos(u)],u[0,Pi], v[0,2*Pi]`

Observe shape of Figure(III)1.1.1 by intuitionistic vision, parts of target parametrized surface above-mentioned show intersected、overlapped、fractured、unclosed states; therefore target parametrized surface belongs to non simply connected orientable closed surface, be independent of 'Poincare Conjecture' and 'Divergence/Curl Theorem at Manifold'.

Figure(III)1.1.2 Non Simply Connected Orientable Closed Surface, Deduced from parametrized expression `[a*sin(u)cos(v), b*sin(u)sin(v), c*cos(u)], u[0,Pi], v[0,2*Pi]'

Suppose discretional pending coefficients

a = sin(u+v)+cos(v), b = cos(v), c = cos(v/2)

(viz. 'a,b,c' are discretional '1th order derivable continuous functions'), then target parametrized surface [a*sin(u)*cos(v), b*sin(u)*sin(v), c*cos(u)], u∈[0, π], v∈[0, 2π] equals

[(sin(u+v)+cos(v))*sin(u)*cos(v), cos(v)*sin(u)*sin(v), cos(v/2)*cos(u)], u∈[0,π], v∈[0,2π]).

Actual shape of target parametrized surface:

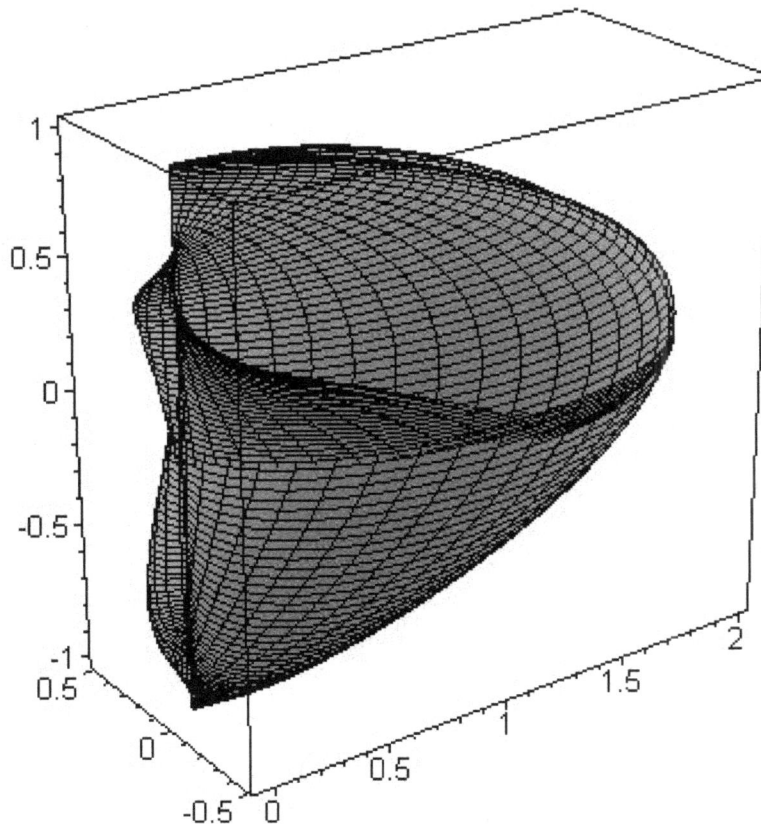

Figure(III)1.1.2 Non Simply Connected Orientable Closed Surface,
Deduced from parametrized expression
`[a*sin(u)cos(v), b*sin(u)sin(v),c*cos(u)], u[0,Pi], v[0,2*Pi]'

Observe shape of Figure(III)1.1.2 by intuitionistic vision, parts of target parametrized surface above-mentioned show intersected、 overlapped、 fractured、 unclosed states; therefore target parametrized surface belongs to non simply connected orientable closed surface, be independent of 'Poincare Conjecture' and 'Divergence/Curl Theorem at Manifold'.

Figure(III)1.1.3 Simply Connected Orientable Closed Surface, Deduced from parametrized expression '[a*sin(u)cos(v), b*sin(u)sin(v), c*cos(u)], u[0,Pi], v[0,2*Pi]'

Suppose discretional pending coefficients

a = sin(u), b = cos(u)+cos(u+3*v)/3, c = cos(u),

(viz. 'a,b,c' are discretional '1th order derivable continuous functions'), then target parametrized surface [a*sin(u)*cos(v), b*sin(u)*sin(v), c*cos(u)], u∈[0,π], v∈[0,2π] equals

[sin(u)*sin(u)*cos(v), (cos(u)+cos(u+3*v)/3)*sin(u)*sin(v), cos(u)*cos(u)], u∈[0,π], v∈[0,2π].

Actual shape of target parametrized surface:

978-1-62265-930-2 (online) 978-1-62265-931-9 (paper)

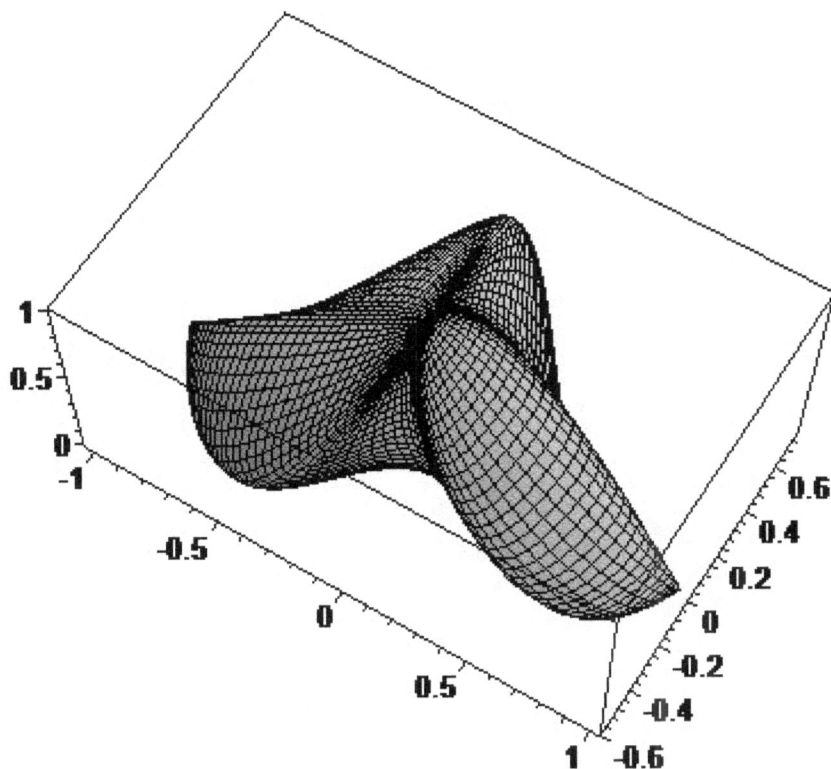

Figure(III)1.1.3 Simply Connected Orientable Closed Surface,

Deduced From parametrized expression

`[a*sin(u)cos(v), b*sin(u)sin(v),c*cos(u)], u[0,Pi], v[0,2*Pi]`

Observe shape of Figure(III)1.1.3 by intuitionistic vision, there aren't intersected、overlapped、fractured、unclosed states about target parametrized surface above-mentioned, and genus value of target parametrized surface is '0'; therefore target parametrized surface belongs to simply connected orientable closed surface, be relational to 'Poincare Conjecture' and 'Divergence/Curl Theorem at Manifold'.

In lay of actual operation, draw certain a parametrized surface by 'Plot3D'(Command of Waterloo Maplesoft), must judge that 'This parametrized surface is simply connected orientable closed surface or not' by intuitionistic vision at first, then decide that 'This parametrized surface is the same with numerical models of Divergence (or curl) theorem at manifold or not' at second. It is impossible to judge that 'Certain a parametrized surface is simply connected orientable closed surface or not' by parametrized expression itself.

Determinant diathesis of 'Parametrized surface is simply connected orientable closed surface or not' is in topology field and not in analytical geometry field; Only method of parametrized expression in analytical geometry is unable to conclude and deduce certain a surface as a simply connected orientable closed surface.

By original phenomenons, experimental data indicate that, similarly belong to parametrized surface '[a*sin(u)cos(v),b*sin(u) sin(v), c*cos(u)], u[0,Pi], v[0,2*Pi]', because of different values of pending coefficients 'a,b,c', a part of surfaces belong to simply connected orientable closed surface, and another part of surfaces are exceptional.

In other words, there are two instances about parametrized surface '[a*sin(u)cos(v),b*sin(u)sin(v),c*cos(u)],u[0,Pi],v[0,2*Pi]':

(1)In the case of pending coefficients 'a,b,c' are discretional nonzero constants, parametrized surface is ellipsoid (Be homeomorphic to 2-Sphere naturally);

(2)In the case of pending coefficients 'a,b,c' are discretional 1th order derivable continuous functions, parametrized surface is simply connected orientable closed surface possibly (Be homeomorphic to 2-Sphere),and parametrized surface isn't simply connected orientable closed surface possibly (Not to be homeomorphic to 2-Sphere).

Naturally, next question is "In mode of parametrized surface [a*sin(u)cos(v),b*sin(u)sin(v),c*cos(u)],u[0,Pi],v[0,2*Pi], it is possible to exclude instance of 'Parametrized surface is not simply connected orientable closed surface' by certain logical defining ?"

1.2 Base on Poincare Conjecture,Define Abstract Simply Connected Orientable Closed Surface and Its Unclosed Surface

Suppose 'Discretional Surfaces' as a set, then 'Discretional Simply Connected Orientable Closed Surfaces' is subset of former, Poincare Conjecture is attribute of this subset, current thesis 'Proof of Divergence or Curl Theorem at Manifold' discuss that Divergence or

Curl Theorem is the same with this subset or not.

Poincare Conjecture supplies the realizing route of describing 'Discretional simply connected orientable closed surface's certain attribute (Viz. 'Be homeomorphic to 2-Sphere' this attribute) by parametrized expression.

Base on situation above-mentioned, abstract unlimited quantitative idiographic simply connected orientable closed surfaces as an uniform expression:

[a*sin(u)cos(v),b*sin(u)sin(v),c*cos(u)], u[0,Pi],v[0,2*Pi]

(thereinto, pending coefficients 'a,b,c' can't be designated discretionarily, 'a,b,c' must obey topology attribute of simply connected orientable closed surface)

In other words, if pending coefficients 'a,b,c' can be designated discretionarily, then target surface '[a*sin(u)cos(v),b*sin(u)sin(v), c*cos(u)],u[0,Pi],v[0,2*Pi]' may be simply connected orientable closed surface, or may be not.

If preestablish that 'Target surface [a*sin(u)cos(v), b*sin(u) sin(v),c*cos(u)], u[0,Pi], v[0,2*Pi] itself is simply connected orientable closed surface', then pending coefficients 'a,b,c' can't be designated discretionarily.

Explain phenomenons above-mentioned on geometrical meaning, in 3-Dimensional Cartesian coordinates, spherical surface (Viz. [sin(u)cos(v), sin(u)sin(v), cos(u)], u[0,Pi], v[0,2*Pi]) itself extends continuously discretionarily along three directions of axis 'x,y,z' (Viz.[a*sin(u)cos(v), b*sin(u)sin(v),c*cos(u)], u[0,Pi], v[0,2*Pi], thereinto, 'a,b,c' are discretional 1th order derivable continuous functions), not always comes into being simply connected orientable closed surface;

Contrarily, in 3-Dimensional Cartesian coordinates, discretional simply connected orientable closed surface,consequentially, based on spherical surface (Viz.[sin(u)cos(v),sin(u)sin(v),cos(u)],u[0,Pi], v[0,2*Pi]), extended continuously along three directions of axis 'x,y,z', came into being (Likewise consequentially, discretional simply connected orientable closed surface extends continuously along three directions of axis 'x,y,z', comes into being spherical surface)

-- Base on Poincare Conjecture.

For example, surface of hexahedron can be regarded as simply connected orientable closed surface -- But surface of hexahedron is difficult or can't be described by parametrized expression -- But it is incontestable: Base on Poincare Conjecture, surface of hexahedron is homeomorphic to 2-Sphere consequentially, surface of hexahedron based on spherical surface (Viz.[sin(u)cos(v),sin(u)sin(v),cos(u)], u[0,Pi],v[0,2*Pi]),extended continuously along three directions of axis 'x,y,z', came into being (Likewise consequentially, surface of hexahedron extends continuously along three directions of axis 'x,y,z', comes into being spherical surface); Base on Poincare Conjecture, similarly, surface of hexahedron can be described by'[a*sin(u)cos(v), b*sin(u)sin(v),c*cos(u)],u[0,Pi],v[0,2*Pi]' mode.

Describe abstract universal simply connected orientable closed surface by '[a*sin(u)cos(v), b*sin(u)sin(v), c*cos(u)], u[0,Pi], v[0,2*Pi]' Mode -- Actually, describe certain inner configuration and attribute (Viz. 'Be homeomorphic to 2-Sphere' this inner configuration and attribute) of simply connected orientable closed surface by Poincare Conjecture, define a felicitous precondition for proofs.

It is necessary to indicate expressly, in the case of 'Idiographic surfaces are multiply connected orientable closed surfaces'(for example, torus and its homeomorphous surfaces), abstraction is impossible yet, correlative theory is absence.

Due to topology classifying of open surface hasn't realized, have to describe unclosed surface by method of "Curtail range of closed surface's parameters" (Viz. u[0,Pi/n-theta],v[0,2*Pi] , thereinto, 'n' is discretional constant, and n ≥ 1; 'θ' is discretional constant or continuous function, and π/n - θ < π);

  [a*sin(u)cos(v), b*sin(u)sin(v), c*cos(u)],

  u[0,Pi/n-theta],v[0,2*Pi] is parametrized expression of abstract simply connected orientable unclosed surface.

1.3 Constitute Abstract Simply Connected Orientable Closed Surface Coordinates

Base on Poincare Conjecture, '[a*sin(u)cos(v),b*sin(u)sin(v), c*cos(u)],u[0,Pi],v[0,2*Pi]'is just abstract universal expression of simply connected orientable closed surface, but it isn't coordinates; abstract simply connected orientable closed surface coordinates is '[r*a*sin(u)cos(v),r*b*sin(u)sin(v),r*c*cos(u)], r[0,∞], u[0,Pi], v[0,2*Pi]', thereinto, 'r' is radius, 'a,b,c' is pending coefficients (Because of 'a,b,c' maybe nonzero constants, maybe 1th order derivable continuous functions),be uncertain.

Actually, expression of abstract simply connected orientable closed surface '[a*sin(u)cos(v),b*sin(u)sin(v),c*cos(u)],u[0,Pi], v[0,2*Pi]' is the same as expression of ellipsoid '[a*sin(u)cos(v), b*sin(u)sin(v), c*cos(u)], u[0,Pi], v[0,2*Pi]' formally, only explanations of pending coefficients 'a,b,c' are different: Former explains 'a,b,c' as 'Discretional nonzero constants or 1th order derivable continuous functions (Non Discretional, be restricted by attribute of simply connected orientable closed surface)'; and latter explains 'a,b,c' as 'Discretional nonzero constants' -- So that corresponding relation of abstract simply connected orientable closed surface coordinates and Cartesian coordinates is 'x=r*a*sin(u)cos(v), y=r*b*sin(u) sin(v),z=r*c cos(u)'(Be the same as transform expression of ellipsoid coordinates and Cartesian coordinates).

(See also Chapter 1 (Part I) 'Precondition of Proof: Constitute Abstract Simply Connected Orientable Closed Parametrized Surface Coordinates')

Current thesis 'Proof of Curl Theorem at Manifold', applies abstract simply connected orientable closed surface coordinates, Viz.

[r*a*sin(u)cos(v), r*b*sin(u)sin(v), r*c*cos(u)],

r[0,∞], u[0,Pi], v[0,2*Pi].

1.4 Base on Poincare Conjecture, Define Abstract Plane Simply Connected Closed Curve

Depart from traditional Cartesian coordinates, how to depict abstract universal 'space closed curve'?

Poincare Conjecture[19] concludes that 'Every closed n-Manifold which is homotopy equivalent to n-Sphere is homeomorphic to n-Sphere'. In 2-Dimensional Euclidean space, corresponding opinion is that: 'Every simply connected closed 1-Manifold is homeomorphic to 1-Sphere (viz. Circle)'.

In other words, according as Poincare Conjecture, in 2-Dimensional Euclidean space, every simply connected closed curve, no matter its Protean geometrical shape, 'Be homeomorphic to 1-Sphere (Circle)' is universal attribute.

Naturally, next question is 'In 2-Dimensional Euclidean space, base on Poincare Conjecture, it is possible to define an abstract universal expression of simply connected closed curve ?' -- It is two of kernel content in this chapter.

In plane analytic geometry, parametrized expression of '1-Sphere' (circle) above-mentioned is '[cos(t),sin(t)]', thereinto, t[0,2*Pi] (In proper meaning, this parametrized expression is transform expression of '1-Sphere' between '2-Dimensional Cartesian coordinates' and 'Polar coordinates', expression of '1-Sphere' in Polar coordinates is constant 1)

In topology field, define 'Homeomorphism' as 'Two manifolds, if it is possible to change one to another by operations of bending、extending、cutting etc., then recognise that two manifolds above-mentioned are homeomorphous'.

Reconsider Poincare Conjecture by visual angle of topology and analytic geometry, since parametrized expression of '1-Sphere' is '[cos(t),sin(t)],t[0,2*Pi]', then its transfiguration '[a*(cos(t), b*sin(t)], t[0,2*Pi]' (Thereinto, 'a,b' are discretional nonzero constants) is parametrized expression of discretional ellipse. In 2-Dimensional Euclidean space, discretional ellipse is homeomorphic to 1-Sphere, this is general knowledge of topology, dispense with discussion;

If 'a,b' are discretional '1th order derivable continuous functions', what instance appears possibly ?

See also below figures:

Figure(III)1.4.1 Non Simply Connected Closed Curve, Deduced from
  parametrized expression '[a*cos(t), b*sin(t)], t∈[0,2π]'

  Suppose discretional pending coefficients
   a = cos(2*t)+2*sin(t)/3, b = sin(3*t)/3
  (viz. 'a,b' are discretional '1th order derivable continuous
functions'), then target parametrized curve '[a*cos(t),b*sin(t)],
  t∈[0,2π]' equals '[(cos(2*t)+2*sin(t)/3)*cos(t),
  sin(3*t)/3*sin(t)], t∈[0,2π]'.

  Actual shape of target parametrized curve:

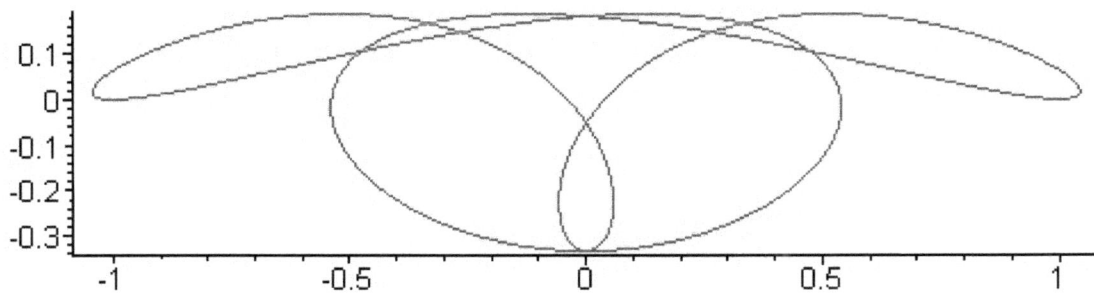

Figure(III)1.4.1 Non Simply Connected Closed Curve, Deduced from
       expression'[a*cos(t),b*sin(t)],t∈[0,2π]'

  Observe shape of Figure(III)1.4.1 by intuitionistic vision, parts
of target parametrized curve above-mentioned show intersected、
overlapped states; therefore target parametrized surface belongs to
non simply connected closed curve, be independent of 'Poincare
Conjecture' and 'Curl Theorem at Manifold'.

Figure(III)1.4.2 Simply Connected Closed Curve, Deduced from
  parametrized expression '[a*cos(t), b*sin(t)], t∈[0,2π]'

  Suppose discretional pending coefficients

a = cos(t-1)+sin(9*t-2)/12, b = sin(t-1)-cos(t)
(viz. 'a,b' are discretional '1th order derivable continuous functions'), then target parametrized curve '[a*cos(t),b*sin(t)], t∈[0,2π]' equals '[(cos(t-1)+sin(9*t-2)/12)*cos(t), (sin(t-1)-cos(t))*sin(t)], t∈[0,2π]'.

Actual shape of target parametrized curve:

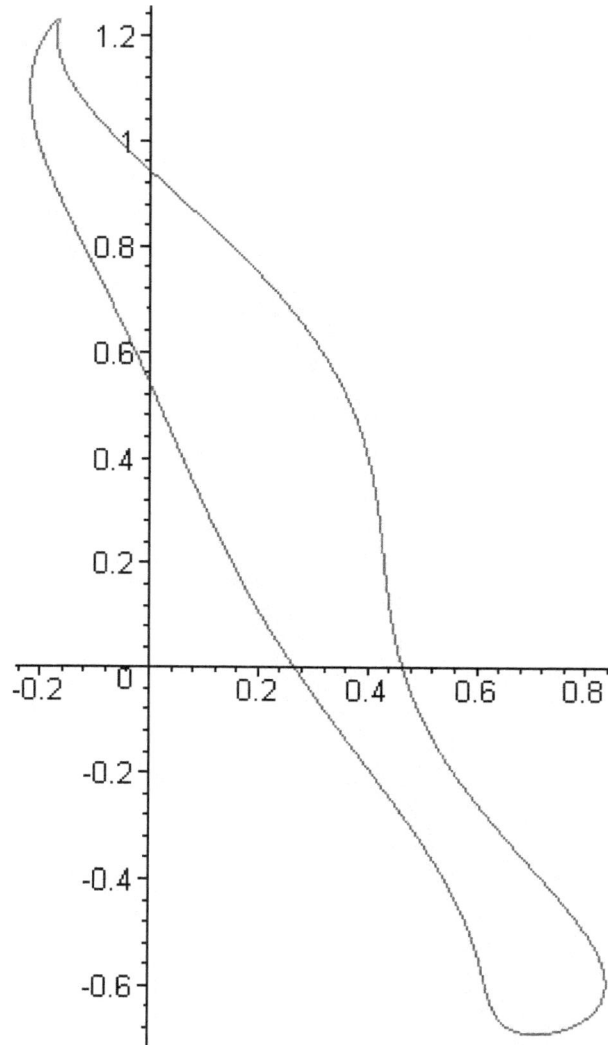

Figure(III)1.4.2 Simply Connected Closed Curve, Deduced from expression'[a*cos(t),b*sin(t)],t∈[0,2π]'

Observe shape of Figure(III)1.4.2 by intuitionistic vision, there aren't intersected、overlapped states about target parametrized curve above-mentioned, therefore target parametrized curve belongs to simply connected closed curve, be relational to 'Poincare Conjecture' and 'Curl Theorem at Manifold'.

978-1-62265-930-2 (online)   978-1-62265-931-9 (paper)

In lay of actual operation, draw certain a parametrized curve by 'Plot'(Command of Waterloo Maplesoft), must judge that 'This parametrized curve is simply connected closed curve or not' by intuitionistic vision; It is impossible to judge that 'Certain a parametrized curve is simply connected closed curve or not' by parametrized expression itself.

Determinant diathesis of 'Parametrized curve (plane) is simply connected closed curve or not' is in topology field and not in analytical geometry field; Only method of parametrized expression in analytical geometry is unable to conclude and deduce certain a curve as a simply connected closed curve.

By original phenomenons, experimental data above-meitioned indicate that, Similarly belong to parametrized curve '[a*cos(t), b*sin(t)],t[0,2*Pi]', because of different values of pending coefficients 'a,b', a part of curves belong to simply connected closed curve, and another part of curves are exceptional.

In other words, there are two instances about parametrized curve '[a*cos(t),b*sin(t)],t[0,2*Pi]':

(1) In the case of pending coefficients 'a,b' are discretional nonzero constants, parametrized curve is ellipse (Be homeomorphic to 1-Sphere naturally);

(2) In the case of pending coefficients 'a,b' are discretional 1th order derivable continuous functions, parametrized curve is simply connected closed curve possibly (Be homeomorphic to 1-Sphere), and parametrized curve isn't simply connected closed curve possibly (Not to be homeomorphic to 1-Sphere).

Naturally, next question is "In the mode of parametrized curve [a*cos(t),b*sin(t)],t[0,2*Pi], it is possible to exclude instance of 'Parametrized curve is not simply connected closed curve' by certain logical defining ?"

Suppose 'Discretional Curves' as a set, then 'Discretional Simply Connected Closed Curves' is the subset of former, Poincare Conjecture is attribute of this subset.

Poincare Conjecture supplies realizing route for describing
`Discretional simply connected closed curve's certain attribute
(Viz. `Be homeomorphic to 1-Sphere' this attribute) by parametrized
expression.

Base on situation above-mentioned, abstract unlimited quantitative
idiographic simply connected closed curves as an uniform expression:
   [a*cos(t),b*sin(t)],t[0,2*Pi]

   (Thereinto, pending coefficients `a,b' can't be designated
discretionarily, `a,b' must obey topology attribute of simply
connected closed curve)

In other words, if pending coefficients `a,b' can be designated
discretionarily, then target curve `[a*cos(t),b*sin(t)],t[0,2*Pi]'
may be simply connected closed curve possibly, or may be not.

If preestablish that `Target curve [a*cos(t),b*sin(t)],t[0,2*Pi]
itself is simply connected closed curve', then pending coefficients
`a,b' can't be designated discretionarily.

Explain phenomenons above-mentioned on geometrical meaning, in
2-Dimensional Cartesian coordinates, circle (Viz. [cos(t),sin(t)],
t[0,2*Pi]) itself extends continuously discretionarily along two
directions of axis `x,y' (Viz. [a*cos(t), b*sin(t)], t[0,2*Pi],
thereinto, `a,b' are discretional 1th order derivable continuous
functions, not always comes into being simply connected closed curve;

Contrarily, in 2-Dimensional Cartesian coordinates, discretional
simply connected closed curve, consequentially, based on circle
(Viz. [cos(t),sin(t)],t[0,2*Pi]), extended continuously along two
directions of axis `x,y', came into being (Likewise consequentially,
discretional simply connected closed curve extends continuously
along two directions of axis `x,y', comes into being circle)-- Base
on Poincare Conjecture.

For example, side lines of square or triangle can be regarded as
simply connected closed curve -- But side lines of square or triangle
is difficult or can't be described by parametrized expression -- But
it is incontestable: Base on Poincare Conjecture, side lines of square
or triangle is homeomorphic to 1-Sphere consequentially, side lines
of square or triangle based on circle (Viz.[cos(t),sin(t)], t[0,2*Pi]),

extended continuously along two directions of axis 'x,y', came into being (Likewise consequentially, side lines of square or triangle extends continuously along two directions of axis 'x,y', comes into being circle); Base on Poincare Conjecture, similarly, side lines of square or triangle can be described by '[a*cos(t), b*sin(t)], t[0,2*Pi]' mode.

Describe abstract universal simply connected closed curve by '[a*cos(t),b*sin(t)],t[0,2*Pi]' Mode -- Actually, describe certain inner configuration and attribute (Viz.'Be homeomorphic to 1-Sphere' this inner configuration and attribute) of simply connected closed curve by Poincare Conjecture, define a felicitous precondition for proof.

(See also Chapter 1 (Part II) 'Precondition of Proof: Constitute Abstract Simply Connected Closed Parametrized Curve Coordinates')

1.5 From Abstract Plane Simply Connected Closed Curve,

To Abstract Space Closed Curve

Deform abstract simply connected closed curve
'[a*cos(t),b*sin(t)],t[0,2*Pi]'

to '[α cos(v), β sin(v), γ], v[0,2*Pi]'

(Viz. Append component 'γ'), then obtain abstract space closed curve.

It is also understood that abstract space closed curve
'[α cos(v), β sin(v), γ], v[0,2*Pi]'

takes abstract plane simply connected closed curve
'[a*cos(t), b*sin(t)], t[0,2*Pi]' (one or more) as plane projection.

Chapter 2 Prove Curl Theorem at Manifold

2.1 Prove Curl Theorem at Manifold

Curl Theorem Suppose positive directional boundary 'L+' of smooth or piecewise smooth orientational surface 'S' is smooth or piecewise smooth closed curve, Right Hand Rule is associated basis between positive direction of boundary curve 'L' and outer side of orientational surface 'S'. If there are 1th order continuous partial derivatives about functions 'P(x,y,z),Q(x,y,z),R(x,y,z)' [Structure vector field 'A'] at orientational surface 'S', then:

$$\int_{L+} A \cdot dL = \iint_{S} rotA \cdot n \, dS \qquad (1)$$

thereinto, 'rotA' is curl of vector field 'A', 'n' is unit normal vector of orientational surface 'S's outer side.

Proof:

Define parametrized expression of abstract simply connected orientable closed surface 'S':

  [a*sin(u)cos(v),b*sin(u)sin(v),c*cos(u)]   (2)

thereinto, 'a,b,c' are nonzero constants or 1th order derivable continuous functions, simply connected orientable closed surface 'S' determines value of 'a,b,c';

Set range of parameters 'u,v' in [0, π/n - θ],[0,2π], thereinto, 'n' is discretional constant, and n ≥ 1; 'θ'is discretional constant or continuous function, and π/n - θ < π, make surface 'S' unclosed. (See also Poincare Conjecture: 'Every closed n-Manifold which is homotopy equivalent to n-Sphere is homeomorphic to n-Sphere' [19], and Chapter 1 (This Part) Section 1.2 'Define Abstract Simply Connected Orientable Closed Surface and Its Unclosed Surface')

Define parametrized expression of boundary curve 'L'; Set Right Hand Rule as associated basis between positive direction of boundary curve 'L' and normal vector of unclosed surface 'S'.

  [α cos(v), β sin(v), γ]   (3)

In parametrized expression '[a*sin(u)cos(v), b*sin(u)sin(v), c*cos(u)]', owing to pending coefficients 'a,b,c' can't be parsed renewedly (Viz. Embedded variables of pending coefficients 'a,b,c' are unidentifiable),it is impossible that "Directly import boundary value of 'u' to expression of surface 'S', and directly obtain expression of boundary curve 'L'" (Method of 'Proof of Curl Theorem In Torus Coordinates' or 'Numerical Models of Curl Theorem at Manifold'), and have to define expression of boundary curve

'[α cos(v), β sin(v), γ]' that depends on surface '[a*sin(u)cos(v), b*sin(u)sin(v), c*cos(u)]' for existence

thereinto, 'α, β, γ' are constants($\alpha \neq 0$, $\beta \neq 0$) or 1th order derivable continuous functions depend on 'a,b,c'; Because of range of parameter 'v' in [0,2π], boundary curve 'L' is closed.

(See also Chapter 1 (This Part) Section 1.5 'From Abstract Plane Simply Connected Closed Curve, To Abstract Space Closed Curve')

Spacial Closed Curve Integral [Vector Field 'A' at Boundary Curve 'L']:

$$\int_0^{2\pi} P(x,y,z)\left(\frac{\partial}{\partial v}(\alpha\cos(v))\right) + Q(x,y,z)\left(\frac{\partial}{\partial v}(\beta\sin(v))\right) + R(x,y,z)\left(\frac{\partial}{\partial v}\gamma\right) dv =$$

$$\int_0^{2\pi} -P(x,y,z)\,\alpha\sin(v) + Q(x,y,z)\,\beta\cos(v)\,dv = 0$$

(4)

//Relative to Protean given 3-Dimensional space vector fields (Composed of idiographic given 3-variable functions), abstract 3-Dimensional space vector field'[P(x,y,z),Q(x,y,z),R(x,y,z)]'is a sort of balanced symmetrical data structure;

In proof of Curl Theorem at Manifold, objectively require a sort of balanced symmetrical abstract orientable unclosed surface(Include its boundary curve)'s expression, match with abstract 3-Dimensional space vector field '[P(x,y,z),Q(x,y,z),R(x,y,z)]' and its curl;

In meaning of data structure, expression of abstract simply connected   orientable   unclosed   surface   '[a*sin(u)*cos(v), b*sin(u)*sin(v), c*cos(u)],u[0,Pi/n-theta],v[0,2*Pi]'also possesses balanced symmetrical attribute.

//Owing to universality and homogeneity of abstract vector field [P(x,y,z),Q(x,y,z),R(x,y,z)], Integral by Embedded Sub-Variable 'v' of Variable 'x,y,z', Result of Integral can be described as 'P(x,y,z) [or Q(x,y,z), or R(x,y,z)]'.

In other words, Integral by Embedded Sub-Variable 'v' of Variable 'x,y,z', can't change structure of abstract function P(x,y,z) [or Q(x,y,z), or R(x,y,z)], so that abstract function structure P(x,y,z) [or Q(x,y,z), or R(x,y,z)] can preserve primary form after integral.

(See also Chapter 4 (This Part) Section 4.1 'Prove Curl Theorem Finite Sums Limits at Manifold' and Chapter 4 (Part I) Section 4.3 'Discuss About Abstract Vector Field、Finite Sums Limit、Integral and Riemann Sums')

//'Integral value equals 0' possesses specific mathematical and physical meaning:

In mathematical meaning, 'Integral value equals 0' is inevitable result of logic illation, incarnates logic balanced state between various integral elements;

In physical meaning, 'Integral value equals 0' implies that circumfluent value of 'Abstract Vector Field at Abstract Spacial Closed Curve(Route)' is constant for '0' (immobile and pending) identically; If integral value equals certain a positive/negative number or an expression, it implies that always there is positive/reverse directional flux or unknown flux about 'Abstract Spacial Vector Field at Abstract Spacial Closed Curve(Route)', this instance will be unaccountable.

About 'Integral value is unequal to 0', See also Chapter 6 (This Part) Section 6.1 'Curl Theorem at Manifold and Mobius Strip'.

According to surface parametrized expression (2), define and calculate matrix of partial derivatives, obtain 'tangent plane normal vector' of surface 'S':

$$\begin{bmatrix} i & j & k \\ \dfrac{\partial}{\partial u} a\sin(u)\cos(v) & \dfrac{\partial}{\partial u} b\sin(u)\sin(v) & \dfrac{\partial}{\partial u} c\cos(u) \\ \dfrac{\partial}{\partial v} a\sin(u)\cos(v) & \dfrac{\partial}{\partial v} b\sin(u)\sin(v) & \dfrac{\partial}{\partial v} c\cos(u) \end{bmatrix} =$$

$$i\, c\sin(u)^2\, b\cos(v) + a\cos(u)\cos(v)^2\, k\, b\sin(u) + a\sin(u)^2\sin(v)\, j\, c$$
$$+ a\sin(u)\sin(v)^2\, k\, b\cos(u) \tag{5}$$

From expression (5), respectively pick up coefficient of item 'i,j,k', obtain 'tangent plane normal vector' of surface 'S':

$$[\, c\sin(u)^2 b\cos(v)\,,\ \sin(u)^2 a\sin(v)c\,,\ \sin(u)ab\cos(u)\,] \tag{6}$$

Calculate curl of vector field 'A', and convert curl from Cartesian Coordinates expression(7) to Surface 'S' Coordinates expression(8):

$$\left[\left(\left(\frac{\partial}{\partial y} R(x,y,z)\right) - \left(\frac{\partial}{\partial z} Q(x,y,z)\right)\right), \left(\left(\frac{\partial}{\partial z} P(x,y,z)\right) - \left(\frac{\partial}{\partial x} R(x,y,z)\right)\right),\right.$$
$$\left. \left(\left(\frac{\partial}{\partial x} Q(x,y,z)\right) - \left(\frac{\partial}{\partial y} P(x,y,z)\right)\right)\right] \tag{7}$$

$$\left[\left(\left(\frac{\partial}{\partial y} R(x,y,z)\right)\left(\frac{\partial}{\partial u}(b\sin(u)\sin(v))\right)\left(\frac{\partial}{\partial v}(b\sin(u)\sin(v))\right)\right.\right.$$
$$\left.- \left(\frac{\partial}{\partial z} Q(x,y,z)\right)\left(\frac{\partial}{\partial u}(c\cos(u))\right)\left(\frac{\partial}{\partial v}(c\cos(u))\right)\right),$$
$$\left(\frac{\partial}{\partial z} P(x,y,z)\right)\left(\frac{\partial}{\partial u}(c\cos(u))\right)\left(\frac{\partial}{\partial v}(c\cos(u))\right)$$
$$- \left(\frac{\partial}{\partial x} R(x,y,z)\right)\left(\frac{\partial}{\partial u}(a\sin(u)\cos(v))\right)\left(\frac{\partial}{\partial v}(a\sin(u)\cos(v))\right),$$
$$\left(\frac{\partial}{\partial x} Q(x,y,z)\right)\left(\frac{\partial}{\partial u}(a\sin(u)\cos(v))\right)\left(\frac{\partial}{\partial v}(a\sin(u)\cos(v))\right)$$
$$\left.- \left(\frac{\partial}{\partial y} P(x,y,z)\right)\left(\frac{\partial}{\partial u}(b\sin(u)\sin(v))\right)\left(\frac{\partial}{\partial v}(b\sin(u)\sin(v))\right)\right]$$
$$=$$
$$\left[\left(\left(\frac{\partial}{\partial y} R(x,y,z)\right) b^2\cos(u)\sin(v)\sin(u)\cos(v),\right.\right.$$
$$\left(\frac{\partial}{\partial x} R(x,y,z)\right) a^2\cos(u)\cos(v)\sin(u)\sin(v),$$
$$-\left(\frac{\partial}{\partial x} Q(x,y,z)\right) a^2\cos(u)\cos(v)\sin(u)\sin(v)$$
$$\left.-\left(\frac{\partial}{\partial y} P(x,y,z)\right) b^2\cos(u)\sin(v)\sin(u)\cos(v)\right] \tag{8}$$

- 364 -

978-1-62265-930-2 (online)   978-1-62265-931-9 (paper)   Yang Ke

//In Spacial Cartesian Coordinates, Curl of abstract vector field [P(x,y,z),Q(x,y,z),R(x,y,z)] is

$$\left[\left(\frac{\partial}{\partial y}R(x,y,z)\right)-\left(\frac{\partial}{\partial z}Q(x,y,z)\right),\left(\frac{\partial}{\partial z}P(x,y,z)\right)-\left(\frac{\partial}{\partial x}R(x,y,z)\right),\right.$$
$$\left.\left(\frac{\partial}{\partial x}Q(x,y,z)\right)-\left(\frac{\partial}{\partial y}P(x,y,z)\right)\right]$$

In logic deduction of this Proof, it is necessary to import curl of abstract vector field above-mentioned to abstract simply connected orientable closed surface coordinates.

Six elements of abstract vector field's curl:

$$\frac{\partial}{\partial y}R(x,y,z)\text{ , }\frac{\partial}{\partial z}Q(x,y,z)\text{ , }\frac{\partial}{\partial z}P(x,y,z)\text{ , }\frac{\partial}{\partial x}R(x,y,z)\text{ , }\frac{\partial}{\partial x}Q(x,y,z)\text{ , }\frac{\partial}{\partial y}P(x,y,z)$$

are abstract differential function, and their variables 'x,y,z' contain sub-variables 'u,v'.

Surface Parametric Transform Expression between 'Spacial Cartesian coordinates' and 'Abstract simply connected orientable closed surface coordinates' is:

x = a*sin(u)cos(v), y = b*sin(u) sin(v), z = c*cos(u).

-"Coordinates Transform Differential Functions" (Correspond to

Three Differential Variables '$\frac{\partial}{\partial x}$, $\frac{\partial}{\partial y}$, $\frac{\partial}{\partial z}$' of Differential Functions

$$\frac{\partial}{\partial x}Q(x,y,z)\text{ , }\frac{\partial}{\partial x}R(x,y,z)\text{ , }\frac{\partial}{\partial y}P(x,y,z)\text{ , }\frac{\partial}{\partial y}R(x,y,z)\text{ , }\frac{\partial}{\partial z}Q(x,y,z)\text{ , }\frac{\partial}{\partial z}P(x,y,z)\text{)}$$

are $\frac{\partial}{\partial u}a\sin(u)\cos(v)\frac{\partial}{\partial v}a\sin(u)\cos(v)$, $\frac{\partial}{\partial u}b\sin(u)\sin(v)\frac{\partial}{\partial v}b\sin(u)\sin(v)$ and

$\frac{\partial}{\partial u}c\cos(u)\frac{\partial}{\partial v}c\cos(u)$.

Product of "Differential Function '$\frac{\partial}{\partial y}R(x,y,z)$, $\frac{\partial}{\partial z}Q(x,y,z)$,

$\frac{\partial}{\partial z}P(x,y,z)$, $\frac{\partial}{\partial x}R(x,y,z)$, $\frac{\partial}{\partial x}Q(x,y,z)$, $\frac{\partial}{\partial y}P(x,y,z)$ '" and "Coordinates

Transform Differential Functions" (Viz. Product of Two Differential functions) constitute "Curl in Abstract Simply Connected Orientable Closed Surface Coordinates".

//'Chain differentiation' or 'Coordinates conversion' ?

If it is 'Chain differentiation', according to principle of 'multiply in identical chain, plus in subdivision chain', it ought to be:

$$\left[\left(\left(\frac{\partial}{\partial y}R(x,y,z)\right)\left(\left(\frac{\partial}{\partial u}(b\sin(u)\sin(v))\right)+\left(\frac{\partial}{\partial v}(b\sin(u)\sin(v))\right)\right)\right.\right.$$

$$-\left(\frac{\partial}{\partial z}Q(x,y,z)\right)\left(\left(\frac{\partial}{\partial u}(c\cos(u))\right)+\left(\frac{\partial}{\partial v}(c\cos(u))\right)\right),$$

$$\left(\frac{\partial}{\partial z}P(x,y,z)\right)\left(\left(\frac{\partial}{\partial u}(c\cos(u))\right)+\left(\frac{\partial}{\partial v}(c\cos(u))\right)\right)$$

$$-\left(\frac{\partial}{\partial x}R(x,y,z)\right)\left(\left(\frac{\partial}{\partial u}(a\sin(u)\cos(v))\right)+\left(\frac{\partial}{\partial v}(a\sin(u)\cos(v))\right)\right),$$

$$\left(\frac{\partial}{\partial x}Q(x,y,z)\right)\left(\left(\frac{\partial}{\partial u}(a\sin(u)\cos(v))\right)+\left(\frac{\partial}{\partial v}(a\sin(u)\cos(v))\right)\right)$$

$$\left.\left.-\left(\frac{\partial}{\partial y}P(x,y,z)\right)\left(\left(\frac{\partial}{\partial u}(b\sin(u)\sin(v))\right)+\left(\frac{\partial}{\partial v}(b\sin(u)\sin(v))\right)\right)\right]\right]$$

No matter 'Chain differentiation' or 'Solve divergence/curl', question for discussion is 'How to solve differential coefficient' or 'method of differentiation' about abstract vector field 'P[x,y,z], Q[x,y,z],R[x,y,z]'; and this step, question for discussion is 'How to convert result of solving curl from a coordinates to another'; character and hiberarchy of 'two questions' above-mentioned are disparate, it is 'multiply', not 'plus' – Be determined by 3-Dimensional spacial attribute of coordinates

Surface Integral of Dot Product [Curl (8) and Tangent Plane Normal Vector (6)] in range of parameters 'u,v' (9):

$$\int_0^{2\pi}\int_0^{\frac{\pi}{n}-\theta}\left(\frac{\partial}{\partial y}R(x,y,z)\right)b^3\cos(u)\sin(v)\sin(u)^3\cos(v)^2 c$$

$$+\left(\frac{\partial}{\partial x}R(x,y,z)\right)a^3\cos(u)\cos(v)\sin(u)^3\sin(v)^2 c+\left(\right.$$

$$-\left(\frac{\partial}{\partial x}Q(x,y,z)\right)a^2\cos(u)\cos(v)\sin(u)\sin(v)$$

$$\left.-\left(\frac{\partial}{\partial y}P(x,y,z)\right)b^2\cos(u)\sin(v)\sin(u)\cos(v)\right)\sin(u)\,a\,b\cos(u)\,du\,dv$$

$$=$$

$$\int_0^{2\pi} \frac{1}{8} \sin(v) \cos(v) \left(2\left(\frac{\partial}{\partial y} R(x,y,z) \right) b^3 \sin\left(\frac{-\pi + \theta\, n}{n} \right)^4 \cos(v)\, c\, n \right.$$

$$+ 2\left(\frac{\partial}{\partial x} R(x,y,z) \right) a^3 \sin\left(\frac{-\pi + \theta\, n}{n} \right)^4 \sin(v)\, c\, n$$

$$- 2\, a^3\, b \sin\left(\frac{-\pi + \theta\, n}{n} \right) \cos\left(\frac{-\pi + \theta\, n}{n} \right)^3 n \left(\frac{\partial}{\partial x} Q(x,y,z) \right)$$

$$- 2\, a\, b^3 \sin\left(\frac{-\pi + \theta\, n}{n} \right) \cos\left(\frac{-\pi + \theta\, n}{n} \right)^3 n \left(\frac{\partial}{\partial y} P(x,y,z) \right)$$

$$+ a^3\, b \cos\left(\frac{-\pi + \theta\, n}{n} \right) \sin\left(\frac{-\pi + \theta\, n}{n} \right) n \left(\frac{\partial}{\partial x} Q(x,y,z) \right)$$

$$+ a\, b^3 \cos\left(\frac{-\pi + \theta\, n}{n} \right) \sin\left(\frac{-\pi + \theta\, n}{n} \right) n \left(\frac{\partial}{\partial y} P(x,y,z) \right) - a^3\, b\, \pi \left(\frac{\partial}{\partial x} Q(x,y,z) \right)$$

$$- a\, b^3\, \pi \left(\frac{\partial}{\partial y} P(x,y,z) \right) + a^3\, b\, \theta\, n \left(\frac{\partial}{\partial x} Q(x,y,z) \right) + a\, b^3\, \theta\, n \left(\frac{\partial}{\partial y} P(x,y,z) \right) \Bigg) / n\; dv$$

$$= 0$$

// Owing to universality and homogeneity of Four Differential

Function Elements ‘$\frac{\partial}{\partial y} R(x,y,z)$, $\frac{\partial}{\partial x} R(x,y,z)$, $\frac{\partial}{\partial x} Q(x,y,z)$, $\frac{\partial}{\partial y} P(x,y,z)$ ’

(Belong to Curl of Abstract Vector Field:

$$\left[\left(\frac{\partial}{\partial y} R(x,y,z) \right) b^2 \cos(u) \sin(v) \sin(u) \cos(v), \right.$$

$$\left(\frac{\partial}{\partial x} R(x,y,z) \right) a^2 \cos(u) \cos(v) \sin(u) \sin(v),$$

$$-\left(\frac{\partial}{\partial x} Q(x,y,z) \right) a^2 \cos(u) \cos(v) \sin(u) \sin(v)$$

$$\left. -\left(\frac{\partial}{\partial y} P(x,y,z) \right) b^2 \cos(u) \sin(v) \sin(u) \cos(v) \right]),$$

In formula deduction above-mentioned, integral by Embedded Sub-Variable ‘u,v’ of Variable ‘x,y,z’, character of integral can be regarded as “Integral about Dot Product (‘Tangent Plane Normal Vector’ and ‘Product between Four Differential Function Elements of Curl

(Viz. $\frac{\partial}{\partial y} R(x,y,z)$, $\frac{\partial}{\partial x} R(x,y,z)$, $\frac{\partial}{\partial x} Q(x,y,z)$, $\frac{\partial}{\partial y} P(x,y,z)$) and Coordinates

Transform Differential Functions’ ”.

Integral by Embedded Sub-Variable 'u,v' of Variable 'x,y,z', Can't change structure of Four Differential Function Elements (Viz.

$\dfrac{\partial}{\partial y}R(x,y,z)$, or $\dfrac{\partial}{\partial x}R(x,y,z)$, or $\dfrac{\partial}{\partial x}Q(x,y,z)$, or $\dfrac{\partial}{\partial y}P(x,y,z)$), they can

preserve primary form after integral;

Three Coordinates Transform Differential Functions (Correspond to

Three Differential Variables '$\dfrac{\partial}{\partial x},\dfrac{\partial}{\partial y},\dfrac{\partial}{\partial z}$'), Viz:

$\dfrac{\partial}{\partial u}a\sin(u)\cos(v)\dfrac{\partial}{\partial v}a\sin(u)\cos(v)$, $\dfrac{\partial}{\partial u}b\sin(u)\sin(v)\dfrac{\partial}{\partial v}b\sin(u)\sin(v)$ and

$\dfrac{\partial}{\partial u}c\cos(u)\dfrac{\partial}{\partial v}c\cos(u)$

will be changed after integral.

(See also Chapter 4 (This Part) Section 4.1 'Prove Curl Theorem Finite Sums Limits at Manifold' and Chapter 4 (Part I) Section 4.3 'Discuss About Abstract Vector Field、Finite Sums Limit、Integral and Riemann Sums')

// About 'Integral value is unequal to 0', See also Chapter 6 (This Part) Section 6.1 'Curl Theorem at Manifold and Mobius Strip'

Viz. expression(4) = expression(9):

$$\int_0^{2\pi} P(x,y,z)\left(\dfrac{\partial}{\partial v}(\alpha\cos(v))\right)+Q(x,y,z)\left(\dfrac{\partial}{\partial v}(\beta\sin(v))\right)+R(x,y,z)\left(\dfrac{\partial}{\partial v}\gamma\right)dv=$$

$$\int_0^{2\pi}\int_0^{\frac{\pi}{n}-\theta}\left(\dfrac{\partial}{\partial y}R(x,y,z)\right)b^3\cos(u)\sin(v)\sin(u)^3\cos(v)^2 c$$

$$+\left(\dfrac{\partial}{\partial x}R(x,y,z)\right)a^3\cos(u)\cos(v)\sin(u)^3\sin(v)^2 c+\Big($$

$$-\left(\dfrac{\partial}{\partial x}Q(x,y,z)\right)a^2\cos(u)\cos(v)\sin(u)\sin(v)$$

$$-\left(\dfrac{\partial}{\partial y}P(x,y,z)\right)b^2\cos(u)\sin(v)\sin(u)\cos(v)\Big)\sin(u)\,a\,b\,\cos(u)\,du\,dv$$

Above mentioned equation can be described as:

$$\int_{L+} A \cdot dL = \iint_S rotA \cdot n \, dS$$ (1), Complete Proof.

2.2 Prove Curl Theorem at Manifold

[Program Template of Waterloo Maplesoft, Optional]

Curl Theorem Suppose positive directional boundary 'L+' of smooth or piecewise smooth orientational surface 'S' is smooth or piecewise smooth closed curve, Right Hand Rule is associated basis between positive direction of boundary curve 'L' and outer side of orientational surface 'S'. If there are 1th order continuous partial derivatives about functions 'P(x,y,z),Q(x,y,z),R(x,y,z)' [Structure vector field 'A'] at orientational surface 'S', then:

$$\int_{L+} A \cdot dL = \iint_S rotA \cdot n \, dS$$ (1)

thereinto, 'rotA' is curl of vector field 'A', 'n' is unit normal vector of orientational surface 'S's outer side.

Symbol System:

Abstract Vector Field 'V',

Curl 'cV1,cV2' of Abstract Vector Field 'V';

Abstract simply connected orientable closed parametrized surface 'CS' [Set it unclosed],

Closed Boundary Curve 'CL' of Surface 'CS',

Tangent Plane Normal Vector '[A,B,C]' of Surface 'CS'

> restart; # Reset computer algebraic system
> with(linalg): # Load 'linear algebra' package
> CS:=[a*sin(u)*cos(v),b*sin(u)*sin(v),c*cos(u)];

Define parametrized expression of abstract simply connected、 orientable closed surface 'CS'; 'a,b,c' are nonzero constants or 1th order derivable continuous functions, simply connected orientable closed surface 'CS' determines value of 'a,b,c'. (See also Poincare Conjecture: 'Every closed n-Manifold which is homotopy equivalent to n-Sphere is homeomorphic to n-Sphere'[19], and Chapter 1 (This Part)

Section 1.2 'Define Abstract Simply Connected Orientable Closed
Surface and Its Unclosed Surface')

$$CS := [\, a \sin(u) \cos(v), \, b \sin(u) \sin(v), \, c \cos(u) \,]$$

>rgu:=[0,Pi/n-theta];

'n' is discretional constant, and n ≥ 1; 'theta'is discretional
constant or continuous function, and Pi/n - theta < Pi; Curtail range
of parameter 'u' (Relative to abstract simply connected orientable
closed parametrized surface)

$$rgu := \left[\, 0, \frac{\pi}{n} - \theta \,\right]$$

> rgv:=[0,2*Pi];

Set range of parameters 'u,v', make parametrized surface 'CS'
unclosed

$$rgv := [\, 0, 2\,\pi \,]$$

> CL:=[alpha*cos(v),beta*sin(v),gamma];

Define parametrized expression of boundary curve 'CL' (it belongs
to abstract simply connected orientable unclosed surface 'CS')

   Set Right Hand Rule as associated basis between positive direction
of boundary curve 'CL' and normal vector of unclosed surface 'CS'

   'alpha, beta, gamma' are constants('alpha' <>0, 'beta' <>0) or 1th
order derivable continuous functions depend on 'a,b,c'; because of
range of parameter 'v' in [0,2π], boundary curve 'L' is closed.

   (See also Chapter 1 (This Part) Section 1.5 'From Abstract Plane
Simply Connected Closed Curve, To Abstract Space Closed Curve')

$$CL := [\, \alpha \cos(v), \, \beta \sin(v), \, \gamma \,]$$

> V:=[(P)(x,y,z),(Q)(x,y,z),(R)(x,y,z)];

Define abstract spacial vector field 'V'(Suppose Vector Field 'V'
possesses 1th order continuous partial derivatives at surface 'CS')

$$V := [\, \mathrm{P}(x,y,z), \, \mathrm{Q}(x,y,z), \, \mathrm{R}(x,y,z) \,]$$

> [Diff(V[3],y)-Diff(V[2],z),Diff(V[1],z)-Diff(V[3],x),
Diff(V[2],x)-Diff(V[1],y)]
=[diff(V[3],y)-diff(V[2],z),diff(V[1],z)-diff(V[3],x),
diff(V[2],x)-diff(V[1],y)];cV1:=rhs(%);

Calculate 'cV1' - Curl of abstract spacial vector field 'V'

$$\left[\left(\left(\frac{\partial}{\partial y}R(x,y,z)\right)-\left(\frac{\partial}{\partial z}Q(x,y,z)\right),\left(\frac{\partial}{\partial z}P(x,y,z)\right)-\left(\frac{\partial}{\partial x}R(x,y,z)\right),\right.\right.$$
$$\left.\left(\frac{\partial}{\partial x}Q(x,y,z)\right)-\left(\frac{\partial}{\partial y}P(x,y,z)\right)\right]=\left[\left(\frac{\partial}{\partial y}R(x,y,z)\right)-\left(\frac{\partial}{\partial z}Q(x,y,z)\right),\right.$$
$$\left.\left(\frac{\partial}{\partial z}P(x,y,z)\right)-\left(\frac{\partial}{\partial x}R(x,y,z)\right),\left(\frac{\partial}{\partial x}Q(x,y,z)\right)-\left(\frac{\partial}{\partial y}P(x,y,z)\right)\right]$$
$$cV1:=\left[\left(\frac{\partial}{\partial y}R(x,y,z)\right)-\left(\frac{\partial}{\partial z}Q(x,y,z)\right),\left(\frac{\partial}{\partial z}P(x,y,z)\right)-\left(\frac{\partial}{\partial x}R(x,y,z)\right),\right.$$
$$\left.\left(\frac{\partial}{\partial x}Q(x,y,z)\right)-\left(\frac{\partial}{\partial y}P(x,y,z)\right)\right]$$

```
> x:=CL[1]:y:=CL[2]:z:=CL[3]:
> Int(V[1]*Diff(CL[1],t)+V[2]*Diff(CL[2],t)+V[3]*Diff(CL[3],t),
t=rgt[1]..rgt[2]);
# Spacial Closed Curve Integral - Vector Field 'V' at Closed Boundary
Curve 'CL'
```

$$\int_0^{2\pi}P(\alpha\cos(v),\beta\sin(v),\gamma)\left(\frac{\partial}{\partial v}(\alpha\cos(v))\right)$$
$$+Q(\alpha\cos(v),\beta\sin(v),\gamma)\left(\frac{\partial}{\partial v}(\beta\sin(v))\right)+R(\alpha\cos(v),\beta\sin(v),\gamma)\left(\frac{\partial}{\partial v}\gamma\right)dv$$

```
> value(%);
```

$$\int_0^{2\pi}-P(\alpha\cos(v),\beta\sin(v),\gamma)\,\alpha\sin(v)+Q(\alpha\cos(v),\beta\sin(v),\gamma)\,\beta\cos(v)\,dv$$

```
# Import 'x = alpha*cos(v), y = beta*sin(v), z = gamma' to abstract
vector field 'P(x,y,z),Q(x,y,z),R(x,y,z)', obtain insoluble result,
abnegate it
> x:='x':y:='y':z:='z':
> Int(V[1]*Diff(CL[1],t)+V[2]*Diff(CL[2],t)+V[3]*Diff(CL[3],t),
t=rgt[1]..rgt[2])
=Int(V[1]*diff(CL[1],t)+V[2]*diff(CL[2],t)+V[3]*diff(CL[3],t),
t=rgt[1]..rgt[2]);
# Spacial Closed Curve Integral (Reserve original form of Abstract
Vector Field '[P(x,y,z),Q(x,y,z),R(x,y,z)]')
```

$$\int_0^{2\pi}P(x,y,z)\left(\frac{\partial}{\partial v}(\alpha\cos(v))\right)+Q(x,y,z)\left(\frac{\partial}{\partial v}(\beta\sin(v))\right)+R(x,y,z)\left(\frac{\partial}{\partial v}\gamma\right)dv=$$

$$\int_{0}^{2\pi} -P(x, y, z)\,\alpha\,\sin(v) + Q(x, y, z)\,\beta\,\cos(v)\;dv$$

```
> delta:=rhs(value(%));
```

Calculate value of spacial closed curve integral

$$\delta := 0$$ # Obtain a constant '0'

Owing to universality and homogeneity of abstract vector field [P(x,y,z),Q(x,y,z),R(x,y,z)], Integral by Embedded Sub-Variable 'v' of Variable 'x,y,z', Result of Integral can be described as 'P(x,y,z) [or Q(x,y,z), or R(x,y,z)]'.

In other words, Integral by Embedded Sub-Variable 'v' of Variable 'x,y,z', can't change structure of abstract function P(x,y,z) [or Q(x,y,z), or R(x,y,z)], so that abstract function structure P(x,y,z) [or Q(x,y,z), or R(x,y,z)] can preserve primary form after integral.

[See also Chapter 4 (This Part) Section 4.1 'Prove Curl Theorem Finite Sums Limits at Manifold' and Chapter 4 (Part I) Section 4.3 'Discuss About Abstract Vector Field、Finite Sums Limit、Integral and Riemann Sums']

```
> x:='x':y:='y':z:='z':
> x:=CS[1]:y:=CS[2]:z:=CS[3]:
> matrix(3,3,[i,j,k,Diff(CS[1],u),Diff(CS[2],u),Diff(CS[3],u),
Diff(CS[1],v),Diff(CS[2],v),Diff(CS[3],v)])=
matrix(3,3,[i,j,k,diff(CS[1],u),diff(CS[2],u),diff(CS[3],u),
diff(CS[1],v),diff(CS[2],v),diff(CS[3],v)]);m:=rhs(%);
```

Define and calculate matrix of partial derivatives 'm', obtain 'tangent plane normal vector' of surface 'CS':

$$\begin{bmatrix} i & j & k \\ \dfrac{\partial}{\partial u}(a\sin(u)\cos(v)) & \dfrac{\partial}{\partial u}(b\sin(u)\sin(v)) & \dfrac{\partial}{\partial u}(c\cos(u)) \\ \dfrac{\partial}{\partial v}(a\sin(u)\cos(v)) & \dfrac{\partial}{\partial v}(b\sin(u)\sin(v)) & \dfrac{\partial}{\partial v}(c\cos(u)) \end{bmatrix} =$$

$$\begin{bmatrix} i & j & k \\ a\cos(u)\cos(v) & b\cos(u)\sin(v) & -c\sin(u) \\ -a\sin(u)\sin(v) & b\sin(u)\cos(v) & 0 \end{bmatrix}$$

$$m := \begin{bmatrix} i & j & k \\ a\cos(u)\cos(v) & b\cos(u)\sin(v) & -c\sin(u) \\ -a\sin(u)\sin(v) & b\sin(u)\cos(v) & 0 \end{bmatrix}$$

```
> det(m);
```

$$i\,c\,\sin(u)^2\,b\,\cos(v) + a\,\cos(u)\,\cos(v)^2\,k\,b\,\sin(u) + a\,\sin(u)^2\,\sin(v)\,j\,c$$
$$+ a\,\sin(u)\,\sin(v)^2\,k\,b\,\cos(u)$$

```
> mn:=simplify(%);
```

$$mn := \sin(u)\,(i\,c\,\sin(u)\,b\,\cos(v) + a\,\sin(u)\,\sin(v)\,j\,c + a\,k\,b\,\cos(u))$$

```
> A:=coeff(mn,i);  # Obtain coefficient of 'i'
```

$$A := c\,\sin(u)^2\,b\,\cos(v)$$

```
> B:=coeff(mn,j); # Obtain coefficient of 'j'
```

$$B := \sin(u)^2\,a\,\sin(v)\,c$$

```
> C:=coeff(mn,k);
```

```
# Obtain coefficient of 'k'
```

$$C := \sin(u)\,a\,b\,\cos(u)$$

```
> [A,B,C]; # [A,B,C] structure 'tangent plane normal vector'
```

$$[c\,\sin(u)^2\,b\,\cos(v),\ \sin(u)^2\,\sin(v)\,a\,c,\ \sin(u)\,a\,\cos(u)\,b]$$

```
> x:='x':y:='y':z:='z':
> [Diff(V[3],y)*Diff(CS[2],u)*Diff(CS[2],v)
-Diff(V[2],z)*Diff(CS[3],u)*Diff(CS[3],v),
Diff(V[1],z)*Diff(CS[3],u)*Diff(CS[3],v)
-Diff(V[3],x)*Diff(CS[1],u)*Diff(CS[1],v),
Diff(V[2],x)*Diff(CS[1],u)*Diff(CS[1],v)
-Diff(V[1],y)*Diff(CS[2],u)*Diff(CS[2],v)]
=[diff(V[3],y)*diff(CS[2],u)*diff(CS[2],v)
-diff(V[2],z)*diff(CS[3],u)*diff(CS[3],v),
diff(V[1],z)*diff(CS[3],u)*diff(CS[3],v)
-diff(V[3],x)*diff(CS[1],u)*diff(CS[1],v),
diff(V[2],x)*diff(CS[1],u)*diff(CS[1],v)
-diff(V[1],y)*diff(CS[2],u)*diff(CS[2],v)];cV2:=rhs(%);
# Convert 'cV1' from Cartesian Coordinates expression to Surface 'CS'
Coordinates expression
# Reserve original form of abstract vector field '[P(x,y,z),Q(x,y,z),
```

R(x,y,z)]' and its partial derivatives (insoluble), Calculate partial

derivatives of 'a*sin(u)*cos(v),b*sin(u)*sin(v),c*cos(u)'

(computable), obtain a new expression 'cV2'(Curl) of vector field

$$\left[\left(\left(\frac{\partial}{\partial y}R(x,y,z)\right)\left(\frac{\partial}{\partial u}(b\sin(u)\sin(v))\right)\left(\frac{\partial}{\partial v}(b\sin(u)\sin(v))\right)\right.\right.$$

$$-\left(\frac{\partial}{\partial z}Q(x,y,z)\right)\left(\frac{\partial}{\partial u}(c\cos(u))\right)\left(\frac{\partial}{\partial v}(c\cos(u))\right),$$

$$\left(\frac{\partial}{\partial z}P(x,y,z)\right)\left(\frac{\partial}{\partial u}(c\cos(u))\right)\left(\frac{\partial}{\partial v}(c\cos(u))\right)$$

$$-\left(\frac{\partial}{\partial x}R(x,y,z)\right)\left(\frac{\partial}{\partial u}(a\sin(u)\cos(v))\right)\left(\frac{\partial}{\partial v}(a\sin(u)\cos(v))\right),$$

$$\left(\frac{\partial}{\partial x}Q(x,y,z)\right)\left(\frac{\partial}{\partial u}(a\sin(u)\cos(v))\right)\left(\frac{\partial}{\partial v}(a\sin(u)\cos(v))\right)$$

$$\left.\left.-\left(\frac{\partial}{\partial y}P(x,y,z)\right)\left(\frac{\partial}{\partial u}(b\sin(u)\sin(v))\right)\left(\frac{\partial}{\partial v}(b\sin(u)\sin(v))\right)\right]=\right[$$

$$\left(\frac{\partial}{\partial y}R(x,y,z)\right)b^2\cos(u)\sin(v)\sin(u)\cos(v),$$

$$\left(\frac{\partial}{\partial x}R(x,y,z)\right)a^2\cos(u)\cos(v)\sin(u)\sin(v),$$

$$-\left(\frac{\partial}{\partial x}Q(x,y,z)\right)a^2\cos(u)\cos(v)\sin(u)\sin(v)$$

$$\left.-\left(\frac{\partial}{\partial y}P(x,y,z)\right)b^2\cos(u)\sin(v)\sin(u)\cos(v)\right]$$

$$cV2:=\left[\left(\frac{\partial}{\partial y}R(x,y,z)\right)b^2\cos(u)\sin(v)\sin(u)\cos(v),\right.$$

$$\left(\frac{\partial}{\partial x}R(x,y,z)\right)a^2\cos(u)\cos(v)\sin(u)\sin(v),$$

$$-\left(\frac{\partial}{\partial x}Q(x,y,z)\right)a^2\cos(u)\cos(v)\sin(u)\sin(v)$$

$$\left.-\left(\frac{\partial}{\partial y}P(x,y,z)\right)b^2\cos(u)\sin(v)\sin(u)\cos(v)\right]$$

```
>Int(Int(cV2[1]*A+cV2[2]*B+cV2[3]*C,u=rgu[1]..rgu[2]),v=rgv[1]..
rgv[2]);
```

Surface Integral of Dot Product (Curl 'cV2' and Tangent Plane Normal

Vector '[A,B,C]') in range of 'u,v'

$$\int_0^{2\pi}\int_0^{\frac{\pi}{n}-\theta}\left(\frac{\partial}{\partial y}R(x,y,z)\right)b^3\cos(u)\sin(v)\sin(u)^3\cos(v)^2c$$

$$+\left(\frac{\partial}{\partial x}\mathrm{R}(x,y,z)\right)a^3\cos(u)\cos(v)\sin(u)^3\sin(v)^2\,c+\Bigg($$

$$-\left(\frac{\partial}{\partial x}\mathrm{Q}(x,y,z)\right)a^2\cos(u)\cos(v)\sin(u)\sin(v)$$

$$-\left(\frac{\partial}{\partial y}\mathrm{P}(x,y,z)\right)b^2\cos(u)\sin(v)\sin(u)\cos(v)\Bigg)\sin(u)\,a\,b\,\cos(u)\;du\,dv$$

```
> epsilon:=value(%);# Calculate value of surface integral
```

$$\varepsilon := 0 \quad\text{\# Obtain a constant '0'}$$

\# Owing to universality and homogeneity of Four Differential Function

Elements' $\dfrac{\partial}{\partial y}R(x,y,z)$, $\dfrac{\partial}{\partial x}R(x,y,z)$, $\dfrac{\partial}{\partial x}Q(x,y,z)$, $\dfrac{\partial}{\partial y}P(x,y,z)$ ' (Belong to

Curl of Abstract Vector Field:

$$\Bigg[\left(\frac{\partial}{\partial y}\mathrm{R}(x,y,z)\right)b^2\cos(u)\sin(v)\sin(u)\cos(v),$$

$$\left(\frac{\partial}{\partial x}\mathrm{R}(x,y,z)\right)a^2\cos(u)\cos(v)\sin(u)\sin(v),$$

$$-\left(\frac{\partial}{\partial x}\mathrm{Q}(x,y,z)\right)a^2\cos(u)\cos(v)\sin(u)\sin(v)$$

$$-\left(\frac{\partial}{\partial y}\mathrm{P}(x,y,z)\right)b^2\cos(u)\sin(v)\sin(u)\cos(v)\Bigg],$$

In formula deduction above-mentioned, integral by Embedded Sub-Variable 'u,v' of Variable 'x,y,z', character of integral can be regarded as "Integral about Dot Product ('Tangent Plane Normal Vector' and 'Product between Four Differential Function Elements of Curl (Viz.

$\dfrac{\partial}{\partial y}R(x,y,z)$, $\dfrac{\partial}{\partial x}R(x,y,z)$, $\dfrac{\partial}{\partial x}Q(x,y,z)$, $\dfrac{\partial}{\partial y}P(x,y,z)$) and Coordinates

Transform Differential Functions'".

Integral by Embedded Sub-Variable 'u,v' of Variable 'x,y,z', Can't change structure of Four Differential Function Elements (Viz.

$\dfrac{\partial}{\partial y}R(x,y,z)$, or $\dfrac{\partial}{\partial x}R(x,y,z)$, or $\dfrac{\partial}{\partial x}Q(x,y,z)$, or $\dfrac{\partial}{\partial y}P(x,y,z)$), they can

preserve primary form after integral;

Three Coordinates Transform Differential Functions (Correspond to

Three Differential Variables $`\dfrac{\partial}{\partial x}, \dfrac{\partial}{\partial y}, \dfrac{\partial}{\partial z}`$), Viz:

$$\frac{\partial}{\partial u}a\sin(u)\cos(v)\frac{\partial}{\partial v}a\sin(u)\cos(v),\ \frac{\partial}{\partial u}b\sin(u)\sin(v)\frac{\partial}{\partial v}b\sin(u)\sin(v)\ \text{and}$$

$\dfrac{\partial}{\partial u}c\cos(u)\dfrac{\partial}{\partial v}c\cos(u)$ will be changed after integral.

[See also Chapter 4 (This Part) Section 4.1 'Prove Curl Theorem Finite Sums Limits at Manifold' and Chapter 4 (Part I) Section 4.3 'Discuss About Abstract Vector Field、Finite Sums Limit、Integral and Riemann Sums']

Viz:

$$\int_0^{2\pi} P(x,y,z)\left(\frac{\partial}{\partial v}(\alpha\cos(v))\right)+Q(x,y,z)\left(\frac{\partial}{\partial v}(\beta\sin(v))\right)+R(x,y,z)\left(\frac{\partial}{\partial v}\gamma\right)dv=$$

$$\int_0^{2\pi}\int_0^{\frac{\pi}{n}-\theta}\left(\frac{\partial}{\partial y}R(x,y,z)\right)b^3\cos(u)\sin(v)\sin(u)^3\cos(v)^2 c$$

$$+\left(\frac{\partial}{\partial x}R(x,y,z)\right)a^3\cos(u)\cos(v)\sin(u)^3\sin(v)^2 c+\bigg($$

$$-\left(\frac{\partial}{\partial x}Q(x,y,z)\right)a^2\cos(u)\cos(v)\sin(u)\sin(v)$$

$$-\left(\frac{\partial}{\partial y}P(x,y,z)\right)b^2\cos(u)\sin(v)\sin(u)\cos(v)\bigg)\sin(u)\,a\,b\,\cos(u)\,du\,dv$$

Value of 'spacial closed curve integral' is equal to 'unclosed surface integral', Complete Proof

2.3 Prove Curl Theorem In Torus Coordinates
(Multiple Connected Orientable Closed Surface Coordinates)

Curl Theorem Suppose positive directional boundary 'L+' of smooth or piecewise smooth orientational surface 'S' is smooth or piecewise smooth closed curve, Right Hand Rule is associated basis between positive direction of boundary curve 'L' and outer side of orientational surface 'S'. If there are 1th order continuous partial derivatives about functions 'P(x,y,z),Q(x,y,z),R(x,y,z)' [Structure vector field 'A'] at orientational surface 'S', then:

$$\int_{L+} A \cdot dL = \iint_S rotA \cdot n \, dS \qquad (1)$$

thereinto, 'rotA' is curl of vector field 'A', 'n' is unit normal vector of orientational surface 'S's outer side.

Proof:

Define parametrized expression of torus surface 'S':

[(2+cos(u))cos(v),(2+cos(u))sin(v),sin(u)]   (2)

Set range of parameters 'u,v' in [0,2π/n - θ],[0,2π], thereinto, 'n' is discretional constant, and n ≥ 1, 'θ' is discretional constant or continuous function, and 2π/n- θ <2π, make torus surface 'S' unclosed.

Import two boundary values of parameter 'u' to expression of torus surface 'S',obtain expressions of boundary curves 'L1' and 'L2';

Set Right Hand Rule as associated basis between positive direction of boundary curves 'L1,L2' and normal vector of unclosed torus surface 'S'

L1: $[3\cos(v), 3\sin(v), 0]$ (3)

L2: $[(2+\cos(-\frac{2\pi}{n}+\theta))\cos(v),(2+\cos(-\frac{2\pi}{n}+\theta))\sin(v),-\sin(-\frac{2\pi}{n}+\theta)]$ (4)

Because of range of parameter 'v' in [0,2π], boundary curves 'L1' and 'L2' are closed.

Spacial Closed Curve Integral -- Vector Field 'A' at Boundary Curves 'L1' and 'L2':

$$\int_0^{2\pi} P(x,y,z)\left(\frac{d}{dv}(3\cos(v))\right) + Q(x,y,z)\left(\frac{d}{dv}(3\sin(v))\right) + R(x,y,z)\left(\frac{d}{dv}0\right) dv$$

$$-\int_0^{2\pi} P(x,y,z)\left(\frac{\partial}{\partial v}\left(\left(2+\cos\left(-\frac{2\pi}{n}+\theta\right)\right)\cos(v)\right)\right)$$

$$+ Q(x, y, z) \left(\frac{\partial}{\partial v} \left(\left(2 + \cos\left(-\frac{2\pi}{n} + \theta \right) \right) \sin(v) \right) \right)$$

$$+ R(x, y, z) \left(\frac{\partial}{\partial v} \left(-\sin\left(-\frac{2\pi}{n} + \theta \right) \right) \right) dv = 0 \qquad (5)$$

// About 'Integral value is unequal to 0', See also Chapter 6 (This Part) Section 6.1 'Curl Theorem at Manifold and Mobius Strip'

According to Torus parametrized expression (2), define and calculate matrix of partial derivatives, obtain 'Tangent plane normal vector' of torus surface 'S' (6):

$$\begin{bmatrix} i & j & k \\ \frac{\partial}{\partial u}(2+\cos(u))\cos(v) & \frac{\partial}{\partial u}(2+\cos(u))\sin(v) & \frac{\partial}{\partial u}\sin(u) \\ \frac{\partial}{\partial v}(2+\cos(u))\cos(v) & \frac{\partial}{\partial v}(2+\cos(u))\sin(v) & \frac{\partial}{\partial v}\sin(u) \end{bmatrix} =$$

$$-2\,i\cos(u)\cos(v) - i\cos(u)^2\cos(v) - 2\sin(u)\cos(v)^2\,k - \sin(u)\cos(v)^2\,k\cos(u)$$
$$- 2\sin(v)\,j\cos(u) - 2\,k\sin(u)\sin(v)^2 - \sin(v)\,j\cos(u)^2 - \cos(u)\,k\sin(u)\sin(v)^2$$

From expression (6), respectively pick up coefficient of item 'i,j,k', obtain tangent plane normal vector of torus surface 'S' (7):

$$[-2\cos(u)\cos(v) - \cos(u)^2\cos(v), -2\sin(v)\cos(u) - \sin(v)\cos(u)^2, -2\sin(u) - \cos(u)\sin(u)]$$

Calculate curl of vector field 'A', and convert curl from Cartesian Coordinates expression(8) to Torus Surface Coordinates expression(9):

$$\left[\left(\left(\frac{\partial}{\partial y} R(x,y,z) \right) - \left(\frac{\partial}{\partial z} Q(x,y,z) \right) \right), \left(\left(\frac{\partial}{\partial z} P(x,y,z) \right) - \left(\frac{\partial}{\partial x} R(x,y,z) \right) \right), \right.$$
$$\left. \left(\left(\frac{\partial}{\partial x} Q(x,y,z) \right) - \left(\frac{\partial}{\partial y} P(x,y,z) \right) \right) \right] \qquad (8)$$

$$\left[\left(\left(\frac{\partial}{\partial y} R(x,y,z) \right) \left(\frac{\partial}{\partial u} ((2+\cos(u))\sin(v)) \right) \left(\frac{\partial}{\partial v} ((2+\cos(u))\sin(v)) \right) \right. \right.$$
$$- \left(\frac{\partial}{\partial z} Q(x,y,z) \right) \left(\frac{d}{du} \sin(u) \right) \left(\frac{\partial}{\partial v} \sin(u) \right), \left(\frac{\partial}{\partial z} P(x,y,z) \right) \left(\frac{d}{du} \sin(u) \right) \left(\frac{\partial}{\partial v} \sin(u) \right)$$
$$- \left(\frac{\partial}{\partial x} R(x,y,z) \right) \left(\frac{\partial}{\partial u} ((2+\cos(u))\cos(v)) \right) \left(\frac{\partial}{\partial v} ((2+\cos(u))\cos(v)) \right),$$
$$\left(\frac{\partial}{\partial x} Q(x,y,z) \right) \left(\frac{\partial}{\partial u} ((2+\cos(u))\cos(v)) \right) \left(\frac{\partial}{\partial v} ((2+\cos(u))\cos(v)) \right)$$

$$-\left(\frac{\partial}{\partial y}P(x,y,z)\right)\left(\frac{\partial}{\partial u}\left((2+\cos(u))\sin(v)\right)\right)\left(\frac{\partial}{\partial v}\left((2+\cos(u))\sin(v)\right)\right)\Bigg]=\Bigg[$$

$$-\left(\frac{\partial}{\partial y}R(x,y,z)\right)\sin(u)\sin(v)\,(2+\cos(u))\cos(v),$$

$$-\left(\frac{\partial}{\partial x}R(x,y,z)\right)\sin(u)\cos(v)\,(2+\cos(u))\sin(v),$$

$$\left(\frac{\partial}{\partial x}Q(x,y,z)\right)\sin(u)\cos(v)\,(2+\cos(u))\sin(v)$$

$$+\left(\frac{\partial}{\partial y}P(x,y,z)\right)\sin(u)\sin(v)\,(2+\cos(u))\cos(v)\Bigg]$$

(9)

Surface Integral of Dot Product [Curl (9) and Tangent Plane Normal
Vector (7)] in range of parameters 'u,v' (10):

$$\int_0^{2\pi}\int_0^{\frac{2\pi}{n}-\theta}-\left(\frac{\partial}{\partial y}R(x,y,z)\right)\sin(u)\sin(v)\,(2+\cos(u))\cos(v)$$

$$(-2\cos(u)\cos(v)-\cos(u)^2\cos(v))-\left(\frac{\partial}{\partial x}R(x,y,z)\right)\sin(u)\cos(v)\,(2+\cos(u))$$

$$\sin(v)\,(-2\sin(v)\cos(u)-\sin(v)\cos(u)^2)+\Bigg($$

$$\left(\frac{\partial}{\partial x}Q(x,y,z)\right)\sin(u)\cos(v)\,(2+\cos(u))\sin(v)$$

$$+\left(\frac{\partial}{\partial y}P(x,y,z)\right)\sin(u)\sin(v)\,(2+\cos(u))\cos(v)\Bigg)(-2\sin(u)-\cos(u)\sin(u))\,d$$
$u\;dv$

$$=$$

$$\int_0^{2\pi}\frac{1}{24}\cos(v)\sin(v)\Bigg(86\cos(v)\left(\frac{\partial}{\partial y}R(x,y,z)\right)n+86\sin(v)\left(\frac{\partial}{\partial x}R(x,y,z)\right)n$$

$$-6\left(\frac{\partial}{\partial y}R(x,y,z)\right)\cos(v)\cos\left(\frac{-2\pi+\theta\,n}{n}\right)^4 n$$

$$-32\left(\frac{\partial}{\partial y}R(x,y,z)\right)\cos(v)\cos\left(\frac{-2\pi+\theta\,n}{n}\right)^3 n$$

$$-48\left(\frac{\partial}{\partial y}R(x,y,z)\right)\cos(v)\cos\left(\frac{-2\pi+\theta\,n}{n}\right)^2 n$$

$$-6\left(\frac{\partial}{\partial x}R(x,y,z)\right)\sin(v)\cos\left(\frac{-2\pi+\theta\,n}{n}\right)^4 n$$

$$- 32 \left(\frac{\partial}{\partial x} R(x, y, z) \right) \sin(v) \cos\left(\frac{-2 \pi + \theta n}{n} \right)^3 n$$

$$- 48 \left(\frac{\partial}{\partial x} R(x, y, z) \right) \sin(v) \cos\left(\frac{-2 \pi + \theta n}{n} \right)^2 n$$

$$- 45 \left(\frac{\partial}{\partial x} Q(x, y, z) \right) \cos\left(\frac{-2 \pi + \theta n}{n} \right) \sin\left(\frac{-2 \pi + \theta n}{n} \right) n - 102 \left(\frac{\partial}{\partial x} Q(x, y, z) \right) \pi$$

$$+ 51 \left(\frac{\partial}{\partial x} Q(x, y, z) \right) \theta\, n - 45 \left(\frac{\partial}{\partial y} P(x, y, z) \right) \cos\left(\frac{-2 \pi + \theta n}{n} \right) \sin\left(\frac{-2 \pi + \theta n}{n} \right) n$$

$$- 102 \left(\frac{\partial}{\partial y} P(x, y, z) \right) \pi + 51 \left(\frac{\partial}{\partial y} P(x, y, z) \right) \theta\, n$$

$$+ 32 \left(\frac{\partial}{\partial x} Q(x, y, z) \right) \sin\left(\frac{-2 \pi + \theta n}{n} \right)^3 n + 32 \left(\frac{\partial}{\partial y} P(x, y, z) \right) \sin\left(\frac{-2 \pi + \theta n}{n} \right)^3 n$$

$$- 6 \left(\frac{\partial}{\partial x} Q(x, y, z) \right) \sin\left(\frac{-2 \pi + \theta n}{n} \right) \cos\left(\frac{-2 \pi + \theta n}{n} \right)^3 n$$

$$- 6 \left(\frac{\partial}{\partial y} P(x, y, z) \right) \sin\left(\frac{-2 \pi + \theta n}{n} \right) \cos\left(\frac{-2 \pi + \theta n}{n} \right)^3 n \Bigg) / n \; dv = 0 \tag{10}$$

// About 'Integral value is unequal to 0', See also Chapter 6 (This Part) Section 6.1 'Curl Theorem at Manifold and Mobius Strip'

Viz. expression (5) = expression (10):

$$\int_0^{2\pi} P(x, y, z) \left(\frac{d}{dv} (3 \cos(v)) \right) + Q(x, y, z) \left(\frac{d}{dv} (3 \sin(v)) \right) + R(x, y, z) \left(\frac{d}{dv} 0 \right) dv$$

$$- \int_0^{2\pi} P(x, y, z) \left(\frac{\partial}{\partial v} \left(\left(2 + \cos\left(-\frac{2 \pi}{n} + \theta \right) \right) \cos(v) \right) \right)$$

$$+ Q(x, y, z) \left(\frac{\partial}{\partial v} \left(\left(2 + \cos\left(-\frac{2 \pi}{n} + \theta \right) \right) \sin(v) \right) \right)$$

$$+ R(x, y, z) \left(\frac{\partial}{\partial v} \left(-\sin\left(-\frac{2 \pi}{n} + \theta \right) \right) \right) dv$$

$$=$$

$$\int_0^{2\pi} \int_0^{\frac{2\pi}{n} - \theta} - \left(\frac{\partial}{\partial y} R(x, y, z) \right) \sin(u) \sin(v) (2 + \cos(u)) \cos(v)$$

$$(-2\cos(u)\cos(v) - \cos(u)^2\cos(v)) - \left(\frac{\partial}{\partial x}R(x,y,z)\right)\sin(u)\cos(v)(2+\cos(u))$$

$$\sin(v)(-2\sin(v)\cos(u) - \sin(v)\cos(u)^2) + \Bigg($$

$$\left(\frac{\partial}{\partial x}Q(x,y,z)\right)\sin(u)\cos(v)(2+\cos(u))\sin(v)$$

$$+\left(\frac{\partial}{\partial y}P(x,y,z)\right)\sin(u)\sin(v)(2+\cos(u))\cos(v)\Bigg)(-2\sin(u) - \cos(u)\sin(u))\,d$$

$u\,dv$

Above mentioned equation can be described as:

$$\int_{L+} A\cdot dL = \iint_S rotA\cdot n\, dS \quad (1),\ \text{Complete Proof.}$$

2.4 Prove Curl Theorem In Torus Coordinates
(Multiple Connected Orientable Closed Surface Coordinates)
[Program Template of Waterloo Maplesoft, Optional]

Curl Theorem Suppose positive directional boundary 'L+' of smooth or piecewise smooth orientational surface 'S' is smooth or piecewise smooth closed curve, Right Hand Rule is associated basis between positive direction of boundary curve 'L' and outer side of orientational surface 'S'. If there are 1th order continuous partial derivatives about functions 'P(x,y,z),Q(x,y,z),R(x,y,z)' [Structure vector field 'A'] at orientational surface 'S', then:

$$\int_{L+} A\cdot dL = \iint_S rotA\cdot n\, dS \quad (1)$$

thereinto, 'rotA' is curl of vector field 'A', 'n' is unit normal vector of orientational surface 'S's outer side.

Symbol System:

Abstract Vector Field 'V',

Curl 'cV1,cV2' of Abstract Vector Field 'V';

Torus surface 'CS'[Set it unclosed],

Closed Boundary Curve 'CL1' and 'CL2' of Torus Surface 'CS',

Tangent Plane Normal Vector '[A,B,C]' of Torus Surface 'CS'

```
> restart;
```

```
> with(linalg):
> CS:=[(2+cos(u))*cos(v),(2+cos(u))*sin(v),sin(u)];
# Define parametrized expression of torus surface
```

$$CS := [(2 + \cos(u)) \cos(v), (2 + \cos(u)) \sin(v), \sin(u)]$$

```
> rgu:=[0,2*Pi/n-theta];
# Set range of parameters 'u,v' in [0,2π/n - θ],[0,2π]
# Thereinto, 'n' is discretional constant, and n ≥ 1, 'θ'is
```

discretional constant or continuous function, and $2\pi/n - \theta < 2\pi$, make torus surface 'CS' unclosed.

$$rgu := \left[0, \frac{2\pi}{n} - \theta \right]$$

```
> rgv:=[0,2*Pi];
```

$$rgv := [0, 2\pi]$$

```
> subs(u=rgu[1],CS)=eval(subs(u=rgu[1],CS));CL1:=rhs(%);
```

Import left boundary values of parameter 'u' to expression of torus surface 'CS', obtain expressions of boundary curves 'CL1'; Set Right Hand Rule as associated basis between positive direction of boundary curves 'CL1' and normal vector of unclosed torus surface 'CS'.

$$[(2 + \cos(0)) \cos(v), (2 + \cos(0)) \sin(v), \sin(0)] = [3 \cos(v), 3 \sin(v), 0]$$

$$CL1 := [3 \cos(v), 3 \sin(v), 0]$$

```
> subs(u=rgu[2],CS)=eval(subs(u=rgu[2],CS));CL2:=rhs(%);
```

Import right boundary values of parameter 'u' to expression of torus surface 'CS', obtain expressions of boundary curves 'CL2'; Set Right Hand Rule as associated basis between positive direction of boundary curves 'CL2' and normal vector of unclosed torus surface 'CS'.

$$\left[\left(2 + \cos\left(\frac{2\pi}{n} - \theta\right)\right) \cos(v), \left(2 + \cos\left(\frac{2\pi}{n} - \theta\right)\right) \sin(v), \sin\left(\frac{2\pi}{n} - \theta\right)\right] =$$

$$\left[\left(2 + \cos\left(-\frac{2\pi}{n} + \theta\right)\right) \cos(v), \left(2 + \cos\left(-\frac{2\pi}{n} + \theta\right)\right) \sin(v), -\sin\left(-\frac{2\pi}{n} + \theta\right)\right]$$

$$CL2 := \left[\left(2 + \cos\left(-\frac{2\pi}{n} + \theta\right)\right) \cos(v), \left(2 + \cos\left(-\frac{2\pi}{n} + \theta\right)\right) \sin(v), -\sin\left(-\frac{2\pi}{n} + \theta\right)\right]$$

```
> V:=[(P)(x,y,z),(Q)(x,y,z),(R)(x,y,z)];
```

Define abstract spacial vector field 'V' (Suppose Vector Field 'V' possesses 1th order continuous partial derivatives at torus surface)

$$V := [\mathrm{P}(x, y, z), \mathrm{Q}(x, y, z), \mathrm{R}(x, y, z)]$$

```
> [Diff(V[3],y)-Diff(V[2],z),Diff(V[1],z)-Diff(V[3],x),
Diff(V[2],x)-Diff(V[1],y)]
```

```
=[diff(V[3],y)-diff(V[2],z),diff(V[1],z)-diff(V[3],x),
diff(V[2],x)-diff(V[1],y)];cV1:=rhs(%);
# Calculate 'cV1' - curl of abstract spacial vector field 'V'
```

$$\left[\left(\frac{\partial}{\partial y}R(x,y,z)\right)-\left(\frac{\partial}{\partial z}Q(x,y,z)\right),\left(\frac{\partial}{\partial z}P(x,y,z)\right)-\left(\frac{\partial}{\partial x}R(x,y,z)\right),\right.$$
$$\left.\left(\frac{\partial}{\partial x}Q(x,y,z)\right)-\left(\frac{\partial}{\partial y}P(x,y,z)\right)\right]=\left[\left(\frac{\partial}{\partial y}R(x,y,z)\right)-\left(\frac{\partial}{\partial z}Q(x,y,z)\right),\right.$$
$$\left.\left(\frac{\partial}{\partial z}P(x,y,z)\right)-\left(\frac{\partial}{\partial x}R(x,y,z)\right),\left(\frac{\partial}{\partial x}Q(x,y,z)\right)-\left(\frac{\partial}{\partial y}P(x,y,z)\right)\right]$$
$$cV1:=\left[\left(\frac{\partial}{\partial y}R(x,y,z)\right)-\left(\frac{\partial}{\partial z}Q(x,y,z)\right),\left(\frac{\partial}{\partial z}P(x,y,z)\right)-\left(\frac{\partial}{\partial x}R(x,y,z)\right),\right.$$
$$\left.\left(\frac{\partial}{\partial x}Q(x,y,z)\right)-\left(\frac{\partial}{\partial y}P(x,y,z)\right)\right]$$

```
> x:='x':y:='y':z:='z':
> Int(V[1]*Diff(CL1[1],v)+V[2]*Diff(CL1[2],v)+V[3]*Diff(CL1[3],v),
v=rgv[1]..rgv[2]);
# Spacial Closed Curve Integral [Vector field 'V' at Boundary Curve
'CL1']
```

$$\int_0^{2\pi}P(x,y,z)\left(\frac{d}{dv}(3\cos(v))\right)+Q(x,y,z)\left(\frac{d}{dv}(3\sin(v))\right)+R(x,y,z)\left(\frac{d}{dv}0\right)dv$$

```
> v1:=value(%);
```
$$v1:=0$$

```
> x:='x':y:='y':z:='z':
> Int(V[1]*Diff(CL2[1],v)+V[2]*Diff(CL2[2],v)+V[3]*Diff(CL2[3],v),
v=rgv[1]..rgv[2]);
# Spacial Closed Curve Integral [Vector Field 'V' at Boundary Curve
'CL2']
```

$$\int_0^{2\pi}P(x,y,z)\left(\frac{\partial}{\partial v}\left(\left(2+\cos\left(-\frac{2\pi}{n}+\theta\right)\right)\cos(v)\right)\right)$$
$$+Q(x,y,z)\left(\frac{\partial}{\partial v}\left(\left(2+\cos\left(-\frac{2\pi}{n}+\theta\right)\right)\sin(v)\right)\right)$$
$$+R(x,y,z)\left(\frac{\partial}{\partial v}\left(-\sin\left(-\frac{2\pi}{n}+\theta\right)\right)\right)dv$$

```
> v2:=value(%);
```
$$v2:=0$$

```
> alpha:=v1-v2;
```

```
# Spacial Closed Curve Integral [Vector Field 'V' at Boundary Curves
'CL1'and'CL2']
```

$$\alpha := 0$$

```
> x:='x':y:='y':z:='z':
> matrix(3,3,[i,j,k,Diff(CS[1],u),Diff(CS[2],u),Diff(CS[3],u),
Diff(CS[1],v),Diff(CS[2],v),Diff(CS[3],v)])=
matrix(3,3,[i,j,k,diff(CS[1],u),diff(CS[2],u),diff(CS[3],u),
diff(CS[1],v),diff(CS[2],v),diff(CS[3],v)]);m:=rhs(%);
# Define and calculate matrix of partial derivatives, obtain 'tangent
plane normal vector' of torus surface 'S'
```

$$\begin{bmatrix} i & j & k \\ \dfrac{\partial}{\partial u}((2+\cos(u))\cos(v)) & \dfrac{\partial}{\partial u}((2+\cos(u))\sin(v)) & \dfrac{d}{du}\sin(u) \\ \dfrac{\partial}{\partial v}((2+\cos(u))\cos(v)) & \dfrac{\partial}{\partial v}((2+\cos(u))\sin(v)) & \dfrac{\partial}{\partial v}\sin(u) \end{bmatrix} =$$

$$\begin{bmatrix} i & j & k \\ -\sin(u)\cos(v) & -\sin(u)\sin(v) & \cos(u) \\ -(2+\cos(u))\sin(v) & (2+\cos(u))\cos(v) & 0 \end{bmatrix}$$

$$m := \begin{bmatrix} i & j & k \\ -\sin(u)\cos(v) & -\sin(u)\sin(v) & \cos(u) \\ -(2+\cos(u))\sin(v) & (2+\cos(u))\cos(v) & 0 \end{bmatrix}$$

```
> det(m);
```

$$-2\,i\cos(u)\cos(v) - i\cos(u)^2\cos(v) - 2\sin(u)\cos(v)^2\,k - \sin(u)\cos(v)^2\,k\cos(u)$$
$$- 2\sin(v)\,j\cos(u) - 2\,k\sin(u)\sin(v)^2 - \sin(v)\,j\cos(u)^2 - \cos(u)\,k\sin(u)\sin(v)^2$$

```
> mn:=simplify(%);
```

$$mn := -2\,i\cos(u)\cos(v) - i\cos(u)^2\cos(v) - 2\sin(v)\,j\cos(u) - 2\,k\sin(u)$$
$$- \sin(v)\,j\cos(u)^2 - \cos(u)\,k\sin(u)$$

```
> A:=coeff(mn,i); # Obtain coefficient of 'i'
```

$$A := -2\cos(u)\cos(v) - \cos(u)^2\cos(v)$$

```
> B:=coeff(mn,j); # Obtain coefficient of 'j'
```

$$B := -2\sin(v)\cos(u) - \sin(v)\cos(u)^2$$

```
> C:=coeff(mn,k); # Obtain coefficient of 'k'
```

$$C := -2\sin(u) - \cos(u)\sin(u)$$

```
> [A,B,C]; # [A,B,C] structure 'tangent plane normal vector'
```

$$[-\cos(u)^2\cos(v) - 2\cos(u)\cos(v), -\cos(u)^2\sin(v) - 2\cos(u)\sin(v),$$
$$-\sin(u)\cos(u) - 2\sin(u)]$$

```
> x:='x':y:='y':z:='z':
> [Diff(V[3],y)*Diff(CS[2],u)*Diff(CS[2],v)
-Diff(V[2],z)*Diff(CS[3],u)*Diff(CS[3],v),
Diff(V[1],z)*Diff(CS[3],u)*Diff(CS[3],v)
-Diff(V[3],x)*Diff(CS[1],u)*Diff(CS[1],v),
Diff(V[2],x)*Diff(CS[1],u)*Diff(CS[1],v)
-Diff(V[1],y)*Diff(CS[2],u)*Diff(CS[2],v)]
=[diff(V[3],y)*diff(CS[2],u)*diff(CS[2],v)
-diff(V[2],z)*diff(CS[3],u)*diff(CS[3],v),
diff(V[1],z)*diff(CS[3],u)*diff(CS[3],v)
-diff(V[3],x)*diff(CS[1],u)*diff(CS[1],v),
diff(V[2],x)*diff(CS[1],u)*diff(CS[1],v)
-diff(V[1],y)*diff(CS[2],u)*diff(CS[2],v)];cV2:=rhs(%);
# Convert `cV1` from Cartesian Coordinates expression to Torus Surface
Coordinates expression
# Reserve original form of abstract vector field `[P(x,y,z),Q(x,y,z),
R(x,y,z)]` and its partial derivatives (insoluble),Calculate partial
derivatives of `a*sin(u)*cos(v), b*sin(u)*sin(v),c*cos(u)`
(computable), obtain a new expression `cV2`(Curl) of vector field
```

$$\left[\left(\frac{\partial}{\partial y}R(x,y,z)\right)\left(\frac{\partial}{\partial u}((2+\cos(u))\sin(v))\right)\left(\frac{\partial}{\partial v}((2+\cos(u))\sin(v))\right)\right.$$
$$-\left(\frac{\partial}{\partial z}Q(x,y,z)\right)\left(\frac{d}{du}\sin(u)\right)\left(\frac{\partial}{\partial v}\sin(u)\right),\left(\frac{\partial}{\partial z}P(x,y,z)\right)\left(\frac{d}{du}\sin(u)\right)\left(\frac{\partial}{\partial v}\sin(u)\right)$$
$$-\left(\frac{\partial}{\partial x}R(x,y,z)\right)\left(\frac{\partial}{\partial u}((2+\cos(u))\cos(v))\right)\left(\frac{\partial}{\partial v}((2+\cos(u))\cos(v))\right),$$
$$\left(\frac{\partial}{\partial x}Q(x,y,z)\right)\left(\frac{\partial}{\partial u}((2+\cos(u))\cos(v))\right)\left(\frac{\partial}{\partial v}((2+\cos(u))\cos(v))\right)$$
$$\left.-\left(\frac{\partial}{\partial y}P(x,y,z)\right)\left(\frac{\partial}{\partial u}((2+\cos(u))\sin(v))\right)\left(\frac{\partial}{\partial v}((2+\cos(u))\sin(v))\right)\right]=\left[\vphantom{\frac{\partial}{\partial y}}\right.$$
$$-\left(\frac{\partial}{\partial y}R(x,y,z)\right)\sin(u)\sin(v)(2+\cos(u))\cos(v),$$
$$-\left(\frac{\partial}{\partial x}R(x,y,z)\right)\sin(u)\cos(v)(2+\cos(u))\sin(v),$$
$$\left(\frac{\partial}{\partial x}Q(x,y,z)\right)\sin(u)\cos(v)(2+\cos(u))\sin(v)$$
$$\left.+\left(\frac{\partial}{\partial y}P(x,y,z)\right)\sin(u)\sin(v)(2+\cos(u))\cos(v)\right]$$

$$cV2 := \left[-\left(\frac{\partial}{\partial y} R(x,y,z) \right) \sin(u) \sin(v) (2 + \cos(u)) \cos(v), \right.$$

$$-\left(\frac{\partial}{\partial x} R(x,y,z) \right) \sin(u) \cos(v) (2 + \cos(u)) \sin(v),$$

$$\left(\frac{\partial}{\partial x} Q(x,y,z) \right) \sin(u) \cos(v) (2 + \cos(u)) \sin(v)$$

$$\left. + \left(\frac{\partial}{\partial y} P(x,y,z) \right) \sin(u) \sin(v) (2 + \cos(u)) \cos(v) \right]$$

```
> Int(Int(cV2[1]*A+cV2[2]*B+cV2[3]*C,u=rgu[1]..rgu[2]),
v=rgv[1]..rgv[2]);
```

Surface Integral of Dot product (Curl 'cV2' and Tangent Plane Normal Vector '[A,B,C]') in range of 'u,v'

$$\int_0^{2\pi} \int_0^{\frac{2\pi}{n} - \theta} -\left(\frac{\partial}{\partial y} R(x,y,z) \right) \sin(u) \sin(v) (2 + \cos(u)) \cos(v)$$

$$(-2\cos(u)\cos(v) - \cos(u)^2\cos(v)) - \left(\frac{\partial}{\partial x} R(x,y,z) \right) \sin(u) \cos(v) (2 + \cos(u))$$

$$\sin(v) (-2\sin(v)\cos(u) - \sin(v)\cos(u)^2) + \Bigg($$

$$\left(\frac{\partial}{\partial x} Q(x,y,z) \right) \sin(u) \cos(v) (2 + \cos(u)) \sin(v)$$

$$+ \left(\frac{\partial}{\partial y} P(x,y,z) \right) \sin(u) \sin(v) (2 + \cos(u)) \cos(v) \Bigg)(-2\sin(u) - \cos(u)\sin(u))\, du$$

$$dv$$

```
> Beta:=value(%);
```

$$\varepsilon := 0$$

Viz.

$$\int_0^{2\pi} P(x,y,z)\left(\frac{d}{dv}(3\cos(v)) \right) + Q(x,y,z)\left(\frac{d}{dv}(3\sin(v)) \right) + R(x,y,z)\left(\frac{d}{dv}0 \right) dv$$

$$- \int_0^{2\pi} P(x,y,z)\left(\frac{\partial}{\partial v}\left(\left(2 + \cos\left(-\frac{2\pi}{n} + \theta\right)\right)\cos(v) \right) \right)$$

$$+ Q(x,y,z)\left(\frac{\partial}{\partial v}\left(\left(2 + \cos\left(-\frac{2\pi}{n} + \theta\right)\right)\sin(v) \right) \right)$$

$$+ R(x,y,z)\left(\frac{\partial}{\partial v}\left(-\sin\left(-\frac{2\pi}{n} + \theta\right) \right) \right) dv$$

$$=$$

$$\int_0^{2\pi} \int_0^{\frac{2\pi}{n}-\theta} -\left(\frac{\partial}{\partial y}R(x,y,z)\right)\sin(u)\sin(v)\,(2+\cos(u))\cos(v)$$

$$(-2\cos(u)\cos(v)-\cos(u)^2\cos(v))-\left(\frac{\partial}{\partial x}R(x,y,z)\right)\sin(u)\cos(v)\,(2+\cos(u))$$

$$\sin(v)\,(-2\sin(v)\cos(u)-\sin(v)\cos(u)^2)+\Bigg($$

$$\left(\frac{\partial}{\partial x}Q(x,y,z)\right)\sin(u)\cos(v)\,(2+\cos(u))\sin(v)$$

$$+\left(\frac{\partial}{\partial y}P(x,y,z)\right)\sin(u)\sin(v)\,(2+\cos(u))\cos(v)\Bigg)(-2\sin(u)-\cos(u)\sin(u))\,d$$

$$u\,dv$$

Above mentioned equation can be described as:

$$\int_{L+} A\cdot dL = \iint_S rotA\cdot n\,dS \quad (1),\; \text{Complete Proof.}$$

Chapter 3   Numerical Models of Curl Theorem at Manifold

3.1 Numerical Model of Curl Theorem at Manifold (I)

Known: Parametrized expression of Simply connected、 Orientable、 Unclosed Surface(Irregular、Asymmetrical)

$$\left[2\sin(u)\cos(v)+\frac{2}{7}\sin(u)\cos(v-2)\cos(7v)-\sin(u),\right.$$

$$2\sin(u)\sin(v)+\frac{2}{7}\sin(u)\sin(v-2)\cos(8v)-\cos(u),$$

$$\left.2\cos(u)-\frac{1}{7}\cos\left(\frac{u}{2}\right)\cos(12u-v)\right] \quad (1)$$

thereinto, $u\in[0,\frac{\pi}{2}]$, $v\in[0,2\pi]$; and Integral Vector Field

$$\left[\frac{\left(\frac{x}{3}+\frac{y}{4}-\frac{z}{5}\right)^2}{2},\frac{y^2}{3}+\frac{xz}{3},\frac{x^2}{3}+\frac{yz}{3}\right] \quad (2)$$

Calculate and Validate Curl Theorem at Manifold.

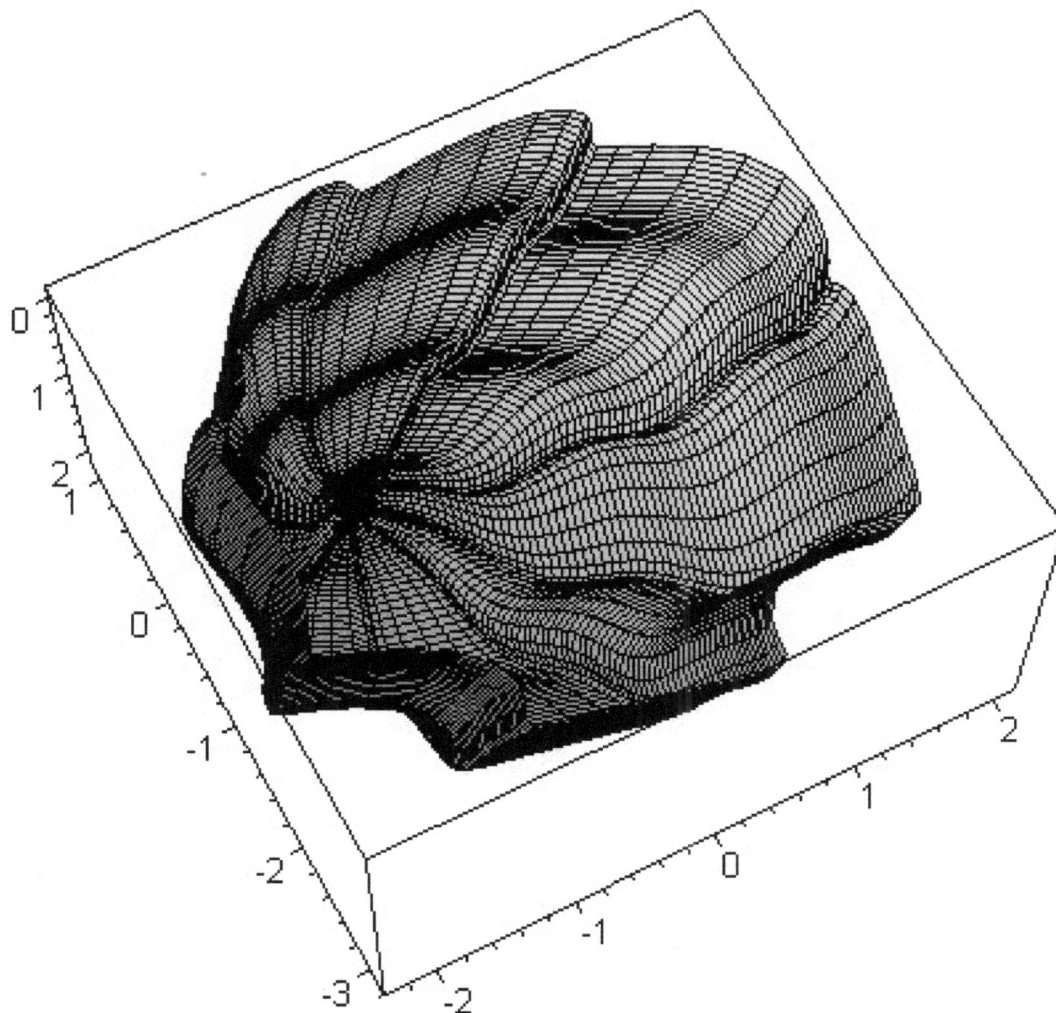

Figure (III) 3.1.1 Simply connected、 Orientable、 Unclosed Surface (1)
[Irregular、 Asymmetrical]

Solution:

First, Free Spacial Closed Curve Integral:

Import right boundary value of 'u'(viz. $\dfrac{\pi}{2}$) to parametrized expression (1) of surface, obtain parametrized expression (3) of Unclosed Surface' Boundary Curve; Set Right Hand Rule as associated basis between positive direction of boundary curve and normal vector of unclosed surface

// Be different to define expression of boundary curve `[$\alpha$ cos(v), $\beta$ sin(v), $\gamma$ ]` that depends on surface `[a*sin(u)cos(v),b*sin(u) sin(v),c*cos(u)]` for existence in `Proof`, in `Numerical Models`, it is possible to import boundary value of `u` to expression of surface(1), directly obtain expression of closed boundary curve

$$
\begin{aligned}
&\left[2\sin\left(\frac{\pi}{2}\right)\cos(v) + \frac{2}{7}\sin\left(\frac{\pi}{2}\right)\cos(v-2)\cos(7v) - \sin\left(\frac{\pi}{2}\right), \right. \\
&\quad 2\sin\left(\frac{\pi}{2}\right)\sin(v) + \frac{2}{7}\sin\left(\frac{\pi}{2}\right)\sin(v-2)\cos(8v) - \cos\left(\frac{\pi}{2}\right), \\
&\quad \left. 2\cos\left(\frac{\pi}{2}\right) - \frac{1}{7}\cos\left(\frac{\pi}{4}\right)\cos(6\pi - v)\right] = \left[2\cos(v) + \frac{2}{7}\cos(v-2)\cos(7v) - 1, \right. \\
&\quad \left. 2\sin(v) + \frac{2}{7}\sin(v-2)\cos(8v), -\frac{1}{14}\sqrt{2}\cos(v)\right]
\end{aligned}
$$

(3)

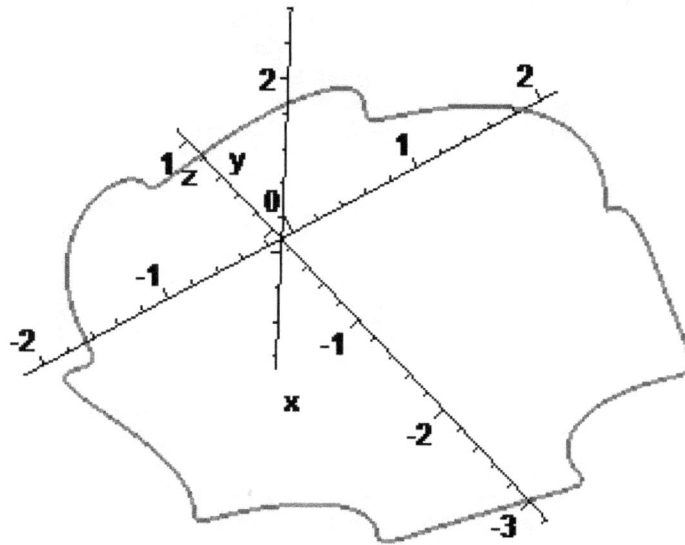

Figure(III)3.1.2 Closed Boundary Curve (3) of Simply connected Orientable、Unclosed Surface (1)

Input expression (3) of Boundary Curve to Vector Field (2); and calculate Spacial Closed Curve Integral [Vetor Field (2) at Boundary Curve(3)] in range of `v` (4):

$$\int_0^{2\pi} \frac{1}{2}\left(\frac{2}{3}\cos(v) + \frac{2}{21}\cos(v-2)\cos(7v) - \frac{1}{3} + \frac{1}{2}\sin(v) + \frac{1}{14}\sin(v-2)\cos(8v)\right.$$

$$\left. + \frac{1}{70}\sqrt{2}\cos(v)\right)^2\left(\frac{d}{dv}\left(2\cos(v) + \frac{2}{7}\cos(v-2)\cos(7v) - 1\right)\right) + \left(\rule{0pt}{20pt}\right.$$

$$\frac{1}{3}\left(2\sin(v) + \frac{2}{7}\sin(v-2)\cos(8v)\right)^2$$

$$-\frac{1}{42}\left(2\cos(v) + \frac{2}{7}\cos(v-2)\cos(7v) - 1\right)\sqrt{2}\cos(v)\rule{0pt}{14pt}\bigg)$$

$$\left(\frac{d}{dv}\left(2\sin(v) + \frac{2}{7}\sin(v-2)\cos(8v)\right)\right) + \left(\frac{1}{3}\left(2\cos(v) + \frac{2}{7}\cos(v-2)\cos(7v) - 1\right)^2\right.$$

$$\left. -\frac{1}{42}\left(2\sin(v) + \frac{2}{7}\sin(v-2)\cos(8v)\right)\sqrt{2}\cos(v)\right)\left(\frac{d}{dv}\left(-\frac{1}{14}\sqrt{2}\cos(v)\right)\right)\,dv =$$

$$\frac{97\pi}{294} + \frac{41}{20580}\sqrt{2}\cos(4)\pi + \frac{1937\sqrt{2}\,\pi}{41160} + \frac{1}{98}\sin(4)\pi + \frac{1}{84}\cos(4)\pi$$

Second, Free (Unclosed) Surface Integral:

According to parametric expression (1) of unclosed surface, define and calculate Matrix of Partial Derivatives, obtain tangent plane normal vector of unclosed surface (5):

$$\begin{bmatrix} i & j & k \\[6pt] \frac{\partial}{\partial u}(2\sin(u)\cos(v) + \frac{2}{7}\sin(u)\cos(v-2)\cos(7v) - \sin(u)) & \frac{\partial}{\partial u}(2\sin(u)\sin(v) + \frac{2}{7}\sin(u)\sin(v-2)\cos(8v) - \cos(u)) & \frac{\partial}{\partial u}(2\cos(u) - \frac{1}{7}\cos(\frac{u}{2})\cos(12u-v)) \\[6pt] \frac{\partial}{\partial v}(2\sin(u)\cos(v) + \frac{2}{7}\sin(u)\cos(v-2)\cos(7v) - \sin(u)) & \frac{\partial}{\partial v}(2\sin(u)\sin(v) + \frac{2}{7}\sin(u)\sin(v-2)\cos(8v) - \cos(u)) & \frac{\partial}{\partial v}(2\cos(u) - \frac{1}{7}\cos(\frac{u}{2})\cos(12u-v)) \end{bmatrix}$$

$$=$$

$$\frac{192}{49}\,i\cos\left(\frac{u}{2}\right)\sin(12u-v)\sin(u)\sin(v-2)\sin(8v)$$

$$+\frac{8}{49}\,i\sin\left(\frac{u}{2}\right)\cos(12u-v)\sin(u)\sin(v-2)\sin(8v)$$

$$-\frac{24}{49}\,i\cos\left(\frac{u}{2}\right)\sin(12u-v)\sin(u)\cos(v-2)\cos(8v)$$

$$-\frac{32}{49}\cos(u)\cos(v-2)\cos(7v)\,k\sin(u)\sin(v-2)\sin(8v)$$

$$+\frac{4}{7}\sin(u)\cos(v-2)\sin(7v)\,k\cos(u)\sin(v-2)\cos(8v)$$

$$-\frac{32}{7}\cos(u)\cos(v)\,k\sin(u)\sin(v-2)\sin(8v)$$

$$-\frac{1}{49}\sin(u)\sin(v-2)\cos(7\,v)\,j\sin\!\left(\frac{u}{2}\right)\cos(12\,u-v)$$

$$+4\sin(u)\cos(v-2)\sin(7\,v)\,k\cos(u)\sin(v)$$

$$-\frac{1}{7}\sin(u)\cos(v-2)\sin(7\,v)\,j\sin\!\left(\frac{u}{2}\right)\cos(12\,u-v)$$

$$+\frac{4}{49}\sin(u)\sin(v-2)^2\cos(7\,v)\,k\cos(u)\cos(8\,v)$$

$$+\frac{4}{7}\sin(u)\sin(v-2)\cos(7\,v)\,k\cos(u)\sin(v)$$

$$-\frac{24}{49}\sin(u)\sin(v-2)\cos(7\,v)\,j\cos\!\left(\frac{u}{2}\right)\sin(12\,u-v)$$

$$-\frac{24}{7}\sin(u)\cos(v-2)\sin(7\,v)\,j\cos\!\left(\frac{u}{2}\right)\sin(12\,u-v)$$

$$+\frac{4}{7}\cos(u)\cos(v)\,k\sin(u)\cos(v-2)\cos(8\,v)$$

$$+\frac{4}{7}\sin(u)\sin(v)\,k\cos(u)\sin(v-2)\cos(8\,v)$$

$$-\frac{1}{49}\,i\sin\!\left(\frac{u}{2}\right)\cos(12\,u-v)\sin(u)\cos(v-2)\cos(8\,v)$$

$$-\frac{2}{49}\,i\cos\!\left(\frac{u}{2}\right)\sin(12\,u-v)\cos(u)\sin(v-2)\cos(8\,v)$$

$$-\frac{2}{7}\,i\cos\!\left(\frac{u}{2}\right)\sin(12\,u-v)\cos(u)\sin(v)-\frac{1}{7}\,i\sin\!\left(\frac{u}{2}\right)\cos(12\,u-v)\sin(u)\cos(v)$$

$$-\frac{24}{7}\,i\cos\!\left(\frac{u}{2}\right)\sin(12\,u-v)\sin(u)\cos(v)+\frac{2}{7}\cos(u)\cos(v)\,j\cos\!\left(\frac{u}{2}\right)\sin(12\,u-v)$$

$$-\frac{2}{7}\cos(u)\,k\sin(u)\cos(v-2)\cos(8\,v)+\frac{16}{7}\cos(u)\,k\sin(u)\sin(v-2)\sin(8\,v)$$

$$-\frac{1}{7}\sin(u)\sin(v)\,j\sin\!\left(\frac{u}{2}\right)\cos(12\,u-v)-\frac{24}{7}\sin(u)\sin(v)\,j\cos\!\left(\frac{u}{2}\right)\sin(12\,u-v)$$

$$+\frac{2}{49}\cos(u)\cos(v-2)\cos(7\,v)\,j\cos\!\left(\frac{u}{2}\right)\sin(12\,u-v)$$

$$+\frac{4}{7}\cos(u)\cos(v-2)\cos(7\,v)\,k\sin(u)\cos(v)$$

$$+\frac{4}{49}\cos(u)\cos(v-2)^2\cos(7\,v)\,k\sin(u)\cos(8\,v)+4\,i\sin(u)^2\cos(v)$$

$$+4\sin(u)^2\sin(v)\,j+2\sin(u)^2\sin(v)\,k-\frac{1}{7}\,i\cos\!\left(\frac{u}{2}\right)\sin(12\,u-v)\sin(u)$$

$$+\frac{4}{7}\,i\sin(u)^2\cos(v-2)\cos(8\,v)-\frac{32}{7}\,i\sin(u)^2\sin(v-2)\sin(8\,v)$$

$$+4\cos(u)\cos(v)^2\,k\sin(u)-\frac{1}{7}\cos(u)\,j\cos\!\left(\frac{u}{2}\right)\sin(12\,u-v)$$

$$-2\cos(u)\,k\sin(u)\cos(v)+4\sin(u)\sin(v)^2\,k\cos(u)$$

$$+ \frac{4}{7} \sin(u)^2 \sin(v-2) \cos(7v) j + \frac{2}{7} \sin(u)^2 \sin(v-2) \cos(7v) k$$

$$+ 4 \sin(u)^2 \cos(v-2) \sin(7v) j + 2 \sin(u)^2 \cos(v-2) \sin(7v) k \qquad (5)$$

From expression (5), respectively pick up coefficient of item 'i,j,k', obtain tangent plane normal vector (6) of surface (1):

$$\left[\frac{192}{49} \cos\left(\frac{u}{2}\right) \sin(12u-v) \sin(u) \sin(v-2) \sin(8v) \right.$$

$$+ \frac{8}{49} \sin\left(\frac{u}{2}\right) \cos(12u-v) \sin(u) \sin(v-2) \sin(8v)$$

$$- \frac{24}{49} \cos\left(\frac{u}{2}\right) \sin(12u-v) \sin(u) \cos(v-2) \cos(8v)$$

$$- \frac{1}{49} \sin\left(\frac{u}{2}\right) \cos(12u-v) \sin(u) \cos(v-2) \cos(8v)$$

$$- \frac{2}{49} \cos\left(\frac{u}{2}\right) \sin(12u-v) \cos(u) \sin(v-2) \cos(8v)$$

$$- \frac{2}{7} \cos\left(\frac{u}{2}\right) \sin(12u-v) \cos(u) \sin(v) - \frac{1}{7} \sin\left(\frac{u}{2}\right) \cos(12u-v) \sin(u) \cos(v)$$

$$- \frac{24}{7} \cos\left(\frac{u}{2}\right) \sin(12u-v) \sin(u) \cos(v) + 4 \cos(v) - 4 \cos(v) \cos(u)^2$$

$$- \frac{4}{7} \cos(v-2) \cos(8v) \cos(u)^2 + \frac{32}{7} \sin(v-2) \sin(8v) \cos(u)^2$$

$$+ \frac{4}{7} \cos(v-2) \cos(8v) - \frac{32}{7} \sin(v-2) \sin(8v) - \frac{1}{7} \cos\left(\frac{u}{2}\right) \sin(12u-v) \sin(u),$$

$$- \frac{1}{49} \sin(u) \sin(v-2) \cos(7v) \sin\left(\frac{u}{2}\right) \cos(12u-v)$$

$$- \frac{1}{7} \sin(u) \cos(v-2) \sin(7v) \sin\left(\frac{u}{2}\right) \cos(12u-v)$$

$$- \frac{24}{49} \sin(u) \sin(v-2) \cos(7v) \cos\left(\frac{u}{2}\right) \sin(12u-v)$$

$$- \frac{24}{7} \sin(u) \cos(v-2) \sin(7v) \cos\left(\frac{u}{2}\right) \sin(12u-v)$$

$$+ \frac{2}{7} \cos(u) \cos(v) \cos\left(\frac{u}{2}\right) \sin(12u-v) - \frac{1}{7} \sin(u) \sin(v) \sin\left(\frac{u}{2}\right) \cos(12u-v)$$

$$- \frac{24}{7} \sin(u) \sin(v) \cos\left(\frac{u}{2}\right) \sin(12u-v)$$

$$+ \frac{2}{49} \cos(u) \cos(v-2) \cos(7v) \cos\left(\frac{u}{2}\right) \sin(12u-v) - 4 \sin(v) \cos(u)^2 + 4 \sin(v)$$

$$+ \frac{4}{7} \sin(v-2) \cos(7v) + 4 \cos(v-2) \sin(7v) - \frac{4}{7} \sin(v-2) \cos(7v) \cos(u)^2$$

$$- 4 \cos(v-2) \sin(7v) \cos(u)^2 - \frac{1}{7} \cos(u) \cos\left(\frac{u}{2}\right) \sin(12u-v),$$

$$-\frac{32}{49}\cos(u)\cos(v-2)\cos(7\,v)\sin(u)\sin(v-2)\sin(8\,v)$$

$$+\frac{4}{7}\sin(u)\cos(v-2)\sin(7\,v)\cos(u)\sin(v-2)\cos(8\,v)$$

$$-\frac{32}{7}\cos(u)\cos(v)\sin(u)\sin(v-2)\sin(8\,v)$$

$$+4\sin(u)\cos(v-2)\sin(7\,v)\cos(u)\sin(v)$$

$$+\frac{4}{7}\sin(u)\sin(v-2)\cos(7\,v)\cos(u)\sin(v)$$

$$+\frac{4}{7}\cos(u)\cos(v)\sin(u)\cos(v-2)\cos(8\,v)$$

$$+\frac{4}{7}\sin(u)\sin(v)\cos(u)\sin(v-2)\cos(8\,v)-\frac{2}{7}\cos(u)\sin(u)\cos(v-2)\cos(8\,v)$$

$$+\frac{16}{7}\cos(u)\sin(u)\sin(v-2)\sin(8\,v)+\frac{4}{7}\cos(u)\cos(v-2)\cos(7\,v)\sin(u)\cos(v)$$

$$+2\sin(v)-2\sin(v)\cos(u)^2+\frac{4}{49}\sin(u)\cos(7\,v)\cos(u)\cos(8\,v)$$

$$+2\cos(v-2)\sin(7\,v)+\frac{2}{7}\sin(v-2)\cos(7\,v)-\frac{2}{7}\sin(v-2)\cos(7\,v)\cos(u)^2$$

$$-2\cos(v-2)\sin(7\,v)\cos(u)^2+4\cos(u)\sin(u)-2\cos(u)\sin(u)\cos(v)\Big]$$

Calculate Curl (7) of Vector Field (2):

$$\left[\left(\frac{\partial}{\partial y}\left(\frac{x^2}{3}+\frac{y\,z}{3}\right)\right)-\left(\frac{\partial}{\partial z}\left(\frac{y^2}{3}+\frac{x\,z}{3}\right)\right),\left(\frac{\partial}{\partial z}\left(\frac{\left(\frac{x}{3}+\frac{y}{4}-\frac{z}{5}\right)^2}{2}\right)\right)-\left(\frac{\partial}{\partial x}\left(\frac{x^2}{3}+\frac{y\,z}{3}\right)\right),\right.$$

$$\left.\left(\frac{\partial}{\partial x}\left(\frac{y^2}{3}+\frac{x\,z}{3}\right)\right)-\left(\frac{\partial}{\partial y}\left(\frac{\left(\frac{x}{3}+\frac{y}{4}-\frac{z}{5}\right)^2}{2}\right)\right)\right]=$$

$$\left[\frac{z}{3}-\frac{x}{3},-\frac{11\,x}{15}-\frac{y}{20}+\frac{z}{25},\frac{23\,z}{60}-\frac{x}{12}-\frac{y}{16}\right] \tag{7}$$

Input expression(1) of Unclosed Surface to Curl (7); and Calculate Integral of Dot Product [Curl (7) and tangent plane normal vector (6)] in range of 'u,v' (8):

// Be different to abstract curl 'cV1,cV2' in 'Proof',in 'Numerical Models', it is possible to input idiographic expression (1) of unclosed surface to idiographic curl (7), then calculate dot product [curl (7) and tangent plane normal vector (6)], complete surface integral

$$\int_0^{2\pi}\int_0^{\frac{\pi}{2}} \left(\frac{2}{3}\cos(u) - \frac{1}{21}\cos\left(\frac{u}{2}\right)\cos(12u-v) - \frac{2}{3}\sin(u)\cos(v)\right.$$

$$\left. -\frac{2}{21}\sin(u)\cos(v-2)\cos(7v) + \frac{1}{3}\sin(u)\right)\left(\right.$$

$$\frac{192}{49}\cos\left(\frac{u}{2}\right)\sin(12u-v)\sin(u)\sin(v-2)\sin(8v)$$

$$+\frac{8}{49}\sin\left(\frac{u}{2}\right)\cos(12u-v)\sin(u)\sin(v-2)\sin(8v)$$

$$-\frac{24}{49}\cos\left(\frac{u}{2}\right)\sin(12u-v)\sin(u)\cos(v-2)\cos(8v)$$

$$-\frac{1}{49}\sin\left(\frac{u}{2}\right)\cos(12u-v)\sin(u)\cos(v-2)\cos(8v)$$

$$-\frac{2}{49}\cos\left(\frac{u}{2}\right)\sin(12u-v)\cos(u)\sin(v-2)\cos(8v)$$

$$-\frac{2}{7}\cos\left(\frac{u}{2}\right)\sin(12u-v)\cos(u)\sin(v) - \frac{1}{7}\sin\left(\frac{u}{2}\right)\cos(12u-v)\sin(u)\cos(v)$$

$$-\frac{24}{7}\cos\left(\frac{u}{2}\right)\sin(12u-v)\sin(u)\cos(v) + 4\cos(v) - 4\cos(v)\cos(u)^2$$

$$-\frac{4}{7}\cos(v-2)\cos(8v)\cos(u)^2 + \frac{32}{7}\sin(v-2)\sin(8v)\cos(u)^2$$

$$+\frac{4}{7}\cos(v-2)\cos(8v) - \frac{32}{7}\sin(v-2)\sin(8v) - \frac{1}{7}\cos\left(\frac{u}{2}\right)\sin(12u-v)\sin(u)\right) +$$

$$\left(-\frac{22}{15}\sin(u)\cos(v) - \frac{22}{105}\sin(u)\cos(v-2)\cos(7v) + \frac{11}{15}\sin(u) - \frac{1}{10}\sin(u)\sin(v)\right.$$

$$\left. -\frac{1}{70}\sin(u)\sin(v-2)\cos(8v) + \frac{13}{100}\cos(u) - \frac{1}{175}\cos\left(\frac{u}{2}\right)\cos(12u-v)\right)\left(\right.$$

$$-\frac{1}{49}\sin(u)\sin(v-2)\cos(7v)\sin\left(\frac{u}{2}\right)\cos(12u-v)$$

$$-\frac{1}{7}\sin(u)\cos(v-2)\sin(7v)\sin\left(\frac{u}{2}\right)\cos(12u-v)$$

$$-\frac{24}{49}\sin(u)\sin(v-2)\cos(7v)\cos\left(\frac{u}{2}\right)\sin(12u-v)$$

$$-\frac{24}{7}\sin(u)\cos(v-2)\sin(7v)\cos\left(\frac{u}{2}\right)\sin(12u-v)$$

$$+\frac{2}{7}\cos(u)\cos(v)\cos\left(\frac{u}{2}\right)\sin(12u-v) - \frac{1}{7}\sin(u)\sin(v)\sin\left(\frac{u}{2}\right)\cos(12u-v)$$

$$-\frac{24}{7}\sin(u)\sin(v)\cos\left(\frac{u}{2}\right)\sin(12u-v)$$

$$+\frac{2}{49}\cos(u)\cos(v-2)\cos(7v)\cos\left(\frac{u}{2}\right)\sin(12u-v) - 4\sin(v)\cos(u)^2 + 4\sin(v)$$

$$+\frac{4}{7}\sin(v-2)\cos(7v) + 4\cos(v-2)\sin(7v) - \frac{4}{7}\sin(v-2)\cos(7v)\cos(u)^2$$

$$- 4 \cos(v - 2) \sin(7\,v) \cos(u)^2 - \frac{1}{7} \cos(u) \cos\left(\frac{u}{2}\right) \sin(12\,u - v)\Big) + \left(\frac{199}{240} \cos(u)\right.$$

$$- \frac{23}{420} \cos\left(\frac{u}{2}\right) \cos(12\,u - v) - \frac{1}{6} \sin(u) \cos(v) - \frac{1}{42} \sin(u) \cos(v - 2) \cos(7\,v)$$

$$+ \frac{1}{12} \sin(u) - \frac{1}{8} \sin(u) \sin(v) - \frac{1}{56} \sin(u) \sin(v - 2) \cos(8\,v)\Big)\Bigg($$

$$- \frac{32}{49} \cos(u) \cos(v - 2) \cos(7\,v) \sin(u) \sin(v - 2) \sin(8\,v)$$

$$+ \frac{4}{7} \sin(u) \cos(v - 2) \sin(7\,v) \cos(u) \sin(v - 2) \cos(8\,v)$$

$$- \frac{32}{7} \cos(u) \cos(v) \sin(u) \sin(v - 2) \sin(8\,v)$$

$$+ 4 \sin(u) \cos(v - 2) \sin(7\,v) \cos(u) \sin(v)$$

$$+ \frac{4}{7} \sin(u) \sin(v - 2) \cos(7\,v) \cos(u) \sin(v)$$

$$+ \frac{4}{7} \cos(u) \cos(v) \sin(u) \cos(v - 2) \cos(8\,v)$$

$$+ \frac{4}{7} \sin(u) \sin(v) \cos(u) \sin(v - 2) \cos(8\,v) - \frac{2}{7} \cos(u) \sin(u) \cos(v - 2) \cos(8\,v)$$

$$+ \frac{16}{7} \cos(u) \sin(u) \sin(v - 2) \sin(8\,v) + \frac{4}{7} \cos(u) \cos(v - 2) \cos(7\,v) \sin(u) \cos(v)$$

$$+ 2 \sin(v) - 2 \sin(v) \cos(u)^2 + \frac{4}{49} \sin(u) \cos(7\,v) \cos(u) \cos(8\,v)$$

$$+ 2 \cos(v - 2) \sin(7\,v) + \frac{2}{7} \sin(v - 2) \cos(7\,v) - \frac{2}{7} \sin(v - 2) \cos(7\,v) \cos(u)^2$$

$$- 2 \cos(v - 2) \sin(7\,v) \cos(u)^2 + 4 \cos(u) \sin(u) - 2 \cos(u) \sin(u) \cos(v)\Bigg) du\ dv$$

$$= \frac{97\,\pi}{294} + \frac{41}{20580} \sqrt{2} \cos(4) \pi + \frac{1937 \sqrt{2}\,\pi}{41160} + \frac{1}{98} \sin(4) \pi + \frac{1}{84} \cos(4) \pi$$

Closed Curve Integral Precision Value (4) [Vector Field (2) at Boundary Curve(3) of Unclosed Surface (1)], is equal to Surface Integral Precision Value (8) [Curl (7) at Unclosed Surface (1)], Complete Calculation and Validation of Curl Theorem at Manifold;

Third, Free Cubage Integral:

Entirely multiply radius vector 'r' (suppose r>0) by each item of surface parametrized expression (1), convert 'Parametrized expression of surface' to 'Parametrized expression of surface coordinates' (9):

$$\left[r\left(2\sin(u)\cos(v) + \frac{2}{7}\sin(u)\cos(v-2)\cos(7v) - \sin(u) \right), \right.$$

$$r\left(2\sin(u)\sin(v) + \frac{2}{7}\sin(u)\sin(v-2)\cos(8v) - \cos(u) \right),$$

$$\left. r\left(2\cos(u) - \frac{1}{7}\cos\left(\frac{u}{2}\right)\cos(12u-v) \right) \right]$$

$$(9)$$

According to parametrized expression of surface coordinates (9), define and calculate Second Matrix of Partial Derivatives, obtain general expression of surface coordinates volume element Coefficient (10):

$$\begin{bmatrix} \frac{\partial}{\partial r}(r(2\sin(u)\cos(v)+\frac{2}{7}\sin(u)\cos(v-2)\cos(7v)-\sin(u))) & \frac{\partial}{\partial u}(r(2\sin(u)\cos(v)+\frac{2}{7}\sin(u)\cos(v-2)\cos(7v)-\sin(u))) & \frac{\partial}{\partial v}(r(2\sin(u)\cos(v)+\frac{2}{7}\sin(u)\cos(v-2)\cos(7v)-\sin(u))) \\ \frac{\partial}{\partial r}(r(2\sin(u)\sin(v)+\frac{2}{7}\sin(u)\sin(v-2)\cos(8v)-\cos(u))) & \frac{\partial}{\partial u}(r(2\sin(u)\sin(v)+\frac{2}{7}\sin(u)\sin(v-2)\cos(8v)-\cos(u))) & \frac{\partial}{\partial v}(r(2\sin(u)\sin(v)+\frac{2}{7}\sin(u)\sin(v-2)\cos(8v)-\cos(u))) \\ \frac{\partial}{\partial r}(r(2\cos(u)-\frac{1}{7}\cos(\frac{u}{2})\cos(12u-v))) & \frac{\partial}{\partial u}(r(2\cos(u)-\frac{1}{7}\cos(\frac{u}{2})\cos(12u-v))) & \frac{\partial}{\partial v}(r(2\cos(u)-\frac{1}{7}\cos(\frac{u}{2})\cos(12u-v))) \end{bmatrix}$$

$$=$$

$$\frac{1}{343}r^2\left(-196\sin(u)\cos(v-2)\cos(8v) + 1568\sin(u)\sin(v-2)\sin(8v) \right.$$

$$-1372\sin(u)\cos(v) - 448\cos(v-2)\cos(7v)\sin(v-2)\sin(8v)\sin(u)$$

$$+32\cos\left(\frac{u}{2}\right)\cos(12u-v)\cos(u)\cos(v-2)\cos(7v)\sin(u)\sin(v-2)\sin(8v)$$

$$-28\cos\left(\frac{u}{2}\right)\cos(12u-v)\sin(u)\cos(v-2)\sin(7v)\cos(u)\sin(v-2)\cos(8v)$$

$$+2744\sin(u) + 392\cos(v)\sin(u)\cos(v-2)\cos(8v)$$

$$-1176\cos\left(\frac{u}{2}\right)\sin(12u-v)\cos(u)^2\cos(v)$$

$$+168\cos(v-2)\cos(8v)\cos\left(\frac{u}{2}\right)\sin(12u-v)$$

$$+7\cos(v-2)\cos(8v)\sin\left(\frac{u}{2}\right)\cos(12u-v)$$

$$-14\cos(v-2)\cos(7v)\cos\left(\frac{u}{2}\right)\sin(12u-v)$$

$$-16\cos(u)^2\sin(v-2)\cos(v-2)\sin(8v)\sin\left(\frac{u}{2}\right)\cos(12u-v)\cos(7v)$$

$$+14\cos(u)^2\sin(v-2)\cos(8v)\cos(v-2)\sin\left(\frac{u}{2}\right)\cos(12u-v)\sin(7v)$$

$$+336\cos(u)^2\sin(v-2)\cos(8v)\cos\left(\frac{u}{2}\right)\sin(12u-v)\cos(v-2)\sin(7v)$$

$$-384\cos(u)^2\sin(v-2)\cos\left(\frac{u}{2}\right)\sin(12\,u-v)\cos(v-2)\sin(8\,v)\cos(7\,v)$$

$$+56\cos(7\,v)\sin(u)\cos(8\,v)-56\sin(v-2)\sin(8\,v)\sin\left(\frac{u}{2}\right)\cos(12\,u-v)$$

$$-1344\sin(v-2)\sin(8\,v)\cos\left(\frac{u}{2}\right)\sin(12\,u-v)$$

$$-98\cos\left(\frac{u}{2}\right)\cos(12\,u-v)\cos(v-2)\sin(7\,v)$$

$$-14\cos\left(\frac{u}{2}\right)\cos(12\,u-v)\sin(v-2)\cos(7\,v)$$

$$-196\cos\left(\frac{u}{2}\right)\cos(12\,u-v)\cos(u)\sin(u)+2744\sin(u)\cos(v-2)\sin(7\,v)\sin(v)$$

$$+14\cos(u)^2\sin(v-2)\cos\left(\frac{u}{2}\right)\cos(12\,u-v)\cos(7\,v)$$

$$+98\cos(u)^2\cos\left(\frac{u}{2}\right)\cos(v-2)\cos(12\,u-v)\sin(7\,v)$$

$$-7\cos(u)^2\cos(8\,v)\cos(v-2)\sin\left(\frac{u}{2}\right)\cos(12\,u-v)$$

$$-168\cos(u)^2\cos(8\,v)\cos\left(\frac{u}{2}\right)\sin(12\,u-v)\cos(v-2)$$

$$+56\cos(u)^2\sin(v-2)\sin(8\,v)\sin\left(\frac{u}{2}\right)\cos(12\,u-v)$$

$$+1344\cos(u)^2\sin(v-2)\cos\left(\frac{u}{2}\right)\sin(12\,u-v)\sin(8\,v)$$

$$+2\cos(u)^2\cos(8\,v)\sin\left(\frac{u}{2}\right)\cos(12\,u-v)\cos(7\,v)$$

$$+48\cos(u)^2\cos(8\,v)\cos\left(\frac{u}{2}\right)\sin(12\,u-v)\cos(7\,v)$$

$$-14\sin(v)\sin(v-2)\cos(7\,v)\sin\left(\frac{u}{2}\right)\cos(12\,u-v)$$

$$-336\sin(v)\sin(v-2)\cos(7\,v)\cos\left(\frac{u}{2}\right)\sin(12\,u-v)$$

$$-98\sin(v)\cos(v-2)\sin(7\,v)\sin\left(\frac{u}{2}\right)\cos(12\,u-v)$$

$$-2352\sin(v)\cos(v-2)\sin(7\,v)\cos\left(\frac{u}{2}\right)\sin(12\,u-v)$$

$$-14\sin(v-2)\cos(8\,v)\sin(v)\sin\left(\frac{u}{2}\right)\cos(12\,u-v)$$

$$-336\sin(v-2)\cos(8\,v)\sin(v)\cos\left(\frac{u}{2}\right)\sin(12\,u-v)$$

$$-14\cos(v)\cos(v-2)\cos(8\,v)\sin\left(\frac{u}{2}\right)\cos(12\,u-v)$$

$$-336\cos(v)\cos(v-2)\cos(8\,v)\cos\left(\frac{u}{2}\right)\sin(12\,u-v)$$

$$+ 112 \cos(v) \sin(v-2) \sin(8\,v) \sin\left(\frac{u}{2}\right) \cos(12\,u-v)$$

$$+ 2688 \cos(v) \sin(v-2) \sin(8\,v) \cos\left(\frac{u}{2}\right) \sin(12\,u-v)$$

$$- 14 \cos(v-2) \cos(7\,v) \cos(v) \sin\left(\frac{u}{2}\right) \cos(12\,u-v)$$

$$- 336 \cos(v-2) \cos(7\,v) \cos(v) \cos\left(\frac{u}{2}\right) \sin(12\,u-v)$$

$$+ 7 \cos(u) \sin(u) \sin(v-2) \cos(7\,v) \sin\left(\frac{u}{2}\right) \cos(12\,u-v)$$

$$+ 168 \cos(u) \sin(u) \sin(v-2) \cos(7\,v) \cos\left(\frac{u}{2}\right) \sin(12\,u-v)$$

$$+ 49 \cos(u) \sin(u) \cos(v-2) \sin(7\,v) \sin\left(\frac{u}{2}\right) \cos(12\,u-v)$$

$$+ 1176 \cos(u) \sin(u) \cos(v-2) \sin(7\,v) \cos\left(\frac{u}{2}\right) \sin(12\,u-v)$$

$$+ 14 \cos\left(\frac{u}{2}\right) \cos(12\,u-v) \cos(u) \sin(u) \cos(v-2) \cos(8\,v)$$

$$- 112 \cos\left(\frac{u}{2}\right) \cos(12\,u-v) \cos(u) \sin(u) \sin(v-2) \sin(8\,v)$$

$$+ 392 \sin(u) \sin(v-2) \cos(7\,v) \sin(v) - 49 \cos(u)^2 \cos(v) \sin\left(\frac{u}{2}\right) \cos(12\,u-v)$$

$$+ 98 \cos(u)^2 \sin(v) \cos\left(\frac{u}{2}\right) \cos(12\,u-v)$$

$$- 2 \cos(8\,v) \cos(7\,v) \sin\left(\frac{u}{2}\right) \cos(12\,u-v)$$

$$- 48 \cos(8\,v) \cos(7\,v) \cos\left(\frac{u}{2}\right) \sin(12\,u-v) + 392 \sin(u) \sin(v) \sin(v-2) \cos(8\,v)$$

$$+ 392 \cos(v-2) \cos(7\,v) \sin(u) \cos(v) - 3136 \cos(v) \sin(u) \sin(v-2) \sin(8\,v)$$

$$+ 392 \sin(v-2) \cos(8\,v) \cos(v-2) \sin(7\,v) \sin(u) - 98 \sin\left(\frac{u}{2}\right) \cos(12\,u-v)$$

$$- 2303 \cos\left(\frac{u}{2}\right) \sin(12\,u-v) + 1078 \cos(v) \cos\left(\frac{u}{2}\right) \sin(12\,u-v)$$

$$+ 49 \cos(v) \sin\left(\frac{u}{2}\right) \cos(12\,u-v) - 98 \cos\left(\frac{u}{2}\right) \cos(12\,u-v) \sin(v)$$

$$+ 98 \cos(u)^2 \sin\left(\frac{u}{2}\right) \cos(12\,u-v)$$

$$- 28 \cos\left(\frac{u}{2}\right) \cos(12\,u-v) \cos(u) \cos(v) \sin(u) \cos(v-2) \cos(8\,v)$$

$$+ 224 \cos\left(\frac{u}{2}\right) \cos(12\,u-v) \cos(u) \cos(v) \sin(u) \sin(v-2) \sin(8\,v)$$

$$-28\cos\left(\frac{u}{2}\right)\cos(12\,u-v)\cos(u)\cos(v-2)\cos(7\,v)\sin(u)\cos(v)$$

$$-28\cos\left(\frac{u}{2}\right)\cos(12\,u-v)\sin(u)\sin(v)\cos(u)\sin(v-2)\cos(8\,v)$$

$$-28\cos\left(\frac{u}{2}\right)\cos(12\,u-v)\sin(u)\sin(v-2)\cos(7\,v)\cos(u)\sin(v)$$

$$-196\cos\left(\frac{u}{2}\right)\cos(12\,u-v)\sin(u)\cos(v-2)\sin(7\,v)\cos(u)\sin(v)$$

$$+2352\cos\left(\frac{u}{2}\right)\sin(12\,u-v)\cos(u)^2$$

$$+49\cos(u)\sin(u)\sin(v)\sin\left(\frac{u}{2}\right)\cos(12\,u-v)$$

$$+98\cos\left(\frac{u}{2}\right)\cos(12\,u-v)\cos(u)\sin(u)\cos(v)$$

$$+1176\sin(u)\cos\left(\frac{u}{2}\right)\sin(12\,u-v)\cos(u)\sin(v)$$

$$-4\cos\left(\frac{u}{2}\right)\cos(12\,u-v)\sin(u)\cos(7\,v)\cos(u)\cos(8\,v)$$

$$+384\cos(v-2)\cos(7\,v)\sin(v-2)\sin(8\,v)\cos\left(\frac{u}{2}\right)\sin(12\,u-v)$$

$$+16\cos(v-2)\cos(7\,v)\sin(v-2)\sin(8\,v)\sin\left(\frac{u}{2}\right)\cos(12\,u-v)$$

$$-14\sin(v-2)\cos(8\,v)\cos(v-2)\sin(7\,v)\sin\left(\frac{u}{2}\right)\cos(12\,u-v)$$

$$-336\sin(v-2)\cos(8\,v)\cos(v-2)\sin(7\,v)\cos\left(\frac{u}{2}\right)\sin(12\,u-v)$$

$$+14\cos(u)^2\sin(v)\sin(v-2)\sin\left(\frac{u}{2}\right)\cos(12\,u-v)\cos(7\,v)$$

$$+336\cos(u)^2\sin(v)\sin(v-2)\cos\left(\frac{u}{2}\right)\sin(12\,u-v)\cos(7\,v)$$

$$+98\cos(u)^2\sin(v)\cos(v-2)\sin\left(\frac{u}{2}\right)\cos(12\,u-v)\sin(7\,v)$$

$$+2352\cos(u)^2\sin(v)\cos\left(\frac{u}{2}\right)\sin(12\,u-v)\cos(v-2)\sin(7\,v)$$

$$+14\cos(u)^2\sin(v)\sin(v-2)\cos(8\,v)\sin\left(\frac{u}{2}\right)\cos(12\,u-v)$$

$$+336\cos(u)^2\sin(v)\sin(v-2)\cos(8\,v)\cos\left(\frac{u}{2}\right)\sin(12\,u-v)$$

$$+14\cos(u)^2\cos(8\,v)\cos(v)\cos(v-2)\sin\left(\frac{u}{2}\right)\cos(12\,u-v)$$

$$+336\cos(u)^2\cos(8\,v)\cos\left(\frac{u}{2}\right)\sin(12\,u-v)\cos(v)\cos(v-2)$$

$$-112\cos(u)^2\sin(v-2)\cos(v)\sin(8\,v)\sin\left(\frac{u}{2}\right)\cos(12\,u-v)$$

$$- 2688 \cos(u)^2 \sin(v-2) \cos\!\left(\frac{u}{2}\right) \sin(12\,u-v) \cos(v) \sin(8\,v)$$

$$+ 14 \cos(u)^2 \cos(v) \cos(v-2) \sin\!\left(\frac{u}{2}\right) \cos(12\,u-v) \cos(7\,v)$$

$$+ 336 \cos(u)^2 \cos\!\left(\frac{u}{2}\right) \sin(12\,u-v) \cos(v) \cos(v-2) \cos(7\,v) \Bigg)$$

(10)

Triple Integrals of 'Surface Coordinates Volume Element' in range of 'r,u,v', absolute value of result (11) is cubage value of unclosed surface (1):

$$\int_0^{2\pi} \int_0^{\frac{\pi}{2}} \int_0^1 \frac{1}{343} r^2 \Big(-196 \sin(u) \cos(v-2) \cos(8\,v) + 1568 \sin(u) \sin(v-2) \sin(8\,v)$$

$$- 1372 \sin(u) \cos(v) - 448 \cos(v-2) \cos(7\,v) \sin(v-2) \sin(8\,v) \sin(u)$$

$$+ 32 \cos\!\left(\frac{u}{2}\right) \cos(12\,u-v) \cos(u) \cos(v-2) \cos(7\,v) \sin(u) \sin(v-2) \sin(8\,v)$$

$$- 28 \cos\!\left(\frac{u}{2}\right) \cos(12\,u-v) \sin(u) \cos(v-2) \sin(7\,v) \cos(u) \sin(v-2) \cos(8\,v)$$

$$+ 2744 \sin(u) + 392 \cos(v) \sin(u) \cos(v-2) \cos(8\,v)$$

$$- 1176 \cos\!\left(\frac{u}{2}\right) \sin(12\,u-v) \cos(u)^2 \cos(v)$$

$$+ 168 \cos(v-2) \cos(8\,v) \cos\!\left(\frac{u}{2}\right) \sin(12\,u-v)$$

$$+ 7 \cos(v-2) \cos(8\,v) \sin\!\left(\frac{u}{2}\right) \cos(12\,u-v)$$

$$- 14 \cos(v-2) \cos(7\,v) \cos\!\left(\frac{u}{2}\right) \sin(12\,u-v)$$

$$- 16 \cos(u)^2 \sin(v-2) \cos(v-2) \sin(8\,v) \sin\!\left(\frac{u}{2}\right) \cos(12\,u-v) \cos(7\,v)$$

$$+ 14 \cos(u)^2 \sin(v-2) \cos(8\,v) \cos(v-2) \sin\!\left(\frac{u}{2}\right) \cos(12\,u-v) \sin(7\,v)$$

$$+ 336 \cos(u)^2 \sin(v-2) \cos(8\,v) \cos\!\left(\frac{u}{2}\right) \sin(12\,u-v) \cos(v-2) \sin(7\,v)$$

$$- 384 \cos(u)^2 \sin(v-2) \cos\!\left(\frac{u}{2}\right) \sin(12\,u-v) \cos(v-2) \sin(8\,v) \cos(7\,v)$$

$$+ 56 \cos(7\,v) \sin(u) \cos(8\,v) - 56 \sin(v-2) \sin(8\,v) \sin\!\left(\frac{u}{2}\right) \cos(12\,u-v)$$

$$- 1344 \sin(v-2) \sin(8\,v) \cos\!\left(\frac{u}{2}\right) \sin(12\,u-v)$$

$$- 98 \cos\!\left(\frac{u}{2}\right) \cos(12\,u-v) \cos(v-2) \sin(7\,v)$$

$$- 14 \cos\!\left(\frac{u}{2}\right) \cos(12\,u-v) \sin(v-2) \cos(7\,v)$$

$$-196 \cos\left(\frac{u}{2}\right) \cos(12u - v) \cos(u) \sin(u) + 2744 \sin(u) \cos(v - 2) \sin(7v) \sin(v)$$

$$+14 \cos(u)^2 \sin(v - 2) \cos\left(\frac{u}{2}\right) \cos(12u - v) \cos(7v)$$

$$+98 \cos(u)^2 \cos\left(\frac{u}{2}\right) \cos(v - 2) \cos(12u - v) \sin(7v)$$

$$-7 \cos(u)^2 \cos(8v) \cos(v - 2) \sin\left(\frac{u}{2}\right) \cos(12u - v)$$

$$-168 \cos(u)^2 \cos(8v) \cos\left(\frac{u}{2}\right) \sin(12u - v) \cos(v - 2)$$

$$+56 \cos(u)^2 \sin(v - 2) \sin(8v) \sin\left(\frac{u}{2}\right) \cos(12u - v)$$

$$+1344 \cos(u)^2 \sin(v - 2) \cos\left(\frac{u}{2}\right) \sin(12u - v) \sin(8v)$$

$$+2 \cos(u)^2 \cos(8v) \sin\left(\frac{u}{2}\right) \cos(12u - v) \cos(7v)$$

$$+48 \cos(u)^2 \cos(8v) \cos\left(\frac{u}{2}\right) \sin(12u - v) \cos(7v)$$

$$-14 \sin(v) \sin(v - 2) \cos(7v) \sin\left(\frac{u}{2}\right) \cos(12u - v)$$

$$-336 \sin(v) \sin(v - 2) \cos(7v) \cos\left(\frac{u}{2}\right) \sin(12u - v)$$

$$-98 \sin(v) \cos(v - 2) \sin(7v) \sin\left(\frac{u}{2}\right) \cos(12u - v)$$

$$-2352 \sin(v) \cos(v - 2) \sin(7v) \cos\left(\frac{u}{2}\right) \sin(12u - v)$$

$$-14 \sin(v - 2) \cos(8v) \sin(v) \sin\left(\frac{u}{2}\right) \cos(12u - v)$$

$$-336 \sin(v - 2) \cos(8v) \sin(v) \cos\left(\frac{u}{2}\right) \sin(12u - v)$$

$$-14 \cos(v) \cos(v - 2) \cos(8v) \sin\left(\frac{u}{2}\right) \cos(12u - v)$$

$$-336 \cos(v) \cos(v - 2) \cos(8v) \cos\left(\frac{u}{2}\right) \sin(12u - v)$$

$$+112 \cos(v) \sin(v - 2) \sin(8v) \sin\left(\frac{u}{2}\right) \cos(12u - v)$$

$$+2688 \cos(v) \sin(v - 2) \sin(8v) \cos\left(\frac{u}{2}\right) \sin(12u - v)$$

$$-14 \cos(v - 2) \cos(7v) \cos(v) \sin\left(\frac{u}{2}\right) \cos(12u - v)$$

$$-336 \cos(v - 2) \cos(7v) \cos(v) \cos\left(\frac{u}{2}\right) \sin(12u - v)$$

$$+7 \cos(u) \sin(u) \sin(v - 2) \cos(7v) \sin\left(\frac{u}{2}\right) \cos(12u - v)$$

$$+ 168 \cos(u) \sin(u) \sin(v-2) \cos(7v) \cos\left(\frac{u}{2}\right) \sin(12u-v)$$

$$+ 49 \cos(u) \sin(u) \cos(v-2) \sin(7v) \sin\left(\frac{u}{2}\right) \cos(12u-v)$$

$$+ 1176 \cos(u) \sin(u) \cos(v-2) \sin(7v) \cos\left(\frac{u}{2}\right) \sin(12u-v)$$

$$+ 14 \cos\left(\frac{u}{2}\right) \cos(12u-v) \cos(u) \sin(u) \cos(v-2) \cos(8v)$$

$$- 112 \cos\left(\frac{u}{2}\right) \cos(12u-v) \cos(u) \sin(u) \sin(v-2) \sin(8v)$$

$$+ 392 \sin(u) \sin(v-2) \cos(7v) \sin(v) - 49 \cos(u)^2 \cos(v) \sin\left(\frac{u}{2}\right) \cos(12u-v)$$

$$+ 98 \cos(u)^2 \sin(v) \cos\left(\frac{u}{2}\right) \cos(12u-v)$$

$$- 2 \cos(8v) \cos(7v) \sin\left(\frac{u}{2}\right) \cos(12u-v)$$

$$- 48 \cos(8v) \cos(7v) \cos\left(\frac{u}{2}\right) \sin(12u-v) + 392 \sin(u) \sin(v) \sin(v-2) \cos(8v)$$

$$+ 392 \cos(v-2) \cos(7v) \sin(u) \cos(v) - 3136 \cos(v) \sin(u) \sin(v-2) \sin(8v)$$

$$+ 392 \sin(v-2) \cos(8v) \cos(v-2) \sin(7v) \sin(u) - 98 \sin\left(\frac{u}{2}\right) \cos(12u-v)$$

$$- 2303 \cos\left(\frac{u}{2}\right) \sin(12u-v) + 1078 \cos(v) \cos\left(\frac{u}{2}\right) \sin(12u-v)$$

$$+ 49 \cos(v) \sin\left(\frac{u}{2}\right) \cos(12u-v) - 98 \cos\left(\frac{u}{2}\right) \cos(12u-v) \sin(v)$$

$$+ 98 \cos(u)^2 \sin\left(\frac{u}{2}\right) \cos(12u-v)$$

$$- 28 \cos\left(\frac{u}{2}\right) \cos(12u-v) \cos(u) \cos(v) \sin(u) \cos(v-2) \cos(8v)$$

$$+ 224 \cos\left(\frac{u}{2}\right) \cos(12u-v) \cos(u) \cos(v) \sin(u) \sin(v-2) \sin(8v)$$

$$- 28 \cos\left(\frac{u}{2}\right) \cos(12u-v) \cos(u) \cos(v-2) \cos(7v) \sin(u) \cos(v)$$

$$- 28 \cos\left(\frac{u}{2}\right) \cos(12u-v) \sin(u) \sin(v) \cos(u) \sin(v-2) \cos(8v)$$

$$- 28 \cos\left(\frac{u}{2}\right) \cos(12u-v) \sin(u) \sin(v-2) \cos(7v) \cos(u) \sin(v)$$

$$- 196 \cos\left(\frac{u}{2}\right) \cos(12u-v) \sin(u) \cos(v-2) \sin(7v) \cos(u) \sin(v)$$

$$+ 2352 \cos\left(\frac{u}{2}\right) \sin(12u-v) \cos(u)^2$$

$$+ 49 \cos(u) \sin(u) \sin(v) \sin\left(\frac{u}{2}\right) \cos(12\,u - v)$$

$$+ 98 \cos\left(\frac{u}{2}\right) \cos(12\,u - v) \cos(u) \sin(u) \cos(v)$$

$$+ 1176 \sin(u) \cos\left(\frac{u}{2}\right) \sin(12\,u - v) \cos(u) \sin(v)$$

$$- 4 \cos\left(\frac{u}{2}\right) \cos(12\,u - v) \sin(u) \cos(7\,v) \cos(u) \cos(8\,v)$$

$$+ 384 \cos(v - 2) \cos(7\,v) \sin(v - 2) \sin(8\,v) \cos\left(\frac{u}{2}\right) \sin(12\,u - v)$$

$$+ 16 \cos(v - 2) \cos(7\,v) \sin(v - 2) \sin(8\,v) \sin\left(\frac{u}{2}\right) \cos(12\,u - v)$$

$$- 14 \sin(v - 2) \cos(8\,v) \cos(v - 2) \sin(7\,v) \sin\left(\frac{u}{2}\right) \cos(12\,u - v)$$

$$- 336 \sin(v - 2) \cos(8\,v) \cos(v - 2) \sin(7\,v) \cos\left(\frac{u}{2}\right) \sin(12\,u - v)$$

$$+ 14 \cos(u)^2 \sin(v) \sin(v - 2) \sin\left(\frac{u}{2}\right) \cos(12\,u - v) \cos(7\,v)$$

$$+ 336 \cos(u)^2 \sin(v) \sin(v - 2) \cos\left(\frac{u}{2}\right) \sin(12\,u - v) \cos(7\,v)$$

$$+ 98 \cos(u)^2 \sin(v) \cos(v - 2) \sin\left(\frac{u}{2}\right) \cos(12\,u - v) \sin(7\,v)$$

$$+ 2352 \cos(u)^2 \sin(v) \cos\left(\frac{u}{2}\right) \sin(12\,u - v) \cos(v - 2) \sin(7\,v)$$

$$+ 14 \cos(u)^2 \sin(v) \sin(v - 2) \cos(8\,v) \sin\left(\frac{u}{2}\right) \cos(12\,u - v)$$

$$+ 336 \cos(u)^2 \sin(v) \sin(v - 2) \cos(8\,v) \cos\left(\frac{u}{2}\right) \sin(12\,u - v)$$

$$+ 14 \cos(u)^2 \cos(8\,v) \cos(v) \cos(v - 2) \sin\left(\frac{u}{2}\right) \cos(12\,u - v)$$

$$+ 336 \cos(u)^2 \cos(8\,v) \cos\left(\frac{u}{2}\right) \sin(12\,u - v) \cos(v) \cos(v - 2)$$

$$- 112 \cos(u)^2 \sin(v - 2) \cos(v) \sin(8\,v) \sin\left(\frac{u}{2}\right) \cos(12\,u - v)$$

$$- 2688 \cos(u)^2 \sin(v - 2) \cos\left(\frac{u}{2}\right) \sin(12\,u - v) \cos(v) \sin(8\,v)$$

$$+ 14 \cos(u)^2 \cos(v) \cos(v - 2) \sin\left(\frac{u}{2}\right) \cos(12\,u - v) \cos(7\,v)$$

$$+ 336 \cos(u)^2 \cos\left(\frac{u}{2}\right) \sin(12\,u - v) \cos(v) \cos(v - 2) \cos(7\,v) \Bigg) dr\,du\,dv$$

$$=$$

$$\frac{109506115472\,\pi}{20538814275} - \frac{1870}{11906559}\cos(4)\,\pi - \frac{137641001\,\sqrt{2}\,\pi}{4323960900}$$
$$+ \frac{160}{11906559}\sin(2)\cos(2)\sqrt{2}\,\pi - \frac{177305}{23813118}\cos(2)^2\,\sqrt{2}\,\pi$$

(11)

3.2 Numerical Model of Curl Theorem at Manifold (I)
[Program Template of Waterloo Maplesoft, Optional]

```
> restart; # Reset computer algebraic system
> with(plots):with(linalg): # Load 'plots' and 'linalg' package
> CS:=[2*sin(u)*cos(v)+2*sin(u)*cos(v-2)*cos(7*v)/7-sin(u),
2*sin(u)*sin(v)+2*sin(u)*sin(v-2)*cos(8*v)/7-cos(u),2*cos(u)
-cos(u/2)*cos(12*u-v)/7];
```

Define parametrized expression of discretional simply connected orientable closed surface 'CS'

$$CS := \left[\, 2\sin(u)\cos(v) + \frac{2}{7}\sin(u)\cos(v-2)\cos(7\,v) - \sin(u), \right.$$
$$2\sin(u)\sin(v) + \frac{2}{7}\sin(u)\sin(v-2)\cos(8\,v) - \cos(u),$$
$$\left. 2\cos(u) - \frac{1}{7}\cos\!\left(\frac{u}{2}\right)\cos(-12\,u+v) \right]$$

```
> rgu:=[0,Pi/2];
```

Halve range of 'u' (relative to closed parametrized surface)

$$rgu := \left[\, 0, \frac{\pi}{2} \,\right]$$

```
> rgv:=[0,2*Pi];
```

Set range of parameters 'u,v', make parametrized surface 'CS' unclosed

$$rgv := [\, 0, 2\,\pi \,]$$

```
> plot3d(CS,u=rgu[1]..rgu[2],v=rgv[1]..rgv[2],scaling=constrained,
projection=0.9,numpoints=5000);g1:=%:
```

Draw unclosed parameterized surface 'CS'

```
> subs(u=rgu[2],CS)=eval(subs(u=rgu[2],CS));CL:=rhs(%);
```

Import boundary value of 'u' to expression of surface 'CS', obtain expression of boundary curve 'CL'; Set Right Hand Rule as associated basis between positive direction of boundary curve 'CL' and normal vector of unclosed surface 'CS'

```
// Be different to define expression of boundary curve '[α cos(v),
```

β sin(v), γ]' that depends on surface '[a*sin(u)cos(v), b*sin(u)

sin(v), c*cos(u)]' for existence in 'Proof', in 'Numerical Models',

it is possible to "Import the boundary value of 'u' to expression of

surface 'CS', directly obtain expression of boundary curve 'CL'"

$$\left[2\sin\left(\frac{\pi}{2}\right)\cos(v) + \frac{2}{7}\sin\left(\frac{\pi}{2}\right)\cos(v-2)\cos(7\,v) - \sin\left(\frac{\pi}{2}\right), \right.$$

$$2\sin\left(\frac{\pi}{2}\right)\sin(v) + \frac{2}{7}\sin\left(\frac{\pi}{2}\right)\sin(v-2)\cos(8\,v) - \cos\left(\frac{\pi}{2}\right),$$

$$2\cos\left(\frac{\pi}{2}\right) - \frac{1}{7}\cos\left(\frac{\pi}{4}\right)\cos(-6\,\pi+v) \right] = \left[2\cos(v) + \frac{2}{7}\cos(v-2)\cos(7\,v) - 1, \right.$$

$$\left. 2\sin(v) + \frac{2}{7}\sin(v-2)\cos(8\,v), -\frac{1}{14}\sqrt{2}\,\cos(v) \right]$$

$$CL := \left[2\cos(v) + \frac{2}{7}\cos(v-2)\cos(7\,v) - 1, 2\sin(v) + \frac{2}{7}\sin(v-2)\cos(8\,v), \right.$$

$$\left. -\frac{1}{14}\sqrt{2}\,\cos(v) \right]$$

```
> spacecurve(CL,v=rgv[1]..rgv[2],scaling=constrained,
  projection=0.9,numpoints=3000,thickness=2,color=red):g2:=%:
# Draw closed boundary curve'CL'
> V:=[(x/3+y/4-z/5)^2/2,y^2/3+x*z/3,x^2/3+y*z/3];
# Define discretional spacial vector field 'V'
```

(Suppose Vector Field 'V' possesses 1th order continuous partial

 derivatives at surface 'CS')

$$V := \left[\frac{\left(\dfrac{x}{3}+\dfrac{y}{4}-\dfrac{z}{5}\right)^2}{2}, \frac{y^2}{3}+\frac{x\,z}{3}, \frac{x^2}{3}+\frac{y\,z}{3} \right]$$

```
> [Diff(V[3],y)-Diff(V[2],z),Diff(V[1],z)-Diff(V[3],x),
Diff(V[2],x)-Diff(V[1],y)]
=[diff(V[3],y)-diff(V[2],z),diff(V[1],z)-diff(V[3],x),
diff(V[2],x)-diff(V[1],y)];cV:=rhs(%);
# Calculate 'cV' -- Curl of spacial vector field 'V'
```

$$\left[\left(\frac{\partial}{\partial y}\left(\frac{x^2}{3}+\frac{y\,z}{3}\right)\right) - \left(\frac{\partial}{\partial z}\left(\frac{y^2}{3}+\frac{x\,z}{3}\right)\right), \left(\frac{\partial}{\partial z}\left(\frac{\left(\dfrac{x}{3}+\dfrac{y}{4}-\dfrac{z}{5}\right)^2}{2}\right)\right) - \left(\frac{\partial}{\partial x}\left(\frac{x^2}{3}+\frac{y\,z}{3}\right)\right), \right.$$

$$\left(\frac{\partial}{\partial x}\left(\frac{y^2}{3}+\frac{x\,z}{3}\right)\right)-\left(\frac{\partial}{\partial y}\left(\frac{\left(\frac{x}{3}+\frac{y}{4}-\frac{z}{5}\right)^2}{2}\right)\right)\Bigg]=$$

$$\left[\frac{z}{3}-\frac{x}{3},\,-\frac{11\,x}{15}-\frac{y}{20}+\frac{z}{25},\frac{23\,z}{60}-\frac{x}{12}-\frac{y}{16}\right]$$

$$cV:=\left[\frac{z}{3}-\frac{x}{3},\,-\frac{11\,x}{15}-\frac{y}{20}+\frac{z}{25},\frac{23\,z}{60}-\frac{x}{12}-\frac{y}{16}\right]$$

```
> rgx:=[-3,1];
```
$$rgx:=[-3,1]$$

```
> rgy:=[-2,2];
```
$$rgy:=[-2,2]$$

```
> rgz:=[-1.25,2.75];
```
$$rgz:=[-1.25,2.75]$$

```
> fieldplot3d(V,x=rgx[1]..rgx[2],y=rgy[1]..rgy[2],z=rgz[1]..
rgz[2],arrows=SLIM,color=red,thickness=1,grid=[6,6,6]):g3:=%:
# Draw spacial vector field 'V'
> fieldplot3d(cV,x=rgx[1]..rgx[2],y=rgy[1]..rgy[2],z=rgz[1]..
rgz[2],arrows=SLIM,color=blue,thickness=1,grid=[6,6,6]):g4:=%:
# Draw curl 'cV'
> display(g1,g2,g3,g4); # Synthesize figures
```

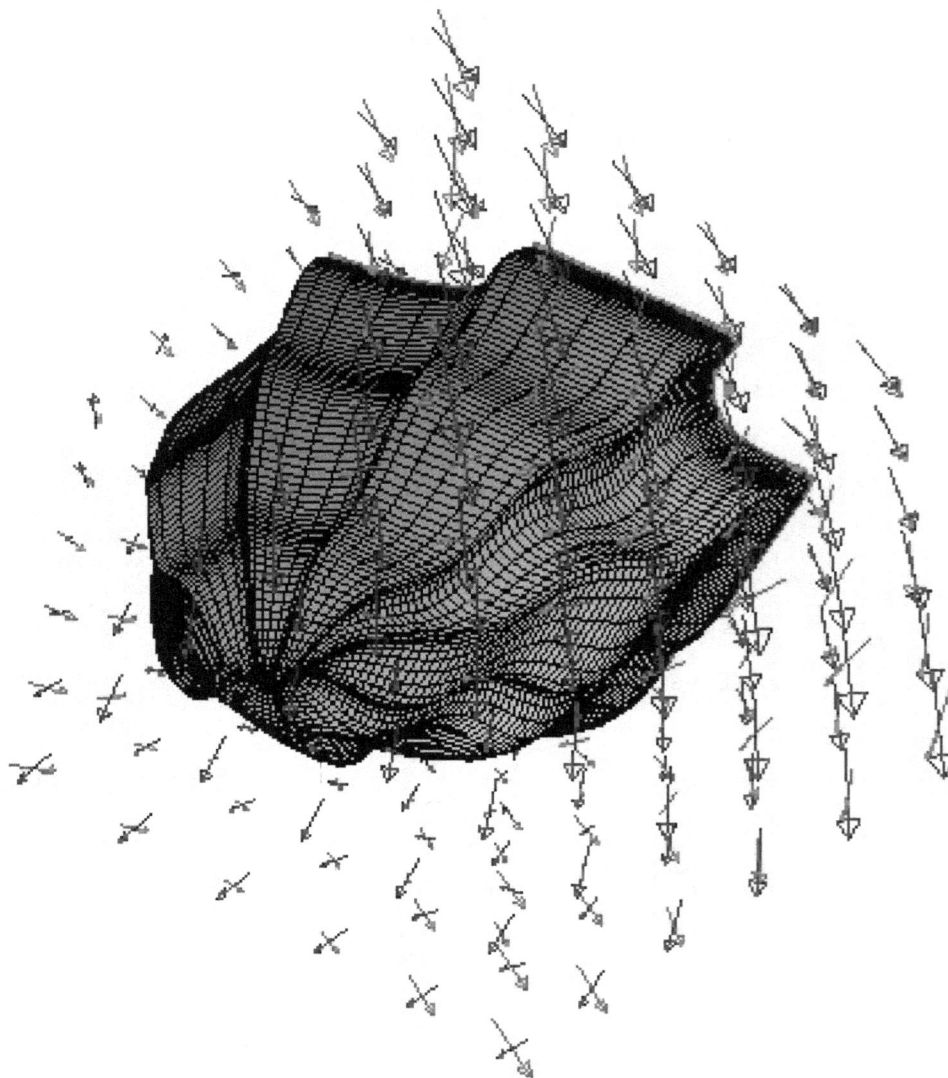

Figure(III)3.2.1 Surface[unclosed] 'CS' and its Boundary' Curve 'CL'
Vector Field 'V' (Red arrows) and its Curl 'cV' (Blue arrows)

```
> display(g2,g3,g4);
```

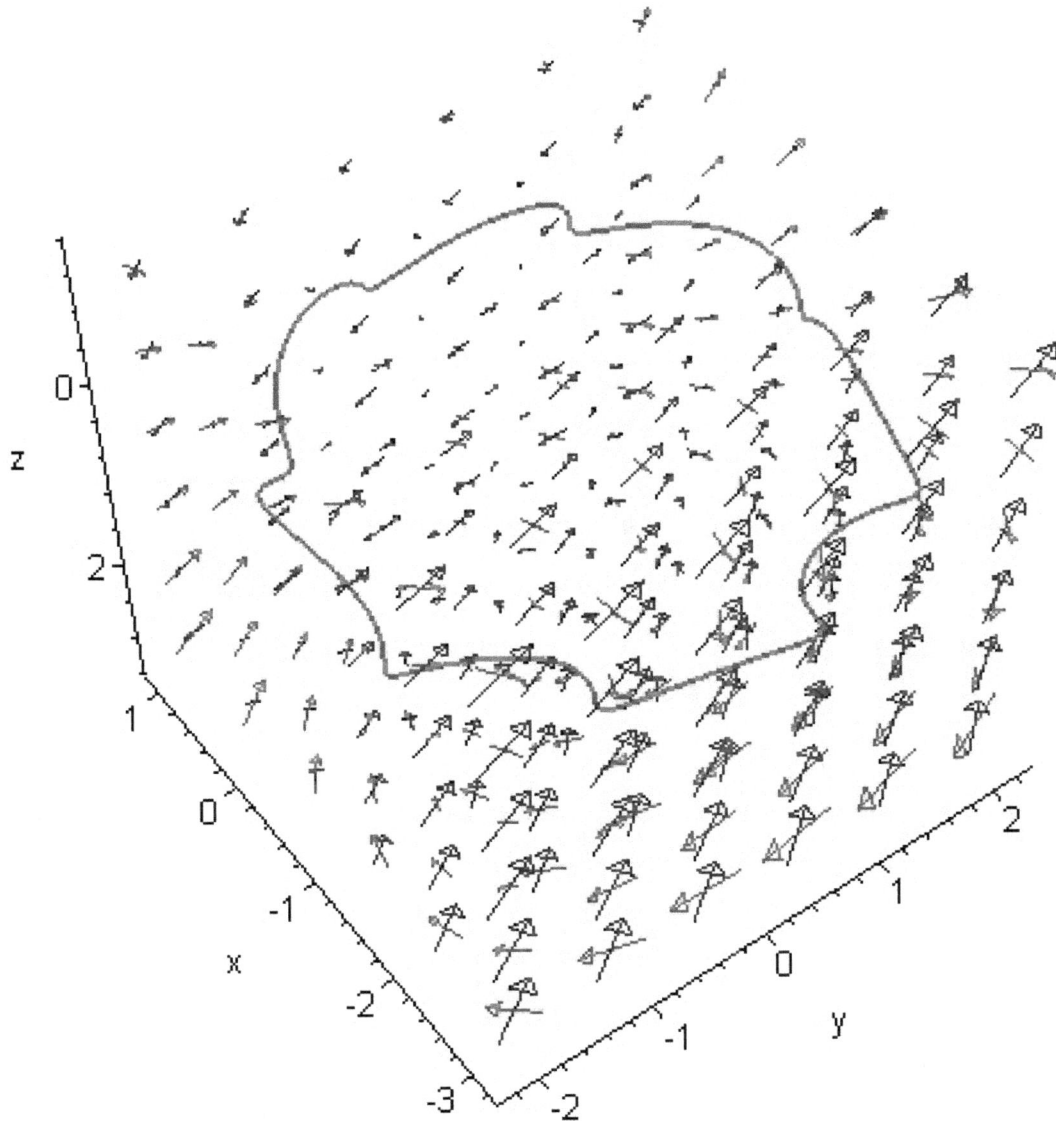

Figure(III)3.2.2 Closed Boundary Curve 'CL';
Vector Field 'V'(Red arrows)and its Curl'cV'(Blue arrows)

```
> x:=CL[1]:y:=CL[2]:z:=CL[3]:
>Int(V[1]*Diff(CL[1],v)+V[2]*Diff(CL[2],v)+V[3]*Diff(CL[3],v),
v=rgv[1]..rgv[2]);
# Closed Curve Integral (Vector Field 'V' at Closed Curve 'CL')
```

$$\int_0^{2\pi} \frac{1}{2}\left(\frac{2}{3}\cos(v)+\frac{2}{21}\cos(v-2)\cos(7v)-\frac{1}{3}+\frac{1}{2}\sin(v)+\frac{1}{14}\sin(v-2)\cos(8v)\right)$$

$$+ \frac{1}{70}\sqrt{2}\,\cos(v)\Bigg)^2 \left(\frac{d}{dv}\left(2\,\cos(v)+\frac{2}{7}\cos(v-2)\cos(7\,v)-1\right)\right)+\Bigg($$

$$\frac{1}{3}\left(2\,\sin(v)+\frac{2}{7}\sin(v-2)\cos(8\,v)\right)^2$$

$$-\frac{1}{42}\left(2\,\cos(v)+\frac{2}{7}\cos(v-2)\cos(7\,v)-1\right)\sqrt{2}\,\cos(v)\Bigg)$$

$$\left(\frac{d}{dv}\left(2\,\sin(v)+\frac{2}{7}\sin(v-2)\cos(8\,v)\right)\right)+\Bigg($$

$$\frac{1}{3}\left(2\,\cos(v)+\frac{2}{7}\cos(v-2)\cos(7\,v)-1\right)^2$$

$$-\frac{1}{42}\left(2\,\sin(v)+\frac{2}{7}\sin(v-2)\cos(8\,v)\right)\sqrt{2}\,\cos(v)\Bigg)\left(\frac{d}{dv}\left(-\frac{1}{14}\sqrt{2}\,\cos(v)\right)\right)\Bigg]dv$$

```
> alpha:=value(%);delta:=evalf(alpha);
# Calculate value of spacial closed curve integral
```

$$\alpha := \frac{1937\sqrt{2}\,\pi}{41160}+\frac{1}{84}\cos(4)\,\pi+\frac{1}{98}\sin(4)\,\pi+\frac{41}{20580}\cos(4)\sqrt{2}\,\pi+\frac{97\,\pi}{294}$$

$$\delta := 1.191102426$$

```
> x:='x':y:='y':z:='z': # Reset variable 'x','y','z'
> x:=CS[1]:y:=CS[2]:z:=CS[3]:
> matrix(3,3,[i,j,k,Diff(CS[1],u),Diff(CS[2],u),Diff(CS[3],u),
Diff(CS[1],v),Diff(CS[2],v),Diff(CS[3],v)])=
matrix(3,3,[i,j,k,diff(CS[1],u),diff(CS[2],u),diff(CS[3],u),
diff(CS[1],v),diff(CS[2],v),diff(CS[3],v)]);m1:=rhs(%);
# Define and calculate matrix of partial derivatives 'm1'; obtain
'tangent plane normal vector' of surface 'CS'
```

$$[i,j,k]$$

$$\left[\frac{\partial}{\partial u}\left(2\,\sin(u)\cos(v)+\frac{2}{7}\sin(u)\cos(v-2)\cos(7\,v)-\sin(u)\right),\right.$$

$$\frac{\partial}{\partial u}\left(2\,\sin(u)\sin(v)+\frac{2}{7}\sin(u)\sin(v-2)\cos(8\,v)-\cos(u)\right),$$

$$\left.\frac{\partial}{\partial u}\left(2\,\cos(u)-\frac{1}{7}\cos\left(\frac{u}{2}\right)\cos(-12\,u+v)\right)\right]$$

$$\left[\frac{\partial}{\partial v}\left(2\,\sin(u)\cos(v)+\frac{2}{7}\sin(u)\cos(v-2)\cos(7\,v)-\sin(u)\right),\right.$$

$$\frac{\partial}{\partial v}\left(2\sin(u)\sin(v)+\frac{2}{7}\sin(u)\sin(v-2)\cos(8v)-\cos(u)\right),$$

$$\frac{\partial}{\partial v}\left(2\cos(u)-\frac{1}{7}\cos\left(\frac{u}{2}\right)\cos(-12u+v)\right)\Bigg]=$$

$$[i,j,k]$$

$$\left[2\cos(u)\cos(v)+\frac{2}{7}\cos(u)\cos(v-2)\cos(7v)-\cos(u),\right.$$

$$2\cos(u)\sin(v)+\frac{2}{7}\cos(u)\sin(v-2)\cos(8v)+\sin(u),$$

$$\left.-2\sin(u)+\frac{1}{14}\sin\left(\frac{u}{2}\right)\cos(-12u+v)-\frac{12}{7}\cos\left(\frac{u}{2}\right)\sin(-12u+v)\right]$$

$$\left[-2\sin(u)\sin(v)-\frac{2}{7}\sin(u)\sin(v-2)\cos(7v)-2\sin(u)\cos(v-2)\sin(7v),\right.$$

$$2\sin(u)\cos(v)+\frac{2}{7}\sin(u)\cos(v-2)\cos(8v)-\frac{16}{7}\sin(u)\sin(v-2)\sin(8v),$$

$$\left.\frac{1}{7}\cos\left(\frac{u}{2}\right)\sin(-12u+v)\right]$$

$$m1:=$$

$$[i,j,k]$$

$$\left[2\cos(u)\cos(v)+\frac{2}{7}\cos(u)\cos(v-2)\cos(7v)-\cos(u),\right.$$

$$2\cos(u)\sin(v)+\frac{2}{7}\cos(u)\sin(v-2)\cos(8v)+\sin(u),$$

$$\left.-2\sin(u)+\frac{1}{14}\sin\left(\frac{u}{2}\right)\cos(12u-v)+\frac{12}{7}\cos\left(\frac{u}{2}\right)\sin(12u-v)\right]$$

$$\left[-2\sin(u)\sin(v)-\frac{2}{7}\sin(u)\sin(v-2)\cos(7v)-2\sin(u)\cos(v-2)\sin(7v),\right.$$

$$2\sin(u)\cos(v)+\frac{2}{7}\sin(u)\cos(v-2)\cos(8v)-\frac{16}{7}\sin(u)\sin(v-2)\sin(8v),$$

$$\left.-\frac{1}{7}\cos\left(\frac{u}{2}\right)\sin(12u-v)\right]$$

```
> det(m1):
> mn:=simplify(%);
```

$$mn:=\frac{1}{7}i\cos\left(\frac{u}{2}\right)\sin(-12u+v)\sin(u)+\frac{1}{7}\cos(u)j\cos\left(\frac{u}{2}\right)\sin(-12u+v)$$

$$-2\cos(u)k\sin(u)\cos(v)+\frac{4}{49}\sin(u)\cos(7v)k\cos(u)\cos(8v)$$

$$-2\cos(v-2)\sin(7v)k\cos(u)^2-\frac{4}{7}i\cos(v-2)\cos(8v)\cos(u)^2$$

$$+\frac{32}{7}i\sin(v-2)\sin(8v)\cos(u)^2-\frac{4}{7}\sin(v-2)\cos(7v)j\cos(u)^2$$

$$-\frac{2}{7}\sin(v-2)\cos(7\,v)\,k\cos(u)^2-4\cos(v-2)\sin(7\,v)\,j\cos(u)^2+2\sin(v)\,k$$

$$+4\,i\cos(v)+\frac{4}{7}\sin(v-2)\cos(7\,v)\,j+\frac{2}{7}\sin(v-2)\cos(7\,v)\,k$$

$$+4\cos(v-2)\sin(7\,v)\,j+4\sin(u)\,k\cos(u)-\frac{32}{7}\,i\sin(v-2)\sin(8\,v)$$

$$+\frac{4}{7}\,i\cos(v-2)\cos(8\,v)-4\sin(v)\,j\cos(u)^2-2\sin(v)\,k\cos(u)^2$$

$$-4\,i\cos(v)\cos(u)^2-\frac{2}{7}\cos(u)\cos(v)\,j\cos\!\left(\frac{u}{2}\right)\sin(-12\,u+v)$$

$$+\frac{4}{7}\cos(u)\cos(v)\,k\sin(u)\cos(v-2)\cos(8\,v)$$

$$-\frac{32}{7}\cos(u)\cos(v)\,k\sin(u)\sin(v-2)\sin(8\,v)$$

$$-\frac{2}{49}\cos(u)\cos(v-2)\cos(7\,v)\,j\cos\!\left(\frac{u}{2}\right)\sin(-12\,u+v)$$

$$+\frac{4}{7}\cos(u)\cos(v-2)\cos(7\,v)\,k\sin(u)\cos(v)$$

$$-\frac{32}{49}\cos(u)\cos(v-2)\cos(7\,v)\,k\sin(u)\sin(v-2)\sin(8\,v)$$

$$-\frac{2}{7}\cos(u)\,k\sin(u)\cos(v-2)\cos(8\,v)+\frac{16}{7}\cos(u)\,k\sin(u)\sin(v-2)\sin(8\,v)$$

$$-\frac{1}{7}\sin(u)\sin(v)\,j\sin\!\left(\frac{u}{2}\right)\cos(-12\,u+v)$$

$$+\frac{24}{7}\sin(u)\sin(v)\,j\cos\!\left(\frac{u}{2}\right)\sin(-12\,u+v)$$

$$+\frac{4}{7}\sin(u)\sin(v)\,k\cos(u)\sin(v-2)\cos(8\,v)$$

$$-\frac{1}{49}\sin(u)\sin(v-2)\cos(7\,v)\,j\sin\!\left(\frac{u}{2}\right)\cos(-12\,u+v)$$

$$+\frac{24}{49}\sin(u)\sin(v-2)\cos(7\,v)\,j\cos\!\left(\frac{u}{2}\right)\sin(-12\,u+v)+4\sin(v)\,j$$

$$+2\cos(v-2)\sin(7\,v)\,k+\frac{4}{7}\sin(u)\sin(v-2)\cos(7\,v)\,k\cos(u)\sin(v)$$

$$-\frac{1}{7}\sin(u)\cos(v-2)\sin(7\,v)\,j\sin\!\left(\frac{u}{2}\right)\cos(-12\,u+v)$$

$$+\frac{24}{7}\sin(u)\cos(v-2)\sin(7\,v)\,j\cos\!\left(\frac{u}{2}\right)\sin(-12\,u+v)$$

$$+4\sin(u)\cos(v-2)\sin(7\,v)\,k\cos(u)\sin(v)$$

$$+\frac{4}{7}\sin(u)\cos(v-2)\sin(7\,v)\,k\cos(u)\sin(v-2)\cos(8\,v)$$

$$+ \frac{2}{7} i \cos\left(\frac{u}{2}\right) \sin(-12\,u + v) \cos(u) \sin(v)$$

$$+ \frac{2}{49} i \cos\left(\frac{u}{2}\right) \sin(-12\,u + v) \cos(u) \sin(v - 2) \cos(8\,v)$$

$$- \frac{1}{7} i \sin\left(\frac{u}{2}\right) \cos(-12\,u + v) \sin(u) \cos(v)$$

$$- \frac{1}{49} i \sin\left(\frac{u}{2}\right) \cos(-12\,u + v) \sin(u) \cos(v - 2) \cos(8\,v)$$

$$+ \frac{8}{49} i \sin\left(\frac{u}{2}\right) \cos(-12\,u + v) \sin(u) \sin(v - 2) \sin(8\,v)$$

$$+ \frac{24}{7} i \cos\left(\frac{u}{2}\right) \sin(-12\,u + v) \sin(u) \cos(v)$$

$$+ \frac{24}{49} i \cos\left(\frac{u}{2}\right) \sin(-12\,u + v) \sin(u) \cos(v - 2) \cos(8\,v)$$

$$- \frac{192}{49} i \cos\left(\frac{u}{2}\right) \sin(-12\,u + v) \sin(u) \sin(v - 2) \sin(8\,v)$$

```
> A:=coeff(mn,i); # Obtain coefficient of 'i'
```

$$A := \frac{1}{7} \cos\left(\frac{u}{2}\right) \sin(-12\,u + v) \sin(u) - \frac{4}{7} \cos(v - 2) \cos(8\,v) \cos(u)^2$$

$$+ \frac{32}{7} \sin(v - 2) \sin(8\,v) \cos(u)^2 + 4 \cos(v) - \frac{32}{7} \sin(v - 2) \sin(8\,v)$$

$$+ \frac{4}{7} \cos(v - 2) \cos(8\,v) - 4 \cos(v) \cos(u)^2$$

$$+ \frac{2}{7} \cos\left(\frac{u}{2}\right) \sin(-12\,u + v) \cos(u) \sin(v)$$

$$+ \frac{2}{49} \cos\left(\frac{u}{2}\right) \sin(-12\,u + v) \cos(u) \sin(v - 2) \cos(8\,v)$$

$$- \frac{1}{7} \sin\left(\frac{u}{2}\right) \cos(-12\,u + v) \sin(u) \cos(v)$$

$$- \frac{1}{49} \sin\left(\frac{u}{2}\right) \cos(-12\,u + v) \sin(u) \cos(v - 2) \cos(8\,v)$$

$$+ \frac{8}{49} \sin\left(\frac{u}{2}\right) \cos(-12\,u + v) \sin(u) \sin(v - 2) \sin(8\,v)$$

$$+ \frac{24}{7} \cos\left(\frac{u}{2}\right) \sin(-12\,u + v) \sin(u) \cos(v)$$

$$+ \frac{24}{49} \cos\left(\frac{u}{2}\right) \sin(-12\,u + v) \sin(u) \cos(v - 2) \cos(8\,v)$$

$$- \frac{192}{49} \cos\left(\frac{u}{2}\right) \sin(-12\,u + v) \sin(u) \sin(v - 2) \sin(8\,v)$$

```
> B:=coeff(mn,j); # Obtain coefficient of 'j'
```

$$B := \frac{1}{7} \cos(u) \cos\left(\frac{u}{2}\right) \sin(-12\,u + v) - \frac{4}{7} \sin(v - 2) \cos(7\,v) \cos(u)^2$$

$$- 4 \cos(v - 2) \sin(7\,v) \cos(u)^2 + \frac{4}{7} \sin(v - 2) \cos(7\,v) + 4 \cos(v - 2) \sin(7\,v)$$

$$-4\sin(v)\cos(u)^2 - \frac{2}{7}\cos(u)\cos(v)\cos\left(\frac{u}{2}\right)\sin(-12\,u+v)$$

$$-\frac{2}{49}\cos(u)\cos(v-2)\cos(7\,v)\cos\left(\frac{u}{2}\right)\sin(-12\,u+v)$$

$$-\frac{1}{7}\sin(u)\sin(v)\sin\left(\frac{u}{2}\right)\cos(-12\,u+v)+\frac{24}{7}\sin(u)\sin(v)\cos\left(\frac{u}{2}\right)\sin(-12\,u+v)$$

$$-\frac{1}{49}\sin(u)\sin(v-2)\cos(7\,v)\sin\left(\frac{u}{2}\right)\cos(-12\,u+v)$$

$$+\frac{24}{49}\sin(u)\sin(v-2)\cos(7\,v)\cos\left(\frac{u}{2}\right)\sin(-12\,u+v)+4\sin(v)$$

$$-\frac{1}{7}\sin(u)\cos(v-2)\sin(7\,v)\sin\left(\frac{u}{2}\right)\cos(-12\,u+v)$$

$$+\frac{24}{7}\sin(u)\cos(v-2)\sin(7\,v)\cos\left(\frac{u}{2}\right)\sin(-12\,u+v)$$

```
> C:=coeff(mn,k); # Obtain coefficient of 'k'
```

$$C := -2\cos(u)\sin(u)\cos(v)+\frac{4}{49}\sin(u)\cos(7\,v)\cos(u)\cos(8\,v)$$

$$-2\cos(v-2)\sin(7\,v)\cos(u)^2-\frac{2}{7}\sin(v-2)\cos(7\,v)\cos(u)^2+2\sin(v)$$

$$+\frac{2}{7}\sin(v-2)\cos(7\,v)+4\sin(u)\cos(u)-2\sin(v)\cos(u)^2$$

$$+\frac{4}{7}\cos(u)\cos(v)\sin(u)\cos(v-2)\cos(8\,v)$$

$$-\frac{32}{7}\cos(u)\cos(v)\sin(u)\sin(v-2)\sin(8\,v)$$

$$+\frac{4}{7}\cos(u)\cos(v-2)\cos(7\,v)\sin(u)\cos(v)$$

$$-\frac{32}{49}\cos(u)\cos(v-2)\cos(7\,v)\sin(u)\sin(v-2)\sin(8\,v)$$

$$-\frac{2}{7}\cos(u)\sin(u)\cos(v-2)\cos(8\,v)+\frac{16}{7}\cos(u)\sin(u)\sin(v-2)\sin(8\,v)$$

$$+\frac{4}{7}\sin(u)\sin(v)\cos(u)\sin(v-2)\cos(8\,v)+2\cos(v-2)\sin(7\,v)$$

$$+\frac{4}{7}\sin(u)\sin(v-2)\cos(7\,v)\cos(u)\sin(v)$$

$$+4\sin(u)\cos(v-2)\sin(7\,v)\cos(u)\sin(v)$$

$$+\frac{4}{7}\sin(u)\cos(v-2)\sin(7\,v)\cos(u)\sin(v-2)\cos(8\,v)$$

```
> [A,B,C]:  # [A,B,C] structure 'tangent plane normal vector'
> Int(Int(cV[1]*A+cV[2]*B+cV[3]*C,u=rgu[1]..rgu[2]),v=rgv[1]..
  rgv[2]);
# Integral of Dot Product (Curl 'cV' and Tangent Plane Normal Vector
'[A,B,C]') in range of 'u,v'
```

```
// Be different to abstract curl 'cV1,cV2' that 'Proof' refers to,
in 'Numerical Models', be able to input idiographic values 'x=CS[1],
y=CS[2],z=CS[3]' to idiographic curl 'cV', then calculate dot product
(Curl 'cV' and tangent plane normal vector '[A,B,C]'), complete surface
integral
```

$$\int_0^{2\pi}\int_0^{\frac{\pi}{2}}\left(\frac{2}{3}\cos(u)-\frac{1}{21}\cos\left(\frac{u}{2}\right)\cos(-12\,u+v)-\frac{2}{3}\sin(u)\cos(v)\right.$$

$$-\frac{2}{21}\sin(u)\cos(v-2)\cos(7\,v)+\frac{1}{3}\sin(u)\right)\left(\frac{1}{7}\cos\left(\frac{u}{2}\right)\sin(-12\,u+v)\sin(u)\right.$$

$$-\frac{4}{7}\cos(v-2)\cos(8\,v)\cos(u)^2+\frac{32}{7}\sin(v-2)\sin(8\,v)\cos(u)^2+4\cos(v)$$

$$-\frac{32}{7}\sin(v-2)\sin(8\,v)+\frac{4}{7}\cos(v-2)\cos(8\,v)-4\cos(v)\cos(u)^2$$

$$+\frac{2}{7}\cos\left(\frac{u}{2}\right)\sin(-12\,u+v)\cos(u)\sin(v)$$

$$+\frac{2}{49}\cos\left(\frac{u}{2}\right)\sin(-12\,u+v)\cos(u)\sin(v-2)\cos(8\,v)$$

$$-\frac{1}{7}\sin\left(\frac{u}{2}\right)\cos(-12\,u+v)\sin(u)\cos(v)$$

$$-\frac{1}{49}\sin\left(\frac{u}{2}\right)\cos(-12\,u+v)\sin(u)\cos(v-2)\cos(8\,v)$$

$$+\frac{8}{49}\sin\left(\frac{u}{2}\right)\cos(-12\,u+v)\sin(u)\sin(v-2)\sin(8\,v)$$

$$+\frac{24}{7}\cos\left(\frac{u}{2}\right)\sin(-12\,u+v)\sin(u)\cos(v)$$

$$+\frac{24}{49}\cos\left(\frac{u}{2}\right)\sin(-12\,u+v)\sin(u)\cos(v-2)\cos(8\,v)$$

$$-\frac{192}{49}\cos\left(\frac{u}{2}\right)\sin(-12\,u+v)\sin(u)\sin(v-2)\sin(8\,v)\right)+\left(-\frac{22}{15}\sin(u)\cos(v)\right.$$

$$-\frac{22}{105}\sin(u)\cos(v-2)\cos(7\,v)+\frac{11}{15}\sin(u)-\frac{1}{10}\sin(u)\sin(v)$$

$$-\frac{1}{70}\sin(u)\sin(v-2)\cos(8\,v)+\frac{13}{100}\cos(u)-\frac{1}{175}\cos\left(\frac{u}{2}\right)\cos(-12\,u+v)\right)\left(\vphantom{\frac{1}{7}}\right.$$

$$\frac{1}{7}\cos(u)\cos\left(\frac{u}{2}\right)\sin(-12\,u+v)-\frac{4}{7}\sin(v-2)\cos(7\,v)\cos(u)^2$$

$$-4\cos(v-2)\sin(7\,v)\cos(u)^2+\frac{4}{7}\sin(v-2)\cos(7\,v)+4\cos(v-2)\sin(7\,v)$$

$$-4\sin(v)\cos(u)^2-\frac{2}{7}\cos(u)\cos(v)\cos\left(\frac{u}{2}\right)\sin(-12\,u+v)$$

$$-\frac{2}{49}\cos(u)\cos(v-2)\cos(7v)\cos\left(\frac{u}{2}\right)\sin(-12u+v)$$

$$-\frac{1}{7}\sin(u)\sin(v)\sin\left(\frac{u}{2}\right)\cos(-12u+v)+\frac{24}{7}\sin(u)\sin(v)\cos\left(\frac{u}{2}\right)\sin(-12u+v)$$

$$-\frac{1}{49}\sin(u)\sin(v-2)\cos(7v)\sin\left(\frac{u}{2}\right)\cos(-12u+v)$$

$$+\frac{24}{49}\sin(u)\sin(v-2)\cos(7v)\cos\left(\frac{u}{2}\right)\sin(-12u+v)+4\sin(v)$$

$$-\frac{1}{7}\sin(u)\cos(v-2)\sin(7v)\sin\left(\frac{u}{2}\right)\cos(-12u+v)$$

$$+\frac{24}{7}\sin(u)\cos(v-2)\sin(7v)\cos\left(\frac{u}{2}\right)\sin(-12u+v)\bigg)+\left(\frac{199}{240}\cos(u)\right.$$

$$-\frac{23}{420}\cos\left(\frac{u}{2}\right)\cos(-12u+v)-\frac{1}{6}\sin(u)\cos(v)-\frac{1}{42}\sin(u)\cos(v-2)\cos(7v)$$

$$+\frac{1}{12}\sin(u)-\frac{1}{8}\sin(u)\sin(v)-\frac{1}{56}\sin(u)\sin(v-2)\cos(8v)\bigg)\bigg($$

$$-2\cos(u)\sin(u)\cos(v)+\frac{4}{49}\sin(u)\cos(7v)\cos(u)\cos(8v)$$

$$-2\cos(v-2)\sin(7v)\cos(u)^2-\frac{2}{7}\sin(v-2)\cos(7v)\cos(u)^2+2\sin(v)$$

$$+\frac{2}{7}\sin(v-2)\cos(7v)+4\sin(u)\cos(u)-2\sin(v)\cos(u)^2$$

$$+\frac{4}{7}\cos(u)\cos(v)\sin(u)\cos(v-2)\cos(8v)$$

$$-\frac{32}{7}\cos(u)\cos(v)\sin(u)\sin(v-2)\sin(8v)$$

$$+\frac{4}{7}\cos(u)\cos(v-2)\cos(7v)\sin(u)\cos(v)$$

$$-\frac{32}{49}\cos(u)\cos(v-2)\cos(7v)\sin(u)\sin(v-2)\sin(8v)$$

$$-\frac{2}{7}\cos(u)\sin(u)\cos(v-2)\cos(8v)+\frac{16}{7}\cos(u)\sin(u)\sin(v-2)\sin(8v)$$

$$+\frac{4}{7}\sin(u)\sin(v)\cos(u)\sin(v-2)\cos(8v)+2\cos(v-2)\sin(7v)$$

$$+\frac{4}{7}\sin(u)\sin(v-2)\cos(7v)\cos(u)\sin(v)$$

$$+4\sin(u)\cos(v-2)\sin(7v)\cos(u)\sin(v)$$

$$+\frac{4}{7}\sin(u)\cos(v-2)\sin(7v)\cos(u)\sin(v-2)\cos(8v)\bigg)du\,dv$$

```
> beta:=value(%);epsilon:=evalf(beta);
# Calculate value of surface integral
```

$$\beta:=\frac{1937\sqrt{2}\,\pi}{41160}+\frac{1}{84}\cos(4)\,\pi+\frac{1}{98}\sin(4)\,\pi+\frac{41}{20580}\cos(4)\sqrt{2}\,\pi+\frac{97\,\pi}{294}$$

$$\varepsilon:=1.191102426$$

```
> alpha;beta; # Two analytic values are equal
```

$$\frac{1937\sqrt{2}\,\pi}{41160} + \frac{1}{84}\cos(4)\,\pi + \frac{1}{98}\sin(4)\,\pi + \frac{41}{20580}\cos(4)\,\sqrt{2}\,\pi + \frac{97\,\pi}{294}$$

$$\frac{1937\sqrt{2}\,\pi}{41160} + \frac{1}{84}\cos(4)\,\pi + \frac{1}{98}\sin(4)\,\pi + \frac{41}{20580}\cos(4)\,\sqrt{2}\,\pi + \frac{97\,\pi}{294}$$

```
> delta;epsilon; # Two float values are equal
```

$$1.191102426$$

$$1.191102426$$

```
> x:='x':y:='y':z:='z':
> [r*CS[1],r*CS[2],r*CS[3]];
```

\# Entirely multiply radius vector 'r' (suppose r>0) by each item of surface 'CS' parametrized expression, Convert "Parametrized expression of surface 'CS'" to "Parametrized expression of surface 'CS' coordinates"

$$\left[r\left(2\sin(u)\cos(v) + \frac{2}{7}\sin(u)\cos(v-2)\cos(7v) - \sin(u)\right),\right.$$
$$r\left(2\sin(u)\sin(v) + \frac{2}{7}\sin(u)\sin(v-2)\cos(8v) - \cos(u)\right),$$
$$\left. r\left(2\cos(u) - \frac{1}{7}\cos\left(\frac{u}{2}\right)\cos(12u-v)\right)\right]$$

```
> matrix(3,3,[Diff(r*CS[1],r),Diff(r*CS[1],u),Diff(r*CS[1],v),
  Diff(r*CS[2],r),Diff(r*CS[2],u),Diff(r*CS[2],v),
  Diff(r*CS[3],r),Diff(r*CS[3],u),Diff(r*CS[3],v)])=
  matrix(3,3,[diff(r*CS[1],r),diff(r*CS[1],u),diff(r*CS[1],v),
  diff(r*CS[2],r),diff(r*CS[2],u),diff(r*CS[2],v),
  diff(r*CS[3],r),diff(r*CS[3],u),diff(r*CS[3],v)]);m2:=rhs(%);
```

\# Define and calculate matrix of partial derivatives 'm2', obtain general expression of surface coordinates volume element coefficient

$$\left[\frac{\partial}{\partial r}\left(r\left(2\sin(u)\cos(v) + \frac{2}{7}\sin(u)\cos(v-2)\cos(7v) - \sin(u)\right)\right),\right.$$
$$\frac{\partial}{\partial u}\left(r\left(2\sin(u)\cos(v) + \frac{2}{7}\sin(u)\cos(v-2)\cos(7v) - \sin(u)\right)\right),$$
$$\left.\frac{\partial}{\partial v}\left(r\left(2\sin(u)\cos(v) + \frac{2}{7}\sin(u)\cos(v-2)\cos(7v) - \sin(u)\right)\right)\right]$$

$$\left[\frac{\partial}{\partial r}\left(r\left(2\sin(u)\sin(v) + \frac{2}{7}\sin(u)\sin(v-2)\cos(8v) - \cos(u)\right)\right),\right.$$
$$\frac{\partial}{\partial u}\left(r\left(2\sin(u)\sin(v) + \frac{2}{7}\sin(u)\sin(v-2)\cos(8v) - \cos(u)\right)\right),$$
$$\left.\frac{\partial}{\partial v}\left(r\left(2\sin(u)\sin(v) + \frac{2}{7}\sin(u)\sin(v-2)\cos(8v) - \cos(u)\right)\right)\right]$$

$$\left[\frac{\partial}{\partial r}\left(r\left(2\cos(u)-\frac{1}{7}\cos\left(\frac{u}{2}\right)\cos(12\,u-v)\right)\right),\right.$$

$$\frac{\partial}{\partial u}\left(r\left(2\cos(u)-\frac{1}{7}\cos\left(\frac{u}{2}\right)\cos(12\,u-v)\right)\right),$$

$$\left.\frac{\partial}{\partial v}\left(r\left(2\cos(u)-\frac{1}{7}\cos\left(\frac{u}{2}\right)\cos(12\,u-v)\right)\right)\right]=$$

$$\left[2\sin(u)\cos(v)+\frac{2}{7}\sin(u)\cos(v-2)\cos(7\,v)-\sin(u),\right.$$

$$r\left(2\cos(u)\cos(v)+\frac{2}{7}\cos(u)\cos(v-2)\cos(7\,v)-\cos(u)\right),$$

$$\left.r\left(-2\sin(u)\sin(v)-\frac{2}{7}\sin(u)\sin(v-2)\cos(7\,v)-2\sin(u)\cos(v-2)\sin(7\,v)\right)\right]$$

$$\left[2\sin(u)\sin(v)+\frac{2}{7}\sin(u)\sin(v-2)\cos(8\,v)-\cos(u),\right.$$

$$r\left(2\cos(u)\sin(v)+\frac{2}{7}\cos(u)\sin(v-2)\cos(8\,v)+\sin(u)\right),$$

$$\left.r\left(2\sin(u)\cos(v)+\frac{2}{7}\sin(u)\cos(v-2)\cos(8\,v)-\frac{16}{7}\sin(u)\sin(v-2)\sin(8\,v)\right)\right]$$

$$\left[2\cos(u)-\frac{1}{7}\cos\left(\frac{u}{2}\right)\cos(12\,u-v),\right.$$

$$r\left(-2\sin(u)+\frac{1}{14}\sin\left(\frac{u}{2}\right)\cos(12\,u-v)+\frac{12}{7}\cos\left(\frac{u}{2}\right)\sin(12\,u-v)\right),$$

$$\left.-\frac{1}{7}r\cos\left(\frac{u}{2}\right)\sin(12\,u-v)\right]$$

m2 :=

$$\left[2\sin(u)\cos(v)+\frac{2}{7}\sin(u)\cos(v-2)\cos(7\,v)-\sin(u),\right.$$

$$r\left(2\cos(u)\cos(v)+\frac{2}{7}\cos(u)\cos(v-2)\cos(7\,v)-\cos(u)\right),$$

$$\left.r\left(-2\sin(u)\sin(v)-\frac{2}{7}\sin(u)\sin(v-2)\cos(7\,v)-2\sin(u)\cos(v-2)\sin(7\,v)\right)\right]$$

$$\left[2\sin(u)\sin(v)+\frac{2}{7}\sin(u)\sin(v-2)\cos(8\,v)-\cos(u),\right.$$

$$r\left(2\cos(u)\sin(v)+\frac{2}{7}\cos(u)\sin(v-2)\cos(8\,v)+\sin(u)\right),$$

$$\left.r\left(2\sin(u)\cos(v)+\frac{2}{7}\sin(u)\cos(v-2)\cos(8\,v)-\frac{16}{7}\sin(u)\sin(v-2)\sin(8\,v)\right)\right]$$

$$\left[2\cos(u)-\frac{1}{7}\cos\left(\frac{u}{2}\right)\cos(12\,u-v),\right.$$

$$r\left(-2\sin(u)+\frac{1}{14}\sin\left(\frac{u}{2}\right)\cos(12\,u-v)+\frac{12}{7}\cos\left(\frac{u}{2}\right)\sin(12\,u-v)\right),$$

$$\left.-\frac{1}{7}r\cos\left(\frac{u}{2}\right)\sin(12\,u-v)\right]$$

```
> det(m2):
```

```
> J:=simplify(%); # General expression of surface coordinates volume
element coefficient
```

$$J := \frac{1}{343} r^2 \Big(-196 \sin(u) \cos(v-2) \cos(8v) + 1568 \sin(u) \sin(v-2) \sin(8v)$$

$$- 1372 \sin(u) \cos(v) - 448 \cos(v-2) \cos(7v) \sin(v-2) \sin(8v) \sin(u)$$

$$+ 32 \cos\!\left(\frac{u}{2}\right) \cos(12u-v) \cos(u) \cos(v-2) \cos(7v) \sin(u) \sin(v-2) \sin(8v)$$

$$- 28 \cos\!\left(\frac{u}{2}\right) \cos(12u-v) \sin(u) \cos(v-2) \sin(7v) \cos(u) \sin(v-2) \cos(8v)$$

$$+ 2744 \sin(u) + 392 \cos(v) \sin(u) \cos(v-2) \cos(8v)$$

$$- 1176 \cos\!\left(\frac{u}{2}\right) \sin(12u-v) \cos(u)^2 \cos(v)$$

$$+ 168 \cos(v-2) \cos(8v) \cos\!\left(\frac{u}{2}\right) \sin(12u-v)$$

$$+ 7 \cos(v-2) \cos(8v) \sin\!\left(\frac{u}{2}\right) \cos(12u-v)$$

$$- 14 \cos(v-2) \cos(7v) \cos\!\left(\frac{u}{2}\right) \sin(12u-v)$$

$$- 16 \cos(u)^2 \sin(v-2) \cos(v-2) \sin(8v) \sin\!\left(\frac{u}{2}\right) \cos(12u-v) \cos(7v)$$

$$+ 14 \cos(u)^2 \sin(v-2) \cos(8v) \cos(v-2) \sin\!\left(\frac{u}{2}\right) \cos(12u-v) \sin(7v)$$

$$+ 336 \cos(u)^2 \sin(v-2) \cos(8v) \cos\!\left(\frac{u}{2}\right) \sin(12u-v) \cos(v-2) \sin(7v)$$

$$- 384 \cos(u)^2 \sin(v-2) \cos\!\left(\frac{u}{2}\right) \sin(12u-v) \cos(v-2) \sin(8v) \cos(7v)$$

$$+ 56 \cos(7v) \sin(u) \cos(8v) - 56 \sin(v-2) \sin(8v) \sin\!\left(\frac{u}{2}\right) \cos(12u-v)$$

$$- 1344 \sin(v-2) \sin(8v) \cos\!\left(\frac{u}{2}\right) \sin(12u-v)$$

$$- 98 \cos\!\left(\frac{u}{2}\right) \cos(12u-v) \cos(v-2) \sin(7v)$$

$$- 14 \cos\!\left(\frac{u}{2}\right) \cos(12u-v) \sin(v-2) \cos(7v)$$

$$- 196 \cos\!\left(\frac{u}{2}\right) \cos(12u-v) \cos(u) \sin(u) + 2744 \sin(u) \cos(v-2) \sin(7v) \sin(v)$$

$$+ 14 \cos(u)^2 \sin(v-2) \cos\!\left(\frac{u}{2}\right) \cos(12u-v) \cos(7v)$$

$$+ 98 \cos(u)^2 \cos\!\left(\frac{u}{2}\right) \cos(v-2) \cos(12u-v) \sin(7v)$$

$$- 7 \cos(u)^2 \cos(8v) \cos(v-2) \sin\!\left(\frac{u}{2}\right) \cos(12u-v)$$

$$- 168 \cos(u)^2 \cos(8v) \cos\!\left(\frac{u}{2}\right) \sin(12u-v) \cos(v-2)$$

$$+ 56 \cos(u)^2 \sin(v-2) \sin(8v) \sin\left(\frac{u}{2}\right) \cos(12u-v)$$

$$+ 1344 \cos(u)^2 \sin(v-2) \cos\left(\frac{u}{2}\right) \sin(12u-v) \sin(8v)$$

$$+ 2 \cos(u)^2 \cos(8v) \sin\left(\frac{u}{2}\right) \cos(12u-v) \cos(7v)$$

$$+ 48 \cos(u)^2 \cos(8v) \cos\left(\frac{u}{2}\right) \sin(12u-v) \cos(7v)$$

$$- 14 \sin(v) \sin(v-2) \cos(7v) \sin\left(\frac{u}{2}\right) \cos(12u-v)$$

$$- 336 \sin(v) \sin(v-2) \cos(7v) \cos\left(\frac{u}{2}\right) \sin(12u-v)$$

$$- 98 \sin(v) \cos(v-2) \sin(7v) \sin\left(\frac{u}{2}\right) \cos(12u-v)$$

$$- 2352 \sin(v) \cos(v-2) \sin(7v) \cos\left(\frac{u}{2}\right) \sin(12u-v)$$

$$- 14 \sin(v-2) \cos(8v) \sin(v) \sin\left(\frac{u}{2}\right) \cos(12u-v)$$

$$- 336 \sin(v-2) \cos(8v) \sin(v) \cos\left(\frac{u}{2}\right) \sin(12u-v)$$

$$- 14 \cos(v) \cos(v-2) \cos(8v) \sin\left(\frac{u}{2}\right) \cos(12u-v)$$

$$- 336 \cos(v) \cos(v-2) \cos(8v) \cos\left(\frac{u}{2}\right) \sin(12u-v)$$

$$+ 112 \cos(v) \sin(v-2) \sin(8v) \sin\left(\frac{u}{2}\right) \cos(12u-v)$$

$$+ 2688 \cos(v) \sin(v-2) \sin(8v) \cos\left(\frac{u}{2}\right) \sin(12u-v)$$

$$- 14 \cos(v-2) \cos(7v) \cos(v) \sin\left(\frac{u}{2}\right) \cos(12u-v)$$

$$- 336 \cos(v-2) \cos(7v) \cos(v) \cos\left(\frac{u}{2}\right) \sin(12u-v)$$

$$+ 7 \cos(u) \sin(u) \sin(v-2) \cos(7v) \sin\left(\frac{u}{2}\right) \cos(12u-v)$$

$$+ 168 \cos(u) \sin(u) \sin(v-2) \cos(7v) \cos\left(\frac{u}{2}\right) \sin(12u-v)$$

$$+ 49 \cos(u) \sin(u) \cos(v-2) \sin(7v) \sin\left(\frac{u}{2}\right) \cos(12u-v)$$

$$+ 1176 \cos(u) \sin(u) \cos(v-2) \sin(7v) \cos\left(\frac{u}{2}\right) \sin(12u-v)$$

$$+ 14 \cos\left(\frac{u}{2}\right) \cos(12u-v) \cos(u) \sin(u) \cos(v-2) \cos(8v)$$

$$- 112 \cos\left(\frac{u}{2}\right) \cos(12u-v) \cos(u) \sin(u) \sin(v-2) \sin(8v)$$

$$+\ 392 \sin(u) \sin(v-2) \cos(7\,v) \sin(v) - 49 \cos(u)^2 \cos(v) \sin\!\left(\frac{u}{2}\right) \cos(12\,u-v)$$

$$+\ 98 \cos(u)^2 \sin(v) \cos\!\left(\frac{u}{2}\right) \cos(12\,u-v)$$

$$-\ 2 \cos(8\,v) \cos(7\,v) \sin\!\left(\frac{u}{2}\right) \cos(12\,u-v)$$

$$-\ 48 \cos(8\,v) \cos(7\,v) \cos\!\left(\frac{u}{2}\right) \sin(12\,u-v) + 392 \sin(u) \sin(v) \sin(v-2) \cos(8\,v)$$

$$+\ 392 \cos(v-2) \cos(7\,v) \sin(u) \cos(v) - 3136 \cos(v) \sin(u) \sin(v-2) \sin(8\,v)$$

$$+\ 392 \sin(v-2) \cos(8\,v) \cos(v-2) \sin(7\,v) \sin(u) - 98 \sin\!\left(\frac{u}{2}\right) \cos(12\,u-v)$$

$$-\ 2303 \cos\!\left(\frac{u}{2}\right) \sin(12\,u-v) + 1078 \cos(v) \cos\!\left(\frac{u}{2}\right) \sin(12\,u-v)$$

$$+\ 49 \cos(v) \sin\!\left(\frac{u}{2}\right) \cos(12\,u-v) - 98 \cos\!\left(\frac{u}{2}\right) \cos(12\,u-v) \sin(v)$$

$$+\ 98 \cos(u)^2 \sin\!\left(\frac{u}{2}\right) \cos(12\,u-v)$$

$$-\ 28 \cos\!\left(\frac{u}{2}\right) \cos(12\,u-v) \cos(u) \cos(v) \sin(u) \cos(v-2) \cos(8\,v)$$

$$+\ 224 \cos\!\left(\frac{u}{2}\right) \cos(12\,u-v) \cos(u) \cos(v) \sin(u) \sin(v-2) \sin(8\,v)$$

$$-\ 28 \cos\!\left(\frac{u}{2}\right) \cos(12\,u-v) \cos(u) \cos(v-2) \cos(7\,v) \sin(u) \cos(v)$$

$$-\ 28 \cos\!\left(\frac{u}{2}\right) \cos(12\,u-v) \sin(u) \sin(v) \cos(u) \sin(v-2) \cos(8\,v)$$

$$-\ 28 \cos\!\left(\frac{u}{2}\right) \cos(12\,u-v) \sin(u) \sin(v-2) \cos(7\,v) \cos(u) \sin(v)$$

$$-\ 196 \cos\!\left(\frac{u}{2}\right) \cos(12\,u-v) \sin(u) \cos(v-2) \sin(7\,v) \cos(u) \sin(v)$$

$$+\ 2352 \cos\!\left(\frac{u}{2}\right) \sin(12\,u-v) \cos(u)^2$$

$$+\ 49 \cos(u) \sin(u) \sin(v) \sin\!\left(\frac{u}{2}\right) \cos(12\,u-v)$$

$$+\ 98 \cos\!\left(\frac{u}{2}\right) \cos(12\,u-v) \cos(u) \sin(u) \cos(v)$$

$$+\ 1176 \sin(u) \cos\!\left(\frac{u}{2}\right) \sin(12\,u-v) \cos(u) \sin(v)$$

$$-\ 4 \cos\!\left(\frac{u}{2}\right) \cos(12\,u-v) \sin(u) \cos(7\,v) \cos(u) \cos(8\,v)$$

$$+\ 384 \cos(v-2) \cos(7\,v) \sin(v-2) \sin(8\,v) \cos\!\left(\frac{u}{2}\right) \sin(12\,u-v)$$

$$+ 16 \cos(v-2) \cos(7v) \sin(v-2) \sin(8v) \sin\left(\frac{u}{2}\right) \cos(12u-v)$$

$$- 14 \sin(v-2) \cos(8v) \cos(v-2) \sin(7v) \sin\left(\frac{u}{2}\right) \cos(12u-v)$$

$$- 336 \sin(v-2) \cos(8v) \cos(v-2) \sin(7v) \cos\left(\frac{u}{2}\right) \sin(12u-v)$$

$$+ 14 \cos(u)^2 \sin(v) \sin(v-2) \sin\left(\frac{u}{2}\right) \cos(12u-v) \cos(7v)$$

$$+ 336 \cos(u)^2 \sin(v) \sin(v-2) \cos\left(\frac{u}{2}\right) \sin(12u-v) \cos(7v)$$

$$+ 98 \cos(u)^2 \sin(v) \cos(v-2) \sin\left(\frac{u}{2}\right) \cos(12u-v) \sin(7v)$$

$$+ 2352 \cos(u)^2 \sin(v) \cos\left(\frac{u}{2}\right) \sin(12u-v) \cos(v-2) \sin(7v)$$

$$+ 14 \cos(u)^2 \sin(v) \sin(v-2) \cos(8v) \sin\left(\frac{u}{2}\right) \cos(12u-v)$$

$$+ 336 \cos(u)^2 \sin(v) \sin(v-2) \cos(8v) \cos\left(\frac{u}{2}\right) \sin(12u-v)$$

$$+ 14 \cos(u)^2 \cos(8v) \cos(v) \cos(v-2) \sin\left(\frac{u}{2}\right) \cos(12u-v)$$

$$+ 336 \cos(u)^2 \cos(8v) \cos\left(\frac{u}{2}\right) \sin(12u-v) \cos(v) \cos(v-2)$$

$$- 112 \cos(u)^2 \sin(v-2) \cos(v) \sin(8v) \sin\left(\frac{u}{2}\right) \cos(12u-v)$$

$$- 2688 \cos(u)^2 \sin(v-2) \cos\left(\frac{u}{2}\right) \sin(12u-v) \cos(v) \sin(8v)$$

$$+ 14 \cos(u)^2 \cos(v) \cos(v-2) \sin\left(\frac{u}{2}\right) \cos(12u-v) \cos(7v)$$

$$+ 336 \cos(u)^2 \cos\left(\frac{u}{2}\right) \sin(12u-v) \cos(v) \cos(v-2) \cos(7v) \Bigg)$$

```
> Int(Int(Int(J,r=0..1),u=rgu[1]..rgu[2]),v=rgv[1]..rgv[2]);
```

Triple Integrals of "Surface 'CS' Coordinates Volume Element" in range of 'r,u,v', absolute value of result is cubage value of unclosed surface 'CS':

$$\int_0^{2\pi} \int_0^{\frac{\pi}{2}} \int_0^1 \frac{1}{343} r^2 \Big(-196 \sin(u) \cos(v-2) \cos(8v) + 1568 \sin(u) \sin(v-2) \sin(8v)$$

$$- 1372 \sin(u) \cos(v) - 448 \cos(v-2) \cos(7v) \sin(v-2) \sin(8v) \sin(u)$$

978-1-62265-930-2 (online)   978-1-62265-931-9 (paper)

$$+ 32 \cos\left(\frac{u}{2}\right) \cos(12\,u - v) \cos(u) \cos(v - 2) \cos(7\,v) \sin(u) \sin(v - 2) \sin(8\,v)$$

$$- 28 \cos\left(\frac{u}{2}\right) \cos(12\,u - v) \sin(u) \cos(v - 2) \sin(7\,v) \cos(u) \sin(v - 2) \cos(8\,v)$$

$$+ 2744 \sin(u) + 392 \cos(v) \sin(u) \cos(v - 2) \cos(8\,v)$$

$$- 1176 \cos\left(\frac{u}{2}\right) \sin(12\,u - v) \cos(u)^2 \cos(v)$$

$$+ 168 \cos(v - 2) \cos(8\,v) \cos\left(\frac{u}{2}\right) \sin(12\,u - v)$$

$$+ 7 \cos(v - 2) \cos(8\,v) \sin\left(\frac{u}{2}\right) \cos(12\,u - v)$$

$$- 14 \cos(v - 2) \cos(7\,v) \cos\left(\frac{u}{2}\right) \sin(12\,u - v)$$

$$- 16 \cos(u)^2 \sin(v - 2) \cos(v - 2) \sin(8\,v) \sin\left(\frac{u}{2}\right) \cos(12\,u - v) \cos(7\,v)$$

$$+ 14 \cos(u)^2 \sin(v - 2) \cos(8\,v) \cos(v - 2) \sin\left(\frac{u}{2}\right) \cos(12\,u - v) \sin(7\,v)$$

$$+ 336 \cos(u)^2 \sin(v - 2) \cos(8\,v) \cos\left(\frac{u}{2}\right) \sin(12\,u - v) \cos(v - 2) \sin(7\,v)$$

$$- 384 \cos(u)^2 \sin(v - 2) \cos\left(\frac{u}{2}\right) \sin(12\,u - v) \cos(v - 2) \sin(8\,v) \cos(7\,v)$$

$$+ 56 \cos(7\,v) \sin(u) \cos(8\,v) - 56 \sin(v - 2) \sin(8\,v) \sin\left(\frac{u}{2}\right) \cos(12\,u - v)$$

$$- 1344 \sin(v - 2) \sin(8\,v) \cos\left(\frac{u}{2}\right) \sin(12\,u - v)$$

$$- 98 \cos\left(\frac{u}{2}\right) \cos(12\,u - v) \cos(v - 2) \sin(7\,v)$$

$$- 14 \cos\left(\frac{u}{2}\right) \cos(12\,u - v) \sin(v - 2) \cos(7\,v)$$

$$- 196 \cos\left(\frac{u}{2}\right) \cos(12\,u - v) \cos(u) \sin(u) + 2744 \sin(u) \cos(v - 2) \sin(7\,v) \sin(v)$$

$$+ 14 \cos(u)^2 \sin(v - 2) \cos\left(\frac{u}{2}\right) \cos(12\,u - v) \cos(7\,v)$$

$$+ 98 \cos(u)^2 \cos\left(\frac{u}{2}\right) \cos(v - 2) \cos(12\,u - v) \sin(7\,v)$$

$$- 7 \cos(u)^2 \cos(8\,v) \cos(v - 2) \sin\left(\frac{u}{2}\right) \cos(12\,u - v)$$

$$- 168 \cos(u)^2 \cos(8\,v) \cos\left(\frac{u}{2}\right) \sin(12\,u - v) \cos(v - 2)$$

$$+ 56 \cos(u)^2 \sin(v - 2) \sin(8\,v) \sin\left(\frac{u}{2}\right) \cos(12\,u - v)$$

$$+ 1344 \cos(u)^2 \sin(v-2) \cos\left(\frac{u}{2}\right) \sin(12\,u-v) \sin(8\,v)$$

$$+ 2 \cos(u)^2 \cos(8\,v) \sin\left(\frac{u}{2}\right) \cos(12\,u-v) \cos(7\,v)$$

$$+ 48 \cos(u)^2 \cos(8\,v) \cos\left(\frac{u}{2}\right) \sin(12\,u-v) \cos(7\,v)$$

$$- 14 \sin(v) \sin(v-2) \cos(7\,v) \sin\left(\frac{u}{2}\right) \cos(12\,u-v)$$

$$- 336 \sin(v) \sin(v-2) \cos(7\,v) \cos\left(\frac{u}{2}\right) \sin(12\,u-v)$$

$$- 98 \sin(v) \cos(v-2) \sin(7\,v) \sin\left(\frac{u}{2}\right) \cos(12\,u-v)$$

$$- 2352 \sin(v) \cos(v-2) \sin(7\,v) \cos\left(\frac{u}{2}\right) \sin(12\,u-v)$$

$$- 14 \sin(v-2) \cos(8\,v) \sin(v) \sin\left(\frac{u}{2}\right) \cos(12\,u-v)$$

$$- 336 \sin(v-2) \cos(8\,v) \sin(v) \cos\left(\frac{u}{2}\right) \sin(12\,u-v)$$

$$- 14 \cos(v) \cos(v-2) \cos(8\,v) \sin\left(\frac{u}{2}\right) \cos(12\,u-v)$$

$$- 336 \cos(v) \cos(v-2) \cos(8\,v) \cos\left(\frac{u}{2}\right) \sin(12\,u-v)$$

$$+ 112 \cos(v) \sin(v-2) \sin(8\,v) \sin\left(\frac{u}{2}\right) \cos(12\,u-v)$$

$$+ 2688 \cos(v) \sin(v-2) \sin(8\,v) \cos\left(\frac{u}{2}\right) \sin(12\,u-v)$$

$$- 14 \cos(v-2) \cos(7\,v) \cos(v) \sin\left(\frac{u}{2}\right) \cos(12\,u-v)$$

$$- 336 \cos(v-2) \cos(7\,v) \cos(v) \cos\left(\frac{u}{2}\right) \sin(12\,u-v)$$

$$+ 7 \cos(u) \sin(u) \sin(v-2) \cos(7\,v) \sin\left(\frac{u}{2}\right) \cos(12\,u-v)$$

$$+ 168 \cos(u) \sin(u) \sin(v-2) \cos(7\,v) \cos\left(\frac{u}{2}\right) \sin(12\,u-v)$$

$$+ 49 \cos(u) \sin(u) \cos(v-2) \sin(7\,v) \sin\left(\frac{u}{2}\right) \cos(12\,u-v)$$

$$+ 1176 \cos(u) \sin(u) \cos(v-2) \sin(7\,v) \cos\left(\frac{u}{2}\right) \sin(12\,u-v)$$

$$+ 14 \cos\left(\frac{u}{2}\right) \cos(12\,u-v) \cos(u) \sin(u) \cos(v-2) \cos(8\,v)$$

$$- 112 \cos\left(\frac{u}{2}\right) \cos(12\,u-v) \cos(u) \sin(u) \sin(v-2) \sin(8\,v)$$

$$+ 392 \sin(u) \sin(v-2) \cos(7\,v) \sin(v) - 49 \cos(u)^2 \cos(v) \sin\left(\frac{u}{2}\right) \cos(12\,u-v)$$

$$+ 98 \cos(u)^2 \sin(v) \cos\left(\frac{u}{2}\right) \cos(12\,u - v)$$

$$- 2 \cos(8\,v) \cos(7\,v) \sin\left(\frac{u}{2}\right) \cos(12\,u - v)$$

$$- 48 \cos(8\,v) \cos(7\,v) \cos\left(\frac{u}{2}\right) \sin(12\,u - v) + 392 \sin(u) \sin(v) \sin(v - 2) \cos(8\,v)$$

$$+ 392 \cos(v - 2) \cos(7\,v) \sin(u) \cos(v) - 3136 \cos(v) \sin(u) \sin(v - 2) \sin(8\,v)$$

$$+ 392 \sin(v - 2) \cos(8\,v) \cos(v - 2) \sin(7\,v) \sin(u) - 98 \sin\left(\frac{u}{2}\right) \cos(12\,u - v)$$

$$- 2303 \cos\left(\frac{u}{2}\right) \sin(12\,u - v) + 1078 \cos(v) \cos\left(\frac{u}{2}\right) \sin(12\,u - v)$$

$$+ 49 \cos(v) \sin\left(\frac{u}{2}\right) \cos(12\,u - v) - 98 \cos\left(\frac{u}{2}\right) \cos(12\,u - v) \sin(v)$$

$$+ 98 \cos(u)^2 \sin\left(\frac{u}{2}\right) \cos(12\,u - v)$$

$$- 28 \cos\left(\frac{u}{2}\right) \cos(12\,u - v) \cos(u) \cos(v) \sin(u) \cos(v - 2) \cos(8\,v)$$

$$+ 224 \cos\left(\frac{u}{2}\right) \cos(12\,u - v) \cos(u) \cos(v) \sin(u) \sin(v - 2) \sin(8\,v)$$

$$- 28 \cos\left(\frac{u}{2}\right) \cos(12\,u - v) \cos(u) \cos(v - 2) \cos(7\,v) \sin(u) \cos(v)$$

$$- 28 \cos\left(\frac{u}{2}\right) \cos(12\,u - v) \sin(u) \sin(v) \cos(u) \sin(v - 2) \cos(8\,v)$$

$$- 28 \cos\left(\frac{u}{2}\right) \cos(12\,u - v) \sin(u) \sin(v - 2) \cos(7\,v) \cos(u) \sin(v)$$

$$- 196 \cos\left(\frac{u}{2}\right) \cos(12\,u - v) \sin(u) \cos(v - 2) \sin(7\,v) \cos(u) \sin(v)$$

$$+ 2352 \cos\left(\frac{u}{2}\right) \sin(12\,u - v) \cos(u)^2$$

$$+ 49 \cos(u) \sin(u) \sin(v) \sin\left(\frac{u}{2}\right) \cos(12\,u - v)$$

$$+ 98 \cos\left(\frac{u}{2}\right) \cos(12\,u - v) \cos(u) \sin(u) \cos(v)$$

$$+ 1176 \sin(u) \cos\left(\frac{u}{2}\right) \sin(12\,u - v) \cos(u) \sin(v)$$

$$- 4 \cos\left(\frac{u}{2}\right) \cos(12\,u - v) \sin(u) \cos(7\,v) \cos(u) \cos(8\,v)$$

$$+ 384 \cos(v - 2) \cos(7\,v) \sin(v - 2) \sin(8\,v) \cos\left(\frac{u}{2}\right) \sin(12\,u - v)$$

$$+ 16 \cos(v - 2) \cos(7\,v) \sin(v - 2) \sin(8\,v) \sin\left(\frac{u}{2}\right) \cos(12\,u - v)$$

$$-14 \sin(v-2) \cos(8v) \cos(v-2) \sin(7v) \sin\left(\frac{u}{2}\right) \cos(12u-v)$$

$$-336 \sin(v-2) \cos(8v) \cos(v-2) \sin(7v) \cos\left(\frac{u}{2}\right) \sin(12u-v)$$

$$+14 \cos(u)^2 \sin(v) \sin(v-2) \sin\left(\frac{u}{2}\right) \cos(12u-v) \cos(7v)$$

$$+336 \cos(u)^2 \sin(v) \sin(v-2) \cos\left(\frac{u}{2}\right) \sin(12u-v) \cos(7v)$$

$$+98 \cos(u)^2 \sin(v) \cos(v-2) \sin\left(\frac{u}{2}\right) \cos(12u-v) \sin(7v)$$

$$+2352 \cos(u)^2 \sin(v) \cos\left(\frac{u}{2}\right) \sin(12u-v) \cos(v-2) \sin(7v)$$

$$+14 \cos(u)^2 \sin(v) \sin(v-2) \cos(8v) \sin\left(\frac{u}{2}\right) \cos(12u-v)$$

$$+336 \cos(u)^2 \sin(v) \sin(v-2) \cos(8v) \cos\left(\frac{u}{2}\right) \sin(12u-v)$$

$$+14 \cos(u)^2 \cos(8v) \cos(v) \cos(v-2) \sin\left(\frac{u}{2}\right) \cos(12u-v)$$

$$+336 \cos(u)^2 \cos(8v) \cos\left(\frac{u}{2}\right) \sin(12u-v) \cos(v) \cos(v-2)$$

$$-112 \cos(u)^2 \sin(v-2) \cos(v) \sin(8v) \sin\left(\frac{u}{2}\right) \cos(12u-v)$$

$$-2688 \cos(u)^2 \sin(v-2) \cos\left(\frac{u}{2}\right) \sin(12u-v) \cos(v) \sin(8v)$$

$$+14 \cos(u)^2 \cos(v) \cos(v-2) \sin\left(\frac{u}{2}\right) \cos(12u-v) \cos(7v)$$

$$\left. +336 \cos(u)^2 \cos\left(\frac{u}{2}\right) \sin(12u-v) \cos(v) \cos(v-2) \cos(7v) \right) dr\, du\, dv$$

```
> value(%);evalf(%); # Cubage Integral Value of Unclosed Surface
```

$$\frac{109506115472\,\pi}{20538814275} - \frac{1870}{11906559} \cos(4)\,\pi - \frac{137641001\sqrt{2}\,\pi}{4323960900}$$

$$+ \frac{160}{11906559} \sin(2) \cos(2) \sqrt{2}\,\pi - \frac{177305}{23813118} \cos(2)^2 \sqrt{2}\,\pi$$

$$16.60307003$$

3.3 Numerical Model of Curl Theorem at Manifold (II)

Known: Parametric expression (1) of Simply connected、 Orientable、 Unclosed Surface(Irregular、Asymmetrical)

$$\left[\sin(u) \cos(v) - \frac{2}{7} \sin(u) \cos(3u-v) + \cos(u), \ \sin(u) \sin(v) - \frac{1}{6} \sin(u) \cos(6v), \right.$$

$$\left. \sin(u) - 2 \cos(u) \right]$$

978-1-62265-930-2 (online)   978-1-62265-931-9 (paper)   Yang Ke

thereinto, $u \in [0, \dfrac{\pi}{2}\text{-}\sin(7v)\text{-}\dfrac{1}{3}\cos(7v-1)]$, $v \in [0, 2\pi]$;

and Integral Vector Field

$$\left[\frac{\left(\dfrac{x}{4}-\dfrac{y}{6}+\dfrac{z}{5}\right)^2}{3}, \frac{z^2}{6}-\frac{xy}{5}, \frac{x^2}{5}-\frac{yz}{6} \right] \quad (2)$$

Calculate and Validate Curl Theorem at Manifold.

Figure(III)3.3.1 Simply connected、 Orientable、 Unclosed Surface (1)

[Irregular、Asymmetrical]

Solution:

First, Free Spacial Closed Curve Integral:

Import right boundary value of 'u'[viz. $\frac{\pi}{2}-\sin(7v)-\frac{1}{3}\cos(7v-1)$] to expression (1) of unclosed surface, obtain expression (3) of Unclosed Surface' Boundary Curve; Set Right Hand Rule as associated basis between positive direction of boundary curve and normal vector of unclosed surface

$$
\left[\sin\!\left(\frac{\pi}{2}-\sin(7\,v)-\frac{1}{3}\cos(7\,v-1)\right)\cos(v) \right.
$$

$$
-\frac{2}{7}\sin\!\left(\frac{\pi}{2}-\sin(7\,v)-\frac{1}{3}\cos(7\,v-1)\right)\cos\!\left(\frac{3\,\pi}{2}-3\sin(7\,v)-\cos(7\,v-1)-v\right)
$$

$$
+\cos\!\left(\frac{\pi}{2}-\sin(7\,v)-\frac{1}{3}\cos(7\,v-1)\right),\ \sin\!\left(\frac{\pi}{2}-\sin(7\,v)-\frac{1}{3}\cos(7\,v-1)\right)\sin(v)
$$

$$
-\frac{1}{6}\sin\!\left(\frac{\pi}{2}-\sin(7\,v)-\frac{1}{3}\cos(7\,v-1)\right)\cos(6\,v),
$$

$$
\left.\sin\!\left(\frac{\pi}{2}-\sin(7\,v)-\frac{1}{3}\cos(7\,v-1)\right)-2\cos\!\left(\frac{\pi}{2}-\sin(7\,v)-\frac{1}{3}\cos(7\,v-1)\right)\right]=\left[\vphantom{\frac12}\right.
$$

$$
\cos\!\left(\sin(7\,v)+\frac{1}{3}\cos(7\,v-1)\right)\cos(v)
$$

$$
+\frac{2}{7}\cos\!\left(\sin(7\,v)+\frac{1}{3}\cos(7\,v-1)\right)\sin(3\sin(7\,v)+\cos(7\,v-1)+v)
$$

$$
+\sin\!\left(\sin(7\,v)+\frac{1}{3}\cos(7\,v-1)\right),
$$

$$
\cos\!\left(\sin(7\,v)+\frac{1}{3}\cos(7\,v-1)\right)\sin(v)-\frac{1}{6}\cos\!\left(\sin(7\,v)+\frac{1}{3}\cos(7\,v-1)\right)\cos(6\,v),
$$

$$
\left.\cos\!\left(\sin(7\,v)+\frac{1}{3}\cos(7\,v-1)\right)-2\sin\!\left(\sin(7\,v)+\frac{1}{3}\cos(7\,v-1)\right)\right]
$$

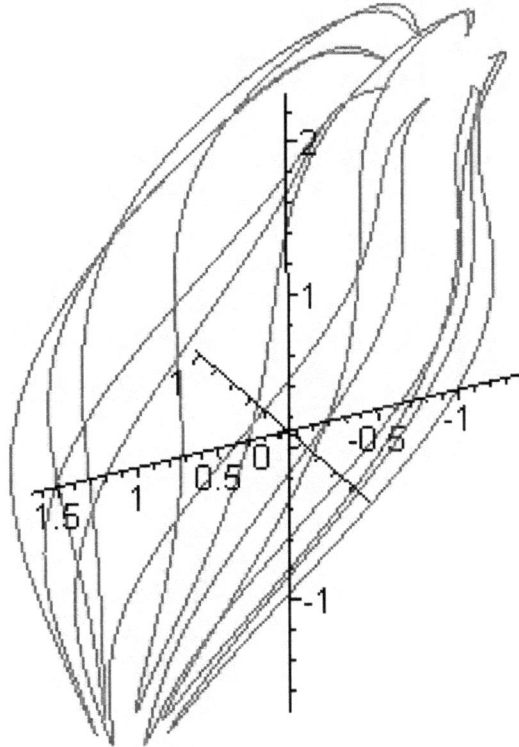

Figure(III)3.3.2 Boundary Curve (3) of

Simply connected、Orientable、Unclosed Surface (1)

Input expression (3) of Boundary Curve to Vector Field (2); and
realize Spacial Closed Curve Integral [Vetor Field (2) at Boundary
Curve(3)] in range of 'v' (4):

$$\int_0^{2\pi} \frac{1}{3}\left(\frac{1}{4}\cos\left(\sin(7\,v) + \frac{1}{3}\cos(7\,v - 1)\right)\right)\cos(v)$$

$$+ \frac{1}{14} \cos\left(\sin(7\,v) + \frac{1}{3} \cos(7\,v - 1) \right) \sin(3\,\sin(7\,v) + \cos(7\,v - 1) + v)$$

$$- \frac{3}{20} \sin\left(\sin(7\,v) + \frac{1}{3} \cos(7\,v - 1) \right) - \frac{1}{6} \cos\left(\sin(7\,v) + \frac{1}{3} \cos(7\,v - 1) \right) \sin(v)$$

$$+ \frac{1}{36} \cos\left(\sin(7\,v) + \frac{1}{3} \cos(7\,v - 1) \right) \cos(6\,v) + \frac{1}{5} \cos\left(\sin(7\,v) + \frac{1}{3} \cos(7\,v - 1) \right)\right)^2 \bigg($$

$$-\sin\left(\sin(7\,v) + \frac{1}{3} \cos(7\,v - 1) \right) \left(7\,\cos(7\,v) - \frac{7}{3} \sin(7\,v - 1) \right) \cos(v)$$

$$- \cos\left(\sin(7\,v) + \frac{1}{3} \cos(7\,v - 1) \right) \sin(v) - \frac{2}{7} \sin\left(\sin(7\,v) + \frac{1}{3} \cos(7\,v - 1) \right)$$

$$\left(7\,\cos(7\,v) - \frac{7}{3} \sin(7\,v - 1) \right) \sin(3\,\sin(7\,v) + \cos(7\,v - 1) + v) + \frac{2}{7}$$

$$\cos\left(\sin(7\,v) + \frac{1}{3} \cos(7\,v - 1) \right) \cos(3\,\sin(7\,v) + \cos(7\,v - 1) + v)$$

$$(21\,\cos(7\,v) - 7\,\sin(7\,v - 1) + 1)$$

$$+ \cos\left(\sin(7\,v) + \frac{1}{3} \cos(7\,v - 1) \right) \left(7\,\cos(7\,v) - \frac{7}{3} \sin(7\,v - 1) \right)\bigg) + \bigg($$

$$\frac{1}{6}\left(\cos\left(\sin(7\,v) + \frac{1}{3} \cos(7\,v - 1) \right) - 2\,\sin\left(\sin(7\,v) + \frac{1}{3} \cos(7\,v - 1) \right)\right)^2 - \frac{1}{5}\bigg($$

$$\cos\left(\sin(7\,v) + \frac{1}{3} \cos(7\,v - 1) \right) \cos(v)$$

$$+ \frac{2}{7} \cos\left(\sin(7\,v) + \frac{1}{3} \cos(7\,v - 1) \right) \sin(3\,\sin(7\,v) + \cos(7\,v - 1) + v)$$

$$+ \sin\left(\sin(7\,v) + \frac{1}{3} \cos(7\,v - 1) \right)\bigg)\bigg)$$

$$\left(\cos\left(\sin(7\,v) + \frac{1}{3} \cos(7\,v - 1) \right) \sin(v) - \frac{1}{6} \cos\left(\sin(7\,v) + \frac{1}{3} \cos(7\,v - 1) \right) \cos(6\,v)\right)\bigg)$$

$$\left(-\sin\left(\sin(7\,v) + \frac{1}{3} \cos(7\,v - 1) \right) \left(7\,\cos(7\,v) - \frac{7}{3} \sin(7\,v - 1) \right) \sin(v)\right.$$

$$+ \cos\left(\sin(7\,v) + \frac{1}{3} \cos(7\,v - 1) \right) \cos(v)$$

$$+ \frac{1}{6} \sin\left(\sin(7\,v) + \frac{1}{3} \cos(7\,v - 1) \right) \left(7\,\cos(7\,v) - \frac{7}{3} \sin(7\,v - 1) \right) \cos(6\,v)$$

$$+ \cos\left(\sin(7\,v) + \frac{1}{3} \cos(7\,v - 1) \right) \sin(6\,v)\bigg) + \bigg(\frac{1}{5}\bigg($$

$$\cos\left(\sin(7\,v) + \frac{1}{3} \cos(7\,v - 1) \right) \cos(v)$$

- 429 -

$$+ \frac{2}{7}\cos\left(\sin(7\,v)+\frac{1}{3}\cos(7\,v-1)\right)\sin(3\,\sin(7\,v)+\cos(7\,v-1)+v)$$

$$+\sin\left(\sin(7\,v)+\frac{1}{3}\cos(7\,v-1)\right)\Bigg)^2-\frac{1}{6}$$

$$\left(\cos\left(\sin(7\,v)+\frac{1}{3}\cos(7\,v-1)\right)\sin(v)-\frac{1}{6}\cos\left(\sin(7\,v)+\frac{1}{3}\cos(7\,v-1)\right)\cos(6\,v)\right)$$

$$\left(\cos\left(\sin(7\,v)+\frac{1}{3}\cos(7\,v-1)\right)-2\,\sin\left(\sin(7\,v)+\frac{1}{3}\cos(7\,v-1)\right)\right)\Bigg)\Bigg)\Bigg($$

$$-\sin\left(\sin(7\,v)+\frac{1}{3}\cos(7\,v-1)\right)\left(7\,\cos(7\,v)-\frac{7}{3}\sin(7\,v-1)\right)$$

$$-2\,\cos\left(\sin(7\,v)+\frac{1}{3}\cos(7\,v-1)\right)\left(7\,\cos(7\,v)-\frac{7}{3}\sin(7\,v-1)\right)\Bigg)dv$$

$$=\ 0.1213635571 \tag{4}$$

Because of similar structures such as 'cos(cos(...))', 'sin(sin(...))', there is no analytical value(Expressed by Elementary Function) about integral result, but only arbitrary precision floating value

Second, Free (Unclosed) Surface Integral:

According to parametric expression (1) of unclosed surface, Define and calculate Matrix of Partial Derivatives, obtain tangent plane normal vector of unclosed surface (5):

$$\begin{bmatrix} i & j & k \\ \frac{\partial}{\partial u}(\sin(u)\cos(v)-\frac{2}{7}\sin(u)\cos(3u-v)+\cos(u)) & \frac{\partial}{\partial u}(\sin(u)\sin(v)-\frac{1}{6}\sin(u)\cos(6v)) & \frac{\partial}{\partial u}(\sin(u)-2\cos(u)) \\ \frac{\partial}{\partial v}(\sin(u)\cos(v)-\frac{2}{7}\sin(u)\cos(3u-v)+\cos(u)) & \frac{\partial}{\partial v}(\sin(u)\sin(v)-\frac{1}{6}\sin(u)\cos(6v)) & \frac{\partial}{\partial v}(\sin(u)-2\cos(u)) \end{bmatrix}$$

$$=$$

$$-i\cos(u)\sin(u)\cos(v)-i\cos(u)\sin(u)\sin(6\,v)-2\,i\sin(u)^2\cos(v)$$

$$-2\,i\sin(u)^2\sin(6\,v)+k\cos(u)\cos(v)^2\sin(u)+k\cos(u)\cos(v)\sin(u)\sin(6\,v)$$

$$-\frac{2}{7}k\cos(u)\cos(3\,u-v)\sin(u)\cos(v)-\frac{2}{7}k\cos(u)\cos(3\,u-v)\sin(u)\sin(6\,v)$$

$$+\frac{6}{7}k\sin(u)^2\sin(3\,u-v)\cos(v)+\frac{6}{7}k\sin(u)^2\sin(3\,u-v)\sin(6\,v)$$

$$-k\sin(u)^2\cos(v)-k\sin(u)^2\sin(6\,v)-\sin(u)\sin(v)j\cos(u)-2\sin(u)^2\sin(v)j$$

$$+\sin(u)\sin(v)^2 k\cos(u)-\frac{1}{6}\sin(u)\sin(v)k\cos(u)\cos(6\,v)$$

$$-\frac{2}{7}\sin(u)\sin(3\,u-v)\,j\cos(u)-\frac{4}{7}\sin(u)^2\sin(3\,u-v)\,j$$

$$+\frac{2}{7}\sin(u)\sin(3\,u-v)\,k\cos(u)\sin(v)-\frac{1}{21}\sin(u)\sin(3\,u-v)\,k\cos(u)\cos(6\,v)$$

From expression (5), respectively pick up coefficient of item 'i,j,k', obtain tangent plane normal vector (6) of surface (1):

$$\left[\frac{1}{42}\sin(u)\right.$$

$$(-42\cos(u)\cos(v)-42\cos(u)\sin(6\,v)-84\sin(u)\cos(v)-84\sin(u)\sin(6\,v)),$$

$$\frac{1}{42}\sin(u)\,(-42\cos(u)\sin(v)-84\sin(u)\sin(v)-12\sin(3\,u-v)\cos(u)$$

$$-24\sin(u)\sin(3\,u-v)),\frac{1}{42}\sin(u)\,(42\cos(u)\cos(v)\sin(6\,v)$$

$$-12\cos(u)\cos(3\,u-v)\cos(v)-12\cos(u)\cos(3\,u-v)\sin(6\,v)$$

$$+36\sin(u)\sin(3\,u-v)\cos(v)+36\sin(u)\sin(3\,u-v)\sin(6\,v)-42\sin(u)\cos(v)$$

$$-42\sin(u)\sin(6\,v)+42\cos(u)-7\sin(v)\cos(u)\cos(6\,v)$$

$$\left.+12\sin(3\,u-v)\cos(u)\sin(v)-2\sin(3\,u-v)\cos(u)\cos(6\,v))\right] \tag{6}$$

Calculate Curl (7) of Vector Field (2):

$$\left[\left(\frac{\partial}{\partial y}\left(\frac{x^2}{5}-\frac{yz}{6}\right)\right)-\left(\frac{\partial}{\partial z}\left(\frac{z^2}{6}-\frac{xy}{5}\right)\right),\left(\frac{\partial}{\partial z}\left(\frac{\left(\frac{x}{4}-\frac{y}{6}+\frac{z}{5}\right)^2}{3}\right)\right)-\left(\frac{\partial}{\partial x}\left(\frac{x^2}{5}-\frac{yz}{6}\right)\right),\right.$$

$$\left.\left(\frac{\partial}{\partial x}\left(\frac{z^2}{6}-\frac{xy}{5}\right)\right)-\left(\frac{\partial}{\partial y}\left(\frac{\left(\frac{x}{4}-\frac{y}{6}+\frac{z}{5}\right)^2}{3}\right)\right)\right]=$$

$$\left[-\frac{z}{2},-\frac{11\,x}{30}-\frac{y}{45}+\frac{2\,z}{75},-\frac{59\,y}{270}+\frac{x}{36}+\frac{z}{45}\right] \tag{7}$$

Input expression(1) of Unclosed Surface to Curl (7); and Calculate Integral of Dot Product [Curl (7) and tangent plane normal vector (6)] in range of 'u,v' (8):

$$\int_0^{2\pi}\int_0^{\frac{\pi}{2}-\sin(7\,v)-1/3\cos(7\,v-1)}\frac{1}{42}\left(-\frac{1}{2}\sin(u)+\cos(u)\right)\sin(u)$$

$$(-42\cos(u)\cos(v)-42\cos(u)\sin(6\,v)-84\sin(u)\cos(v)-84\sin(u)\sin(6\,v))+$$

$$\frac{1}{42}\left(-\frac{11}{30}\sin(u)\cos(v)+\frac{11}{105}\sin(u)\cos(3\,u-v)-\frac{21}{50}\cos(u)-\frac{1}{45}\sin(u)\sin(v)\right.$$

$$\left.+\frac{1}{270}\sin(u)\cos(6\,v)+\frac{2}{75}\sin(u)\right)\sin(u)\,(-42\cos(u)\sin(v)-84\sin(u)\sin(v)$$

$$- 12 \sin(3\,u - v)\cos(u) - 24 \sin(u)\sin(3\,u - v)) + \frac{1}{42}\left(-\frac{59}{270}\sin(u)\sin(v)\right.$$

$$+ \frac{59}{1620}\sin(u)\cos(6\,v) + \frac{1}{36}\sin(u)\cos(v) - \frac{1}{126}\sin(u)\cos(3\,u - v) - \frac{1}{60}\cos(u)$$

$$+ \left.\frac{1}{45}\sin(u)\right)\sin(u)\,(42\cos(u)\cos(v)\sin(6\,v) - 12\cos(u)\cos(3\,u - v)\cos(v)$$

$$- 12\cos(u)\cos(3\,u - v)\sin(6\,v) + 36\sin(u)\sin(3\,u - v)\cos(v)$$

$$+ 36\sin(u)\sin(3\,u - v)\sin(6\,v) - 42\sin(u)\cos(v) - 42\sin(u)\sin(6\,v)$$

$$+ 42\cos(u) - 7\sin(v)\cos(u)\cos(6\,v) + 12\sin(3\,u - v)\cos(u)\sin(v)$$

$$- 2\sin(3\,u - v)\cos(u)\cos(6\,v))\,du\,dv$$

$$= 0.1213635571 \tag{8}$$

Since right boundary value of 'u' is trigonometric function expression, import it to integral expression above-mentioned, it will come into being similar structures such as 'cos(cos(...))', 'sin(sin(...))', there is no analytical value(Expressed by Elementary Function) about integral result, but only arbitrary precision floating value

Closed Curve Integral Precision Value (4) [Vector Field (2) at Boundary Curve (3) of Unclosed Surface (1)], is equal to Surface Integral Precision Value (8) [Curl (7) at Unclosed Surface (1)], Complete Calculation and Validation of Curl Theorem at Manifold.

3.4 Numerical Model of Curl Theorem at Manifold (II)
[Program Template of Waterloo Maplesoft, Optional]

```
> restart;
> with(plots):with(linalg):
>CS:=[sin(u)*cos(v)-2*sin(u)*cos(3*u-v)/7+cos(u),sin(u)*sin(v)
-sin(u)*cos(6*v)/6,sin(u)-2*cos(u)];
# Define parametrized expression of discretional simply connected
orientable closed surface 'CS'
```

$$CS := \left[\sin(u)\cos(v) - \frac{2}{7}\sin(u)\cos(-3\,u + v) + \cos(u),\right.$$
$$\left.\sin(u)\sin(v) - \frac{1}{6}\sin(u)\cos(6\,v),\ \sin(u) - 2\cos(u)\right]$$

```
> rgu:=[0,Pi/2-sin(7*v)-cos(7*v-1)/3];
```

\# Curtail range of parameter 'u' (Relative to closed parametrized surface)

$$rgu := \left[0, \frac{\pi}{2} - \sin(7\,v) - \frac{1}{3}\cos(7\,v - 1)\right]$$

> rgv:=[0,2*Pi];

\# Set range of 'u,v', make parametrized surface 'CS' unclosed

$$rgv := [0, 2\,\pi]$$

> plot3d(CS,u=rgu[1]..rgu[2],v=rgv[1]..rgv[2],scaling=constrained,
projection=0.9,numpoints=30000):g1:=%:

\# Draw unclosed parameterized surface 'CS'

> subs(u=rgu[2],CS)=eval(subs(u=rgu[2],CS));CL:=rhs(%);

\# Import boundary value of 'u' to expression of surface 'CS', obtain expression of boundary curve 'CL'; Set Right Hand Rule as associated basis between positive direction of boundary curve 'CL' and normal vector of unclosed surface 'CS'

$$\left[\sin\left(\frac{\pi}{2} - \sin(7\,v) - \frac{1}{3}\cos(7\,v-1)\right)\cos(v)\right.$$

$$-\frac{2}{7}\sin\left(\frac{\pi}{2} - \sin(7\,v) - \frac{1}{3}\cos(7\,v-1)\right)\cos\left(-\frac{3\,\pi}{2} + 3\sin(7\,v) + \cos(7\,v-1) + v\right)$$

$$+\cos\left(\frac{\pi}{2} - \sin(7\,v) - \frac{1}{3}\cos(7\,v-1)\right), \sin\left(\frac{\pi}{2} - \sin(7\,v) - \frac{1}{3}\cos(7\,v-1)\right)\sin(v)$$

$$-\frac{1}{6}\sin\left(\frac{\pi}{2} - \sin(7\,v) - \frac{1}{3}\cos(7\,v-1)\right)\cos(6\,v),$$

$$\sin\left(\frac{\pi}{2} - \sin(7\,v) - \frac{1}{3}\cos(7\,v-1)\right) - 2\cos\left(\frac{\pi}{2} - \sin(7\,v) - \frac{1}{3}\cos(7\,v-1)\right)\right] = \left[\right.$$

$$\cos\left(\sin(7\,v) + \frac{1}{3}\cos(7\,v-1)\right)\cos(v)$$

$$+\frac{2}{7}\cos\left(\sin(7\,v) + \frac{1}{3}\cos(7\,v-1)\right)\sin(3\sin(7\,v) + \cos(7\,v-1) + v)$$

$$+\sin\left(\sin(7\,v) + \frac{1}{3}\cos(7\,v-1)\right),$$

$$\cos\left(\sin(7\,v) + \frac{1}{3}\cos(7\,v-1)\right)\sin(v) - \frac{1}{6}\cos\left(\sin(7\,v) + \frac{1}{3}\cos(7\,v-1)\right)\cos(6\,v),$$

$$\cos\left(\sin(7\,v) + \frac{1}{3}\cos(7\,v-1)\right) - 2\sin\left(\sin(7\,v) + \frac{1}{3}\cos(7\,v-1)\right)\right]$$

$$CL := \left[\cos\left(\sin(7\,v) + \frac{1}{3}\cos(7\,v-1)\right)\cos(v)\right.$$

$$+\frac{2}{7}\cos\left(\sin(7\,v) + \frac{1}{3}\cos(7\,v-1)\right)\sin(3\sin(7\,v) + \cos(7\,v-1) + v)$$

$$+ \sin\left(\sin(7\,v) + \frac{1}{3}\cos(7\,v - 1) \right),$$

$$\cos\left(\sin(7\,v) + \frac{1}{3}\cos(7\,v - 1) \right)\sin(v) - \frac{1}{6}\cos\left(\sin(7\,v) + \frac{1}{3}\cos(7\,v - 1) \right)\cos(6\,v),$$

$$\cos\left(\sin(7\,v) + \frac{1}{3}\cos(7\,v - 1) \right) - 2\sin\left(\sin(7\,v) + \frac{1}{3}\cos(7\,v - 1) \right)\Bigg]$$

```
> spacecurve(CL,v=rgv[1]..rgv[2],scaling=constrained,
projection=0.9,numpoints=3000,thickness=2,color=red):g2:=%:
# Draw closed boundary curve'CL'
> V:=[(x/4-y/6+z/5)^2/3,z^2/6-x*y/5,x^2/5-y*z/6];
```

Define discretional spacial vector field 'V' (Suppose Vector Field 'V' possesses 1th order continuous partial derivatives at surface 'CS')

$$V := \left[\frac{\left(\dfrac{x}{4} - \dfrac{y}{6} + \dfrac{z}{5} \right)^2}{3},\ \frac{z^2}{6} - \frac{x\,y}{5},\ \frac{x^2}{5} - \frac{y\,z}{6} \right]$$

```
> [Diff(V[3],y)-Diff(V[2],z),Diff(V[1],z)-Diff(V[3],x),
Diff(V[2],x)-Diff(V[1],y)]
=[diff(V[3],y)-diff(V[2],z),diff(V[1],z)-diff(V[3],x),
diff(V[2],x)-diff(V[1],y)];cV:=rhs(%);
```

Calculate 'cV'-- Curl of spacial vector field 'V'

$$\Bigg[\left(\frac{\partial}{\partial y}\left(\frac{x^2}{5} - \frac{y\,z}{6} \right) \right) - \left(\frac{\partial}{\partial z}\left(\frac{z^2}{6} - \frac{x\,y}{5} \right) \right),\ \left(\frac{\partial}{\partial z}\left(\frac{\left(\dfrac{x}{4} - \dfrac{y}{6} + \dfrac{z}{5} \right)^2}{3} \right) \right) - \left(\frac{\partial}{\partial x}\left(\frac{x^2}{5} - \frac{y\,z}{6} \right) \right),$$

$$\left(\frac{\partial}{\partial x}\left(\frac{z^2}{6} - \frac{x\,y}{5} \right) \right) - \left(\frac{\partial}{\partial y}\left(\frac{\left(\dfrac{x}{4} - \dfrac{y}{6} + \dfrac{z}{5} \right)^2}{3} \right) \right)\Bigg] =$$

$$\left[-\frac{z}{2},\ -\frac{11\,x}{30} - \frac{y}{45} + \frac{2\,z}{75},\ -\frac{59\,y}{270} + \frac{x}{36} + \frac{z}{45} \right]$$

$$cV := \left[-\frac{z}{2},\ -\frac{11\,x}{30} - \frac{y}{45} + \frac{2\,z}{75},\ -\frac{59\,y}{270} + \frac{x}{36} + \frac{z}{45} \right]$$

```
> rgx:=[-2,2];
```

$$rgx := [-2, 2]$$

```
> rgy:=[-2,2];
```

$$rgy := [-2, 2]$$

```
> rgz:=[-1.9,2.1];
```

$$rgz := [-1.9, 2.1]$$

```
> fieldplot3d(V,x=rgx[1]..rgx[2],y=rgy[1]..rgy[2],z=rgz[1]..
```

```
rgz[2],arrows=SLIM,color=red,thickness=1,grid=[6,6,6]):g3:=%:
# Draw spacial vector field 'V'
> fieldplot3d(cV,x=rgx[1]..rgx[2],y=rgy[1]..rgy[2],z=rgz[1]..
rgz[2],arrows=SLIM,color=blue,thickness=1,grid=[6,6,6]):g4:=%:
# Draw curl 'cV'
> display(g1,g2,g3,g4); # Synthesize figures
```

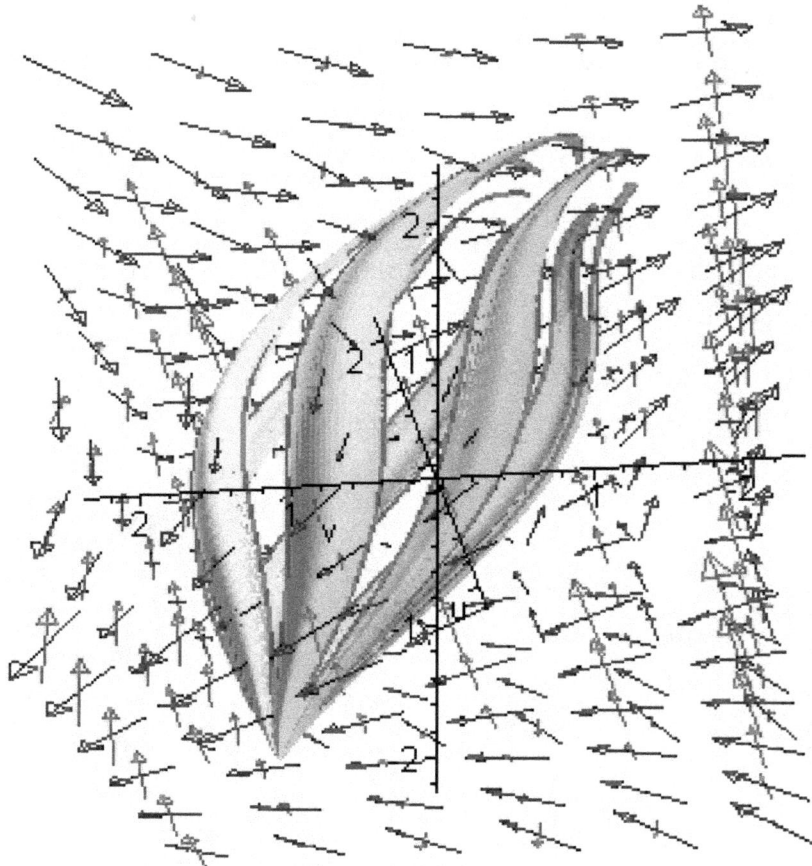

Figure(III)3.4.1 Surface[unclosed] 'CS' and its Boundary' Curve 'CL'
   Vector Field 'V'(Red arrows) and its Curl 'cV' (Blue arrows)

```
> display(g2); # Boundary Curve 'CL'
```

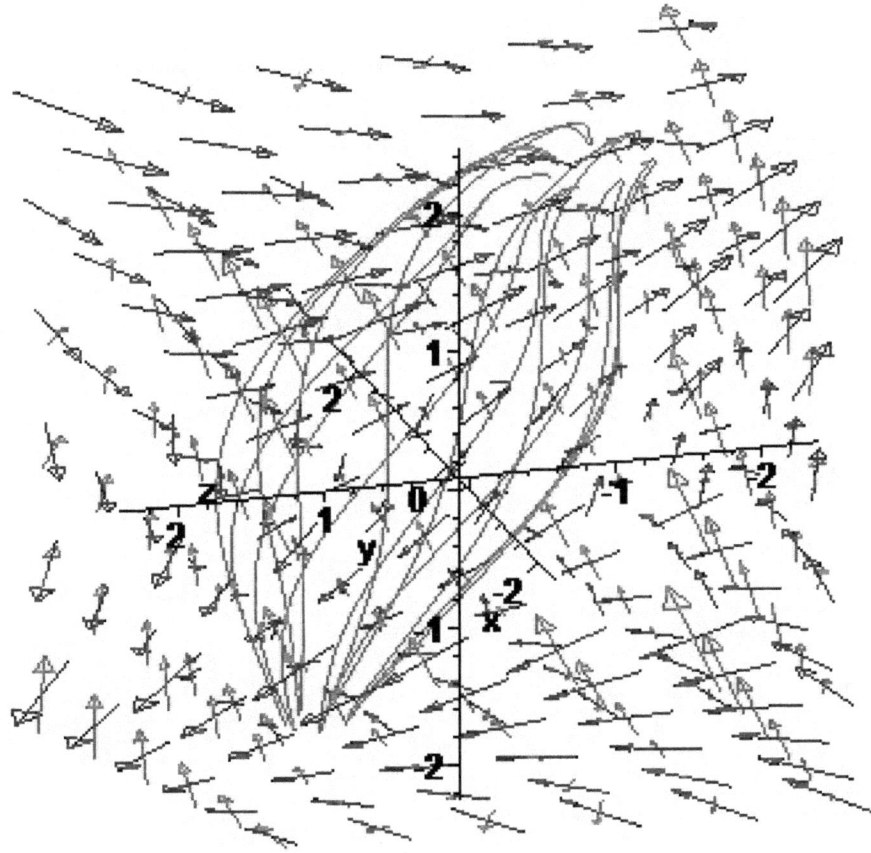

Figure(III)3.4.2 Closed Boundary Curve 'CL'

Vector Field 'V' and its Curl 'cV'

```
> x:=CL[1]:y:=CL[2]:z:=CL[3]:
> Int(V[1]*Diff(CL[1],v)+V[2]*Diff(CL[2],v)+V[3]*Diff(CL[3],v),
v=rgv[1]..rgv[2]);
# Closed Curve Integral(Vector Field 'V' at Closed Curve 'CL')
```

$$\int_0^{2\pi} \frac{1}{3}\left(\frac{1}{4}\cos\left(\sin(7\,v)+\frac{1}{3}\cos(7\,v-1)\right)\cos(v)\right.$$

$$+ \frac{1}{14} \cos\left(\sin(7\,v) + \frac{1}{3} \cos(7\,v - 1) \right) \sin(3 \sin(7\,v) + \cos(7\,v - 1) + v)$$

$$- \frac{3}{20} \sin\left(\sin(7\,v) + \frac{1}{3} \cos(7\,v - 1) \right) - \frac{1}{6} \cos\left(\sin(7\,v) + \frac{1}{3} \cos(7\,v - 1) \right) \sin(v)$$

$$+ \frac{1}{36} \cos\left(\sin(7\,v) + \frac{1}{3} \cos(7\,v - 1) \right) \cos(6\,v) + \frac{1}{5} \cos\left(\sin(7\,v) + \frac{1}{3} \cos(7\,v - 1) \right) \Big)\Big)^2 \Big($$

$$-\sin\left(\sin(7\,v) + \frac{1}{3} \cos(7\,v - 1) \right)\left(7 \cos(7\,v) - \frac{7}{3} \sin(7\,v - 1) \right) \cos(v)$$

$$- \cos\left(\sin(7\,v) + \frac{1}{3} \cos(7\,v - 1) \right) \sin(v) - \frac{2}{7} \sin\left(\sin(7\,v) + \frac{1}{3} \cos(7\,v - 1) \right)$$

$$\left(7 \cos(7\,v) - \frac{7}{3} \sin(7\,v - 1) \right) \sin(3 \sin(7\,v) + \cos(7\,v - 1) + v) + \frac{2}{7}$$

$$\cos\left(\sin(7\,v) + \frac{1}{3} \cos(7\,v - 1) \right) \cos(3 \sin(7\,v) + \cos(7\,v - 1) + v)$$

$$(21 \cos(7\,v) - 7 \sin(7\,v - 1) + 1)$$

$$+ \cos\left(\sin(7\,v) + \frac{1}{3} \cos(7\,v - 1) \right)\left(7 \cos(7\,v) - \frac{7}{3} \sin(7\,v - 1) \right)\Big) + \Big($$

$$\frac{1}{6}\left(\cos\left(\sin(7\,v) + \frac{1}{3} \cos(7\,v - 1) \right) - 2 \sin\left(\sin(7\,v) + \frac{1}{3} \cos(7\,v - 1) \right)\Big)\right)^2 - \frac{1}{5}\Big($$

$$\cos\left(\sin(7\,v) + \frac{1}{3} \cos(7\,v - 1) \right) \cos(v)$$

$$+ \frac{2}{7} \cos\left(\sin(7\,v) + \frac{1}{3} \cos(7\,v - 1) \right) \sin(3 \sin(7\,v) + \cos(7\,v - 1) + v)$$

$$+ \sin\left(\sin(7\,v) + \frac{1}{3} \cos(7\,v - 1) \right)\Big)\Big)$$

$$\left(\cos\left(\sin(7\,v) + \frac{1}{3} \cos(7\,v - 1) \right) \sin(v) - \frac{1}{6} \cos\left(\sin(7\,v) + \frac{1}{3} \cos(7\,v - 1) \right) \cos(6\,v) \right)$$

$$\Big)\left(-\sin\left(\sin(7\,v) + \frac{1}{3} \cos(7\,v - 1) \right)\left(7 \cos(7\,v) - \frac{7}{3} \sin(7\,v - 1) \right) \sin(v) \right)$$

$$+ \cos\left(\sin(7\,v) + \frac{1}{3} \cos(7\,v - 1) \right) \cos(v)$$

$$+ \frac{1}{6} \sin\left(\sin(7\,v) + \frac{1}{3} \cos(7\,v - 1) \right)\left(7 \cos(7\,v) - \frac{7}{3} \sin(7\,v - 1) \right) \cos(6\,v)$$

$$+ \cos\left(\sin(7\,v) + \frac{1}{3} \cos(7\,v - 1) \right) \sin(6\,v) \right) + \left(\frac{1}{5}\Big(\right.$$

$$\cos\left(\sin(7\,v) + \frac{1}{3} \cos(7\,v - 1) \right) \cos(v)$$

$$+ \frac{2}{7} \cos\left(\sin(7\,v) + \frac{1}{3} \cos(7\,v - 1) \right) \sin(3 \sin(7\,v) + \cos(7\,v - 1) + v)$$

$$+ \sin\left(\sin(7\,v) + \frac{1}{3} \cos(7\,v - 1) \right)\Bigg)^{2} - \frac{1}{6}$$

$$\left(\cos\left(\sin(7\,v) + \frac{1}{3} \cos(7\,v - 1) \right) \sin(v) - \frac{1}{6} \cos\left(\sin(7\,v) + \frac{1}{3} \cos(7\,v - 1) \right) \cos(6\,v) \right)$$

$$\left(\cos\left(\sin(7\,v) + \frac{1}{3} \cos(7\,v - 1) \right) - 2 \sin\left(\sin(7\,v) + \frac{1}{3} \cos(7\,v - 1) \right) \right)\Bigg)\Bigg($$

$$-\sin\left(\sin(7\,v) + \frac{1}{3} \cos(7\,v - 1) \right)\left(7 \cos(7\,v) - \frac{7}{3} \sin(7\,v - 1) \right)$$

$$- 2 \cos\left(\sin(7\,v) + \frac{1}{3} \cos(7\,v - 1) \right)\left(7 \cos(7\,v) - \frac{7}{3} \sin(7\,v - 1) \right)\Bigg) dv$$

```
> alpha:=evalf(%);
```

Calculate value of spacial closed curve integral

$$\alpha := 0.1213635571$$

```
> x:='x':y:='y':z:='z':
> x:=CS[1]:y:=CS[2]:z:=CS[3]:
> matrix(3,3,[i,j,k,Diff(CS[1],u),Diff(CS[2],u),Diff(CS[3],u),
Diff(CS[1],v),Diff(CS[2],v),Diff(CS[3],v)])=
matrix(3,3,[i,j,k,diff(CS[1],u),diff(CS[2],u),diff(CS[3],u),
diff(CS[1],v),diff(CS[2],v),diff(CS[3],v)]);m:=rhs(%);
```

Define and calculate matrix of partial derivatives 'm'; obtain
'tangent plane normal vector' of surface 'CS'

$[i, j, k]$

$$\left[\frac{\partial}{\partial u}\left(\sin(u) \cos(v) - \frac{2}{7} \sin(u) \cos(-3\,u + v) + \cos(u) \right),\right.$$

$$\frac{\partial}{\partial u}\left(\sin(u) \sin(v) - \frac{1}{6} \sin(u) \cos(6\,v) \right), \frac{d}{du}\left(\sin(u) - 2 \cos(u) \right)\right]$$

$$\left[\frac{\partial}{\partial v}\left(\sin(u) \cos(v) - \frac{2}{7} \sin(u) \cos(-3\,u + v) + \cos(u) \right),\right.$$

$$\left.\frac{\partial}{\partial v}\left(\sin(u) \sin(v) - \frac{1}{6} \sin(u) \cos(6\,v) \right), \frac{\partial}{\partial v}\left(\sin(u) - 2 \cos(u) \right)\right] =$$

$[i, j, k]$

$$\left[\cos(u) \cos(v) - \frac{2}{7} \cos(u) \cos(-3\,u + v) - \frac{6}{7} \sin(u) \sin(-3\,u + v) - \sin(u),\right.$$

$$\left.\cos(u) \sin(v) - \frac{1}{6} \cos(u) \cos(6\,v), \cos(u) + 2 \sin(u)\right]$$

$$\left[-\sin(u) \sin(v) + \frac{2}{7} \sin(u) \sin(-3\,u + v), \sin(u) \cos(v) + \sin(u) \sin(6\,v), 0\right]$$

$m :=$

$[i, j, k]$

$$\left[\cos(u)\cos(v) - \frac{2}{7}\cos(u)\cos(-3u+v) - \frac{6}{7}\sin(u)\sin(-3u+v) - \sin(u),\right.$$

$$\left.\cos(u)\sin(v) - \frac{1}{6}\cos(u)\cos(6v), \cos(u) + 2\sin(u)\right]$$

$$\left[-\sin(u)\sin(v) + \frac{2}{7}\sin(u)\sin(-3u+v), \sin(u)\cos(v) + \sin(u)\sin(6v), 0\right]$$

```
> det(m):
> mn:=simplify(%);
```

$$mn := -\frac{1}{42}\sin(u)\,(42\,i\cos(u)\cos(v) + 42\,i\cos(u)\sin(6v) + 84\,i\sin(u)\cos(v)$$

$$+ 84\,i\sin(u)\sin(6v) - 42\,k\cos(u)\cos(v)\sin(6v)$$

$$+ 12\,k\cos(u)\cos(-3u+v)\cos(v) + 12\,k\cos(u)\cos(-3u+v)\sin(6v)$$

$$+ 36\,k\sin(u)\sin(-3u+v)\cos(v) + 36\,k\sin(u)\sin(-3u+v)\sin(6v)$$

$$+ 42\,k\sin(u)\cos(v) + 42\,k\sin(u)\sin(6v) + 42\sin(v)\,j\cos(u) + 84\sin(u)\sin(v)\,j$$

$$- 42\,k\cos(u) + 7\sin(v)\,k\cos(u)\cos(6v) - 12\sin(-3u+v)\,j\cos(u)$$

$$- 24\sin(u)\sin(-3u+v)\,j + 12\sin(-3u+v)\,k\cos(u)\sin(v)$$

$$- 2\sin(-3u+v)\,k\cos(u)\cos(6v))$$

```
> A:=coeff(mn,i); # Obtain coefficient of 'i'
```

$$A := -\frac{1}{42}\sin(u)$$

$$(42\cos(u)\cos(v) + 42\cos(u)\sin(6v) + 84\sin(u)\cos(v) + 84\sin(u)\sin(6v))$$

```
> B:=coeff(mn,j); # Obtain coefficient of 'j'
```

$$B := -\frac{1}{42}\sin(u)\,(42\cos(u)\sin(v) + 84\sin(u)\sin(v) - 12\sin(-3u+v)\cos(u)$$

$$- 24\sin(u)\sin(-3u+v))$$

```
> C:=coeff(mn,k);
```

```
# Obtain coefficient of 'k'
```

$$C := -\frac{1}{42}\sin(u)\,(-42\cos(u)\cos(v)\sin(6v) + 12\cos(u)\cos(-3u+v)\cos(v)$$

$$+ 12\cos(u)\cos(-3u+v)\sin(6v) + 36\sin(u)\sin(-3u+v)\cos(v)$$

$$+ 36\sin(u)\sin(-3u+v)\sin(6v) + 42\sin(u)\cos(v) + 42\sin(u)\sin(6v)$$

$$- 42\cos(u) + 7\sin(v)\cos(u)\cos(6v) + 12\sin(-3u+v)\cos(u)\sin(v)$$

$$- 2\sin(-3u+v)\cos(u)\cos(6v))$$

```
> [A,B,C]: # [A,B,C] structure 'tangent plane normal vector'
> Int(Int(cV[1]*A+cV[2]*B+cV[3]*C,u=rgu[1]..rgu[2]),v=rgv[1]..
rgv[2]);
```

```
# Integral of Dot Product (Curl 'cV' and Tangent Plane Normal Vector
'[A,B,C]') in range of 'u,v'
```

$$\int_0^{2\pi} \int_0^{\frac{\pi}{2} - \sin(7v) - 1/3\cos(7v-1)} -\frac{1}{42}\left(-\frac{1}{2}\sin(u) + \cos(u)\right)\sin(u)$$

$$(42\cos(u)\cos(v) + 42\cos(u)\sin(6v) + 84\sin(u)\cos(v) + 84\sin(u)\sin(6v)) -$$
$$\frac{1}{42}\left(-\frac{11}{30}\sin(u)\cos(v) + \frac{11}{105}\sin(u)\cos(-3u+v) - \frac{21}{50}\cos(u) - \frac{1}{45}\sin(u)\sin(v)\right.$$
$$\left. + \frac{1}{270}\sin(u)\cos(6v) + \frac{2}{75}\sin(u)\right)\sin(u)(42\cos(u)\sin(v) + 84\sin(u)\sin(v)$$
$$- 12\sin(-3u+v)\cos(u) - 24\sin(u)\sin(-3u+v)) - \frac{1}{42}\left(-\frac{59}{270}\sin(u)\sin(v)\right.$$
$$+ \frac{59}{1620}\sin(u)\cos(6v) + \frac{1}{36}\sin(u)\cos(v) - \frac{1}{126}\sin(u)\cos(-3u+v) - \frac{1}{60}\cos(u)$$
$$\left. + \frac{1}{45}\sin(u)\right)\sin(u)(-42\cos(u)\cos(v)\sin(6v) + 12\cos(u)\cos(-3u+v)\cos(v)$$
$$+ 12\cos(u)\cos(-3u+v)\sin(6v) + 36\sin(u)\sin(-3u+v)\cos(v)$$
$$+ 36\sin(u)\sin(-3u+v)\sin(6v) + 42\sin(u)\cos(v) + 42\sin(u)\sin(6v)$$
$$- 42\cos(u) + 7\sin(v)\cos(u)\cos(6v) + 12\sin(-3u+v)\cos(u)\sin(v)$$
$$- 2\sin(-3u+v)\cos(u)\cos(6v))\,du\,dv$$

```
> beta:=evalf(%); # Calculate value of surface integral
```
$$\beta := 0.1213635571$$

```
> alpha;beta; # Two float values are equal
```
$$0.1213635571$$
$$0.1213635571$$

3.5 Numerical Model of Curl Theorem in Torus Coordinates

```
    Known: Parametric Expression of Torus:
    (Multiply connected、Orientable、Unclosed Surface)
```
$$[(2 + \cos(u))\cos(v), (2 + \cos(u))\sin(v), \sin(u)] \qquad (1)$$

```
    thereinto, u∈
```
$[0, 2\pi]$, v∈ $\left[0, \frac{5\pi}{3}\right]$; and Integral Vector Field

$$\left[\frac{\left(\frac{x}{3} + \frac{y}{5} - \frac{z}{7}\right)^3}{9} - \frac{z^2}{5}, \frac{x^2}{5} + \frac{yz}{6}, \frac{x^2}{3} + \frac{yz}{5}\right] \qquad (2)$$

```
    Calculate and Validate Curl Theorem at Torus.
```

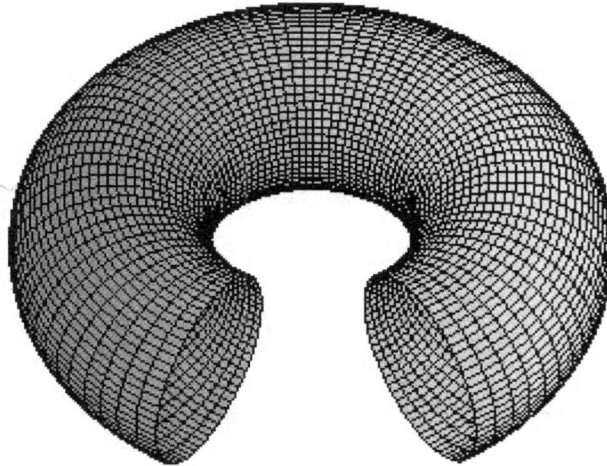

Figure(III)3.5.1 Unclosed Torus (1)

Solution:

First, Spacial Closed Curve Integral about Torus:

Import Left Boundary Value of 'v' (Viz. 0) to Expression of Torus (1), obtain Expression of Boundary Curve 'CL1'(3); Set Right Hand Rule as associated basis between positive direction of boundary curve 'CL1' and normal vector of unclosed torus surface

$$[(2+\cos(u))\cos(0), (2+\cos(u))\sin(0), \sin(u)] = [2+\cos(u), 0, \sin(u)]$$

$$CL1 := [2+\cos(u), 0, \sin(u)] \qquad (3)$$

Import Right Boundary Value of 'v' (Viz. $\dfrac{5\pi}{3}$) to Expression of Torus (1), obtain Expression of Boundary Curve 'CL2'(4); Set Right Hand Rule as associated basis between positive direction of boundary curve 'CL2' and normal vector of unclosed torus surface.

$$\left[(2+\cos(u))\cos\left(\frac{5\pi}{3}\right), (2+\cos(u))\sin\left(\frac{5\pi}{3}\right), \sin(u)\right] =$$

$$\left[1+\frac{1}{2}\cos(u), -\frac{1}{2}(2+\cos(u))\sqrt{3}, \sin(u)\right]$$

$$CL2 := \left[1+\frac{1}{2}\cos(u), -\frac{1}{2}(2+\cos(u))\sqrt{3}, \sin(u)\right] \qquad (4)$$

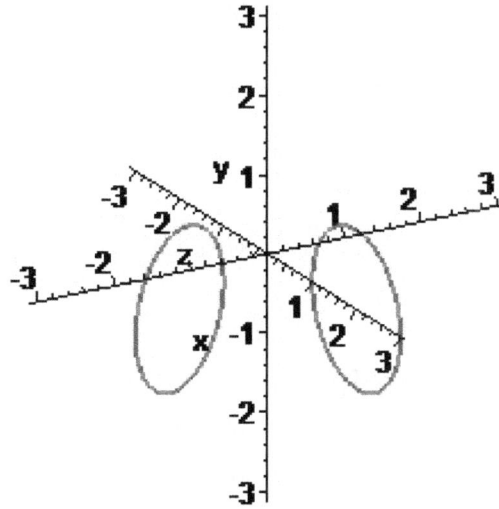

Figure(III)3.5.2 Closed Boundary Curves (3) and (4)

of Unclosed Torus (1)

Input Expression of Boundary Curve 'CL1'(3) to Vector Field(2); and Curve Integral [Vector Field(2) at Boundary Curve 'CL1'(3)]:

$$\int_0^{2\pi} \left(\left(\frac{1}{9} \left(\frac{2}{3} + \frac{1}{3}\cos(u) - \frac{1}{7}\sin(u) \right)^3 - \frac{1}{5}\sin(u)^2 \right) \left(\frac{d}{du}(2 + \cos(u)) \right) \right.$$

$$\left. + \frac{1}{5}(2 + \cos(u))^2 \left(\frac{d}{du} 0 \right) + \frac{1}{3}(2 + \cos(u))^2 \left(\frac{d}{du}\sin(u) \right) \right) du$$

$$V1 = \frac{25117\,\pi}{18522}$$

(5)

Input Expression of Boundary Curve 'CL2'(4) to Vector Field(2); and Curve Integral [Vector Field(2) at Boundary Curve 'CL2'(4)]:

$$\int_0^{2\pi} \left(\left(\frac{1}{9} \left(\frac{1}{3} + \frac{1}{6}\cos(u) - \frac{1}{10}(2 + \cos(u))\sqrt{3} - \frac{1}{7}\sin(u) \right)^3 - \frac{1}{5}\sin(u)^2 \right) \right.$$

$$\left(\frac{d}{du} \left(1 + \frac{1}{2}\cos(u) \right) \right)$$

$$+ \left(\frac{1}{5} \left(1 + \frac{1}{2}\cos(u) \right)^2 - \frac{1}{12}(2 + \cos(u))\sqrt{3}\,\sin(u) \right) \left(\frac{d}{du} \left(-\frac{1}{2}(2 + \cos(u))\sqrt{3} \right) \right)$$

$$+\left(\frac{1}{3}\left(1+\frac{1}{2}\cos(u)\right)^2-\frac{1}{10}(2+\cos(u))\sqrt{3}\sin(u)\right)\left(\frac{d}{du}\sin(u)\right)du$$

$$V2 = -\frac{17}{5040}\sqrt{3}\,\pi+\frac{41351}{463050}\,\pi \tag{6}$$

Curve Integral Value (Vector Field (2) at Closed Boundary Curve 'CL1' and 'CL2') (7):

$$V1 - V2 = \frac{293287}{231525}\pi+\frac{17}{5040}\sqrt{3}\,\pi \tag{7}$$

Second, Surface Integral at Torus:

According to parametric expression of Torus (1), define and calculate Matrix of Partial Derivatives, obtain tangent plane normal vector of Torus (9):

$$\begin{bmatrix} i & j & k \\ \frac{\partial}{\partial u}((2+\cos(u))\cos(v)) & \frac{\partial}{\partial u}((2+\cos(u))\sin(v)) & \frac{d}{du}\sin(u) \\ \frac{\partial}{\partial v}((2+\cos(u))\cos(v)) & \frac{\partial}{\partial v}((2+\cos(u))\sin(v)) & \frac{\partial}{\partial v}\sin(u) \end{bmatrix}$$

$$=$$

$$-i\cos(u)^2\cos(v)-\cos(u)^2\sin(v)j-2i\cos(u)\cos(v)-\sin(u)k\cos(u)$$
$$-2\cos(u)\sin(v)j-2\sin(u)k \tag{8}$$

From Expression (8), Respectively pick up coefficient of item 'i,j,k', obtain Tangent Plane Normal Vector of Torus (9):

$$[-\cos(u)^2\cos(v)-2\cos(u)\cos(v),-\cos(u)^2\sin(v)-2\cos(u)\sin(v),$$
$$-\sin(u)\cos(u)-2\sin(u)] \tag{9}$$

Calculate Curl of Vector Field (10):

$$\left[\left(\frac{\partial}{\partial y}\left(\frac{x^2}{3}+\frac{yz}{5}\right)\right)-\left(\frac{\partial}{\partial z}\left(\frac{x^2}{5}+\frac{yz}{6}\right)\right),\left(\frac{\partial}{\partial z}\left(\frac{\left(\frac{x}{3}+\frac{y}{5}-\frac{z}{7}\right)^3}{9}-\frac{z^2}{5}\right)\right)-\left(\frac{\partial}{\partial x}\left(\frac{x^2}{3}+\frac{yz}{5}\right)\right),\right.$$

$$\left.\left(\frac{\partial}{\partial x}\left(\frac{x^2}{5}+\frac{yz}{6}\right)\right)-\left(\frac{\partial}{\partial y}\left(\frac{\left(\frac{x}{3}+\frac{y}{5}-\frac{z}{7}\right)^3}{9}-\frac{z^2}{5}\right)\right)\right]=$$

$$\left[\frac{z}{5}-\frac{y}{6},\ -\frac{\left(\frac{x}{3}+\frac{y}{5}-\frac{z}{7}\right)^2}{21}-\frac{2z}{5}-\frac{2x}{3},\ \frac{2x}{5}-\frac{\left(\frac{x}{3}+\frac{y}{5}-\frac{z}{7}\right)^2}{15}\right] \tag{10}$$

Input Expression of Torus(1) to Curl(10); and Surface Integral
[Curl(10) at Torus(1)] (11):

$$\int_{0}^{\frac{5\pi}{3}}\int_{0}^{2\pi}(-\cos(u)^2\cos(v)-2\cos(u)\cos(v))\left(\frac{1}{5}\sin(u)-\frac{1}{6}(2+\cos(u))\sin(v)\right)+$$

$$(-\cos(u)^2\sin(v)-2\cos(u)\sin(v))\Bigg($$

$$-\frac{1}{21}\left(\frac{1}{3}(2+\cos(u))\cos(v)+\frac{1}{5}(2+\cos(u))\sin(v)-\frac{1}{7}\sin(u)\right)^2-\frac{2}{5}\sin(u)$$

$$-\frac{2}{3}(2+\cos(u))\cos(v)\Bigg)+(-\sin(u)\cos(u)-2\sin(u))\left(\frac{2}{5}(2+\cos(u))\cos(v)\right.$$

$$-\frac{1}{15}\left(\frac{1}{3}(2+\cos(u))\cos(v)+\frac{1}{5}(2+\cos(u))\sin(v)-\frac{1}{7}\sin(u)\right)^2\Bigg)du\ dv$$

$$=\frac{293287}{231525}\pi+\frac{17}{5040}\sqrt{3}\ \pi \tag{11}$$

Curve Integral Precision Value (7) [Vector Field (2) at Closed
Boundary Curves (3) and (4) of Unclosed Torus Surface (1)], is equal
to Surface Integral Precision Value(11) [Curl (10) at Unclosed Torus
Surface (1)], Complete Calculation and Validation of Curl Theorem at
Torus.

3.6 Numerical Model of Curl Theorem in Torus Coordinates
[Program Template of Waterloo Maplesoft, Optional]

```
> restart;
> with(plots):with(linalg):
> CS:=[(2+cos(u))*cos(v),(2+cos(u))*sin(v),sin(u)];
# Define parametrized expression of torus 'CS'
```

$$CS:=[(2+\cos(u))\cos(v),(2+\cos(u))\sin(v),\sin(u)]$$

```
> rgu:=[0,2*Pi];
```

Curtail range of parameter 'u' (Relative to closed torus parametrized surface)

$$rgu := [0, 2\pi]$$

```
> rgv:=[0, 5*Pi/3];
```

Set range of parameters 'u,v', make parametrized torus 'CS' unclosed

$$rgv := \left[0, \frac{5\pi}{3} \right]$$

```
>plot3d(CS,u=rgu[1]..rgu[2],v=rgv[1]..rgv[2],scaling=constrained,
projection=0.9,numpoints=5000):g1:=%:
> subs(v=rgv[1],CS)=eval(subs(v=rgv[1],CS));CL1:=rhs(%);
```

Import left boundary value of 'v' to expression of torus 'CS', obtain expression of boundary curve 'CL1'; Set Right Hand Rule as associated basis between positive direction of boundary curve 'CL1' and normal vector of unclosed torus surface 'CS'

$$[(2 + \cos(u))\cos(0), (2 + \cos(u))\sin(0), \sin(u)] = [2 + \cos(u), 0, \sin(u)]$$

$$CL1 := [2 + \cos(u), 0, \sin(u)]$$

```
> subs(v=rgv[2],CS)=eval(subs(v=rgv[2],CS));CL2:=rhs(%);
```

Import right boundary value of parameter 'v' to expression of surface 'CS', obtain expression of boundary curve 'CL2'; Set Right Hand Rule as associated basis between positive direction of boundary curve 'CL2' and normal vector of unclosed torus surface 'CS'

$$\left[(2 + \cos(u))\cos\left(\frac{5\pi}{3}\right), (2 + \cos(u))\sin\left(\frac{5\pi}{3}\right), \sin(u) \right] =$$
$$\left[1 + \frac{1}{2}\cos(u), -\frac{1}{2}(2 + \cos(u))\sqrt{3}, \sin(u) \right]$$

$$CL2 := \left[1 + \frac{1}{2}\cos(u), -\frac{1}{2}(2 + \cos(u))\sqrt{3}, \sin(u) \right]$$

```
> spacecurve(CL1, u=rgu[1]..rgu[2],thickness=2,color=red):g2:=%:
> spacecurve(CL2, u=rgu[1]..rgu[2],thickness=2,color=red):g3:=%:
```

Draw closed boundary curve 'CL1,CL2'

```
> V:=[ (x/3+y/5-z/7)^3/9-z^2/5,x^2/5+y*z/6,x^2/3+y*z/5];
```

Define discretional spacial vector field 'V' (Suppose Vector Field 'V' possesses 1th order continuous partial derivatives at surface 'CS')

$$V := \left[\frac{\left(\dfrac{x}{3} + \dfrac{y}{5} - \dfrac{z}{7} \right)^3}{9} - \frac{z^2}{5}, \frac{x^2}{5} + \frac{yz}{6}, \frac{x^2}{3} + \frac{yz}{5} \right]$$

```
[Diff(V[3],y)-Diff(V[2],z),Diff(V[1],z)-Diff(V[3],x),
 Diff(V[2],x)-Diff(V[1],y)]
=[diff(V[3],y)-diff(V[2],z),diff(V[1],z)-diff(V[3],x),
 diff(V[2],x)-diff(V[1],y)];cV:=rhs(%);
# Calculate 'cV'-- Curl of vector field 'V'
```

$$\left[\left(\frac{\partial}{\partial y}\left(\frac{x^2}{3}+\frac{yz}{5}\right)\right)-\left(\frac{\partial}{\partial z}\left(\frac{x^2}{5}+\frac{yz}{6}\right)\right),\left(\frac{\partial}{\partial z}\left(\frac{\left(\frac{x}{3}+\frac{y}{5}-\frac{z}{7}\right)^3}{9}-\frac{z^2}{5}\right)\right)-\left(\frac{\partial}{\partial x}\left(\frac{x^2}{3}+\frac{yz}{5}\right)\right),\right.$$

$$\left.\left(\frac{\partial}{\partial x}\left(\frac{x^2}{5}+\frac{yz}{6}\right)\right)-\left(\frac{\partial}{\partial y}\left(\frac{\left(\frac{x}{3}+\frac{y}{5}-\frac{z}{7}\right)^3}{9}-\frac{z^2}{5}\right)\right)\right]=$$

$$\left[\frac{z}{5}-\frac{y}{6},-\frac{\left(\frac{x}{3}+\frac{y}{5}-\frac{z}{7}\right)^2}{21}-\frac{2z}{5}-\frac{2x}{3},\frac{2x}{5}-\frac{\left(\frac{x}{3}+\frac{y}{5}-\frac{z}{7}\right)^2}{15}\right]$$

$$cV:=\left[\frac{z}{5}-\frac{y}{6},-\frac{\left(\frac{x}{3}+\frac{y}{5}-\frac{z}{7}\right)^2}{21}-\frac{2z}{5}-\frac{2x}{3},\frac{2x}{5}-\frac{\left(\frac{x}{3}+\frac{y}{5}-\frac{z}{7}\right)^2}{15}\right]$$

```
> rgx:=[-3,3];
```
$$rgx := [-3, 3]$$

```
> rgy:=[-3,3];
```
$$rgy := [-3, 3]$$

```
> rgz:=[-3,3];
```
$$rgz := [-3, 3]$$

```
> fieldplot3d(V,x=rgx[1]..rgx[2],y=rgy[1]..rgy[2],z=rgz[1]..
rgz[2],projection=0.9,arrows=SLIM,color=red,thickness=1,
grid=[6,6,6]):g4:=%:
> fieldplot3d(cV,x=rgx[1]..rgx[2],y=rgy[1]..rgy[2],z=rgz[1]..
rgz[2],arrows=SLIM,color=blue,thickness=1,grid=[6,6,6]):g5:=%:
> display(g1,g2,g3,g4,g5);
```

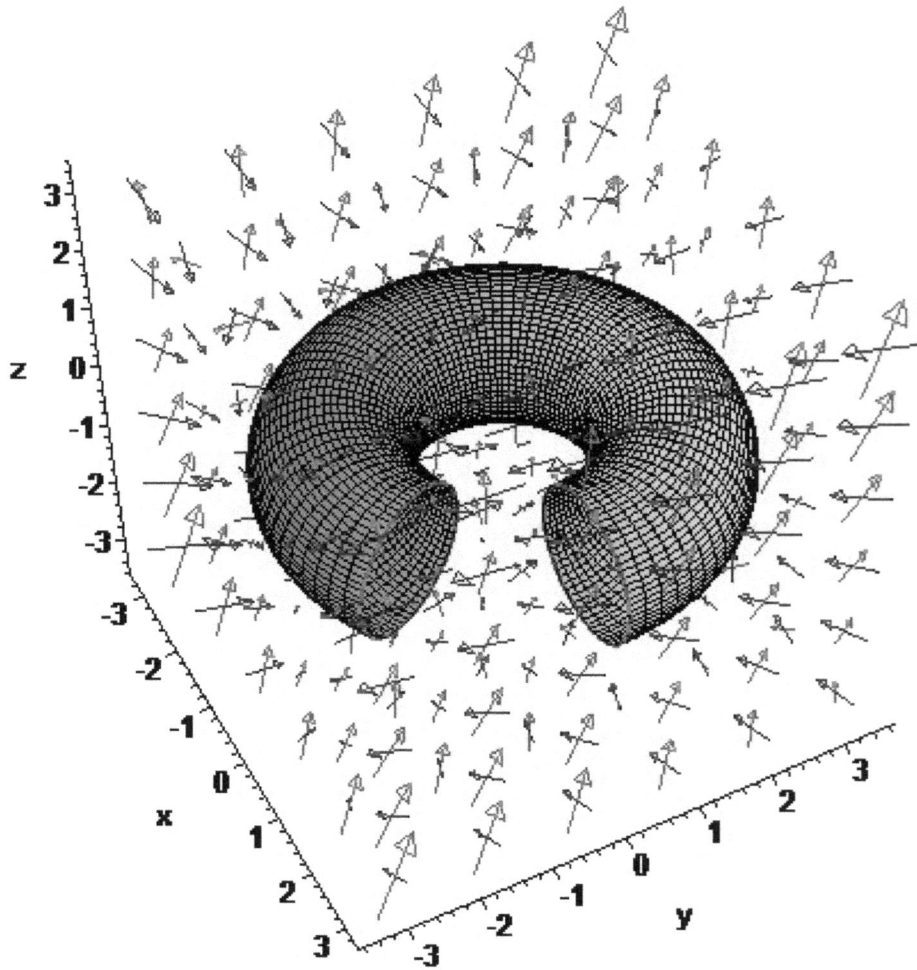

Figure(III)3.6.1 Torus Surface[unclosed] 'CS'
and its Closed Boundary Curves 'CL1' and 'CL2';
Vector Field 'V' (Red Arrows) and its Curl 'cV' (Blue Arrows)

```
> display(g2,g3,g4,g5,scaling=constrained,projection=0.9);
```

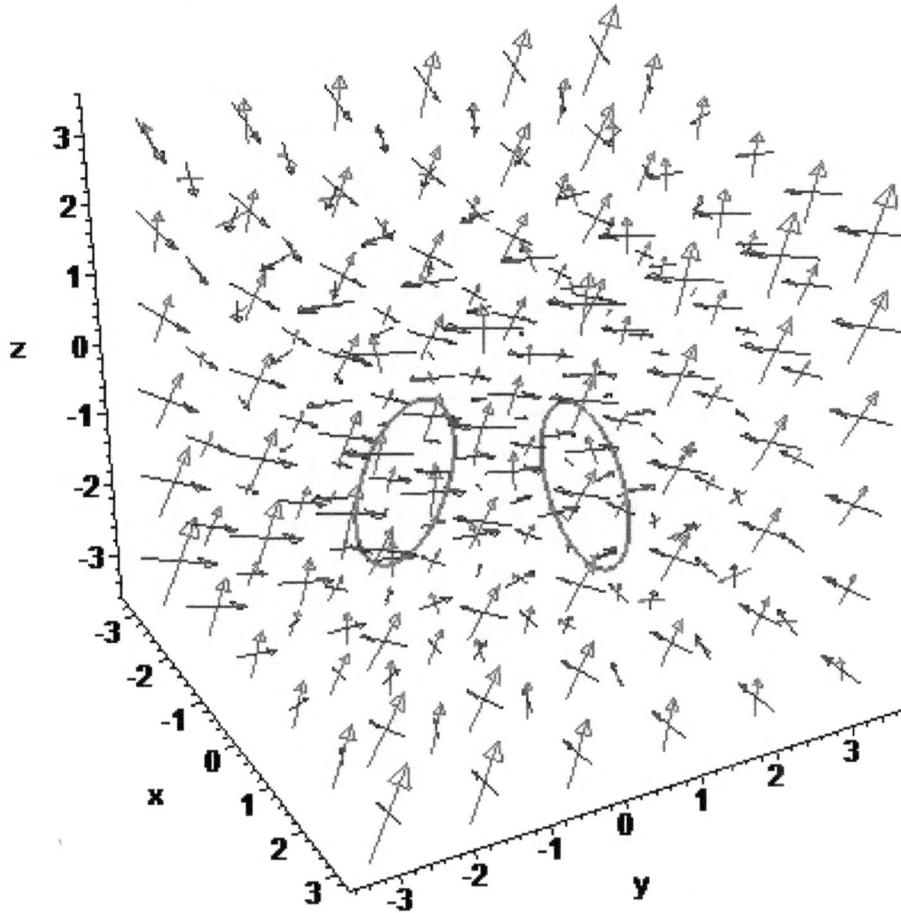

Figure(III)3.6.2 Closed Boundary Curves 'CL1' and 'CL2';
Vector Field 'V' and its Curl 'cV'

```
> x:=CL1[1]:y:=CL1[2]:z:=CL1[3]:
>Int(V[1]*Diff(CL1[1],u)+V[2]*Diff(CL1[2],u)+V[3]*Diff(CL1[3],u),
u=rgu[1]..rgu[2]); # Closed Curve Integral of Vector Field 'V' at
Closed Boundary Curve 'CL1'
```

$$\int_0^{2\pi} \left(\left(\frac{1}{9}\left(\frac{2}{3} + \frac{1}{3}\cos(u) - \frac{1}{7}\sin(u) \right)^3 - \frac{1}{5}\sin(u)^2 \right)\left(\frac{d}{du}(2+\cos(u)) \right) \right.$$

$$\left. + \frac{1}{5}(2+\cos(u))^2\left(\frac{d}{du}0 \right) + \frac{1}{3}(2+\cos(u))^2\left(\frac{d}{du}\sin(u) \right) \right) du$$

```
> v1:=value(%);
```

$$v1 := \frac{25117\,\pi}{18522}$$

```
> x:='x':y:='y':z:='z':
> x:=CL2[1]:y:=CL2[2]:z:=CL2[3]:
>Int(V[1]*Diff(CL2[1],u)+V[2]*Diff(CL2[2],u)+V[3]*Diff(CL2[3],u),
u=rgu[1]..rgu[2]); # Closed Curve Integral of Vector Field 'V' at
Closed Boundary Curve 'CL2'
```

$$\int_0^{2\pi} \left(\frac{1}{9}\left(\frac{1}{3} + \frac{1}{6}\cos(u) - \frac{1}{10}(2+\cos(u))\sqrt{3} - \frac{1}{7}\sin(u) \right)^3 - \frac{1}{5}\sin(u)^2 \right)$$

$$\left(\frac{d}{du}\left(1 + \frac{1}{2}\cos(u) \right) \right)$$

$$+ \left(\frac{1}{5}\left(1 + \frac{1}{2}\cos(u) \right)^2 - \frac{1}{12}(2+\cos(u))\sqrt{3}\,\sin(u) \right)\left(\frac{d}{du}\left(-\frac{1}{2}(2+\cos(u))\sqrt{3} \right) \right)$$

$$+ \left(\frac{1}{3}\left(1 + \frac{1}{2}\cos(u) \right)^2 - \frac{1}{10}(2+\cos(u))\sqrt{3}\,\sin(u) \right)\left(\frac{d}{du}\sin(u) \right) du$$

```
> v2:=value(%);
```

$$v2 := -\frac{17}{5040}\sqrt{3}\,\pi + \frac{41351}{463050}\,\pi$$

```
> alpha:=v1-v2;delta:=evalf(alpha);
# Closed Curve Integral of Vector Field 'V' at Closed Boundary Curves
'CL1,CL2'
```

$$\alpha := \frac{293287}{231525}\,\pi + \frac{17}{5040}\sqrt{3}\,\pi$$

$$\delta := 3.998003139$$

```
> x:='x':y:='y':z:='z':
> x:=CS[1]:y:=CS[2]:z:=CS[3]:
> matrix(3,3,[i,j,k,Diff(CS[1],u),Diff(CS[2],u),Diff(CS[3],u),
Diff(CS[1],v),Diff(CS[2],v),Diff(CS[3],v)])=
matrix(3,3,[i,j,k,diff(CS[1],u),diff(CS[2],u),diff(CS[3],u),
diff(CS[1],v),diff(CS[2],v),diff(CS[3],v)]);m:=rhs(%);
# Define and calculate matrix of partial derivatives 'm'; obtain
'tangent plane normal vector' of unclosed torus surface 'CS'
```

$$
\begin{bmatrix}
i & j & k \\
\dfrac{\partial}{\partial u}\left((2+\cos(u))\cos(v)\right) & \dfrac{\partial}{\partial u}\left((2+\cos(u))\sin(v)\right) & \dfrac{d}{du}\sin(u) \\
\dfrac{\partial}{\partial v}\left((2+\cos(u))\cos(v)\right) & \dfrac{\partial}{\partial v}\left((2+\cos(u))\sin(v)\right) & \dfrac{\partial}{\partial v}\sin(u)
\end{bmatrix} =
$$

$$
\begin{bmatrix}
i & j & k \\
-\sin(u)\cos(v) & -\sin(u)\sin(v) & \cos(u) \\
-(2+\cos(u))\sin(v) & (2+\cos(u))\cos(v) & 0
\end{bmatrix}
$$

$$
m := \begin{bmatrix}
i & j & k \\
-\sin(u)\cos(v) & -\sin(u)\sin(v) & \cos(u) \\
-(2+\cos(u))\sin(v) & (2+\cos(u))\cos(v) & 0
\end{bmatrix}
$$

```
> det(m):
> mn:=simplify(%);
```

$$
mn := -2\,i\cos(u)\cos(v) - i\cos(u)^2\cos(v) - 2\sin(v)\,j\cos(u) - 2\,k\sin(u)
$$
$$
- \sin(v)\,j\cos(u)^2 - \cos(u)\,k\sin(u)
$$

```
> A:=coeff(mn,i); # Obtain coefficient of 'i'
```

$$
A := -2\cos(u)\cos(v) - \cos(u)^2\cos(v)
$$

```
> B:=coeff(mn,j); # Obtain coefficient of 'j'
```

$$
B := -2\sin(v)\cos(u) - \sin(v)\cos(u)^2
$$

```
> C:=coeff(mn,k); # Obtain coefficient of 'k'
```

$$
C := -2\sin(u) - \cos(u)\sin(u)
$$

```
> [A,B,C]:  # [A,B,C] constitute 'tangent plane normal vector'
> -Int(Int(cV[1]*A+cV[2]*B+cV[3]*C,u=rgu[1]..rgu[2]),v=rgv[1]..
rgv[2]);
# Integral of dot product (curl 'cV' and tangent plane normal vector
'[A,B,C]') in range of 'u,v'
```

$$
\int_{0}^{\frac{5\pi}{3}}\int_{0}^{2\pi}\left(-\cos(u)^2\cos(v)-2\cos(u)\cos(v)\right)\left(\frac{1}{5}\sin(u)-\frac{1}{6}(2+\cos(u))\sin(v)\right)+
$$

$$
\left(-\cos(u)^2\sin(v)-2\cos(u)\sin(v)\right)\Bigg(
$$

$$
-\frac{1}{21}\left(\frac{1}{3}(2+\cos(u))\cos(v)+\frac{1}{5}(2+\cos(u))\sin(v)-\frac{1}{7}\sin(u)\right)^2-\frac{2}{5}\sin(u)
$$

$$-\frac{2}{3}\,(2+\cos(u))\cos(v)\Bigg) + (-\sin(u)\cos(u)-2\sin(u))\Bigg(\frac{2}{5}\,(2+\cos(u))\cos(v)$$

$$-\frac{1}{15}\left(\frac{1}{3}\,(2+\cos(u))\cos(v)+\frac{1}{5}\,(2+\cos(u))\sin(v)-\frac{1}{7}\sin(u)\right)^{2}\Bigg]\,du\;dv$$

```
> beta:=value(%);epsilon:=evalf(beta);
# Calculate value of surface integral
```

$$\beta := \frac{293287}{231525}\,\pi + \frac{17}{5040}\,\sqrt{3}\;\pi$$

$$\varepsilon := 3.998003139$$

```
> alpha;beta; # Two analytic values are equal
```

$$\frac{293287}{231525}\,\pi + \frac{17}{5040}\,\sqrt{3}\;\pi$$

$$\frac{293287}{231525}\,\pi + \frac{17}{5040}\,\sqrt{3}\;\pi$$

```
> delta;epsilon; # Two float values are equal
```

$$3.998003139$$

$$3.998003139$$

Chapter 4 Other Typical Numerical Model of Spacial Closed Curve Integral and Surface Integral

4.1 Other Typical Numerical Model of Spacial Closed Curve Integral and Surface Integral

Known: Parametrized Expression of Discretional Unclosed Surface

$$\left[\frac{2}{3}\,v\,(1+2\cos(u))\cos(v),\frac{2}{3}\,v\,(1+2\cos(u))\sin(v),2\,v\,(2-\sin(u))\right] \quad (1)$$

thereinto,u∈[0,2π],v∈[0,7π]; and Integral Vector Field

$$\left[\frac{\left(\dfrac{x}{3}-\dfrac{y}{3}-\dfrac{z}{6}\right)^{2}}{6},\frac{x^{2}}{5}+\frac{yz}{6},\frac{x^{2}}{3}-\frac{yz}{5}\right] \quad (2)$$

Calculate Discretional Unclosed Surface Integral、Cubage Integral and Associated Spacial Closed Curve Integral.

978-1-62265-930-2 (online)   978-1-62265-931-9 (paper)   Yang Ke

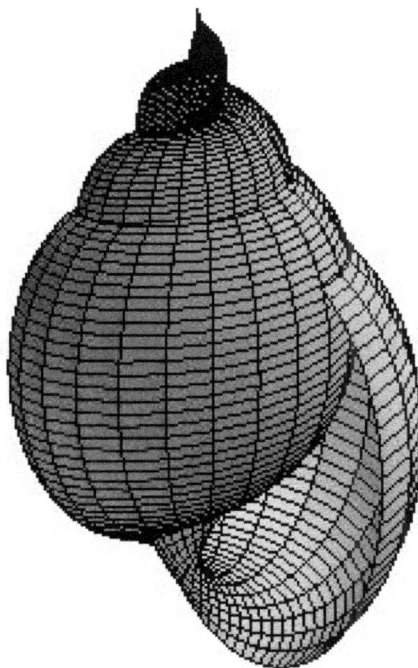

Figure(III)4.1.1 Discretional Unclosed Surface (1)

[Irregular、Asymmetrical]

Solution:

Observe shape of unclosed surface(1) by intuitionistic vision, be orientable admittedly, but parts of surface show intersected state; therefore unclosed surface(1) does not accord with precondition of proof and numerical models - Curl Theorem at Manifold, [See also Chapter 1 (this Part) Section 1.2 'Base on Poincare Conjecture, Define Abstract Simply Connected Orientable Closed Surface and Its Unclosed Surface'], does not belong to research scope of Curl Theorem at Manifold;

but science explorers might as well calculate surface integral by 'Parameterized Dot Product Method', calculate cubage integral by 'Idiographic Volume Element Coefficient Method', and associated spacial closed curve integral, in individualized unclosed surface (1) coordinates, about whelk in ocean of unknown and mysterious world.

This numerical model belongs to extended application of primary integral methods (Viz. Surface Integral bases on Parameterized Dot Product Method and Volume/Cubage Integral bases on Idiographic Volume Element Coefficient Method in Individualized Surface Coordinates) in fields of 'Curl Theorem at Manifold' and 'Divergence Theorem at Manifold'.

First, Free Spacial Closed Curve Integral:

Import right boundary value of 'v' (Viz. 7π) to parametrized expression of unclosed surface (1), obtain parametrized expression of boundary curve (3);

Set Right Hand Rule as associated basis between positive direction of boundary curve (3) and normal vector of unclosed surface (1).

$$\left[\frac{14}{3}\pi\,(1+2\cos(u))\cos(7\,\pi),\frac{14}{3}\pi\,(1+2\cos(u))\sin(7\,\pi),14\,\pi\,(2-\sin(u))\right]=$$
$$\left[-\frac{14}{3}\pi\,(1+2\cos(u)),0,14\,\pi\,(2-\sin(u))\right]$$

(3)

Figure(III)4.1.2 Boundary Curve (3) of Unclosed Surface(1)

Input expression (3) of Boundary Curve to Vector Field (2); and realize Spacial Closed Curve Integral [Vetor Field (2) at Boundary Curve (3)] in range of 'u' (4):

$$\int_0^{2\pi} \frac{1}{6}\left(-\frac{14}{9}\pi\,(1+2\cos(u))-\frac{7}{3}\pi\,(2-\sin(u))\right)^2\left(\frac{\partial}{\partial u}\left(-\frac{14}{3}\pi\,(1+2\cos(u))\right)\right)$$

$$+\frac{196}{45}\pi^2\,(1+2\cos(u))^2\left(\frac{d}{du}\,0\right)+\frac{196}{27}\pi^2\,(1+2\cos(u))^2\left(\frac{\partial}{\partial u}\,(14\,\pi\,(2-\sin(u)))\right)$$

$$du$$

$$=-\frac{109760\,\pi^4}{243}$$

(4)

Second, Free (Unclosed) Surface Integral:

According to parametric expression (1) of unclosed surface, define and calculate Matrix of Partial Derivatives, obtain tangent plane normal vector of unclosed surface (5):

$$\begin{bmatrix} i, & j, & k \\ \frac{\partial}{\partial u}\left(\frac{2}{3}v\,(1+2\cos(u))\cos(v)\right), & \frac{\partial}{\partial u}\left(\frac{2}{3}v\,(1+2\cos(u))\sin(v)\right), & \frac{\partial}{\partial u}\,(2\,v\,(2-\sin(u))) \\ \frac{\partial}{\partial v}\left(\frac{2}{3}v\,(1+2\cos(u))\cos(v)\right), & \frac{\partial}{\partial v}\left(\frac{2}{3}v\,(1+2\cos(u))\sin(v)\right), & \frac{\partial}{\partial v}\,(2\,v\,(2-\sin(u))) \end{bmatrix}$$

$$=$$

$$-\frac{4}{9}v\,(-6\,v\cos(u)^2\sin(v)\,j-6\,i\,v\cos(u)^2\cos(v)+4\,v\cos(u)\,k\sin(u)$$

$$-3\,v\sin(v)\,j\cos(u)-3\,i\,v\cos(u)\cos(v)+12\,i\sin(u)\sin(v)-12\sin(u)\cos(v)\,j$$

$$+2\,v\,k\sin(u)-3\,i\cos(u)\sin(v)+3\cos(v)\,j\cos(u)-6\,i\sin(v)+6\cos(v)\,j)$$

(5)

From expression(5), respectively pick up coefficient of item 'i,j,k', obtain tangent plane normal vector(6) of unclosed surface (1):

$$\left[-\frac{4}{9}v\,(-6\,v\cos(u)^2\cos(v)-3\,v\cos(u)\cos(v)+12\sin(u)\sin(v)-3\sin(v)\cos(u)\right.$$

$$-6\sin(v)),\,-\frac{4}{9}v\,(-6\,v\cos(u)^2\sin(v)-3\,v\cos(u)\sin(v)-12\sin(u)\cos(v)$$

$$\left.+3\cos(u)\cos(v)+6\cos(v)),\,-\frac{4}{9}v\,(4\,v\cos(u)\sin(u)+2\,v\sin(u))\right]$$

(6)

Calculate Curl(7) of Vector Field(2):

$$\left[\left(\frac{\partial}{\partial y}\left(\frac{x^2}{3}-\frac{y\,z}{5}\right)\right)-\left(\frac{\partial}{\partial z}\left(\frac{x^2}{5}+\frac{y\,z}{6}\right)\right),\left(\frac{\partial}{\partial z}\left(\frac{\left(\frac{x}{3}-\frac{y}{3}-\frac{z}{6}\right)^2}{6}\right)\right)-\left(\frac{\partial}{\partial x}\left(\frac{x^2}{3}-\frac{y\,z}{5}\right)\right),\right.$$

$$\left(\frac{\partial}{\partial x}\left(\frac{x^2}{5}+\frac{yz}{6}\right)\right)-\left(\frac{\partial}{\partial y}\left(\frac{\left(\frac{x}{3}-\frac{y}{3}-\frac{z}{6}\right)^2}{6}\right)\right)\right]=$$

$$\left[-\frac{z}{5}-\frac{y}{6},\ -\frac{37x}{54}+\frac{y}{54}+\frac{z}{108},\ \frac{59x}{135}-\frac{y}{27}-\frac{z}{54}\right] \qquad (7)$$

Input expression(1) of Unclosed Surface to Curl(7); and Calculate Integral of Dot Product [Curl(7) and tangent plane normal vector(6)] in range of 'u,v'(8):

$$-\int_0^{7\pi}\int_0^{2\pi}-\frac{4}{9}v\left(-6v\cos(u)^2\cos(v)-3v\cos(u)\cos(v)+12\sin(u)\sin(v)\right.$$

$$-3\sin(v)\cos(u)-6\sin(v)\right)\left(-\frac{2}{5}v(2-\sin(u))-\frac{1}{9}v(1+2\cos(u))\sin(v)\right)-\frac{4}{9}v\,($$

$$-6v\cos(u)^2\sin(v)-3v\cos(u)\sin(v)-12\sin(u)\cos(v)+3\cos(u)\cos(v)$$
$$+6\cos(v))$$

$$\left(-\frac{37}{81}v(1+2\cos(u))\cos(v)+\frac{1}{81}v(1+2\cos(u))\sin(v)+\frac{1}{54}v(2-\sin(u))\right)-\frac{4}{9}$$
$$v\,(4v\cos(u)\sin(u)+2v\sin(u))$$

$$\left(\frac{118}{405}v(1+2\cos(u))\cos(v)-\frac{2}{81}v(1+2\cos(u))\sin(v)-\frac{1}{27}v(2-\sin(u))\right)du\,d$$
$$v$$

$$=-\frac{109760\,\pi^4}{243} \qquad (8)$$

Closed Curve Integral Precision Value (4) [Vector Field (2) at Boundary Curve (3) of Unclosed Surface (1)], is equal to Surface Integral Precision Value (8) [Curl (7) at Unclosed Surface (1)], Complete Calculation of Discretional Unclosed Surface Integral and Associated Spacial Closed Curve Integral.

Surface Integral of 'Modulus of Tangent Plane Normal Vector [A,B,C]'in range of 'u,v', absolute value of result is area value of unclosed surface (1):

$$\int_0^{7\pi}\int_0^{2\pi}\frac{4}{9}\left(v^2\left(-6v\cos(u)^2\cos(v)-3v\cos(u)\cos(v)+12\sin(u)\sin(v)\right.\right.$$

$$-3\sin(v)\cos(u)-6\sin(v)\right)^2+v^2\left(-6v\cos(u)^2\sin(v)-3v\cos(u)\sin(v)\right.$$

$$- 12 \sin(u) \cos(v) + 3 \cos(u) \cos(v) + 6 \cos(v))^2$$

$$+ v^2 (4 v \cos(u) \sin(u) + 2 v \sin(u))^2)^{(1/2)} \quad du \ dv$$

$$= \quad 38705.23446 \qquad (9)$$

Third, Free Cubage Integral:

Entirely multiply radius vector 'r' (suppose r>0) by each item of surface parametrized expression (1), convert 'Parametrized expression of surface' to 'Parametrized expression of surface coordinates' (10):

$$\left[\frac{2}{3} r v (1 + 2 \cos(u)) \cos(v), \frac{2}{3} r v (1 + 2 \cos(u)) \sin(v), 2 r v (2 - \sin(u)) \right]_{(10)}$$

According to parametrized expression of surface coordinates (10), define and calculate Second Matrix of Partial Derivatives, obtain general expression of surface coordinates volume element Coefficient (11):

$$\begin{bmatrix} \frac{\partial}{\partial r}(\frac{2}{3}rv(1+2\cos(u))\cos(v)) & \frac{\partial}{\partial u}(\frac{2}{3}rv(1+2\cos(u))\cos(v)) & \frac{\partial}{\partial v}(\frac{2}{3}rv(1+2\cos(u))\cos(v)) \\ \frac{\partial}{\partial r}(\frac{2}{3}rv(1+2\cos(u))\sin(v)) & \frac{\partial}{\partial u}(\frac{2}{3}rv(1+2\cos(u))\sin(v)) & \frac{\partial}{\partial v}(\frac{2}{3}rv(1+2\cos(u))\sin(v)) \\ \frac{\partial}{\partial r}(2rv(2-\sin(u))) & \frac{\partial}{\partial u}(2rv(2-\sin(u))) & \frac{\partial}{\partial v}(2rv(2-\sin(u))) \end{bmatrix}$$

$$=$$

$$- \frac{8}{9} r^2 v^3 (8 \cos(u) \sin(u) - 2 \cos(u)^2 + 4 \sin(u) - 5 \cos(u) - 2) \qquad (11)$$

Triple Integrals of 'Surface Coordinates Volume Element' in range of 'r,u,v', absolute value of result (12) is cubage value of unclosed surface (1):

$$\int_0^{7\pi} \int_0^{2\pi} \int_0^1 - \frac{8}{9} r^2 v^3 (8 \cos(u) \sin(u) - 2 \cos(u)^2 + 4 \sin(u) - 5 \cos(u) - 2) \, dr \, du \, dv$$

$$= \frac{9604 \, \pi^5}{9} \qquad (12)$$

4.2 Other Typical Numerical Model of Spacial Closed Curve Integral
and Surface Integral
[Program Template of Waterloo Maplesoft, Optional]

```
> restart;
> with(plots):with(linalg):
> CS:=[2*v*(1+2*cos(u))*cos(v)/3,2*v*(1+2*cos(u))*sin(v)/3,
  2*v*(2-sin(u))];
```

Define parametrized expression of discretional unclosed surface 'CS'

$$CS := \left[\frac{2}{3} v \left(1 + 2 \cos(u) \right) \cos(v), \frac{2}{3} v \left(1 + 2 \cos(u) \right) \sin(v), 2 v \left(2 - \sin(u) \right) \right]$$

```
> rgu:=[0,2*Pi]; # Set range of parameters 'u,v'
```

$$rgu := [0, 2\pi]$$

```
> rgv:=[0,7*Pi];
```

$$rgv := [0, 7\pi]$$

```
> plot3d(CS,u=rgu[1]..rgu[2],v=rgv[1]..rgv[2],scaling=constrained,
projection=0.9,numpoints=7000):g1:=%:
```

Draw unclosed parameterized surface 'CS'

```
> subs(v=rgv[2],CS)=eval(subs(v=rgv[2],CS));CL:=rhs(%);
```

Import boundary value of 'v' to expression of surface 'CS', obtain
expression of boundary curve 'CL'; Set Right Hand Rule as associated
basis between positive direction of boundary curve 'CL' and normal
vector of unclosed surface 'CS'

$$\left[\frac{14}{3} \pi \left(1 + 2 \cos(u) \right) \cos(7\pi), \frac{14}{3} \pi \left(1 + 2 \cos(u) \right) \sin(7\pi), 14 \pi \left(2 - \sin(u) \right) \right] =$$
$$\left[-\frac{14}{3} \pi \left(1 + 2 \cos(u) \right), 0, 14 \pi \left(2 - \sin(u) \right) \right]$$
$$CL := \left[-\frac{14}{3} \pi \left(1 + 2 \cos(u) \right), 0, 14 \pi \left(2 - \sin(u) \right) \right]$$

```
> rgu:=[rgu[1],rgu[2]];
```

$$rgu := [0, 2\pi]$$

```
> spacecurve(CL,u=rgu[1]..rgu[2],scaling=constrained,
 projection=0.9,numpoints=3000,thickness=2,color=red):g2:=%:
```

Draw closed boundary curve 'CL'

```
> V:=[(x/3-y/3-z/6)^2/6,x^2/5+y*z/6,x^2/3-y*z/5];
```

Define discretional spacial vector field 'V'
 (Suppose Vector Field 'V' possesses 1th order continuous partial

derivatives at surface 'CS')

$$V := \left[\frac{\left(\dfrac{x}{3} - \dfrac{y}{3} - \dfrac{z}{6} \right)^2}{6}, \frac{x^2}{5} + \frac{y\,z}{6}, \frac{x^2}{3} - \frac{y\,z}{5} \right]$$

```
> [Diff(V[3],y)-Diff(V[2],z),Diff(V[1],z)-Diff(V[3],x),
Diff(V[2],x)-Diff(V[1],y)]
=[diff(V[3],y)-diff(V[2],z),diff(V[1],z)-diff(V[3],x),
diff(V[2],x)-diff(V[1],y)];cV:=rhs(%);
# Calculate 'cV' - Curl of spacial vector field 'V'
```

$$\left[\left(\frac{\partial}{\partial y}\left(\frac{x^2}{3} - \frac{y\,z}{5} \right) \right) - \left(\frac{\partial}{\partial z}\left(\frac{x^2}{5} + \frac{y\,z}{6} \right) \right), \left(\frac{\partial}{\partial z}\left(\frac{\left(\dfrac{x}{3} - \dfrac{y}{3} - \dfrac{z}{6} \right)^2}{6} \right) \right) - \left(\frac{\partial}{\partial x}\left(\frac{x^2}{3} - \frac{y\,z}{5} \right) \right), \right.$$

$$\left. \left(\frac{\partial}{\partial x}\left(\frac{x^2}{5} + \frac{y\,z}{6} \right) \right) - \left(\frac{\partial}{\partial y}\left(\frac{\left(\dfrac{x}{3} - \dfrac{y}{3} - \dfrac{z}{6} \right)^2}{6} \right) \right) \right] =$$

$$\left[-\frac{z}{5} - \frac{y}{6}, -\frac{37\,x}{54} + \frac{y}{54} + \frac{z}{108}, \frac{59\,x}{135} - \frac{y}{27} - \frac{z}{54} \right]$$

$$cV := \left[-\frac{z}{5} - \frac{y}{6}, -\frac{37\,x}{54} + \frac{y}{54} + \frac{z}{108}, \frac{59\,x}{135} - \frac{y}{27} - \frac{z}{54} \right]$$

```
> rgx:=[-66,66];
```

$$rgx := [-66, 66]$$

```
> rgy:=[-66,66];
```

$$rgy := [-66, 66]$$

```
> rgz:=[0,132];
```

$$rgz := [0, 132]$$

```
> fieldplot3d(V,x=rgx[1]..rgx[2],y=rgy[1]..rgy[2],z=rgz[1]..
rgz[2],arrows=SLIM,color=red,thickness=1,grid=[6,6,6]):g3:=%:
# Draw spacial vector field 'V'
>fieldplot3d(cV,x=rgx[1]..rgx[2],y=rgy[1]..rgy[2],z=rgz[1]..
rgz[2],arrows=SLIM,color=blue,thickness=1,grid=[6,6,6]):g4:=%:
# Draw curl 'cV'
> display(g1,g2,g3,g4); # Synthesize figures
```

978-1-62265-930-2 (online)   978-1-62265-931-9 (paper)

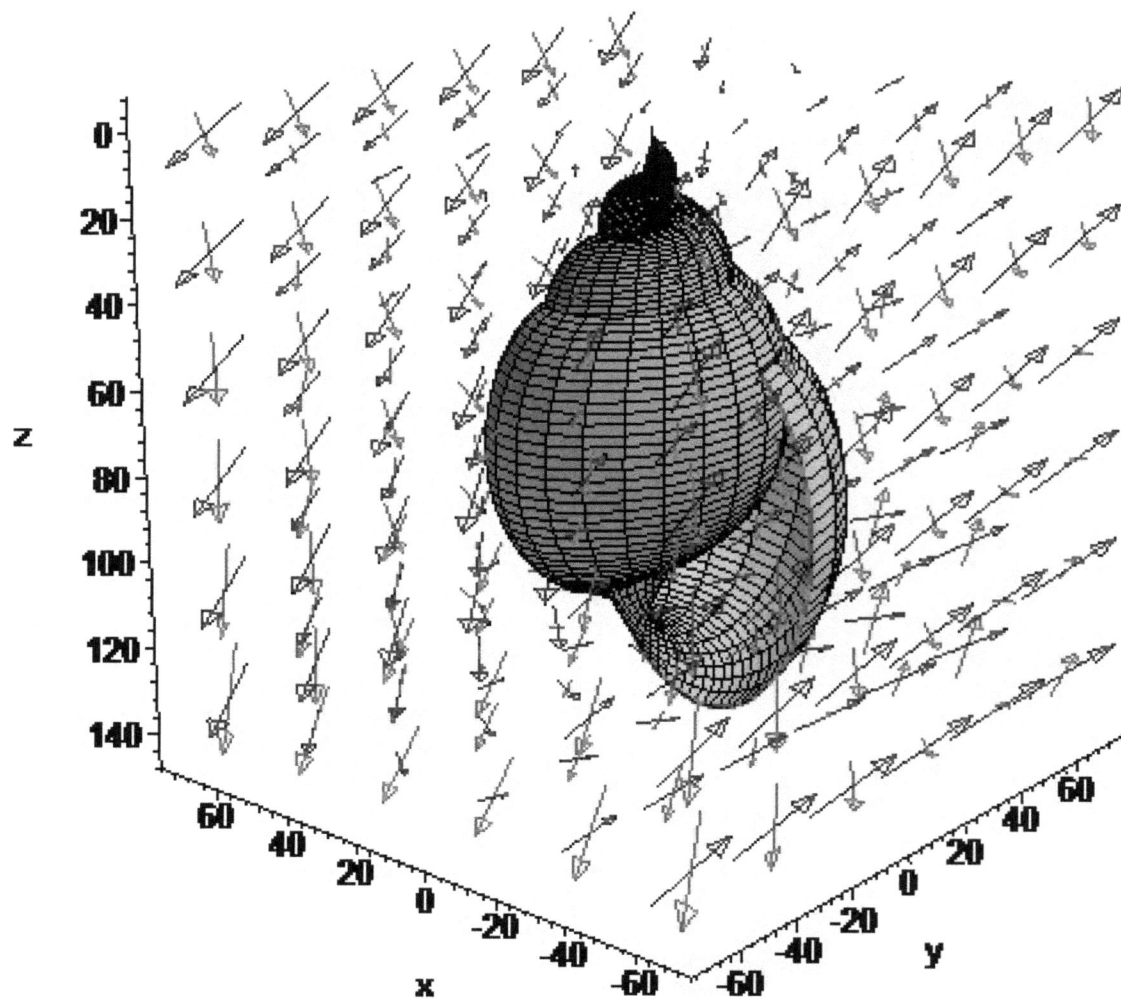

Figure(III)4.2.1 Discretional Unclosed Surface 'CS'
and its Boundary Curve 'CL'
Vector Field 'V'(Red arrows) and its Curl 'cV'(Blue arrows)

```
> display(g2,g3,g4);
```

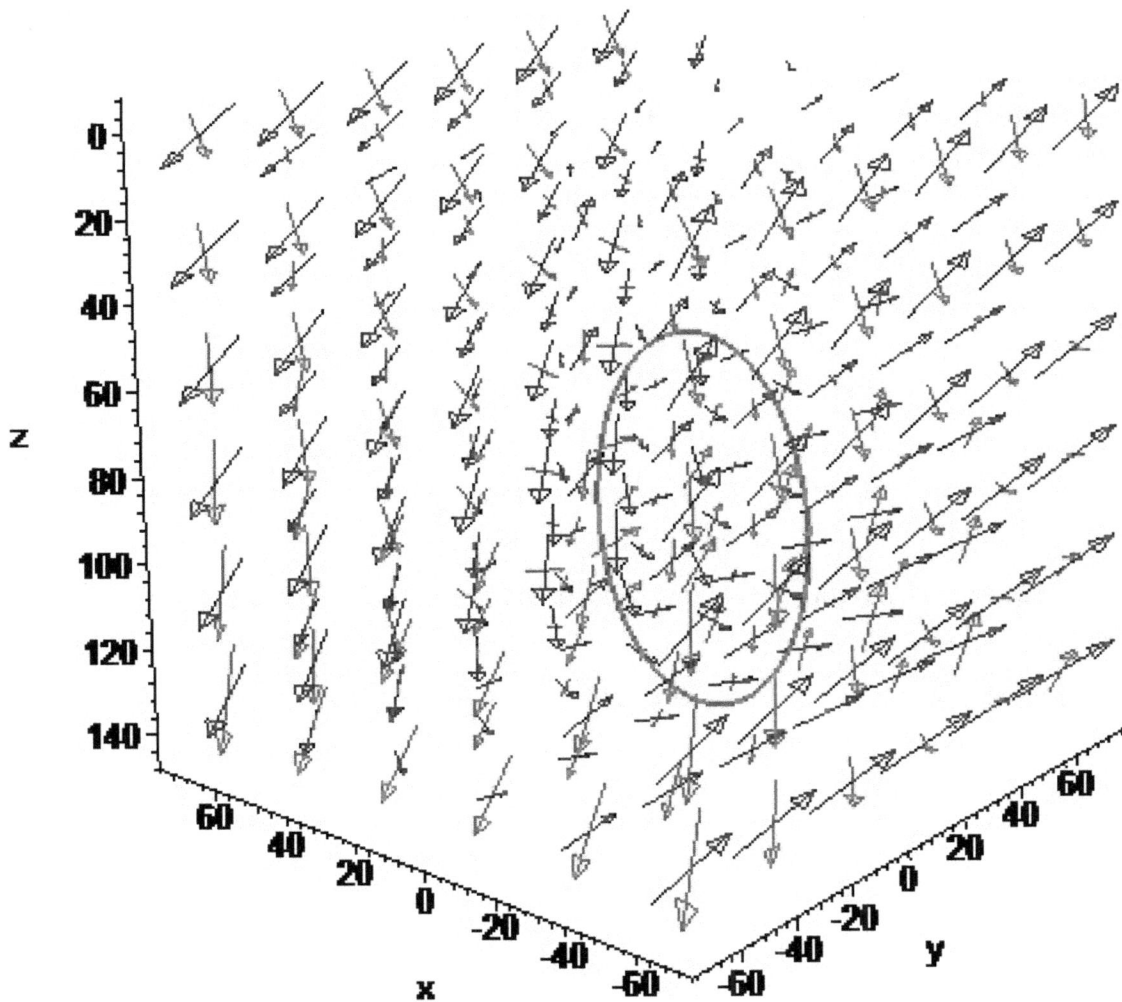

Figure(III)4.2.2 Closed Boundary Curve 'CL'
Vector Field 'V'(Red arrows) and its Curl'cV'(Blue arrows)

```
> x:=CL[1]:y:=CL[2]:z:=CL[3]:
> Int(V[1]*Diff(CL[1],u)+V[2]*Diff(CL[2],u)+V[3]*Diff(CL[3],u),
u=rgu[1]..rgu[2]);
# Closed Curve Integral (Vector Field 'V' at Closed Boundary
Curve 'CL')
```

$$\int_0^{2\pi} \frac{1}{6} \left(-\frac{14}{9}\pi\,(1+2\cos(u)) - \frac{7}{3}\pi\,(2-\sin(u)) \right)^2 \left(\frac{\partial}{\partial u}\left(-\frac{14}{3}\pi\,(1+2\cos(u)) \right) \right)$$

$$+ \frac{196}{45}\pi^2\,(1+2\cos(u))^2 \left(\frac{d}{du}\,0 \right) + \frac{196}{27}\pi^2\,(1+2\cos(u))^2 \left(\frac{\partial}{\partial u}\,(14\pi\,(2-\sin(u))) \right)$$

$$du$$

```
> alpha:=value(%);delta:=evalf(alpha);
```

Calculate value of spacial closed curve integral

$$\alpha := -\frac{109760\,\pi^4}{243}$$

$$\delta := -43998.44377$$

```
> x:='x':y:='y':z:='z':
> x:=CS[1]:y:=CS[2]:z:=CS[3]:
> matrix(3,3,[i,j,k,Diff(CS[1],u),Diff(CS[2],u),Diff(CS[3],u),
Diff(CS[1],v),Diff(CS[2],v),Diff(CS[3],v)])=
matrix(3,3,[i,j,k,diff(CS[1],u),diff(CS[2],u),diff(CS[3],u),
diff(CS[1],v),diff(CS[2],v),diff(CS[3],v)]); m1:=rhs(%);
```

Define and calculate matrix of partial derivatives 'm1'; obtain
'tangent plane normal vector' of unclosed surface 'CS'

$$\begin{bmatrix} i, & j, & k \\ \dfrac{\partial}{\partial u}\left(\dfrac{2}{3}v\,(1+2\cos(u))\cos(v) \right), & \dfrac{\partial}{\partial u}\left(\dfrac{2}{3}v\,(1+2\cos(u))\sin(v) \right), & \dfrac{\partial}{\partial u}\,(2v\,(2-\sin(u))) \\ \dfrac{\partial}{\partial v}\left(\dfrac{2}{3}v\,(1+2\cos(u))\cos(v) \right), & \dfrac{\partial}{\partial v}\left(\dfrac{2}{3}v\,(1+2\cos(u))\sin(v) \right), & \dfrac{\partial}{\partial v}\,(2v\,(2-\sin(u))) \end{bmatrix}$$

$$=$$

$$[i,j,k]$$

$$\left[-\frac{4}{3}v\sin(u)\cos(v), -\frac{4}{3}v\sin(u)\sin(v), -2v\cos(u) \right]$$

$$\left[\frac{2}{3}(1+2\cos(u))\cos(v) - \frac{2}{3}v\,(1+2\cos(u))\sin(v), \right.$$

$$\left. \frac{2}{3}(1+2\cos(u))\sin(v) + \frac{2}{3}v\,(1+2\cos(u))\cos(v), 4-2\sin(u) \right]$$

$$m1 :=$$

$$[i,j,k]$$

$$\left[-\frac{4}{3}v\sin(u)\cos(v), -\frac{4}{3}v\sin(u)\sin(v), -2v\cos(u) \right]$$

$$\left[\frac{2}{3}(1+2\cos(u))\cos(v) - \frac{2}{3}v\,(1+2\cos(u))\sin(v), \right.$$

$$\left. \frac{2}{3}(1+2\cos(u))\sin(v) + \frac{2}{3}v\,(1+2\cos(u))\cos(v), 4-2\sin(u) \right]$$

```
> det(m1):

> mn:=simplify(%);
```

$$mn := -\frac{4}{9}v\,(-6\,v\cos(u)^2\sin(v)\,j - 6\,i\,v\cos(u)^2\cos(v) + 4\,v\cos(u)\,k\sin(u)$$
$$-3\,v\sin(v)\,j\cos(u) - 3\,i\,v\cos(u)\cos(v) + 12\,i\sin(u)\sin(v) - 12\sin(u)\cos(v)\,j$$
$$+2\,v\,k\sin(u) - 3\,i\cos(u)\sin(v) + 3\cos(v)\,j\cos(u) - 6\,i\sin(v) + 6\cos(v)\,j)$$

```
> A:=coeff(mn,i); # Obtain coefficient of 'i'
```

$$A := -\frac{4}{9}v\,(-6\,v\cos(u)^2\cos(v) - 3\,v\cos(u)\cos(v) + 12\sin(u)\sin(v) - 3\sin(v)\cos(u)$$
$$-6\sin(v))$$

```
> B:=coeff(mn,j); # Obtain coefficient of 'j'
```

$$B := -\frac{4}{9}v\,(-6\,v\cos(u)^2\sin(v) - 3\,v\cos(u)\sin(v) - 12\sin(u)\cos(v) + 3\cos(u)\cos(v)$$
$$+6\cos(v))$$

```
> C:=coeff(mn,k); # Obtain coefficient of 'k'
```

$$C := -\frac{4}{9}v\,(4\,v\cos(u)\sin(u) + 2\,v\sin(u))$$

```
> [A,B,C]: # [A,B,C] structure 'tangent plane normal vector'

> -Int(Int(cV[1]*A+cV[2]*B+cV[3]*C,u=rgu[1]..rgu[2]),
  v=rgv[1]..rgv[2]);

# Integral of dot product (Curl 'cV' and tangent plane normal vector
'[A,B,C]') in range of 'u,v'
```

$$-\int_0^{7\pi}\int_0^{2\pi} -\frac{4}{9}v\,(-6\,v\cos(u)^2\cos(v) - 3\,v\cos(u)\cos(v) + 12\sin(u)\sin(v)$$
$$-3\sin(v)\cos(u) - 6\sin(v))\left(-\frac{2}{5}v\,(2-\sin(u)) - \frac{1}{9}v\,(1+2\cos(u))\sin(v)\right) - \frac{4}{9}v\,($$
$$-6\,v\cos(u)^2\sin(v) - 3\,v\cos(u)\sin(v) - 12\sin(u)\cos(v) + 3\cos(u)\cos(v)$$
$$+6\cos(v))$$
$$\left(-\frac{37}{81}v\,(1+2\cos(u))\cos(v) + \frac{1}{81}v\,(1+2\cos(u))\sin(v) + \frac{1}{54}v\,(2-\sin(u))\right) - \frac{4}{9}$$
$$v\,(4\,v\cos(u)\sin(u) + 2\,v\sin(u))$$
$$\left(\frac{118}{405}v\,(1+2\cos(u))\cos(v) - \frac{2}{81}v\,(1+2\cos(u))\sin(v) - \frac{1}{27}v\,(2-\sin(u))\right)du\,d$$
$$v$$

```
> beta:=value(%);epsilon:=evalf(beta);

# Calculate value of surface integral
```

$$\beta := -\frac{109760\,\pi^4}{243}$$

$$\varepsilon := -43998.44377$$

```
> alpha;beta; # Two analytic values are equal
```

$$-\frac{109760\,\pi^4}{243}$$

$$-\frac{109760\,\pi^4}{243}$$

```
> delta;epsilon; # Two float values are equal
```

$$-43998.44377$$

$$-43998.44377$$

```
> Int(Int(sqrt(A^2+B^2+C^2),u=rgu[1]..rgu[2]),v=rgv[1]..rgv[2]);
```

Surface Integral of 'Modulus of Tangent Plane Normal Vector [A,B,C]' in range of 'u,v', absolute value of result is area value of unclosed surface (1):

$$\int_0^{7\pi}\int_0^{2\pi} \frac{4}{9}\left(v^2\left(-6\,v\cos(u)^2\cos(v)-3\,v\cos(u)\cos(v)+12\sin(u)\sin(v)\right.\right.$$

$$\left.\left.-3\sin(v)\cos(u)-6\sin(v)\right)^2 + v^2\left(-6\,v\cos(u)^2\sin(v)-3\,v\cos(u)\sin(v)\right.\right.$$

$$\left.\left.-12\sin(u)\cos(v)+3\cos(u)\cos(v)+6\cos(v)\right)^2\right.$$

$$\left.+v^2\left(4\,v\cos(u)\sin(u)+2\,v\sin(u)\right)^2\right)^{(1/2)}\quad du\ dv$$

```
> evalf(%);
```

$$38705.23446$$

```
> x:='x':y:='y':z:='z':
> [r*CS[1],r*CS[2],r*CS[3]];
```

Entirely multiply radius vector 'r' (suppose r>0) by each item of surface 'CS' parametrized expression, Convert "Parametrized expression of surface 'CS'" to "Parametrized expression of surface 'CS' coordinates"

$$\left[\frac{2}{3}r\,v\,(1+2\cos(u))\cos(v),\frac{2}{3}r\,v\,(1+2\cos(u))\sin(v),2\,r\,v\,(2-\sin(u))\right]$$

```
> x:=r*CS[1]:y:=r*CS[2]:z:=r*CS[3]:
> matrix(3,3,[Diff(r*CS[1],r),Diff(r*CS[1],u),Diff(r*CS[1],v),
   Diff(r*CS[2],r),Diff(r*CS[2],u),Diff(r*CS[2],v),
   Diff(r*CS[3],r),Diff(r*CS[3],u),Diff(r*CS[3],v)])=
   matrix(3,3,[diff(r*CS[1],r),diff(r*CS[1],u),diff(r*CS[1],v),
   diff(r*CS[2],r),diff(r*CS[2],u),diff(r*CS[2],v),
```

```
diff(r*CS[3],r),diff(r*CS[3],u),diff(r*CS[3],v)]);m2:=rhs(%);
```

Define and calculate matrix of partial derivatives 'm2', obtain general expression of surface 'CS' coordinates volume element coefficient

$$
\left[\frac{\partial}{\partial r}\left(\frac{2}{3}\,r\,v\,(1+2\cos(u))\cos(v)\right), \frac{\partial}{\partial u}\left(\frac{2}{3}\,r\,v\,(1+2\cos(u))\cos(v)\right),\right.
$$

$$
\left.\frac{\partial}{\partial v}\left(\frac{2}{3}\,r\,v\,(1+2\cos(u))\cos(v)\right)\right]
$$

$$
\left[\frac{\partial}{\partial r}\left(\frac{2}{3}\,r\,v\,(1+2\cos(u))\sin(v)\right), \frac{\partial}{\partial u}\left(\frac{2}{3}\,r\,v\,(1+2\cos(u))\sin(v)\right),\right.
$$

$$
\left.\frac{\partial}{\partial v}\left(\frac{2}{3}\,r\,v\,(1+2\cos(u))\sin(v)\right)\right]
$$

$$
\left[\frac{\partial}{\partial r}\left(2\,r\,v\,(2-\sin(u))\right), \frac{\partial}{\partial u}\left(2\,r\,v\,(2-\sin(u))\right), \frac{\partial}{\partial v}\left(2\,r\,v\,(2-\sin(u))\right)\right] =
$$

$$
\left[\frac{2}{3}\,v\,(1+2\cos(u))\cos(v), -\frac{4}{3}\,r\,v\,\sin(u)\cos(v),\right.
$$

$$
\left.\frac{2}{3}\,r\,(1+2\cos(u))\cos(v) - \frac{2}{3}\,r\,v\,(1+2\cos(u))\sin(v)\right]
$$

$$
\left[\frac{2}{3}\,v\,(1+2\cos(u))\sin(v), -\frac{4}{3}\,r\,v\,\sin(u)\sin(v),\right.
$$

$$
\left.\frac{2}{3}\,r\,(1+2\cos(u))\sin(v) + \frac{2}{3}\,r\,v\,(1+2\cos(u))\cos(v)\right]
$$

$$
\left[2\,v\,(2-\sin(u)), -2\,r\,v\,\cos(u), 2\,r\,(2-\sin(u))\right]
$$

$$
m2 :=
$$

$$
\left[\frac{2}{3}\,v\,(1+2\cos(u))\cos(v), -\frac{4}{3}\,r\,v\,\sin(u)\cos(v),\right.
$$

$$
\left.\frac{2}{3}\,r\,(1+2\cos(u))\cos(v) - \frac{2}{3}\,r\,v\,(1+2\cos(u))\sin(v)\right]
$$

$$
\left[\frac{2}{3}\,v\,(1+2\cos(u))\sin(v), -\frac{4}{3}\,r\,v\,\sin(u)\sin(v),\right.
$$

$$
\left.\frac{2}{3}\,r\,(1+2\cos(u))\sin(v) + \frac{2}{3}\,r\,v\,(1+2\cos(u))\cos(v)\right]
$$

$$
\left[2\,v\,(2-\sin(u)), -2\,r\,v\,\cos(u), 2\,r\,(2-\sin(u))\right]
$$

```
> det(m2):
```

```
> J:=simplify(%);
```

$$
J := -\frac{8}{9}\,r^2\,v^3\,(8\cos(u)\sin(u) - 2\cos(u)^2 + 4\sin(u) - 5\cos(u) - 2)
$$

```
> Int(Int(Int(J,r=0..1),u=rgu[1]..rgu[2]),v=rgv[1]..rgv[2]);
```

Triple Integrals of "Surface 'CS' Coordinates Volume Element" in range of 'r,u,v', absolute value of result is cubage value of unclosed surface 'CS':

$$\int_{0}^{7\pi}\int_{0}^{2\pi}\int_{0}^{1} -\frac{8}{9}r^2 v^3 \left(8\cos(u)\sin(u) - 2\cos(u)^2 + 4\sin(u) - 5\cos(u) - 2\right) dr\, du\, dv$$

> value(%);evalf(%); # Cubage Integral Value of Unclosed Surface'CS'

$$\frac{9604\,\pi^5}{9}$$

$$326557.0060$$

Chapter 5 Prove Curl Theorem Finite Sums Limits at Manifold.

5.1 Prove Curl Theorem Finite Sums Limits at Manifold

Curl Theorem Suppose positive directional boundary 'L+' of smooth or piecewise smooth orientational surface 'S' is smooth or piecewise smooth closed curve, Right Hand Rule is associated basis between positive direction of boundary curve 'L' and outer side of orientational surface 'S'.If there are 1th order continuous partial derivatives about functions 'P(x,y,z),Q(x,y,z),R(x,y,z)' [Structure vector field 'A'] at orientational surface 'S', then:

$$\int_{L+} A\cdot dL = \iint_{S} rotA\cdot n\, dS \qquad (1)$$

thereinto, 'rotA' is curl of vector field 'A', 'n' is unit normal vector of orientational surface 'S's outer side.

Proof (Finite Sums Limits):

Define parametrized expression of abstract simply connected orientable closed surface 'S':

  [a*sin(u)cos(v),b*sin(u)sin(v),c*cos(u)]   (2)

thereinto,'a,b,c' are nonzero constants or 1th order derivable continuous functions, simply connected orientable closed surface 'S' determines value of 'a,b,c';

Set Range of parameters 'u,v' in $[0, \pi/n - \theta]$, $[0, 2\pi]$, thereinto, 'n' is discretional constant, and n ≥ 1; 'θ'is discretional constant or continuous function, and $\pi/n - \theta < \pi$, make surface 'S' unclosed. (See also Poincare Conjecture: 'Every closed n-Manifold which is homotopy equivalent to n-Sphere is homeomorphic to n-Sphere'[19], and

Chapter 1 (This Part) Section 1.2 'Define Abstract Simply Connected Orientable Closed Surface and Its Unclosed Surface')

Define parametrized expression of boundary curve 'L'; Set Right Hand Rule as associated basis between positive direction of boundary curve 'L' and normal vector of unclosed surface 'S'

$[\alpha \cos(v), \beta \sin(v), \gamma]$ (3)

thereinto, 'α, β, γ' are constants($\alpha \neq 0$, $\beta \neq 0$) or 1th order derivable continuous functions depend on 'a,b,c'; owing to range of parameter 'v' in $[0, 2\pi]$, boundary curve 'L' is closed.

(See also Chapter 1 (This Part) Section 1.5 'From Abstract Plane Simply Connected Closed Curve, To Abstract Space Closed Curve')

Calculate tangent vector of closed boundary curve 'L' (4):

$$\left[\frac{\partial}{\partial v}(\alpha \cos(v)), \frac{\partial}{\partial v}(\beta \sin(v)), \frac{\partial}{\partial v}\gamma\right] = [-\alpha \sin(v), \beta \cos(v), 0]$$
(4)

Set Amount of closed boundary curve 'L's parametrized segmental cells as 50 (It can be discretional natural number) (5)

1.Microcosmic Curve Integral Course at Closed Boundary Curve 'L's first segmental cell:

Segment range $[0, 2\pi]$ of 'v': dv $= \dfrac{2\pi}{50} = \dfrac{\pi}{25}$ (6)

Segment Tangent Vector (7): [Viz. Input (6) to (4)]

$$\left[-\alpha \sin\left(\frac{\pi}{25}\right), \beta \cos\left(\frac{\pi}{25}\right), 0\right]$$
(7)

Segment Abstract Vector Field [P(x,y,z),Q(x,y,z),R(x,y,z)]:

[P(x,y,z),Q(x,y,z),R(x,y,z)]   (8)

// Owing to universality and homogeneity of abstract vector field [P(x,y,z),Q(x,y,z),R(x,y,z)], its value at first segment cell is [P(x,y,z),Q(x,y,z),R(x,y,z)]; See also Section 4.3 Chapter 4 (Part I) 'Discuss About Abstract Vector Field、Finite Sums Limit、Integral and Riemann Sums'

Calculate Microcosmic Curve Integral Value at Closed Boundary
Curve 'L's first segmental cell (9):

Base on integral median theorem, dot product of abstract vector
field (8) and tangent vector (7), times segment unit of 'v'(6), that
is Microcosmic Curve Integral Value at first segmental cell (9)

$$\frac{1}{25}\pi\left(-P(x,y,z)\,\alpha\,\sin\left(\frac{\pi}{25}\right)+Q(x,y,z)\,\beta\,\cos\left(\frac{\pi}{25}\right)\right) \quad (9)$$

2.Microcosmic Curve Integral Course at Closed Boundary Curve 'L's
all (Viz. 50) segmental cells:

Segment range [0,2π] of 'v': dv = $\frac{i2\pi}{50}=\frac{i\pi}{25}$

(thereinto, 'i' stand for natural number 1~50)   (10)

Segment Tangent Vector (11): [Viz. Input (10) to (4)]

$$\left[-\alpha\,\sin\left(\frac{\pi i}{25}\right),\,\beta\,\cos\left(\frac{\pi i}{25}\right),\,0\right] \quad (11)$$

// Expression (11) isn't that 'single vector value', but that 'Set
of finite quantitative (viz. 50) vector values'

Segment Abstract Vector Field [P(x,y,z),Q(x,y,z),R(x,y,z)]:
        [P(x,y,z),Q(x,y,z),R(x,y,z)]   (12)
// Owing to universality and homogeneity of abstract vector field
[P(x,y,z),Q(x,y,z),R(x,y,z)], its values at umpty segment cells are
[P(x,y,z),Q(x,y,z),R(x,y,z)]; See also Section 4.3 Chapter 4 (Part I)
'Discuss About Abstract Vector Field、Finite Sums Limit、Integral and
Riemann Sums'

Calculate Microcosmic Curve Integral Values at Closed Boundary
Curve 'L's all segmental cells (13):

Base on Integral Median Theorem, dot product of abstract vector
field (12) and tangent vector (11), times segment unit of 'v'(10),
these are Microcosmic Curve Integral Values at all segmental
cells (13)

$$\frac{1}{25}\pi\left(-P(x,y,z)\,\alpha\,\sin\left(\frac{i\pi}{25}\right)+Q(x,y,z)\,\beta\,\cos\left(\frac{i\pi}{25}\right)\right) \quad (13)$$

// Expression (13) isn't that 'single value', but that 'set of finite quantitative (viz.50) values'

Structure Sequence that is composed of finite quantitative (viz. 50) Microcosmic Curve Integral Values (14):

[Lengthy expression of sequence is elided;

In program template of Waterloo Maplesoft (see also Section 5.2):

```
sqn:=seq(dv*(idV[1]*idCL[1]+idV[2]*idCL[2]+idV[3]*idCL[3]),
i=1..dus):
```

change ':'(last) to ';', then obtain expression of sequence]

Accumulational sum of sequence (viz. Sum of integral values of dot product [abstract vector field (12) and tangent vector (11) at all (Viz. 50) segmental cells], obtain Curve Integral Value at manifold (15):

[Lengthy expression of accumulational sum result is elided;

In program template of Waterloo Maplesoft (see also Section 5.2):

```
add(k,k=sqn):xi:=evalf(%);
```

change ':'(middle) to ';', then obtain expression of result]

Change expression of accumulational sum result to float value:

$$0.9 \ 10^{-9} \ P(x,y,z) \ \alpha - 0.610 \ 10^{-9} \ Q(x,y,z) \ \beta$$

// Amount of Closed Boundary Curve 'L's segmental cells increases infinitely, sum of integral values above-mentioned tends to '0' infinitely

Set Amount of closed boundary curve 'L's parametrized segmental cells as natural number 'w'   (16)

3.Microcosmic Curve Integral Course at Closed Boundary Curve 'L's first segmental cell:

Segment range $[0,2\pi]$ of 'v': $dv = \dfrac{2\pi}{w}$ (17)

Segment Tangent Vector (18): [Viz. Input (17) to (4)]

$$\left[-\alpha \sin\left(\frac{2\pi}{w}\right), \beta \cos\left(\frac{2\pi}{w}\right), 0 \right]$$

                                                        (18)

Segment Abstract Vector Field [P(x,y,z),Q(x,y,z),R(x,y,z)]:

$$[P(x,y,z),Q(x,y,z),R(x,y,z)] \qquad (19)$$

// Owing to universality and homogeneity of abstract vector field [P(x,y,z),Q(x,y,z),R(x,y,z)], its value at first segment cell is [P(x,y,z),Q(x,y,z),R(x,y,z)]; See also Section 4.3 Chapter 4 (Part I) 'Discuss About Abstract Vector Field、Finite Sums Limit、Integral and Riemann Sums'

Calculate Microcosmic Curve Integral Value at Closed Boundary Curve 'L's first segmental cell (20):

Base on Integral Median Theorem, dot product of abstract vector field (19) and tangent vector (18), times segment unit of 'v'(17), that is Microcosmic Curve Integral Value at first segmental cell (20)

$$\frac{2\pi\left(-P(x,y,z)\,\alpha\,\sin\left(\dfrac{2\pi}{w}\right)+Q(x,y,z)\,\beta\,\cos\left(\dfrac{2\pi}{w}\right)\right)}{w} \qquad (20)$$

4.Microcosmic Curve Integral Course at Closed Boundary Curve 'L's all (Viz.50) segmental cells:

Segment range [0,2π] of 'v': dv = $\dfrac{2\pi i}{w}$ \qquad (21)

(thereinto, 'i' stand for natural number 1~w)

Segment Tangent Vector (22): [Viz. Input (21) to (4)]

$$\left[-\alpha\,\sin\left(\frac{2\pi i}{w}\right),\ \beta\,\cos\left(\frac{2\pi i}{w}\right),\ 0\right] \qquad (22)$$

// Expression (22) isn't that 'single vector value', but that 'Set of finite quantitative (viz. w) vector values'

Segment Abstract Vector Field [P(x,y,z),Q(x,y,z),R(x,y,z)]:

$$[P(x,y,z),Q(x,y,z),R(x,y,z)] \qquad (23)$$

// Owing to universality and homogeneity of abstract vector field [P(x,y,z),Q(x,y,z),R(x,y,z)], its values at umpty segment cells are [P(x,y,z),Q(x,y,z),R(x,y,z)]; See also Section 4.3 Chapter 4 (Part I) 'Discuss About Abstract Vector Field、Finite Sums Limit、Integral and Riemann Sums'

Calculate Microcosmic Curve Integral Values at Closed Boundary Curve 'L's all segmental cells (24):

Base on Integral Median Theorem, dot product of abstract vector field (23) and tangent vector (22), times segment unit of 'v' (21), these are Microcosmic Curve Integral Values at all segmental cells (24)

$$\frac{2\pi\left(-P(x,y,z)\,\alpha\,\sin\left(\frac{2\,i\,\pi}{w}\right)+Q(x,y,z)\,\beta\,\cos\left(\frac{2\,i\,\pi}{w}\right)\right)}{w} \tag{24}$$

// Expression (24) isn't that 'single value', but that 'set of finite quantitative (viz. w) values'

Structure finite sums (25):

(In the case of "Amount 'w' of closed boundary curve 'L's segmental cells is uncertain", sum of integral values [dot product about absract vector field (23) and tangent vector (22)] at all segmental cells)

$$\sum_{i=1}^{w}\left(\frac{2\pi\left(-P(x,y,z)\,\alpha\,\sin\left(\frac{2\,\pi\,i}{w}\right)+Q(x,y,z)\,\beta\,\cos\left(\frac{2\,\pi\,i}{w}\right)\right)}{w}\right) \tag{25}$$

Infinitize finite sums, its limit operational value is 'Curl Integral Value at Manifold' (26):

(In the case of "Amount 'w' of closed boundary curve 'L's segmental cells tends to infinity", limit value of integral values' sums [dot product about abstract vector field (23) and tangent vector (22)] at all segmental cells)

$$\lim_{w\to\infty}\sum_{i=1}^{w}\left(\frac{2\pi\left(-P(x,y,z)\,\alpha\,\sin\left(\frac{2\,\pi\,i}{w}\right)+Q(x,y,z)\,\beta\,\cos\left(\frac{2\,\pi\,i}{w}\right)\right)}{w}\right)$$

$$=$$

$$\lim_{w\to\infty}-2\pi\cos\left(\frac{\pi}{w}\right)^2\left(P(x,y,z)\,\alpha\,\sin\left(\frac{\pi}{w}\right)\cos\left(\frac{\pi}{w}\right)^3-P(x,y,z)\,\alpha\,\sin\left(\frac{\pi}{w}\right)\cos\left(\frac{\pi}{w}\right)\right.$$
$$\left.+Q(x,y,z)\,\beta\,\cos\left(\frac{\pi}{w}\right)^4-2Q(x,y,z)\,\beta\,\cos\left(\frac{\pi}{w}\right)^2+Q(x,y,z)\,\beta\right.$$

$$+ \sin\left(\frac{\pi}{w}\right)^3 \cos\left(\frac{\pi}{w}\right) P(x,y,z)\,\alpha + \sin\left(\frac{\pi}{w}\right)^2 \cos\left(\frac{\pi}{w}\right)^2 Q(x,y,z)\,\beta$$

$$\left. - \sin\left(\frac{\pi}{w}\right)^2 Q(x,y,z)\,\beta\right)(w+1)\ \Bigg/\ \left(\left(\left(\cos\left(\frac{\pi}{w}\right)^2 - 1\right)^2 w\right) - 2\pi\left(\vphantom{\frac{\pi}{w}}\right.\right.$$

$$\sin\left(\frac{\pi}{w}\right)^2 \cos\left(\frac{\pi}{w}\right)^2 P(x,y,z)\,\alpha + \sin\left(\frac{\pi}{w}\right)\cos\left(\frac{\pi}{w}\right)^3 Q(x,y,z)\,\beta$$

$$\left. - \sin\left(\frac{\pi}{w}\right)\cos\left(\frac{\pi}{w}\right)Q(x,y,z)\,\beta + P(x,y,z)\,\alpha\cos\left(\frac{\pi}{w}\right)^2 - P(x,y,z)\,\alpha\right)\sin\left(\frac{(w+1)\pi}{w}\right)$$

$$\cos\left(\frac{(w+1)\pi}{w}\right)\ \Bigg/\ \left(\left(\cos\left(\frac{\pi}{w}\right)^2 - 1\right)^2 w\right) - \frac{2\pi\left(P(x,y,z)\,\alpha\sin\left(\frac{\pi}{w}\right)\cos\left(\frac{\pi}{w}\right) + Q(x,y,z\right.}{w\left(\cos\left(\vphantom{\frac{\pi}{w}}\right.\right.}$$

$$+ 2\pi\cos\left(\frac{\pi}{w}\right)^2\left(P(x,y,z)\,\alpha\sin\left(\frac{\pi}{w}\right)\cos\left(\frac{\pi}{w}\right)^3 - P(x,y,z)\,\alpha\sin\left(\frac{\pi}{w}\right)\cos\left(\frac{\pi}{w}\right)\right.$$

$$+ Q(x,y,z)\,\beta\cos\left(\frac{\pi}{w}\right)^4 - 2\,Q(x,y,z)\,\beta\cos\left(\frac{\pi}{w}\right)^2 + Q(x,y,z)\,\beta$$

$$+ \sin\left(\frac{\pi}{w}\right)^3 \cos\left(\frac{\pi}{w}\right) P(x,y,z)\,\alpha + \sin\left(\frac{\pi}{w}\right)^2 \cos\left(\frac{\pi}{w}\right)^2 Q(x,y,z)\,\beta$$

$$\left. - \sin\left(\frac{\pi}{w}\right)^2 Q(x,y,z)\,\beta\right)\ \Bigg/\ \left(\left(\left(\cos\left(\frac{\pi}{w}\right)^2 - 1\right)^2 w\right) + 2\pi\left(\vphantom{\frac{\pi}{w}}\right.\right.$$

$$\sin\left(\frac{\pi}{w}\right)^2 \cos\left(\frac{\pi}{w}\right)^2 P(x,y,z)\,\alpha + \sin\left(\frac{\pi}{w}\right)\cos\left(\frac{\pi}{w}\right)^3 Q(x,y,z)\,\beta$$

$$\left. - \sin\left(\frac{\pi}{w}\right)\cos\left(\frac{\pi}{w}\right)Q(x,y,z)\,\beta + P(x,y,z)\,\alpha\cos\left(\frac{\pi}{w}\right)^2 - P(x,y,z)\,\alpha\right)\sin\left(\frac{\pi}{w}\right)$$

$$\cos\left(\frac{\pi}{w}\right)\ \Bigg/\ \left(\left(\cos\left(\frac{\pi}{w}\right)^2 - 1\right)^2 w\right)$$

$$+ \frac{2\pi\left(P(x,y,z)\,\alpha\sin\left(\frac{\pi}{w}\right)\cos\left(\frac{\pi}{w}\right) + Q(x,y,z)\,\beta\cos\left(\frac{\pi}{w}\right)^2 - Q(x,y,z)\,\beta\right)\cos\left(\frac{\pi}{w}\right)^2}{w\left(\cos\left(\frac{\pi}{w}\right)^2 - 1\right)}$$

$$= 0 \tag{26}$$

//'Integral value equals 0' possesses specific mathematical and physical meaning:

In mathematical meaning, 'Integral value equals 0' is inevitable result of logic illation, incarnates logic balanced state between various integral elements;

In physical meaning, 'Integral value equals 0' implies that circumfluent value of 'Abstract Vector Field at Abstract Spacial Closed Curve(Route)' is constant for '0'(immobile and pending) identically; If integral value equals certain a positive/negative number or an expression, it implies that always there is positive/reverse directional flux or unknown flux about 'Abstract Spacial Vector Field at Abstract Spacial Closed Curve(Route)', this instance will be unaccountable

About 'Integral value is unequal to 0', See also Chapter 6 (Part III) Section 6.1 'Curl Theorem at Manifold and Mobius Strip'

According to surface parametrized expression (2), define and calculate matrix of partial derivatives (27), obtain 'tangent plane normal vector' of surface 'S'

$$
\begin{bmatrix}
i & j & k \\
\dfrac{\partial}{\partial u} a\sin(u)\cos(v) & \dfrac{\partial}{\partial u} b\sin(u)\sin(v) & \dfrac{\partial}{\partial u} c\cos(u) \\
\dfrac{\partial}{\partial v} a\sin(u)\cos(v) & \dfrac{\partial}{\partial v} b\sin(u)\sin(v) & \dfrac{\partial}{\partial v} c\cos(u)
\end{bmatrix} =
$$

$$
i\,c\sin(u)^2 b\cos(v) + a\cos(u)\cos(v)^2 k\,b\sin(u) + a\sin(u)^2\sin(v)\,j\,c
$$
$$
+ a\sin(u)\sin(v)^2 k\,b\cos(u) \tag{27}
$$

From expression (27), respectively pick up coefficient of item 'i,j,k', obtain tangent plane normal vector of surface 'S' (28):

$$
[\,c\sin(u)^2 b\cos(v)\,,\ \sin(u)^2 a\sin(v)c\,,\ \sin(u)ab\cos(u)\,] \tag{28}
$$

Calculate Curl of Abstract Vector Field [P(x,y,z),Q(x,y,z), R(x,y,z)], and convert it from Cartesian Coordinates expression (29) to Surface 'S's Coordinates expression (30):

$$
\left[\left(\frac{\partial}{\partial y}R(x,y,z)\right)-\left(\frac{\partial}{\partial z}Q(x,y,z)\right),\left(\frac{\partial}{\partial z}P(x,y,z)\right)-\left(\frac{\partial}{\partial x}R(x,y,z)\right),\right.
$$
$$
\left.\left(\frac{\partial}{\partial x}Q(x,y,z)\right)-\left(\frac{\partial}{\partial y}P(x,y,z)\right)\right] \tag{29}
$$

$$\left[\left(\frac{\partial}{\partial y}R(x,y,z)\right)\left(\frac{\partial}{\partial u}(b\,\sin(u)\,\sin(v))\right)\left(\frac{\partial}{\partial v}(b\,\sin(u)\,\sin(v))\right)\right.$$

$$-\left(\frac{\partial}{\partial z}Q(x,y,z)\right)\left(\frac{\partial}{\partial u}(c\,\cos(u))\right)\left(\frac{\partial}{\partial v}(c\,\cos(u))\right),$$

$$\left(\frac{\partial}{\partial z}P(x,y,z)\right)\left(\frac{\partial}{\partial u}(c\,\cos(u))\right)\left(\frac{\partial}{\partial v}(c\,\cos(u))\right)$$

$$-\left(\frac{\partial}{\partial x}R(x,y,z)\right)\left(\frac{\partial}{\partial u}(a\,\sin(u)\,\cos(v))\right)\left(\frac{\partial}{\partial v}(a\,\sin(u)\,\cos(v))\right),$$

$$\left(\frac{\partial}{\partial x}Q(x,y,z)\right)\left(\frac{\partial}{\partial u}(a\,\sin(u)\,\cos(v))\right)\left(\frac{\partial}{\partial v}(a\,\sin(u)\,\cos(v))\right)$$

$$\left.-\left(\frac{\partial}{\partial y}P(x,y,z)\right)\left(\frac{\partial}{\partial u}(b\,\sin(u)\,\sin(v))\right)\left(\frac{\partial}{\partial v}(b\,\sin(u)\,\sin(v))\right)\right]=\left[\right.$$

$$\left(\frac{\partial}{\partial y}R(x,y,z)\right)b^2\cos(u)\sin(v)\sin(u)\cos(v),$$

$$\left(\frac{\partial}{\partial x}R(x,y,z)\right)a^2\cos(u)\cos(v)\sin(u)\sin(v),$$

$$-\left(\frac{\partial}{\partial x}Q(x,y,z)\right)a^2\cos(u)\cos(v)\sin(u)\sin(v)$$

$$\left.-\left(\frac{\partial}{\partial y}P(x,y,z)\right)b^2\cos(u)\sin(v)\sin(u)\cos(v)\right]$$

(30)

// In Spacial Cartesian Coordinates, Curl of abstract vector field
[P(x,y,z),Q(x,y,z),R(x,y,z)] is

$$\left[\left(\frac{\partial}{\partial y}R(x,y,z)\right)-\left(\frac{\partial}{\partial z}Q(x,y,z)\right),\left(\frac{\partial}{\partial z}P(x,y,z)\right)-\left(\frac{\partial}{\partial x}R(x,y,z)\right),\right.$$

$$\left.\left(\frac{\partial}{\partial x}Q(x,y,z)\right)-\left(\frac{\partial}{\partial y}P(x,y,z)\right)\right]$$

In logic deduction of this Proof, it is necessary to import curl
of abstract vector field above-mentioned to abstract simply connected
orientable closed surface coordinates.

Six elements of abstract vector field's curl:

$$\frac{\partial}{\partial y}R(x,y,z)\ ,\ \frac{\partial}{\partial z}Q(x,y,z)\ ,\ \frac{\partial}{\partial z}P(x,y,z)\ ,\ \frac{\partial}{\partial x}R(x,y,z)\ ,\ \frac{\partial}{\partial x}Q(x,y,z)\ ,\ \frac{\partial}{\partial y}P(x,y,z)$$

are abstract differential function, and their variables 'x,y,z'
contain sub-variables 'u,v'.

Surface Parametric Transform Expression between 'Spacial
Cartesian coordinates' and 'Abstract simply connected orientable
closed surface coordinates' is:

x = a*sin(u)cos(v), y = b*sin(u) sin(v), z = c*cos(u)

–'Coordinates Transform Differential Functions' (Correspond to

Three Differential Variables ' $\dfrac{\partial}{\partial x}$, $\dfrac{\partial}{\partial y}$, $\dfrac{\partial}{\partial z}$ ' of Differential Functions

$\dfrac{\partial}{\partial x} Q(x,y,z)$, $\dfrac{\partial}{\partial x} R(x,y,z)$, $\dfrac{\partial}{\partial y} P(x,y,z)$, $\dfrac{\partial}{\partial y} R(x,y,z)$, $\dfrac{\partial}{\partial z} Q(x,y,z)$, $\dfrac{\partial}{\partial z} P(x,y,z)$)

are $\dfrac{\partial}{\partial u} a\sin(u)\cos(v)\dfrac{\partial}{\partial v} a\sin(u)\cos(v)$, $\dfrac{\partial}{\partial u} b\sin(u)\sin(v)\dfrac{\partial}{\partial v} b\sin(u)\sin(v)$ and

$\dfrac{\partial}{\partial u} c\cos(u)\dfrac{\partial}{\partial v} c\cos(u)$.

Product of "Differential Function ' $\dfrac{\partial}{\partial y} R(x,y,z)$, $\dfrac{\partial}{\partial z} Q(x,y,z)$,

$\dfrac{\partial}{\partial z} P(x,y,z)$, $\dfrac{\partial}{\partial x} R(x,y,z)$, $\dfrac{\partial}{\partial x} Q(x,y,z)$, $\dfrac{\partial}{\partial y} P(x,y,z)$ ' " and "Coordinates

Transform Differential Functions" (Viz. Product of Two Differential functions) constitute "Curl in Abstract Simply Connected Orientable Closed Surface Coordinates".

//'Chain differentiation' or 'Coordinates conversion' ?

   If it is 'Chain differentiation', according to principle of 'multiply in identical chain, plus in subdivision chain', it ought to be:

$$\left[\left(\frac{\partial}{\partial y}R(x,y,z)\right)\left(\left(\frac{\partial}{\partial u}(b\sin(u)\sin(v))\right)+\left(\frac{\partial}{\partial v}(b\sin(u)\sin(v))\right)\right)\right.$$
$$-\left(\frac{\partial}{\partial z}Q(x,y,z)\right)\left(\left(\frac{\partial}{\partial u}(c\cos(u))\right)+\left(\frac{\partial}{\partial v}(c\cos(u))\right)\right),$$
$$\left(\frac{\partial}{\partial z}P(x,y,z)\right)\left(\left(\frac{\partial}{\partial u}(c\cos(u))\right)+\left(\frac{\partial}{\partial v}(c\cos(u))\right)\right)$$
$$-\left(\frac{\partial}{\partial x}R(x,y,z)\right)\left(\left(\frac{\partial}{\partial u}(a\sin(u)\cos(v))\right)+\left(\frac{\partial}{\partial v}(a\sin(u)\cos(v))\right)\right),$$
$$\left(\frac{\partial}{\partial x}Q(x,y,z)\right)\left(\left(\frac{\partial}{\partial u}(a\sin(u)\cos(v))\right)+\left(\frac{\partial}{\partial v}(a\sin(u)\cos(v))\right)\right)$$
$$\left.-\left(\frac{\partial}{\partial y}P(x,y,z)\right)\left(\left(\frac{\partial}{\partial u}(b\sin(u)\sin(v))\right)+\left(\frac{\partial}{\partial v}(b\sin(u)\sin(v))\right)\right)\right]$$

No matter 'Chain differentiation' or 'Solve divergence/curl', question for discussion is 'How to solve differential coefficient' or 'method of differentiation' about abstract vector field 'P[x,y,z], Q[x,y,z],R[x,y,z]'; and this step, question for discussion is 'How to convert result of curl from a coordinates to another'; character and hiberarchy of 'two questions' above-mentioned are disparate, it is 'multiply', not 'plus' — be determined by 3-Dimensional spacial attribute of coordinates

Set Amount of surface 'S's segmental cells as 50

(This amount can be discretional natural number)   (31)

5.Microcosmic surface integral course at surface 'S's first segmental cell:

Segment range $[0, \pi/n - \theta]$ of 'u': du = $\dfrac{\pi}{50n} - \dfrac{\theta}{50}$

Segment range $[0, 2\pi]$ of 'v': dv = $\dfrac{2\pi}{50} = \dfrac{\pi}{25}$ (32)

Segment 'tangent plane normal vector'(33):[viz. Input (32) to (28)]

$$\left[c\sin\left(-\frac{\pi}{50\,n}+\frac{\theta}{50}\right)^2 b\cos\left(\frac{\pi}{25}\right), \sin\left(-\frac{\pi}{50\,n}+\frac{\theta}{50}\right)^2 \sin\left(\frac{\pi}{25}\right) a\,c, \right.$$
$$\left. -\sin\left(-\frac{\pi}{50\,n}+\frac{\theta}{50}\right) a\cos\left(-\frac{\pi}{50\,n}+\frac{\theta}{50}\right) b \right]$$

(33)

Segment abstract Curl (34): [Viz. Input (32) to (30)]

$$\left[\left(\frac{\partial}{\partial y}R(x,y,z)\right) b^2 \cos\left(\frac{\pi}{50\,n}-\frac{\theta}{50}\right)\sin\left(\frac{\pi}{25}\right)\sin\left(\frac{\pi}{50\,n}-\frac{\theta}{50}\right)\cos\left(\frac{\pi}{25}\right), \right.$$
$$\left(\frac{\partial}{\partial x}R(x,y,z)\right) a^2 \cos\left(\frac{\pi}{50\,n}-\frac{\theta}{50}\right)\cos\left(\frac{\pi}{25}\right)\sin\left(\frac{\pi}{50\,n}-\frac{\theta}{50}\right)\sin\left(\frac{\pi}{25}\right),$$
$$-\left(\frac{\partial}{\partial x}Q(x,y,z)\right) a^2 \cos\left(\frac{\pi}{50\,n}-\frac{\theta}{50}\right)\cos\left(\frac{\pi}{25}\right)\sin\left(\frac{\pi}{50\,n}-\frac{\theta}{50}\right)\sin\left(\frac{\pi}{25}\right)$$
$$\left. -\left(\frac{\partial}{\partial y}P(x,y,z)\right) b^2 \cos\left(\frac{\pi}{50\,n}-\frac{\theta}{50}\right)\sin\left(\frac{\pi}{25}\right)\sin\left(\frac{\pi}{50\,n}-\frac{\theta}{50}\right)\cos\left(\frac{\pi}{25}\right) \right]$$

(34)

// Owing to universality and homogeneity of abstract Curl

$$\left[\left(\frac{\partial}{\partial y}R(x,y,z)\right)-\left(\frac{\partial}{\partial z}Q(x,y,z)\right), \left(\frac{\partial}{\partial z}P(x,y,z)\right)-\left(\frac{\partial}{\partial x}R(x,y,z)\right), \right.$$
$$\left. \left(\frac{\partial}{\partial x}Q(x,y,z)\right)-\left(\frac{\partial}{\partial y}P(x,y,z)\right) \right]$$

if this abstract Curl is continuous at surface 'S', then this abstract curl still can be described as

$$\left[\left(\frac{\partial}{\partial y}R(x,y,z)\right)-\left(\frac{\partial}{\partial z}Q(x,y,z)\right), \left(\frac{\partial}{\partial z}P(x,y,z)\right)-\left(\frac{\partial}{\partial x}R(x,y,z)\right), \right.$$
$$\left. \left(\frac{\partial}{\partial x}Q(x,y,z)\right)-\left(\frac{\partial}{\partial y}P(x,y,z)\right) \right]$$

in one or more segment cells of surface'S'.

(See also Section 4.3 Chapter 4 (Part I)'Discuss About Abstract Vector Field、Finite Sums Limit、Integral and Riemann Sums')

Calculate microcosmic surface integral value at surface 'S's first segmental cell (35):

Base on integral median theorem, dot product of abstract curl(34) and tangent plane normal vector (33), times segment unit of 'u,v' (32), that is microcosmic surface integral value at first segmental Cell (35)

$$\frac{1}{25}\left(-c\sin\left(-\frac{\pi}{50\,n}+\frac{\theta}{50}\right)^3 b^3\cos\left(\frac{\pi}{25}\right)^2\left(\frac{\partial}{\partial y}R(x,y,z)\right)\cos\left(-\frac{\pi}{50\,n}+\frac{\theta}{50}\right)\sin\left(\frac{\pi}{25}\right)\right.$$

$$-\sin\left(-\frac{\pi}{50\,n}+\frac{\theta}{50}\right)^3\sin\left(\frac{\pi}{25}\right)^2 a^3 c\left(\frac{\partial}{\partial x}R(x,y,z)\right)\cos\left(-\frac{\pi}{50\,n}+\frac{\theta}{50}\right)\cos\left(\frac{\pi}{25}\right)-$$

$$\sin\left(-\frac{\pi}{50\,n}+\frac{\theta}{50}\right)a\cos\left(-\frac{\pi}{50\,n}+\frac{\theta}{50}\right)b\left(\vphantom{\frac{\pi}{50}}\right.$$

$$\left(\frac{\partial}{\partial x}Q(x,y,z)\right)a^2\cos\left(-\frac{\pi}{50\,n}+\frac{\theta}{50}\right)\cos\left(\frac{\pi}{25}\right)\sin\left(-\frac{\pi}{50\,n}+\frac{\theta}{50}\right)\sin\left(\frac{\pi}{25}\right)$$

$$\left.+\left(\frac{\partial}{\partial y}P(x,y,z)\right)b^2\cos\left(-\frac{\pi}{50\,n}+\frac{\theta}{50}\right)\sin\left(\frac{\pi}{25}\right)\sin\left(-\frac{\pi}{50\,n}+\frac{\theta}{50}\right)\cos\left(\frac{\pi}{25}\right)\right)\right)$$

$$\left(\frac{\pi}{50\,n}-\frac{\theta}{50}\right)\pi$$

6.Microcosmic surface integral course at surface 'S's all (viz.50) segmental cells:

Segment range [0, π/n $-\theta$] of 'u': du $= \dfrac{s\pi}{50n}-\dfrac{s\theta}{50}$

Segment range [0,2π] of 'v': dv $= \dfrac{2t\pi}{50}=\dfrac{t\pi}{25}$ (36)

(thereinto, 's' and 't' stand for natural number 1~50)

Segment 'tangent plane normal vector' (37):[viz. Input (36) to (28)]

$$\left[c\sin\left(\left(\frac{\pi}{50\,n}-\frac{\theta}{50}\right)s\right)^2 b\cos\left(\frac{\pi\,t}{25}\right),\sin\left(\left(\frac{\pi}{50\,n}-\frac{\theta}{50}\right)s\right)^2\sin\left(\frac{\pi\,t}{25}\right)a\,c,\right.$$

$$\left.\sin\left(\left(\frac{\pi}{50\,n}-\frac{\theta}{50}\right)s\right)a\cos\left(\left(\frac{\pi}{50\,n}-\frac{\theta}{50}\right)s\right)b\right]$$

(37)

// Expression (37) isn't that 'single vector value',but that 'Set of finite quantitative (viz. n) vector values'

Segment abstract Curl (38): [Viz. Input (36) to (30)]

$$\left[\left(\frac{\partial}{\partial y}R(x,y,z)\right)b^2\cos\left(\left(\frac{\pi}{50\,n}-\frac{\theta}{50}\right)s\right)\sin\left(\frac{\pi\,t}{25}\right)\sin\left(\left(\frac{\pi}{50\,n}-\frac{\theta}{50}\right)s\right)\cos\left(\frac{\pi\,t}{25}\right),\right.$$

$$\left(\frac{\partial}{\partial x}R(x,y,z)\right)a^2\cos\left(\left(\frac{\pi}{50\,n}-\frac{\theta}{50}\right)s\right)\cos\left(\frac{\pi\,t}{25}\right)\sin\left(\left(\frac{\pi}{50\,n}-\frac{\theta}{50}\right)s\right)\sin\left(\frac{\pi\,t}{25}\right),$$

$$-\left(\frac{\partial}{\partial x}Q(x,y,z)\right)a^2\cos\left(\left(\frac{\pi}{50\,n}-\frac{\theta}{50}\right)s\right)\cos\left(\frac{\pi\,t}{25}\right)\sin\left(\left(\frac{\pi}{50\,n}-\frac{\theta}{50}\right)s\right)\sin\left(\frac{\pi\,t}{25}\right)$$

$$\left.-\left(\frac{\partial}{\partial y}P(x,y,z)\right)b^2\cos\left(\left(\frac{\pi}{50\,n}-\frac{\theta}{50}\right)s\right)\sin\left(\frac{\pi\,t}{25}\right)\sin\left(\left(\frac{\pi}{50\,n}-\frac{\theta}{50}\right)s\right)\cos\left(\frac{\pi\,t}{25}\right)\right]$$

// Expression (38) isn't that 'single vector value',but that 'Set of finite quantitative (viz. 50) vector values'

// Owing to universality and homogeneity of abstract Curl

$$\left[\left(\frac{\partial}{\partial y}R(x,y,z)\right)-\left(\frac{\partial}{\partial z}Q(x,y,z)\right),\left(\frac{\partial}{\partial z}P(x,y,z)\right)-\left(\frac{\partial}{\partial x}R(x,y,z)\right),\right.$$

$$\left.\left(\frac{\partial}{\partial x}Q(x,y,z)\right)-\left(\frac{\partial}{\partial y}P(x,y,z)\right)\right]$$

if this abstract Curl is continuous at surface 'S', then this abstract curl still can be described as

$$\left[\left(\frac{\partial}{\partial y}R(x,y,z)\right)-\left(\frac{\partial}{\partial z}Q(x,y,z)\right),\left(\frac{\partial}{\partial z}P(x,y,z)\right)-\left(\frac{\partial}{\partial x}R(x,y,z)\right),\right.$$

$$\left.\left(\frac{\partial}{\partial x}Q(x,y,z)\right)-\left(\frac{\partial}{\partial y}P(x,y,z)\right)\right]$$

in one or more segment cells of surface 'S'.

(See also Section 4.3 Chapter 4 (Part I) 'Discuss About Abstract Vector Field、Finite Sums Limit、Integral and Riemann Sums')

Calculate microcosmic surface integral value at surface 'S's all segmental cells (39):

Base on integral median theorem, dot product of abstract curl (38) and tangent plane normal vector (37), times segment unit of 'u,v' (36), these are microcosmic surface integral values at all segmental cells(39)

$$\frac{1}{25}\left(c\sin\left(\left(\frac{\pi}{50\,n}-\frac{\theta}{50}\right)s\right)^3 b^3\cos\left(\frac{\pi\,t}{25}\right)^2\left(\frac{\partial}{\partial y}R(x,y,z)\right)\cos\left(\left(\frac{\pi}{50\,n}-\frac{\theta}{50}\right)s\right)\sin\left(\frac{\pi\,t}{25}\right)\right.$$

$$+\sin\left(\left(\frac{\pi}{50\,n}-\frac{\theta}{50}\right)s\right)^3\sin\left(\frac{\pi\,t}{25}\right)^2 a^3\,c\left(\frac{\partial}{\partial x}R(x,y,z)\right)\cos\left(\left(\frac{\pi}{50\,n}-\frac{\theta}{50}\right)s\right)\cos\left(\frac{\pi\,t}{25}\right)$$

$$+ \sin\left(\left(\frac{\pi}{50\,n} - \frac{\theta}{50}\right)s\right)a\cos\left(\left(\frac{\pi}{50\,n} - \frac{\theta}{50}\right)s\right)b\Bigg($$

$$-\left(\frac{\partial}{\partial x}Q(x,y,z)\right)a^2\cos\left(\left(\frac{\pi}{50\,n} - \frac{\theta}{50}\right)s\right)\cos\left(\frac{\pi\,t}{25}\right)\sin\left(\left(\frac{\pi}{50\,n} - \frac{\theta}{50}\right)s\right)\sin\left(\frac{\pi\,t}{25}\right)$$

$$-\left(\frac{\partial}{\partial y}P(x,y,z)\right)b^2\cos\left(\left(\frac{\pi}{50\,n} - \frac{\theta}{50}\right)s\right)\sin\left(\frac{\pi\,t}{25}\right)\sin\left(\left(\frac{\pi}{50\,n} - \frac{\theta}{50}\right)s\right)\cos\left(\frac{\pi\,t}{25}\right)\Bigg)\Bigg)$$

$$\left(\frac{\pi}{50\,n} - \frac{\theta}{50}\right)\pi$$

// Expression (39) isn't that 'single value', but that 'set of finite quantitative (viz.50) values'

Structure sequence that is composed of finite quantitative (viz. 50) microcomosic surface integral values (40):
  [Lengthy expression of sequence is elided;
  In program template of Waterloo Maplesoft (see also Section 4.2):
  sqn:=seq(seq((stdV[1]*stdA+stdV[2]*stdB+stdV[3]*stdC)*du*dv,
  s=1..dus),t=1..dus):
change ':'(last) to ';', then obtain expression of sequence]

Accumulational sum of sequence (viz. Sum of integral values of dot product [abstract curl(38) and tangent plane normal vector(37)] at all segmental cells, obtain surface integral value at manifold (41):
  [Lengthy expression of accumulational sum result is elided;
  In program template of Waterloo Maplesoft (see also Section 5.2):
  add(k,k=sqn):xi:=evalf(%);
change ':'(middle) to ';', then obtain expression of result]

Set Amount of surface 'S's segmental cells as w
(viz. uncertain natural number)   (42)

7.Microcosmic surface integral course at surface 'S's first segmental cell:

Segment range [0, π/n $-\theta$] of 'u': du = $\dfrac{\dfrac{\pi}{n} - \theta}{w}$

Segment range [0,2π] of 'v': dv = $\dfrac{2\pi}{w}$ (43)

Segment 'tangent plane normal vector' (44):[viz. Input (43) to (28)]

$$\left[c\sin\left(\frac{\frac{\pi}{n}-\theta}{w}\right)^2 b\cos\left(\frac{2\pi}{w}\right),\ \sin\left(\frac{\frac{\pi}{n}-\theta}{w}\right)^2\sin\left(\frac{2\pi}{w}\right)a\,c,\ \sin\left(\frac{\frac{\pi}{n}-\theta}{w}\right)a\cos\left(\frac{\frac{\pi}{n}-\theta}{w}\right)b \right]$$

Segment abstract curl (45): [Viz. Input (43) to (30)]

$$\left[\left(\frac{\partial}{\partial y}R(x,y,z)\right)b^2\cos\left(\frac{\frac{\pi}{n}-\theta}{w}\right)\sin\left(\frac{2\pi}{w}\right)\sin\left(\frac{\frac{\pi}{n}-\theta}{w}\right)\cos\left(\frac{2\pi}{w}\right), \right.$$

$$\left(\frac{\partial}{\partial x}R(x,y,z)\right)a^2\cos\left(\frac{\frac{\pi}{n}-\theta}{w}\right)\cos\left(\frac{2\pi}{w}\right)\sin\left(\frac{\frac{\pi}{n}-\theta}{w}\right)\sin\left(\frac{2\pi}{w}\right),$$

$$-\left(\frac{\partial}{\partial x}Q(x,y,z)\right)a^2\cos\left(\frac{\frac{\pi}{n}-\theta}{w}\right)\cos\left(\frac{2\pi}{w}\right)\sin\left(\frac{\frac{\pi}{n}-\theta}{w}\right)\sin\left(\frac{2\pi}{w}\right)$$

$$\left. -\left(\frac{\partial}{\partial y}P(x,y,z)\right)b^2\cos\left(\frac{\frac{\pi}{n}-\theta}{w}\right)\sin\left(\frac{2\pi}{w}\right)\sin\left(\frac{\frac{\pi}{n}-\theta}{w}\right)\cos\left(\frac{2\pi}{w}\right) \right] \qquad (45)$$

// Owing to universality and homogeneity of abstract Curl

$$\left[\left(\frac{\partial}{\partial y}R(x,y,z)\right)-\left(\frac{\partial}{\partial z}Q(x,y,z)\right), \left(\frac{\partial}{\partial z}P(x,y,z)\right)-\left(\frac{\partial}{\partial x}R(x,y,z)\right), \right.$$

$$\left. \left(\frac{\partial}{\partial x}Q(x,y,z)\right)-\left(\frac{\partial}{\partial y}P(x,y,z)\right) \right]$$

if this abstract Curl is continuous at surface 'S', then this abstract curl still can be described as

$$\left[\left(\frac{\partial}{\partial y}R(x,y,z)\right)-\left(\frac{\partial}{\partial z}Q(x,y,z)\right), \left(\frac{\partial}{\partial z}P(x,y,z)\right)-\left(\frac{\partial}{\partial x}R(x,y,z)\right), \right.$$

$$\left. \left(\frac{\partial}{\partial x}Q(x,y,z)\right)-\left(\frac{\partial}{\partial y}P(x,y,z)\right) \right]$$

in one or more segment cells of surface 'S'.

(See also Section 4.3 Chapter 4 (Part I) 'Discuss About Abstract Vector Field、Finite Sums Limit、Integral and Riemann Sums')

Calculate microcosmic surface integral value at surface 'S's first segmental cell (46):

Base on integral median theorem, dot product of abstract curl (45) and tangent plane normal vector(44), times segment unit of 'u,v' (43), that is microcosmic surface integral value at first segmental cell(46)

$$2\left(c\sin\left(\frac{\frac{\pi}{n}-\theta}{w}\right)^3 b^3 \cos\left(\frac{2\pi}{w}\right)^2 \left(\frac{\partial}{\partial y}R(x,y,z)\right)\cos\left(\frac{\frac{\pi}{n}-\theta}{w}\right)\sin\left(\frac{2\pi}{w}\right)\right.$$

$$+\sin\left(\frac{\frac{\pi}{n}-\theta}{w}\right)^3\sin\left(\frac{2\pi}{w}\right)^2 a^3 c\left(\frac{\partial}{\partial x}R(x,y,z)\right)\cos\left(\frac{\frac{\pi}{n}-\theta}{w}\right)\cos\left(\frac{2\pi}{w}\right)+\sin\left(\frac{\frac{\pi}{n}-\theta}{w}\right)a$$

$$\cos\left(\frac{\frac{\pi}{n}-\theta}{w}\right)b\left(-\left(\frac{\partial}{\partial x}Q(x,y,z)\right)a^2\cos\left(\frac{\frac{\pi}{n}-\theta}{w}\right)\cos\left(\frac{2\pi}{w}\right)\sin\left(\frac{\frac{\pi}{n}-\theta}{w}\right)\sin\left(\frac{2\pi}{w}\right)\right.$$

$$\left.\left.-\left(\frac{\partial}{\partial y}P(x,y,z)\right)b^2\cos\left(\frac{\frac{\pi}{n}-\theta}{w}\right)\sin\left(\frac{2\pi}{w}\right)\sin\left(\frac{\frac{\pi}{n}-\theta}{w}\right)\cos\left(\frac{2\pi}{w}\right)\right)\right)\left(\frac{\pi}{n}-\theta\right)\pi\ /\ w^2$$

$$2\left(c\sin\left(\frac{\frac{\pi}{n}-\theta}{w}\right)^3 b^3 \cos\left(\frac{2\pi}{w}\right)^2 \left(\frac{\partial}{\partial y}R(x,y,z)\right)\cos\left(\frac{\frac{\pi}{n}-\theta}{w}\right)\sin\left(\frac{2\pi}{w}\right)\right.$$

$$+\sin\left(\frac{\frac{\pi}{n}-\theta}{w}\right)^3\sin\left(\frac{2\pi}{w}\right)^2 a^3 c\left(\frac{\partial}{\partial x}R(x,y,z)\right)\cos\left(\frac{\frac{\pi}{n}-\theta}{w}\right)\cos\left(\frac{2\pi}{w}\right)+\sin\left(\frac{\frac{\pi}{n}-\theta}{w}\right)a$$

$$\cos\left(\frac{\frac{\pi}{n}-\theta}{w}\right)b\left(-\left(\frac{\partial}{\partial x}Q(x,y,z)\right)a^2\cos\left(\frac{\frac{\pi}{n}-\theta}{w}\right)\cos\left(\frac{2\pi}{w}\right)\sin\left(\frac{\frac{\pi}{n}-\theta}{w}\right)\sin\left(\frac{2\pi}{w}\right)\right.$$

$$\left.\left.-\left(\frac{\partial}{\partial y}P(x,y,z)\right)b^2\cos\left(\frac{\frac{\pi}{n}-\theta}{w}\right)\sin\left(\frac{2\pi}{w}\right)\sin\left(\frac{\frac{\pi}{n}-\theta}{w}\right)\cos\left(\frac{2\pi}{w}\right)\right)\right)\left(\frac{\pi}{n}-\theta\right)\pi\ /\ w^2$$

$$2\left(c\sin\left(\frac{\frac{\pi}{n}-\theta}{w}\right)^3 b^3 \cos\left(\frac{2\pi}{w}\right)^2 \left(\frac{\partial}{\partial y}R(x,y,z)\right)\cos\left(\frac{\frac{\pi}{n}-\theta}{w}\right)\sin\left(\frac{2\pi}{w}\right)\right.$$

$$+\sin\left(\frac{\frac{\pi}{n}-\theta}{w}\right)^3\sin\left(\frac{2\pi}{w}\right)^2 a^3 c\left(\frac{\partial}{\partial x}R(x,y,z)\right)\cos\left(\frac{\frac{\pi}{n}-\theta}{w}\right)\cos\left(\frac{2\pi}{w}\right)+\sin\left(\frac{\frac{\pi}{n}-\theta}{w}\right)a$$

$$\cos\left(\frac{\frac{\pi}{n}-\theta}{w}\right)b\left(-\left(\frac{\partial}{\partial x}Q(x,y,z)\right)a^2\cos\left(\frac{\frac{\pi}{n}-\theta}{w}\right)\cos\left(\frac{2\pi}{w}\right)\sin\left(\frac{\frac{\pi}{n}-\theta}{w}\right)\sin\left(\frac{2\pi}{w}\right)\right.$$

$$\left.\left.-\left(\frac{\partial}{\partial y}P(x,y,z)\right)b^2\cos\left(\frac{\frac{\pi}{n}-\theta}{w}\right)\sin\left(\frac{2\pi}{w}\right)\sin\left(\frac{\frac{\pi}{n}-\theta}{w}\right)\cos\left(\frac{2\pi}{w}\right)\right)\right)\left(\frac{\pi}{n}-\theta\right)\pi\ /\ w^2$$

8.Microcosmic surface integral course at surface 'S's all (viz. n) segmental cells:

Segment range $[0, \pi/n - \theta]$ of 'u': $du = \dfrac{(\frac{\pi}{n} - \theta)s}{w}$

Segment range $[0, 2\pi]$ of 'v': $dv = \dfrac{2t\pi}{w}$

(thereinto, 's' and 't' stand for natural number 1~w)   (47)

Segment 'tangent plane normal vector' (48):[viz. Input (47) to (28)]

$$\left[c\sin\left(\frac{\left(\frac{\pi}{n}-\theta\right)s}{w}\right)^2 b\cos\left(\frac{2\pi t}{w}\right),\ \sin\left(\frac{\left(\frac{\pi}{n}-\theta\right)s}{w}\right)^2 \sin\left(\frac{2\pi t}{w}\right)a\,c, \right.$$

$$\left. \sin\left(\frac{\left(\frac{\pi}{n}-\theta\right)s}{w}\right)a\cos\left(\frac{\left(\frac{\pi}{n}-\theta\right)s}{w}\right)b \right]$$

(48)

// Expression (48) isn't that 'single vector value', but that 'Set of finite quantitative (viz. w) vector values'

Segment abstract curl (49): [Viz. Input (47) to (30)]

$$\left[\left(\frac{\partial}{\partial y}\text{R}(x,y,z)\right)b^2\cos\left(\frac{\left(\frac{\pi}{n}-\theta\right)s}{w}\right)\sin\left(\frac{2\pi t}{w}\right)\sin\left(\frac{\left(\frac{\pi}{n}-\theta\right)s}{w}\right)\cos\left(\frac{2\pi t}{w}\right), \right.$$

$$\left(\frac{\partial}{\partial x}\text{R}(x,y,z)\right)a^2\cos\left(\frac{\left(\frac{\pi}{n}-\theta\right)s}{w}\right)\cos\left(\frac{2\pi t}{w}\right)\sin\left(\frac{\left(\frac{\pi}{n}-\theta\right)s}{w}\right)\sin\left(\frac{2\pi t}{w}\right),$$

$$-\left(\frac{\partial}{\partial x}\text{Q}(x,y,z)\right)a^2\cos\left(\frac{\left(\frac{\pi}{n}-\theta\right)s}{w}\right)\cos\left(\frac{2\pi t}{w}\right)\sin\left(\frac{\left(\frac{\pi}{n}-\theta\right)s}{w}\right)\sin\left(\frac{2\pi t}{w}\right)$$

$$\left. -\left(\frac{\partial}{\partial y}\text{P}(x,y,z)\right)b^2\cos\left(\frac{\left(\frac{\pi}{n}-\theta\right)s}{w}\right)\sin\left(\frac{2\pi t}{w}\right)\sin\left(\frac{\left(\frac{\pi}{n}-\theta\right)s}{w}\right)\cos\left(\frac{2\pi t}{w}\right) \right]$$

// Expression (49) isn't that 'single vector value', but that 'Set of finite quantitative (viz. w) vector values'
// Owing to universality and homogeneity of abstract Curl

$$\left[\left(\frac{\partial}{\partial y}\text{R}(x,y,z)\right)-\left(\frac{\partial}{\partial z}\text{Q}(x,y,z)\right),\ \left(\frac{\partial}{\partial z}\text{P}(x,y,z)\right)-\left(\frac{\partial}{\partial x}\text{R}(x,y,z)\right), \right.$$

$$\left. \left(\frac{\partial}{\partial x}\text{Q}(x,y,z)\right)-\left(\frac{\partial}{\partial y}\text{P}(x,y,z)\right) \right]$$

if this abstract Curl is continuous at surface'S', then this abstract
curl still can be described as

$$\left[\left(\frac{\partial}{\partial y}R(x,y,z)\right)-\left(\frac{\partial}{\partial z}Q(x,y,z)\right),\left(\frac{\partial}{\partial z}P(x,y,z)\right)-\left(\frac{\partial}{\partial x}R(x,y,z)\right),\right.$$
$$\left.\left(\frac{\partial}{\partial x}Q(x,y,z)\right)-\left(\frac{\partial}{\partial y}P(x,y,z)\right)\right]$$

in one or more segment cells of surface'S'.

(See also Section 4.3 Chapter 4 (Part I) 'Discuss About Abstract
Vector Field、Finite Sums Limit、Integral and Riemann Sums')

Calculate microcosmic surface integral values at surface'S's all
segmental cells (50):

Base on integral median theorem, dot product of abstract curl(49)
and tangent plane normal vector (48), times segment unit of 'u,v' (47),
these are microcosmic surface integral values at all segmental
cells (50)

$$2\left(c\sin\left(\frac{\left(\frac{\pi}{n}-\theta\right)s}{w}\right)^3 b^3\cos\left(\frac{2\pi t}{w}\right)^2\left(\frac{\partial}{\partial y}R(x,y,z)\right)\cos\left(\frac{\left(\frac{\pi}{n}-\theta\right)s}{w}\right)\sin\left(\frac{2\pi t}{w}\right)\right.$$

$$+\sin\left(\frac{\left(\frac{\pi}{n}-\theta\right)s}{w}\right)^3\sin\left(\frac{2\pi t}{w}\right)^2 a^3 c\left(\frac{\partial}{\partial x}R(x,y,z)\right)\cos\left(\frac{\left(\frac{\pi}{n}-\theta\right)s}{w}\right)\cos\left(\frac{2\pi t}{w}\right)+$$

$$\sin\left(\frac{\left(\frac{\pi}{n}-\theta\right)s}{w}\right)a\cos\left(\frac{\left(\frac{\pi}{n}-\theta\right)s}{w}\right)b\left(\right.$$

$$-\left(\frac{\partial}{\partial x}Q(x,y,z)\right)a^2\cos\left(\frac{\left(\frac{\pi}{n}-\theta\right)s}{w}\right)\cos\left(\frac{2\pi t}{w}\right)\sin\left(\frac{\left(\frac{\pi}{n}-\theta\right)s}{w}\right)\sin\left(\frac{2\pi t}{w}\right)$$

$$\left.\left.-\left(\frac{\partial}{\partial y}P(x,y,z)\right)b^2\cos\left(\frac{\left(\frac{\pi}{n}-\theta\right)s}{w}\right)\sin\left(\frac{2\pi t}{w}\right)\sin\left(\frac{\left(\frac{\pi}{n}-\theta\right)s}{w}\right)\cos\left(\frac{2\pi t}{w}\right)\right)\right)$$

$$\left(\frac{\pi}{n}-\theta\right)\pi\,/\,w^2$$

// Expression (50) isn't that 'single value', but that 'set of finite
quantitative (viz. w) values'

Structure finite sums (51):

(In the case of "Amount 'w' of surface 'S's segmental cells is uncertain", sum of integral values [dot product about absract curl(49) and tangent plane normal vector (48)] at all segmental cells)

$$\sum_{t=1}^{w}\left(\sum_{s=1}^{w}\left(2\left(\right.\right.\right.$$

$$\left(\frac{\partial}{\partial y}R(x,y,z)\right)b^3\cos\left(\frac{s\left(\frac{\pi}{n}-\theta\right)}{w}\right)\sin\left(\frac{2t\pi}{w}\right)\sin\left(\frac{s\left(\frac{\pi}{n}-\theta\right)}{w}\right)^3\cos\left(\frac{2t\pi}{w}\right)^2 c$$

$$+\left(\frac{\partial}{\partial x}R(x,y,z)\right)a^3\cos\left(\frac{s\left(\frac{\pi}{n}-\theta\right)}{w}\right)\cos\left(\frac{2t\pi}{w}\right)\sin\left(\frac{s\left(\frac{\pi}{n}-\theta\right)}{w}\right)^3\sin\left(\frac{2t\pi}{w}\right)^2 c+\left(\right.$$

$$-\left(\frac{\partial}{\partial x}Q(x,y,z)\right)a^2\cos\left(\frac{s\left(\frac{\pi}{n}-\theta\right)}{w}\right)\cos\left(\frac{2t\pi}{w}\right)\sin\left(\frac{s\left(\frac{\pi}{n}-\theta\right)}{w}\right)\sin\left(\frac{2t\pi}{w}\right)$$

$$-\left(\frac{\partial}{\partial y}P(x,y,z)\right)b^2\cos\left(\frac{s\left(\frac{\pi}{n}-\theta\right)}{w}\right)\sin\left(\frac{2t\pi}{w}\right)\sin\left(\frac{s\left(\frac{\pi}{n}-\theta\right)}{w}\right)\cos\left(\frac{2t\pi}{w}\right)\right)$$

$$\sin\left(\frac{s\left(\frac{\pi}{n}-\theta\right)}{w}\right)a\,b\cos\left(\frac{s\left(\frac{\pi}{n}-\theta\right)}{w}\right)\right)\left(\frac{\pi}{n}-\theta\right)\pi\,/\,w^2\right)\right)$$

$$(51)$$

Launch by Intel Celeron CPU E3400 2.6G (32 bit Processor) 1GB Memory, Operation of Sum can't be completed, Await to enhance computer hardware capability;

Simplify sum expression, change range '[0,Pi/n-theta]' of 'u' to '[0,Pi/2]' (Elide two variables: 'n' and 'theta') (52), calculate renewedly (Include closed curve integral aforementioned) (53) 、

$$\lim_{w\to\infty}\sum_{t=1}^{w}\left(\sum_{s=1}^{w}\left(c\sin\left(\frac{\pi s}{2w}\right)^3 b^3\cos\left(\frac{2\pi t}{w}\right)^2\left(\frac{\partial}{\partial y}R(x,y,z)\right)\cos\left(\frac{\pi s}{2w}\right)\sin\left(\frac{2\pi t}{w}\right)\right.\right.$$

$$+\sin\left(\frac{\pi s}{2w}\right)^3\sin\left(\frac{2\pi t}{w}\right)^2 a^3 c\left(\frac{\partial}{\partial x}R(x,y,z)\right)\cos\left(\frac{\pi s}{2w}\right)\cos\left(\frac{2\pi t}{w}\right)+\sin\left(\frac{\pi s}{2w}\right)a$$

$$\cos\left(\frac{\pi s}{2w}\right)b\left(-\left(\frac{\partial}{\partial x}Q(x,y,z)\right)a^2\cos\left(\frac{\pi s}{w}\right)\cos\left(\frac{2\pi t}{w}\right)\sin\left(\frac{\pi s}{2w}\right)\sin\left(\frac{2\pi t}{w}\right)\right.$$

$$\left.\left.\left.-\left(\frac{\partial}{\partial y}P(x,y,z)\right)b^2\cos\left(\frac{\pi s}{2w}\right)\sin\left(\frac{2\pi t}{w}\right)\sin\left(\frac{\pi s}{2w}\right)\cos\left(\frac{2\pi t}{w}\right)\right)\right)\pi^2\,/\,w^2\right)$$

= 0   (53)

// About 'Integral value is unequal to 0', See also Chapter 6 (This Part) Section 6.1 'Curl Theorem at Manifold and Mobius Strip'

Viz. In the case of 'n → ∞', (25)=(52):

$$\sum_{i=1}^{w}\left(\frac{2\pi\left(-P(x,y,z)\,\alpha\,\sin\left(\frac{2\,i\,\pi}{w}\right)+Q(x,y,z)\,\beta\,\cos\left(\frac{2\,i\,\pi}{w}\right)\right)}{w}\right)$$

$$=$$

$$\sum_{t=1}^{w}\left(\sum_{s=1}^{w}\left(\left(\left(\frac{\partial}{\partial y}R(x,y,z)\right)b^3\cos\left(\frac{s\,\pi}{2\,w}\right)\sin\left(\frac{2\,t\,\pi}{w}\right)\sin\left(\frac{s\,\pi}{2\,w}\right)^3\cos\left(\frac{2\,t\,\pi}{w}\right)^2 c\right.\right.\right.$$

$$+\left(\frac{\partial}{\partial x}R(x,y,z)\right)a^3\cos\left(\frac{s\,\pi}{2\,w}\right)\cos\left(\frac{2\,t\,\pi}{w}\right)\sin\left(\frac{s\,\pi}{2\,w}\right)^3\sin\left(\frac{2\,t\,\pi}{w}\right)^2 c+\left(\right.$$

$$-\left(\frac{\partial}{\partial x}Q(x,y,z)\right)a^2\cos\left(\frac{s\,\pi}{2\,w}\right)\cos\left(\frac{2\,t\,\pi}{w}\right)\sin\left(\frac{s\,\pi}{2\,w}\right)\sin\left(\frac{2\,t\,\pi}{w}\right)$$

$$-\left(\frac{\partial}{\partial y}P(x,y,z)\right)b^2\cos\left(\frac{s\,\pi}{2\,w}\right)\sin\left(\frac{2\,t\,\pi}{w}\right)\sin\left(\frac{s\,\pi}{2\,w}\right)\cos\left(\frac{2\,t\,\pi}{w}\right)\right)\sin\left(\frac{s\,\pi}{2\,w}\right)a\,b$$

$$\left.\left.\cos\left(\frac{s\,\pi}{2\,w}\right)\right)\pi^2/w^2\right)$$

Equation above-mentioned can be described as:

$$\int_{L+}A\cdot dL=\iint_{S}rotA\cdot n\,dS \quad \textbf{(1)}, \text{ Complete Proof.}$$

5.2 Prove Curl Theorem Finite Sums Limits at Manifold [Program Template of Waterloo Maplesoft, Optional]

Curl Theorem Suppose positive directional boundary 'L+' of smooth or piecewise smooth orientational surface 'S' is smooth or piecewise smooth closed curve, Right Hand Rule is associated basis between positive direction of boundary curve 'L' and outer side of orientational surface 'S'. If there are 1th order continuous partial derivatives about functions 'P(x,y,z),Q(x,y,z),R(x,y,z)' [Structure vector field 'A'] at orientational surface 'S', then:

$$\int_{L+} A \cdot dL = \iint_{S} rotA \cdot n \, dS \qquad \textbf{(1)}$$

thereinto, 'rotA' is curl of vector field 'A', 'n' is unit normal vector of orientational surface 'S's outer side.

Symbol System:

Abstract Spacial Vector Field 'V',

Curl 'cV1,cV2' of Abstract Spacial Vector Field 'V'

Abstract simply connected orientable closed Parametrized Surface 'CS' [Suppose as Unclosed],

Closed boundary curve 'CL' of surface 'CS',

Tangent Plane Normal Vector '[A,B,C]' of Surface 'CS';

Amount 'dus' of boundary curve 'CL's parameter segmental cells (It can be discretional natural number),

Segmental range 'dv' of parameter 'v',

Tangent vector 'dCL' of boundary curve 'CL';

Spacial vector field 'V's average value 'dV' that corresponds to boundary curve 'CL's first segmental cell,

'Tangent vector dCL's average value 'dCLm' that corresponds to boundary curve 'CL's first segmental cell;

Spacial vector field 'V's average values 'idV' that correspond to boundary curve 'CL's all segmental cells,

'Tangent vector dCL's average values 'idCL' that correspond to boundary curve 'CL's all segmental cells;

(In actual expressions,'i' stands for natural number)

Amount 'dus' of surface 'CS's parameter segmental cells (It can be discretional natural number),

Segmental range 'du' of parameter 'u',

Segmental range 'dv' of parameter 'v';

'Tangent Plane Normal Vector [A,B,C]'s average value '[dA,dB,dC]' that corresponds to surface 'CS's first segmental cell,

Curl 'cV2's average value 'dcV2' that corresponds to surface'CS's first segmental cell;

'Tangent Plane Normal Vector [A,B,C]'s average values '[stdA, stdB,stdC]' that correspond to surface 'CS's all segmental cells,

Curl 'cV2's average values 'stdcV2' that correspond to surface 'CS's all segmental cell

(In actual expressions, 's,t' stand for natural number)

> restart;
> with(linalg):
Define parametrized expression of abstract simply connected orientable closed surface 'CS':
> CS:=[a*sin(u)*cos(v),b*sin(u)*sin(v),c*cos(u)];   **(2)**

$$CS := [a \sin(u) \cos(v), b \sin(u) \sin(v), c \cos(u)]$$

Thereinto, 'a,b,c' are nonzero constants or 1th order derivable continuous functions, simply connected orientable closed surface 'CS' determines value of 'a,b,c'
> rgu:=[0,Pi/n-theta];

$$rgu := \left[0, \frac{\pi}{n} - \theta \right]$$

> rgv:=[0,2*Pi];

$$rgv := [0, 2\pi]$$

Set range of 'u,v' as $[0, \pi/n - \theta], [0, 2\pi]$, thereinto, 'n' is discretional constant, and $n \geq 1$; 'θ' is discretional constant or continuous function, and $\pi/n - \theta < \pi$, make surface 'CS' unclosed. (See also Poincare Conjecture: 'Every closed n-Manifold which is homotopy equivalent to n-Sphere is homeomorphic to n-Sphere'[19] and Chapter 1 (This Part) Section 1.2 'Define Abstract Simply Connected Orientable Closed Surface and Its Unclosed Surface')

Define parametrized expression of boundary curve 'CL'; Set Right Hand Rule as associated basis between positive direction of boundary curve 'CL' and normal vector of unclosed surface 'CS':
> CL:=[alpha*cos(v),beta*sin(v),gamma];   **(3)**

$$CL := [\alpha \cos(v), \beta \sin(v), \gamma]$$

Thereinto, 'α, β, γ' are constants($\alpha \neq 0, \beta \neq 0$) or 1th order derivable continuous functions depend on 'a,b,c'; Because of range of 'v' in $[0, 2\pi]$, boundary curve 'CL' is closed

[See also Chapter 1 (This Part) Section 1.5 'From Abstract Plane Simply Connected Closed Curve, To Abstract Space Closed Curve']

```
> [Diff(CL[1],v),Diff(CL[2],v),Diff(CL[3],v)]=[diff(CL[1],v),
diff(CL[2],v),diff(CL[3],v)];dCL:=rhs(%);        (4)
```

Calculate tangent vector 'dCL' of closed boundary curve 'CL'

$$\left[\frac{\partial}{\partial v}(\alpha\cos(v)),\frac{\partial}{\partial v}(\beta\sin(v)),\frac{\partial}{\partial v}\gamma\right]=[-\alpha\sin(v),\beta\cos(v),0]$$

$$dCL:=[-\alpha\sin(v),\beta\cos(v),0]$$

```
> V:=[(P)(x,y,z),(Q)(x,y,z),(R)(x,y,z)];    (5)
```

Define abstract spacial vector field 'V'(Suppose Vector Field 'V' possesses 1th order continuous partial derivatives at surface 'CS')

$$V:=[\mathrm{P}(x,y,z),\mathrm{Q}(x,y,z),\mathrm{R}(x,y,z)]$$

```
> [Diff(V[3],y)-Diff(V[2],z),Diff(V[1],z)-Diff(V[3],x),
Diff(V[2],x)-Diff(V[1],y)]
=[diff(V[3],y)-diff(V[2],z),diff(V[1],z)-diff(V[3],x),
diff(V[2],x)-diff(V[1],y)];cV1:=rhs(%);    (6)
```

Calculate 'cV1'-- Curl of abstract spacial vector field 'V'

$$\left[\left(\frac{\partial}{\partial y}\mathrm{R}(x,y,z)\right)-\left(\frac{\partial}{\partial z}\mathrm{Q}(x,y,z)\right),\left(\frac{\partial}{\partial z}\mathrm{P}(x,y,z)\right)-\left(\frac{\partial}{\partial x}\mathrm{R}(x,y,z)\right),\right.$$
$$\left.\left(\frac{\partial}{\partial x}\mathrm{Q}(x,y,z)\right)-\left(\frac{\partial}{\partial y}\mathrm{P}(x,y,z)\right)\right]=\left[\left(\frac{\partial}{\partial y}\mathrm{R}(x,y,z)\right)-\left(\frac{\partial}{\partial z}\mathrm{Q}(x,y,z)\right),\right.$$
$$\left.\left(\frac{\partial}{\partial z}\mathrm{P}(x,y,z)\right)-\left(\frac{\partial}{\partial x}\mathrm{R}(x,y,z)\right),\left(\frac{\partial}{\partial x}\mathrm{Q}(x,y,z)\right)-\left(\frac{\partial}{\partial y}\mathrm{P}(x,y,z)\right)\right]$$

$$cV1:=\left[\left(\frac{\partial}{\partial y}\mathrm{R}(x,y,z)\right)-\left(\frac{\partial}{\partial z}\mathrm{Q}(x,y,z)\right),\left(\frac{\partial}{\partial z}\mathrm{P}(x,y,z)\right)-\left(\frac{\partial}{\partial x}\mathrm{R}(x,y,z)\right),\right.$$
$$\left.\left(\frac{\partial}{\partial x}\mathrm{Q}(x,y,z)\right)-\left(\frac{\partial}{\partial y}\mathrm{P}(x,y,z)\right)\right]$$

```
// Segment range '[0,2*Pi]' of 'v':
> dus:=50;
```

Define amount 'dus' of boundary curve 'CL's parameter segmental cells (It can be discretional natural number)

$$dus:=50$$

```
> dv:=(rgv[2]-rgv[1])/dus; # Segment range '[0,2*Pi]' of 'v'
```

$$dv:=\frac{\pi}{25}$$

```
> x:='x':y:='y':z:='z':
```

```
// Transform variables, prevent that 'Import 'x =
```
α `cos(v),y=` β `sin(v),`

`z =` γ `' to abstract vector field '[P(x,y,z),Q(x,y,z),R(x,y,z)]''`

Microcosmic curve integral course of boundary curve 'CL's first segmental cell (50 segmental cells):

```
// Segment special vector field 'V':
```
(This step possesses formal meaning in 'Proof' only, and possesses essential meaning in 'Numerical Models')

```
> dV:=subs(v=rgv[1]+dv,V);
```

Spacial vector field 'V's average value that corresponds to boundary curve 'CL's first segmental cell

$$dV := [\,P(x, y, z),\, Q(x, y, z),\, R(x, y, z)\,]$$

```
// Segment tangent vector 'dCL':
```

```
> dCLm:=subs(v=rgv[1]+dv,dCL);
```

#'Tangent vector dCL's average value that corresponds to closed boundary curve 'CL's first segmental cell

$$dCLm := \left[\, -\alpha \sin\!\left(\frac{\pi}{25}\right),\, \beta \cos\!\left(\frac{\pi}{25}\right),\, 0 \,\right]$$

```
// Calculate microcosmic curve integral value of closed boundary curve
```
'CL's first segmental cell:

```
> dv*(dV[1]*dCLm[1]+dV[2]*dCLm[2]+dV[3]*dCLm[3]);
```

Base on integral median theorem, integral value of dot product (Spacial vector field 'V' and tangent vector 'dCL') at first segmental cell

$$\frac{1}{25}\,\pi\left(\, -P(x, y, z)\, \alpha \sin\!\left(\frac{\pi}{25}\right) + Q(x, y, z)\, \beta \cos\!\left(\frac{\pi}{25}\right)\right)$$

Microcosmic curve integral course of closed boundary curve 'CL's all segmental cells (50 segmental cells):

```
// Segment spacial vector field 'V':
```
(This step possesses formal meaning in 'Proof' only, and possesses essential meaning in 'Numerical Models')

```
> idV:=subs(v=rgv[1]+i*dv,V);
```

Spacial vector field 'V's average values that correspond to boundary curve 'CL's all segmental cells

$$idV := [\,P(x, y, z),\, Q(x, y, z),\, R(x, y, z)\,]$$

```
> seq(([i,idV,evalf(idV)]),i=1..dus);
```
List of spacial vector field 'V's average values that correspond to boundary curve 'CL's all segmental cells, be elided


```
// Segment tangent vector 'dCL':
> idCL:=subs(v=rgv[1]+i*dv,dCL);
```
#'Tangent vector dCL's average values that correspond to boundary curve 'CL's all segmental cells

$$idCL := \left[-\alpha \sin\left(\frac{i\,\pi}{25}\right),\ \beta \cos\left(\frac{i\,\pi}{25}\right),\ 0 \right]$$

```
> seq(([i,idCL,evalf(idCL)]),i=1..dus):
```
List of 'Tangent vector dCL's average values that correspond to boundary curve 'CL's all segmental cells, be elided
// In actual expressions, 'i' stands for natural number 1~50
//'idCL' isn't that 'single vector value', but that 'set of finite quantitative (viz.50) vector values'

// Calculate microcosmic curve integral values of closed boundary curve 'CL's all segmental cells:
```
> dv*(idV[1]*idCL[1]+idV[2]*idCL[2]+idV[3]*idCL[3]);
```
Base on integral median theorem, integral values of dot product (Spacial vector field 'V' and tangent vector 'dCL') at all segmental cells

$$\frac{1}{25}\pi\left(-P(x,y,z)\,\alpha \sin\left(\frac{i\,\pi}{25}\right) + Q(x,y,z)\,\beta \cos\left(\frac{i\,\pi}{25}\right) \right)$$

// Expression above-mentioned isn't that 'single value', but that 'set of finite quantitative (viz.50) values'

// List of finite quantitative microcosmic curl integral values:
```
> seq(([i,dv*(idV[1]*idCL[1]+idV[2]*idCL[2]+idV[3]*idCL[3]),
evalf(dv*(idV[1]*idCL[1]+idV[2]*idCL[2]+idV[3]*idCL[3]))]),
i=1..dus):
```
List of integral values of dot product(Spacial vector field 'V' and tangent vector 'dCL') at all segmental cells, be elided
// Structure sequence that is composed of finite quantitative microcomosic curl integral values:

```
> sqn:=seq(dv*(idV[1]*idCL[1]+idV[2]*idCL[2]+idV[3]*idCL[3]),
i=1..dus):
```

// Accumulational sum of sequence, obtain spacial closed curve integral
value at manifold:

```
> add(k,k=sqn):xi:=evalf(%);     (7)
```

Sum of integral values of dot product (Spacial vector field 'V' and
tangent vector 'dCL')at all segmental cells; Amount 'dus'of closed
boundary curve 'CL's segmental cells increases infinitely, sum of
integral values above-mentioned tends to '0' infinitely (Verbose
analytical expression is elided)

$$\xi := -0.20\ 10^{-9}\ P(x, y, z)\ \alpha$$

// Renewedly segment the range '[0,2*Pi]' of 'v':

```
> dus:=w;
```

Set amount 'dus' of boundary curve 'CL's parameter segmental cells
as natural number 'w'

$$dus := w$$

```
> dv:=(rgv[2]-rgv[1])/dus; # Segment range '[0,2*Pi]' of 'v'
```

$$dv := \frac{2\pi}{w}$$

**Microcosmic curve integral course of boundary curve 'CL's first
segmental cell (w segmental cells):**

// Segment spacial vector field 'V':

(This step possesses formal meaning in 'Proof' only, and possesses
essential meaning in 'Numerical Models')

```
> dV:=subs(v=rgv[1]+dv,V);
```

Spacial vector field 'V's average value that corresponds to closed
boundary curve 'CL's first segmental cell

$$dV := [P(x, y, z), Q(x, y, z), R(x, y, z)]$$

// Segment tangent vector 'dCL':

```
> dCLm:=subs(v=rgv[1]+dv,dCL);
```

#'Tangent vector dCL's average value that corresponds to closed
boundary curve 'CL's first segmental cell

$$dCLm := \left[-\alpha \sin\left(\frac{2\pi}{w}\right), \beta \cos\left(\frac{2\pi}{w}\right), 0 \right]$$

```
// Calculate microcosmic curve integral value of closed boundary
curve 'CL's first segmental cell:
> dv*(dV[1]*dCLm[1]+dV[2]*dCLm[2]+dV[3]*dCLm[3]);
```

Base on integral median theorem, integral value of dot product
(Spacial vector field 'V' and tangent vector 'dCL') at first segmental
cell

$$\frac{2\pi\left(-P(x,y,z)\,\alpha\,\sin\!\left(\dfrac{2\pi}{w}\right)+Q(x,y,z)\,\beta\,\cos\!\left(\dfrac{2\pi}{w}\right)\right)}{w}$$

Microcosmic curve integral course of closed boundary curve 'CL's all segmental cells (w segmental cells):

```
// Segment special vector field 'V':
```

(This step possesses formal meaning in 'Proof' only, and possesses
essential meaning in 'Numerical Models')

```
> idV:=subs(v=rgv[1]+i*dv,V);
```

Spacial vector field 'V's average values that correspond to boundary
curve 'CL's all segmental cells

$$idV := [\,P(x,y,z),\,Q(x,y,z),\,R(x,y,z)\,]$$

```
// Segment tangent vector 'dCL':
> idCL:=subs(v=rgv[1]+i*dv,dCL);
```

#'Tangent vector dCL's average values that correspond to boundary
curve 'CL's all segmental cells

$$idCL := \left[\,-\alpha\,\sin\!\left(\frac{2\,i\,\pi}{w}\right),\,\beta\,\cos\!\left(\frac{2\,i\,\pi}{w}\right),\,0\,\right]$$

```
// In actual expressions, 'i' stands for natural number 1~w
//'idCL' isn't that 'single vector value', but that 'set of finite
quantitative (viz. w) vector values'
```

```
// Calculate microcosmic curve integral values of closed boundary curve
'CL's all segmental cells:
> dv*(idV[1]*idCL[1]+idV[2]*idCL[2]+idV[3]*idCL[3]);
```

Base on integral median theorem, integral values of dot product
(Spacial vector field 'V' and tangent vector 'dCL') at all segmental
cells

978-1-62265-930-2 (online)　978-1-62265-931-9 (paper)

$$\frac{2\pi\left(-P(x,y,z)\,\alpha\,\sin\!\left(\dfrac{2\,i\,\pi}{w}\right)+Q(x,y,z)\,\beta\,\cos\!\left(\dfrac{2\,i\,\pi}{w}\right)\right)}{w}$$

// Expression above-mentioned isn't that 'single value', but that 'set of finite quantitative (viz. w) values'

// Structure finite sums:

> Sum(dv*(idV[1]*idCL[1]+idV[2]*idCL[2]+idV[3]*idCL[3]),i=1..dus);

In the case of 'Amount 'w' of closed boundary curve 'CL's parameter segmental cells is pending', sum of integral values (dot product about spacial vector field 'V' and tangent vector 'dCL')at all segmental cells

$$\sum_{i=1}^{w}\left(\frac{2\pi\left(-P(x,y,z)\,\alpha\,\sin\!\left(\dfrac{2\,i\,\pi}{w}\right)+Q(x,y,z)\,\beta\,\cos\!\left(\dfrac{2\,i\,\pi}{w}\right)\right)}{w}\right) \tag{8}$$

> vs:=value(%):

> Limit(vs,w=infinity);

// Infinitize finite sums, its limit operational value is 'Spacial Closed Curve Integral Value at Manifold'

In the case of "Amount 'w' of closed boundary curve 'CL's parameter segmental cells tends to infinity", limit value of integral values' sums (dot product about Spacial vector field 'V' and tangent vector 'dCL') at all segmental cells

$$\lim_{w\to\infty}-2\pi\cos\!\left(\frac{\pi}{w}\right)^{2}\left(\cos\!\left(\frac{\pi}{w}\right)^{3}\sin\!\left(\frac{\pi}{w}\right)P(x,y,z)\,\alpha-P(x,y,z)\,\alpha\,\sin\!\left(\frac{\pi}{w}\right)\cos\!\left(\frac{\pi}{w}\right)\right.$$

$$+Q(x,y,z)\,\beta\,\cos\!\left(\frac{\pi}{w}\right)^{4}-2\,Q(x,y,z)\,\beta\,\cos\!\left(\frac{\pi}{w}\right)^{2}+Q(x,y,z)\,\beta$$

$$+\cos\!\left(\frac{\pi}{w}\right)\sin\!\left(\frac{\pi}{w}\right)^{3}P(x,y,z)\,\alpha+\cos\!\left(\frac{\pi}{w}\right)^{2}\sin\!\left(\frac{\pi}{w}\right)^{2}\beta\,Q(x,y,z)$$

$$\left.-\sin\!\left(\frac{\pi}{w}\right)^{2}\beta\,Q(x,y,z)\right)(w+1)\Bigg/\left(\left(\cos\!\left(\frac{\pi}{w}\right)^{2}-1\right)^{2}w\right)-2\pi\Bigg($$

$$\cos\!\left(\frac{\pi}{w}\right)^{2}\sin\!\left(\frac{\pi}{w}\right)^{2}P(x,y,z)\,\alpha+\cos\!\left(\frac{\pi}{w}\right)^{3}\sin\!\left(\frac{\pi}{w}\right)Q(x,y,z)\,\beta$$

$$-\cos\!\left(\frac{\pi}{w}\right)\sin\!\left(\frac{\pi}{w}\right)\beta\,Q(x,y,z)+P(x,y,z)\,\alpha\,\cos\!\left(\frac{\pi}{w}\right)^{2}-P(x,y,z)\,\alpha\Bigg)$$

$$\sin\left(\frac{(w+1)\pi}{w}\right)\cos\left(\frac{(w+1)\pi}{w}\right) \Big/ \left(\left(\cos\left(\frac{\pi}{w}\right)^2-1\right)^2 w\right) - 2\pi$$

$$\left(P(x,y,z)\,\alpha\sin\left(\frac{\pi}{w}\right)\cos\left(\frac{\pi}{w}\right)+Q(x,y,z)\,\beta\cos\left(\frac{\pi}{w}\right)^2-Q(x,y,z)\,\beta\right)$$

$$\cos\left(\frac{(w+1)\pi}{w}\right)^2 \Big/ \left(w\left(\cos\left(\frac{\pi}{w}\right)^2-1\right)\right)+2\pi\cos\left(\frac{\pi}{w}\right)^2\Bigg($$

$$\cos\left(\frac{\pi}{w}\right)^3\sin\left(\frac{\pi}{w}\right)P(x,y,z)\,\alpha-P(x,y,z)\,\alpha\sin\left(\frac{\pi}{w}\right)\cos\left(\frac{\pi}{w}\right)+Q(x,y,z)\,\beta\cos\left(\frac{\pi}{w}\right)^4$$

$$-2Q(x,y,z)\,\beta\cos\left(\frac{\pi}{w}\right)^2+Q(x,y,z)\,\beta+\cos\left(\frac{\pi}{w}\right)\sin\left(\frac{\pi}{w}\right)^3 P(x,y,z)\,\alpha$$

$$+\cos\left(\frac{\pi}{w}\right)^2\sin\left(\frac{\pi}{w}\right)^2\beta\,Q(x,y,z)-\sin\left(\frac{\pi}{w}\right)^2\beta\,Q(x,y,z)\Bigg) \Big/ \left(\left(\cos\left(\frac{\pi}{w}\right)^2-1\right)^2 w\right)$$

$$\Bigg)+2\pi\Bigg(\cos\left(\frac{\pi}{w}\right)^2\sin\left(\frac{\pi}{w}\right)^2 P(x,y,z)\,\alpha+\cos\left(\frac{\pi}{w}\right)^3\sin\left(\frac{\pi}{w}\right)Q(x,y,z)\,\beta$$

$$-\cos\left(\frac{\pi}{w}\right)\sin\left(\frac{\pi}{w}\right)\beta\,Q(x,y,z)+P(x,y,z)\,\alpha\cos\left(\frac{\pi}{w}\right)^2-P(x,y,z)\,\alpha\Bigg)\sin\left(\frac{\pi}{w}\right)$$

$$\cos\left(\frac{\pi}{w}\right) \Big/ \left(\left(\cos\left(\frac{\pi}{w}\right)^2-1\right)^2 w\right)+$$

$$\frac{2\pi\left(P(x,y,z)\,\alpha\sin\left(\frac{\pi}{w}\right)\cos\left(\frac{\pi}{w}\right)+Q(x,y,z)\,\beta\cos\left(\frac{\pi}{w}\right)^2-Q(x,y,z)\,\beta\right)\cos\left(\frac{\pi}{w}\right)^2}{w\left(\cos\left(\frac{\pi}{w}\right)^2-1\right)}$$

```
> delta:=value(%);
```

$$\delta := 0$$

```
> x:='x':y:='y':z:='z':
> matrix(3,3,[i,j,k,Diff(CS[1],u),Diff(CS[2],u),Diff(CS[3],u),
Diff(CS[1],v),Diff(CS[2],v),Diff(CS[3],v)])=
matrix(3,3,[i,j,k,diff(CS[1],u),diff(CS[2],u),diff(CS[3],u),
diff(CS[1],v),diff(CS[2],v),diff(CS[3],v)]);m:=rhs(%);
# Define and calculate matrix of partial derivatives 'm', obtain
'tangent plane normal vector' of surface 'CS'
```

$$\begin{bmatrix} i & j & k \\ \dfrac{\partial}{\partial u}(a\sin(u)\cos(v)) & \dfrac{\partial}{\partial u}(b\sin(u)\sin(v)) & \dfrac{\partial}{\partial u}(c\cos(u)) \\ \dfrac{\partial}{\partial v}(a\sin(u)\cos(v)) & \dfrac{\partial}{\partial v}(b\sin(u)\sin(v)) & \dfrac{\partial}{\partial v}(c\cos(u)) \end{bmatrix} =$$

$$\begin{bmatrix} i & j & k \\ a\cos(u)\cos(v) & b\cos(u)\sin(v) & -c\sin(u) \\ -a\sin(u)\sin(v) & b\sin(u)\cos(v) & 0 \end{bmatrix}$$

$$m := \begin{bmatrix} i & j & k \\ a\cos(u)\cos(v) & b\cos(u)\sin(v) & -c\sin(u) \\ -a\sin(u)\sin(v) & b\sin(u)\cos(v) & 0 \end{bmatrix}$$

```
> det(m);
```

$$i\,c\sin(u)^2\,b\cos(v) + a\cos(u)\cos(v)^2\,k\,b\sin(u) + a\sin(u)^2\sin(v)\,j\,c$$
$$+ a\sin(u)\sin(v)^2\,k\,b\cos(u)$$

```
> mn:=simplify(%);
```

$$mn := \sin(u)\,(i\,c\sin(u)\,b\cos(v) + a\sin(u)\sin(v)\,j\,c + a\,k\,b\cos(u))$$

```
> A:=coeff(mn,i); # Obtain coefficient of 'i'
```

$$A := c\sin(u)^2\,b\cos(v)$$

```
> B:=coeff(mn,j); # Obtain coefficient of 'j'
```

$$B := \sin(u)^2\,a\sin(v)\,c$$

```
> C:=coeff(mn,k);  # Obtain coefficient of 'k'
```

$$C := \sin(u)\,a\,b\cos(u)$$

```
> [A,B,C]; # [A,B,C] structure 'tangent plane normal vector'     (9)
```

$$[c\sin(u)^2\,b\cos(v),\ \sin(u)^2\sin(v)\,a\,c,\ \sin(u)\,a\cos(u)\,b]$$

```
> x:='x':y:='y':z:='z':
> [Diff(V[3],y)*Diff(CS[2],u)*Diff(CS[2],v)
-Diff(V[2],z)*Diff(CS[3],u)*Diff(CS[3],v),
Diff(V[1],z)*Diff(CS[3],u)*Diff(CS[3],v)
-Diff(V[3],x)*Diff(CS[1],u)*Diff(CS[1],v),
Diff(V[2],x)*Diff(CS[1],u)*Diff(CS[1],v)
-Diff(V[1],y)*Diff(CS[2],u)*Diff(CS[2],v)]
=[diff(V[3],y)*diff(CS[2],u)*diff(CS[2],v)
-diff(V[2],z)*diff(CS[3],u)*diff(CS[3],v),
diff(V[1],z)*diff(CS[3],u)*diff(CS[3],v)-
```

```
diff(V[3],x)*diff(CS[1],u)*diff(CS[1],v),

diff(V[2],x)*diff(CS[1],u)*diff(CS[1],v)

-diff(V[1],y)*diff(CS[2],u)*diff(CS[2],v)];cV2:=rhs(%);    (10)
```

\# Convert curl 'cV1' from Cartesian coordinates expression to surface 'CS's coordinates expression

$$
\left[\left[\left(\frac{\partial}{\partial y} R(x,y,z) \right) \left(\frac{\partial}{\partial u} (b \sin(u) \sin(v)) \right) \left(\frac{\partial}{\partial v} (b \sin(u) \sin(v)) \right) \right. \right.
$$

$$
- \left(\frac{\partial}{\partial z} Q(x,y,z) \right) \left(\frac{\partial}{\partial u} (c \cos(u)) \right) \left(\frac{\partial}{\partial v} (c \cos(u)) \right),
$$

$$
\left(\frac{\partial}{\partial z} P(x,y,z) \right) \left(\frac{\partial}{\partial u} (c \cos(u)) \right) \left(\frac{\partial}{\partial v} (c \cos(u)) \right)
$$

$$
- \left(\frac{\partial}{\partial x} R(x,y,z) \right) \left(\frac{\partial}{\partial u} (a \sin(u) \cos(v)) \right) \left(\frac{\partial}{\partial v} (a \sin(u) \cos(v)) \right),
$$

$$
\left(\frac{\partial}{\partial x} Q(x,y,z) \right) \left(\frac{\partial}{\partial u} (a \sin(u) \cos(v)) \right) \left(\frac{\partial}{\partial v} (a \sin(u) \cos(v)) \right)
$$

$$
\left. - \left(\frac{\partial}{\partial y} P(x,y,z) \right) \left(\frac{\partial}{\partial u} (b \sin(u) \sin(v)) \right) \left(\frac{\partial}{\partial v} (b \sin(u) \sin(v)) \right) \right] = \left[
$$

$$
\left(\frac{\partial}{\partial y} R(x,y,z) \right) b^2 \cos(u) \sin(v) \sin(u) \cos(v),
$$

$$
\left(\frac{\partial}{\partial x} R(x,y,z) \right) a^2 \cos(u) \cos(v) \sin(u) \sin(v),
$$

$$
- \left(\frac{\partial}{\partial x} Q(x,y,z) \right) a^2 \cos(u) \cos(v) \sin(u) \sin(v)
$$

$$
- \left(\frac{\partial}{\partial y} P(x,y,z) \right) b^2 \cos(u) \sin(v) \sin(u) \cos(v) \right]
$$

$$
cV2 := \left[\left[\left(\frac{\partial}{\partial y} R(x,y,z) \right) b^2 \cos(u) \sin(v) \sin(u) \cos(v), \right. \right.
$$

$$
\left(\frac{\partial}{\partial x} R(x,y,z) \right) a^2 \cos(u) \cos(v) \sin(u) \sin(v),
$$

$$
- \left(\frac{\partial}{\partial x} Q(x,y,z) \right) a^2 \cos(u) \cos(v) \sin(u) \sin(v)
$$

$$
\left. - \left(\frac{\partial}{\partial y} P(x,y,z) \right) b^2 \cos(u) \sin(v) \sin(u) \cos(v) \right]
$$

```
// Segment range '[0,Pi],[0,2*Pi]' of 'u,v':
> dus:=50;
```

\# Set amount of surface 'CS's segmental cells,this amount can be discretional natural number

$$
dus := 50
$$

```
> du:=(rgu[2]-rgu[1])/dus; # Segment range of 'u'
```

$$du := \frac{\pi}{50\,n} - \frac{\theta}{50}$$

```
> dv:=(rgv[2]-rgv[1])/dus; # Segment range of 'v'
```

$$dv := \frac{\pi}{25}$$

Microcosmic surface integral course of surface 'CS's first segmental cell (50 segmental cells):

```
// Segment 'tangent plane normal vector'
> dA:=subs(v=rgv[1]+dv,subs(u=rgu[1]+du,A));
```

$$dA := c\,\sin\!\left(\frac{\pi}{50\,n} - \frac{\theta}{50}\right)^2 b\,\cos\!\left(\frac{\pi}{25}\right)$$

```
> dB:=subs(v=rgv[1]+dv,subs(u=rgu[1]+du,B));
```

$$dB := \sin\!\left(\frac{\pi}{50\,n} - \frac{\theta}{50}\right)^2 a\,\sin\!\left(\frac{\pi}{25}\right) c$$

```
> dC:=subs(v=rgv[1]+dv,subs(u=rgu[1]+du,C));
```

$$dC := \sin\!\left(\frac{\pi}{50\,n} - \frac{\theta}{50}\right) a\,b\,\cos\!\left(\frac{\pi}{50\,n} - \frac{\theta}{50}\right)$$

```
> [dA,dB,dC]; #'[dA,dB,dC]' is 'tangent plane normal vector's average
```
value that corresponds to surface 'CS's first segmental cell

$$\left[c\,\sin\!\left(-\frac{\pi}{50\,n} + \frac{\theta}{50}\right)^2 b\,\cos\!\left(\frac{\pi}{25}\right),\ \sin\!\left(-\frac{\pi}{50\,n} + \frac{\theta}{50}\right)^2 \sin\!\left(\frac{\pi}{25}\right) a\,c, \right.$$
$$\left. -\sin\!\left(-\frac{\pi}{50\,n} + \frac{\theta}{50}\right) a\,\cos\!\left(-\frac{\pi}{50\,n} + \frac{\theta}{50}\right) b \right]$$

```
// Segment curl 'cV2':
> dcV2:=subs(v=rgv[1]+dv,subs(u=rgu[1]+du,cV2));
```

```
# Curl 'cV2's average value that corresponds to surface 'CS's first
segmental cell
```

$$dcV2 := \left[\left(\frac{\partial}{\partial y} R(x,y,z)\right) b^2 \cos\!\left(\frac{\pi}{50\,n} - \frac{\theta}{50}\right) \sin\!\left(\frac{\pi}{25}\right) \sin\!\left(\frac{\pi}{50\,n} - \frac{\theta}{50}\right) \cos\!\left(\frac{\pi}{25}\right), \right.$$
$$\left(\frac{\partial}{\partial x} R(x,y,z)\right) a^2 \cos\!\left(\frac{\pi}{50\,n} - \frac{\theta}{50}\right) \cos\!\left(\frac{\pi}{25}\right) \sin\!\left(\frac{\pi}{50\,n} - \frac{\theta}{50}\right) \sin\!\left(\frac{\pi}{25}\right),$$
$$-\left(\frac{\partial}{\partial x} Q(x,y,z)\right) a^2 \cos\!\left(\frac{\pi}{50\,n} - \frac{\theta}{50}\right) \cos\!\left(\frac{\pi}{25}\right) \sin\!\left(\frac{\pi}{50\,n} - \frac{\theta}{50}\right) \sin\!\left(\frac{\pi}{25}\right)$$
$$\left. -\left(\frac{\partial}{\partial y} P(x,y,z)\right) b^2 \cos\!\left(\frac{\pi}{50\,n} - \frac{\theta}{50}\right) \sin\!\left(\frac{\pi}{25}\right) \sin\!\left(\frac{\pi}{50\,n} - \frac{\theta}{50}\right) \cos\!\left(\frac{\pi}{25}\right) \right]$$

```
// Calculate  microcosmic surface integral value of surface 'CS's
first segmental cell:
```

```
> (dcV2[1]*dA+dcV2[2]*dB+dcV2[3]*dC)*du*dv;
```

```
# Base on integral median theorem, integral value of dot product(Curl
'cV2' and '[dA,dB,dC]') at first segmental cell
```

$$\frac{1}{25}\left(-\left(\frac{\partial}{\partial y}R(x,y,z)\right)b^3\cos\left(-\frac{\pi}{50\,n}+\frac{\theta}{50}\right)\sin\left(\frac{\pi}{25}\right)\sin\left(-\frac{\pi}{50\,n}+\frac{\theta}{50}\right)^3\cos\left(\frac{\pi}{25}\right)^2 c\right.$$

$$-\left(\frac{\partial}{\partial x}R(x,y,z)\right)a^3\cos\left(-\frac{\pi}{50\,n}+\frac{\theta}{50}\right)\cos\left(\frac{\pi}{25}\right)\sin\left(-\frac{\pi}{50\,n}+\frac{\theta}{50}\right)^3\sin\left(\frac{\pi}{25}\right)^2 c-\left($$

$$\left(\frac{\partial}{\partial x}Q(x,y,z)\right)a^2\cos\left(-\frac{\pi}{50\,n}+\frac{\theta}{50}\right)\cos\left(\frac{\pi}{25}\right)\sin\left(-\frac{\pi}{50\,n}+\frac{\theta}{50}\right)\sin\left(\frac{\pi}{25}\right)$$

$$\left.+\left(\frac{\partial}{\partial y}P(x,y,z)\right)b^2\cos\left(-\frac{\pi}{50\,n}+\frac{\theta}{50}\right)\sin\left(\frac{\pi}{25}\right)\sin\left(-\frac{\pi}{50\,n}+\frac{\theta}{50}\right)\cos\left(\frac{\pi}{25}\right)\right)$$

$$\sin\left(-\frac{\pi}{50\,n}+\frac{\theta}{50}\right)a\,b\cos\left(-\frac{\pi}{50\,n}+\frac{\theta}{50}\right)\right)\left(\frac{\pi}{50\,n}-\frac{\theta}{50}\right)\pi$$

Microcosmic surface integral courses of surface 'CS's all segmental cells (50 segmental cells):

```
// Segment 'tangent plane normal vector'
> stdA:=subs(v=rgv[1]+t*dv,subs(u=rgu[1]+s*du,A));
```

$$stdA := c\sin\left(s\left(\frac{\pi}{50\,n}-\frac{\theta}{50}\right)\right)^2 b\cos\left(\frac{t\,\pi}{25}\right)$$

```
> stdB:=subs(v=rgv[1]+t*dv,subs(u=rgu[1]+s*du,B));
```

$$stdB := \sin\left(s\left(\frac{\pi}{50\,n}-\frac{\theta}{50}\right)\right)^2 a\sin\left(\frac{t\,\pi}{25}\right)c$$

```
> stdC:=subs(v=rgv[1]+t*dv,subs(u=rgu[1]+s*du,C));
```

$$stdC := \sin\left(s\left(\frac{\pi}{50\,n}-\frac{\theta}{50}\right)\right)a\,b\cos\left(s\left(\frac{\pi}{50\,n}-\frac{\theta}{50}\right)\right)$$

```
> [stdA,stdB,stdC]; #'[stdA,stdB,stdC]' are 'tangent plane normal
vector's average values that correspond to surface 'CS's all segmental
cells
```

$$\left[c\sin\left(\left(\frac{\pi}{50\,n}-\frac{\theta}{50}\right)s\right)^2 b\cos\left(\frac{\pi\,t}{25}\right),\sin\left(\left(\frac{\pi}{50\,n}-\frac{\theta}{50}\right)s\right)^2\sin\left(\frac{\pi\,t}{25}\right)a\,c,\right.$$

$$\left.\sin\left(\left(\frac{\pi}{50\,n}-\frac{\theta}{50}\right)s\right)a\cos\left(\left(\frac{\pi}{50\,n}-\frac{\theta}{50}\right)s\right)b\right]$$

```
// In actual expressions, 's,t'stand for natural number 1~50
//'[stdA,stdB,stdC]'isn't that 'single vector value', but that 'set
of finite quantitative(viz.50) vector values'
```

```
// Segment curl `cV2':
> stdcV2:=subs(v=rgv[1]+t*dv,subs(u=rgu[1]+s*du,cV2));
# Curl `cV2's average values that correspond to surface`CS's all
segmental cells
```

$$stdcV2 := \Bigg[$$

$$\left(\frac{\partial}{\partial y}R(x,y,z)\right)b^2\cos\left(s\left(\frac{\pi}{50\,n}-\frac{\theta}{50}\right)\right)\sin\left(\frac{t\,\pi}{25}\right)\sin\left(s\left(\frac{\pi}{50\,n}-\frac{\theta}{50}\right)\right)\cos\left(\frac{t\,\pi}{25}\right),$$

$$\left(\frac{\partial}{\partial x}R(x,y,z)\right)a^2\cos\left(s\left(\frac{\pi}{50\,n}-\frac{\theta}{50}\right)\right)\cos\left(\frac{t\,\pi}{25}\right)\sin\left(s\left(\frac{\pi}{50\,n}-\frac{\theta}{50}\right)\right)\sin\left(\frac{t\,\pi}{25}\right),$$

$$-\left(\frac{\partial}{\partial x}Q(x,y,z)\right)a^2\cos\left(s\left(\frac{\pi}{50\,n}-\frac{\theta}{50}\right)\right)\cos\left(\frac{t\,\pi}{25}\right)\sin\left(s\left(\frac{\pi}{50\,n}-\frac{\theta}{50}\right)\right)\sin\left(\frac{t\,\pi}{25}\right)$$

$$-\left(\frac{\partial}{\partial y}P(x,y,z)\right)b^2\cos\left(s\left(\frac{\pi}{50\,n}-\frac{\theta}{50}\right)\right)\sin\left(\frac{t\,\pi}{25}\right)\sin\left(s\left(\frac{\pi}{50\,n}-\frac{\theta}{50}\right)\right)\cos\left(\frac{t\,\pi}{25}\right)\Bigg]$$

```
// In actual expressions, `s,t'stand for natural number 1~50
//`stdcV2'isn't that `single vector value', but that `set of finite
quantitative (viz.50) vector values'


// Calculate microcosmosic surface integral value at all segmental
cells of surface `CS':
> (stdcV2[1]*stdA+stdcV2[2]*stdB+stdcV2[3]*stdC)*du*dv;
# Base on integral median theorem, integral value of dot product
(Curl `cV2' and `[stdA,stdB,stdC]')at all segmental cells
```

$$\frac{1}{25}\Bigg(\left(\frac{\partial}{\partial y}R(x,y,z)\right)b^3\cos\left(s\left(\frac{\pi}{50\,n}-\frac{\theta}{50}\right)\right)\sin\left(\frac{t\,\pi}{25}\right)\sin\left(s\left(\frac{\pi}{50\,n}-\frac{\theta}{50}\right)\right)^3\cos\left(\frac{t\,\pi}{25}\right)^2 c$$

$$+\left(\frac{\partial}{\partial x}R(x,y,z)\right)a^3\cos\left(s\left(\frac{\pi}{50\,n}-\frac{\theta}{50}\right)\right)\cos\left(\frac{t\,\pi}{25}\right)\sin\left(s\left(\frac{\pi}{50\,n}-\frac{\theta}{50}\right)\right)^3\sin\left(\frac{t\,\pi}{25}\right)^2 c$$

$$+\left(-\left(\frac{\partial}{\partial x}Q(x,y,z)\right)a^2\cos\left(s\left(\frac{\pi}{50\,n}-\frac{\theta}{50}\right)\right)\cos\left(\frac{t\,\pi}{25}\right)\sin\left(s\left(\frac{\pi}{50\,n}-\frac{\theta}{50}\right)\right)\sin\left(\frac{t\,\pi}{25}\right)\right.$$

$$\left.-\left(\frac{\partial}{\partial y}P(x,y,z)\right)b^2\cos\left(s\left(\frac{\pi}{50\,n}-\frac{\theta}{50}\right)\right)\sin\left(\frac{t\,\pi}{25}\right)\sin\left(s\left(\frac{\pi}{50\,n}-\frac{\theta}{50}\right)\right)\cos\left(\frac{t\,\pi}{25}\right)\right)$$

$$\sin\left(s\left(\frac{\pi}{50\,n}-\frac{\theta}{50}\right)\right)a\,b\cos\left(s\left(\frac{\pi}{50\,n}-\frac{\theta}{50}\right)\right)\Bigg)\left(\frac{\pi}{50\,n}-\frac{\theta}{50}\right)\pi$$

```
// Expression above-mentioned isn't that `single value', but that `set
of finite quantitative(viz.50) values'


// List of finite quantitative microcosmic surface integral values:
> seq(seq([s*t,(stdcV2[1]*stdA+stdcV2[2]*stdB
```

```
+stdcV2[3]*stdC)*du*dv,evalf((stdcV2[1]*stdA+stdcV2[2]*stdB
+stdcV2[3]*stdC)*du*dv)],s=1..dus),t=1..dus):
# List of integral values of dot product (Curl'cV2' and '[stdA,stdB,
stdC]') at all segmental cells, be elided
// Structure sequence that is composed of finite quantitative
microcomosic surface integral values:
> sqn:=seq(seq((stdcV2[1]*stdA+stdcV2[2]*stdB
+stdcV2[3]*stdC)*du*dv,s=1..dus),t=1..dus):
// Accumulational sum of sequence, obtain surface integral value at
manifold:
> add(k,k=sqn):omega:=value(%):  (11)
# Sum of integral values of dot product (Curl 'cV2' and '[stdA,stdB,
stdC]')at all segmental cells; Verbose analytical and float expression
are elided; Amount of surface 'CS's segmental cells increases
infinitely, sum of integral values above-mentioned tends to '0'
infinitely


// Renewedly segment range '[0,Pi],[0,2*Pi]' of 'u,v':
> dus:=w;
# Set amount of surface 'CS's segmental cells as natural number'w'
```

$$dus := w$$

```
> du:=(rgu[2]-rgu[1])/dus; # Segment range of 'u'
```

$$du := \frac{\dfrac{\pi}{n} - \theta}{w}$$

```
> dv:=(rgv[2]-rgv[1])/dus; # Segment range of 'v'
```

$$dv := \frac{2\,\pi}{w}$$

Microcosmic surface integral course of surface 'CS's first segmental cell (w segmental cells):

```
// Segment 'tangent plane normal vector'
> dA:=subs(v=rgv[1]+dv,subs(u=rgu[1]+du,A));
```

$$dA := c \sin\left(\frac{\dfrac{\pi}{n} - \theta}{w}\right)^{2} b \cos\left(\frac{2\,\pi}{w}\right)$$

```
> dB:=subs(v=rgv[1]+dv,subs(u=rgu[1]+du,B));
```

$$dB := \sin\left(\frac{\frac{\pi}{n} - \theta}{w}\right)^2 a \sin\left(\frac{2\pi}{w}\right) c$$

```
> dC:=subs(v=rgv[1]+dv,subs(u=rgu[1]+du,C));
```

$$dC := \sin\left(\frac{\frac{\pi}{n} - \theta}{w}\right) a \, b \cos\left(\frac{\frac{\pi}{n} - \theta}{w}\right)$$

```
> [dA,dB,dC]; #'[dA,dB,dC]' is 'tangent plane normal vector's average
```
value that corresponds to surface 'CS's first segmental cell

$$\left[c \sin\left(\frac{\frac{\pi}{n}-\theta}{w}\right)^2 b \cos\left(\frac{2\pi}{w}\right), \sin\left(\frac{\frac{\pi}{n}-\theta}{w}\right)^2 \sin\left(\frac{2\pi}{w}\right) a\, c, \sin\left(\frac{\frac{\pi}{n}-\theta}{w}\right) a \cos\left(\frac{\frac{\pi}{n}-\theta}{w}\right) b \right]$$

```
// Segment curl 'cV2':

> dcV2:=subs(v=rgv[1]+dv,subs(u=rgu[1]+du,cV2));
```

Curl 'cV2's average value that corresponds to surface 'CS's first
segmental cell

$$dcV2 := \left[\left(\frac{\partial}{\partial y} R(x,y,z)\right) b^2 \cos\left(\frac{\frac{\pi}{n}-\theta}{w}\right) \sin\left(\frac{2\pi}{w}\right) \sin\left(\frac{\frac{\pi}{n}-\theta}{w}\right) \cos\left(\frac{2\pi}{w}\right), \right.$$

$$\left(\frac{\partial}{\partial x} R(x,y,z)\right) a^2 \cos\left(\frac{\frac{\pi}{n}-\theta}{w}\right) \cos\left(\frac{2\pi}{w}\right) \sin\left(\frac{\frac{\pi}{n}-\theta}{w}\right) \sin\left(\frac{2\pi}{w}\right),$$

$$-\left(\frac{\partial}{\partial x} Q(x,y,z)\right) a^2 \cos\left(\frac{\frac{\pi}{n}-\theta}{w}\right) \cos\left(\frac{2\pi}{w}\right) \sin\left(\frac{\frac{\pi}{n}-\theta}{w}\right) \sin\left(\frac{2\pi}{w}\right)$$

$$\left. -\left(\frac{\partial}{\partial y} P(x,y,z)\right) b^2 \cos\left(\frac{\frac{\pi}{n}-\theta}{w}\right) \sin\left(\frac{2\pi}{w}\right) \sin\left(\frac{\frac{\pi}{n}-\theta}{w}\right) \cos\left(\frac{2\pi}{w}\right) \right]$$

```
// Calculate microcosmic surface integral value of surface 'CS's
first segmental cell:

> (dcV2[1]*dA+dcV2[2]*dB+dcV2[3]*dC)*du*dv;
```

Base on integral median theorem, integral value of dot product (Curl
'cV2' and '[dA,dB,dC]') at first segmental cell

$$2\left(\left(\frac{\partial}{\partial y} R(x,y,z)\right) b^3 \cos\left(\frac{\frac{\pi}{n}-\theta}{w}\right) \sin\left(\frac{2\pi}{w}\right) \sin\left(\frac{\frac{\pi}{n}-\theta}{w}\right)^3 \cos\left(\frac{2\pi}{w}\right)^2 c \right.$$

$$+ \left(\frac{\partial}{\partial x} R(x,y,z) \right) a^3 \cos\left(\frac{\frac{\pi}{n} - \theta}{w} \right) \cos\left(\frac{2\,\pi}{w} \right) \sin\left(\frac{\frac{\pi}{n} - \theta}{w} \right)^3 \sin\left(\frac{2\,\pi}{w} \right)^2 c + \Bigg($$

$$- \left(\frac{\partial}{\partial x} Q(x,y,z) \right) a^2 \cos\left(\frac{\frac{\pi}{n} - \theta}{w} \right) \cos\left(\frac{2\,\pi}{w} \right) \sin\left(\frac{\frac{\pi}{n} - \theta}{w} \right) \sin\left(\frac{2\,\pi}{w} \right)$$

$$- \left(\frac{\partial}{\partial y} P(x,y,z) \right) b^2 \cos\left(\frac{\frac{\pi}{n} - \theta}{w} \right) \sin\left(\frac{2\,\pi}{w} \right) \sin\left(\frac{\frac{\pi}{n} - \theta}{w} \right) \cos\left(\frac{2\,\pi}{w} \right) \sin\left(\frac{\frac{\pi}{n} - \theta}{w} \right) a\,b$$

$$\cos\left(\frac{\frac{\pi}{n} - \theta}{w} \right) \Bigg) \left(\frac{\pi}{n} - \theta \right) \pi \,/\, w^2$$

Microcosmic surface integral course of surface 'CS's all segmental cells (w segmental cells):

// Segment 'tangent plane normal vector'

> stdA:=subs(v=rgv[1]+t*dv,subs(u=rgu[1]+s*du,A));

$$stdA := c \sin\left(\frac{s\left(\frac{\pi}{n} - \theta \right)}{w} \right)^2 b \cos\left(\frac{2\,t\,\pi}{w} \right)$$

> stdB:=subs(v=rgv[1]+t*dv,subs(u=rgu[1]+s*du,B));

$$stdB := \sin\left(\frac{s\left(\frac{\pi}{n} - \theta \right)}{w} \right)^2 a \sin\left(\frac{2\,t\,\pi}{w} \right) c$$

> stdC:=subs(v=rgv[1]+t*dv,subs(u=rgu[1]+s*du,C));

$$stdC := \sin\left(\frac{s\left(\frac{\pi}{n} - \theta \right)}{w} \right) a\,b \cos\left(\frac{s\left(\frac{\pi}{n} - \theta \right)}{w} \right)$$

> [stdA,stdB,stdC]; #'[stdA,stdB,stdC]' are 'tangent plane normal vector's average values that correspond to surface 'CS's all segmental cells

$$\left[c \sin\left(\frac{\left(\frac{\pi}{n} - \theta \right) s}{w} \right)^2 b \cos\left(\frac{2\,\pi\,t}{w} \right), \; \sin\left(\frac{\left(\frac{\pi}{n} - \theta \right) s}{w} \right)^2 \sin\left(\frac{2\,\pi\,t}{w} \right) a\,c, \right.$$

$$\left. \sin\left(\frac{\left(\frac{\pi}{n} - \theta \right) s}{w} \right) a \cos\left(\frac{\left(\frac{\pi}{n} - \theta \right) s}{w} \right) b \right]$$

```
// In actual expressions, 's,t'stand for natural number 1~w
//'[stdA,stdB,stdC]' isn't that 'single vector value', but that 'set
of finite quantitative(viz. w) vector values'


// Segment curl 'cV2':
> stdcV2:=subs(v=rgv[1]+t*dv,subs(u=rgu[1]+s*du,cV2));
# Curl 'cV2's average values that correspond to surface'CS's all
segmental cells
```

$$
stdcV2 := \left[\left(\frac{\partial}{\partial y} R(x,y,z) \right) b^2 \cos\left(\frac{s\left(\frac{\pi}{n} - \theta \right)}{w} \right) \sin\left(\frac{2\,t\,\pi}{w} \right) \sin\left(\frac{s\left(\frac{\pi}{n} - \theta \right)}{w} \right) \cos\left(\frac{2\,t\,\pi}{w} \right), \right.
$$

$$
\left(\frac{\partial}{\partial x} R(x,y,z) \right) a^2 \cos\left(\frac{s\left(\frac{\pi}{n} - \theta \right)}{w} \right) \cos\left(\frac{2\,t\,\pi}{w} \right) \sin\left(\frac{s\left(\frac{\pi}{n} - \theta \right)}{w} \right) \sin\left(\frac{2\,t\,\pi}{w} \right),
$$

$$
-\left(\frac{\partial}{\partial x} Q(x,y,z) \right) a^2 \cos\left(\frac{s\left(\frac{\pi}{n} - \theta \right)}{w} \right) \cos\left(\frac{2\,t\,\pi}{w} \right) \sin\left(\frac{s\left(\frac{\pi}{n} - \theta \right)}{w} \right) \sin\left(\frac{2\,t\,\pi}{w} \right)
$$

$$
\left. -\left(\frac{\partial}{\partial y} P(x,y,z) \right) b^2 \cos\left(\frac{s\left(\frac{\pi}{n} - \theta \right)}{w} \right) \sin\left(\frac{2\,t\,\pi}{w} \right) \sin\left(\frac{s\left(\frac{\pi}{n} - \theta \right)}{w} \right) \cos\left(\frac{2\,t\,\pi}{w} \right) \right]
$$

```
// In actual expressions, 's,t' stand for natural number 1~w
//'stdcV2' isn't that 'single vector value', but that 'set of finite
quantitative (viz. w) vector values'


// Calculate microcosmosic surface integral value at all segmental
cells of surface 'CS':
> (stdcV2[1]*stdA+stdcV2[2]*stdB+stdcV2[3]*stdC)*du*dv;
# Base on integral median theorem, integral value of dot product
(Curl 'cV2' and '[stdA,stdB,stdC]') at all segmental cells
```

$$
2\left(\left(\frac{\partial}{\partial y} R(x,y,z) \right) b^3 \cos\left(\frac{s\left(\frac{\pi}{n} - \theta \right)}{w} \right) \sin\left(\frac{2\,t\,\pi}{w} \right) \sin\left(\frac{s\left(\frac{\pi}{n} - \theta \right)}{w} \right)^3 \cos\left(\frac{2\,t\,\pi}{w} \right)^2 c \right.
$$

$$
+\left(\frac{\partial}{\partial x} R(x,y,z) \right) a^3 \cos\left(\frac{s\left(\frac{\pi}{n} - \theta \right)}{w} \right) \cos\left(\frac{2\,t\,\pi}{w} \right) \sin\left(\frac{s\left(\frac{\pi}{n} - \theta \right)}{w} \right)^3 \sin\left(\frac{2\,t\,\pi}{w} \right)^2 c + \left(
$$

$$
-\left(\frac{\partial}{\partial x} Q(x,y,z) \right) a^2 \cos\left(\frac{s\left(\frac{\pi}{n} - \theta \right)}{w} \right) \cos\left(\frac{2\,t\,\pi}{w} \right) \sin\left(\frac{s\left(\frac{\pi}{n} - \theta \right)}{w} \right) \sin\left(\frac{2\,t\,\pi}{w} \right)
$$

$$-\left(\frac{\partial}{\partial y}\,\mathrm{P}(x,y,z)\right)b^2\cos\!\left(\frac{s\left(\frac{\pi}{n}-\theta\right)}{w}\right)\sin\!\left(\frac{2\,t\,\pi}{w}\right)\sin\!\left(\frac{s\left(\frac{\pi}{n}-\theta\right)}{w}\right)\cos\!\left(\frac{2\,t\,\pi}{w}\right)\Bigg)$$

$$\sin\!\left(\frac{s\left(\frac{\pi}{n}-\theta\right)}{w}\right)a\,b\,\cos\!\left(\frac{s\left(\frac{\pi}{n}-\theta\right)}{w}\right)\Bigg)\left(\frac{\pi}{n}-\theta\right)\pi\,/\,w^2$$

// Expression above-mentioned isn't that 'single value', but that 'set
of finite quantitative (viz. w) values'

// Structure finite sums:

```
> Sum(Sum((stdcV2[1]*stdA+stdcV2[2]*stdB+stdcV2[3]*stdC)*du*dv,
s=1..dus),t=1..dus);
```

In the case of "Amount 'w' of surface 'CS's parameter segmental cells
is pending", sum of integral values (dot product about curl 'cV2' and
'[stdA,stdB,stdC]') at all segmental cells

$$\sum_{t=1}^{w}\left(\sum_{s=1}^{w}\left(2\right.\right.$$

$$\left(\frac{\partial}{\partial y}\,\mathrm{R}(x,y,z)\right)b^3\cos\!\left(\frac{s\left(\frac{\pi}{n}-\theta\right)}{w}\right)\sin\!\left(\frac{2\,t\,\pi}{w}\right)\sin\!\left(\frac{s\left(\frac{\pi}{n}-\theta\right)}{w}\right)^3\cos\!\left(\frac{2\,t\,\pi}{w}\right)^2c$$

$$+\left(\frac{\partial}{\partial x}\,\mathrm{R}(x,y,z)\right)a^3\cos\!\left(\frac{s\left(\frac{\pi}{n}-\theta\right)}{w}\right)\cos\!\left(\frac{2\,t\,\pi}{w}\right)\sin\!\left(\frac{s\left(\frac{\pi}{n}-\theta\right)}{w}\right)^3\sin\!\left(\frac{2\,t\,\pi}{w}\right)^2c+\Bigg($$

$$-\left(\frac{\partial}{\partial x}\,\mathrm{Q}(x,y,z)\right)a^2\cos\!\left(\frac{s\left(\frac{\pi}{n}-\theta\right)}{w}\right)\cos\!\left(\frac{2\,t\,\pi}{w}\right)\sin\!\left(\frac{s\left(\frac{\pi}{n}-\theta\right)}{w}\right)\sin\!\left(\frac{2\,t\,\pi}{w}\right)$$

$$-\left(\frac{\partial}{\partial y}\,\mathrm{P}(x,y,z)\right)b^2\cos\!\left(\frac{s\left(\frac{\pi}{n}-\theta\right)}{w}\right)\sin\!\left(\frac{2\,t\,\pi}{w}\right)\sin\!\left(\frac{s\left(\frac{\pi}{n}-\theta\right)}{w}\right)\cos\!\left(\frac{2\,t\,\pi}{w}\right)\Bigg)$$

$$\sin\!\left(\frac{s\left(\frac{\pi}{n}-\theta\right)}{w}\right)a\,b\,\cos\!\left(\frac{s\left(\frac{\pi}{n}-\theta\right)}{w}\right)\Bigg)\left(\frac{\pi}{n}-\theta\right)\pi\,/\,w^2\Bigg)\Bigg)$$

```
> vs:=value(%);
```

Launch by Intel Celeron CPU E3400 2.6G (32 bit Processor) 1GB Memory,
operation of sum can't be completed, Await to enhance computer

```
hardware capability;
# Simplify sum expression,change range '[0,Pi/n-theta]' of 'u' to
'[0,Pi/2]' (Elide two variables:'n' and 'theta'),calculate renewedly
(include closed curve integral aforementioned):
> Sum(Sum((stdcV2[1]*stdA+stdcV2[2]*stdB+stdcV2[3]*stdC)*du*dv,
s=1..dus),t=1..dus);    (12)
```

$$\sum_{t=1}^{w}\left(\sum_{s=1}^{w}\left(\left(\left(\frac{\partial}{\partial y}R(x,y,z)\right)b^3\cos\left(\frac{s\,\pi}{2\,w}\right)\sin\left(\frac{2\,t\,\pi}{w}\right)\sin\left(\frac{s\,\pi}{2\,w}\right)^3\cos\left(\frac{2\,t\,\pi}{w}\right)^2 c\right.\right.\right.$$

$$+\left(\frac{\partial}{\partial x}R(x,y,z)\right)a^3\cos\left(\frac{s\,\pi}{2\,w}\right)\cos\left(\frac{2\,t\,\pi}{w}\right)\sin\left(\frac{s\,\pi}{2\,w}\right)^3\sin\left(\frac{2\,t\,\pi}{w}\right)^2 c+\left(\right.$$

$$-\left(\frac{\partial}{\partial x}Q(x,y,z)\right)a^2\cos\left(\frac{s\,\pi}{2\,w}\right)\cos\left(\frac{2\,t\,\pi}{w}\right)\sin\left(\frac{s\,\pi}{2\,w}\right)\sin\left(\frac{2\,t\,\pi}{w}\right)$$

$$-\left(\frac{\partial}{\partial y}P(x,y,z)\right)b^2\cos\left(\frac{s\,\pi}{2\,w}\right)\sin\left(\frac{2\,t\,\pi}{w}\right)\sin\left(\frac{s\,\pi}{2\,w}\right)\cos\left(\frac{2\,t\,\pi}{w}\right)\right)\sin\left(\frac{s\,\pi}{2\,w}\right)a\,b$$

$$\left.\left.\left.\cos\left(\frac{s\,\pi}{2\,w}\right)\right)\pi^2/w^2\right)\right)$$

```
> vs:=value(%): # Extremely lengthy sum expression is elided
> Limit(vs,w=infinity):
// Infinitize finite sums, its limit operational value is 'Surface
Integral Value at Manifold'
# In the case of "Amount 'n' of surface 'CS's segmental cells tends
to infinity", limit value of integral values' sums (dot product about
curl 'cV2' and '[stdA,stdB,stdC]') at all segmental cells
> epsilon:=value(%);
```

$$\varepsilon:=0$$

Viz. In the case of 'w → ∞', (8) = (12):

$$\sum_{i=1}^{w}\left(\frac{2\,\pi\left(-P(x,y,z)\,\alpha\,\sin\left(\frac{2\,i\,\pi}{w}\right)+Q(x,y,z)\,\beta\,\cos\left(\frac{2\,i\,\pi}{w}\right)\right)}{w}\right)$$

$$=$$

$$\sum_{t=1}^{w}\left(\sum_{s=1}^{w}\left(\left(\left(\frac{\partial}{\partial y}R(x,y,z)\right)b^3\cos\left(\frac{s\,\pi}{2\,w}\right)\sin\left(\frac{2\,t\,\pi}{w}\right)\sin\left(\frac{s\,\pi}{2\,w}\right)^3\cos\left(\frac{2\,t\,\pi}{w}\right)^2 c\right.\right.\right.$$

$$+\left(\frac{\partial}{\partial x}R(x,y,z)\right)a^3\cos\left(\frac{s\,\pi}{2\,w}\right)\cos\left(\frac{2\,t\,\pi}{w}\right)\sin\left(\frac{s\,\pi}{2\,w}\right)^3\sin\left(\frac{2\,t\,\pi}{w}\right)^2 c+\left(\right.$$

$$-\left(\frac{\partial}{\partial x} Q(x,y,z)\right) a^2 \cos\left(\frac{s\,\pi}{2\,w}\right) \cos\left(\frac{2\,t\,\pi}{w}\right) \sin\left(\frac{s\,\pi}{2\,w}\right) \sin\left(\frac{2\,t\,\pi}{w}\right)$$

$$-\left(\frac{\partial}{\partial y} P(x,y,z)\right) b^2 \cos\left(\frac{s\,\pi}{2\,w}\right) \sin\left(\frac{2\,t\,\pi}{w}\right) \sin\left(\frac{s\,\pi}{2\,w}\right) \cos\left(\frac{2\,t\,\pi}{w}\right)\right) \sin\left(\frac{s\,\pi}{2\,w}\right) a\,b$$

$$\cos\left(\frac{s\,\pi}{2\,w}\right)\right)\pi^2 \big/ w^2\bigg)$$

Above mentioned equation can be described as:

$$\int_{L+} A \cdot dL = \iint_{S} rotA \cdot n\, dS \quad \textbf{(1)}, \text{ Complete Proof.}$$

Chapter 6 Numerical Model of Curl Theorem Finite Sums Limits at Manifold

6.1 Numerical Model of Curl Theorem Finite Sums Limits at Manifold

Known: Parametrized expression of Simply connected、 Orientable、 Unclosed Surface(Irregular、Asymmetrical)

$$\left[\sin(u)\cos(v),\ \sin(u)\sin(v)\sin\left(\frac{v}{3}\right),\ \cos(u)\right] \tag{1}$$

thereinto, $u \in [0, \frac{\pi}{2}]$, $v \in [0, 2\pi]$;

and Integral Vector Field $\left[\dfrac{y}{2}, \dfrac{z^2}{3}, \dfrac{x}{3}\right]$ (2)

Calculate and Validate Curl Theorem at Manifold
(Finite Sums Limits)

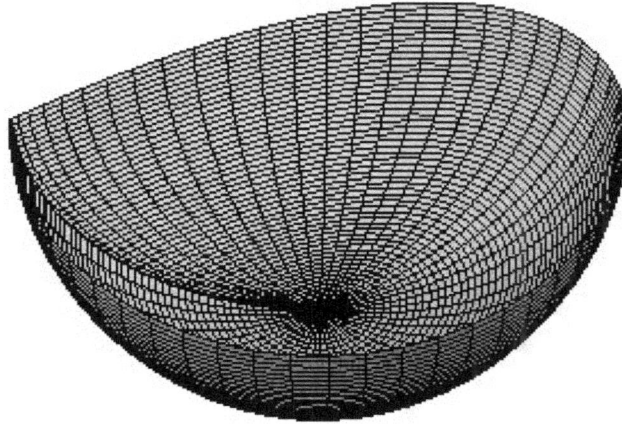

Figure(III)6.1.1 Simply connected、 Orientable、 Unclosed Surface (1)

[Irregular、 Asymmetrical]

Solution:

First, Free Spacial Closed Curve Integral (Finite Sums Limits):

Import right boundary value of 'u' (viz. $\dfrac{\pi}{2}$) to parametrized expression(1) of surface, obtain parametrized expression(3) of Unclosed Surface' Closed Boundary Curve; Set Right Hand Rule as associated basis between positive direction of boundary curve and normal vector of unclosed surface.

Be different to define expression of boundary curve'$[\alpha$ cos(v), β sin(v),γ]' that depends on surface '[a*sin(u)cos(v), b*sin(u) sin(v),c*cos(u)]' for existence in 'Proof', in 'Numerical Models', it is possible to import boundary value of 'u' to expression of surface (1), directly obtain expression of closed boundary curve

$$\left[\sin\left(\frac{\pi}{2}\right)\cos(v), \sin\left(\frac{\pi}{2}\right)\sin(v)\sin\left(\frac{v}{3}\right), \cos\left(\frac{\pi}{2}\right) \right] = \left[\cos(v), \sin(v)\sin\left(\frac{v}{3}\right), 0 \right] \quad (3)$$

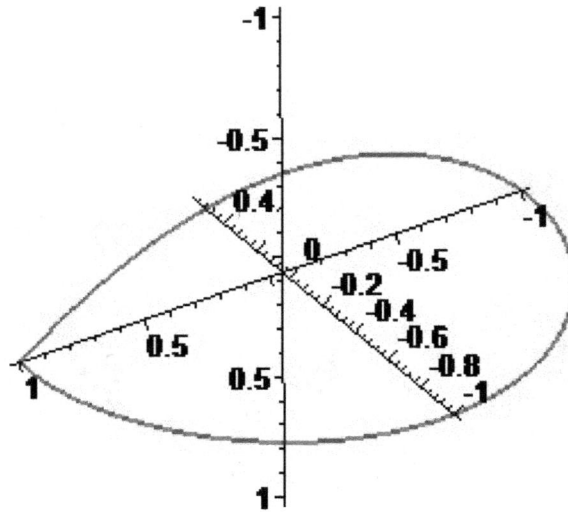

Figure(III)6.1.2 Closed Boundary Curve(3) of Simply connected Orientable、 Unclosed Surface(1)

Calculate tangent vector of closed boundary curve 'L' (4):

$$\left[\frac{d}{dv}\cos(v),\frac{d}{dv}\left(\sin(v)\sin\left(\frac{v}{3}\right)\right),\frac{d}{dv}0\right]=\left[-\sin(v),\cos(v)\sin\left(\frac{v}{3}\right)+\frac{1}{3}\sin(v)\cos\left(\frac{v}{3}\right),0\right]$$

Set Amount of closed boundary curve 'L's parametrized segmental cells as 50 (It can be discretional natural number)   (5)

1.Microcosmic Curve Integral Course at Closed Boundary Curve 'L's first segmental cell:

Segment range [0,2π] of 'v': dv = $\dfrac{2\pi}{50}=\dfrac{\pi}{25}$ (6)

Segment Tangent Vector (7): [Viz. Input (6) to (4)]

$$\left[-\sin\left(\frac{\pi}{25}\right),\cos\left(\frac{\pi}{25}\right)\sin\left(\frac{\pi}{75}\right)+\frac{1}{3}\sin\left(\frac{\pi}{25}\right)\cos\left(\frac{\pi}{75}\right),0\right]_{(7)}$$

Segment Vector Field: [Viz. Input (3) to (2), then input (6)]

$$\left[\frac{1}{2}\sin\left(\frac{\pi}{25}\right)\sin\left(\frac{\pi}{75}\right),0,\frac{1}{3}\cos\left(\frac{\pi}{25}\right)\right]\qquad(8)$$

Calculate Microcosmic Curve Integral Value at Closed Boundary Curve'L's first segmental cell (9):

Base on integral median theorem, dot product of vector field (8) and tangent vector (7), times segment unit of 'v'(6), that is Microcosmic Curve Integral Value at first segmental cell (9)

$$-\frac{1}{50}\pi\sin\left(\frac{\pi}{25}\right)^2\sin\left(\frac{\pi}{75}\right)$$

(9)

2.Microcosmic Curve Integral Course at Closed Boundary Curve 'L's all (Viz. 50) segmental cells:

Segment range $[0,2\pi]$ of 'v': $dv = \dfrac{i2\pi}{50}=\dfrac{i\pi}{25}$ (10)

(thereinto, 'i' stand for natural number 1~50)

Segment Tangent Vector (11): [Viz. Input (10) to (4)]

$$\left[-\sin\left(\frac{\pi i}{25}\right), \cos\left(\frac{\pi i}{25}\right)\sin\left(\frac{\pi i}{75}\right)+\frac{1}{3}\sin\left(\frac{\pi i}{25}\right)\cos\left(\frac{\pi i}{75}\right), 0\right]$$

(11)

//Expression (11) isn't that 'Single vector value', but that 'Set of finite quantitative (Viz.50) vector values'

Segment Vector Field: [Viz. Input (3) to (2), then input (10)]

$$\left[\frac{1}{2}\sin\left(\frac{\pi i}{25}\right)\sin\left(\frac{\pi i}{75}\right), 0, \frac{1}{3}\cos\left(\frac{\pi i}{25}\right)\right]$$

(12)

//Expression (12) isn't that 'Single vector value', but that 'Set of finite quantitative (Viz.50) vector values'

Calculate Microcosmic Curve Integral Values at Closed Boundary Curve 'L's all segmental cells (13):

Base on Integral Median Theorem, dot product of vector field (12) and tangent vector (11), times segment unit of 'v'(10), these are Microcosmic Curve Integral Values at all segmental cells (13)

$$-\frac{1}{50}\pi\sin\left(\frac{\pi i}{25}\right)^2\sin\left(\frac{\pi i}{75}\right)$$

(13)

//Expression (13) isn't that 'Single value', but that 'Set of finite quantitative (Viz.50) values'

Structure Sequence that is composed of finite quantitative (viz. 50) Microcosmic Curve Integral Values (14):

[Lengthy expression of sequence is elided;

In program template of Waterloo Maplesoft (see also Section 5.2):

sqn:=seq(dv*(idV[1]*idCL[1]+idV[2]*idCL[2]+idV[3]*idCL[3]),

i=1..dus):

change ':'(last) to ';', then obtain expression of sequence]

Accumulational sum of sequence (viz. Sum of integral values of dot product [vector field(12) and tangent vector(11)] at all (Viz.50) segmental cells, obtain Curve Integral Value at manifold(15):

[Lengthy expression of accumulational sum result is elided;

In program template of Waterloo Maplesoft (See also Section 5.2):

add(k,k=sqn):xi:=evalf(%);

change ':'(middle) to ';', then obtain expression of result]

Change expression of accumulational sum result to float value:
$$-1.157143379$$

Set Amount of closed boundary curve 'L's parametrized segmental cells as natural number 'n'   (16)

3.Microcosmic Curve Integral Course at Closed Boundary Curve 'L's first segmental cell:

Segment range [0,2π] of 'v': dv = $\dfrac{2\pi}{n}$ (17)

Segment Tangent Vector (18): [Viz. Input (17) to (4)]

$$\left[-\sin\left(\frac{2\pi}{n}\right),\cos\left(\frac{2\pi}{n}\right)\sin\left(\frac{2\pi}{3n}\right)+\frac{1}{3}\sin\left(\frac{2\pi}{n}\right)\cos\left(\frac{2\pi}{3n}\right),0\right]_{(18)}$$

Segment Vector Field (19): [Viz. Input (3) to (2), then input (17)]

$$\left[\frac{1}{2}\sin\left(\frac{2\pi}{n}\right)\sin\left(\frac{2\pi}{3n}\right),0,\frac{1}{3}\cos\left(\frac{2\pi}{n}\right)\right]$$ (19)

Calculate Microcosmic Curve Integral Value at Closed Boundary Curve 'L's first segmental cell (20):

Base on Integral Median Theorem, dot product of vector field (19) and tangent vector(18), times segment unit of 'v'(17), that is Microcosmic Curve Integral Value at first segmental cell (20)

$$-\frac{\pi \sin\left(\dfrac{2\pi}{n}\right)^2 \sin\left(\dfrac{2\pi}{3n}\right)}{n}$$

(20)

4.Microcosmic Curve Integral Course at Closed Boundary Curve 'L's all (Viz. 50) segmental cells:

Segment range $[0, 2\pi]$ of 'v': dv = $\dfrac{2\pi i}{n}$

(thereinto, 'i' stand for natural number 1~n)   (21)

Segment Tangent Vector (22): [Viz. Input (21) to (4)]

$$\left[-\sin\left(\frac{2\pi i}{n}\right), \cos\left(\frac{2\pi i}{n}\right)\sin\left(\frac{2\pi i}{3n}\right) + \frac{1}{3}\sin\left(\frac{2\pi i}{n}\right)\cos\left(\frac{2\pi i}{3n}\right), 0\right]$$ (22)

// Expression (22) isn't that 'Single vector value', but that 'Set of finite quantitative (Viz.n) vector values'

Segment Vector Field (23): [Viz. Input (3) to (2), then input (21)]

$$\left[\frac{1}{2}\sin\left(\frac{2\pi i}{n}\right)\sin\left(\frac{2\pi i}{3n}\right), 0, \frac{1}{3}\cos\left(\frac{2\pi i}{n}\right)\right]$$ (23)

// Expression (23) isn't that 'Single vector value', but that 'Set of finite quantitative (Viz.n) vector values'

Calculate Microcosmic Curve Integral Values at Closed Boundary Curve 'L's all segmental cells (24):

Base on Integral Median Theorem, dot product of vector field (23) and tangent vector(22), times segment unit of 'v'(21), these are Microcosmic Curve Integral Values at all segmental cells (24)

$$-\frac{\pi \sin\left(\dfrac{2\pi i}{n}\right)^2 \sin\left(\dfrac{2\pi i}{3n}\right)}{n}$$

(24)

// Expression (24) isn't that 'Single value', but that 'Set of finite quantitative (Viz.n) values'

Structure finite sums (25):

(In the case of "Amount 'n' of closed boundary curve 'L's segmental cells is uncertain", sum of integral values [dot product about vector field(23) and tangent vector(22)] at all segmental cells)

$$\sum_{i=1}^{n} \left(- \frac{\pi \sin\left(\frac{2\pi i}{n}\right)^2 \sin\left(\frac{2\pi i}{3n}\right)}{n} \right) \tag{25}$$

Infinitize finite sums, its limit operational value is 'Curl Integral Value at Manifold'(26):

(In the case of "Amount 'n' of closed boundary curve 'L's segmental cells tends to infinity", limit value of integral values' sums [dot product about vector field(23) and tangent vector(22)] at all segmental cells)

$$\lim_{n \to \infty} \sum_{i=1}^{n} \left(- \frac{\pi \sin\left(\frac{2\pi i}{n}\right)^2 \sin\left(\frac{2\pi i}{3n}\right)}{n} \right)$$

$$=$$

$$\lim_{n \to \infty} \frac{2\pi \sin\left(\frac{2\pi}{3n}\right) \cos\left(\frac{\pi(n+1)}{n}\right)^2 \cos\left(\frac{2\pi(n+1)}{3n}\right)}{n\left(8\cos\left(\frac{\pi}{n}\right)^4 - 8\cos\left(\frac{\pi}{n}\right)^2 - \cos\left(\frac{2\pi}{3n}\right) + 1\right)}$$

$$- \frac{2\pi \sin\left(\frac{2\pi}{3n}\right) \cos\left(\frac{\pi(n+1)}{n}\right)^4 \cos\left(\frac{2\pi(n+1)}{3n}\right)}{n\left(8\cos\left(\frac{\pi}{n}\right)^4 - 8\cos\left(\frac{\pi}{n}\right)^2 - \cos\left(\frac{2\pi}{3n}\right) + 1\right)}$$

$$+ \frac{2\pi \sin\left(\frac{2\pi(n+1)}{3n}\right) \cos\left(\frac{\pi(n+1)}{n}\right)^2}{n} - \frac{2\pi \sin\left(\frac{2\pi(n+1)}{3n}\right) \cos\left(\frac{\pi(n+1)}{n}\right)^4}{n}$$

$$+ \frac{4\pi\left(2\cos\left(\frac{\pi}{n}\right)^2 - 1\right) \sin\left(\frac{\pi}{n}\right) \cos\left(\frac{\pi}{n}\right) \sin\left(\frac{\pi(n+1)}{n}\right) \cos\left(\frac{\pi(n+1)}{n}\right) \sin\left(\frac{2\pi(n+1)}{3n}\right)}{n\left(8\cos\left(\frac{\pi}{n}\right)^4 - 8\cos\left(\frac{\pi}{n}\right)^2 - \cos\left(\frac{2\pi}{3n}\right) + 1\right)}$$

$$- \frac{8\pi\left(2\cos\left(\frac{\pi}{n}\right)^2 - 1\right) \sin\left(\frac{\pi}{n}\right) \cos\left(\frac{\pi}{n}\right) \sin\left(\frac{\pi(n+1)}{n}\right) \cos\left(\frac{\pi(n+1)}{n}\right)^3 \sin\left(\frac{2\pi(n+1)}{3n}\right)}{n\left(8\cos\left(\frac{\pi}{n}\right)^4 - 8\cos\left(\frac{\pi}{n}\right)^2 - \cos\left(\frac{2\pi}{3n}\right) + 1\right)}$$

$$-2\pi\left(\cos\left(\frac{\pi}{n}\right)^2-1\right)\cos\left(\frac{\pi}{n}\right)^2\sin\left(\frac{2\pi}{3n}\right)\cos\left(\frac{2\pi(n+1)}{3n}\right)\right)\ \Bigg/\ \left(n\left(\right.\right.$$

$$8\cos\left(\frac{\pi}{n}\right)^4\cos\left(\frac{2\pi}{3n}\right)-8\cos\left(\frac{\pi}{n}\right)^4-8\cos\left(\frac{\pi}{n}\right)^2\cos\left(\frac{2\pi}{3n}\right)+8\cos\left(\frac{\pi}{n}\right)^2-\cos\left(\frac{2\pi}{3n}\right)^2$$

$$+2\cos\left(\frac{2\pi}{3n}\right)-1\Bigg)\Bigg)-\frac{2\pi\sin\left(\frac{2\pi}{3n}\right)\cos\left(\frac{\pi}{n}\right)^2\cos\left(\frac{2\pi}{3n}\right)}{n\left(8\cos\left(\frac{\pi}{n}\right)^4-8\cos\left(\frac{\pi}{n}\right)^2-\cos\left(\frac{2\pi}{3n}\right)+1\right)}$$

$$+\frac{2\pi\sin\left(\frac{2\pi}{3n}\right)\cos\left(\frac{\pi}{n}\right)^4\cos\left(\frac{2\pi}{3n}\right)}{n\left(8\cos\left(\frac{\pi}{n}\right)^4-8\cos\left(\frac{\pi}{n}\right)^2-\cos\left(\frac{2\pi}{3n}\right)+1\right)}-\frac{2\pi\sin\left(\frac{2\pi}{3n}\right)\cos\left(\frac{\pi}{n}\right)^2}{n}$$

$$+\frac{2\pi\sin\left(\frac{2\pi}{3n}\right)\cos\left(\frac{\pi}{n}\right)^4}{n}-\frac{4\pi\left(2\cos\left(\frac{\pi}{n}\right)^2-1\right)\sin\left(\frac{\pi}{n}\right)^2\cos\left(\frac{\pi}{n}\right)^2\sin\left(\frac{2\pi}{3n}\right)}{n\left(8\cos\left(\frac{\pi}{n}\right)^4-8\cos\left(\frac{\pi}{n}\right)^2-\cos\left(\frac{2\pi}{3n}\right)+1\right)}$$

$$+\frac{8\pi\left(2\cos\left(\frac{\pi}{n}\right)^2-1\right)\sin\left(\frac{\pi}{n}\right)^2\cos\left(\frac{\pi}{n}\right)^4\sin\left(\frac{2\pi}{3n}\right)}{n\left(8\cos\left(\frac{\pi}{n}\right)^4-8\cos\left(\frac{\pi}{n}\right)^2-\cos\left(\frac{2\pi}{3n}\right)+1\right)}+2\pi\left(\cos\left(\frac{\pi}{n}\right)^2-1\right)\cos\left(\frac{\pi}{n}\right)^2$$

$$\sin\left(\frac{2\pi}{3n}\right)\cos\left(\frac{2\pi}{3n}\right)\ \Bigg/\ \left(n\left(8\cos\left(\frac{\pi}{n}\right)^4\cos\left(\frac{2\pi}{3n}\right)-8\cos\left(\frac{\pi}{n}\right)^4\right.\right.$$

$$-8\cos\left(\frac{\pi}{n}\right)^2\cos\left(\frac{2\pi}{3n}\right)+8\cos\left(\frac{\pi}{n}\right)^2-\cos\left(\frac{2\pi}{3n}\right)^2+2\cos\left(\frac{2\pi}{3n}\right)-1\Bigg)\Bigg)$$

$$=\frac{-81}{70}\qquad(26)$$

Second, Free (Unclosed) Surface Integral (Finite Sums Limits):

According to surface parametrized expression(1), define and calculate matrix of partial derivatives(27), obtain 'tangent plane normal vector' of surface 'S':

$$\begin{bmatrix} i & j & k \\[4pt] \dfrac{\partial}{\partial u}(\sin(u)\cos(v)) & \dfrac{\partial}{\partial u}\left(\sin(u)\sin(v)\sin\left(\dfrac{v}{3}\right)\right) & \dfrac{d}{du}\cos(u) \\[10pt] \dfrac{\partial}{\partial v}(\sin(u)\cos(v)) & \dfrac{\partial}{\partial v}\left(\sin(u)\sin(v)\sin\left(\dfrac{v}{3}\right)\right) & \dfrac{\partial}{\partial v}\cos(u) \end{bmatrix}=$$

$$\frac{1}{3}\sin(u)\left(\cos(u)\cos(v)\,k\,\sin(v)\cos\left(\frac{v}{3}\right)+3\,i\,\sin(u)\cos(v)\sin\left(\frac{v}{3}\right)\right.$$
$$\left.+\,i\,\sin(u)\sin(v)\cos\left(\frac{v}{3}\right)+3\,\sin(u)\sin(v)\,j+3\,\cos(u)\,k\,\sin\left(\frac{v}{3}\right)\right) \tag{27}$$

From expression(27), respectively pick up coefficient of item 'i,j,k',obtain tangent plane normal vector of surface 'S' (28):

$$\left[\frac{1}{3}\sin(u)\left(3\,\sin(u)\cos(v)\sin\left(\frac{v}{3}\right)+\sin(u)\sin(v)\cos\left(\frac{v}{3}\right)\right),\ \sin(u)^2\sin(v),\right.$$
$$\left.\frac{1}{3}\sin(u)\left(\cos(u)\cos(v)\sin(v)\cos\left(\frac{v}{3}\right)+3\,\cos(u)\sin\left(\frac{v}{3}\right)\right)\right] \tag{28}$$

Calculate Curl of Vector Field (29):

$$\left[\left(\frac{\partial}{\partial y}\left(\frac{x}{3}\right)\right)-\left(\frac{d}{dz}\left(\frac{z^2}{3}\right)\right),\ \left(\frac{\partial}{\partial z}\left(\frac{y}{2}\right)\right)-\left(\frac{d}{dx}\left(\frac{x}{3}\right)\right),\ \left(\frac{\partial}{\partial x}\left(\frac{z^2}{3}\right)\right)-\left(\frac{d}{dy}\left(\frac{y}{2}\right)\right)\right]=\left[-\frac{2z}{3},\frac{-1}{3},\frac{-1}{2}\right]$$

Set Amount of surface'S's segmental cells as 50

(This amount can be discretional natural number)　　　(30)

5.Microcosmic surface integral course at surface'S's first segmental cell:

Segment range [0,π/2] of 'u': du $=\dfrac{\pi}{2}/50=\dfrac{\pi}{100}$

Segment range [0,2π] of 'v': dv $=\dfrac{2\pi}{50}=\dfrac{\pi}{25}$　　　(31)

Segment 'tangent plane normal vector'(32):[viz. Input (31) to (28)]

$$\left[\frac{1}{3}\sin\left(\frac{\pi}{100}\right)\left(3\,\sin\left(\frac{\pi}{100}\right)\cos\left(\frac{\pi}{25}\right)\sin\left(\frac{\pi}{75}\right)+\sin\left(\frac{\pi}{100}\right)\sin\left(\frac{\pi}{25}\right)\cos\left(\frac{\pi}{75}\right)\right),\right.$$
$$\sin\left(\frac{\pi}{100}\right)^2\sin\left(\frac{\pi}{25}\right),$$
$$\left.\frac{1}{3}\sin\left(\frac{\pi}{100}\right)\left(\cos\left(\frac{\pi}{100}\right)\cos\left(\frac{\pi}{25}\right)\sin\left(\frac{\pi}{25}\right)\cos\left(\frac{\pi}{75}\right)+3\,\cos\left(\frac{\pi}{100}\right)\sin\left(\frac{\pi}{75}\right)\right)\right]$$

Segment Curl (33): [Viz. Input (1) to (29), then input (31)]

$$\left[-\frac{2}{3}\cos\left(\frac{\pi}{100}\right),\frac{-1}{3},\frac{-1}{2}\right] \tag{33}$$

Calculate microcosmic surface integral value at surface'S's first segmental cell(34):

Base on integral median theorem, dot product of curl(33) and tangent plane normal vector(32), times segment unit of 'u,v'(31), that is microcosmic surface integral value at first segmental Cell(34)

$$\frac{1}{2500}\left(-\frac{2}{9}\right.$$

$$\sin\left(\frac{\pi}{100}\right)\left(3\sin\left(\frac{\pi}{100}\right)\cos\left(\frac{\pi}{25}\right)\sin\left(\frac{\pi}{75}\right)+\sin\left(\frac{\pi}{100}\right)\sin\left(\frac{\pi}{25}\right)\cos\left(\frac{\pi}{75}\right)\right)\cos\left(\frac{\pi}{100}\right)$$

$$-\frac{1}{3}\sin\left(\frac{\pi}{100}\right)^2\sin\left(\frac{\pi}{25}\right)$$

$$-\frac{1}{6}\sin\left(\frac{\pi}{100}\right)\left(\cos\left(\frac{\pi}{100}\right)\cos\left(\frac{\pi}{25}\right)\sin\left(\frac{\pi}{25}\right)\cos\left(\frac{\pi}{75}\right)+3\cos\left(\frac{\pi}{100}\right)\sin\left(\frac{\pi}{75}\right)\right)\right)\pi^2$$

6.Microcosmic surface integral course at surface'S's all (viz.50) segmental cells:

Segment range $[0,\pi/2]$ of 'u': du $=\dfrac{s\pi}{2}/50=\dfrac{s\pi}{100}$

Segment range $[0,2\pi]$ of 'v': dv $=\dfrac{2t\pi}{50}=\dfrac{t\pi}{25}$

(thereinto, 's' and 't' stand for natural number 1~50)   (35)

Segment 'tangent plane normal vector'(36):[viz. Input (35) to (28)]

$$\left[\frac{1}{3}\sin\left(\frac{\pi s}{100}\right)\left(3\sin\left(\frac{\pi s}{100}\right)\cos\left(\frac{\pi t}{25}\right)\sin\left(\frac{\pi t}{75}\right)+\sin\left(\frac{\pi s}{100}\right)\sin\left(\frac{\pi t}{25}\right)\cos\left(\frac{\pi t}{75}\right)\right),\right.$$

$$\sin\left(\frac{\pi s}{100}\right)^2\sin\left(\frac{\pi t}{25}\right),$$

$$\left.\frac{1}{3}\sin\left(\frac{\pi s}{100}\right)\left(\cos\left(\frac{\pi s}{100}\right)\cos\left(\frac{\pi t}{25}\right)\sin\left(\frac{\pi t}{25}\right)\cos\left(\frac{\pi t}{75}\right)+3\cos\left(\frac{\pi s}{100}\right)\sin\left(\frac{\pi t}{75}\right)\right)\right]$$

// Expression (36) isn't that 'Single vector value', but that 'Set of finite quantitative (Viz.50) vector values'

Segment Curl (37): [Viz. Input (1) to (29), then input (35)]

$$\left[-\frac{2}{3}\cos\left(\frac{\pi s}{100}\right),\frac{-1}{3},\frac{-1}{2}\right]$$

(37)

// Expression (37) isn't that 'Single vector value', but that 'Set of finite quantitative (Viz.50) vector values'

Calculate microcosmic surface integral value at surface'S's all segmental cells (38):

Base on integral median theorem, dot product of curl (37) and tangent plane normal vector(36), times segment unit of 'u,v'(35), these are microcosmic surface integral values at all segmental Cells (38)

$$
\frac{1}{2500}\left(-\frac{2}{9}\right.
$$

$$
\sin\left(\frac{\pi s}{100}\right)\left(3\sin\left(\frac{\pi s}{100}\right)\cos\left(\frac{\pi t}{25}\right)\sin\left(\frac{\pi t}{75}\right)+\sin\left(\frac{\pi s}{100}\right)\sin\left(\frac{\pi t}{25}\right)\cos\left(\frac{\pi t}{75}\right)\right)\cos\left(\frac{\pi s}{100}\right)
$$

$$
-\frac{1}{3}\sin\left(\frac{\pi s}{100}\right)^2\sin\left(\frac{\pi t}{25}\right)
$$

$$
\left.-\frac{1}{6}\sin\left(\frac{\pi s}{100}\right)\left(\cos\left(\frac{\pi s}{100}\right)\cos\left(\frac{\pi t}{25}\right)\sin\left(\frac{\pi t}{25}\right)\cos\left(\frac{\pi t}{75}\right)+3\cos\left(\frac{\pi s}{100}\right)\sin\left(\frac{\pi t}{75}\right)\right)\right)\pi^2
$$

// Expression (38) isn't that 'Single value', but that 'Set of finite quantitative (Viz.50) values'

Structure sequence that is composed of finite quantitative (viz. 50) microcomosic surface integral values (39):

[Lengthy expression of sequence is elided;

In program template of Waterloo Maplesoft (see also Section 5.2):

```
sqn:=seq(seq(((stdV[1]*stdA+stdV[2]*stdB+stdV[3]*stdC)*du*dv,
s=1..dus),t=1..dus):
```

change ':'(last) to ';', then obtain expression of sequence]

Accumulational sum of sequence (viz. Sum of integral values of dot product [curl (37) and tangent plane normal vector (36)] at all segmental cells), obtain surface integral value at manifold (40):

[Lengthy expression of accumulational sum result is elided;

In program template of Waterloo Maplesoft (See also Section 5.2):

```
add(k,k=sqn):xi:=evalf(%);
```

change ':'(middle) to ';', then obtain expression of result]

Change expression of accumulational sum result to float value:

$$-1.181828580$$

Set Amount of surface 'S's segmental cells as n

(viz. uncertain natural number)   (41)

7.Microcosmic surface integral course at surface 'S's first segmental cell:

Segment range $[0, \pi/2]$ of 'u': $du = \dfrac{\pi}{2n}$

Segment range $[0, 2\pi]$ of 'v': $dv = \dfrac{2\pi}{n}$ (42)

Segment 'tangent plane normal vector' (43): [viz. Input (42) to (28)]

$$\left[\frac{1}{3}\sin\left(\frac{\pi}{2n}\right)\left(3\sin\left(\frac{\pi}{2n}\right)\cos\left(\frac{2\pi}{n}\right)\sin\left(\frac{2\pi}{3n}\right)+\sin\left(\frac{\pi}{2n}\right)\sin\left(\frac{2\pi}{n}\right)\cos\left(\frac{2\pi}{3n}\right)\right),\right.$$

$$\sin\left(\frac{\pi}{2n}\right)^2\sin\left(\frac{2\pi}{n}\right),$$

$$\left.\frac{1}{3}\sin\left(\frac{\pi}{2n}\right)\left(\cos\left(\frac{\pi}{2n}\right)\cos\left(\frac{2\pi}{n}\right)\sin\left(\frac{2\pi}{n}\right)\cos\left(\frac{2\pi}{3n}\right)+3\cos\left(\frac{\pi}{2n}\right)\sin\left(\frac{2\pi}{3n}\right)\right)\right]$$

Segment Curl(44): [Viz. Input (1) to (29), then input (42)]

$$\left[-\frac{2}{3}\cos\left(\frac{\pi}{2n}\right),\frac{-1}{3},\frac{-1}{2}\right]$$ (44)

Calculate microcosmic surface integral value at surface 'S's first segmental cell(45):

Base on integral median theorem, dot product of curl(44) and tangent plane normal vector(43), times segment unit of 'u,v'(42), that is microcosmic surface integral value at first segmental cell (45)

$$\left(-\frac{2}{9}\right.$$

$$\sin\left(\frac{\pi}{2n}\right)\left(3\sin\left(\frac{\pi}{2n}\right)\cos\left(\frac{2\pi}{n}\right)\sin\left(\frac{2\pi}{3n}\right)+\sin\left(\frac{\pi}{2n}\right)\sin\left(\frac{2\pi}{n}\right)\cos\left(\frac{2\pi}{3n}\right)\right)\cos\left(\frac{\pi}{2n}\right)$$

$$-\frac{1}{3}\sin\left(\frac{\pi}{2n}\right)^2\sin\left(\frac{2\pi}{n}\right)$$

$$\left.-\frac{1}{6}\sin\left(\frac{\pi}{2n}\right)\left(\cos\left(\frac{\pi}{2n}\right)\cos\left(\frac{2\pi}{n}\right)\sin\left(\frac{2\pi}{n}\right)\cos\left(\frac{2\pi}{3n}\right)+3\cos\left(\frac{\pi}{2n}\right)\sin\left(\frac{2\pi}{3n}\right)\right)\right)\pi^2$$

$$/\,n^2$$ (45)

8.Microcosmic surface integral course at surface 'S's all (viz. n) segmental cells:

978-1-62265-930-2 (online)   978-1-62265-931-9 (paper)   Yang Ke

Segment range $[0, \pi/2]$ of 'u': du $= \dfrac{\pi s}{2n}$

Segment range $[0, 2\pi]$ of 'v': dv $= \dfrac{2t\pi}{n}$

(thereinto, 's' and 't' stand for natural number 1~n)   (46)

Segment 'tangent plane normal vector' (47): [viz. Input (46) to (28)]

$$\left[\frac{1}{3}\sin\left(\frac{\pi s}{2n}\right)\left(3\sin\left(\frac{\pi s}{2n}\right)\cos\left(\frac{2\pi t}{n}\right)\sin\left(\frac{2\pi t}{3n}\right)+\sin\left(\frac{\pi s}{2n}\right)\sin\left(\frac{2\pi t}{n}\right)\cos\left(\frac{2\pi t}{3n}\right)\right),\right.$$

$$\sin\left(\frac{\pi s}{2n}\right)^2\sin\left(\frac{2\pi t}{n}\right),$$

$$\left.\frac{1}{3}\sin\left(\frac{\pi s}{2n}\right)\left(\cos\left(\frac{\pi s}{2n}\right)\cos\left(\frac{2\pi t}{n}\right)\sin\left(\frac{2\pi t}{n}\right)\cos\left(\frac{2\pi t}{3n}\right)+3\cos\left(\frac{\pi s}{2n}\right)\sin\left(\frac{2\pi t}{3n}\right)\right)\right]$$

// Expression (47) isn't that 'Single vector value', but that 'Set of finite quantitative (Viz.n) vector values'

Segment Curl (48): [Viz. Input (1) to (29), then input (46)]

$$\left[-\frac{2}{3}\cos\left(\frac{\pi s}{2n}\right),\frac{-1}{3},\frac{-1}{2}\right]$$

(48)

// Expression (48) isn't that 'Single vector value', but that 'Set of finite quantitative (Viz.n) vector values'

Calculate microcosmic surface integral values at surface'S's all segmental cells (49):

Base on integral median theorem, dot product of curl(48) and tangent plane normal vector(47), times segment unit of 'u,v'(46), these are microcosmic surface integral values at all segmental cells(49)

$$\left(-\frac{2}{9}\sin\left(\frac{\pi s}{2n}\right)\left(3\sin\left(\frac{\pi s}{2n}\right)\cos\left(\frac{2\pi t}{n}\right)\sin\left(\frac{2\pi t}{3n}\right)+\sin\left(\frac{\pi s}{2n}\right)\sin\left(\frac{2\pi t}{n}\right)\cos\left(\frac{2\pi t}{3n}\right)\right)\right.$$

$$\cos\left(\frac{\pi s}{2n}\right)-\frac{1}{3}\sin\left(\frac{\pi s}{2n}\right)^2\sin\left(\frac{2\pi t}{n}\right)$$

$$-\frac{1}{6}\sin\left(\frac{\pi s}{2n}\right)\left(\cos\left(\frac{\pi s}{2n}\right)\cos\left(\frac{2\pi t}{n}\right)\sin\left(\frac{2\pi t}{n}\right)\cos\left(\frac{2\pi t}{3n}\right)+3\cos\left(\frac{\pi s}{2n}\right)\sin\left(\frac{2\pi t}{3n}\right)\right)$$

$$\left.\right)\pi^2/n^2$$

// Expression (49) isn't that 'Single value', but that 'Set of finite
quantitative (Viz.n) values'

Structure finite sums (50):

(In the case of "Amount 'n' of surface 'S's segmental cells is
uncertain", sum of integral values [dot product about curl(48) and
tangent plane normal vector(47)] at all segmental cells)

$$
\sum_{t=1}^{n}\left(\sum_{s=1}^{n}\left(-\frac{2}{9}\sin\left(\frac{\pi s}{2n}\right)\right.\right.
$$

$$
\left(3\sin\left(\frac{\pi s}{2n}\right)\cos\left(\frac{2\pi t}{n}\right)\sin\left(\frac{2\pi t}{3n}\right)+\sin\left(\frac{\pi s}{2n}\right)\sin\left(\frac{2\pi t}{n}\right)\cos\left(\frac{2\pi t}{3n}\right)\right)\cos\left(\frac{\pi s}{2n}\right)
$$

$$
-\frac{1}{3}\sin\left(\frac{\pi s}{2n}\right)^2\sin\left(\frac{2\pi t}{n}\right)
$$

$$
\left.-\frac{1}{6}\sin\left(\frac{\pi s}{2n}\right)\left(\cos\left(\frac{\pi s}{2n}\right)\cos\left(\frac{2\pi t}{n}\right)\sin\left(\frac{2\pi t}{n}\right)\cos\left(\frac{2\pi t}{3n}\right)+3\cos\left(\frac{\pi s}{2n}\right)\sin\left(\frac{2\pi t}{3n}\right)\right)\right)
$$

$$
\left.\bigg)\pi^2/n^2\right)
$$

(50)

Infinitize finite sums, its limit operational value is 'Surface
Integral Value at Manifold' (51):

(In the case of "Amount 'n' of surface 'S's segmental cells tends
to infinity", limit value of integral values' sums [dot product about
curl(48) and tangent plane normal vector (47)] at all segmental cells)

$$
\lim_{n\to\infty}\sum_{t=1}^{n}\left(\sum_{s=1}^{n}\left(-\frac{2}{9}\sin\left(\frac{\pi s}{2n}\right)\right.\right.
$$

$$
\left(3\sin\left(\frac{\pi s}{2n}\right)\cos\left(\frac{2\pi t}{n}\right)\sin\left(\frac{2\pi t}{3n}\right)+\sin\left(\frac{\pi s}{2n}\right)\sin\left(\frac{2\pi t}{n}\right)\cos\left(\frac{2\pi t}{3n}\right)\right)\cos\left(\frac{\pi s}{2n}\right)
$$

$$
-\frac{1}{3}\sin\left(\frac{\pi s}{2n}\right)^2\sin\left(\frac{2\pi t}{n}\right)
$$

$$
\left.-\frac{1}{6}\sin\left(\frac{\pi s}{2n}\right)\left(\cos\left(\frac{\pi s}{2n}\right)\cos\left(\frac{2\pi t}{n}\right)\sin\left(\frac{2\pi t}{n}\right)\cos\left(\frac{2\pi t}{3n}\right)+3\cos\left(\frac{\pi s}{2n}\right)\sin\left(\frac{2\pi t}{3n}\right)\right)\right)
$$

$$
\left.\bigg)\pi^2/n^2\right)
$$

$$
=\frac{-81}{70}
$$

(51)

Viz. In the case of 'n →∞', (26)=(51)

978-1-62265-930-2 (online)   978-1-62265-931-9 (paper)

Complete Calculation and Validation of Curl Theorem at Manifold (Finite Sums Limits)

6.2 Numerical Model of Curl Theorem Finite Sums Limits at Manifold [Program Template of Waterloo Maplesoft, Optional]

```
> restart;
> with(plots):with(linalg):
> CS:=[sin(u)*cos(v),sin(u)*sin(v)*sin(v/3),cos(u)];   (1)
# Define parametrized expression of discretional simply connected
orientable closed surface 'CS'
```

$$CS := \left[\sin(u)\cos(v),\, \sin(u)\sin(v)\sin\left(\frac{v}{3}\right),\, \cos(u) \right]$$

```
> rgu:=[0,Pi/2];
# Halve range of 'u' (relative to closed parametrized surface)
```

$$rgu := \left[0, \frac{\pi}{2} \right]$$

```
> rgv:=[0,2*Pi];
# Set range of parameters 'u,v', make parametrized surface 'CS'
unclosed
```

$$rgv := [0, 2\,\pi]$$

```
> plot3d(CS,u=rgu[1]..rgu[2],v=rgv[1]..rgv[2],scaling=constrained,
projection=0.9,numpoints=3000):g1:=%:
# Draw unclosed parameterized surface 'CS'
> subs(u=rgu[2],CS)=eval(subs(u=rgu[2],CS));CL:=rhs(%);   (2)
# Import  boundary value of 'u' to expression of surface 'CS', obtain
```
expression of boundary curve 'CL'; Set Right Hand Rule as associated basis between positive direction of boundary curve 'CL' and normal vector of unclosed surface 'CS'

// Be different to define expression of boundary curve'[α cos(v), β sin(v),γ]' that depends on surface '[a*sin(u)cos(v), b*sin(u) sin(v), c*cos(u)]' for existence in 'Proof', in 'Numerical Models', it is possible to 'Import boundary value of 'u' to expression of surface 'CS', directly obtain expression of boundary curve 'CL''

$$\left[\sin\left(\frac{\pi}{2}\right)\cos(v),\ \sin\left(\frac{\pi}{2}\right)\sin(v)\sin\left(\frac{v}{3}\right),\ \cos\left(\frac{\pi}{2}\right) \right] = \left[\cos(v),\ \sin(v)\sin\left(\frac{v}{3}\right),\ 0 \right]$$

$$CL := \left[\cos(v),\ \sin(v)\sin\left(\frac{v}{3}\right),\ 0 \right]$$

```
> [Diff(CL[1],v),Diff(CL[2],v),Diff(CL[3],v)]=[diff(CL[1],v),
diff(CL[2],v),diff(CL[3],v)];dCL:=rhs(%);      (3)
```
Calculate tangent vector 'dCL' of closed boundary curve 'CL'

$$\left[\frac{d}{dv}\cos(v),\ \frac{d}{dv}\left(\sin(v)\sin\left(\frac{v}{3}\right)\right),\ \frac{d}{dv}0 \right] = \left[-\sin(v),\ \cos(v)\sin\left(\frac{v}{3}\right) + \frac{1}{3}\sin(v)\cos\left(\frac{v}{3}\right),\ 0 \right]$$

$$dCL := \left[-\sin(v),\ \cos(v)\sin\left(\frac{v}{3}\right) + \frac{1}{3}\sin(v)\cos\left(\frac{v}{3}\right),\ 0 \right]$$

```
> spacecurve(CL,v=rgv[1]..rgv[2],numpoints=2000,thickness=2,
color=red):g2:=%: # Draw closed boundary curve'CL'


> V:=[y/2,z^2/3,x/3];     (4)
```
Define discretional spacial vector field 'V' (Suppose Vector Field
'V' possesses 1th order continuous partial derivatives at surface 'CS')

$$V := \left[\frac{y}{2},\ \frac{z^2}{3},\ \frac{x}{3} \right]$$

```
> [Diff(V[3],y)-Diff(V[2],z),Diff(V[1],z)-Diff(V[3],x),
Diff(V[2],x)-Diff(V[1],y)]
=[diff(V[3],y)-diff(V[2],z),diff(V[1],z)-diff(V[3],x),
diff(V[2],x)-diff(V[1],y)];cV:=rhs(%);     (5)
```
Calculate 'cV'-- Curl of spacial vector field 'V'

$$\left[\left(\frac{\partial}{\partial y}\left(\frac{x}{3}\right)\right) - \left(\frac{d}{dz}\left(\frac{z^2}{3}\right)\right),\ \left(\frac{\partial}{\partial z}\left(\frac{y}{2}\right)\right) - \left(\frac{d}{dx}\left(\frac{x}{3}\right)\right),\ \left(\frac{\partial}{\partial x}\left(\frac{z^2}{3}\right)\right) - \left(\frac{d}{dy}\left(\frac{y}{2}\right)\right) \right] =$$
$$\left[-\frac{2z}{3},\ \frac{-1}{3},\ \frac{-1}{2} \right]$$

$$cV := \left[-\frac{2z}{3},\ \frac{-1}{3},\ \frac{-1}{2} \right]$$

```
> rgx:=[-1,1];
```
$$rgx := [-1,\ 1]$$

```
> rgy:=[-1,1];
```
$$rgy := [-1,\ 1]$$

```
> rgz:=[-0.7,1.3];
```
$$rgz := [-0.7,\ 1.3]$$

```
> fieldplot3d(V,x=rgx[1]..rgx[2],y=rgy[1]..rgy[2],z=rgz[1]..
```

```
rgz[2],arrows=SLIM,color=red,thickness=1,grid=[6,6,6]):g3:=%:
# Draw spacial vector field 'V'
> fieldplot3d(cV,x=rgx[1]..rgx[2],y=rgy[1]..rgy[2],z=rgz[1]..
rgz[2],arrows=SLIM,color=blue,thickness=1,grid=[6,6,6]):g4:=%:
# Draw curl 'cV'
> display(g1,g2,g3,g4); # Synthesize figures
```

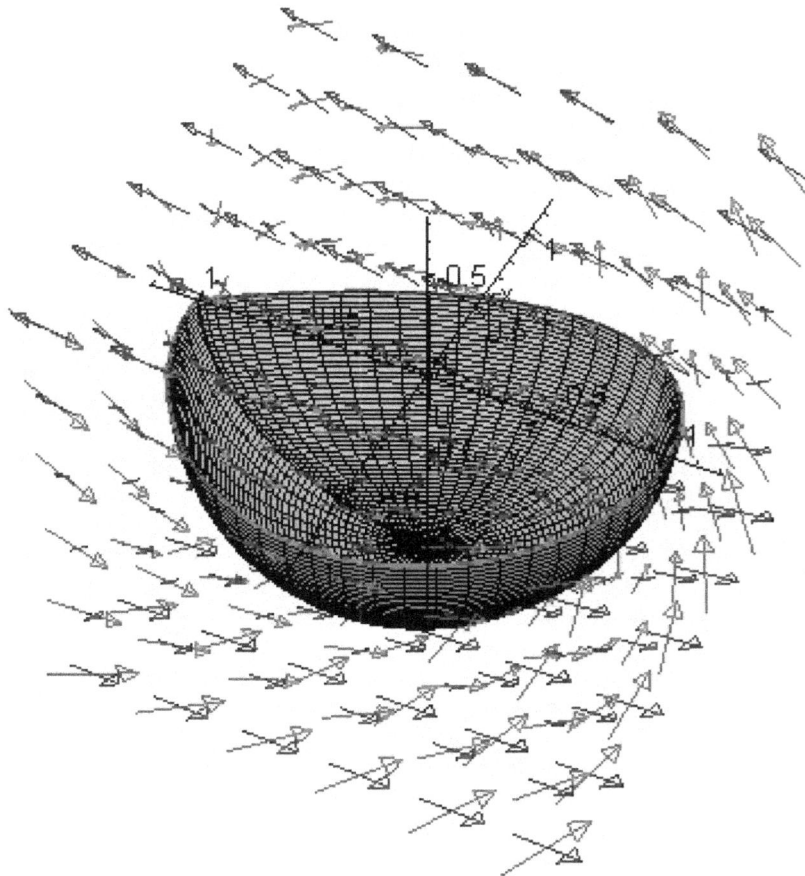

Figure(III)6.2.1 Surface[unclosed] 'CS' and its Closed Boundary' Curve 'CL'; Vector Field 'V'(Red arrows) and its Curl 'cV'(Blue arrows)

```
> display(g2,g3,g4);
```

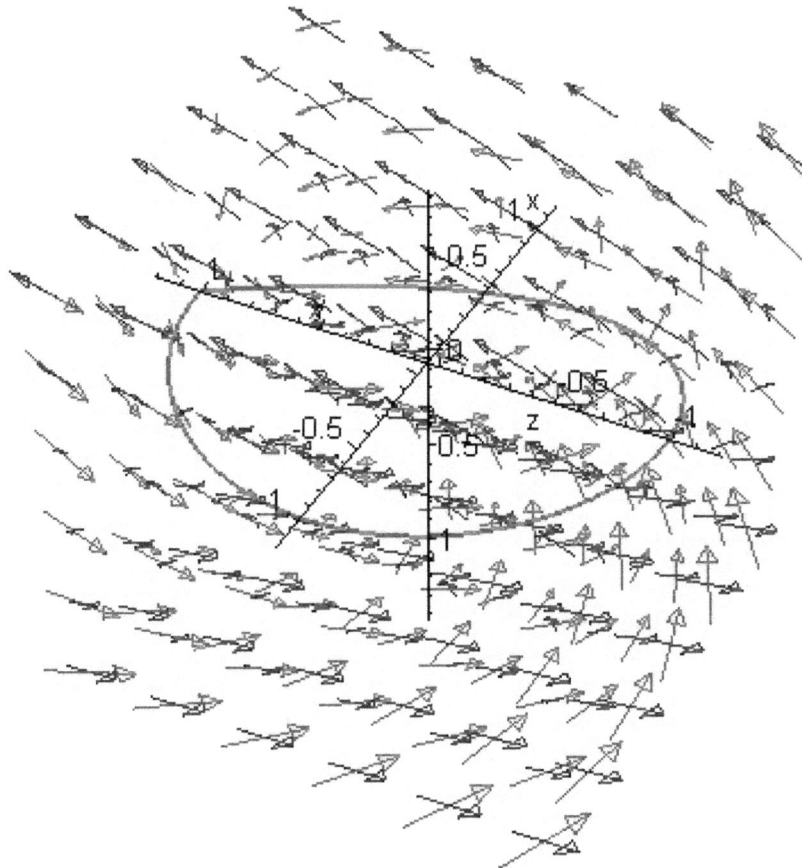

Figure(III)6.2.2 Closed Boundary Curve 'CL';

Vector Field 'V'(Red arrows)and its Curl'cV'(Blue arrows)


```
> x:=CL[1]:y:=CL[2]:z:=CL[3]:
```

// Segment the range `[0,2*Pi]' of `v':

```
> dus:=50;
```

Define amount 'dus' of boundary curve 'CL's parameter segmental cells
(It can be discretional natural number)

$$dus := 50$$

```
> dv:=(rgv[2]-rgv[1])/dus; # Segment range `[0,2*Pi]' of `v'
```

$$dv := \frac{\pi}{25}$$

Microcosmic curve integral course of boundary curve 'CL's first segmental cell (50 segmental cells):

// Segment special vector field 'V':

(This step possesses formal meaning in 'Proof' only, and possesses essential meaning in 'Numerical Models')

> dV:=subs(v=rgv[1]+dv,V);

Spacial vector field 'V's average value that corresponds to boundary curve 'CL's first segmental cell

$$dV := \left[\frac{1}{2} \sin\left(\frac{\pi}{25}\right) \sin\left(\frac{\pi}{75}\right), 0, \frac{1}{3} \cos\left(\frac{\pi}{25}\right) \right]$$

// Segment tangent vector 'dCL':

> dCLm:=subs(v=rgv[1]+dv,dCL);

#'Tangent vector dCL's average value that corresponds to closed boundary curve 'CL's first segmental cell

$$dCLm := \left[-\sin\left(\frac{\pi}{25}\right), \cos\left(\frac{\pi}{25}\right) \sin\left(\frac{\pi}{75}\right) + \frac{1}{3} \sin\left(\frac{\pi}{25}\right) \cos\left(\frac{\pi}{75}\right), 0 \right]$$

// Calculate microcosmic curve integral value of closed boundary curve 'CL's first segmental cell:

> dv*(dV[1]*dCLm[1]+dV[2]*dCLm[2]+dV[3]*dCLm[3]);

Base on integral median theorem, integral value of dot product (Spacial vector field 'V' and tangent vector 'dCL') at first segmental cell

$$-\frac{1}{50} \pi \sin\left(\frac{\pi}{25}\right)^2 \sin\left(\frac{\pi}{75}\right)$$

Microcosmic curve integral course of closed boundary curve 'CL's all segmental cells (50 segmental cells):

// Segment special vector field 'V':

(This step possesses formal meaning in 'Proof' only, and possesses essential meaning in 'Numerical Models')

> idV:=subs(v=rgv[1]+i*dv,V);

Spacial vector field 'V's average values that correspond to boundary curve 'CL's all segmental cells

$$idV := \left[\frac{1}{2} \sin\left(\frac{i\pi}{25}\right) \sin\left(\frac{i\pi}{75}\right), 0, \frac{1}{3} \cos\left(\frac{i\pi}{25}\right) \right]$$

> seq(([i,idV]),i=1..dus):

```
# List of spacial vector field 'V's average values that correspond
to boundary curve 'CL's all segmental cells, be elided
//'idV' isn't that 'Single vector value', but that 'Set of finite
quantitative (Viz.50) vector values'


// Segment tangent vector 'dCL':
> idCL:=subs(v=rgv[1]+i*dv,dCL);
#'Tangent vector dCL's average values that correspond to boundary curve
'CL's all segmental cells
```

$$idCL := \left[-\sin\left(\frac{i\,\pi}{25}\right), \cos\left(\frac{i\,\pi}{25}\right)\sin\left(\frac{i\,\pi}{75}\right) + \frac{1}{3}\sin\left(\frac{i\,\pi}{25}\right)\cos\left(\frac{i\,\pi}{75}\right), 0 \right]$$

```
> seq(([i,idCL]),i=1..dus):
# List of 'Tangent vector dCL's average values that correspond to
boundary curve 'CL's all segmental cells, be elided
//'idCL' isn't that 'Single vector value', but that 'Set of finite
quantitative (Viz.50) vector values'


// Calculate microcosmic curve integral values of closed boundary
curve 'CL's all segmental cells:
> dv*(idV[1]*idCL[1]+idV[2]*idCL[2]+idV[3]*idCL[3]);
# Base on integral median theorem, integral values of dot product
(Spacial vector field 'V' and tangent vector 'dCL') at all segmental
cells
```

$$-\frac{1}{50}\,\pi\,\sin\left(\frac{i\,\pi}{25}\right)^{2}\,\sin\left(\frac{i\,\pi}{75}\right)$$

```
// Expression above-mentioned isn't that 'Single value', but that
'Set of finite quantitative (Viz.50) values'


// List of finite quantitative microcosmic curl integral values:
> seq(([i,dv*(idV[1]*idCL[1]+idV[2]*idCL[2]+idV[3]*idCL[3])]),
i=1..dus):
# List of integral values of dot product(Spacial vector field 'V' and
tangent vector 'dCL')at all segmental cells, be elided
// Structure sequence that is composed of finite quantitative
microcomosic curl integral values:
```

```
> sqn:=seq(dv*(idV[1]*idCL[1]+idV[2]*idCL[2]+idV[3]*idCL[3]),
i=1..dus):
```
// Accumulational sum of sequence, obtain spacial closed curve integral value at manifold:

```
> add(k,k=sqn):xi:=evalf(%);      (6)
```

\# Sum of integral values of dot product (Spacial vector field 'V' and tangent vector 'dCL') at all segmental cells (Verbose analytical expression is elided)

$$\xi := -1.157143379$$

// Renewedly segment range '[0,2*Pi]' of 'v':

```
> dus:=n;
```

\# Set amount 'dus' of boundary curve 'CL's parameter segmental cells as natural number 'n'

$$dus := n$$

```
> dv:=(rgv[2]-rgv[1])/dus; # Segment range '[0,2*Pi]' of 'v'
```

$$dv := \frac{2\pi}{n}$$

\# Microcosmic curve integral course of boundary curve 'CL's first segmental cell (n segmental cells):

// Segment special vector field 'V':

(This step possesses formal meaning in 'Proof' only, and possesses essential meaning in 'Numerical Models')

```
> dV:=subs(v=rgv[1]+dv,V);
```

\# Spacial vector field 'V's average value that corresponds to closed boundary curve 'CL's first segmental cell

$$dV := \left[\frac{1}{2} \sin\left(\frac{2\pi}{n}\right) \sin\left(\frac{2\pi}{3n}\right), \ 0, \ \frac{1}{3} \cos\left(\frac{2\pi}{n}\right) \right]$$

// Segment tangent vector 'dCL':

```
> dCLm:=subs(v=rgv[1]+dv,dCL);
```

\#'Tangent vector dCL's average value that corresponds to closed boundary curve 'CL's first segmental cell

$$dCLm := \left[-\sin\left(\frac{2\pi}{n}\right), \ \cos\left(\frac{2\pi}{n}\right) \sin\left(\frac{2\pi}{3n}\right) + \frac{1}{3} \sin\left(\frac{2\pi}{n}\right) \cos\left(\frac{2\pi}{3n}\right), \ 0 \right]$$

// Calculate microcosmic curve integral value of closed boundary curve 'CL's first segmental cell:

```
> dv*(dV[1]*dCLm[1]+dV[2]*dCLm[2]+dV[3]*dCLm[3]);
```

Base on integral median theorem, integral value of dot product (Spacial vector field 'V' and tangent vector 'dCL') at first segmental cell

$$-\frac{\pi \sin\left(\dfrac{2\,\pi}{n}\right)^{2} \sin\left(\dfrac{2\,\pi}{3\,n}\right)}{n}$$

Microcosmic curve integral course of closed boundary curve 'CL's all segmental cells (n segmental cells):

// Segment special vector field 'V':

(This step possesses formal meaning in 'Proof' only, and possesses essential meaning in 'Numerical Models')

```
> idV:=subs(v=rgv[1]+i*dv,V);
```

Spacial vector field 'V's average values that correspond to boundary curve 'CL's all segmental cells

$$idV := \left[\frac{1}{2} \sin\left(\frac{2\,i\,\pi}{n}\right) \sin\left(\frac{2\,i\,\pi}{3\,n}\right),\, 0,\, \frac{1}{3}\cos\left(\frac{2\,i\,\pi}{n}\right) \right]$$

//'idV' isn't that 'Single vector value', but that 'Set of finite quantitative (Viz.n) vector values'

// Segment tangent vector 'dCL':

```
> idCL:=subs(v=rgv[1]+i*dv,dCL);
```

'Tangent vector dCL's average values that correspond to boundary curve 'CL's all segmental cells

$$idCL := \left[-\sin\left(\frac{2\,i\,\pi}{n}\right),\, \cos\left(\frac{2\,i\,\pi}{n}\right)\sin\left(\frac{2\,i\,\pi}{3\,n}\right) + \frac{1}{3}\sin\left(\frac{2\,i\,\pi}{n}\right)\cos\left(\frac{2\,i\,\pi}{3\,n}\right),\, 0 \right]$$

//'idCL' isn't that 'Single vector value', but that 'Set of finite quantitative (Viz.n) vector values'

// Calculate microcosmic curve integral values of closed boundary curve 'CL's all segmental cells:

```
> dv*(idV[1]*idCL[1]+idV[2]*idCL[2]+idV[3]*idCL[3]);
```

Base on integral median theorem, integral values of dot product (Spacial vector field 'V' and tangent vector 'dCL') at all segmental cells

$$-\frac{\pi \sin\left(\frac{2\,i\,\pi}{n}\right)^2 \sin\left(\frac{2\,i\,\pi}{3\,n}\right)}{n}$$

// Expression above-mentioned isn't that 'Single value', but that 'Set of finite quantitative (Viz.n) vector values'

// Structure finite sums:
> Sum(dv*(idV[1]*idCL[1]+idV[2]*idCL[2]+idV[3]*idCL[3]),i=1..dus);
In the case of "Amount 'n' of closed boundary curve 'CL's parameter segmental cells is pending", sum of integral values (dot product about spacial vector field 'V' and tangent vector 'dCL') at all segmental cells

$$\sum_{i=1}^{n}\left(-\frac{\pi \sin\left(\frac{2\,i\,\pi}{n}\right)^2 \sin\left(\frac{2\,i\,\pi}{3\,n}\right)}{n}\right)$$

> vs:=value(%):
> Limit(vs,n=infinity):   **(7)**
// Infinitize finite sums, its limit operational value is 'Spacial Closed Curve Integral Value at Manifold'
In the case of "Amount 'w' of closed boundary curve 'CL's parameter segmental cells tends to infinity", limit value of integral values' sums (dot product about Spacial vector field 'V' and tangent vector 'dCL') at all segmental cells
> alpha:=value(%);delta:=evalf(alpha);

$$\alpha := \frac{-81}{70}$$

$$\delta := -1.157142857$$

> xi;delta;
There is tiny difference between "Integral values' sum of finite quantitative segmental cells" and "Integral values' sum of infinite quantitative segmental cells"

$$-1.157143379$$
$$-1.157142857$$

> x:='x':y:='y':z:='z':

```
> x:=CS[1]:y:=CS[2]:z:=CS[3]:
> matrix(3,3,[i,j,k,Diff(CS[1],u),Diff(CS[2],u),Diff(CS[3],u),
Diff(CS[1],v),Diff(CS[2],v),Diff(CS[3],v)])=
matrix(3,3,[i,j,k,diff(CS[1],u),diff(CS[2],u),diff(CS[3],u),
diff(CS[1],v),diff(CS[2],v),diff(CS[3],v)]);m:=rhs(%);
# Define and calculate matrix of partial derivatives 'm', obtain
'tangent plane normal vector' of surface 'CS'
```

$$\begin{bmatrix} i & j & k \\ \frac{\partial}{\partial u}(\sin(u)\cos(v)) & \frac{\partial}{\partial u}\left(\sin(u)\sin(v)\sin\left(\frac{v}{3}\right)\right) & \frac{d}{du}\cos(u) \\ \frac{\partial}{\partial v}(\sin(u)\cos(v)) & \frac{\partial}{\partial v}\left(\sin(u)\sin(v)\sin\left(\frac{v}{3}\right)\right) & \frac{\partial}{\partial v}\cos(u) \end{bmatrix} =$$

$$\begin{bmatrix} i & j & k \\ \cos(u)\cos(v) & \cos(u)\sin(v)\sin\left(\frac{v}{3}\right) & -\sin(u) \\ -\sin(u)\sin(v) & \sin(u)\cos(v)\sin\left(\frac{v}{3}\right)+\frac{1}{3}\sin(u)\sin(v)\cos\left(\frac{v}{3}\right) & 0 \end{bmatrix}$$

$$m := \begin{bmatrix} i & j & k \\ \cos(u)\cos(v) & \cos(u)\sin(v)\sin\left(\frac{v}{3}\right) & -\sin(u) \\ -\sin(u)\sin(v) & \sin(u)\cos(v)\sin\left(\frac{v}{3}\right)+\frac{1}{3}\sin(u)\sin(v)\cos\left(\frac{v}{3}\right) & 0 \end{bmatrix}$$

```
> det(m):
> mn:=simplify(%);
```

$$mn := \frac{1}{3}\sin(u)\left(3\,i\,\sin(u)\cos(v)\sin\left(\frac{v}{3}\right)+i\,\sin(u)\sin(v)\cos\left(\frac{v}{3}\right)\right.$$
$$\left.+\cos(u)\cos(v)\,k\,\sin(v)\cos\left(\frac{v}{3}\right)+3\sin(u)\sin(v)\,j+3\,k\,\cos(u)\sin\left(\frac{v}{3}\right)\right)$$

```
> A:=coeff(mn,i); # Obtain coefficient of 'i'
```

$$A := \frac{1}{3}\sin(u)\left(3\sin(u)\cos(v)\sin\left(\frac{v}{3}\right)+\sin(u)\sin(v)\cos\left(\frac{v}{3}\right)\right)$$

```
> B:=coeff(mn,j); # Obtain coefficient of 'j'
```

$$B := \sin(u)^2\sin(v)$$

```
> C:=coeff(mn,k); # Obtain coefficient of 'k'
```

$$C := \frac{1}{3}\sin(u)\left(\cos(u)\cos(v)\sin(v)\cos\left(\frac{v}{3}\right)+3\cos(u)\sin\left(\frac{v}{3}\right)\right)$$

```
> [A,B,C]: # [A,B,C] structure 'tangent plane normal vector'    (8)

// Segment range '[0,Pi],[0,2*Pi]' of 'u,v':
```

```
> dus:=50;
```

\# Set amount of surface 'CS's segmental cells, this amount can be discretional natural number

$$dus := 50$$

```
> du:=(rgu[2]-rgu[1])/dus; # Segment range of 'u'
```

$$du := \frac{\pi}{100}$$

```
> dv:=(rgv[2]-rgv[1])/dus; # Segment range of 'v'
```

$$dv := \frac{\pi}{25}$$

Microcosmic surface integral course of surface 'CS's first segmental cell (50 segmental cells):

// Segment 'tangent plane normal vector'

```
> dA:=subs(v=rgv[1]+dv,subs(u=rgu[1]+du,A));
```

$$dA := \frac{1}{3} \sin\left(\frac{\pi}{100}\right)\left(3 \sin\left(\frac{\pi}{100}\right)\cos\left(\frac{\pi}{25}\right)\sin\left(\frac{\pi}{75}\right) + \sin\left(\frac{\pi}{100}\right)\sin\left(\frac{\pi}{25}\right)\cos\left(\frac{\pi}{75}\right)\right)$$

```
> dB:=subs(v=rgv[1]+dv,subs(u=rgu[1]+du,B));
```

$$dB := \sin\left(\frac{\pi}{100}\right)^2 \sin\left(\frac{\pi}{25}\right)$$

```
> dC:=subs(v=rgv[1]+dv,subs(u=rgu[1]+du,C));
```

$$dC := \frac{1}{3} \sin\left(\frac{\pi}{100}\right)\left(\cos\left(\frac{\pi}{100}\right)\cos\left(\frac{\pi}{25}\right)\sin\left(\frac{\pi}{25}\right)\cos\left(\frac{\pi}{75}\right) + 3 \cos\left(\frac{\pi}{100}\right)\sin\left(\frac{\pi}{75}\right)\right)$$

```
> [dA,dB,dC]: #'[dA,dB,dC]' is 'tangent plane normal vector's average
```
value that corresponds to surface 'CS's first segmental cell

// Segment idiographic curl 'cV':

// Be different to abstract curl 'cV1,cV2' that 'Proof' refers to, in 'Numerical Models', be able to directly input idiographic segmented values of 'u,v' to idiographic Curl 'cV', then segment 'cV' in accordance with amount 'dus' of parameter segmental cells

```
> dcV:=subs(v=rgv[1]+dv,subs(u=rgu[1]+du,cV));
```

\# Curl 'cV's average value that corresponds to surface'CS's first segmental cell

$$dcV := \left[-\frac{2}{3}\cos\left(\frac{\pi}{100}\right), \frac{-1}{3}, \frac{-1}{2}\right]$$

```
// Calculate microcosmic surface integral value of surface 'CS's first
segmental cell:
> (dcV[1]*dA+dcV[2]*dB+dcV[3]*dC)*du*dv;
# Base on integral median theorem, integral value of dot product (Curl
'cV' and '[dA,dB,dC]') at first segmental cell
```

$$\frac{1}{2500}\left(\frac{2}{9} \right.$$

$$\cos\left(\frac{\pi}{100} \right) \sin\left(\frac{\pi}{100} \right) \left(3 \sin\left(\frac{\pi}{100} \right) \cos\left(\frac{\pi}{25} \right) \sin\left(\frac{\pi}{75} \right) + \sin\left(\frac{\pi}{100} \right) \sin\left(\frac{\pi}{25} \right) \cos\left(\frac{\pi}{75} \right) \right)$$

$$-\frac{1}{3} \sin\left(\frac{\pi}{100} \right)^2 \sin\left(\frac{\pi}{25} \right)$$

$$-\frac{1}{6} \sin\left(\frac{\pi}{100} \right) \left(\cos\left(\frac{\pi}{100} \right) \cos\left(\frac{\pi}{25} \right) \sin\left(\frac{\pi}{25} \right) \cos\left(\frac{\pi}{75} \right) + 3 \cos\left(\frac{\pi}{100} \right) \sin\left(\frac{\pi}{75} \right) \right) \left. \right) \pi^2$$

Microcosmic surface integral courses of surface 'CS's all segmental cells (50 segmental cells):

```
// Segment 'tangent plane normal vector'
> stdA:=subs(v=rgv[1]+t*dv,subs(u=rgu[1]+s*du,A));
```

$$stdA := \frac{1}{3} \sin\left(\frac{s\,\pi}{100} \right) \left(3 \sin\left(\frac{s\,\pi}{100} \right) \cos\left(\frac{t\,\pi}{25} \right) \sin\left(\frac{t\,\pi}{75} \right) + \sin\left(\frac{s\,\pi}{100} \right) \sin\left(\frac{t\,\pi}{25} \right) \cos\left(\frac{t\,\pi}{75} \right) \right)$$

```
> stdB:=subs(v=rgv[1]+t*dv,subs(u=rgu[1]+s*du,B));
```

$$stdB := \sin\left(\frac{s\,\pi}{100} \right)^2 \sin\left(\frac{t\,\pi}{25} \right)$$

```
> stdC:=subs(v=rgv[1]+t*dv,subs(u=rgu[1]+s*du,C));
```

$$stdC := \frac{1}{3} \sin\left(\frac{s\,\pi}{100} \right) \left(\cos\left(\frac{s\,\pi}{100} \right) \cos\left(\frac{t\,\pi}{25} \right) \sin\left(\frac{t\,\pi}{25} \right) \cos\left(\frac{t\,\pi}{75} \right) + 3 \cos\left(\frac{s\,\pi}{100} \right) \sin\left(\frac{t\,\pi}{75} \right) \right)$$

```
> [stdA,stdB,stdC]: #'[stdA,stdB,stdC]' are 'tangent plane normal
vector's average values that correspond to surface 'CS's all segmental
cells
//'[stdA,stdB,stdC]' isn't that 'Single vector value', but that 'Set
of finite quantitative (Viz.50) vector values'

// Segment idiographic curl 'cV':
// Be different to abstract  curl 'cV1,cV2' that 'Proof' refers to,
in 'Numerical Models', be able to directly input idiographic segmented
values of 'u,v' to idiographic Curl 'cV', then segment 'cV' in
```

accordance with amount 'dus' of parameter segmental cells

```
> stdcV:=subs(v=rgv[1]+t*dv,subs(u=rgu[1]+s*du,cV));
```

Curl 'cV's average values that correspond to surface 'CS's all segmental cells

$$stdcV := \left[-\frac{2}{3} \cos\left(\frac{s\,\pi}{100} \right), \frac{-1}{3}, \frac{-1}{2} \right]$$

//'stdcV' isn't that 'Single vector value', but that 'Set of finite quantitative (Viz.50) vector values'

// Calculate microcosmosic surface integral value at all segmental cells of surface 'CS':

```
> (stdcV[1]*stdA+stdcV[2]*stdB+stdcV[3]*stdC)*du*dv;
```

Base on integral median theorem, integral value of dot product (Curl 'cV' and '[stdA,stdB,stdC]') at all segmental cells

$$\frac{1}{2500}\left(-\frac{2}{9} \cos\left(\frac{s\,\pi}{100} \right) \sin\left(\frac{s\,\pi}{100} \right) \right.$$

$$\left(3\sin\left(\frac{s\,\pi}{100} \right) \cos\left(\frac{t\,\pi}{25} \right) \sin\left(\frac{t\,\pi}{75} \right) + \sin\left(\frac{s\,\pi}{100} \right) \sin\left(\frac{t\,\pi}{25} \right) \cos\left(\frac{t\,\pi}{75} \right) \right)$$

$$-\frac{1}{3} \sin\left(\frac{s\,\pi}{100} \right)^2 \sin\left(\frac{t\,\pi}{25} \right)$$

$$\left. -\frac{1}{6} \sin\left(\frac{s\,\pi}{100} \right) \left(\cos\left(\frac{s\,\pi}{100} \right) \cos\left(\frac{t\,\pi}{25} \right) \sin\left(\frac{t\,\pi}{25} \right) \cos\left(\frac{t\,\pi}{75} \right) + 3 \cos\left(\frac{s\,\pi}{100} \right) \sin\left(\frac{t\,\pi}{75} \right) \right) \right) \pi^2$$

// Expression above-mentioned isn't that 'Single value', but that 'Set of finite quantitative (Viz.50) values'

// List of finite quantitative microcosmic surface integral values:

```
> seq(seq([s*t,(stdcV[1]*stdA+stdcV[2]*stdB+stdcV[3]*stdC)*du*dv],
s=1..dus),t=1..dus):
```

List of integral values of dot product (Curl 'cV2' and '[stdA,stdB,stdC]') at all segmental cells, be elided

// Structure sequence that is composed of finite quantitative microcomosic surface integral values:

```
> sqn:=seq(seq((stdcV[1]*stdA+stdcV[2]*stdB+stdcV[3]*stdC)*du*dv,
s=1..dus),t=1..dus):
```

// Accumulational sum of sequence, obtain surface integral value at manifold:

```
> add(k,k=sqn):omega:=evalf(%);      (9)
```
Sum of integral values of dot product (Curl 'cV2' and '[stdA,stdB, stdC]') at all segmental cells; Verbose analytical expression is elided

$$\omega := -1.181828581$$

```
// Renewedly segment range '[0,Pi],[0,2*Pi]' of 'u,v':
> dus:=n;
```
Set amount of surface 'CS's segmental cells as natural number 'n'

$$dus := n$$

```
> du:=(rgu[2]-rgu[1])/dus; # Segment range of 'u'
```

$$du := \frac{\pi}{2\,n}$$

```
> dv:=(rgv[2]-rgv[1])/dus; # Segment range of 'v'
```

$$dv := \frac{2\,\pi}{n}$$

Microcosmic surface integral course of surface 'CS's first segmental cell (n segmental cells):

```
// Segment 'tangent plane normal vector'
> dA:=subs(v=rgv[1]+dv,subs(u=rgu[1]+du,A));
```

$$dA := \frac{1}{3}\sin\!\left(\frac{\pi}{2\,n}\right)\left(3\sin\!\left(\frac{\pi}{2\,n}\right)\cos\!\left(\frac{2\,\pi}{n}\right)\sin\!\left(\frac{2\,\pi}{3\,n}\right)+\sin\!\left(\frac{\pi}{2\,n}\right)\sin\!\left(\frac{2\,\pi}{n}\right)\cos\!\left(\frac{2\,\pi}{3\,n}\right)\right)$$

```
> dB:=subs(v=rgv[1]+dv,subs(u=rgu[1]+du,B));
```

$$dB := \sin\!\left(\frac{\pi}{2\,n}\right)^{2}\sin\!\left(\frac{2\,\pi}{n}\right)$$

```
> dC:=subs(v=rgv[1]+dv,subs(u=rgu[1]+du,C));
```

$$dC := \frac{1}{3}\sin\!\left(\frac{\pi}{2\,n}\right)\left(\cos\!\left(\frac{\pi}{2\,n}\right)\cos\!\left(\frac{2\,\pi}{n}\right)\sin\!\left(\frac{2\,\pi}{n}\right)\cos\!\left(\frac{2\,\pi}{3\,n}\right)+3\cos\!\left(\frac{\pi}{2\,n}\right)\sin\!\left(\frac{2\,\pi}{3\,n}\right)\right)$$

```
> [dA,dB,dC]: #'[dA,dB,dC]' is 'tangent plane normal vector's average
```
value that corresponds to surface 'CS's first segmental cell

```
// Segment idiographic curl 'cV':
// Be different to abstract  curl 'cV1,cV2' that 'Proof' refers to,
in 'Numerical Models', be able to directly input idiographic segmented
values of 'u,v' to idiographic Curl 'cV', then segment 'cV' in
accordance with amount 'dus' of parameter segmental cells
> dcV:=subs(v=rgv[1]+dv,subs(u=rgu[1]+du,cV));
```

\# Curl 'cV's average value that corresponds to surface'CS's first segmental cell

$$dcV := \left[-\frac{2}{3} \cos\left(\frac{\pi}{2\,n} \right), \frac{-1}{3}, \frac{-1}{2} \right]$$

// Calculate microcosmic surface integral value of surface 'CS's first segmental cell:

`> (dcV[1]*dA+dcV[2]*dB+dcV[3]*dC)*du*dv;`

\# Base on integral median theorem, integral value of dot product (Curl 'cV2' and '[dA,dB,dC]') at first segmental cell

$$\left(\left(-\frac{2}{9} \cos\left(\frac{\pi}{2\,n} \right) \sin\left(\frac{\pi}{2\,n} \right) \right) \left(3 \sin\left(\frac{\pi}{2\,n} \right) \cos\left(\frac{2\,\pi}{n} \right) \sin\left(\frac{2\,\pi}{3\,n} \right) + \sin\left(\frac{\pi}{2\,n} \right) \sin\left(\frac{2\,\pi}{n} \right) \cos\left(\frac{2\,\pi}{3\,n} \right) \right) \right.$$
$$-\frac{1}{3} \sin\left(\frac{\pi}{2\,n} \right)^2 \sin\left(\frac{2\,\pi}{n} \right)$$
$$\left. -\frac{1}{6} \sin\left(\frac{\pi}{2\,n} \right) \left(\cos\left(\frac{\pi}{2\,n} \right) \cos\left(\frac{2\,\pi}{n} \right) \sin\left(\frac{2\,\pi}{n} \right) \cos\left(\frac{2\,\pi}{3\,n} \right) + 3 \cos\left(\frac{\pi}{2\,n} \right) \sin\left(\frac{2\,\pi}{3\,n} \right) \right) \right) \pi^2$$
$$/\,n^2$$

\# Microcosmic surface integral course of surface 'CS's all segmental cells (n segmental cells):

// Segment 'tangent plane normal vector'

`> stdA:=subs(v=rgv[1]+t*dv,subs(u=rgu[1]+s*du,A));`

$stdA :=$
$$\frac{1}{3} \sin\left(\frac{s\,\pi}{2\,n} \right) \left(3 \sin\left(\frac{s\,\pi}{2\,n} \right) \cos\left(\frac{2\,t\,\pi}{n} \right) \sin\left(\frac{2\,t\,\pi}{3\,n} \right) + \sin\left(\frac{s\,\pi}{2\,n} \right) \sin\left(\frac{2\,t\,\pi}{n} \right) \cos\left(\frac{2\,t\,\pi}{3\,n} \right) \right)$$

`> stdB:=subs(v=rgv[1]+t*dv,subs(u=rgu[1]+s*du,B));`

$$stdB := \sin\left(\frac{s\,\pi}{2\,n} \right)^2 \sin\left(\frac{2\,t\,\pi}{n} \right)$$

`> stdC:=subs(v=rgv[1]+t*dv,subs(u=rgu[1]+s*du,C));`

$stdC :=$
$$\frac{1}{3} \sin\left(\frac{s\,\pi}{2\,n} \right) \left(\cos\left(\frac{s\,\pi}{2\,n} \right) \cos\left(\frac{2\,t\,\pi}{n} \right) \sin\left(\frac{2\,t\,\pi}{n} \right) \cos\left(\frac{2\,t\,\pi}{3\,n} \right) + 3 \cos\left(\frac{s\,\pi}{2\,n} \right) \sin\left(\frac{2\,t\,\pi}{3\,n} \right) \right)$$

`> [stdA,stdB,stdC]:` \# '[stdA,stdB,stdC]' are 'tangent plane normal vector's average values that correspond to surface 'CS's all segmental cells

//'[stdA,stdB,stdC]' isn't that 'Single vector value', but that 'Set of finite quantitative (Viz.n) vector values'

```
// Segment idiographic curl 'cV':
// Be different to abstract  curl 'cV1,cV2' that 'Proof' refers to,
in 'Numerical Models', be able to directly input idiographic segmented
values of 'u,v' to idiographic Curl 'cV', then segment 'cV' in
accordance with amount 'dus' of parameter segmental cells
> stdcV:=subs(v=rgv[1]+t*dv,subs(u=rgu[1]+s*du,cV));
# Curl'cV's average values that correspond to surface'CS's all
segmental cells
```

$$stdcV := \left[-\frac{2}{3} \cos\left(\frac{s\,\pi}{2\,n} \right), \frac{-1}{3}, \frac{-1}{2} \right]$$

```
//'stdcV' isn't that 'Single vector value', but that 'Set of finite
quantitative (Viz.n) vector values'


// Calculate  microcosmosic surface integral value at all segmental
cells of surface 'CS':
> (stdcV[1]*stdA+stdcV[2]*stdB+stdcV[3]*stdC)*du*dv;
# Base on integral median theorem, integral value of dot product (Curl
'cV' and '[stdA,stdB,stdC]') at all segmental cells
```

$$\left(-\frac{2}{9} \cos\left(\frac{s\,\pi}{2\,n} \right) \sin\left(\frac{s\,\pi}{2\,n} \right) \right.$$
$$\left(3 \sin\left(\frac{s\,\pi}{2\,n} \right) \cos\left(\frac{2\,t\,\pi}{n} \right) \sin\left(\frac{2\,t\,\pi}{3\,n} \right) + \sin\left(\frac{s\,\pi}{2\,n} \right) \sin\left(\frac{2\,t\,\pi}{n} \right) \cos\left(\frac{2\,t\,\pi}{3\,n} \right) \right) .$$
$$- \frac{1}{3} \sin\left(\frac{s\,\pi}{2\,n} \right)^2 \sin\left(\frac{2\,t\,\pi}{n} \right) -$$
$$\frac{1}{6} \sin\left(\frac{s\,\pi}{2\,n} \right) \left(\cos\left(\frac{s\,\pi}{2\,n} \right) \cos\left(\frac{2\,t\,\pi}{n} \right) \sin\left(\frac{2\,t\,\pi}{n} \right) \cos\left(\frac{2\,t\,\pi}{3\,n} \right) + 3 \cos\left(\frac{s\,\pi}{2\,n} \right) \sin\left(\frac{2\,t\,\pi}{3\,n} \right) \right) \right)$$
$$\pi^2 / n^2$$

```
// Expression above-mentioned isn't that 'Single value', but that 'Set
of finite quantitative (Viz.n) values'


// Structure finite sums:
> Sum(Sum((stdcV[1]*stdA+stdcV[2]*stdB+stdcV[3]*stdC)*du*dv,
s=1..dus),t=1..dus);
# In the case of "Amount 'n' of surface 'CS's parameter segmental cells
is pending", sum of integral values (Dot product about curl 'cV' and
```

'[stdA,stdB,stdC]') at all segmental cells

$$\sum_{t=1}^{n}\left(\sum_{s=1}^{n}\left(-\frac{2}{9}\cos\left(\frac{s\,\pi}{2\,n}\right)\sin\left(\frac{s\,\pi}{2\,n}\right)\right)\right.$$

$$\left(3\sin\left(\frac{s\,\pi}{2\,n}\right)\cos\left(\frac{2\,t\,\pi}{n}\right)\sin\left(\frac{2\,t\,\pi}{3\,n}\right)+\sin\left(\frac{s\,\pi}{2\,n}\right)\sin\left(\frac{2\,t\,\pi}{n}\right)\cos\left(\frac{2\,t\,\pi}{3\,n}\right)\right)$$

$$-\frac{1}{3}\sin\left(\frac{s\,\pi}{2\,n}\right)^{2}\sin\left(\frac{2\,t\,\pi}{n}\right)-$$

$$\frac{1}{6}\sin\left(\frac{s\,\pi}{2\,n}\right)\left(\cos\left(\frac{s\,\pi}{2\,n}\right)\cos\left(\frac{2\,t\,\pi}{n}\right)\sin\left(\frac{2\,t\,\pi}{n}\right)\cos\left(\frac{2\,t\,\pi}{3\,n}\right)+3\cos\left(\frac{s\,\pi}{2\,n}\right)\sin\left(\frac{2\,t\,\pi}{3\,n}\right)\right)\right)$$

$$\left.\pi^{2}\,/\,n^{2}\right)$$

```
> vs:=value(%):
> Limit(vs,n=infinity):     (10)
// Infinitize finite sums, its limit operational value is 'Surface
Integral Value at Manifold'
# In the case of "Amount 'n' of surface 'CS's segmental cells tends
to infinity", limit value of integral values' sums (dot product about
curl 'cV' and '[stdA,stdB,stdC]') at all segmental cells
> beta:=value(%);epsilon:=evalf(beta);
```

$$\beta := \frac{-81}{70}$$

$$\varepsilon := -1.157142857$$

```
> omega;epsilon;
# There is tiny difference between "Integral values' sum of finite
quantitative segmental cells" and "Integral values' sum of infinite
quantitative segmental cells"
```

$$-1.181828581$$

$$-1.157142857$$

```
> alpha,beta; # Two analytic values are equal
```

$$\frac{-81}{70}, \frac{-81}{70}$$

```
> delta,epsilon; # Two float values are equal
```

$$-1.157142857, -1.157142857$$

Chapter 7 Counterexample of Curl Theorem at Manifold
--Spacial Closed Curve Integral and Surface Integral about Mobius Strip

7.1 Curl Theorem at Manifold and Mobius Strip

Confessedly, Mobius strip is representative unorientable unclosed surface; If try to conclude and deduce Mobius strip by logic method of 'Prove Curl Theorem at Manifold', what instance will appear possibly ?

Curl Theorem Suppose positive directional boundary 'L+' of smooth or piecewise smooth orientational surface 'S' is smooth or piecewise smooth closed curve, Right Hand Rule is associated basis between positive direction of boundary curve 'L' and outer side of orientational surface 'S'. If there are 1th order continuous partial derivatives about functions 'P(x,y,z),Q(x,y,z),R(x,y,z)' [Structure vector field 'A'] at orientational surface 'S', then:

$$\int_{L+} A \cdot dL = \iint_S rotA \cdot n \, dS \qquad (1)$$

thereinto, 'rotA' is curl of vector field 'A', 'n' is unit normal vector of orientational surface 'S's outer side.

Proof (Counterexample):
Define Parameterized Expression of Mobius Strip:
(Unorientable Unclosed)
[(3 + v cos(u/2)) cos(u),(3 + v cos(u/2)) sin(u), v sin(u/2)]   (2)
Set Range of u∈[0,2π] and v∈[-1,1]:

978-1-62265-930-2 (online)   978-1-62265-931-9 (paper)   Yang Ke

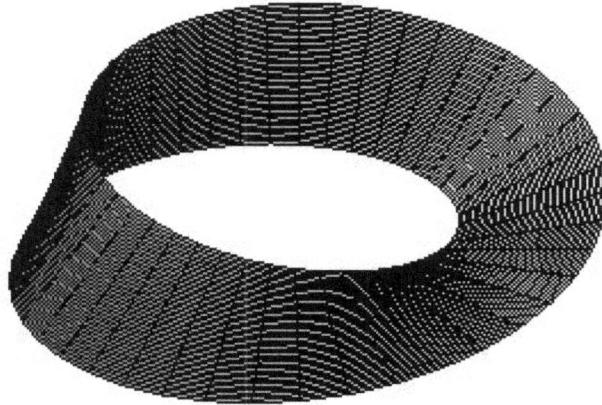

Figure(III)7.1.1 Mobius Strip (Unorientable、Unclosed)

Import Left Boundary Value of 'v' (Viz. -1) to Expression of Mobius Strip (2), obtain Expression of Boundary Curve 'CL1' (3);

Set Right Hand Rule as associated basis between positive direction of boundary curve 'CL1' and normal vector of Mobius Strip

$$\left[\left(3-\cos\left(\frac{u}{2}\right)\right)\cos(u),\left(3-\cos\left(\frac{u}{2}\right)\right)\sin(u),-\sin\left(\frac{u}{2}\right)\right]=$$

$$\left[\left(3-\cos\left(\frac{u}{2}\right)\right)\cos(u),\left(3-\cos\left(\frac{u}{2}\right)\right)\sin(u),-\sin\left(\frac{u}{2}\right)\right]$$

$$CL1:=\left[\left(3-\cos\left(\frac{u}{2}\right)\right)\cos(u),\left(3-\cos\left(\frac{u}{2}\right)\right)\sin(u),-\sin\left(\frac{u}{2}\right)\right]$$

(3)

Import Right Boundary Value of 'v' (Viz. 1) to Expression of Mobius Strip (2), obtain Expression of Boundary Curve 'CL2' (4);

Set Right Hand Rule as associated basis between positive direction of boundary curve 'CL2' and normal vector of Mobius Strip

$$\left[\left(3+\cos\left(\frac{u}{2}\right)\right)\cos(u),\left(3+\cos\left(\frac{u}{2}\right)\right)\sin(u),\sin\left(\frac{u}{2}\right)\right]=$$

$$\left[\left(3+\cos\left(\frac{u}{2}\right)\right)\cos(u),\left(3+\cos\left(\frac{u}{2}\right)\right)\sin(u),\sin\left(\frac{u}{2}\right)\right]$$

$$CL2:=\left[\left(3+\cos\left(\frac{u}{2}\right)\right)\cos(u),\left(3+\cos\left(\frac{u}{2}\right)\right)\sin(u),\sin\left(\frac{u}{2}\right)\right]$$

(4)

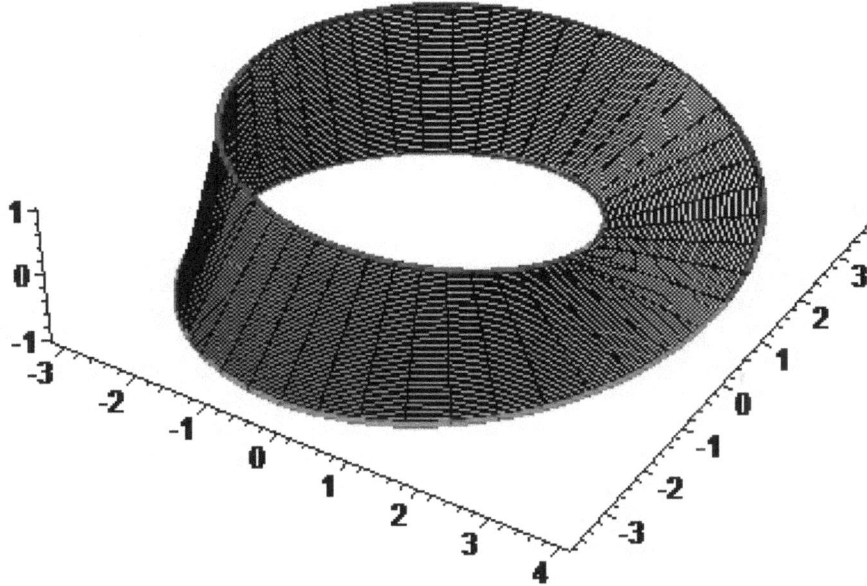

Figure(III)7.1.2 Mobius strip and its boundary curves 'CL1,CL2'
(Closed Boundary Curve of Mobius Strip is composed by 'CL1' and 'CL2')

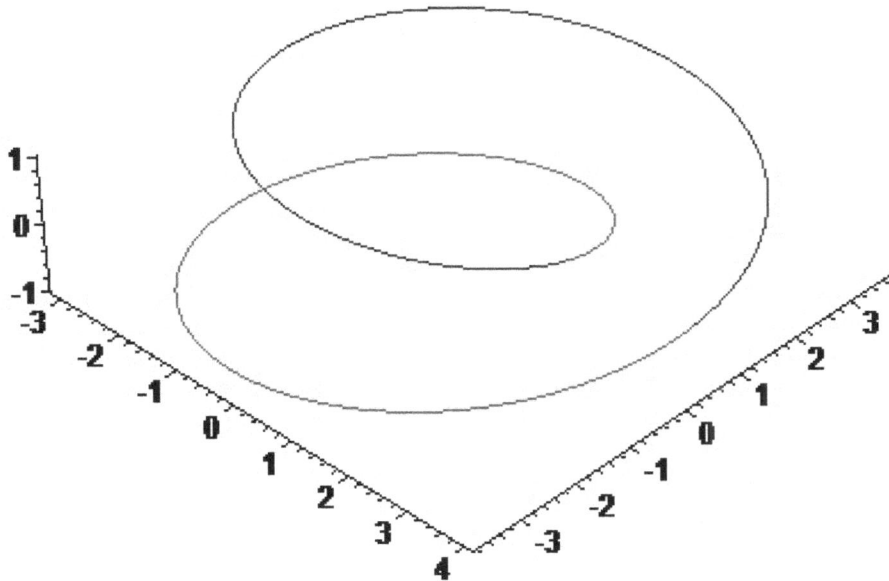

Figture(III)7.1.3 Patchwork of Mobius Strip's
Boundary Curves 'CL1' & 'CL2'

Curve Integral(Abstract Vector Field 'V' at Boundary Curve 'CL1'):

$$\int_0^{2\pi} P(x,y,z)\left(\frac{d}{du}\left(\left(3-\cos\left(\frac{u}{2}\right)\right)\cos(u)\right)\right) + Q(x,y,z)\left(\frac{d}{du}\left(\left(3-\cos\left(\frac{u}{2}\right)\right)\sin(u)\right)\right)$$

$$+ R(x,y,z)\left(\frac{d}{du}\left(-\sin\left(\frac{u}{2}\right)\right)\right) du$$

V1 = 2P(x,y,z)   (5)

Curve Integral(Abstract Vector Field 'V' at Boundary Curve 'CL2'):

$$\int_0^{2\pi} P(x,y,z)\left(\frac{d}{du}\left(\left(3+\cos\left(\frac{u}{2}\right)\right)\cos(u)\right)\right) + Q(x,y,z)\left(\frac{d}{du}\left(\left(3+\cos\left(\frac{u}{2}\right)\right)\sin(u)\right)\right)$$

$$+ R(x,y,z)\left(\frac{d}{du}\sin\left(\frac{u}{2}\right)\right) du$$

V2 = -2P(x,y,z)   (6)

Closed Curve Integral Value (Abstract Vector Field 'V' at Closed
Boundary Curve 'Patchwork CL1/CL2')(7):

V1 - V2 = 4P(x,y,z)   (7)

According to expression(2),Define and calculate matrix of partial
derivatives,obtain'tangent plane normal vector'of'Mobius Strip'(8):

$$\begin{bmatrix} i & j & k \\ \frac{\partial}{\partial u}\left(\left(3+v\cos\left(\frac{u}{2}\right)\right)\cos(u)\right) & \frac{\partial}{\partial u}\left(\left(3+v\cos\left(\frac{u}{2}\right)\right)\sin(u)\right) & \frac{\partial}{\partial u}\left(v\sin\left(\frac{u}{2}\right)\right) \\ \frac{\partial}{\partial v}\left(\left(3+v\cos\left(\frac{u}{2}\right)\right)\cos(u)\right) & \frac{\partial}{\partial v}\left(\left(3+v\cos\left(\frac{u}{2}\right)\right)\sin(u)\right) & \frac{\partial}{\partial v}\left(v\sin\left(\frac{u}{2}\right)\right) \end{bmatrix}$$

$$=$$

$$2\sin\left(\frac{u}{2}\right)\cos\left(\frac{u}{2}\right)^3 iv + 6\sin\left(\frac{u}{2}\right)\cos\left(\frac{u}{2}\right)^2 i - 2\sin\left(\frac{u}{2}\right)\cos\left(\frac{u}{2}\right)iv - 3i\sin\left(\frac{u}{2}\right)$$

$$- 2\cos\left(\frac{u}{2}\right)^4 jv - 6\cos\left(\frac{u}{2}\right)^3 j + 3\cos\left(\frac{u}{2}\right)^2 jv - \cos\left(\frac{u}{2}\right)^2 kv + 6j\cos\left(\frac{u}{2}\right)$$

$$- 3\cos\left(\frac{u}{2}\right)k - \frac{jv}{2}$$

(8)

From expression (8), Respectively pick up coefficient of item
'i,j,k', obtain Tangent Plane Normal Vector of Mobius Strip (9):

$$\left[\, 2\sin\!\left(\frac{u}{2}\right)\cos\!\left(\frac{u}{2}\right)^{3} v + 6\sin\!\left(\frac{u}{2}\right)\cos\!\left(\frac{u}{2}\right)^{2} - 2\sin\!\left(\frac{u}{2}\right)\cos\!\left(\frac{u}{2}\right) v - 3\sin\!\left(\frac{u}{2}\right),\right.$$

$$\left. -2\cos\!\left(\frac{u}{2}\right)^{4} v - 6\cos\!\left(\frac{u}{2}\right)^{3} + 3\cos\!\left(\frac{u}{2}\right)^{2} v + 6\cos\!\left(\frac{u}{2}\right) - \frac{v}{2}, -\cos\!\left(\frac{u}{2}\right)^{2} v - 3\cos\!\left(\frac{u}{2}\right)\right]$$

Calculate Curl of Vector Field 'A', and Convert Curl from Cartesian Coordinates Expression(10) to Mobius Strip Coordinates Expression(11):

$$\left[\left(\frac{\partial}{\partial y}R(x,y,z)\right)-\left(\frac{\partial}{\partial z}Q(x,y,z)\right),\left(\frac{\partial}{\partial z}P(x,y,z)\right)-\left(\frac{\partial}{\partial x}R(x,y,z)\right),\right.$$
$$\left.\left(\frac{\partial}{\partial x}Q(x,y,z)\right)-\left(\frac{\partial}{\partial y}P(x,y,z)\right)\right] \tag{10}$$

$$\left[\left(\frac{\partial}{\partial y}R(x,y,z)\right)\left(\frac{\partial}{\partial u}\left(\left(3+v\cos\!\left(\frac{u}{2}\right)\right)\sin(u)\right)\right)\left(\frac{\partial}{\partial v}\left(\left(3+v\cos\!\left(\frac{u}{2}\right)\right)\sin(u)\right)\right)\right.$$

$$-\left(\frac{\partial}{\partial z}Q(x,y,z)\right)\left(\frac{\partial}{\partial u}\left(v\sin\!\left(\frac{u}{2}\right)\right)\right)\left(\frac{\partial}{\partial v}\left(v\sin\!\left(\frac{u}{2}\right)\right)\right),$$

$$\left(\frac{\partial}{\partial z}P(x,y,z)\right)\left(\frac{\partial}{\partial u}\left(v\sin\!\left(\frac{u}{2}\right)\right)\right)\left(\frac{\partial}{\partial v}\left(v\sin\!\left(\frac{u}{2}\right)\right)\right)$$

$$-\left(\frac{\partial}{\partial x}R(x,y,z)\right)\left(\frac{\partial}{\partial u}\left(\left(3+v\cos\!\left(\frac{u}{2}\right)\right)\cos(u)\right)\right)\left(\frac{\partial}{\partial v}\left(\left(3+v\cos\!\left(\frac{u}{2}\right)\right)\cos(u)\right)\right),$$

$$\left(\frac{\partial}{\partial x}Q(x,y,z)\right)\left(\frac{\partial}{\partial u}\left(\left(3+v\cos\!\left(\frac{u}{2}\right)\right)\cos(u)\right)\right)\left(\frac{\partial}{\partial v}\left(\left(3+v\cos\!\left(\frac{u}{2}\right)\right)\cos(u)\right)\right)$$

$$\left.-\left(\frac{\partial}{\partial y}P(x,y,z)\right)\left(\frac{\partial}{\partial u}\left(\left(3+v\cos\!\left(\frac{u}{2}\right)\right)\sin(u)\right)\right)\left(\frac{\partial}{\partial v}\left(\left(3+v\cos\!\left(\frac{u}{2}\right)\right)\sin(u)\right)\right)\right]$$

$$=$$

$$\left[\left(\frac{\partial}{\partial y}R(x,y,z)\right)\left(-\frac{1}{2}v\sin\!\left(\frac{u}{2}\right)\sin(u)+\left(3+v\cos\!\left(\frac{u}{2}\right)\right)\cos(u)\right)\cos\!\left(\frac{u}{2}\right)\sin(u)\right.$$

$$-\frac{1}{2}\left(\frac{\partial}{\partial z}Q(x,y,z)\right)v\cos\!\left(\frac{u}{2}\right)\sin\!\left(\frac{u}{2}\right), \frac{1}{2}\left(\frac{\partial}{\partial z}P(x,y,z)\right)v\cos\!\left(\frac{u}{2}\right)\sin\!\left(\frac{u}{2}\right)$$

$$-\left(\frac{\partial}{\partial x}R(x,y,z)\right)\left(-\frac{1}{2}v\sin\!\left(\frac{u}{2}\right)\cos(u)-\left(3+v\cos\!\left(\frac{u}{2}\right)\right)\sin(u)\right)\cos\!\left(\frac{u}{2}\right)\cos(u),$$

$$\left(\frac{\partial}{\partial x}Q(x,y,z)\right)\left(-\frac{1}{2}v\sin\!\left(\frac{u}{2}\right)\cos(u)-\left(3+v\cos\!\left(\frac{u}{2}\right)\right)\sin(u)\right)\cos\!\left(\frac{u}{2}\right)\cos(u)$$

$$\left.-\left(\frac{\partial}{\partial y}P(x,y,z)\right)\left(-\frac{1}{2}v\sin\!\left(\frac{u}{2}\right)\sin(u)+\left(3+v\cos\!\left(\frac{u}{2}\right)\right)\cos(u)\right)\cos\!\left(\frac{u}{2}\right)\sin(u)\right]$$

Surface Integral of Dot Product [Curl (11) and Tangent Plane Normal Vector of Mobius Strip (9)] in Range of 'u,v' (12):

$$\int_{-1}^{1}\int_{0}^{2\pi}\left(2\sin\!\left(\frac{u}{2}\right)\cos\!\left(\frac{u}{2}\right)^{3} v + 6\sin\!\left(\frac{u}{2}\right)\cos\!\left(\frac{u}{2}\right)^{2} - 2\sin\!\left(\frac{u}{2}\right)\cos\!\left(\frac{u}{2}\right) v - 3\sin\!\left(\frac{u}{2}\right)\right)\Big($$

$$\left(\frac{\partial}{\partial y} R(x,y,z)\right)\left(-\frac{1}{2}v\sin\left(\frac{u}{2}\right)\sin(u)+\left(3+v\cos\left(\frac{u}{2}\right)\right)\cos(u)\right)\cos\left(\frac{u}{2}\right)\sin(u)$$

$$-\frac{1}{2}\left(\frac{\partial}{\partial z}Q(x,y,z)\right)v\cos\left(\frac{u}{2}\right)\sin\left(\frac{u}{2}\right)\right)+$$

$$\left(-2\cos\left(\frac{u}{2}\right)^4 v-6\cos\left(\frac{u}{2}\right)^3+3\cos\left(\frac{u}{2}\right)^2 v+6\cos\left(\frac{u}{2}\right)-\frac{v}{2}\right)\left(\right.$$

$$\frac{1}{2}\left(\frac{\partial}{\partial z}P(x,y,z)\right)v\cos\left(\frac{u}{2}\right)\sin\left(\frac{u}{2}\right)$$

$$-\left(\frac{\partial}{\partial x}R(x,y,z)\right)\left(-\frac{1}{2}v\sin\left(\frac{u}{2}\right)\cos(u)-\left(3+v\cos\left(\frac{u}{2}\right)\right)\sin(u)\right)\cos\left(\frac{u}{2}\right)\cos(u)\right)+$$

$$\left(-\cos\left(\frac{u}{2}\right)^2 v-3\cos\left(\frac{u}{2}\right)\right)\left(\right.$$

$$\left(\frac{\partial}{\partial x}Q(x,y,z)\right)\left(-\frac{1}{2}v\sin\left(\frac{u}{2}\right)\cos(u)-\left(3+v\cos\left(\frac{u}{2}\right)\right)\sin(u)\right)\cos\left(\frac{u}{2}\right)\cos(u)$$

$$-\left(\frac{\partial}{\partial y}P(x,y,z)\right)\left(-\frac{1}{2}v\sin\left(\frac{u}{2}\right)\sin(u)+\left(3+v\cos\left(\frac{u}{2}\right)\right)\cos(u)\right)\cos\left(\frac{u}{2}\right)\sin(u)\right)du$$

$$dv$$

$$=$$

$$\frac{1}{12}\left(\frac{\partial}{\partial z}Q(x,y,z)\right)\pi+\frac{37}{16}\left(\frac{\partial}{\partial y}R(x,y,z)\right)\pi \tag{12}$$

Viz. Expression (7) ≠ Expression (12)

Conclude and deduce 'Mobius Strip' by logic method of 'Prove Curl Theorem at Manifold', Surface Integral at 'Mobius Strip' and Closed Curve Integral at Mobius Strip's Closed Boundary Curve (Patchwork) are unequal in logic.

Complete Proof (Counterexample).

7.2 Curl Theorem at Manifold and Mobius Strip
[Program Template of Waterloo Maplesoft, Optional]

Curl Theorem Suppose positive directional boundary 'L+' of smooth or piecewise smooth orientational surface 'S' is smooth or piecewise smooth closed curve, Right Hand Rule is associated basis between positive direction of boundary curve 'L' and outer side of orientational surface 'S'. If there are 1th order continuous partial derivatives about functions 'P(x,y,z),Q(x,y,z),R(x,y,z)' [Structure vector field 'A'] at orientational surface 'S', then:

$$\int_{L+} A \cdot dL = \iint_{S} rotA \cdot n \, dS \qquad (1)$$

thereinto, 'rotA' is curl of vector field 'A', 'n' is unit normal vector of orientational surface 'S's outer side.

Symbol System:

Abstract Spacial Vector Field 'V',

Curl of Abstract Vector Field 'V' as 'cV1,cV2',

Unorientable unclosed surface 'CS'(Viz.'Mobius strip'),

Boundary curves 'CL1,CL2' of 'Mobius strip'

'Tangent plane normal vector [A,B,C]' of 'Mobius strip',

> restart;

> with(plots):with(linalg):

> CS:=[(3+v*cos(u/2))*cos(u),(3+v*cos(u/2))*sin(u),v*sin(u/2)];

Define unorientable unclosed parameterized surface 'CS' (Viz. 'Mobius strip')

$$CS := \left[\left(3 + v \cos\left(\frac{u}{2}\right) \right) \cos(u), \left(3 + v \cos\left(\frac{u}{2}\right) \right) \sin(u), v \sin\left(\frac{u}{2}\right) \right]$$

> rgu:=[0,2*Pi];

$$rgu := [0, 2\pi]$$

> rgv:=[-1,1]; # Define range of 'u,v'

$$rgv := [-1, 1]$$

> plot3d(CS,u=rgu[1]..rgu[2],v=rgv[1]..rgv[2],scaling=constrained, projection=0.9,numpoints=1500):g1:=%:

Draw 'Mobius strip', See also Figure(III)7.1.1

> subs(v=rgv[1],CS)=eval(subs(v=rgv[1],CS));CL1:=rhs(%);

Import left boundary value of 'v' to expression of ' Mobius strip, obtain expression of boundary curve 'CL1'; Set Right Hand Rule as associated basis between positive direction of boundary curve 'CL1' and normal vector of Mobius strip

$$\left[\left(3 - \cos\left(\frac{u}{2}\right) \right) \cos(u), \left(3 - \cos\left(\frac{u}{2}\right) \right) \sin(u), -\sin\left(\frac{u}{2}\right) \right] =$$
$$\left[\left(3 - \cos\left(\frac{u}{2}\right) \right) \cos(u), \left(3 - \cos\left(\frac{u}{2}\right) \right) \sin(u), -\sin\left(\frac{u}{2}\right) \right]$$

$$CL1 := \left[\left(3 - \cos\left(\frac{u}{2}\right)\right)\cos(u), \left(3 - \cos\left(\frac{u}{2}\right)\right)\sin(u), -\sin\left(\frac{u}{2}\right)\right]$$

```
> subs(v=rgv[2],CS)=eval(subs(v=rgv[2],CS));CL2:=rhs(%);
```

\# Import right boundary value of 'v' to expression of ' Mobius strip', obtain expression of boundary curve 'CL2'; Set Right Hand Rule as associated basis between positive direction of boundary curve 'CL2' and normal vector of Mobius strip

$$\left[\left(3 + \cos\left(\frac{u}{2}\right)\right)\cos(u), \left(3 + \cos\left(\frac{u}{2}\right)\right)\sin(u), \sin\left(\frac{u}{2}\right)\right] =$$
$$\left[\left(3 + \cos\left(\frac{u}{2}\right)\right)\cos(u), \left(3 + \cos\left(\frac{u}{2}\right)\right)\sin(u), \sin\left(\frac{u}{2}\right)\right]$$
$$CL2 := \left[\left(3 + \cos\left(\frac{u}{2}\right)\right)\cos(u), \left(3 + \cos\left(\frac{u}{2}\right)\right)\sin(u), \sin\left(\frac{u}{2}\right)\right]$$

```
> spacecurve(CL1,u=rgu[1]..rgu[2],numpoints=2000,thickness=2,
color=red):g2:=%: # Draw boundary curve'CL1'
> spacecurve(CL2,u=rgu[1]..rgu[2],numpoints=2000,thickness=2,
color=blue):g3:=%: # Draw boundary curve'CL2'
> display(g1,g2,g3): # See also Figure(III)7.1.2
> display(g2,g3,scaling=constrained,projection=0.9):
```

\# See also Figure(III)7.1.3

```
> V:=[(P)(x,y,z),(Q)(x,y,z),(R)(x,y,z)];
```

\# Define abstract spacial vector field 'V' (Suppose spacial vector field 'V' possesses 1th order continuous partial derivatives at 'Mobius strip')

$$V := [\mathrm{P}(x, y, z), \mathrm{Q}(x, y, z), \mathrm{R}(x, y, z)]$$

```
> [Diff(V[3],y)-Diff(V[2],z),Diff(V[1],z)-Diff(V[3],x),
Diff(V[2],x)-Diff(V[1],y)]
=[diff(V[3],y)-diff(V[2],z),diff(V[1],z)-diff(V[3],x),
diff(V[2],x)-diff(V[1],y)];cV1:=rhs(%);
```

\# Calculate curl 'cV1' of abstract spacial vector field 'V'

$$\left[\left(\frac{\partial}{\partial y}\mathrm{R}(x,y,z)\right) - \left(\frac{\partial}{\partial z}\mathrm{Q}(x,y,z)\right), \left(\frac{\partial}{\partial z}\mathrm{P}(x,y,z)\right) - \left(\frac{\partial}{\partial x}\mathrm{R}(x,y,z)\right),\right.$$
$$\left(\frac{\partial}{\partial x}\mathrm{Q}(x,y,z)\right) - \left(\frac{\partial}{\partial y}\mathrm{P}(x,y,z)\right)\right] = \left[\left(\frac{\partial}{\partial y}\mathrm{R}(x,y,z)\right) - \left(\frac{\partial}{\partial z}\mathrm{Q}(x,y,z)\right),\right.$$
$$\left(\frac{\partial}{\partial z}\mathrm{P}(x,y,z)\right) - \left(\frac{\partial}{\partial x}\mathrm{R}(x,y,z)\right), \left(\frac{\partial}{\partial x}\mathrm{Q}(x,y,z)\right) - \left(\frac{\partial}{\partial y}\mathrm{P}(x,y,z)\right)\right]$$

$$cV1 := \left[\left(\left(\frac{\partial}{\partial y} R(x,y,z)\right) - \left(\frac{\partial}{\partial z} Q(x,y,z)\right)\right), \left(\left(\frac{\partial}{\partial z} P(x,y,z)\right) - \left(\frac{\partial}{\partial x} R(x,y,z)\right)\right),\right.$$
$$\left.\left(\frac{\partial}{\partial x} Q(x,y,z)\right) - \left(\frac{\partial}{\partial y} P(x,y,z)\right)\right]$$

```
> x:='x':y:='y':z:='z':
> Int(V[1]*Diff(CL1[1],u)+V[2]*Diff(CL1[2],u)+V[3]*Diff(CL1[3],u),
u=rgu[1]..rgu[2]);
# Curve Integral (Spacial Vector Field 'V' at boundary curve 'CL1')
```

$$\int_0^{2\pi} P(x,y,z)\left(\frac{d}{du}\left(\left(3-\cos\left(\frac{u}{2}\right)\right)\cos(u)\right)\right) + Q(x,y,z)\left(\frac{d}{du}\left(\left(3-\cos\left(\frac{u}{2}\right)\right)\sin(u)\right)\right)$$
$$+ R(x,y,z)\left(\frac{d}{du}\left(-\sin\left(\frac{u}{2}\right)\right)\right) du$$

```
> v1:=value(%);
```

$$v1 := 2\,P(x,y,z)$$

```
> x:='x':y:='y':z:='z':
> Int(V[1]*Diff(CL2[1],u)+V[2]*Diff(CL2[2],u)+V[3]*Diff(CL2[3],u),
u=rgu[1]..rgu[2]);
# Curve Integral (Spacial Vector Field 'V' at boundary curve 'CL2')
```

$$\int_0^{2\pi} P(x,y,z)\left(\frac{d}{du}\left(\left(3+\cos\left(\frac{u}{2}\right)\right)\cos(u)\right)\right) + Q(x,y,z)\left(\frac{d}{du}\left(\left(3+\cos\left(\frac{u}{2}\right)\right)\sin(u)\right)\right)$$
$$+ R(x,y,z)\left(\frac{d}{du}\,\sin\left(\frac{u}{2}\right)\right) du$$

```
> v2:=value(%);
```

$$v2 := -2\,P(x,y,z)$$

```
> alpha:=v1-v2;
# Closed Curve Integral (Spacial Vector Field 'V' at closed boundary
curve 'Patchwork CL1/CL2')
```

$$\alpha := 4\,P(x,y,z)$$

```
> x:='x':y:='y':z:='z':
> matrix(3,3,[i,j,k,Diff(CS[1],u),Diff(CS[2],u),Diff(CS[3],u),
Diff(CS[1],v),Diff(CS[2],v),Diff(CS[3],v)])=
matrix(3,3,[i,j,k,diff(CS[1],u),diff(CS[2],u),diff(CS[3],u),
diff(CS[1],v),diff(CS[2],v),diff(CS[3],v)]);m:=rhs(%);
# Define and calculate matrix of partial derivatives 'm', obtain
```

'tangent plane normal vector' of 'Mobius strip'

$$\begin{bmatrix} i & j & k \\ \dfrac{\partial}{\partial u}\left(\left(3+v\cos\left(\dfrac{u}{2}\right)\right)\cos(u)\right) & \dfrac{\partial}{\partial u}\left(\left(3+v\cos\left(\dfrac{u}{2}\right)\right)\sin(u)\right) & \dfrac{\partial}{\partial u}\left(v\sin\left(\dfrac{u}{2}\right)\right) \\ \dfrac{\partial}{\partial v}\left(\left(3+v\cos\left(\dfrac{u}{2}\right)\right)\cos(u)\right) & \dfrac{\partial}{\partial v}\left(\left(3+v\cos\left(\dfrac{u}{2}\right)\right)\sin(u)\right) & \dfrac{\partial}{\partial v}\left(v\sin\left(\dfrac{u}{2}\right)\right) \end{bmatrix}=$$

$$[i,j,k]$$

$$\left[-\frac{1}{2}v\sin\left(\frac{u}{2}\right)\cos(u)-\left(3+v\cos\left(\frac{u}{2}\right)\right)\sin(u)\,,\right.$$

$$\left.-\frac{1}{2}v\sin\left(\frac{u}{2}\right)\sin(u)+\left(3+v\cos\left(\frac{u}{2}\right)\right)\cos(u)\,,\frac{1}{2}v\cos\left(\frac{u}{2}\right)\right]$$

$$\left[\cos\left(\frac{u}{2}\right)\cos(u)\,,\cos\left(\frac{u}{2}\right)\sin(u)\,,\sin\left(\frac{u}{2}\right)\right]$$

$m :=$

$$[i,j,k]$$

$$\left[-\frac{1}{2}v\sin\left(\frac{u}{2}\right)\cos(u)-\left(3+v\cos\left(\frac{u}{2}\right)\right)\sin(u)\,,\right.$$

$$\left.-\frac{1}{2}v\sin\left(\frac{u}{2}\right)\sin(u)+\left(3+v\cos\left(\frac{u}{2}\right)\right)\cos(u)\,,\frac{1}{2}v\cos\left(\frac{u}{2}\right)\right]$$

$$\left[\cos\left(\frac{u}{2}\right)\cos(u)\,,\cos\left(\frac{u}{2}\right)\sin(u)\,,\sin\left(\frac{u}{2}\right)\right]$$

```
> det(m);
```

$$-\frac{1}{2}i\,v\sin\left(\frac{u}{2}\right)^2\sin(u)+3\,i\sin\left(\frac{u}{2}\right)\cos(u)+i\sin\left(\frac{u}{2}\right)\cos(u)\,v\cos\left(\frac{u}{2}\right)$$

$$-\frac{1}{2}i\,v\cos\left(\frac{u}{2}\right)^2\sin(u)+\frac{1}{2}v\sin\left(\frac{u}{2}\right)^2\cos(u)\,j+3\sin(u)\,j\sin\left(\frac{u}{2}\right)$$

$$-3\,k\cos\left(\frac{u}{2}\right)\sin(u)^2+\sin(u)\,v\cos\left(\frac{u}{2}\right)j\sin\left(\frac{u}{2}\right)-\sin(u)^2\,v\cos\left(\frac{u}{2}\right)^2k$$

$$+\frac{1}{2}\cos\left(\frac{u}{2}\right)^2\cos(u)\,j\,v-3\cos\left(\frac{u}{2}\right)\cos(u)^2\,k-\cos\left(\frac{u}{2}\right)^2\cos(u)^2\,k\,v$$

```
> mn:=simplify(%);
```

$$mn:=-3\sin\left(\frac{u}{2}\right)i+2\sin\left(\frac{u}{2}\right)i\,v\cos\left(\frac{u}{2}\right)^3+6\sin\left(\frac{u}{2}\right)i\cos\left(\frac{u}{2}\right)^2-2\sin\left(\frac{u}{2}\right)i\,v\cos\left(\frac{u}{2}\right)$$

$$-2\cos\left(\frac{u}{2}\right)^4v\,j+3\cos\left(\frac{u}{2}\right)^2v\,j-\cos\left(\frac{u}{2}\right)^2v\,k-3\,k\cos\left(\frac{u}{2}\right)+6\cos\left(\frac{u}{2}\right)j-\frac{v\,j}{2}$$

$$-6\cos\left(\frac{u}{2}\right)^3j$$

```
> A:=coeff(mn,i); # Obtain coefficient of 'i'
```

$$A := -3\sin\left(\frac{u}{2}\right) + 2\sin\left(\frac{u}{2}\right)v\cos\left(\frac{u}{2}\right)^3 + 6\sin\left(\frac{u}{2}\right)\cos\left(\frac{u}{2}\right)^2 - 2\sin\left(\frac{u}{2}\right)v\cos\left(\frac{u}{2}\right)$$

```
> B:=coeff(mn,j); # Obtain coefficient of 'j'
```

$$B := -2\cos\left(\frac{u}{2}\right)^4 v + 3\cos\left(\frac{u}{2}\right)^2 v + 6\cos\left(\frac{u}{2}\right) - \frac{v}{2} - 6\cos\left(\frac{u}{2}\right)^3$$

```
> C:=coeff(mn,k);
# Obtain coefficient of 'k'
```

$$C := -\cos\left(\frac{u}{2}\right)^2 v - 3\cos\left(\frac{u}{2}\right)$$

```
> [A,B,C];  # [A,B,C] structure 'tangent plane normal vector'
```

$$\left[2\sin\left(\frac{u}{2}\right)\cos\left(\frac{u}{2}\right)^3 v + 6\sin\left(\frac{u}{2}\right)\cos\left(\frac{u}{2}\right)^2 - 2\sin\left(\frac{u}{2}\right)\cos\left(\frac{u}{2}\right)v - 3\sin\left(\frac{u}{2}\right),\right.$$
$$\left. -2\cos\left(\frac{u}{2}\right)^4 v - 6\cos\left(\frac{u}{2}\right)^3 + 3\cos\left(\frac{u}{2}\right)^2 v + 6\cos\left(\frac{u}{2}\right) - \frac{v}{2}, -\cos\left(\frac{u}{2}\right)^2 v - 3\cos\left(\frac{u}{2}\right)\right]$$

```
> x:='x':y:='y':z:='z':
> [Diff(V[3],y)*Diff(CS[2],u)*Diff(CS[2],v)
-Diff(V[2],z)*Diff(CS[3],u)*Diff(CS[3],v),
Diff(V[1],z)*Diff(CS[3],u)*Diff(CS[3],v)
-Diff(V[3],x)*Diff(CS[1],u)*Diff(CS[1],v),
Diff(V[2],x)*Diff(CS[1],u)*Diff(CS[1],v)
-Diff(V[1],y)*Diff(CS[2],u)*Diff(CS[2],v)]
=[diff(V[3],y)*diff(CS[2],u)*diff(CS[2],v)
-diff(V[2],z)*diff(CS[3],u)*diff(CS[3],v),
diff(V[1],z)*diff(CS[3],u)*diff(CS[3],v)
-diff(V[3],x)*diff(CS[1],u)*diff(CS[1],v),
diff(V[2],x)*diff(CS[1],u)*diff(CS[1],v)
-diff(V[1],y)*diff(CS[2],u)*diff(CS[2],v)];cV2:=rhs(%);
```
Convert 'cV1' from Cartesian Coordinates expression to Mobius strip Coordinates expression
Reserve original form of abstract vector field '[P(x,y,z),Q(x,y,z), R(x,y,z)]' and its partial derivatives (insoluble), Calculate partial derivatives of '[(3+v*cos(u/2))*cos(u), (3+v*cos(u/2))*sin(u), v*sin(u/2)]'(Computable), obtain a new expression 'cV2' (Curl) of vector field

$$\left[\left(\frac{\partial}{\partial y}R(x,y,z)\right)\left(\frac{\partial}{\partial u}\left(\left(3+v\cos\left(\frac{u}{2}\right)\right)\sin(u)\right)\right)\left(\frac{\partial}{\partial v}\left(\left(3+v\cos\left(\frac{u}{2}\right)\right)\sin(u)\right)\right)\right.$$

$$-\left(\frac{\partial}{\partial z}Q(x,y,z)\right)\left(\frac{\partial}{\partial u}\left(v\sin\left(\frac{u}{2}\right)\right)\right)\left(\frac{\partial}{\partial v}\left(v\sin\left(\frac{u}{2}\right)\right)\right),$$

$$\left(\frac{\partial}{\partial z}P(x,y,z)\right)\left(\frac{\partial}{\partial u}\left(v\sin\left(\frac{u}{2}\right)\right)\right)\left(\frac{\partial}{\partial v}\left(v\sin\left(\frac{u}{2}\right)\right)\right)$$

$$-\left(\frac{\partial}{\partial x}R(x,y,z)\right)\left(\frac{\partial}{\partial u}\left(\left(3+v\cos\left(\frac{u}{2}\right)\right)\cos(u)\right)\right)\left(\frac{\partial}{\partial v}\left(\left(3+v\cos\left(\frac{u}{2}\right)\right)\cos(u)\right)\right),$$

$$\left(\frac{\partial}{\partial x}Q(x,y,z)\right)\left(\frac{\partial}{\partial u}\left(\left(3+v\cos\left(\frac{u}{2}\right)\right)\cos(u)\right)\right)\left(\frac{\partial}{\partial v}\left(\left(3+v\cos\left(\frac{u}{2}\right)\right)\cos(u)\right)\right)$$

$$\left.-\left(\frac{\partial}{\partial y}P(x,y,z)\right)\left(\frac{\partial}{\partial u}\left(\left(3+v\cos\left(\frac{u}{2}\right)\right)\sin(u)\right)\right)\left(\frac{\partial}{\partial v}\left(\left(3+v\cos\left(\frac{u}{2}\right)\right)\sin(u)\right)\right)\right]=\left[\right.$$

$$\left(\frac{\partial}{\partial y}R(x,y,z)\right)\left(-\frac{1}{2}v\sin\left(\frac{u}{2}\right)\sin(u)+\left(3+v\cos\left(\frac{u}{2}\right)\right)\cos(u)\right)\cos\left(\frac{u}{2}\right)\sin(u)$$

$$-\frac{1}{2}\left(\frac{\partial}{\partial z}Q(x,y,z)\right)v\cos\left(\frac{u}{2}\right)\sin\left(\frac{u}{2}\right),\frac{1}{2}\left(\frac{\partial}{\partial z}P(x,y,z)\right)v\cos\left(\frac{u}{2}\right)\sin\left(\frac{u}{2}\right)$$

$$-\left(\frac{\partial}{\partial x}R(x,y,z)\right)\left(-\frac{1}{2}v\sin\left(\frac{u}{2}\right)\cos(u)-\left(3+v\cos\left(\frac{u}{2}\right)\right)\sin(u)\right)\cos\left(\frac{u}{2}\right)\cos(u),$$

$$\left(\frac{\partial}{\partial x}Q(x,y,z)\right)\left(-\frac{1}{2}v\sin\left(\frac{u}{2}\right)\cos(u)-\left(3+v\cos\left(\frac{u}{2}\right)\right)\sin(u)\right)\cos\left(\frac{u}{2}\right)\cos(u)$$

$$\left.-\left(\frac{\partial}{\partial y}P(x,y,z)\right)\left(-\frac{1}{2}v\sin\left(\frac{u}{2}\right)\sin(u)+\left(3+v\cos\left(\frac{u}{2}\right)\right)\cos(u)\right)\cos\left(\frac{u}{2}\right)\sin(u)\right]$$

$$cV2:=\left[\left(\frac{\partial}{\partial y}R(x,y,z)\right)\left(-\frac{1}{2}v\sin\left(\frac{u}{2}\right)\sin(u)+\left(3+v\cos\left(\frac{u}{2}\right)\right)\cos(u)\right)\cos\left(\frac{u}{2}\right)\sin(u)\right.$$

$$-\frac{1}{2}\left(\frac{\partial}{\partial z}Q(x,y,z)\right)v\cos\left(\frac{u}{2}\right)\sin\left(\frac{u}{2}\right),\frac{1}{2}\left(\frac{\partial}{\partial z}P(x,y,z)\right)v\cos\left(\frac{u}{2}\right)\sin\left(\frac{u}{2}\right)$$

$$-\left(\frac{\partial}{\partial x}R(x,y,z)\right)\left(-\frac{1}{2}v\sin\left(\frac{u}{2}\right)\cos(u)-\left(3+v\cos\left(\frac{u}{2}\right)\right)\sin(u)\right)\cos\left(\frac{u}{2}\right)\cos(u),$$

$$\left(\frac{\partial}{\partial x}Q(x,y,z)\right)\left(-\frac{1}{2}v\sin\left(\frac{u}{2}\right)\cos(u)-\left(3+v\cos\left(\frac{u}{2}\right)\right)\sin(u)\right)\cos\left(\frac{u}{2}\right)\cos(u)$$

$$\left.-\left(\frac{\partial}{\partial y}P(x,y,z)\right)\left(-\frac{1}{2}v\sin\left(\frac{u}{2}\right)\sin(u)+\left(3+v\cos\left(\frac{u}{2}\right)\right)\cos(u)\right)\cos\left(\frac{u}{2}\right)\sin(u)\right]$$

```
> Int(Int(cV2[1]*A+cV2[2]*B+cV2[3]*C,u=rgu[1]..rgu[2]),v=rgv[1]..
rgv[2]);
```

Surface Integral of Dot Product (Curl 'cV2' and Mobius Strip' Tangent Plane Normal Vector '[A,B,C]') in range of 'u,v'

$$\int_{-1}^{1}\int_{0}^{2\pi}\left(\left(\frac{\partial}{\partial y}R(x,y,z)\right)\left(-\frac{1}{2}v\sin\left(\frac{u}{2}\right)\sin(u)+\left(3+v\cos\left(\frac{u}{2}\right)\right)\cos(u)\right)\cos\left(\frac{u}{2}\right)\sin(u)\right.$$

$$-\frac{1}{2}\left(\frac{\partial}{\partial z}Q(x,y,z)\right)v\cos\left(\frac{u}{2}\right)\sin\left(\frac{u}{2}\right)\Bigg)$$

$$\left(-3\sin\left(\frac{u}{2}\right)+2\sin\left(\frac{u}{2}\right)v\cos\left(\frac{u}{2}\right)^3+6\sin\left(\frac{u}{2}\right)\cos\left(\frac{u}{2}\right)^2-2\sin\left(\frac{u}{2}\right)v\cos\left(\frac{u}{2}\right)\right)+\Bigg($$

$$\frac{1}{2}\left(\frac{\partial}{\partial z}P(x,y,z)\right)v\cos\left(\frac{u}{2}\right)\sin\left(\frac{u}{2}\right)$$

$$-\left(\frac{\partial}{\partial x}R(x,y,z)\right)\left(-\frac{1}{2}v\sin\left(\frac{u}{2}\right)\cos(u)-\left(3+v\cos\left(\frac{u}{2}\right)\right)\sin(u)\right)\cos\left(\frac{u}{2}\right)\cos(u)\Bigg)$$

$$\left(-2\cos\left(\frac{u}{2}\right)^4v+3\cos\left(\frac{u}{2}\right)^2v+6\cos\left(\frac{u}{2}\right)-\frac{v}{2}-6\cos\left(\frac{u}{2}\right)^3\right)+\Bigg($$

$$\left(\frac{\partial}{\partial x}Q(x,y,z)\right)\left(-\frac{1}{2}v\sin\left(\frac{u}{2}\right)\cos(u)-\left(3+v\cos\left(\frac{u}{2}\right)\right)\sin(u)\right)\cos\left(\frac{u}{2}\right)\cos(u)$$

$$-\left(\frac{\partial}{\partial y}P(x,y,z)\right)\left(-\frac{1}{2}v\sin\left(\frac{u}{2}\right)\sin(u)+\left(3+v\cos\left(\frac{u}{2}\right)\right)\cos(u)\right)\cos\left(\frac{u}{2}\right)\sin(u)\Bigg)$$

$$\left(-\cos\left(\frac{u}{2}\right)^2v-3\cos\left(\frac{u}{2}\right)\right)du\,dv$$

```
> beta:=value(%);
# Calculate value of Mobius strip' surface integral
```

$$\beta:=\frac{37}{16}\left(\frac{\partial}{\partial y}R(x,y,z)\right)\pi+\frac{1}{12}\left(\frac{\partial}{\partial z}Q(x,y,z)\right)\pi$$

```
> alpha;beta;
```

$$4\,P(x,y,z)$$

$$\frac{37}{16}\left(\frac{\partial}{\partial y}R(x,y,z)\right)\pi+\frac{1}{12}\left(\frac{\partial}{\partial z}Q(x,y,z)\right)\pi$$

```
// Conclude and deduce 'Mobius strip' by logic method of 'Proving Curl
Theorem at Manifold', surface integral at 'Mobius strip' and spacial
closed curve integral at boundary curves of 'Mobius strip' are unequal
in logic
```

7.3 Spacial Closed Curve Integral and Surface Integral about Mobius Strip, Numerical Model (I)

```
Known: Parametric Expression of Mobius Strip:
(Unorientable、Unclosed Surface)
[(3 + v cos(u/2))cos(u),(3 + v  cos(u/2))sin(u), v sin(u/2)] (1)
thereinto, u∈[0,2π], v∈[-1,1];
```

and Integral Vector Field $\left[-z^2, xz, -\left(x-y+\dfrac{z}{3}\right)^2 - \dfrac{xz}{5} - y\right]$ (2)

Calculate and Validate Curl Theorem at Manifold (Counterexample).

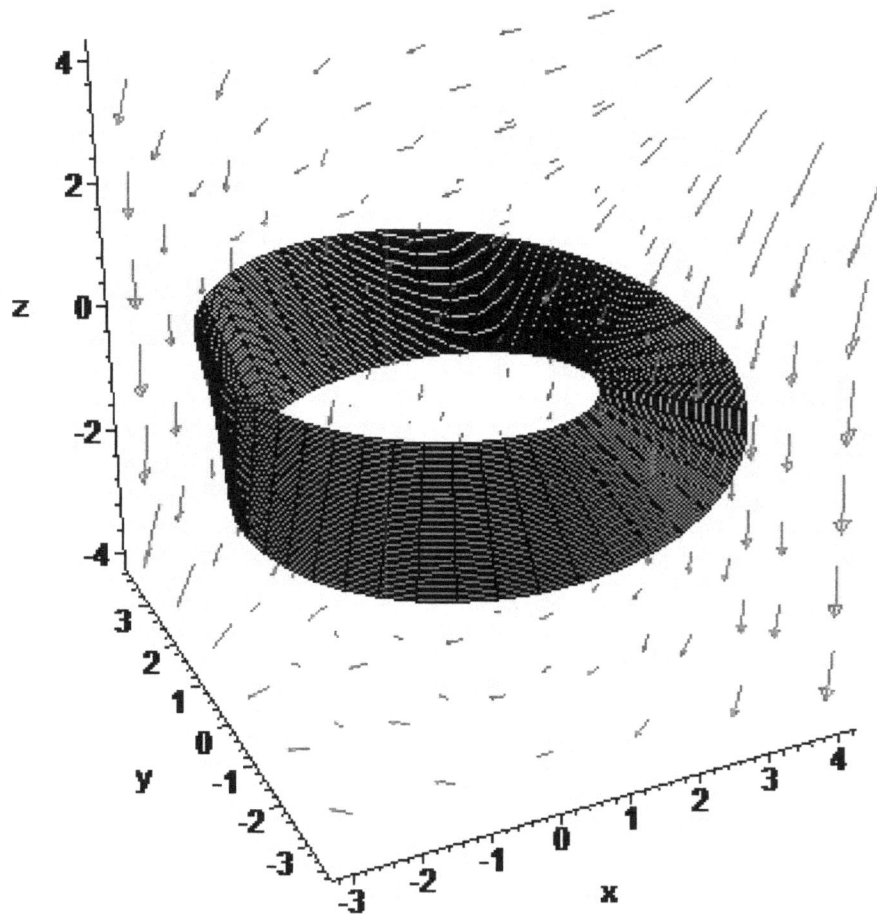

Figure(III)7.3 Mobius Strip and Vector Field

Solution:

First, Spacial Closed Curve Integral about Mobius Strip:

Import Left Boundary Value of 'v' (Viz. -1) to Expression of Mobius Strip (1), obtain Expression of Boundary Curve 'CL1' (3);

Set Right Hand Rule as associated basis between positive direction of boundary curve 'CL1' and normal vector of Mobius Strip

$$\left[\left(3-\cos\left(\frac{u}{2}\right)\right)\cos(u),\left(3-\cos\left(\frac{u}{2}\right)\right)\sin(u),-\sin\left(\frac{u}{2}\right)\right]=$$

$$\left[\left(3-\cos\left(\frac{u}{2}\right)\right)\cos(u),\left(3-\cos\left(\frac{u}{2}\right)\right)\sin(u),-\sin\left(\frac{u}{2}\right)\right]$$

$$CL1:=\left[\left(3-\cos\left(\frac{u}{2}\right)\right)\cos(u),\left(3-\cos\left(\frac{u}{2}\right)\right)\sin(u),-\sin\left(\frac{u}{2}\right)\right] \tag{3}$$

Import Right Boundary Value of 'v'(Viz. 1) to Expression of Mobius Strip (1), obtain Expression of Boundary Curve 'CL2' (4);

Set Right Hand Rule as associated basis between positive direction of boundary curve 'CL2' and normal vector of Mobius Strip

$$\left[\left(3+\cos\left(\frac{u}{2}\right)\right)\cos(u),\left(3+\cos\left(\frac{u}{2}\right)\right)\sin(u),\sin\left(\frac{u}{2}\right)\right]=$$

$$\left[\left(3+\cos\left(\frac{u}{2}\right)\right)\cos(u),\left(3+\cos\left(\frac{u}{2}\right)\right)\sin(u),\sin\left(\frac{u}{2}\right)\right]$$

$$CL2:=\left[\left(3+\cos\left(\frac{u}{2}\right)\right)\cos(u),\left(3+\cos\left(\frac{u}{2}\right)\right)\sin(u),\sin\left(\frac{u}{2}\right)\right] \tag{4}$$

Input Expression of Boundary Curve 'CL1' (3) to Vector Field (2); and Curve Integral [Vector Field (2) at Boundary Curve 'CL1' (3)]:

$$\int_0^{2\pi} -\sin\left(\frac{u}{2}\right)^2\left(\frac{d}{du}\left(\left(3-\cos\left(\frac{u}{2}\right)\right)\cos(u)\right)\right)$$

$$-\left(3-\cos\left(\frac{u}{2}\right)\right)\cos(u)\sin\left(\frac{u}{2}\right)\left(\frac{d}{du}\left(\left(3-\cos\left(\frac{u}{2}\right)\right)\sin(u)\right)\right)+\left(\right.$$

$$-\left(\left(3-\cos\left(\frac{u}{2}\right)\right)\cos(u)-\left(3-\cos\left(\frac{u}{2}\right)\right)\sin(u)-\frac{1}{3}\sin\left(\frac{u}{2}\right)\right)^2$$

$$+\frac{1}{5}\left(3-\cos\left(\frac{u}{2}\right)\right)\cos(u)\sin\left(\frac{u}{2}\right)-\left(3-\cos\left(\frac{u}{2}\right)\right)\sin(u)\left)\left(\frac{d}{du}\left(-\sin\left(\frac{u}{2}\right)\right)\right)\right)du$$

$$V1 = -\frac{29818}{1575}-\frac{5\pi}{2} \tag{5}$$

Input Expression of Boundary Curve 'CL2' (4) to Vector Field(2); and Curve Integral [Vector Field(2) at Boundary Curve 'CL2'(4)]:

$$\int_0^{2\pi} -\sin\left(\frac{t}{2}\right)^2\left(\frac{d}{dt}\left(\left(3+\cos\left(\frac{t}{2}\right)\right)\cos(t)\right)\right)$$

$$+ \left(3 + \cos\left(\frac{t}{2}\right)\right) \cos(t) \sin\left(\frac{t}{2}\right) \left(\frac{d}{dt}\left(\left(3 + \cos\left(\frac{t}{2}\right)\right) \sin(t)\right)\right) + \Bigg($$

$$- \left(\left(3 + \cos\left(\frac{t}{2}\right)\right) \cos(t) - \left(3 + \cos\left(\frac{t}{2}\right)\right) \sin(t) + \frac{1}{3} \sin\left(\frac{t}{2}\right)\right)^2$$

$$- \frac{1}{5}\left(3 + \cos\left(\frac{t}{2}\right)\right) \cos(t) \sin\left(\frac{t}{2}\right) - \left(3 + \cos\left(\frac{t}{2}\right)\right) \sin(t)\right) \left(\frac{d}{dt} \sin\left(\frac{t}{2}\right)\right) dt$$

$$V2 = \frac{29818}{1575} - \frac{5\pi}{2} \tag{6}$$

Closed Curve Integral Value (Vector Field (2) at Closed Boundary Curve 'Patchwork CL1/CL2') (7):

$$V1 - V2 = \frac{-59636}{1575} \tag{7}$$

Second, Surface Integral about Mobius Strip:

According to parametric expression (1) of Mobius Strip, define and calculate Matrix of Partial Derivatives, obtain tangent plane normal vector of Mobius Strip (8):

$$\begin{bmatrix} i & j & k \\ \frac{\partial}{\partial u}\left(\left(3 + v\cos\left(\frac{u}{2}\right)\right)\cos(u)\right) & \frac{\partial}{\partial u}\left(\left(3 + v\cos\left(\frac{u}{2}\right)\right)\sin(u)\right) & \frac{\partial}{\partial u}\left(v\sin\left(\frac{u}{2}\right)\right) \\ \frac{\partial}{\partial v}\left(\left(3 + v\cos\left(\frac{u}{2}\right)\right)\cos(u)\right) & \frac{\partial}{\partial v}\left(\left(3 + v\cos\left(\frac{u}{2}\right)\right)\sin(u)\right) & \frac{\partial}{\partial v}\left(v\sin\left(\frac{u}{2}\right)\right) \end{bmatrix}$$

$$=$$

$$2\sin\left(\frac{u}{2}\right)\cos\left(\frac{u}{2}\right)^3 i\,v + 6\sin\left(\frac{u}{2}\right)\cos\left(\frac{u}{2}\right)^2 i - 2\sin\left(\frac{u}{2}\right)\cos\left(\frac{u}{2}\right) i\,v - 3\,i\sin\left(\frac{u}{2}\right)$$

$$- 2\cos\left(\frac{u}{2}\right)^4 j\,v - 6\cos\left(\frac{u}{2}\right)^3 j + 3\cos\left(\frac{u}{2}\right)^2 j\,v - \cos\left(\frac{u}{2}\right)^2 k\,v + 6\,j\cos\left(\frac{u}{2}\right)$$

$$- 3\cos\left(\frac{u}{2}\right)k - \frac{j\,v}{2} \tag{8}$$

From Expression (8), Respectively pick up coefficient of item 'i,j,k',obtain Tangent Plane Normal Vector of Mobius Strip (9):

$$\Bigg[2\sin\left(\frac{u}{2}\right)\cos\left(\frac{u}{2}\right)^3 v + 6\sin\left(\frac{u}{2}\right)\cos\left(\frac{u}{2}\right)^2 - 2\sin\left(\frac{u}{2}\right)\cos\left(\frac{u}{2}\right) v - 3\sin\left(\frac{u}{2}\right),$$

$$- 2\cos\left(\frac{u}{2}\right)^4 v - 6\cos\left(\frac{u}{2}\right)^3 + 3\cos\left(\frac{u}{2}\right)^2 v + 6\cos\left(\frac{u}{2}\right) - \frac{v}{2}, -\cos\left(\frac{u}{2}\right)^2 v - 3\cos\left(\frac{u}{2}\right) \Bigg]$$

Calculate Curl of Vector Field (10):

$$\left[\left(\frac{\partial}{\partial y}\left(-\left(x-y+\frac{z}{3}\right)^2-\frac{xz}{5}-y\right)\right)-\left(\frac{\partial}{\partial z}(xz)\right),\right.$$

$$\left.\left(\frac{d}{dz}(-z^2)\right)-\left(\frac{\partial}{\partial x}\left(-\left(x-y+\frac{z}{3}\right)^2-\frac{xz}{5}-y\right)\right),\left(\frac{\partial}{\partial x}(xz)\right)-\left(\frac{\partial}{\partial y}(-z^2)\right)\right]$$

$$=$$

$$\left[x-2y+\frac{2z}{3}-1,-\frac{17z}{15}+2x-2y,z\right] \quad (10)$$

Input Expression of Mobius Strip(1) to Curl(10); and Surface Integral [Curl(10) at Mobius Strip(1)] (11):

$$\int_{-1}^{1}\int_{0}^{2\pi}\left(2\sin\left(\frac{u}{2}\right)\cos\left(\frac{u}{2}\right)^3 v+6\sin\left(\frac{u}{2}\right)\cos\left(\frac{u}{2}\right)^2-2\sin\left(\frac{u}{2}\right)\cos\left(\frac{u}{2}\right)v-3\sin\left(\frac{u}{2}\right)\right)$$

$$\left(\left(3+v\cos\left(\frac{u}{2}\right)\right)\cos(u)-2\left(3+v\cos\left(\frac{u}{2}\right)\right)\sin(u)+\frac{2}{3}v\sin\left(\frac{u}{2}\right)-1\right)+$$

$$\left(-2\cos\left(\frac{u}{2}\right)^4 v-6\cos\left(\frac{u}{2}\right)^3+3\cos\left(\frac{u}{2}\right)^2 v+6\cos\left(\frac{u}{2}\right)-\frac{v}{2}\right)$$

$$\left(-\frac{17}{15}v\sin\left(\frac{u}{2}\right)+2\left(3+v\cos\left(\frac{u}{2}\right)\right)\cos(u)-2\left(3+v\cos\left(\frac{u}{2}\right)\right)\sin(u)\right)$$

$$+\left(-\cos\left(\frac{u}{2}\right)^2 v-3\cos\left(\frac{u}{2}\right)\right)v\sin\left(\frac{u}{2}\right)\,du\,dv$$

$$=\frac{-59636}{1575} \quad (11)$$

In the case of idiographic vector field, Surface Integral about Mobius Strip is equal to Spacial Closed Curve Intergral about Its Boundary Curves.

7.4 Spacial Closed Curve Integral and Surface Integral about Mobius Strip, Numerical Model (I) [Program Template of Waterloo Maplesoft, Optional]

```
> restart;
> with(plots):with(linalg):
> CS:=[(3+v*cos(u/2))*cos(u),(3+v*cos(u/2))*sin(u),v*sin(u/2)];
```

\# Define unorientable unclosed parameterized surface 'CS' (Viz. Mobius strip)

$$CS := \left[\left(3 + v \cos\left(\frac{u}{2}\right) \right) \cos(u), \left(3 + v \cos\left(\frac{u}{2}\right) \right) \sin(u), v \sin\left(\frac{u}{2}\right) \right]$$

```
> rgu:=[0,2*Pi];
```

$$rgu := [0, 2\pi]$$

```
> rgv:=[-1,1]; # Define the range of 'u,v'
```

$$rgv := [-1, 1]$$

```
> plot3d(CS,u=rgu[1]..rgu[2],v=rgv[1]..rgv[2],scaling=constrained,
projection=0.9,numpoints=1500):g1:=%: # Draw Mobius strip, be elided
> subs(v=rgv[1],CS)=eval(subs(v=rgv[1],CS));CL1:=rhs(%);
```

\# Import left boundary value of 'v' to expression of Mobius strip, obtain expression of boundary curve 'CL1'; Set Right Hand Rule as associated basis between positive direction of boundary curve 'CL1' and normal vector of Mobius strip

$$\left[\left(3 - \cos\left(\frac{u}{2}\right) \right) \cos(u), \left(3 - \cos\left(\frac{u}{2}\right) \right) \sin(u), -\sin\left(\frac{u}{2}\right) \right] =$$
$$\left[\left(3 - \cos\left(\frac{u}{2}\right) \right) \cos(u), \left(3 - \cos\left(\frac{u}{2}\right) \right) \sin(u), -\sin\left(\frac{u}{2}\right) \right]$$
$$CL1 := \left[\left(3 - \cos\left(\frac{u}{2}\right) \right) \cos(u), \left(3 - \cos\left(\frac{u}{2}\right) \right) \sin(u), -\sin\left(\frac{u}{2}\right) \right]$$

```
> subs(v=rgv[2],CS)=eval(subs(v=rgv[2],CS));CL2:=rhs(%);
```

\# Import right boundary value of 'v' to expression of Mobius strip, obtain expression of boundary curve 'CL2'; Set Right Hand Rule as associated basis between positive direction of boundary curve 'CL2' and normal vector of Mobius strip

$$\left[\left(3 + \cos\left(\frac{u}{2}\right) \right) \cos(u), \left(3 + \cos\left(\frac{u}{2}\right) \right) \sin(u), \sin\left(\frac{u}{2}\right) \right] =$$
$$\left[\left(3 + \cos\left(\frac{u}{2}\right) \right) \cos(u), \left(3 + \cos\left(\frac{u}{2}\right) \right) \sin(u), \sin\left(\frac{u}{2}\right) \right]$$
$$CL2 := \left[\left(3 + \cos\left(\frac{u}{2}\right) \right) \cos(u), \left(3 + \cos\left(\frac{u}{2}\right) \right) \sin(u), \sin\left(\frac{u}{2}\right) \right]$$

```
> spacecurve(CL1,u=rgu[1]..rgu[2],numpoints=2000,thickness=2,
color=red):g2:=%: # Draw boundary curve'CL1'
> spacecurve(CL2,u=rgu[1]..rgu[2],numpoints=2000,thickness=2,
color=blue):g3:=%: # Draw boundary curve'CL2'
> V:=[-z^2,x*z,-(x-y+z/3)^2-x*z/5-y];
```

\# Define discretional spacial vector field 'V' (Suppose spacial vector field 'V' possesses 1th order continuous partial derivatives at 'Mobius strip')

$$V := \left[-z^2, x\,z, -\left(x - y + \frac{z}{3} \right)^2 - \frac{x\,z}{5} - y \right]$$

```
> [Diff(V[3],y)-Diff(V[2],z),Diff(V[1],z)-Diff(V[3],x),
Diff(V[2],x)-Diff(V[1],y)]
=[diff(V[3],y)-diff(V[2],z),diff(V[1],z)-diff(V[3],x),
diff(V[2],x)-diff(V[1],y)];cV:=rhs(%);
# Calculate curl 'cV' of spacial vector field 'V'
```

$$\left[\left(\frac{\partial}{\partial y} \left(-\left(x - y + \frac{z}{3} \right)^2 - \frac{x\,z}{5} - y \right) \right) - \left(\frac{\partial}{\partial z} (x\,z) \right), \right.$$
$$\left(\frac{d}{dz}(-z^2) \right) - \left(\frac{\partial}{\partial x} \left(-\left(x - y + \frac{z}{3} \right)^2 - \frac{x\,z}{5} - y \right) \right), \left(\frac{\partial}{\partial x}(x\,z) \right) - \left(\frac{\partial}{\partial y}(-z^2) \right) \right] =$$
$$\left[x - 2\,y + \frac{2\,z}{3} - 1, -\frac{17\,z}{15} + 2\,x - 2\,y, z \right]$$
$$cV := \left[x - 2\,y + \frac{2\,z}{3} - 1, -\frac{17\,z}{15} + 2\,x - 2\,y, z \right]$$

```
> rgx:=[-3,4];
```

$$rgx := [-3, 4]$$

```
> rgy:=[-7/2,7/2];
```

$$rgy := \left[\frac{-7}{2}, \frac{7}{2} \right]$$

```
> rgz:=[-7/2,7/2];
```

$$rgz := \left[\frac{-7}{2}, \frac{7}{2} \right]$$

```
> fieldplot3d(V,x=rgx[1]..rgx[2],y=rgy[1]..rgy[2],z=rgz[1]..
rgz[2],arrows=SLIM,color=red,thickness=1,grid=[6,6,6]):g4:=%:
# Draw spacial vector field 'V'
> fieldplot3d(cV,x=rgx[1]..rgx[2],y=rgy[1]..rgy[2],z=rgz[1]..
rgz[2],arrows=SLIM,color=blue,thickness=1,grid=[6,6,6]):g5:=%:
# Draw curl 'cV'
> display(g1,g2,g3,g4,g5);
```

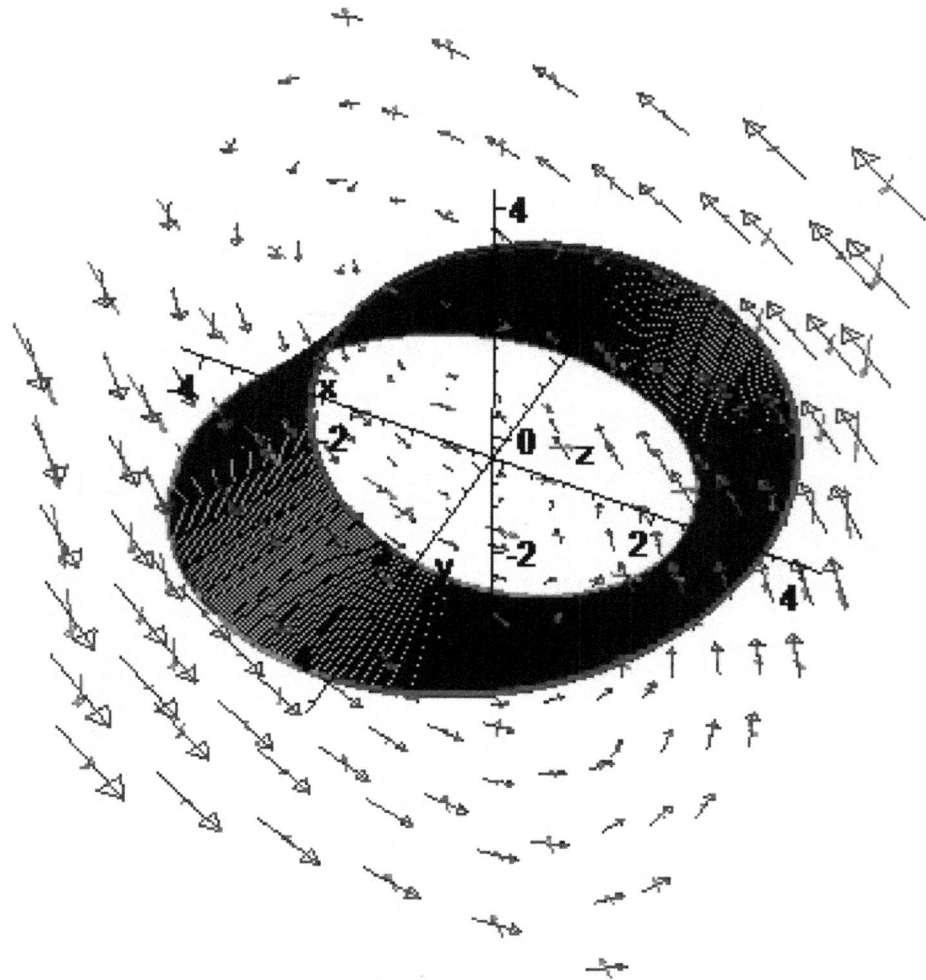

Figure(III)7.4.1 Mobius strip and its boundary curves 'CL1,CL2'
Vector Field 'V'(Red arrows) and its Curl 'cV' (Blue arrows)

```
> display(g2,g3,g4,g5);
```

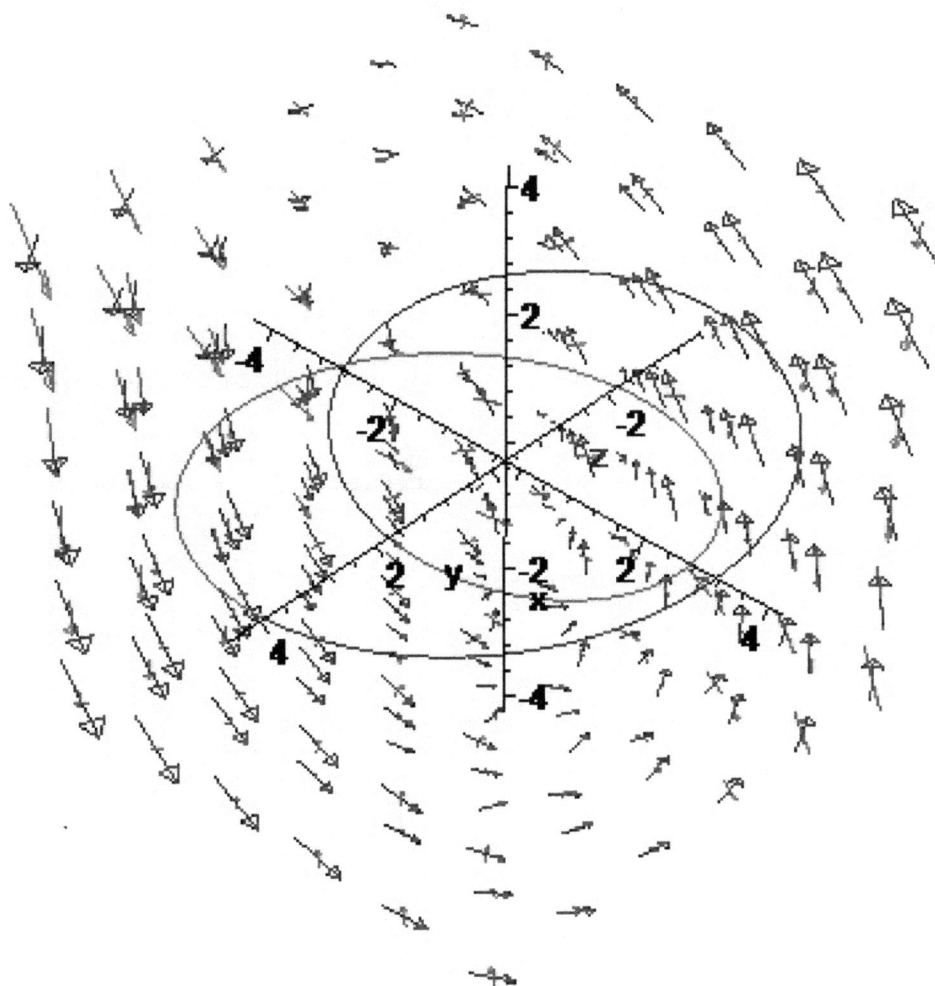

Figure(III)7.4.2 Mobius strip's boundary curves 'CL1,CL2'
Vector Field 'V'(Red arrows) and its Curl 'cV' (Blue arrows)

```
> x:=CL1[1]:y:=CL1[2]:z:=CL1[3]:
> Int(V[1]*Diff(x,u)+V[2]*Diff(y,u)+V[3]*Diff(z,u),u=rgu[1]..
rgu[2]);
# Curve Integral (Spacial Vector Field 'V' at Boundary Curve 'CL1')
```

$$\int_0^{2\pi} -\sin\left(\frac{u}{2}\right)^2 \left(\frac{d}{du}\left(\left(3-\cos\left(\frac{u}{2}\right)\right)\cos(u)\right)\right)$$

$$-\left(3-\cos\left(\frac{u}{2}\right)\right)\cos(u)\sin\left(\frac{u}{2}\right)\left(\frac{d}{du}\left(\left(3-\cos\left(\frac{u}{2}\right)\right)\sin(u)\right)\right)+\Bigg($$

$$-\left(\left(3-\cos\left(\frac{u}{2}\right)\right)\cos(u)-\left(3-\cos\left(\frac{u}{2}\right)\right)\sin(u)-\frac{1}{3}\sin\left(\frac{u}{2}\right)\right)^2$$

$$+\frac{1}{5}\left(3-\cos\left(\frac{u}{2}\right)\right)\cos(u)\sin\left(\frac{u}{2}\right)-\left(3-\cos\left(\frac{u}{2}\right)\right)\sin(u)\Bigg)\left(\frac{d}{du}\left(-\sin\left(\frac{u}{2}\right)\right)\right)du$$

```
> v1:=value(%);
```

$$v1:=-\frac{29818}{1575}-\frac{5\,\pi}{2}$$

```
> x:=CL2[1]:y:=CL2[2]:z:=CL2[3]:
> Int(V[1]*Diff(x,u)+V[2]*Diff(y,u)+V[3]*Diff(z,u),u=rgu[1]..
rgu[2]);
# Curve Integral (Spacial Vector Field 'V' at Boundary Curve 'CL2')
```

$$\int_0^{2\pi}-\sin\left(\frac{u}{2}\right)^2\left(\frac{d}{du}\left(\left(3+\cos\left(\frac{u}{2}\right)\right)\cos(u)\right)\right)$$

$$+\left(3+\cos\left(\frac{u}{2}\right)\right)\cos(u)\sin\left(\frac{u}{2}\right)\left(\frac{d}{du}\left(\left(3+\cos\left(\frac{u}{2}\right)\right)\sin(u)\right)\right)+\Bigg($$

$$-\left(\left(3+\cos\left(\frac{u}{2}\right)\right)\cos(u)-\left(3+\cos\left(\frac{u}{2}\right)\right)\sin(u)+\frac{1}{3}\sin\left(\frac{u}{2}\right)\right)^2$$

$$-\frac{1}{5}\left(3+\cos\left(\frac{u}{2}\right)\right)\cos(u)\sin\left(\frac{u}{2}\right)-\left(3+\cos\left(\frac{u}{2}\right)\right)\sin(u)\Bigg)\left(\frac{d}{du}\sin\left(\frac{u}{2}\right)\right)du$$

```
> v2:=value(%);
```

$$v2:=\frac{29818}{1575}-\frac{5\,\pi}{2}$$

```
> alpha:=v1-v2;delta:=evalf(alpha);
# Closed Curve Integral (Spacial Vector Field 'V' at Closed Boundary
Curve 'Patchwork CL1/CL2')
```

$$\alpha:=\frac{-59636}{1575}$$

$$\delta:=-37.86412698$$

```
> x:='x':y:='y':z:='z':
> x:=CS[1]:y:=CS[2]:z:=CS[3]:
> matrix(3,3,[i,j,k,Diff(CS[1],u),Diff(CS[2],u),Diff(CS[3],u),
Diff(CS[1],v),Diff(CS[2],v),Diff(CS[3],v)])=
```

```
matrix(3,3,[i,j,k,diff(CS[1],u),diff(CS[2],u),diff(CS[3],u),
diff(CS[1],v),diff(CS[2],v),diff(CS[3],v)]);m:=rhs(%);
```

\# Define and calculate matrix of partial derivatives 'm', obtain
'tangent plane normal vector' of 'Mobius strip'

$$
\begin{bmatrix} i & j & k \\ \dfrac{\partial}{\partial u}\left(\left(3+v\cos\left(\dfrac{u}{2}\right)\right)\cos(u)\right) & \dfrac{\partial}{\partial u}\left(\left(3+v\cos\left(\dfrac{u}{2}\right)\right)\sin(u)\right) & \dfrac{\partial}{\partial u}\left(v\sin\left(\dfrac{u}{2}\right)\right) \\ \dfrac{\partial}{\partial v}\left(\left(3+v\cos\left(\dfrac{u}{2}\right)\right)\cos(u)\right) & \dfrac{\partial}{\partial v}\left(\left(3+v\cos\left(\dfrac{u}{2}\right)\right)\sin(u)\right) & \dfrac{\partial}{\partial v}\left(v\sin\left(\dfrac{u}{2}\right)\right) \end{bmatrix} =
$$

$$
[i,j,k]
$$
$$
\left[-\frac{1}{2}v\sin\left(\frac{u}{2}\right)\cos(u)-\left(3+v\cos\left(\frac{u}{2}\right)\right)\sin(u),\right.
$$
$$
\left.-\frac{1}{2}v\sin\left(\frac{u}{2}\right)\sin(u)+\left(3+v\cos\left(\frac{u}{2}\right)\right)\cos(u),\frac{1}{2}v\cos\left(\frac{u}{2}\right)\right]
$$
$$
\left[\cos\left(\frac{u}{2}\right)\cos(u),\cos\left(\frac{u}{2}\right)\sin(u),\sin\left(\frac{u}{2}\right)\right]
$$

$m :=$

$$
[i,j,k]
$$
$$
\left[-\frac{1}{2}v\sin\left(\frac{u}{2}\right)\cos(u)-\left(3+v\cos\left(\frac{u}{2}\right)\right)\sin(u),\right.
$$
$$
\left.-\frac{1}{2}v\sin\left(\frac{u}{2}\right)\sin(u)+\left(3+v\cos\left(\frac{u}{2}\right)\right)\cos(u),\frac{1}{2}v\cos\left(\frac{u}{2}\right)\right]
$$
$$
\left[\cos\left(\frac{u}{2}\right)\cos(u),\cos\left(\frac{u}{2}\right)\sin(u),\sin\left(\frac{u}{2}\right)\right]
$$

```
> det(m):
> mn:=simplify(%);
```

$$
mn := -3\sin\left(\frac{u}{2}\right)i+2\sin\left(\frac{u}{2}\right)iv\cos\left(\frac{u}{2}\right)^3+6\sin\left(\frac{u}{2}\right)i\cos\left(\frac{u}{2}\right)^2-2\sin\left(\frac{u}{2}\right)iv\cos\left(\frac{u}{2}\right)
$$
$$
-2\cos\left(\frac{u}{2}\right)^4vj+3\cos\left(\frac{u}{2}\right)^2vj-\cos\left(\frac{u}{2}\right)^2vk-3k\cos\left(\frac{u}{2}\right)+6\cos\left(\frac{u}{2}\right)j-\frac{vj}{2}
$$
$$
-6\cos\left(\frac{u}{2}\right)^3j
$$

```
> A:=coeff(mn,i); # Obtain coefficient of 'i'
```

$$
A := -3\sin\left(\frac{u}{2}\right)+2\sin\left(\frac{u}{2}\right)v\cos\left(\frac{u}{2}\right)^3+6\sin\left(\frac{u}{2}\right)\cos\left(\frac{u}{2}\right)^2-2\sin\left(\frac{u}{2}\right)v\cos\left(\frac{u}{2}\right)
$$

```
> B:=coeff(mn,j); # Obtain coefficient of 'j'
```

$$
B := -2\cos\left(\frac{u}{2}\right)^4v+3\cos\left(\frac{u}{2}\right)^2v+6\cos\left(\frac{u}{2}\right)-\frac{v}{2}-6\cos\left(\frac{u}{2}\right)^3
$$

```
> C:=coeff(mn,k);
# Obtain coefficient of 'k'
```

$$C := -\cos\left(\frac{u}{2}\right)^2 v - 3\cos\left(\frac{u}{2}\right)$$

```
> [A,B,C]: # [A,B,C] structure 'tangent plane normal vector'
> Int(Int(cV[1]*A+cV[2]*B+cV[3]*C,u=rgu[1]..rgu[2]),v=rgv[1]..
rgv[2]);
# Surface Integral of Dot Product (Curl 'cV' and Mobius Strip' Tangent
Plane Normal Vector '[A,B,C]') in range of 'u,v'
```

$$\int_{-1}^{1}\int_{0}^{2\pi} \left(\left(\left(3+v\cos\left(\frac{u}{2}\right)\right)\cos(u) - 2\left(3+v\cos\left(\frac{u}{2}\right)\right)\sin(u) + \frac{2}{3}v\sin\left(\frac{u}{2}\right) - 1\right)\right.$$

$$\left(-3\sin\left(\frac{u}{2}\right) + 2\sin\left(\frac{u}{2}\right)v\cos\left(\frac{u}{2}\right)^3 + 6\sin\left(\frac{u}{2}\right)\cos\left(\frac{u}{2}\right)^2 - 2\sin\left(\frac{u}{2}\right)v\cos\left(\frac{u}{2}\right)\right) +$$

$$\left(-\frac{17}{15}v\sin\left(\frac{u}{2}\right) + 2\left(3+v\cos\left(\frac{u}{2}\right)\right)\cos(u) - 2\left(3+v\cos\left(\frac{u}{2}\right)\right)\sin(u)\right)$$

$$\left(-2\cos\left(\frac{u}{2}\right)^4 v + 3\cos\left(\frac{u}{2}\right)^2 v + 6\cos\left(\frac{u}{2}\right) - \frac{v}{2} - 6\cos\left(\frac{u}{2}\right)^3\right)$$

$$\left. + v\sin\left(\frac{u}{2}\right)\left(-\cos\left(\frac{u}{2}\right)^2 v - 3\cos\left(\frac{u}{2}\right)\right)\right) du\, dv$$

```
> beta:=value(%);epsilon:=evalf(beta);
```

$$\beta := \frac{-59636}{1575}$$

$$\varepsilon := -37.86412698$$

```
> alpha;beta; # Two analytic values are equal
```

$$\frac{-59636}{1575}$$

$$\frac{-59636}{1575}$$

```
> delta;epsilon; # Two float values are equal
```

$$-37.86412698$$

$$-37.86412698$$

// In the case of idiographic spacial vector field, surface integral
about Mobius strip is equal to spacial closed curve intergral about
its boundary curves

978-1-62265-930-2 (online)   978-1-62265-931-9 (paper)

7.5 Spacial Closed Curve Integral and Surface Integral about Mobius Strip, Numerical Models (II)

Known: Parametric Expression of Mobius Strip:

(Unorientable、Unclosed Surface)

$[(3 + v \cos(u/2))\cos(u),(3 + v \cos(u/2))\sin(u), v \sin(u/2)]$ (1)

thereinto, $u \in [0,2\pi]$, $v \in [-1,1]$;

and Integral Vector Field

$$\left[\frac{1}{2}x^2 + \frac{1}{3}yz - \frac{1}{5}z^2, \frac{\left(\frac{x}{3} - \frac{z}{5}\right)^3}{7} - \frac{xy}{3}, \left(\frac{y}{2} - \frac{z}{3}\right)^2 + \frac{x^2}{5} \right]$$ (2)

Calculate and Validate Curl Theorem at Manifold (Counterexample).

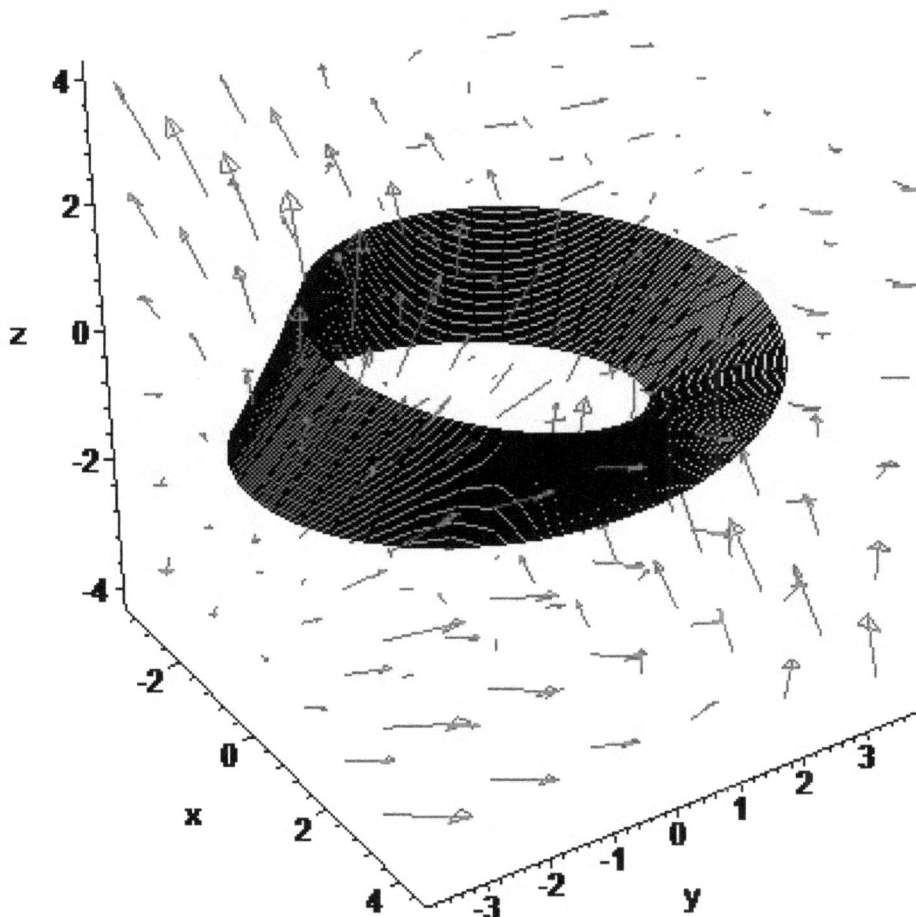

Figure(III)7.5 Mobius Strip and Vector Field

Solution:

First, Spacial Closed Curve Integral about Mobius Strip:

Import Left Boundary Value of 'v' (Viz. -1) to Expression of Mobius Strip (1), obtain Expression of Boundary Curve 'CL1' (3);

Set Right Hand Rule as associated basis between positive direction of boundary curve 'CL1' and normal vector of Mobius Strip

$$\left[\left(3-\cos\left(\frac{u}{2}\right)\right)\cos(u), \left(3-\cos\left(\frac{u}{2}\right)\right)\sin(u), -\sin\left(\frac{u}{2}\right)\right]=$$

$$\left[\left(3-\cos\left(\frac{u}{2}\right)\right)\cos(u), \left(3-\cos\left(\frac{u}{2}\right)\right)\sin(u), -\sin\left(\frac{u}{2}\right)\right]$$

$$CL1:=\left[\left(3-\cos\left(\frac{u}{2}\right)\right)\cos(u), \left(3-\cos\left(\frac{u}{2}\right)\right)\sin(u), -\sin\left(\frac{u}{2}\right)\right] \quad (3)$$

Import Right Boundary Value of 'v' (Viz. 1) to Expression of Mobius Strip (1), obtain Expression of Boundary Curve 'CL2' (4);

Set Right Hand Rule as associated basis between positive direction of boundary curve 'CL2' and normal vector of Mobius Strip

$$\left[\left(3+\cos\left(\frac{u}{2}\right)\right)\cos(u), \left(3+\cos\left(\frac{u}{2}\right)\right)\sin(u), \sin\left(\frac{u}{2}\right)\right]=$$

$$\left[\left(3+\cos\left(\frac{u}{2}\right)\right)\cos(u), \left(3+\cos\left(\frac{u}{2}\right)\right)\sin(u), \sin\left(\frac{u}{2}\right)\right]$$

$$CL2:=\left[\left(3+\cos\left(\frac{u}{2}\right)\right)\cos(u), \left(3+\cos\left(\frac{u}{2}\right)\right)\sin(u), \sin\left(\frac{u}{2}\right)\right] \quad (4)$$

Input Expression of Boundary Curve 'CL1' (3) to Vector Field (2); and Curve Integral [Vector Field(2) at Boundary Curve 'CL1'(3)]:

$$\int_0^{2\pi} \left(\frac{1}{2}\left(3-\cos\left(\frac{t}{2}\right)\right)^2\cos(t)^2 - \frac{1}{3}\left(3-\cos\left(\frac{t}{2}\right)\right)\sin(t)\sin\left(\frac{t}{2}\right) - \frac{1}{5}\sin\left(\frac{t}{2}\right)^2\right)$$

$$\left(\frac{d}{dt}\left(\left(3-\cos\left(\frac{t}{2}\right)\right)\cos(t)\right)\right)+$$

$$\left(\frac{1}{7}\left(\frac{1}{3}\left(3-\cos\left(\frac{t}{2}\right)\right)\cos(t)+\frac{1}{5}\sin\left(\frac{t}{2}\right)\right)^3 - \frac{1}{3}\left(3-\cos\left(\frac{t}{2}\right)\right)^2\cos(t)\sin(t)\right)$$

$$\left(\frac{d}{dt}\left(\left(3-\cos\left(\frac{t}{2}\right)\right)\sin(t)\right)\right)+$$

$$\left(\left(\frac{1}{2}\left(3-\cos\left(\frac{t}{2}\right)\right)\sin(t)+\frac{1}{3}\sin\left(\frac{t}{2}\right)\right)^2 + \frac{1}{5}\left(3-\cos\left(\frac{t}{2}\right)\right)^2\cos(t)^2\right)\left(\frac{d}{dt}\left(-\sin\left(\frac{t}{2}\right)\right)\right)$$

$$dt$$

$$V1 = \frac{6580288}{275625} + \frac{88861\,\pi}{100800} \tag{5}$$

Input Expression of Boundary Curve 'CL2'(4) to Vector Field(2); and Curve Integral [Vector Field(2) at Boundary Curve 'CL2'(4)]:

$$\int_0^{2\pi} \left(\frac{1}{2}\left(3 + \cos\left(\frac{t}{2}\right)\right)^2 \cos(t)^2 + \frac{1}{3}\left(3 + \cos\left(\frac{t}{2}\right)\right)\sin(t)\sin\left(\frac{t}{2}\right) - \frac{1}{5}\sin\left(\frac{t}{2}\right)^2 \right)$$

$$\left(\frac{d}{dt}\left(\left(3 + \cos\left(\frac{t}{2}\right)\right)\cos(t)\right)\right) +$$

$$\left(\frac{1}{7}\left(\frac{1}{3}\left(3 + \cos\left(\frac{t}{2}\right)\right)\cos(t) - \frac{1}{5}\sin\left(\frac{t}{2}\right)\right)^3 - \frac{1}{3}\left(3 + \cos\left(\frac{t}{2}\right)\right)^2\cos(t)\sin(t)\right)$$

$$\left(\frac{d}{dt}\left(\left(3 + \cos\left(\frac{t}{2}\right)\right)\sin(t)\right)\right)$$

$$+ \left(\left(\frac{1}{2}\left(3 + \cos\left(\frac{t}{2}\right)\right)\sin(t) - \frac{1}{3}\sin\left(\frac{t}{2}\right)\right)^2 + \frac{1}{5}\left(3 + \cos\left(\frac{t}{2}\right)\right)^2\cos(t)^2\right)\left(\frac{d}{dt}\sin\left(\frac{t}{2}\right)\right) dt$$

$$V2 = -\frac{6580288}{275625} + \frac{88861\,\pi}{100800} \tag{6}$$

Closed Curve Integral Value (Vector Field (2) at Closed Boundary Curve 'Patchwork CL1/CL2') (7):

$$V1 - V2 = \frac{13160576}{275625} \tag{7}$$

Second, Surface Integral about Mobius Strip:

According to parametric expression of Mobius Strip (1), define and calculate Matrix of Partial Derivatives, obtain tangent plane normal vector of Mobius Strip (8):

$$\begin{bmatrix} i & j & k \\ \frac{\partial}{\partial u}\left(\left(3 + v\cos\left(\frac{u}{2}\right)\right)\cos(u)\right) & \frac{\partial}{\partial u}\left(\left(3 + v\cos\left(\frac{u}{2}\right)\right)\sin(u)\right) & \frac{\partial}{\partial u}\left(v\sin\left(\frac{u}{2}\right)\right) \\ \frac{\partial}{\partial v}\left(\left(3 + v\cos\left(\frac{u}{2}\right)\right)\cos(u)\right) & \frac{\partial}{\partial v}\left(\left(3 + v\cos\left(\frac{u}{2}\right)\right)\sin(u)\right) & \frac{\partial}{\partial v}\left(v\sin\left(\frac{u}{2}\right)\right) \end{bmatrix}$$

$$=$$

$$2 \sin\left(\frac{u}{2}\right) \cos\left(\frac{u}{2}\right)^3 i\,v + 6 \sin\left(\frac{u}{2}\right) \cos\left(\frac{u}{2}\right)^2 i - 2 \sin\left(\frac{u}{2}\right) \cos\left(\frac{u}{2}\right) i\,v - 3\,i \sin\left(\frac{u}{2}\right)$$

$$- 2 \cos\left(\frac{u}{2}\right)^4 j\,v - 6 \cos\left(\frac{u}{2}\right)^3 j + 3 \cos\left(\frac{u}{2}\right)^2 j\,v - \cos\left(\frac{u}{2}\right)^2 k\,v + 6\,j \cos\left(\frac{u}{2}\right)$$

$$- 3 \cos\left(\frac{u}{2}\right) k - \frac{j\,v}{2}$$

(8)

From expression (8), Respectively pick up coefficient of item 'i,j,k', obtain Tangent Plane Normal Vector of Mobius Strip (9):

$$\left[2 \sin\left(\frac{u}{2}\right) \cos\left(\frac{u}{2}\right)^3 v + 6 \sin\left(\frac{u}{2}\right) \cos\left(\frac{u}{2}\right)^2 - 2 \sin\left(\frac{u}{2}\right) \cos\left(\frac{u}{2}\right) v - 3 \sin\left(\frac{u}{2}\right), \right.$$

$$\left. -2 \cos\left(\frac{u}{2}\right)^4 v - 6 \cos\left(\frac{u}{2}\right)^3 + 3 \cos\left(\frac{u}{2}\right)^2 v + 6 \cos\left(\frac{u}{2}\right) - \frac{v}{2}, -\cos\left(\frac{u}{2}\right)^2 v - 3 \cos\left(\frac{u}{2}\right) \right]$$

Calculate Curl of Vector Field (10):

$$\left[\left(\frac{\partial}{\partial y}\left(\left(\frac{y}{2} - \frac{z}{3}\right)^2 + \frac{x^2}{5} \right) \right) - \left(\frac{\partial}{\partial z}\left(\frac{\left(\frac{x}{3} - \frac{z}{5}\right)^3}{7} - \frac{x\,y}{3} \right) \right), \right.$$

$$\left(\frac{\partial}{\partial z}\left(\frac{1}{2}x^2 + \frac{1}{3}y\,z - \frac{1}{5}z^2 \right) \right) - \left(\frac{\partial}{\partial x}\left(\left(\frac{y}{2} - \frac{z}{3}\right)^2 + \frac{x^2}{5} \right) \right),$$

$$\left. \left(\frac{\partial}{\partial x}\left(\frac{\left(\frac{x}{3} - \frac{z}{5}\right)^3}{7} - \frac{x\,y}{3} \right) \right) - \left(\frac{\partial}{\partial y}\left(\frac{1}{2}x^2 + \frac{1}{3}y\,z - \frac{1}{5}z^2 \right) \right) \right]$$

$$=$$

$$\left[\frac{y}{2} - \frac{z}{3} + \frac{3\left(\frac{x}{3} - \frac{z}{5}\right)^2}{35}, \frac{y}{3} - \frac{2z}{5} - \frac{2x}{5}, \frac{\left(\frac{x}{3} - \frac{z}{5}\right)^2}{7} - \frac{y}{3} - \frac{z}{3} \right]$$

(10)

Input Expression of Mobius Strip (1) to Curl (10); and Surface Integral [Curl (10) at Mobius Strip (1)] (11):

$$\int_{-1}^{1}\int_{0}^{2\pi} \left(2 \sin\left(\frac{u}{2}\right) \cos\left(\frac{u}{2}\right)^3 v + 6 \sin\left(\frac{u}{2}\right) \cos\left(\frac{u}{2}\right)^2 - 2 \sin\left(\frac{u}{2}\right) \cos\left(\frac{u}{2}\right) v - 3 \sin\left(\frac{u}{2}\right) \right) \Bigg($$

$$\frac{1}{2}\left(3 + v \cos\left(\frac{u}{2}\right) \right) \sin(u) - \frac{1}{3}v \sin\left(\frac{u}{2}\right) + \frac{3}{35}\left(\frac{1}{3}\left(3 + v \cos\left(\frac{u}{2}\right) \right) \cos(u) - \frac{1}{5}v \sin\left(\frac{u}{2}\right) \right)^2$$

$$\Bigg) + \left(-2 \cos\left(\frac{u}{2}\right)^4 v - 6 \cos\left(\frac{u}{2}\right)^3 + 3 \cos\left(\frac{u}{2}\right)^2 v + 6 \cos\left(\frac{u}{2}\right) - \frac{v}{2} \right)$$

$$\left(\frac{1}{3}\left(3+v\cos\left(\frac{u}{2}\right)\right)\sin(u)-\frac{2}{5}v\sin\left(\frac{u}{2}\right)-\frac{2}{5}\left(3+v\cos\left(\frac{u}{2}\right)\right)\cos(u)\right)+$$

$$\left(-\cos\left(\frac{u}{2}\right)^2 v-3\cos\left(\frac{u}{2}\right)\right)\Bigg($$

$$\frac{1}{7}\left(\frac{1}{3}\left(3+v\cos\left(\frac{u}{2}\right)\right)\cos(u)-\frac{1}{5}v\sin\left(\frac{u}{2}\right)\right)^2-\frac{1}{3}\left(3+v\cos\left(\frac{u}{2}\right)\right)\sin(u)-\frac{1}{3}v\sin\left(\frac{u}{2}\right)\Bigg)$$

du dv

$$=\frac{8015576}{275625}$$

(11)

In the case of idiographic vector field, Surface Integral about Mobius Strip is unequal to Spacial Closed Curve Intergral about Its Boundary Curves.

In the case of abstract spacial vector field '[P(x,y,z),Q(x,y,z), R(x,y,z)]', Mobius strip' surface integral is unequal to spacial closed curve intergral of its boundary curves in logic;

Owing to various values of idiographic vector field, surface integral about Mobius strip and spacial closed curve intergral about its boundary curves are equal or unequal possibly.

See also Chapter 6 (Part I) 'Counterexample of Divergence Theorem at Manifold--Surface Integral and Triple Integrals about Klein Bottle'

And Chapter 2 (This Part) Section 2.3 'Prove Curl Theorem In Torus Coordinates (Multiple Connected Orientable Closed Surface Coordinates)'

7.6 Spacial Closed Curve Integral and Surface Integral about Mobius Strip, Numerical Model (II) [Program Template of Waterloo Maplesoft, Optional]

```
> restart;
> with(plots):with(linalg):
> CS:=[(3+v*cos(u/2))*cos(u),(3+v*cos(u/2))*sin(u),v*sin(u/2)];
# Define unorientable unclosed parameterized surface 'CS' (Viz.
```

'Mobius strip')

$$CS := \left[\left(3 + v \cos\left(\frac{u}{2}\right) \right) \cos(u), \left(3 + v \cos\left(\frac{u}{2}\right) \right) \sin(u), v \sin\left(\frac{u}{2}\right) \right]$$

```
> rgu:=[0,2*Pi];
```

$$rgu := [0, 2\pi]$$

```
> rgv:=[-1,1]; # Define range of 'u,v'
```

$$rgv := [-1, 1]$$

```
> plot3d(CS,u=rgu[1]..rgu[2],v=rgv[1]..rgv[2],scaling=constrained,
projection=0.9,numpoints=1500):g1:=%: # Draw Mobius strip
> subs(v=rgv[1],CS)=eval(subs(v=rgv[1],CS));CL1:=rhs(%);
```

Import left boundary value of 'v' to expression of Mobius strip, obtain expression of boundary curve 'CL1'; Set Right Hand Rule as associated basis between positive direction of boundary curve 'CL1' and normal vector of Mobius strip

$$\left[\left(3 - \cos\left(\frac{u}{2}\right) \right) \cos(u), \left(3 - \cos\left(\frac{u}{2}\right) \right) \sin(u), -\sin\left(\frac{u}{2}\right) \right] =$$
$$\left[\left(3 - \cos\left(\frac{u}{2}\right) \right) \cos(u), \left(3 - \cos\left(\frac{u}{2}\right) \right) \sin(u), -\sin\left(\frac{u}{2}\right) \right]$$
$$CL1 := \left[\left(3 - \cos\left(\frac{u}{2}\right) \right) \cos(u), \left(3 - \cos\left(\frac{u}{2}\right) \right) \sin(u), -\sin\left(\frac{u}{2}\right) \right]$$

```
> subs(v=rgv[2],CS)=eval(subs(v=rgv[2],CS));CL2:=rhs(%);
```

Import right boundary value of 'v' to expression of Mobius strip, obtain expression of boundary curve 'CL2'; Set Right Hand Rule as associated basis between positive direction of boundary curve 'CL2' and normal vector of Mobius strip

$$\left[\left(3 + \cos\left(\frac{u}{2}\right) \right) \cos(u), \left(3 + \cos\left(\frac{u}{2}\right) \right) \sin(u), \sin\left(\frac{u}{2}\right) \right] =$$
$$\left[\left(3 + \cos\left(\frac{u}{2}\right) \right) \cos(u), \left(3 + \cos\left(\frac{u}{2}\right) \right) \sin(u), \sin\left(\frac{u}{2}\right) \right]$$
$$CL2 := \left[\left(3 + \cos\left(\frac{u}{2}\right) \right) \cos(u), \left(3 + \cos\left(\frac{u}{2}\right) \right) \sin(u), \sin\left(\frac{u}{2}\right) \right]$$

```
> spacecurve(CL1,u=rgu[1]..rgu[2],numpoints=2000,thickness=2,
color=red):g2:=%: # Draw boundary curve 'CL1'
> spacecurve(CL2,u=rgu[1]..rgu[2],numpoints=2000,thickness=2,
color=blue):g3:=%: # Draw boundary curve 'CL2'
> V:=[x^2/2+y*z/3-z^2/5,(x/3-z/5)^3/7-x*y/3,(y/2-z/3)^2+x^2/5];
```

Define discretional spacial vector field 'V' (Suppose spacial vector field 'V' possesses 1th order continuous partial derivatives at

'Mobius strip')

$$V := \left[\frac{1}{2}x^2 + \frac{1}{3}yz - \frac{1}{5}z^2, \frac{\left(\frac{x}{3}-\frac{z}{5}\right)^3}{7} - \frac{xy}{3}, \left(\frac{y}{2}-\frac{z}{3}\right)^2 + \frac{x^2}{5} \right]$$

```
> [Diff(V[3],y)-Diff(V[2],z),Diff(V[1],z)-Diff(V[3],x),
Diff(V[2],x)-Diff(V[1],y)]
=[diff(V[3],y)-diff(V[2],z),diff(V[1],z)-diff(V[3],x),
diff(V[2],x)-diff(V[1],y)];cV:=rhs(%);
# Calculate curl 'cV' of spacial vector field 'V'
```

$$\left[\left(\frac{\partial}{\partial y}\left(\left(\frac{y}{2}-\frac{z}{3}\right)^2 + \frac{x^2}{5}\right)\right) - \left(\frac{\partial}{\partial z}\left(\frac{\left(\frac{x}{3}-\frac{z}{5}\right)^3}{7} - \frac{xy}{3}\right)\right), \right.$$

$$\left(\frac{\partial}{\partial z}\left(\frac{1}{2}x^2 + \frac{1}{3}yz - \frac{1}{5}z^2\right)\right) - \left(\frac{\partial}{\partial x}\left(\left(\frac{y}{2}-\frac{z}{3}\right)^2 + \frac{x^2}{5}\right)\right),$$

$$\left.\left(\frac{\partial}{\partial x}\left(\frac{\left(\frac{x}{3}-\frac{z}{5}\right)^3}{7} - \frac{xy}{3}\right)\right) - \left(\frac{\partial}{\partial y}\left(\frac{1}{2}x^2 + \frac{1}{3}yz - \frac{1}{5}z^2\right)\right) \right] =$$

$$\left[\frac{y}{2} - \frac{z}{3} + \frac{3\left(\frac{x}{3}-\frac{z}{5}\right)^2}{35}, \frac{y}{3} - \frac{2z}{5} - \frac{2x}{5}, \frac{\left(\frac{x}{3}-\frac{z}{5}\right)^2}{7} - \frac{y}{3} - \frac{z}{3} \right]$$

$$cV := \left[\frac{y}{2} - \frac{z}{3} + \frac{3\left(\frac{x}{3}-\frac{z}{5}\right)^2}{35}, \frac{y}{3} - \frac{2z}{5} - \frac{2x}{5}, \frac{\left(\frac{x}{3}-\frac{z}{5}\right)^2}{7} - \frac{y}{3} - \frac{z}{3} \right]$$

```
> rgx:=[-3,4];
```

$$rgx := [-3, 4]$$

```
> rgy:=[-7/2,7/2];
```

$$rgy := \left[\frac{-7}{2}, \frac{7}{2}\right]$$

```
> rgz:=[-7/2,7/2];
```

$$rgz := \left[\frac{-7}{2}, \frac{7}{2}\right]$$

```
> fieldplot3d(V,x=rgx[1]..rgx[2],y=rgy[1]..rgy[2],z=rgz[1]..
rgz[2],arrows=SLIM,color=red,thickness=1,grid=[6,6,6]):g4:=%:
# Draw spacial vector field 'V'
> fieldplot3d(cV,x=rgx[1]..rgx[2],y=rgy[1]..rgy[2],z=rgz[1]..
```

```
rgz[2],arrows=SLIM,color=blue,thickness=1,grid=[6,6,6]):g5:=%:
# Draw curl `cV'
> display(g1,g2,g3,g4,g5);
```

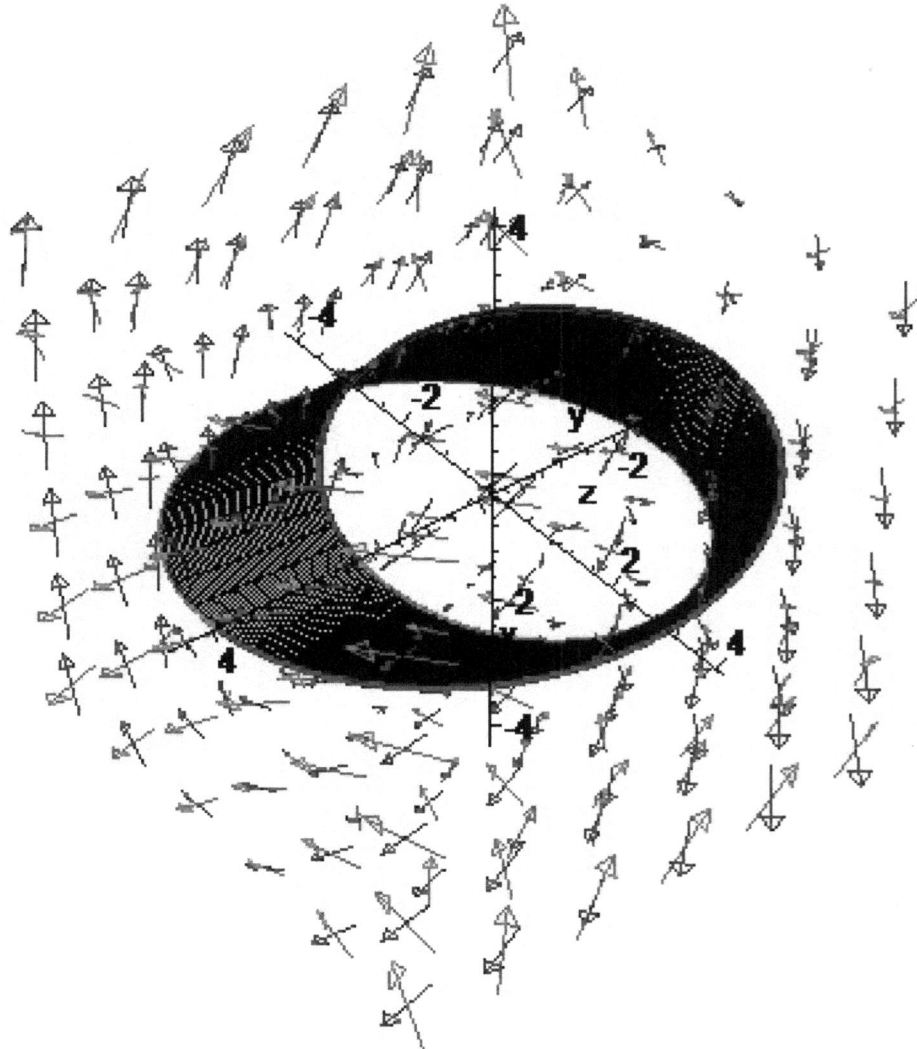

Figure(III)7.6.1 Mobius Strip and its Boundary Curves 'CL1,CL2'
   Vector Field `V'(Red arrows) and its Curl `cV (Blue arrows)

```
> display(g2,g3,g4,g5);
```

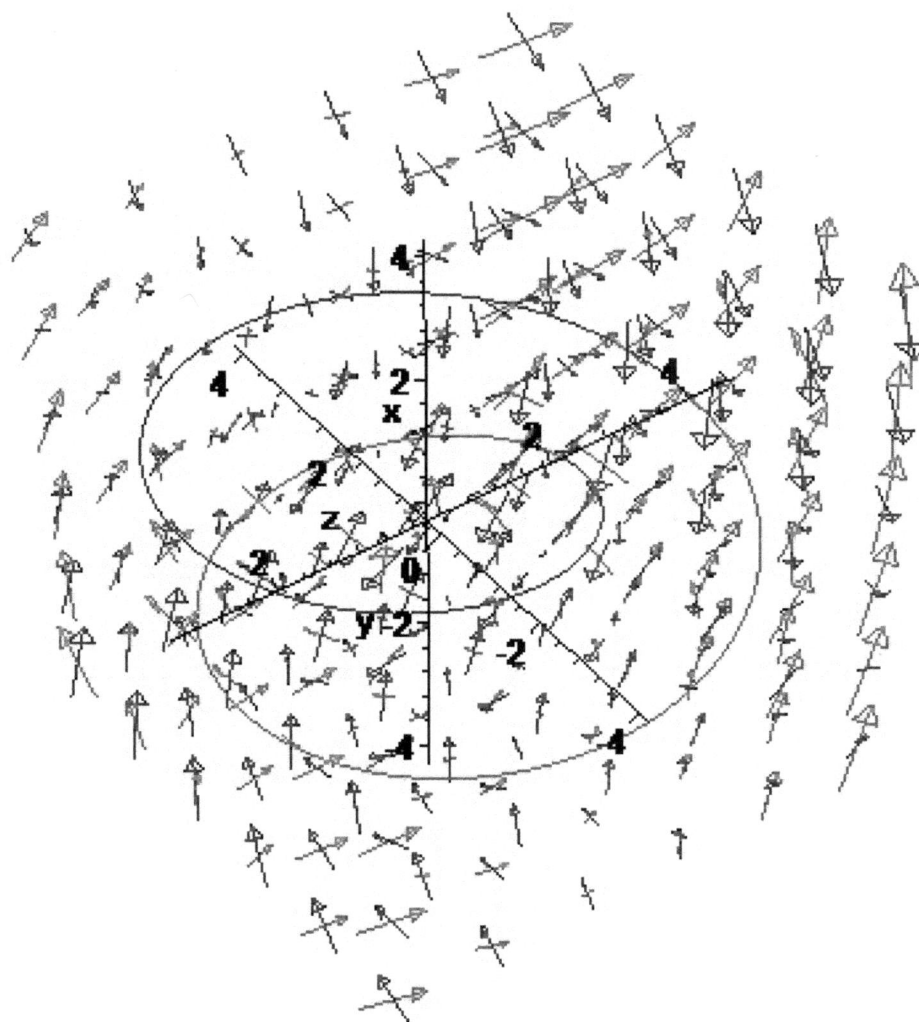

Figure(III)7.6.2 Mobius Strip's Boundary Curves 'CL1,CL2'
Vector Field 'V'(Red arrows) and its Curl 'cV (Blue arrows)

```
> x:=CL1[1]:y:=CL1[2]:z:=CL1[3]:
> Int(V[1]*Diff(x,u)+V[2]*Diff(y,u)+V[3]*Diff(z,u),u=rgu[1]..
rgu[2]);
# Curve Integral (Spacial Vector Field 'V' at Boundary Curve 'CL1')
```

$$\int_0^{2\pi} \left(\frac{1}{2}\left(3 - \cos\left(\frac{u}{2}\right)\right)^2 \cos(u)^2 - \frac{1}{3}\left(3 - \cos\left(\frac{u}{2}\right)\right)\sin(u)\sin\left(\frac{u}{2}\right) - \frac{1}{5}\sin\left(\frac{u}{2}\right)^2 \right)$$

$$\left(\frac{d}{du}\left(\left(3-\cos\left(\frac{u}{2}\right)\right)\cos(u)\right)\right)+$$

$$\left(\frac{1}{7}\left(\frac{1}{3}\left(3-\cos\left(\frac{u}{2}\right)\right)\cos(u)+\frac{1}{5}\sin\left(\frac{u}{2}\right)\right)^{3}-\frac{1}{3}\left(3-\cos\left(\frac{u}{2}\right)\right)^{2}\cos(u)\sin(u)\right)$$

$$\left(\frac{d}{du}\left(\left(3-\cos\left(\frac{u}{2}\right)\right)\sin(u)\right)\right)+$$

$$\left(\left(\frac{1}{2}\left(3-\cos\left(\frac{u}{2}\right)\right)\sin(u)+\frac{1}{3}\sin\left(\frac{u}{2}\right)\right)^{2}+\frac{1}{5}\left(3-\cos\left(\frac{u}{2}\right)\right)^{2}\cos(u)^{2}\right)$$

$$\left(\frac{d}{du}\left(-\sin\left(\frac{u}{2}\right)\right)\right)du$$

```
> v1:=value(%);
```

$$v1:=\frac{6580288}{275625}+\frac{88861\,\pi}{100800}$$

```
> x:=CL2[1]:y:=CL2[2]:z:=CL2[3]:
> Int(V[1]*Diff(x,u)+V[2]*Diff(y,u)+V[3]*Diff(z,u),u=rgu[1]..
rgu[2]);
```

\# Curve Integral (Spacial Vector Field 'V' at Boundary Curve 'CL2')

$$\int_{0}^{2\pi}\left(\frac{1}{2}\left(3+\cos\left(\frac{u}{2}\right)\right)^{2}\cos(u)^{2}+\frac{1}{3}\left(3+\cos\left(\frac{u}{2}\right)\right)\sin(u)\sin\left(\frac{u}{2}\right)-\frac{1}{5}\sin\left(\frac{u}{2}\right)^{2}\right)$$

$$\left(\frac{d}{du}\left(\left(3+\cos\left(\frac{u}{2}\right)\right)\cos(u)\right)\right)+$$

$$\left(\frac{1}{7}\left(\frac{1}{3}\left(3+\cos\left(\frac{u}{2}\right)\right)\cos(u)-\frac{1}{5}\sin\left(\frac{u}{2}\right)\right)^{3}-\frac{1}{3}\left(3+\cos\left(\frac{u}{2}\right)\right)^{2}\cos(u)\sin(u)\right)$$

$$\left(\frac{d}{du}\left(\left(3+\cos\left(\frac{u}{2}\right)\right)\sin(u)\right)\right)$$

$$+\left(\left(\frac{1}{2}\left(3+\cos\left(\frac{u}{2}\right)\right)\sin(u)-\frac{1}{3}\sin\left(\frac{u}{2}\right)\right)^{2}+\frac{1}{5}\left(3+\cos\left(\frac{u}{2}\right)\right)^{2}\cos(u)^{2}\right)\left(\frac{d}{du}\sin\left(\frac{u}{2}\right)\right)$$

$$du$$

```
> v2:=value(%);
```

$$v2:=-\frac{6580288}{275625}+\frac{88861\,\pi}{100800}$$

```
> alpha:=v1-v2;delta:=evalf(alpha);
```

\# Closed Curve Integral (Spacial Vector Field 'V' at Closed Boundary Curve 'Patchwork CL1/CL2')

$$\alpha:=\frac{13160576}{275625}$$

$$\delta := 47.74812154$$

```
> x:='x':y:='y':z:='z':
> x:=CS[1]:y:=CS[2]:z:=CS[3]:
> matrix(3,3,[i,j,k,Diff(CS[1],u),Diff(CS[2],u),Diff(CS[3],u),
Diff(CS[1],v),Diff(CS[2],v),Diff(CS[3],v)])=
matrix(3,3,[i,j,k,diff(CS[1],u),diff(CS[2],u),diff(CS[3],u),
diff(CS[1],v),diff(CS[2],v),diff(CS[3],v)]);m:=rhs(%);
# Define and calculate matrix of partial derivatives 'm', obtain
'Tangent Plane Normal Vector' of 'Mobius Strip'
```

$$
\begin{bmatrix}
i & j & k \\
\dfrac{\partial}{\partial u}\left(\left(3+v\cos\left(\dfrac{u}{2}\right)\right)\cos(u)\right) & \dfrac{\partial}{\partial u}\left(\left(3+v\cos\left(\dfrac{u}{2}\right)\right)\sin(u)\right) & \dfrac{\partial}{\partial u}\left(v\sin\left(\dfrac{u}{2}\right)\right) \\
\dfrac{\partial}{\partial v}\left(\left(3+v\cos\left(\dfrac{u}{2}\right)\right)\cos(u)\right) & \dfrac{\partial}{\partial v}\left(\left(3+v\cos\left(\dfrac{u}{2}\right)\right)\sin(u)\right) & \dfrac{\partial}{\partial v}\left(v\sin\left(\dfrac{u}{2}\right)\right)
\end{bmatrix}=
$$

$$[i,j,k]$$

$$\left[-\frac{1}{2}v\sin\left(\frac{u}{2}\right)\cos(u)-\left(3+v\cos\left(\frac{u}{2}\right)\right)\sin(u),\right.$$

$$\left.-\frac{1}{2}v\sin\left(\frac{u}{2}\right)\sin(u)+\left(3+v\cos\left(\frac{u}{2}\right)\right)\cos(u),\frac{1}{2}v\cos\left(\frac{u}{2}\right)\right]$$

$$\left[\cos\left(\frac{u}{2}\right)\cos(u),\cos\left(\frac{u}{2}\right)\sin(u),\sin\left(\frac{u}{2}\right)\right]$$

$$m :=$$

$$[i,j,k]$$

$$\left[-\frac{1}{2}v\sin\left(\frac{u}{2}\right)\cos(u)-\left(3+v\cos\left(\frac{u}{2}\right)\right)\sin(u),\right.$$

$$\left.-\frac{1}{2}v\sin\left(\frac{u}{2}\right)\sin(u)+\left(3+v\cos\left(\frac{u}{2}\right)\right)\cos(u),\frac{1}{2}v\cos\left(\frac{u}{2}\right)\right]$$

$$\left[\cos\left(\frac{u}{2}\right)\cos(u),\cos\left(\frac{u}{2}\right)\sin(u),\sin\left(\frac{u}{2}\right)\right]$$

```
> det(m):
> mn:=simplify(%);
```

$$mn := 2\sin\left(\frac{u}{2}\right)i\,v\cos\left(\frac{u}{2}\right)^3+6\sin\left(\frac{u}{2}\right)i\cos\left(\frac{u}{2}\right)^2-2\sin\left(\frac{u}{2}\right)i\,v\cos\left(\frac{u}{2}\right)-3\sin\left(\frac{u}{2}\right)i$$

$$+3\cos\left(\frac{u}{2}\right)^2v\,j-\cos\left(\frac{u}{2}\right)^2v\,k-3\,k\cos\left(\frac{u}{2}\right)-2\cos\left(\frac{u}{2}\right)^4v\,j-6\cos\left(\frac{u}{2}\right)^3j$$

$$+6\cos\left(\frac{u}{2}\right)j-\frac{v\,j}{2}$$

```
> A:=coeff(mn,i); # Obtain coefficient of 'i'
```

$$A := 2 \sin\left(\frac{u}{2}\right) v \cos\left(\frac{u}{2}\right)^3 + 6 \sin\left(\frac{u}{2}\right) \cos\left(\frac{u}{2}\right)^2 - 2 \sin\left(\frac{u}{2}\right) v \cos\left(\frac{u}{2}\right) - 3 \sin\left(\frac{u}{2}\right)$$

```
> B:=coeff(mn,j); # Obtain coefficient of 'j'
```

$$B := 3 \cos\left(\frac{u}{2}\right)^2 v - 2 \cos\left(\frac{u}{2}\right)^4 v - 6 \cos\left(\frac{u}{2}\right)^3 + 6 \cos\left(\frac{u}{2}\right) - \frac{v}{2}$$

```
> C:=coeff(mn,k);
# Obtain coefficient of 'k'
```

$$C := -\cos\left(\frac{u}{2}\right)^2 v - 3 \cos\left(\frac{u}{2}\right)$$

```
> [A,B,C]: # [A,B,C] structure 'tangent plane normal vector'
> Int(Int(cV[1]*A+cV[2]*B+cV[3]*C,u=rgu[1]..rgu[2]),v=rgv[1]..
rgv[2]);
# Surface Integral of Dot Product (Curl 'cV' and  Mobius Strip' Tangent
Plane Normal Vector '[A,B,C]') in range of 'u,v'
```

$$\int_{-1}^{1} \int_{0}^{2\pi} \left(\left(\frac{1}{2}\left(3 + v \cos\left(\frac{u}{2}\right)\right) \right) \sin(u) - \frac{1}{3} v \sin\left(\frac{u}{2}\right) \right.$$

$$+ \frac{3}{35}\left(\frac{1}{3}\left(3 + v \cos\left(\frac{u}{2}\right)\right) \cos(u) - \frac{1}{5} v \sin\left(\frac{u}{2}\right) \right)^2 \right)$$

$$\left(2 \sin\left(\frac{u}{2}\right) v \cos\left(\frac{u}{2}\right)^3 + 6 \sin\left(\frac{u}{2}\right) \cos\left(\frac{u}{2}\right)^2 - 2 \sin\left(\frac{u}{2}\right) v \cos\left(\frac{u}{2}\right) - 3 \sin\left(\frac{u}{2}\right) \right) +$$

$$\left(\left(\frac{1}{3}\left(3 + v \cos\left(\frac{u}{2}\right)\right) \right) \sin(u) - \frac{2}{5} v \sin\left(\frac{u}{2}\right) - \frac{2}{5}\left(3 + v \cos\left(\frac{u}{2}\right)\right) \cos(u) \right)$$

$$\left(3 \cos\left(\frac{u}{2}\right)^2 v - 2 \cos\left(\frac{u}{2}\right)^4 v - 6 \cos\left(\frac{u}{2}\right)^3 + 6 \cos\left(\frac{u}{2}\right) - \frac{v}{2} \right) + \left(\vphantom{\frac{1}{7}} \right.$$

$$\frac{1}{7}\left(\frac{1}{3}\left(3 + v \cos\left(\frac{u}{2}\right)\right) \cos(u) - \frac{1}{5} v \sin\left(\frac{u}{2}\right) \right)^2 - \frac{1}{3}\left(3 + v \cos\left(\frac{u}{2}\right)\right) \sin(u) - \frac{1}{3} v \sin\left(\frac{u}{2}\right)$$

$$\left. \right)\left(-\cos\left(\frac{u}{2}\right)^2 v - 3 \cos\left(\frac{u}{2}\right) \right) du \, dv$$

```
> beta:=value(%);epsilon:=evalf(beta);
```

$$\beta := \frac{8015576}{275625}$$

$$\varepsilon := 29.08145488$$

```
> alpha;beta; # Two analytic values are unequal
```

$$\frac{13160576}{275625}$$

$$\frac{8015576}{275625}$$

> delta;epsilon; # Two float values are unequal

47.74812154

29.08145488

// In the case of idiographic spacial vector field, surface integral of Mobius strip is unequal to spacial closed curve intergral of its boundary curves

// In the case of abstract spacial vector field '[P(x,y,z),Q(x,y,z), R(x,y,z)]', Mobius strip' surface integral is unequal to spacial closed curve intergral of its boundary curves in logic; Because of various values of idiographic spacial vector field, Mobius strip' surface integral and spacial closed curve intergral of its boundary curves are equal or unequal possibly

Conclusion A in Part III

Constitute individualized geometric object coordinates that matches with idiographic geometric object [Manifold] (Viz. What idiographic surface, what coordinates of idiographic surface, what volume element coefficient of idiographic surface coordinates; no longer rely on a few existent coordinates: Cartesian coordinates、 Spherical coordinates、Cylindrical coordinates、 Generalized Spherical coordinates and Generalized Cylindrical coordinates etc.), by two methods (Integral、Finite Sums Limits), prove presence of Curl Theorem in unlimited quantitative Individualized Surface coordinates [Summarized and described by abstract simply connected orientable closed surface coordinates (Bases on Poincare Conjecture) and multiple connected orientable closed surface (Torus) coordinates], enable Curl Theorem surpass traditional architecture of 3-Dimensional Cartesian coordinates, in Individualized Surface coordinates, establish new formular association between surface integral (Bases on parameterized dot product method) and spacial closed curve integral, and realize mutual validation between two typical integrals in unlimited

- 572 -

quantitative individualized and gorgeous formular numerical model operations, radicate theoretical logic basis and numerical models of new typical surface integral (Bases on parameterized dot product method in Individualized Surface coordinates).

'Prove Curl Theorem at Manifold' itself is not sole purpose, 'Establish new formular association between surface integral (Bases on parameterized dot product method) and spacial closed curve integral in Individualized Surface coordinates, radicate theoretical logic basis and numerical models of new typical surface integral (Bases on parameterized dot product method in Individualized Surface coordinates)' is prime purpose.

Correlative numerical models of this part have indicated, by surface integral (Bases on parameterized dot product method) and triple integrals (Bases on idiographic volume element coefficient method) in Individualized Surface coordinates, science explorers can obtain analytic integral value and float integral value in discretional precision about complicated geometric objects (Manifold, irregular、asymmetrical、unclosed surface especially). Realize free surface integral、free cubage integral and free closed curve integral, indeed realize artistic integral interval; realize exact integral calculation of vector field [Electric field、Magnetic field、Hydromechanical field、Gravitational field etc.] and scalar field [Electric potential field、temperature field、density field etc.] in discretional free space region (Unclosed surface、space region that unclosed surface contains、unclosed surface boundary closed route), radicate logic associated relationship between unclosed surface integral and unclosed surface boundary closed route integral, seek direct joint points of calculus、topology and physical engineering calculation, realize Curl Theorem at Manifold and manifold integral in physical、engineering meaning, realize much more vast and free physical、mathematical explorations and engineering practices.

Conclusion B Comparison of Different Methods in Part III

Method	Normalization	Application Range	Calculation Precision	Result Verification
Traditional Surface Integral bases on Projective Method in 3-Dimensional Cartesian Coordinates	Tedious & Divergent Calculational Method	Symmetric & Regular Geometric Objects, such as Spherical, Ellipsoidal, Conical, Cylindrical, Hexahedral Surface & their simple combinations	Analytic Value	No Verification
Traditional Partial Differential Equations, such as Finite Difference Method、Finite Element Method、Boundary Element Method etc.	Tedious & Divergent Calculational Method; Method & Geometric Object, One-to-One; 'Algorithm' Design is a huge project & tedious process	Slightly Complex Geometric objects; Mainstream & Research Direction of Numerical Analysis in Contemporary Physical & Engineering Fields	Approximate Value	Convergence & Stability Verification
New Surface Integral bases on Parameterized Dot Product Method & Cubage Integral bases on Idiographic Volume Element Coefficient Method in Individualized Surface Coordinates	Standardized & Matrixing Calculational Method, Simple & Convergent; Method & Geometric Object, One-to-Many	Discretional Complex Geometric Objects	Analytic Value	One to One Verification by Numerical Models of New Integral Theorem

Epilogue

Data and logic are primary tool and ideaistic method for science explorers' observing and cognizing world. In scientific theoretical field, any concept、principle and theorem must accept verification of data and logic, without an exception. Endless Explorations, endless cognitions and practices, endless progress and transcendence for science and technology.

References:

[1] Pangen, Commentative Tutorial of Basal Physics (Science Press, Beijing, 2002.1)
    pp.360-361, pp.363-364, p.385, p.401

[2] Richard Feynman, Robert B. Leighton, Matthew L. Sands, The Feynman Lectures on Physics (Pearson Education, 1989; Shanghai, ed.1 2005.6, 6thPrinting 2009.10)   pp.31-32, pp.36-38, pp.48-54, pp.55-65, pp.159-165, pp.213-242, pp.259-289

[3] Electromagnetics (Higher Education Press, Beijing 2001.1)   pp.33-43, pp.172-175

[4] Electrondynamics and Computer Aided Education (Science Press, Beijing, 2007.8)   pp.1-47

[5] Field Theory (Atomic Energy Press, Beijing, 2006.10) pp.9-17, pp.147-150, pp.184-186

[6] Hydrodynamics (Metallurgical Industry Press, Beijing, 2010.2) pp.37-80

[7] Applied Hydrodynamics (Tsinghua University Press, 2006.3) pp.21-32, p.46

[8] Engineering Hydrodynamics (China Communications Press, 2010.1) pp.28-46, pp.57-59, pp.88-96

[9] Basal Multidimensional Aerodynamics (Beihang University Press, Beijing, ed.2, 2008.6) p.15, pp.36-43

[10] А.Я.Хинчин,Concise Tutorial of Mathematical Analysis (Higher Education Press, Beijing,1956.8)
    vol 2, pp.619-624, pp.644-653

[11] ShuyiXie, Engineering Mathematics:Vector Analysis and Field Theory (Higher Education Press, Beijing, ed.1, 1978.12; ed.2, 1985.3;   23th Printing ,2002.3)   pp.41-45, pp.53-57, p.85, pp.90-91

[12] Tongji University-Department of Applied Mathematics, Calculus (Higher Education Press, Beijing, 2002.1)
    Volume 2, pp.145-151, pp.163-165, pp.208-210, pp.234-248

[13] Higher Mathematics: Multivariable Calculus and Instructional Software (Science Press, Beijing,1999.6)
    pp.249-251, pp.291-294, pp.312-313, pp.348-351, pp.388-390, pp.393-412

[14] XiaoqingDing, Engineering Calculus (Science Press, Beijing, 2002.9) Volume 2, pp.235-239, pp.288-303, pp.309-315, pp.317-321

[15] Tongji University--Department of Mathematics, Higher Mathematics [6th Edition] , Higher Education Press, Beijing,ed.1, 1978.10; ed.6, 2007.6;   9th Printing, 2009.8)   pp.142-145, pp.168-170, pp.174-179

[16] Thomas's Calculus [Tenth Edition] (Higher Education Press, Beijing, 2003.8)   pp.1025-1029, pp.1048-1056,pp.1104-1114, pp.1137-1157

[17] M.R.Spiegel, Schaum's Outline of Theory and Problems of Advanced Calculus (McGraw-Hill Companies, Inc, Copyright 1963, 37thPrinting, 1998;   Science Press, Beijing, 2002.1)   pp.162-165, pp.169-170, p.180, p.183, pp.190-201

[18] JijunLiu, Methods of Modern Numerical Calculation (Science Press, Beijing, 2010.3)   pp.62-136

[19] Hamilton-Perelman's Proof of the Poincare Conjecture and the Geometrization Conjecture
    Huai-Dong Cao and Xi-Ping Zhu, arXiv: math/0912069v1 [math.DG] 3 Dec 2006

[20] M.Spivak , Calculus on Manifolds (Posts & Telecom Press, Beijing, 2006.1) pp.114-143

[21] Reference Manual of MapleSoft Instruction (National Defense Industry Press, Beijing, 2002.1)

[22] Yangke, Integral Geometry and Fields–Theorematic Proofs and Numerical Models
    Work Registration Certificate: 川作登字-2015-A-00001712,
    Copyright Bureau of Sichuan Province, China

Brief Introduction of Author:

Yangke (1968.12.6 -), Graduated from University of Electronic Science and Technology of China (UESTC) in July 1990, Free programmer, Email: more2010e@sina.com; Numerical modeling began in September 2005, and formular proofs were successful preliminarily in December 2009; So far, total more than 15,000 Maplesoft program templates.